Plant Biology

Plant Biology

Alison M. Smith

George Coupland

Liam Dolan

Nicholas Harberd

Jonathan Jones

Cathie Martin

Robert Sablowski

Abigail Amey

 Garland Science
Taylor & Francis Group

Garland Science
Vice President: Denise Schanck
Production Editor: Simon Hill
Typesetting: Georgina Lucas
Illustrations: Nigel Orme
Editorial and Production Assistant: Helen Powis
Copyeditor: Linda Strange
Proofreader: Jo Clayton
Design: Matthew McClements, Blink Studio Ltd.
Cover Design: Andy Magee
Permissions: Julene Knox
Indexer: Merrall-Ross International Ltd.

©2010 by Garland Science, Taylor & Francis Group, LLC

ISBN 978-0-8153-4025-6

Front cover image courtesy of Dr. Jean Armstrong, University of Hull, UK. The cover image shows a transverse section of rice root with lysigenous aerenchyma. The root was previously exposed to light and produced chloroplasts, which fluoresced red in blue light.

Library of Congress Cataloging-in-Publication Data

Plant biology / Alison M. Smith ... [et al.].
        p. cm.
    ISBN 978-0-8153-4025-6
    1. Botany. 2. Plants--Development. 3. Plants--Effect of stress on.
    4. Crops. I. Smith, Alison M. (Alison Mary), 1954-
    QK45.2.P575 2009
    580--dc22

                                                    2009013147

Published by Garland Science
Taylor & Francis Group, LLC, an informa business
270 Madison Avenue, New York NY 10016, USA
and 2 Park Square, Milton Park, Abingdon, OX14 4RN, UK.

Printed in the United States of America
15 14 13 12 11 10 9 8 7 6 5 4 3 2 1

Visit our web site at http://www.garlandscience.com

Alison M. Smith is a project leader at the John Innes Centre, and an honorary Professor at the University of East Anglia. She trained in botany and ecology before specializing in plant biochemistry. Her research is on primary carbohydrate metabolism in plants, particularly starch synthesis and degradation and sugar catabolism.

George Coupland trained in Microbiology and Molecular Biology at the Universities of Glasgow, Edinburgh, and Cologne. He was a project leader at the John Innes Centre from 1989 to 2001 and since then has been a Director of the Max Planck Institute for Plant Breeding in Cologne. His research group focuses on the transition from vegetative growth to flowering, particularly in response to seasonal changes in the environment.

Liam Dolan joined the John Innes Centre as a project leader in 1995. He obtained a BSc and MSc in Botany from University College Dublin and a PhD in Biology from the University of Pennsylvania. From October 2009 he is Sherardian Professor of Botany at Oxford University. His research group investigates cellular development and evolution.

Nicholas Harberd is Sibthorpian Professor of Plant Science and Head of Research in the Department of Plant Sciences, and a Fellow of St. John's College, at the University of Oxford. Formerly, he was a project leader at the John Innes Centre, Norwich, UK, and Honorary Professor at the University of East Anglia, Norwich. His research and teaching focuses on the genetics, development, and evolution of plants, and he is the author of the popular science book *Seed to Seed*.

Jonathan Jones was a PhD student in cereal cytogenetics with Dick Flavell at the Plant Breeding Institute in Cambridge, studied symbiotic nitrogen fixation as a postdoc with Fred Ausubel at Harvard, and spent five years with a small agbiotech company. He joined the Sainsbury Laboratory at the John Innes Centre in 1988, and uses molecular genetics and genomics to investigate plant/pathogen interactions.

Cathie Martin joined the John Innes Centre as a project leader in 1983 and has focused her research on cellular specialization in plants including the control of cellular morphogenesis and metabolic specialization, in particular with respect to pollinator attraction. She is currently Editor-in-Chief of The Plant Cell. She holds a Niels Bohr Visiting Professorship at the faculty of Life Science, University of Copenhagen and is Professor at the University of East Anglia.

Robert Sablowski trained in Biology, Biochemistry, and Developmental Genetics in Brazil (University of Campinas and University of Sao Paulo), at the John Innes Centre, and at Caltech. He has been a group leader at the John Innes Centre since 1999, working on meristem and flower development in Arabidopsis, and has been a Visiting Professor at Leeds University (UK) since 2007.

Abigail Amey is a freelance science editor based in London. She received her degree in Biological Sciences from the University of Birmingham, UK, and her PhD in plant biochemistry from Durham University, UK.

PREFACE

The sunlight that is captured by photosynthetic organisms is the energy source that ultimately fuels almost all of the life on earth. Plants are by far the predominant land-based photosynthetic organisms, and provide the energy that supports almost all of the earth's ecosystems. An understanding of the biology of plants is therefore one of the most important goals of contemporary science, a goal that becomes increasingly pressing as human-generated changes in the earth's environment threaten ecosystem stability.

Plant Biology is a 'where we are now' account of plant science. It acknowledges the distinguished history of the subject, but its approach is strongly influenced by the radically new outlook that has emerged in the last twenty years. Historically, the growth, development, metabolism, and environmental responses of plants were largely considered at biochemical, cellular, and whole organismal levels. But then, in the early 1980s, came the first of two waves of change that were to completely alter the way in which plant biology was thought about.

The first of these waves was the realization that genetics and molecular genetics could be used to shed light on general plant biology. Of course, the genetics of plants itself has a venerable history. The discoveries of the gene as the unit of heredity and of transposable DNA elements were plant-based discoveries. What was different about this first wave of change was the realization that genetics and molecular genetics could be used as tools to understand aspects of plant biology that had not previously been considered as part of the territory of plant genetics. Growth, metabolism, development, and many other areas of plant biology began to be analyzed by studying genetically variant (mutant) plants that carried mutant genes which altered normal processes.

And then there followed a second wave of change. The initiation of this second wave can perhaps be pinpointed more precisely than can the first, to the very end of 1999, when the first complete DNA sequence of a plant genome was published. Since then, many more complete plant genome sequences have been determined. We are still living through the 'genome' wave, and its consequences are still being played out. But what is clear is that these two waves of change have completely transformed the way we think about the biology of plants.

We wrote *Plant Biology* because we felt the need of a textbook that reflected these changes in outlook, and the excitement that they have generated. The book begins with a summary of what is known of the origins of modern-day plants: of how the ancestors of the land-plants are thought to have been aquatic algal species, of the conquest of the land, and of how the flowering plants (the angiosperms) came to dominate terrestrial vegetation. Next, given the importance of genetics to the rest of the book, we cover the special features of plant genomes and genetics. Subsequent chapters then provide summaries of current understanding of plant cell biology, plant metabolism, and plant developmental biology. The remaining three chapters outline the interactions of plants with their environments: how plants modulate their growth in response to environmental variables, how plants cope with stressful environments, how plants interact with other organisms. The final chapter outlines the relationship of plants with humans: domestication, agriculture, and crop breeding.

What we know now would have been inconceivable only a few years ago. To give just one example, we now understand how relative spatial distributions and interactions of specific gene transcription factors enable the construction of a flower. Many of the recent advances in plant biology we describe are useful in the understanding of modern biology as a whole.

The highly visual illustration program and features such as learning goals, chapter summaries, enrichment boxes, and suggestions for further reading, help to strengthen the reader's understanding of essential concepts. Instructors who adopt *Plant Biology* for their course will have access to Garland Science Classwire™. The Classwire course management system allows instructors to build websites for their courses easily. It also serves as an online archive for instructors' resources. After registering for Classwire, instructors will be able to download all the figures from *Plant Biology*, which are available in JPEG and PowerPoint formats. Please visit the Garland Science website at www.garlandscience.com or e-mail science@garland.com for additional information on Classwire.

Plant Biology would not have been possible without the work and commitment of many people. We are indebted to our reviewers, whose constructive and insightful feedback was invaluable to us. They are listed below. Special thanks are due to Anil Day, Rob Martiennsen, and Graham Moore for individual contributions to the text. Needless to say, any mistakes that remain are our fault and not theirs. Nigel Orme's outstanding illustrations immeasurably enhance the clarity and quality of our text, and are complemented by some wonderful photographs by Tobias Kieser. We thank Linda Strange for so skilfully applying her editorial precision to the clarification of our text. Keith Roberts provided much needed advice, encouragement, and quality control at various stages in the project. We thank Dick Flavell for his considerable input into the early stages of the development of this book, and Chris Lamb and the John Innes Centre for supporting the project. Finally we thank all at Garland (past and present), especially Matt Day for coordinating the project through its early phases, Dominic Holdsworth for steering it (almost) to completion, Liz Owen, Simon Hill,

Georgina Lucas, and Helen Powis for the final editorial and production phases, and Denise Schanck for being constantly supportive of the project throughout.

We hope you enjoy reading this book!

Alison Smith
George Coupland
Liam Dolan
Nicholas Harberd
Jonathan Jones
Cathie Martin
Robert Sablowski
Abigail Amey

Acknowledgments

In writing this book we have benefited from the advice of many plant biologists. We would like to thank them for their suggestions.

Richard Amasino, University of Wisconsin-Madison
Dorothea Bartels, University of Bonn
David Baulcombe, University of Cambridge
Andrew Bent, University of Wisconsin-Madison
Frederic Berger, Temasek Life Sciences Laboratory, Singapore
Hans Bohnert, University of Illinois
Terry Brown, University of Manchester
Maarten J. Chrispeels, University of California, San Diego
Jeff Dangl, University of North Carolina
Anil Day, University of Manchester
David T. Dennis, Performance Plants Inc.
Allan Downie, John Innes Centre, Norwich
Jeff Ellis, CSIRO Plant Industry, Canberra
Noel Ellis, John Innes Centre, Norwich
Robert Furbank, CSIRO Plant Industry, Canberra
Jeremy Harbinson, Wageningen University
Patrick Hayes, Oregon State University
Elizabeth A. Kellogg, University of Missouri, St. Louis

Paul Kenrick, The Natural History Museum
Ross E. Koning, Eastern Connecticut State University
Jane Langdale, Oxford University
Ottoline Leyser, University of York
Chentao Lin, University of California, Los Angeles
Enrique Lopez-Juez, Royal Holloway, University of London
John Mansfield, Imperial College
Ron Martiennsen, Cold Spring Harbor Laboratory
Graham Moore, John Innes Centre, Norwich
Andy Maule, John Innes Centre, Norwich
Timothy Nelson, Yale University
T. Kaye Peterman, Wellesley College
Eric J. Richards, Boyce Thompson Institute for Plant Research
Fred Sack, University of British Columbia
Peter Shaw, John Innes Centre, Norwich
Jonathan Walton, Michigan State University
Gary Whitelam, Leicester University

CONTENTS IN BRIEF

CONTENTS

ORIGINS

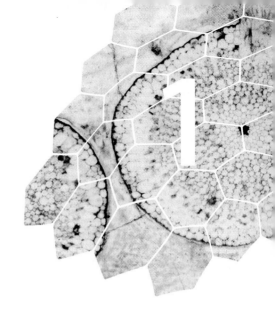

When you have read Chapter 1, you should be able to:

- Describe the atmosphere of the early (abiotic) earth and how it changed with the evolution of oxygen-producing photosynthesis and colonization of the continents by plants.

- Outline the types of direct and indirect evidence used to estimate the timing of important events in plant evolution.

- Describe the role of endosymbiotic events in the formation of the first eukaryotic cells and the first photosynthetic eukaryotes.

- Discuss the transition of plants from water to land, the evidence for descent of land plants from algae, and the features that distinguish plants from their algal ancestor.

- Outline the evolution of the seed plants and list their unique structural features.

- Outline the different types of life cycles in the land plants and give examples of plants using each type.

- Summarize the distinguishing features of the angiosperms and their divergence from the other plant groups.

- Distinguish between the major groups of angiosperms and summarize their phylogenetic relationships.

In the **leaves** of a green plant, a remarkable thing happens: a small, chemical event, but one that opens a gateway from the inorganic to the organic world. The interaction of sunlight with pigments in a leaf cell provides energy that drives the reaction of a small molecule in that cell with carbon dioxide in the air—the process of **photosynthesis**. The organic compound produced by this "fixation" of inorganic carbon becomes the source of all the organic matter of plants and the organisms that feed on them. Plants are not the only organisms that fix carbon—some bacteria do so, too—but our main interest in this book is the plants: their structures and biochemical processes, their growth and development, their diversity, their responses to a changing environment, their interactions with each other and with other types of organisms, and their use and manipulation by humans.

CHAPTER SUMMARY

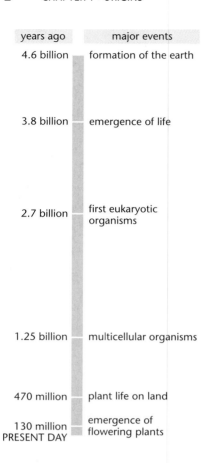

years ago	major events
4.6 billion	formation of the earth
3.8 billion	emergence of life
2.7 billion	first eukaryotic organisms
1.25 billion	multicellular organisms
470 million	plant life on land
130 million PRESENT DAY	emergence of flowering plants

Figure 1–1.
Major events described in this chapter, from formation of the earth 4.6 billion years ago to the present day. More detailed timelines of evolutionary events are provided in Figures 1–2 and 1–11.

A study of the biology of plants must begin with history: the origins and evolution of the ancestors of modern-day plants. Only by learning as much as possible about this history and by studying the relationships among the plants now living—from the macroscopic to the microscopic to the molecular level, using all the tools now at our disposal—can we fully appreciate and understand the diversity of plant life on our planet.

We begin, then, with some of the key events in the evolution of organisms that carry out photosynthesis, with emphasis on the land plants. We describe the development of plant life in the context of the earth system: the web of physical, chemical, and biological interactions involving the earth, the **hydrosphere**, and the **atmosphere**. We start this chapter with the formation of the earth and end with the diversification of the flowering plants. The major events we describe, and the times at which they occurred, are shown in Figure 1–1.

The evolution of photosynthetic organisms had dramatic effects on the earth system. It contributed to a decrease in the level of atmospheric carbon dioxide (CO_2) through increasing rates of **carbon burial** and, later, contributed to the development of an oxidizing atmosphere resulting from the production of oxygen (O_2) by oxygenic photosynthesis. The earliest species were unicellular, but by 1.25 billion (1.25×10^9) years ago, organisms composed of more than one cell type had appeared in the oceans. These earliest multicellular organisms were **algae** (seaweeds), and these fossil algae provide the first evidence of **sexual reproduction**.

Then, around 470 million years ago, plant life was established on the land and the first terrestrial ecosystems formed. This was followed by a pivotal time in the history of plants: the "Devonian explosion." The emergence of simple terrestrial plant forms between 470 and 405 million years ago led to a massive **radiation** of plants—an increase in diversity and geographic spread—over the next 50 million years. In this short, 50-million-year period, many of the plant groups that were to live on land evolved. Over the following 250 million years, the diversity of these groups increased, and then dramatic events, occurring around 130 million years ago, gave rise to the flowering plants—and a further radiation followed. Today, flowering plants dominate all continental surfaces where plants can survive.

This introductory chapter, then, places the flowering plants—whose biology is the major focus of the genetics, molecular biology, and biochemistry described in this book—in their earth system context. Our aim here is to provide a general outline of the major events on the journey from the solar nebula to the complex terrestrial ecosystems of today, as they relate to plant life. We do not intend to be comprehensive. Other textbooks deal with the details of the genesis of the earth, the functioning of the earth system, plant evolution, plant systematics, and environmental science.

1.1 EARTH, CELLS, AND PHOTOSYNTHESIS

We begin with the formation of the earth system and the first cells. This early history shows how evolving life was, of course, dependent on the conditions on the young earth, but also how, with the evolution of photosynthesis, these living organisms came to change those conditions and thus affect all subsequent evolution.

The earth formed 4.6 billion years ago

Our sun formed approximately 5 billion years ago in a rotating disk of dust and gas called a "solar nebula." The solar nebula formed as interstellar dust and gas condensed into a spinning disk, hottest at its center and coolest at its edges. The sun developed at the hot center, and the remaining materials gradually coalesced into planets and asteroids. The planets nearest the sun—Mercury, Venus, Earth, and Mars—formed from the aggregation of rocky material composed largely of metal silicates. Farther from the sun, the aggregation of gases and ice developed into the gas-giant planets: Jupiter, Saturn, Uranus, and Neptune. The silicate-rich origins of the earth are reflected in the chemical composition of its soils, which are complex matrices containing sand (quartz, SiO_2) and clays (plate silicates) combined with organic matter and water. Thus the chemistry of silicates that condensed from the solar nebula is one of the important determinants of soil characteristics on our planet.

The formation of the planets was a gradual process. By 400 million years after the formation of the sun (i.e., 4.6 billion years ago), the earth existed in a primitive form. It continued to be bombarded by meteors and comets for a further 800 million years; during this period, the layered structure of the planet developed: an inner core, outer core, mantle, and crust. The continuous bombardment delivered much of the water that accumulated on the early earth, kept temperatures high, and caused extensive volcanic activity. The high temperatures "sterilized" the environment and probably wiped out any early manifestation of life. The bombardment stage of the earth's development, known as the Hadean Era (Figure 1–2), ended 3.8 billion years ago. With the close of the Hadean, the most tumultuous stage of the earth's history ended, and this is thought to be the time when primitive life emerged.

The atmosphere of the Late Hadean earth consisted of gases released by volcanic activity; these would have included large amounts of CO_2. Water vapor (H_2O) was also present in

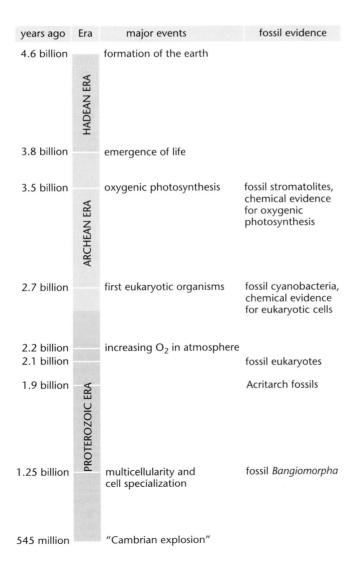

years ago	Era	major events	fossil evidence
4.6 billion	HADEAN ERA	formation of the earth	
3.8 billion		emergence of life	
3.5 billion	ARCHEAN ERA	oxygenic photosynthesis	fossil stromatolites, chemical evidence for oxygenic photosynthesis
2.7 billion		first eukaryotic organisms	fossil cyanobacteria, chemical evidence for eukaryotic cells
2.2 billion		increasing O_2 in atmosphere	
2.1 billion			fossil eukaryotes
1.9 billion	PROTEROZOIC ERA		Acritarch fossils
1.25 billion		multicellularity and cell specialization	fossil *Bangiomorpha*
545 million		"Cambrian explosion"	

Figure 1–2.
Major evolutionary events from formation of the earth until the "Cambrian explosion" of diversity about 0.5 billion years ago. The three eras that encompass this period—the Hadean, Archean, and Proterozoic—are collectively known as the Cryptozoic Eon. Major evolutionary events in the period from 0.5 billion years ago to the present day—the Phanerozoic Eon—are shown in Figure 1–11.

the atmosphere, as oceans had formed from a combination of water from the initial aggregation of the planet and water delivered by comet bombardment. The early atmosphere was very different from that of today. No oxygen gas was present, and there was no outermost protective ozone (O_3) layer. Over the next 2 billion years, two events—the evolution of oxygenic photosynthesis and the gradual oxygenation of the oceans—would result in accumulation of small amounts of oxygen in the atmosphere.

Photosynthesis evolved by approximately 3.5 billion years ago

The earliest fossilized cells are found in rocks laid down in the Archean Era (3.8–2.5 billion years ago; see Figure 1–2), indicating that life had evolved by 3.5 billion years ago. The fossils contain petrified bacteria in **stromatolites** (Figure 1–3). Modern stromatolites are mound-shaped structures that build up in warm, shallow seawater through the activity of photosynthetic bacteria known as **cyanobacteria**. The bacteria live in a gelatinous mat, over which a layer of calcium carbonate precipitates. As the calcium carbonate layer thickens, bacteria migrate to the upper surface of the layer and produce a further gelatinous mat. The cycle repeats, resulting in mounds of alternating layers of gelatinous substance and calcium carbonate. Over time, these structures are petrified to form a type of limestone, thus preserving the enclosed bacteria. The presence of stromatolite fossils suggests not only that life had evolved by 3.5 billion years ago, but also that these early bacterial cells were light-sensitive and probably carried out a form of photosynthesis.

While fossilized organisms provide visual evidence for the flourishing of life 3.5 billion years ago, chemical and isotopic information from rocks provides indirect evidence for the existence of life even earlier. Isotopic data indicate that photosynthetic life may have existed as early as 3.8 billion years ago. If this is true, then life first arose at the end of the Hadean Era.

Carbon exists as several different isotopes, varying in the number of neutrons in the **nucleus.** In the modern atmosphere, about 99% of the carbon in CO_2 is ^{12}C and 1% is the heavier ^{13}C isotope. Both are stable over geologic time—that is, they do not undergo radioactive decay to other isotopes. During photosynthesis, plants assimilate carbon from atmospheric CO_2 into organic compounds. The **enzyme** that catalyzes addition of atmospheric carbon to acceptor molecules in the plant cell has a slight preference for the ^{12}C isotope over the ^{13}C isotope. Therefore, carbon-containing material derived from photosynthetic cells has a lower ratio of ^{13}C to ^{12}C than inorganic carbon-containing material derived directly from atmospheric CO_2. By measuring the ^{13}C content of carbon-containing material (Figure 1–4), it is possible to determine whether the material was derived from living, photosynthetic cells or from precipitation of abiotic (nonliving)

Figure 1–3.
Modern and fossil stromatolites.
(A) Modern stromatolites in Shark Bay, on the west coast of Australia. (B) Section through a fossil stromatolite. This specimen is from rocks about 2.4 billion years old, in Argentina. The alternating layers look very similar to the alternating layers of calcium carbonate and gelatinous material found in modern-day stromatolites.

(A)

(B)

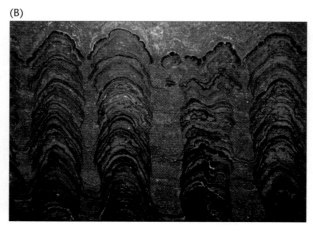

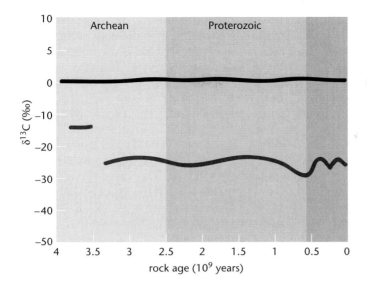

Figure 1–4.
The ratio of ^{12}C to ^{13}C in carbon-containing rocks. In this diagram, the amount of ^{13}C in samples is expressed relative to the amount in a standard rock type: a number known as a $\delta^{13}C$ value. For example, if the proportion of ^{13}C in the sample rock is 3% (i.e., 30 parts per thousand, or 30‰) lower than the proportion in the standard, the sample rock is said to have a $\delta^{13}C$ value of −30‰. The $\delta^{13}C$ value of most plants today is about −27‰. The *black line* shows the $\delta^{13}C$ value of inorganic carbon deposits (e.g., carbonate-containing rocks such as limestones); these values have not changed over geologic time. The $\delta^{13}C$ values of organic carbon-containing deposits, shown by the *red line*, are more negative than those of inorganic carbon deposits from about 3.8 billion years ago, and they drop further at about 3.5 billion years ago. This is consistent with the idea that organic carbon compounds were being formed by biochemical reactions, in living organisms, from about 3.8 billion years ago.

sources (inorganic carbonates precipitate continuously in the oceans, and some organic compounds can be formed by chemical processes rather than biochemical ones).

A low ^{13}C:^{12}C ratio has been found in rocks from the Isua peninsula of Greenland that are 3.8 billion years old; these are among the oldest known rocks on earth. This finding suggests that the carbon in the rocks is derived from photosynthetic organisms, hence that photosynthesis had evolved by 3.8 billion years ago. From what we know about the earth's atmosphere at that time, it is likely that this early photosynthesis was not oxygenic. In the vast majority of photosynthetic organisms on earth today, the process of photosynthesis uses electrons derived from water, and as a consequence generates oxygen—this is referred to as oxygenic photosynthesis (see Section 4.2 for details of the mechanism of photosynthesis). However, photosynthesis in some single-celled organisms uses electrons from compounds other than water. The product is then an oxidized compound other than oxygen. Photosynthetic organisms that evolved 3.8 billion years ago probably fall into this category. They may, for example, have derived electrons from hydrogen sulfide (H_2S), producing sulfur-containing compounds rather than oxygen. Thus although photosynthesis had evolved by 3.8 billion years ago, it did not yet produce oxygen. Oxygenic photosynthesis was a later evolutionary development.

Oxygen-producing photosynthesis was widespread by 2.2 billion years ago

Analyses of 2.2-billion-year-old rocks have provided two main lines of evidence for the timescale of the evolution of oxygen-producing photosynthesis. The first line of evidence derives from a process called "mineral weathering"; this takes place when iron-containing minerals in rock react with atmospheric oxygen. In today's oxygen-rich atmosphere, soluble ferrous iron (Fe^{2+}) in newly exposed rock is oxidized to the insoluble ferric form (Fe^{3+}). This insoluble form remains where it precipitates and is not carried away by water, so soils that develop from the weathering of iron-containing rocks in an oxidizing atmosphere are relatively iron-rich. In contrast, in an atmosphere that lacks oxygen, exposed ferrous iron is not oxidized; the soluble iron is carried away by water and the resulting soils are relatively iron-poor. Notably, fossilized soils (**paleosols**) that are more than 2.2 billion years old are iron-poor, indicating that they were formed in an atmosphere lacking significant levels of oxygen. Paleosols dating from around 2.2 billion years ago are relatively iron-rich, indicating that they were formed in a more oxygen-rich, oxidizing atmosphere. This evidence suggests that oxygen appeared at significant levels in earth's atmosphere around 2.2 billion years ago. Development of this oxidizing atmosphere resulted from the accumulation of O_2 produced by oxygenic photosynthesis.

The second line of chemical evidence for a change in oxygen levels in the atmosphere comes from observed changes in carbon burial over time. Carbon burial takes place when the remains of organisms form sediments that become locked into rock. When this happens, the rate of incorporation of atmospheric carbon into living organisms (by photosynthesis) is different from the rate of loss of carbon from the biosphere back into the atmosphere (during the process of **respiration**). As a result, the global rates of photosynthesis (generating oxygen) and respiration (consuming oxygen) are not in equilibrium. This results in an increase of oxygen in the atmosphere; we explain the process in further detail later in Section 1.3. Examination of rocks formed around 2.2 billion years ago has shown an increase at that time in the amount of buried carbon derived from living organisms (**biomass**). This suggests that oxygen levels in the atmosphere were increasing.

Thus, together, evidence from mineral weathering and carbon burial indicates that oxygen levels had started to increase in the atmosphere by 2.2 billion years ago. Given the great amount of time needed for gas production by bacteria to be reflected in global atmospheric gas composition, the oxygen-producing organisms must have been in existence for a considerable time before significant levels of oxygen appeared in the atmosphere.

Photosynthetic cyanobacteria produced an oxygen-rich atmosphere

The first known organisms to carry out oxygenic photosynthesis were cyanobacteria, the organisms thought to have formed stromatolites from about 3.5 billion years ago. The first definitive evidence for the existence of cyanobacteria has been obtained by chemical techniques in younger rocks, 2.7 billion years old. Cyanobacteria synthesize characteristic compounds, the 2-methylbacteriohopanepolyols, which are not found in any other organism. In sediments, these **polyols** are transformed to 2-methylhopanes. The identification of 2-methylhopanes in 2.7-billion-year-old rocks provides evidence for the existence of cyanobacteria before the observed increase in atmospheric oxygen, 2.2 billion years ago. And, given that cyanobacteria carry out oxygenic photosynthesis, it is likely that they were responsible for the production of the oxygen that was in turn responsible for the formation of the oxidizing atmosphere.

If cyanobacteria existed 2.7 billion years ago and were producing oxygen, why did it take a further 500 million years for the atmosphere to become oxidizing? As noted above, oxygen produced by oxygenic photosynthesis can react with ferrous iron and other elements to form precipitates (such as insoluble ferric oxide). In the oceans, oxygen produced by photosynthetic organisms would have reacted with ferrous iron produced by mineral weathering on the land. This process would prevent release of the oxygen into the atmosphere, so delaying formation of an oxygen-rich atmosphere. The resulting iron-containing precipitates would have petrified over time to form iron-rich rocks on the floors of the oceans. Rocks of this type have been found in sediments laid down between 3.5 and 1.9 billion years ago. This suggests that oxygen was being produced in large amounts during this period. Thus, although the first definitive evidence for the existence of cyanobacteria-like organisms is from sediments 2.7 billion years old, the existence of older iron-rich rocks indicates that oxygen-producing bacteria had existed before this time. Perhaps photosynthetic bacteria were carrying out oxygenic photosynthesis as long as 3.5 billion years ago.

Early life on earth evolved in the absence of a protective atmospheric ozone layer

The formation of an oxygen-rich atmosphere opened the way for the formation of atmospheric ozone. Today, ozone accounts for only 0.00001% of the volume of gas in the atmosphere, and it accumulates high in the stratosphere (in comparison, oxygen constitutes almost 21% of the atmospheric volume). Ozone is important, despite its low abundance, because it absorbs 99% of the incoming ultraviolet (UV) radiation between the wavelengths of 190 and 310 nm. UV radiation is very damaging to living organisms. **DNA (deoxyribonucleic acid)** absorbs UV in the region around 260 nm and, as a result, can undergo chemical changes, or **mutations**. High mutation rates are detrimental for life, and the ozone in today's atmosphere greatly reduces organisms' exposure to DNA-damaging UV radiation (Figure 1–5).

Given that the first life evolved around 3.8 billion years ago and an oxygenated atmosphere did not form until 2.2 billion years ago, life must have existed on earth for a considerable time without the protective effect of an ozone layer. Early life most likely avoided exposure to UV or had protective mechanisms that reduced DNA damage. Some protective mechanisms found in modern bacteria include an extracellular mucilaginous sheath that absorbs UV radiation, gliding movements that allow bacterial cells to escape intense irradiation, and UV-inducible biochemical systems that repair damage to DNA. Analysis of ancient stromatolites suggests that some of the oldest photosynthetic bacteria lived in a UV-shielding mucilaginous milieu (see Figure 1–3). Furthermore, almost all early life was marine, and the ocean environment gave some protection from UV.

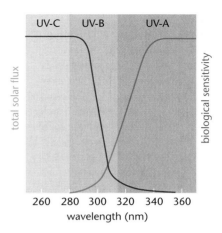

Figure 1–5.
Ultraviolet radiation. The graph shows the flux of radiation of different wavelengths in the ultraviolet range reaching the earth's surface (*blue line*) and the sensitivity of organisms to DNA damage by this radiation (*red line*). The absence of flux below about 300 nm is due to the absorption of UV light by ozone (O_3). These wavelengths are shorter than those visible to the human eye (see Chapter 4, Box 4–1).

1.2 EUKARYOTIC CELLS

The first organisms that flourished on earth were **prokaryotes**. Today, prokaryotes are represented by two broad groups of single-celled organisms, the archaea and the bacteria. Other extant organisms are **eukaryotes**. Prokaryotic organisms are single-celled and have no internal membrane-bounded compartments. Eukaryotic cells are much more complex in structure. They contain several types of internal membrane-bounded compartments (called **organelles**), including **mitochondria**, a **nucleus**, and, in plants and a few other groups, **plastids** (a **chloroplast** is a type of plastid). The biology of these organelles is described in subsequent chapters; here we describe how and when the evolution of eukaryotic from prokaryotic cell types is thought to have occurred, in the context of the evolution of plants.

The mitochondria and plastids of eukaryotic cells arose as a result of the uptake of one single-celled, prokaryotic organism by another and retention of the engulfed organism within the host cell's **cytoplasm**. The symbiotic incorporation of one organism by another is known as an **endosymbiotic** event; after it has taken place, each partner in the **symbiotic interaction** contributes to the survival of the new organism.

Photosynthetic eukaryotic cells arose from two endosymbiotic events

The evolutionary history of plant cells involves at least two independent endosymbiotic events. The first involved the uptake of an organism (referred to as an "α-proteobacterium" into a host cell (Figure 1–6A). The nature of the host cell is not known, but it may have been related to today's archaea. Neither the engulfed organism nor its host were photosynthetic: both they and the new organism created by the endosymbiotic event were **heterotrophic** (that is, they relied for food on organic carbon produced by photosynthetic organisms). The cell resulting from this original endosymbiotic event—the **proto-eukaryotic cell**—gave rise to all major groups of eukaryotic organisms. The symbiont gave rise to the mitochondrion of the eukaryotic cell. Over evolutionary time, most of its **genes** transferred to the **genome** of the host cell, which became enclosed in a membrane to form the nucleus of the cell. In modern eukaryotic cells, the organization of the genes remaining in the mitochondria (the **mitochondrial genome**) resembles that of bacteria (see Section 2.6), while the **nuclear genome** has characteristics that indicate a joint archaeal and bacterial heritage. Formation of the proto-eukaryotic cell was followed by an increase in the number of eukaryotic species, resulting in the diverse morphologies and life histories that we see in the fossil record and around us today.

Figure 1–6.
Events giving rise to photosynthetic eukaryotic organisms. (A) The first endosymbiotic event: engulfment of a bacterium (the α-proteobacterium) by another prokaryotic host cell. The resulting proto-eukaryotic cell underwent transfer of genes from the genome of the engulfed cell to the host-cell genome and conversion of the engulfed cell to a specialized subcellular compartment (organelle)—the mitochondrion. The host-cell genome became enclosed in a membrane, giving rise to the nucleus. (B) The second endosymbiotic event: a eukaryotic cell engulfed a prokaryotic, photosynthetic cyanobacterium. Transfer of genes from the engulfed cell to the host nucleus then took place, together with conversion of the engulfed cell to a specialized organelle—the chloroplast (plastid). (C) The original photosynthetic eukaryotic organism gave rise to three clades of organisms present today: glaucophytes, red algae, and green algae. The glaucophytes are unicellular algae. Their chloroplasts are unlike those of other extant photosynthetic organisms in that they retain a peptidoglycan outer layer. The organism illustrated is *Cyanophora paradoxa*. The red algae occur in both unicellular and multicellular forms. Shown here are (*top*) a unicellular, colonial, freshwater species of the genus *Porphyridium* and (*bottom*) the marine, multicellular species *Rhodymenia palmata*. The green algae also occur in both unicellular and multicellular forms. Shown here is a member of the unicellular freshwater genus *Chlorella*.

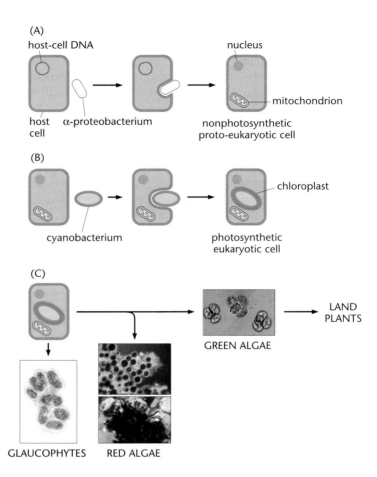

Some groups of eukaryotes later acquired another organelle, the plastid, through a second endosymbiotic event involving the uptake of a photosynthetic cyanobacterium by a mitochondria-containing eukaryote to form an **autotrophic** cell—an organism capable of generating all of its organic carbon from CO_2 through photosynthesis (Figure 1–6B). The symbiotic cyanobacterium gave rise to the plastid. Many of the cyanobacterial genes transferred to the genome of the host cell. The genes that remained—the **plastid genome**—are similar in structure and organization to those of cyanobacteria (see Section 2.6). The nuclear genome of plastid-containing cells is **chimeric**, containing both the elements of the proto-eukaryotic genome and elements derived from the cyanobacterial genome.

Several groups of photosynthetic organisms are derived from the endosymbiotic event that gave rise to plastids

As a result of their endosymbiotic origin, plastids (like mitochondria) are bounded by two membranes: an inner and outer membrane that are derived from the membranes of the ancestral cyanobacterium and the engulfing host cell, respectively. Three **clades** (groups of organisms derived from a single common ancestor) are directly derived from the photosynthetic eukaryotic cell formed by the two endosymbiotic events described above: the glaucophytes, the rhodophytes (red algae), and the chlorophytes (green algae) that are ancestral to today's land plants (Figure 1–6C). Glaucophytes are a small group of freshwater organisms with a unique plastid that has a morphology intermediate between that of a cyanobacterial cell and a plastid of the green or red algae. The glaucophyte plastid retains an outer **peptidoglycan** layer similar to that in the envelope of

(A)

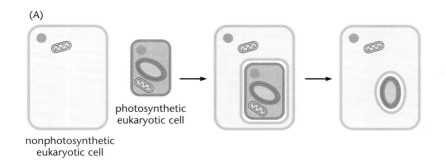

nonphotosynthetic
eukaryotic cell

photosynthetic
eukaryotic cell

(B)

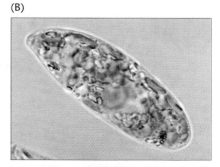

Figure 1–7.
Secondary endosymbiosis. (A) In a secondary endosymbiotic event, an alga—a photosynthetic eukaryotic organism—was engulfed by a nonphotosynthetic host cell. Over evolutionary time, most of the functions of the engulfed cell, apart from its chloroplast, were lost. The chloroplast retained the extra external membranes acquired during engulfment of the endosymbiont. (B) The unicellular euglenophyte *Euglena viridis* (*top*) is a highly motile organism that swims by movement of a whiplike flagellum (not seen). Euglena cells range from 25 to 250 μm in length. The brown alga (*bottom*) is a giant kelp, growing off the coast of California.

cyanobacteria; this layer is missing in plastids of all other groups. Glaucophyte plastids also have other features characteristic of cyanobacteria and absent from the plastids of the green algae and their descendants, including the presence of pigments called **phycobilins** and small particles called **phycobilisomes** on the face of their **thylakoid** membranes. Red algae are microscopic and macroscopic algae found in both freshwater and marine environments. Like glaucophytes, their plastids contain phycobilins and phycobilisomes.

Green algae are microscopic and macroscopic algae found in both freshwater and marine environments. In addition to chlorophyll *a*, the form of **chlorophyll** found in cyanobacteria, glaucophytes, and red algae, green algae also contain a second form of chlorophyll, chlorophyll *b*. The green algae comprise several groups, including the chlorophytes, such as the unicellular alga *Chlorella* and the multicellular seaweed *Ulva* (sea lettuce), and the charophytes, such as the freshwater alga *Chara*. The charophytes are a sister group to the land plants—that is, the charophytes and land plants are derived from a single common ancestor.

Other groups of organisms evolved by further endosymbiotic events in which single-celled species of plastid-containing eukaryotes became endosymbionts in other eukaryotes that lacked plastids (Figure 1–7A). These secondary endosymbiotic events gave rise to several groups of extant photosynthetic organisms. For example, euglenophytes were formed in a secondary endosymbiotic event involving a green alga, and the brown algae (including most of the largest seaweeds) were formed in a secondary event involving a red alga (Figure 1–7B).

Fossil evidence indicates that eukaryotic organisms had evolved by 2.7 billion years ago and multicellular organisms by 1.25 billion years ago

Eukaryotic cells are believed to have come into existence by 2.7 billion years ago. In addition to the possession of internal organelles, eukaryotic cells can be distinguished from prokaryotic cells (archaea and bacteria) by some chemical features. For example, no known prokaryotes synthesize **sterols** such as cholesterol (although some do synthesize simpler, related molecules). Thus sterols can be used as a **molecular marker** specific for eukaryotes. When the remains of a eukaryotic organism are incorporated in sediments, its sterols may be preserved in modified forms, collectively known as "steranes." The oldest rocks in which steranes have been identified are 2.7-billion-year-old rocks from Australia.

The morphologies of these first eukaryotes remain a mystery. Rocks of this great age are rare on earth, and the minerals they contain have usually undergone chemical and physical changes that destroy fossil structure. The oldest fossils that do preserve the morphology of a probable eukaryote —*Grypania spiralis*, found in 2.1-billion-year-old rocks—suggest that some of the earliest eukaryotic organisms resembled algae. On the basis of its relatively large size (2 mm in diameter), *Grypania* is thought to have been both photosynthetic and eukaryotic. Size is often used to distinguish between fossils of prokaryotes and eukaryotes, as most prokaryotic cells are much smaller than eukaryotic cells.

After their first appearance, eukaryotic organisms underwent a slow increase in diversity. Single-celled, probably eukaryotic organisms called acritarchs are found in rocks 1.9 billion years old. They resemble in morphology much younger fossils that have been identified with certainty as eukaryotic, but their fossils have lost the subcellular detail that would reveal the presence of diagnostic organelles.

The earliest known acritarchs were spherical cells, 20 to 200 μm (20–200 × 10⁻⁶ m) in diameter. There was a gradual increase in diversity of eukaryotic cells between 1.9 and 1.25 billion years ago; later acritarchs were up to 2700 μm in diameter and morphologically more intricate, with complex surface decorations (Figure 1–8).

Although the morphologies of the acritarchs indicate that they were photosynthetic eukaryotes, it is not possible to place them in any group, or **taxon**, of extant eukaryotes. The first eukaryote that can be classified in an extant taxon is *Bangiomorpha pubescens*, a fossilized marine red alga preserved in 1.2-billion-year-old rock derived from intertidal sediments in Northern Canada. This species is remarkably similar to *Bangia atropurpurea* (Figure 1–9), an extant alga that grows at the waterline of ocean and freshwater environments worldwide. The two species look alike, and also lived or live in almost identical habitats. Thus we can conclude that eukaryotic organisms that have modern extant counterparts had evolved by 1.2 billion years ago.

Figure 1–8.
Acritarchs. Acritarchs are microfossils that increased in size and complexity during the Proterozoic Era. (A) A simple acritarch in a section of rock 1.5 billion years old. The cell surface is smooth and the cell is 120 μm in diameter. (B) A larger, more complex, spined acritarch microfossil with a diameter of more than 200 μm, in a section of a rock 580 million years old. (A and B, courtesy of Andrew Knoll.)

(A)

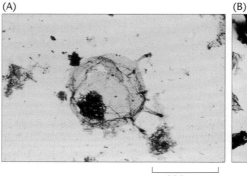

100 μm

(B)

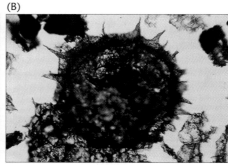

100 μm

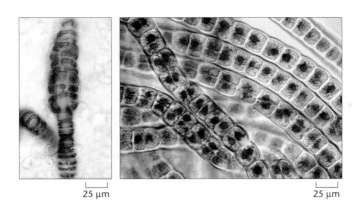

25 µm 25 µm

Figure 1–9.
Ancient and modern algae. A fossil of *Bangiomorpha pubescens* (*left*) and extant *Bangia atropurpurea* from Lake Balaton, Hungary (*right*), showing the similar morphologies of these algae. (Left-hand image courtesy of N. Butterfield; right-hand image courtesy of Lajos Vörös.)

Bangiomorpha shows the first evidence of multicellularity, cellular specialization, and sexual reproduction in a eukaryotic organism. Because of the remarkable state of preservation of these fossils, it has been possible to follow the development of the multicellular body of the alga. The **germination** of a **spore** gave rise to the multicellular body of the alga. Cells at the base of the developing "plant" formed a **holdfast**-like structure that rooted the organism to its substrate (sediment or rock). Cells at the other (upper) end differentiated into a filamentous body or **thallus** (the upper part of the organism in Figure 1–9, *left*). The *Bangiomorpha* body plan allowed the organism to grow vertically, up through the water column, and hence to capture light more effectively. Cells in the thallus differentiated to become **gametes** (sex cells). We describe the evolution of sexual life cycles later in this chapter.

The evolution of multicellularity was soon accompanied by the evolution of larger organisms. Between 1200 and 700 million years ago, many more large, multicellular organisms appear in the fossil record. All major groups of extant macroscopic algae had appeared by 600 million years ago. By 570 million years ago, bilateran (bilaterally symmetric) marine animals had appeared, large animals were soon to make their appearance, and plants were about to move onto the land.

Animals and algae diversified in the Early Cambrian Period

The rate of increase in diversity of living organisms accelerated rapidly immediately before and during the Cambrian Period, which followed the Proterozoic Era from around 550 million years ago. For photosynthetic organisms, this acceleration is well illustrated by the fossil record of the acritarchs. As noted above, the acritarchs diversified slowly between 1.9 and 1 billion years ago. Between 1000 and 600 million years ago, diversity increased more rapidly. Then, between about 600 and 500 million years ago, two periods of dramatic increase in acritarch diversity took place (Figure 1–10). These rapid changes in the diversity of the acritarch fossil record suggest that rates of evolution of new species of photosynthetic eukaryotes had accelerated.

Parallel increases in the rate of evolution of heterotrophic eukaryotic organisms are seen over the same period. There was a striking increase in animal diversity in the period immediately prior to the Cambrian, with the emergence of the soft-bodied animals known as the Vendian or Ediacaran fauna. During the Cambrian, there was a large increase in the numbers and morphologies of animal species with hard external skeletons. This increase in diversity is reflected by the sudden appearance of huge numbers of large, hard-shelled organisms in the fossil record. The increase in diversity of animals was so great that in this relatively short period of time, all but one of the animal phyla appeared. This dramatic phase of evolutionary history is known as the "Cambrian explosion."

It seems likely that environmental changes were at least in part responsible for the increase in diversity of eukaryotic groups immediately prior to and during the Cambrian

Figure 1–10.
The increase in number of acritarch species in rocks formed during the Proterozoic Era. As the histogram shows, the numbers of species of acritarchs were low until approximately 1000 million years ago, after which there were at least three explosions of species numbers, each followed by a decline.

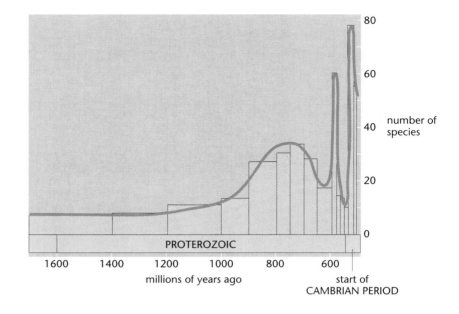

Period. For example, the evolution of the first large animals in the oceans around 570 million years ago is thought to have resulted partly from increasing atmospheric oxygen levels. Whatever the causes, after the Cambrian explosion the seas were filled with predatory animals and animals with armored defenses (shells and skeletons). There was also an abundance of photosynthetic eukaryotes, and one group of these was soon to colonize the wet flanks of the continental verges. The invasion of the land was about to begin.

1.3 LAND PLANTS

The fossil record indicates that plants were growing on the wet areas of the continental surfaces by 470 million years ago. They may have been present for a considerable time before this, but, if so, these early pioneers left no trace in the fossil record. Here we briefly chart some of the major evolutionary events in the history of the land plants. The timescales we refer to are shown in Figure 1–11.

Figure 1–11.
Timescales of major evolutionary events in the Phanerozoic Eon, from 545 million years ago to the present. This eon is divided into three eras, the Paleozoic, Mesozoic, and Cenozoic, which are further divided into several periods.

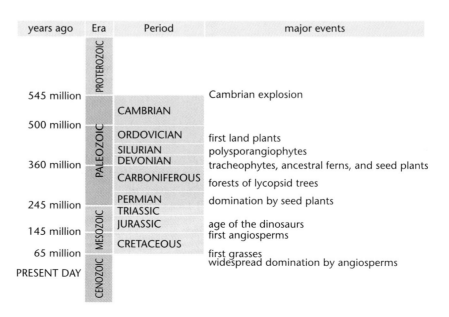

Green plants are monophyletic

The eukaryotic "tree of life" (Figure 1–12) shows that green plants are a **monophyletic group**—that is, they are derived from a single common ancestor. **Fungi** and animals also each constitute a monophyletic group.

Within the green plant–red algae–glaucophyte clade, each group is monophyletic. The green plants constitute two clades: the Chlorophyta (the chlorophytes), which includes algae such as *Chlorella*, *Chlamydomonas*, and *Ulva*; and the Streptophytina, which includes the charophytes and the land plants. Land plants are called **embryophytes**, because, in sexual reproduction, the products of **fertilization** develop as **embryos** inside protective structures. In all but the flowering land plants these structures are called **archegonia** (the flowering plants lost their archegonia in the course of evolution; their embryos develop within a different kind of protective structure; see Chapter 5).

Land plants may be descended from plants related to charophycean (charophyte) algae

Land plants probably evolved from a species of Charophyceae (charophytes), a small class of predominantly freshwater green algae. This is suggested by comparisons of DNA sequences of land plants and charophyte algae, which show that land plants and the Charales, an order within the Charophyceae, are derived from a single common ancestor. (Box 1–1 summarizes how DNA analysis can be used in studies of phylogenetic relationships.) The Charophyceae and the land plants also share some characteristics not found in other groups. These include the presence of **cellulose** in the **cell wall**; the formation of a **phragmoplast**, a structure that mediates cell separation after cell division (**mitosis**); and **plasmodesmata**, channels connecting cells (Figure 1–13).

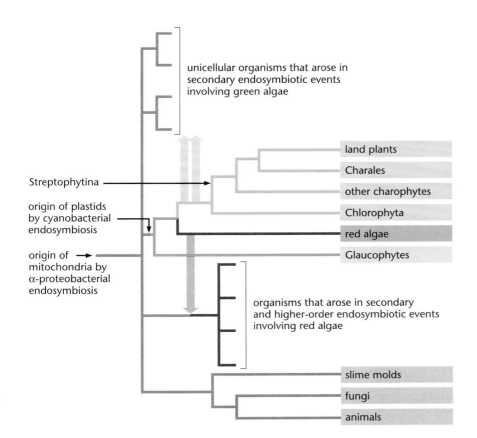

Figure 1–12.
Phylogenetic tree showing the relationships between the major groups of eukaryotic organisms. Together, the green algae, red algae, and glaucophytes form a monophyletic group derived from a common ancestor that acquired a plastid through the primary endosymbiotic event (see Figure 1–6). The *red* and *green arrows* indicate the stages in evolutionary history when red algae or green algae underwent further rounds of endosymbiosis in other eukaryotic hosts.

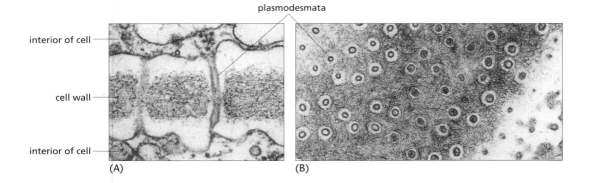

plasmodesmata

interior of cell

cell wall

interior of cell

(A)   (B)

Figure 1–13.
Plasmodesmata in *Chara fibrosa*, an extant alga. Plasmodesmata are membrane-lined pores through the cell wall that connect adjacent cells of charophyte algae and land plants. Note the similarity between these pictures and those of the plasmodesmata of higher plants in Figure 3–57. The structure of plasmodesmata is described in more detail in Section 3.4. (A) Longitudinal view of two plasmodesmata connecting adjacent cells on either side of a wall. The cell wall is approximately 500 nm thick. (B) Cross section through a wall, showing a cluster of plasmodesmata in the wall.

(See Chapter 3 for descriptions of cellular structures and cell division.) The first land plants probably evolved in freshwater pools where they would have been exposed to periodic drying out, so the ability to survive desiccation was essential. This type of habitat is occupied by some desiccation-tolerant charophyte algae today.

Although the evidence for the relationship between the Charales and the ancestral land plants is compelling, it is not supported by the fossil record. The oldest charophyte fossils are only 400 million years old, which is approximately 70 million years younger than the oldest fossils of embryophyte spores. If land plants are indeed descended from early species of charophyte algae, these algae must have existed before 475 million years ago, when the first embryophyte fossils appeared in the record. What might explain the absence of charophyte fossils from the Cambrian, Ordovician, and Silurian Periods (545–409 million years ago), when we suspect these organisms existed? It is

Box 1–1 What DNA Can Reveal about Phylogeny and Evolution

DNA sequences from living organisms can be used to determine the order of speciation events—the formation and divergence of new species—among groups of related species. Changes, or mutations, in the DNA molecule take place from time to time, and these heritable mutations are passed down through lineages. So, by comparing the DNA molecules of different species, we can reconstruct a tree of relatedness, called a **cladogram**. A cladogram is a statistical representation of the hypothetical relationships within a group of organisms and can be interpreted as representative of their **phylogeny,** or historical origins. The relative positioning of species on the branches of the phylogenetic tree provides hypotheses for the order in which speciation events took place. These hypotheses can then be tested as further data become available.

DNA sequences can also be used to give estimates of how long ago the branches of a phylogenetic tree formed. If the mutation rate is constant, it can be used like the ticking of a clock—a **molecular clock**. According to the molecular clock hypothesis, if the rate of mutation is constant in a group of organisms, the difference in DNA sequences among the organisms can be used to determine the time when speciation events occurred. However, we know that the molecular clock runs at different rates in different lineages, so we have to take into account this

variability across a phylogenetic tree. Various analytical methods have been developed to counteract the effects of this variability. In setting a baseline for the rate at which the clock ticks, at least one piece of fossil evidence is usually required. Additional fossil evidence can then be used to test the ages derived from the "gene tree" or to add further constraints on the ages of particular branches. Estimates based on fossil age and molecular clocks often differ because neither technique is perfect. Because the geologic record is incomplete, we know that fossil evidence consistently underestimates the ages of clades. The first appearance of a fossil can only ever provide a *minimum age* constraint for a particular group. The clock has been used to date the times of divergences (speciation events) of lineages that lack a good fossil record. For example, the fossil record indicates that the land plants were present on the continents sometime around 470 million years ago, but it does not tell us when the first land plants evolved from their green algal ancestors. Based on the molecular clock, it has been suggested that these lineages diverged much earlier—perhaps in the Cambrian. It is important to remember that cladograms of relationships and their derivatives, such as calibrated phylogenies, are working hypotheses. They can be tested and modified as new data become available.

possible that the Charales of these periods were tiny organisms and that their fossils have simply not been found, or that they had few resistant tissues and thus failed to form fossils or did so only as fragments.

Microfossils indicate that the first land plants appeared in the Middle Ordovician Period, about 475 million years ago

The appearance of spores with embryophyte characteristics in 475-million-year-old fossils is the first evidence of land plants. In extant spore-forming land plants—bryophytes (including mosses, liverworts, and hornworts) and monilophytes (lycophytes, ferns, and horsetails)—spores are dispersed by water and by wind. Spore walls contain **sporopollenin**, a highly stable substance that renders the spore robust and resistant to decay. Analysis of the occurrence of fossilized spores reveals two separate "bursts" of increased diversity, or radiations, in the early stages of the colonization of the land by plants.

Evidence for the first radiation comes from fossil spores of approximately 475 million years ago (Middle Ordovician). The spores were dispersed as tetrads—groups of four connected spores. Detailed analysis of the spore walls by electron microscopy shows that they are morphologically similar to the spores of some extant bryophytes. This suggests that the first radiation of land plants resulted in the development of a bryophyte-dominated flora by approximately 460 million years ago.

The second radiation, approximately 430 million years ago (in the Silurian), is revealed by the appearance of different spore morphologies in the fossil record. The spores were dispersed individually, but carry scars that indicate they were part of a tetrad during their development. This morphology is similar to that of the spores of some more complex extant plants, including ferns. The increasing complexity of the spore-bearing parts of early plants is discussed in more detail when we consider the evolution of the seed plants (see Section 1.4).

Further evidence on the nature of early land plants comes from the discovery in a few sites of fossilized, spore-containing **sporangia**—the structures in which spores develop on the mother plant. These fossils are much rarer than fossils of spores alone, and the earliest evidence comes from rocks of the Late Ordovician (460 million years ago) found in Oman. Here, spores have been found in saclike structures that resemble the simple sporangia of extant liverworts (Figure 1–14). This evidence is consistent with the idea that bryophyte-like plants were early land colonizers. These earliest of land plants did not leave an extensive fossil record: fossils are found only of spores (referred to as "microfossils") and occasionally of sporangia, but not of other plant parts. This is probably because these early land plants had soft bodies, as do extant

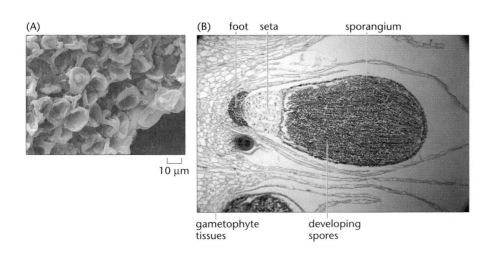

(A)

(B) foot seta sporangium

10 μm

gametophyte tissues developing spores

Figure 1–14.
Fossil and modern-day sporangia.
(A) Fossil spores from 460-million-year-old rocks, clustered on a sporangium. (B) Section through a sporangium of an extant liverwort, *Marchantia*. The sporangium is filled with maturing spores. This structure dominates the sporophyte of the liverwort, which also consists of a stalk (seta) and a foot through which it is in contact with the gametophyte tissues of the liverwort on which it develops. At maturity, the sporangium is approximately 2 mm long. The sporophyte and gametophyte generations of land plants are described in Section 1.3. (A, from C.H. Wellman et al., *Nature* 425:282–285, 2003. With permission from Macmillan Publishers Ltd; B, courtesy of David Polcyn.)

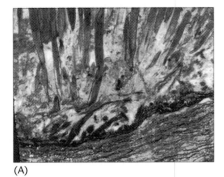

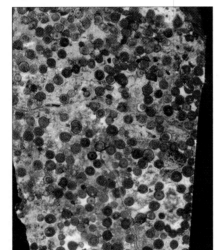

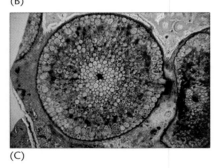

Figure 1–15.
Flora of the Rhynie chert. (A) A polished section through a piece of chert from Rhynie, Scotland, showing vertical axes of plants growing from a soil substrate. (B) A section through the same piece of rock, at right angles to the first, shows cross sections of plant stems. (C) Higher magnification of a stem of *Rhynia gwynne-vaughnii*, showing the preserved cellular details. The stems are approximately 0.5 cm in diameter.
(A and B, courtesy of Hans Kerp and Hagen Hass; C, courtesy of Hans Steur.)

mosses and liverworts, and these were decomposed by microorganisms before rock minerals could replace the plant components and form a durable fossil. Macrofossil evidence for the existence of land plants is found only later, in the Silurian and Devonian Periods, when harder and more resistant plant parts had evolved.

Plant diversity increased in the Silurian and Devonian Periods

As described above, evidence from fossilized spores suggests that the first land plants were related to liverworts and were present by 475 million years ago, after which there was a gradual increase in plant diversity. By approximately 400 million years ago (Early Devonian), more complex ecosystems had developed on land. One of the best examples comes from the small village of Rhynie, Scotland, where an early land-plant environment is preserved in silica (Figure 1–15). This ecosystem was preserved because it was located near hot springs that periodically inundated the area with silica-rich water. The silica permeated the inundated plants, replacing the organic components and preserving the plants in their original growth position. Over time, the silica consolidated and formed "chert," a stone similar to flint, which is composed almost entirely of silica.

What makes the Rhynie chert unique is that the entire ecosystem was fossilized and preserved. This contrasts with the mode of preservation of most other plant fossils from the same period. Fossilization occurs mostly in the beds of lakes and rivers, and because most land plants do not grow in such environments, their fossilization depended on the plant bodies being transported (by streams, for example) to these sites. The plants were broken up during transport, so the resulting fossils are incomplete. For example, another rich source of Devonian plant material is the Senni beds of Wales, where land plants were preserved in the sediments of extensive freshwater lakes (alongside fish and aquatic animals). These fossils provide information on the morphology and size of Devonian land plants but, unlike the Rhynie chert, they tell us little about the other organisms that lived alongside them on the land.

The Rhynie chert ecosystem shows that several different types of ancestral land plants related to extant lycophytes (including club mosses, spike mosses, and quillworts) formed a sward of approximately 20 to 30 cm in height, which was inhabited by arthropods such as springtails, harvestmen, myriapods, and the earliest known winged insect (Figure 1–16A). Similar suites of animals live in this type of environment today, although the plants that comprise these environments are different—much of the Devonian flora is now extinct. Examination of the gut contents and mouthparts of the animal remains in Rhynie chert indicates that these animals fed at least partly, and possibly entirely, on detritus (dead plant material), spores, and the microorganisms that lived off detritus. They do not appear to have eaten living plant material, so it is likely that the herbivorous lifestyle evolved later. The fossil plants of the Rhynie chert contain evidence of associations with fungi. Most extant land plants form symbiotic associations with fungi, called **mycorrhizae** (described in Section 8.5). The fungi receive sugars from the plant and provide the plant with nutrients, especially phosphate, from the soil. The presence inside the stems of fossil plants of structures that look like modern-day mycorrhizal fungi (Figure 1–16B) suggests that this plant-fungal symbiosis arose very early in the evolution of land plants.

The number of sporangia distinguishes the first land plants from their evolutionary descendants

The first land plants, as noted earlier, were probably morphologically similar to extant bryophytes: the spores of 460-million-year-old fossils resemble those of some modern liverworts, and younger fossils also show bryophyte characteristics such as simple spore-bearing structures and rootlike **rhizoids**. Through the Silurian and Devonian

Figure 1–16.
Fossils of other organisms in the Rhynie chert ecosystem. (A) An arthropod, similar to modern-day mites. (B) Mycorrhizal fungus growing inside the stem of an *Aglaophyton* plant. The threadlike structures, thought to be the fungal hyphae, are approximately 10 μm in diameter. (A and B, courtesy of Hans Kerp and Hagen Hass.)

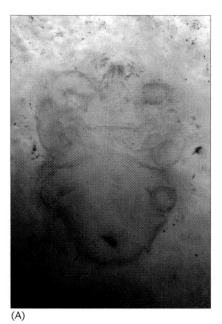

(A)

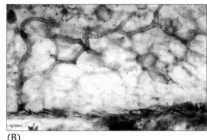

(B)

Periods, between 425 and 350 million years ago, there was an increase in morphological diversity of land plants. By the Carboniferous Period (360–300 million years ago), complex forest ecosystems had developed that contained plants with a great diversity of size and morphology.

Bryophytes are embryophytes characterized by a life cycle in which the **gametophyte** (gamete-producing) generation is larger than the **sporophyte** (spore-producing) generation (further discussed below). The bryophyte gametophyte is the "leafy" part of the plant—it photosynthesizes and absorbs water and nutrients from the substrate, to which it is attached by rhizoids or scalelike outgrowths. Bryophyte gametophytes lack the specialized water-conducting cells—the **tracheary elements**—found in higher plants. They are said to be "nonvascular," whereas higher plants are "vascular." A small number of bryophytes with particularly large gametophytes develop conducting cells called **hydroids**. The sporophyte, the "capsule" of mosses and liverworts, develops directly on the gametophyte and is dependent on the gametophyte for its nutrition. A characteristic shared by all bryophyte sporophytes is that each develops a single sporangium (see Figure 1–14B). This trait distinguishes bryophytes from all other land plants, which develop multiple sporangia on a single sporophyte.

The first plants with multiple sporangia per sporophyte are called the "polysporangiophytes." The oldest polysporangiophyte fossils are 425 million years old, from the Late Silurian Period. The leafless sporophytes of these plants were generally less than 10 cm high and branched by simple bifurcation ("dichotomous branching"). Each branch was the same length, an arrangement known as "isotomous branching." Single sporangia were located at the ends of these branches. *Aglaophyton major* is a good example of an early polysporangiophyte (Figure 1–17). It was attached to the substrate by rhizoids

(A)

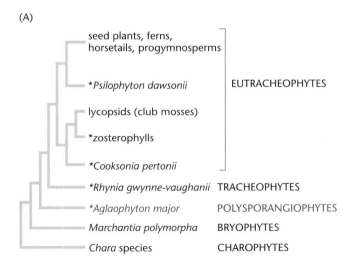

(B)

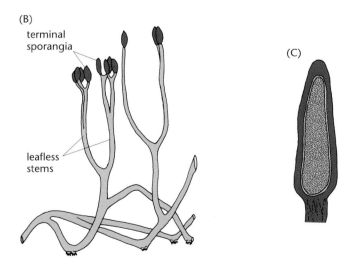

(C)

Figure 1–17.
***Aglaophyton major*, an extinct polysporangiophyte.** (A) Phylogenetic tree showing the likely evolutionary relationship of *Aglaophyton major* to other species and groups of land plants discussed in the text. Species or groups marked with an asterisk are extinct. (B) Morphology of the *Aglaophyton* sporophyte. Note the much greater complexity and size (approximately 10 cm high) of this sporophyte compared with the bryophyte *Marchantia*, shown in Figure 1–14B. (C) An *Aglaophyton* sporangium, with masses of spores at its center.

and bore elongated sporangia at the tips of upright branches. Like bryophytes, the poly-sporangiophytes lacked the specialized water-conducting cells (tracheary elements) found in higher plants, but the branches contained very simple water-conducting tissues similar to the hydroids of some bryophytes. Plants such as *A. major* are thus known as **protracheophytes**.

Increases in plant size were accompanied by evolution of a vascular system

Plant height increased dramatically between the Late Silurian and Late Devonian Periods (425–360 million years ago). Silurian plants were generally less than 10 cm in height; by the end of the Devonian, plants as large as modern-day trees had evolved. This increase in stature was accompanied by the development of specialized cells, with thickened walls, that transport water from the base of the plant to its uppermost regions and provide mechanical support for the upward growth of the plant. These plants are the **tracheophytes** (*trachea* is Latin for "tube"); their water-conducting tubes are composed of thickened cell walls that also provide mechanical support (see Section 3.6). The ancestral tracheophytes, found in sediments from the Silurian Period, include plants such as *Rhynia gwynne-vaughanii* (Figure 1–18) and *Cooksonia pertoni*, which, rather like *Aglaophyton*, had branched, leafless sporophytes with sporangia at the tips of branches. Unlike the protracheophytes, however, *Rhynia* and *Cooksonia* both had thickenings in the walls of water-conducting cells in their branches.

As plants increased in height, changes in the nature of the thickenings in the walls of the water-transporting cells also took place. In the eutracheophytes (true vascular plants), the thickenings of the walls of the tracheary elements (or **tracheids**) became reinforced with a very strong, complex polymer called **lignin** (see Section 3.6). Lignin is water-repellent and resistant to decay; it has an estimated half-life of approximately 500 years! The appearance of lignin resulted in a large increase in the numbers of plants that became fossilized, because the plants were not decomposed before fossilization could take place. Lignin surrounds the cellulose fibers of the tracheid cell walls, stiffening the wall and allowing the tracheids to withstand the large negative pressures to which they are exposed when transporting water through the plant body (see Section 4.10). The thickenings on the tracheid walls of the most ancestral tracheophytes formed either as rings (annular) or as spirals, further reinforcing the cells. (See Sections 3.6 and 5.4 on the structure and development of tracheids and related **vessels** in the **xylem** of modern land plants.)

Figure 1–18.
***Rhynia gwynne-vaughanii*, an early tracheophyte.** (A) Phylogenetic tree showing the likely evolutionary relationship of *Rhynia gwynne-vaughanii* to other land plants discussed in the text. Species or groups marked with an asterisk are extinct.
(B) Morphology of the *Rhynia* sporophyte.
(C) Section through a stem of *Rhynia*, showing the region thought to contain cells responsible for conducting water (tracheary elements) and sugars produced in photosynthesis (phloem). (C, courtesy of Paul Kenrick.)

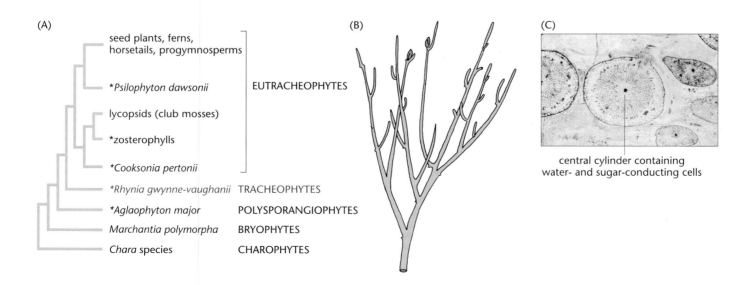

(A)

seed plants, ferns, horsetails, progymnosperms

Psilophyton dawsonii   EUTRACHEOPHYTES

lycopsids (club mosses)

*zosterophylls

Cooksonia pertonii

Rhynia gwynne-vaughanii   TRACHEOPHYTES

Aglaophyton major   POLYSPORANGIOPHYTES

Marchantia polymorpha   BRYOPHYTES

Chara species   CHAROPHYTES

(B)

(C)

central cylinder containing water- and sugar-conducting cells

Some of the earliest vascular plants were related to extant lycophytes

The increase in plant height led to the formation of dense, tall forests. By the end of the Devonian Period (360 million years ago), 50% of the species in these forests were lycophytes (Figure 1–19). These plants are relatives of modern-day club mosses (Lycopodiaceae), quillworts (Isoetaceae), and spike mosses (Selaginellaceae) (Figure 1–20).

The lycophytes comprise two subgroups: the Zosterophyllopsida (zosterophylls) and the Lycopsida (lycopsids) (see Figure 1–19). Zosterophylls are now extinct: they formed swards in damp places or grew as part of the understory below taller plants during the Devonian Period. They superficially resembled rhyniophytes such as *Rhynia* (see Figure 1–18) in that they were short and leafless with horizontal **rhizomes** with erect, sporangia-bearing stems (Figure 1–19B, C). However, whereas rhyniophyte plants bore single sporangia at the tips of stems, the zosterophylls bore sporangia laterally, on branches clustered in a conelike arrangement. This typifies an evolutionary trend toward the clustering of sporangia during the early evolution of the tracheophytes.

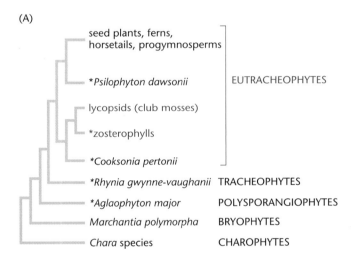

(A)

seed plants, ferns,
horsetails, progymnosperms

Psilophyton dawsonii   EUTRACHEOPHYTES

lycopsids (club mosses)

*zosterophylls

*Cooksonia pertonii

*Rhynia gwynne-vaughanii   TRACHEOPHYTES

*Aglaophyton major   POLYSPORANGIOPHYTES

Marchantia polymorpha   BRYOPHYTES

Chara species   CHAROPHYTES

(C) lateral sporangia

(B)
lateral sporangia

leafless stems

(D)

(E) lateral sporangia

microphyll (leaves)

Figure 1–19.
Lycophytes. (A) Phylogenetic tree showing the likely evolutionary relationship of lycophytes (lycopsids and zosterophylls) to other land plants discussed in the text. Species or groups marked with an asterisk are extinct. (B, C) Morphology of *Zosterophyllum rhenanum*, a leafless plant bearing clustered sporangia pressed close to the stem. (D, E) Morphology of the extinct lycopsid *Asteroxylon mackiei*, showing leafy shoots bearing sporangia.

Figure 1–20.
Extant lycophytes. (A) A spike moss, *Selaginella* sp. (B) A club moss, *Lycopodium annotinum*. (C) A quillwort, *Isoetes lacustris*; this aquatic species grows on lake beds in northern Europe, Canada and the northern United States.

(A) (B) (C)

Many of the tall trees in the Devonian forests and the later Carboniferous swamps were members of the class Lycopsida. Lycopsida also included **herbaceous** forms, such as *Asteroxylon mackiei* (Figure 1–19D, E) preserved in 405-million-year-old rocks. The Lycopsida differed from the Zosterophyllopsida in that they developed **microphylls**, leafy structures that had a central vascular strand, or **leaf trace**. Microphylls resemble the leaves of the later **seed plants** and ferns, but these two sorts of structures are not considered **homologous**. That is, they are not derived from a common ancestral structure, but instead evolved independently.

Some of the Lycopsida grew to the same height as modern trees, and the further development of a water-transporting system and a means of physical support accompanied this increase in stature. The production of vascular strands in the low-growing land plants up to this point, and in the related Zosterophyllopsida, was a result of **primary development**, meaning that the vascular cells developed from cells produced by divisions at the branch tips (the **apical meristems**). **Meristems** are populations of cells that give rise to the tissues of the plant body. In contrast, the strong, tracheid-rich stems of the tree-forming Lycopsida resulted from cell divisions in a secondary meristematic tissue called **cambium**. Cambium develops some distance behind the apical meristem, where it produces vascular cells that increase the girth and rigidity of the stem. This **secondary development** of the stem has evolved numerous times in the land plants. For example, secondary development evolved independently in seed plants and in the lycopsids: the last common ancestor of both groups was herbaceous and did not undergo secondary development. (Primary and secondary development in modern angiosperms are described in Chapters 3 and 5.)

The great height attained by the tree lycopsids involved not only mechanical strengthening of the stem by secondary development, but also strong anchorage mechanisms (Figure 1–21). The tracheophytes and zosterophylls had horizontal rhizomes from which sporangia-bearing aerial stems arose, and typically reached heights of only 25 cm. In contrast, the tree Lycopsida were anchored by large, rootlike modified stems called "rhizomorphs," bearing modified microphylls called "rootlets."

Horsetails, ferns, and seed plants are derived from a group of leafless plants of the Early Devonian Period, 400 million years ago

As well as **woody** lycopsids, the Devonian forests and the Carboniferous swamps also included tree-like species of other groups of plants, including the horsetails, ferns, and progenitors of the seed plants, the progymnosperms. These Paleozoic giants were the source of the coal that has fueled much of the world's industrial development since the beginning of the Industrial Revolution.

The horsetails, ferns, and seed plants are included in a monophyletic group, the Euphyllophytina. Most of these plants have true leaves, or **megaphylls**. Those that lack

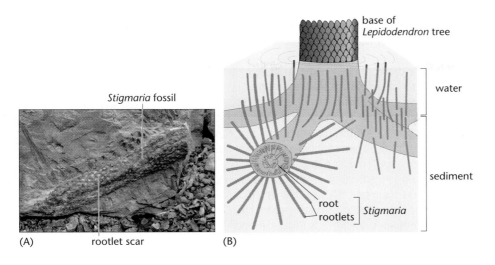

(A) *Stigmaria* fossil

rootlet scar

(B) base of *Lepidodendron* tree

water

sediment

root rootlets } *Stigmaria*

Figure 1–21.
The rooting structures of extinct tree lycopsids. (A) Rock containing a fossil rhizomorph. The surface of the fossil is covered with scars, the sites where the rootlets were attached. (B) Artist's impression of a reconstructed rhizomorph with rootlets. Fossil rhizomorphs and fossils of the trunks and leaves of the same tree lycopsids were originally discovered and named independently. Only later was it realized that these fossils were parts of the same species. This is reflected in the names of the fossils: the rhizomorph fossils of this species of tree lycopsid are named *Stigmaria* and the aboveground parts are named *Lepidodendron*. (A, courtesy of Peter Skelton.)

megaphylls are either the most ancestral, extinct members of the group or species that originally had megaphylls but lost them during subsequent evolution, such as the extant whisk ferns (Psilotales). Megaphylls are thought to have evolved by the gradual amalgamation of leafless stem systems like those of one of the most ancestral, extinct species of this group, *Psilophyton dawsonii* (Figure 1–22). According to this hypothesis, known as the "telome theory," the three-dimensional branch system first lost its

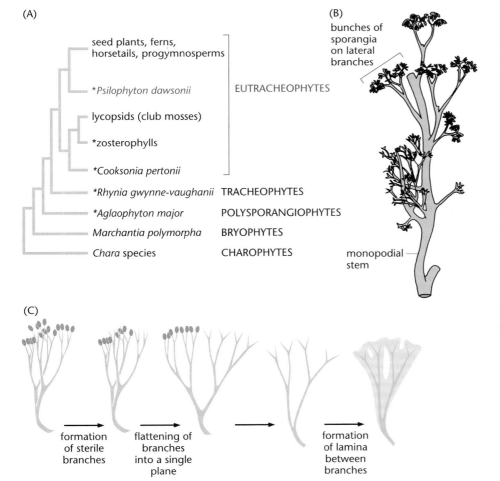

(A)

seed plants, ferns, horsetails, progymnosperms

Psilophyton dawsonii

lycopsids (club mosses)

*zosterophylls

*Cooksonia pertonii

EUTRACHEOPHYTES

*Rhynia gwynne-vaughanii   TRACHEOPHYTES

*Aglaophyton major   POLYSPORANGIOPHYTES

Marchantia polymorpha   BRYOPHYTES

Chara species   CHAROPHYTES

(B)
bunches of sporangia on lateral branches

monopodial stem

(C)

formation of sterile branches

flattening of branches into a single plane

formation of lamina between branches

Figure 1–22.
***Psilophyton dawsonii*, a eutracheophyte.**
(A) Phylogenetic tree showing the presumed evolutionary relationship of *Psilophyton dawsonii* to other land plants discussed in the text. Species or groups marked with an asterisk are extinct. *Psilophyton* and the seed plants, ferns, horsetails, and progymnosperms arose from a common ancestor and together constitute the Euphyllophytina.
(B) Morphology of *P. dawsonii*, a leafless plant with a single main stem and bunches of sporangia produced on lateral branches. Contrast this monopodial growth habit with the isotomous branching of the early land plants such as *Aglaophyton* (see Figure 1–17).
(C) The telome theory of the evolution of megaphylls. An early member of the Euphyllophytina underwent progressive sterilization of some branches, flattening of bunches of branches, then filling of the spaces between branches with tissue to form a lamina, or leaf blade.

sporangia (sterilization) and then became flattened into one plane (planation); a lamina then developed in the interstices between the branches (Figure 1–22C).

Members of the Euphyllophytina share two obvious traits: the branching pattern of the stems and the positioning of the sporangia. The Euphyllophytina branch monopodially (a branching mechanism that forms a "monopodium," a single dominant stem or trunk). Monopodia can develop in two different ways. They may develop from division of an apical meristem into two unequal meristems at the tip of the **shoot**, the larger meristem forming a stem that becomes dominant and forms the monopodium, and the smaller meristem forming a lateral branch. Alternatively, in later-evolved groups, the monopodium forms from an apical meristem and separate, **lateral meristems** give rise to the lateral branches. This results in the development of a plant with a main stem and a system of many smaller lateral branches. These branching patterns contrast with the dichotomous branching generally found among the lycophytes and earlier land plants, which branch by bifurcation: the meristem divides equally into two meristems, each of which gives rise to a stem. Continued bifurcations then result in the formation of ever-finer branches. The monopodial stem morphology is illustrated by the fossil plant *Psilophyton dawsonii* (see Figure 1–22B). In this species, aerial branches arose from horizontally growing stems (rhizomes). Note that *P. dawsonii* did not have leaves (megaphylls).

As described above, early land plants bore single sporangia on the distal ends of stems whereas sporangia in the later lycophytes were arranged laterally, on clusters of branches arising from the sides of stems. The most ancestral group of the Euphyllophytina had a third arrangement of sporangia: many terminal sporangia developed on a single branch. These clusters of sporangia are seen on spirally arranged lateral branches in *P. dawsonii* (see Figure 1–22B). It is clear that the horsetails, ferns, and seed plants all arose from ancestors like *P. dawsonii* and that the seed plants are sister to the clade that contains the horsetails and the ferns. However, because of the incompleteness of the fossil record and our incomplete understanding of the relationships among fossils, we cannot describe each of the steps that led from *Psilophyton*-like morphologies to those seen in the horsetails, ferns, and seed plants.

Ferns and horsetails evolved in the Devonian Period

The extant Euphyllophytina comprise two sister groups: the seed plants and the Moniliformopses (the monilophytes). The monilophytes include the ferns and the horsetails (Figure 1–23).

Figure 1–23.
The euphyllophytes. (A) Phylogenetic tree showing the presumed evolutionary sister relationships among the monilophytes, the seed plants, and some of their common ancestors. Note that the horsetails are a monophyletic group nested within the ferns, which suggests that the ferns are a paraphyletic group, meaning that not all of the species in the group are derived from a common ancestor. Most of the extant families of ferns are leptosporangiates. (B) Some extant monilophytes: *Osmunda regalis*, a leptosporangiate fern (*left*); a species of *Equisetum*, a horsetail (*middle*); and *Ophioglossum vulgatum*, an ophioglossoid fern (*right*). (Bi and ii, courtesy of Tobias Kieser; Biii, courtesy of David J Glaves.)

(A)

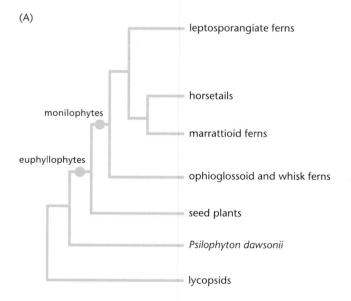

(B)

A comparison of nuclear DNA sequences of living monilophyte species suggests a steady diversification of this group over the last 390 million years. By estimating the timing of speciation events in the monilophyte phylogenetic tree, we know that the major groups formed between 380 and 200 million years ago (i.e., between the Devonian and Jurassic Periods in the Paleozoic and Early Mesozoic Eras). The oldest fern fossils are found in Upper Devonian rocks from around 370 million years ago, so the fern lineage must have evolved early in monilophyte history. However, many of the monilophyte species and genera that are alive today evolved in the last 140 million years. Thus, much of the present species richness of the monilophytes emerged relatively recently, alongside the flowering plants, and is not an evolutionary relic of an earlier time.

Chemical and cellular complexity increased early in the evolution of land plants

We have described how the land plants are derived from a common ancestor shared with algae of the order Charales. The earliest land plants retained some traits from this algal ancestor and acquired novel traits associated with life on the relatively dry land. Here we highlight a few of the features that are thought to have been important as plants emerged from aquatic environments onto the land and into the air.

Some chemical features found among the earliest embryophytes would have contributed to their ability to survive on dry land. Examples include the presence in the cell wall of **waxes**, polymers formed from phenolic compounds, and sporopollenin. These molecules act to reduce water loss and protect the cell from damage. Waxes form a waterproof surface layer— the **cuticle**—on the aboveground parts of land plants. The presence of sporopollenin in the walls of spores of early embryophytes would have protected them against desiccation and microbial decomposition. The biochemical pathways leading to the production of waxes, phenolic compounds (**phenylpropanoids**), and sporopollenin are also found in the Charales, although their products do not contribute in the same way to the waterproofing and strengthening of cell walls. It seems likely that these biochemical pathways were present in the common ancestor of the charophyte algae and the land plants. Thus, biochemical pathways that were crucial in the evolution of land plants probably were not assembled from scratch during their early evolution but were inherited from algal ancestors. Evolutionary modification of these and other ancestral biochemical pathways produced the great chemical diversity observed in the land plants today (see Chapter 4).

Some morphological features of the early land plants were probably associated with survival in drier environments. These include the formation of specialized water-conducting cells. Rather than being restricted to environments in which all of their tissues were in contact with water, plants with water-conducting cells could grow upward into drier air. As we have seen, elaboration of these early transport systems by the Late Devonian and Early Carboniferous Periods was associated with the development of tree-like plant forms in which water was transported from ground level to heights of many meters, as in today's trees.

Another novel cell type in land plants was the **transfer cell**. Transfer cells have extensive invaginations of the cell wall that result in a very large surface area in relation to the cell volume. Such cells are located in regions of the plant in which sugars and other nutrients are exported from one cell and taken up by another: the large surface area allows a high rate of uptake of nutrients into the cell. The appearance of transfer cells was probably important for the changes in the plant life cycle that occurred during the early evolution of land plants. It seems likely that the gametophyte and sporophyte generations of the life cycle were completely separate in the aquatic ancestors of the land plants (described in more detail below). After the emergence onto land, the sporophyte and gametophyte became integrated. The sporophyte was dependent on the gametophyte for nutrition in the early land plants, as in today's bryophytes (see, for example,

Figure 1–14B). Transfer cells at the junction between the sporophyte and the gameto-phyte facilitated the transport of nutrients from the gametophyte to the developing sporophyte.

A major development during the evolution of the land plants was the appearance of meristems. Meristems consist of populations of cells that continue to divide and give rise to the tissues and organs of the plant body (the process of **morphogenesis**) throughout the life of the plant (see Chapter 5). Much of the increase in morphological diversity of land plants that occurred between 470 and 350 million years ago resulted from changes in the activity of meristems. For example, the increasing variety and complexity of meristem activity in plants during the Devonian Period led to the elaboration of complex, branched shoot and **root** systems.

Atmospheric CO_2 and O_2 levels are determined by rates of photosynthesis and carbon burial

The evolution of land plants caused changes in the relative abundance of carbon dioxide and oxygen in the earth's atmosphere. Before going into this in more detail, we need to outline some of the factors that determine the levels of these gases in the atmosphere.

Let's consider a hypothetical abiotic earth—that is, earth in the absence of life. The CO_2 levels in the atmosphere are controlled by geological processes. Conditions such as these existed on earth before 3.8 billion years ago and exist on Venus and Mars today. CO_2 levels in the atmosphere of this abiotic earth are increased by volcanic activity and the release of CO_2 from rocks undergoing metamorphism. Levels are reduced by the chemical weathering of various silicate rocks. This is because atmospheric CO_2 reacts with exposed silicate minerals to form products that remain as minerals in the soil or water. For example, during the chemical weathering of a sodium-containing feldspar ($NaAlSi_3O_8$), atmospheric CO_2 reacts with the mineral and water vapor, $H_2O(g)$, to form a clay mineral ($Al_2Si_2O_5 (OH)_4$), sodium bicarbonate, and dissolved silica:

$$2NaAlSi_3O_8 + 2CO_2 + 3H_2O(g) \rightarrow Al_2Si_2O_5 (OH)_4 + 2NaHCO_3 + 4SiO_2$$

The appearance of life changed the cycling of carbon on earth. The greatest source of CO_2 in the atmosphere is respiration, the process by which organisms generate the energy-rich molecules needed for the maintenance of life processes. Organic mole-cules—often carbohydrates such as sugars—are oxidized to CO_2, and energy-rich mol-ecules are generated in the process (see Chapter 4 for a full description of these path-ways). In simple terms, respiration may be represented by the following equation:

$$nO_2 + (CH_2O)_n \rightarrow nCO_2 + nH_2O$$

where $(CH_2O)_n$ represents carbohydrate. Thus respiration consumes O_2 and generates CO_2.

During photosynthesis, an opposite process occurs: atmospheric CO_2 is converted into organic molecules (biomass), with the release of O_2:

$$nCO_2 + nH_2O \rightarrow (CH_2O)_n + nO_2$$

As we described at the start of this chapter, the large amounts of O_2 in the atmosphere today were formed through photosynthesis over the last 3 billion years.

Carbon contained in the biomass of living organisms can have one of two eventual fates. First, when the organism dies, the biomass may be consumed by other organisms that use the carbon to fuel respiration, consuming oxygen and releasing carbon back into the atmosphere as CO_2. Second, the biomass may be buried in sediments that petrify as coal or other carbon-rich rock. A consequence of carbon burial is that for every mole of carbon buried, a mole of oxygen remains in the atmosphere. Thus there is a direct relationship between the amount of carbon buried in sediments and the increase in O_2 levels in the atmosphere.

To determine the relationship over geological time between atmospheric CO_2 and O_2 levels, we need a means of estimating the abundances of these gases. The relative abundance of atmospheric gases can be measured in ice cores from the polar ice caps, which are up to 700,000 years old. For earlier times there is no means of direct measurement, so atmospheric gas abundances must be estimated. Estimates of atmospheric CO_2 are made either by analyzing the carbonate levels in ancient, fossilized soils or by using complex computer models that incorporate information from all measurable sources that affect atmospheric CO_2, such as rates of organic and inorganic carbon burial, weathering, and volcanic activity. These two independent methods provide very similar values for atmospheric CO_2 levels over the last 550 million years. Similar modeling methods have been used to estimate O_2 levels over the same period.

Figure 1–24 shows the general trends. Atmospheric CO_2 was high about 500 million years ago (the Cambrian Period) and steadily decreased between 450 million years ago (the Middle Ordovician) and 300 million years ago (around the Carboniferous–Permian boundary). It rose again to a lower maximum in the Triassic, then decreased until 1750 and the Industrial Revolution. Oxygen, on the other hand, shows a dramatic increase around 300 million years ago. As atmospheric CO_2 decreased, atmospheric O_2 increased.

The evolution of land plants was at least partly responsible for the decrease in atmospheric CO_2 beginning 450 million years ago

The decrease in atmospheric CO_2 was the result of a combination of land-plant activities, including burial of organic material, and increased chemical weathering. As noted earlier, the first evidence for the existence of land plants comes from sediments about 470 million years old. These plants began to assimilate large amounts of atmospheric CO_2, and some of this carbon was buried when the plants died. The rate of burial increased dramatically, reaching a maximum between 350 and 300 million years ago. This was the time of the Carboniferous coal swamps, when trees that became buried

Figure 1–24.
Change in CO_2 and O_2 levels in the atmosphere over the past 550 million years. (A) Estimated changes in atmospheric CO_2 levels (the present level is around 0.385 parts per thousand, or 385 parts per million, or 0.0385%, by volume). (B) Estimated changes in atmospheric O_2 levels. Abbreviations indicate the geological periods: Ca, Cambrian; O, Ordovician; S, Silurian; D, Devonian; C, Carboniferous; P, Permian; Tr, Triassic; J, Jurassic; K, Cretaceous; T, Tertiary.

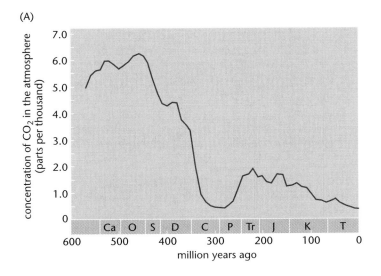

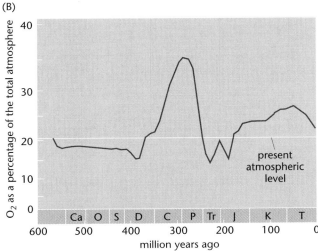

in sediments formed the coal that powered the Industrial Revolution 300 million years later. The increased carbon burial was in part responsible for the decrease in atmospheric CO_2 at this time.

As described above, the other factor that reduced atmospheric CO_2 levels was chemical weathering of rocks, which resulted in conversion of atmospheric carbon either to insoluble products that remained *in situ* and contributed to the development of soils (such as clay minerals) or to soluble substances that were removed by water. Removal of CO_2 from the atmosphere by this process is appropriately called "atmospheric CO_2 drawdown." The growth of plants on the land surface enhanced chemical weathering because it increased the exposed surface area of rocks. Furthermore, the roots of plants secrete acids that dissolve some of the rock minerals and enhance weathering reactions.

An important contributor to this effect of plants on CO_2 drawdown was the evolution of root systems. The short-statured early land plants had no roots and were anchored to substrates by rhizomes bearing single-celled rhizoids, which absorbed nutrients and water and secreted acids into the surrounding substrate (see, for example, *Aglaophyton* in Figure 1–17). By the Carboniferous Period, large trees with true roots and extensive root systems had evolved. As plants increased in stature between 400 and 350 million years ago, the depth of rooting systems also increased, from less than a centimeter to more than a meter. This not only augmented weathering rates but also gave rise to the first soils, rich in organic matter derived from plant biomass. These soils acted as "carbon sinks," in which some of the carbon removed from the atmosphere during photosynthesis became sequestered. One consequence of the sequestration of carbon in soils and its mass burial was an increase in atmospheric O_2 between 390 and 275 million years ago.

Oxygen levels in the atmosphere increased for as long as the rates of carbon burial increased. Toward the end of the Carboniferous Period, the Pangea supercontinent formed and the climate became more arid, resulting in decreased production of plant biomass. Formation of the supercontinent was accompanied by a drop in sea levels, which exposed a large proportion of the carbon-rich sediments on the continental shelf to atmospheric oxygen and thus to **oxidation**. This oxidation, together with the decreased rate of biomass production and carbon burial on land, is thought to have led to the decrease in atmospheric O_2 levels after the end of the Carboniferous Period. Thus a combination of biological and geological events together have determined the atmospheric composition over the past 550 million years.

Today there is concern about the elevation of atmospheric CO_2 levels resulting from the burning of fossil fuels, which has increased CO_2 levels from 270 ppm (parts per million) in the pre-industrial world to more than 380 ppm today. The concern is warranted, because CO_2 is a greenhouse gas: the greater the amount of atmospheric CO_2, the greater the amount of heat retained in the atmosphere. There is evidence for a reverse greenhouse effect between 400 and 250 million years ago, following the steep fall in atmospheric CO_2 levels. This was reflected in the formation of ice caps at the poles and ice ages during the Carboniferous Period, which continued into the Permian Period.

The mid-Paleozoic decrease in atmospheric CO_2 was a driving force in the evolution of big leaves

The decrease in atmospheric CO_2 levels in the late Devonian and Carboniferous Periods had dramatic effects on the evolution of land-plant leaves. We know from physiological experiments on living plants that atmospheric CO_2 levels in part determine the numbers of **stomata** on a given area of leaf (the stomatal density). Stomata are the pores in the surface layer of leaf cells (the **epidermis**) through which gases and water

(A)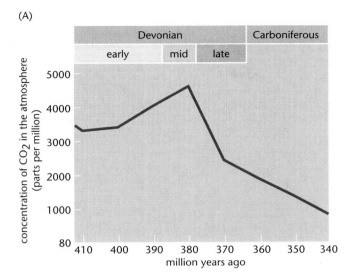

(B)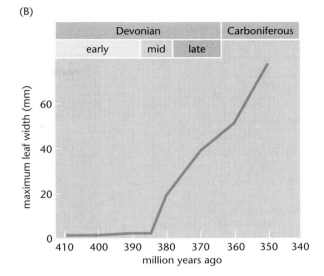

vapor are exchanged between the atmosphere and the air spaces between cells inside the leaf. The lower the atmospheric CO_2 level at which plants are grown, the higher the stomatal density on their leaf surfaces. This developmental response to CO_2 thus allows a more rapid equilibration of CO_2 between the atmosphere and the air spaces of the leaf when CO_2 concentration is more limiting for photosynthesis. Consistent with these observations on modern-day plants, the fossil record shows that stomatal densities were relatively high when atmospheric CO_2 levels were low. (Stomatal functions are discussed in detail in Chapters 3, 4, 6, and 7.)

Figure 1–25.
Drop in atmospheric CO_2 levels and increase in leaf size during the Devonian Period. (A) Estimated levels of atmospheric CO_2, showing the dramatic decrease beginning in the mid-Devonian. (B) Increase in maximum leaf width seen in the fossil record, beginning in the mid-Devonian.

There is evidence that lower atmospheric CO_2 levels were at least partly responsible for the evolution of large leaves around 390 million years ago (Figure 1–25). As well as allowing entry of CO_2 into the leaf, stomata also allow the movement of water vapor out of the leaf into the atmosphere. This process is essential in maintaining a flow of water through the vascular system from the soil to the leaves, and it also provides evaporative cooling of the leaves. This cooling function is important in dissipating the heating effect of solar radiation on the leaf. Because stomatal density is determined largely by atmospheric CO_2 level, changes in atmospheric CO_2 levels can alter the capacity of the leaf for evaporative cooling. If the atmospheric CO_2 level is high, and thus stomatal density is low, the capacity of the leaf for evaporative cooling will be low. These considerations are thought to have limited the size of leaves prior to 390 million years ago, when atmospheric CO_2 levels were high and, thus, stomatal densities were low. Leaves at this time were small—microphylls with small surface areas. Larger surface areas, which would have intercepted more solar radiation, might have resulted in overheating. As atmospheric CO_2 levels decreased and the density of stomata increased, so the capacity for evaporative cooling increased. This in turn is thought to have allowed an increase in leaf size, because the greater evaporative cooling meant that solar radiation could be intercepted over a larger surface area without overheating.

1.4 SEED PLANTS

The **spermatophytes** are a group of plants that share the characteristic of bearing **seeds**. The seed contains the **zygote**—the cell produced by fertilization—and the embryo that develops from it. There are in excess of 223,000 extant species of seed plants, which fall into five monophyletic groups: the orders Cycadales, Gnetales, Ginkgoales, and Coniferales (Pinales) (Figure 1–26), and the **angiosperms**—by far the largest living group. Many other groups of seed plants found as fossils are now extinct.

Figure 1–26.
Modern seed plants (spermatophytes).
(A) A cycad, *Encephalartos altensteinii*, native to South Africa. (B) A conifer, *Pinus contorta* (lodgepole pine), native to North America. (C) *Ginkgo biloba*, a tree native to eastern China. This is the only extant *Ginkgo* species; other species known from the fossil record became extinct by around 2 million years ago. (D) A member of the Gnetales, *Ephedra viridis*, a species native to North America. Members of the fifth group of modern spermatophytes, the angiosperms, are illustrated later in this chapter.

Seeds contain the genetic products of fertilization protected by tissue derived from the sporophyte

Seeds arose just once during the evolution of plants: all seed-bearing plants are derived from a common ancestor. Seeds are the means by which a plant species disperses new genetic individuals. In seed plants, the sporophyte plant body produces two types of spore in its reproductive structures (sporangia): the **megaspore** (which gives rise to the female **megagametophyte**) and the **microspore** (which gives rise to the male **microgametophyte**). The two types of spore are produced in different types of sporangia (the **megasporangium** and the **microsporangium**) and are morphologically different (Figure 1–27). Seed plants are therefore said to be "heterosporous"—they have different types of spores. The megagametophyte forms within the megaspore, inside the megasporangium. The composite structure consisting of the megagametophyte within the megaspore, the surrounding megasporangium (the **nucellus**), and a variable number of sheathing cell layers, the **integuments**, comprises the **ovule**. The ovule remains embedded within the sporophyte. In contrast to the megagametophyte, the microgametophyte is released from the sporangium in which it is formed—it is the **pollen** of the seed plant. Pollen is transported to a receptive surface of the sporophyte close to the ovule. Fertilization then occurs: one of the nuclei of the microgametophyte (the **sperm**) fuses with one of the nuclei of the megagametophyte (the **egg**) to form a zygote, from which the embryo will develop. Following fertilization, the ovule, together with surrounding tissues from the maternal sporophyte, is termed a "seed." The process of gamete formation and fertilization in angiosperms is described in more detail in Chapter 5.

We described earlier how the ancestral land plants had life cycles in which the gametophyte generation was free-living—that is, it developed in isolation from the sporophyte and was autotrophic. This trait is seen in extant groups such as the

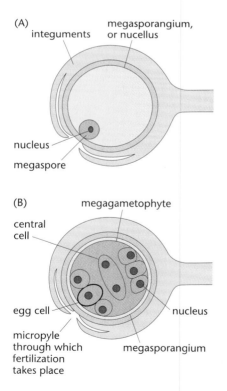

Figure 1–27.
Schematic representation of the spores of seed plants. (A) An early stage in development of the ovule (female reproductive structure). The ovule is embedded in the parent plant (sporophyte). The ovule consists of outer integuments surrounding the megasporangium, which contains the megaspore. (B) Nuclear and cell divisions within the megaspore give rise to the megagametophyte. One of the cells in the megagametophyte, the egg cell, will fuse with a nucleus from a microgametophyte (the pollen) during fertilization, to give rise to the zygote from which the embryo will develop. (C) A mature microgametophyte, or pollen grain (male reproductive structure). The pollen grain consists of two cells. The small generative cell has two nuclei, one of which will fuse with the egg cell during fertilization.

bryophytes, lycophytes, ferns, and horsetails. In seed plants, by contrast, the female gametophyte (the megagametophyte) remains enclosed within the megasporangium on the sporophyte and acquires all of its carbon and nutrients from the sporophyte. The male gametophyte (microgametophyte) of seed plants tends to be smaller than the female gametophyte and undergoes initial development on the sporophyte before it is dispersed as pollen. Pollen is a structure with a hard wall containing sporopollenin, the highly resistant polymer first seen in the spore walls of the earliest plants. Although it is released from the sporophyte, pollen is not autotrophic: it utilizes stores of carbon accumulated from the sporophyte during gametophyte development.

The product of fertilization, the zygote, undergoes cell division to form the embryo. This develops within the gametophyte, embedded in sporophyte tissue (the nucellus and integuments). Thus the ripe seed, ready for dispersal, contains the genetic product of fertilization (the next sporophyte generation) wrapped in layers of tissue derived from the parent sporophyte.

Seed plants evolved in the Devonian and diversified in the Permian, 290 to 250 million years ago

The oldest fossil seed plants are found in rocks between 385 and 365 million years old (Devonian Period), and seed plants form the majority of the world's flora today. By 300 million years ago, seed plants had diverged into two main groups, the Cordaites and the pteridosperms, which coexisted with spore-bearing plants such as ferns, horsetails, and lycophytes (Figure 1–28).

Further groups of seed plants appeared over the next 50 to 80 million years, during the Permian Period, including the Cycadales, the Coniferales (conifers), and the Ginkgoales. By the Triassic Period (230 million years ago), the seed plants had become the dominant plant type in tropical wet forests (the giant lycopsids that had dominated the Devonian and Carboniferous forests had become extinct in the Permian). An

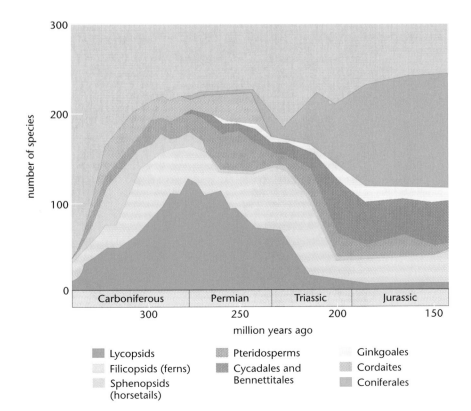

**Figure 1–28.
Diversification of seed plants between 350 and 150 million years ago.** The early groups of seed plants increased in numbers of species during the Carboniferous, then gave rise to new groups of seed plants during the Permian. Spore-bearing plants—lycopsids, ferns, and horsetails—continued to dominate the world flora until the Triassic. During the Triassic, enormous numbers of new species of seed plants arose, especially within the conifers, and the numbers of spore-bearing plant species fell dramatically. All of the seed-bearing plants up to the end of the Jurassic were gymnosperms; angiosperms first appeared about 130 million years ago.

extensive development of seed plants occurred during the Triassic and Jurassic Periods, including a huge diversification of the conifers (Coniferales) and the appearance of groups such as Bennettitales (now extinct) and Gnetales (see Figure 1–26D). All of these groups are collectively referred to as **gymnosperms**. The angiosperms were the last of the seed plant groups to appear. The oldest angiosperm fossils are found in rocks from the Early Cretaceous Period, about 130 million years ago.

The earliest seed plants are thought to be derived from a group of spore-bearing plants known as the progymnosperms. Although these plants had free-living gametophytes, they formed true wood, had root systems similar to those of modern-day seed plants, and in some cases exhibited heterospory. The earliest-known fossil plant with a seed-like structure, in which the female gametophyte remains enclosed within integuments that are part of the sporophyte, is the 385-million-year-old *Runcaria* (Figure 1–29). The megasporangium of *Runcaria* is surrounded by an open, filamentous integument. This is not a true seed, but may well represent an intermediate state in the evolution of seeds. By 360 million years ago, plants with partially fused integuments surrounding the megasporangium had arisen, such as *Archaeosperma*. By the Lower Carboniferous Period, around 350 million years ago, megasporangia completely surrounded by integuments had evolved.

The sporophyte phase became dominant in the land-plant life cycle in the Devonian Period

During the first 100 million years of the evolution of land plants, the sporophyte stage of the life cycle became larger and more complex and the gametophyte stage became smaller. Here we trace the major changes in the life cycle of the land plants since the last shared common ancestor with the charophyte algae.

All land-plant life cycles have two, alternating generations of multicellular organisms. A **haploid**, sexual generation (the gametophyte) alternates with a **diploid**, asexual generation (the sporophyte). The cells of a haploid organism each contain a single set

Figure 1–29.
Evolution of the seed. (A) *Runcaria heinzelinii* (a pre-seed plant), with a filamentous, open integument surrounding the megasporangium (*brown*). The megasporangium has a columnar extension that may have been involved in the reception of pollen. (B) Fossil of *Runcaria* from which the drawing in (A) was made. The specimen is about 7 mm long and is seen in a section of sandstone rock from Belgium. (C) The megasporangium of *Archaeosperma arnoldii* was enclosed within partially fused integuments. (B, courtesy of Philippe Gerrienne.)

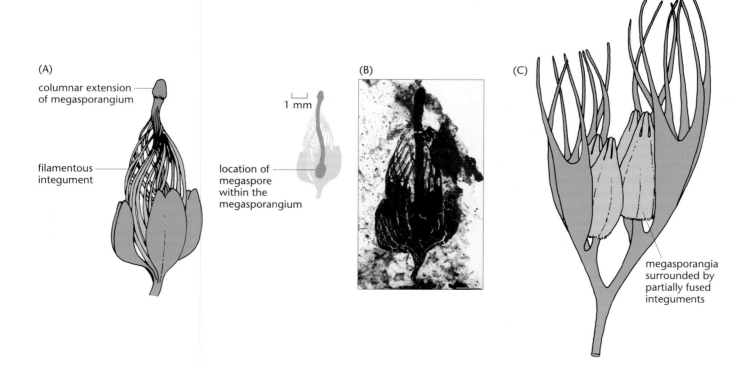

(A)
columnar extension of megasporangium

filamentous integument

(B)
1 mm

location of megaspore within the megasporangium

(C)
megasporangia surrounded by partially fused integuments

of the **chromosomes** that carry its genetic information; the cells of a diploid organism each contain two copies of each of the chromosomes. This type of life cycle, in which both haploid and diploid generations are multicellular, is called "diplobiontic." The charophycean ancestors of the land plants had a different life cycle—one in which the diploid stage was represented only by a single cell—the zygote—and the plant itself was the haploid gametophyte. This is called a "haplobiontic" life cycle. Thus in the alga *Chara*, the haploid plant produces haploid gametes. The fusion of two haploid gametes produces a diploid sporophyte, the zygote. The zygote immediately undergoes a specialized type of division, called **meiosis**, to form four haploid cells, the spores. During meiosis, the two sets of chromosomes of the diploid cell are first replicated to produce four sets, then the four sets are separated to give four (haploid) products, each with one set of chromosomes (this is described in detail in Section 3.2). The spore of *Chara* germinates and develops into the haploid plant (Figure 1–30). Most ancestral land plants, and therefore the land plants most closely related to the charophytes, had the diplobiontic cycle, indicating that the transition from water to the land was accompanied by evolution of the diplobiontic life cycle with its multicellular sporophyte stage.

Further evolution of the land plants was accompanied by a modification of the diplobiontic life cycle. The hypothetical ancestral state among the land plants is still found in extant bryophytes, in which the diploid sporophyte stage is smaller than the haploid gametophyte stage. The life cycle later went through several transitions, including a phase in which both stages were equal in size (as seen in the rhyniophytes, such as *Aglaophyton*), and ultimately leading to the seed plants, in which the female haploid gametophyte is diminished in size and enclosed entirely within the tissue of a very large sporophyte. We can now summarize some details of four types of land-plant life cycles:

1. *Large gametophyte and small sporophyte.* The most ancestral diplobiontic life cycle in the land plants is characterized by a "parasitic" sporophyte developing and maturing in close association with the gametophyte; this is found in the bryophytes (mosses, liverworts, and hornworts). For example, in mosses, on germination of a haploid spore, a filamentous system of cells develops. It comprises two cell types: **chloronema** cells, which contain large chloroplasts, and **caulonema** cells, from which the leafy gametophytes of the moss plant develop (see Figure 1–30C). The archegonia and **antheridia** (female and male gamete-producing organs) form at the top of the leafy gametophyte. Once fertilization has occurred, the diploid sporophyte develops *in situ* at the top of the gametophyte and draws on the gametophyte for carbon and nutrients (similar to the situation in liverworts; see Figure 1–14B). Meiosis and spore formation occur when the sporangium, the capsule at the top of the sporophyte, has matured (see Figure 1–30C). Note the contrast to the situation in the Charales, in which meiosis immediately follows nuclear fusion at fertilization and there is no multicellular diploid stage.

2. *Morphologically similar gametophyte and sporophyte.* In the protracheophytes such as *Aglaophyton*—plants with morphologically distinct water-transporting cells resembling tracheids—the sporophyte and gametophyte stages were independent and morphologically similar. Both the gametophytes and the sporophytes were branched and leafless, bearing at their tips either sporangia, in the case of sporophytes (see Figure 1–17), or specialized gamete-producing organs (antheridia and archegonia, producing male and female gametes, respectively), in the case of gametophytes. The contrast with the relatively simple sporophyte of the bryophytes indicates that elaboration of the sporophyte stage of the life cycle had occurred among the earliest protracheophytes or their immediate precursors, for which we have no record. No extant plants have life cycles in which the gametophyte and sporophyte are similar in size and appearance, so this life cycle has been reconstructed from fossils from between 410 and 390 million years ago (the Upper Silurian and Lower Devonian Periods).

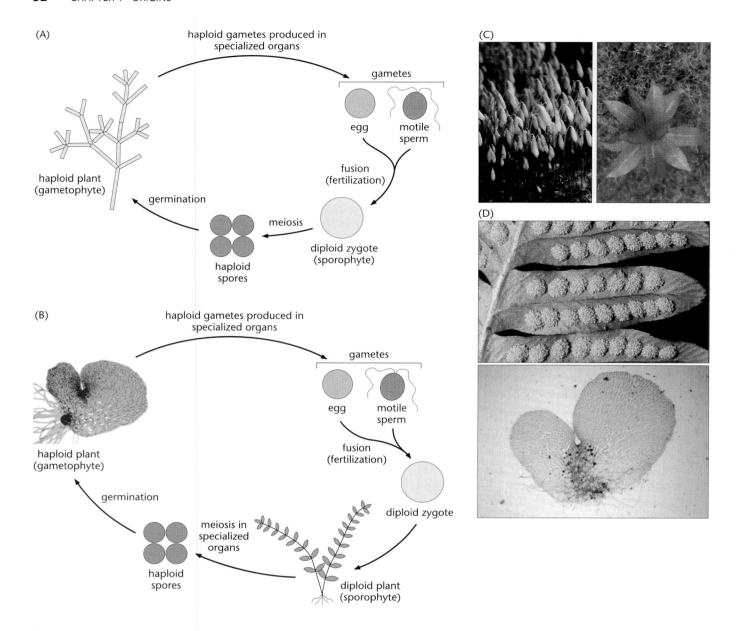

Figure 1–30.

Haplobiontic and diplobiontic life cycles. (A) The haplobiontic life cycle of members of the Charales, the immediate ancestors of the land plants. The haploid plant gives rise to male and female gametophytes (egg and sperm), produced in separate, specialized organs on the plant body. Fusion of the gametes gives a single-celled, diploid zygote. This undergoes meiosis, and a haploid spore germinates and grows into the next generation of the haploid plant. (B) The diplobiontic life cycle of land plants (shown here is a simple version of the fern life cycle). The haploid generation (in ferns, a tiny organism called a prothallus) gives rise to male and female gametophytes (egg and sperm), produced in separate, specialized organs on the plant body. Fusion of the gametes gives a single-celled, diploid zygote. Mitotic cell division of the zygote (not meiosis) gives rise to the multicellular, diploid sporophyte generation (in ferns, the adult fern plant). Meiosis in specialized organs of the sporophyte gives rise to haploid spores, which germinate and grow into the next haploid generation. (C) Maturing sporangia (capsules) of a moss of the genus *Bryum* (*left*); the leafy gametophytes from which the sporophytes are emerging are visible at the bottom of the photograph. A leafy gametophyte of the moss *Physcomitrella* (*right*) emerges from a mass of filaments produced after germination of the haploid spores. (D) The underside of a fern leaf (*top*), showing specialized structures (sori, singular sorus) that contain golden-colored clusters of sporangia. Each sporangium produces many spores. These are released and dispersed by the wind when the mature sporangium bursts open. A fern prothallus (*bottom*), the haploid gametophyte generation, consists of a few hundred cells, anchored by rhizoids (visible in the picture) to the substrate. Male and female gametes (sperm and eggs) are produced in different organs on the surface of the prothallus—antheridia and archegonia, respectively (*not shown*). The motile sperm swim in surface water to the archegonia, where fertilization takes place. (Cii, courtesy of Ralf Reski; Dii, courtesy of Michael Knee.)

3. *Small gametophyte and larger sporophyte.* In the next type of life cycle to evolve, the diploid sporophyte became dominant. This is observed in many fossil species and also in extant monilophytes (such as the lycophytes, ferns, and horsetails). The free-living gametophyte is smaller than the sporophyte, as exemplified by the fern life cycle, in which the "fern plant" (the diploid sporophyte) produces spores that germinate into independent haploid gametophytes of a few hundred cells—a **prothallus** (see Figure 1–30B, D). Antheridia and archegonia may develop on the same or on different prothalli. After fertilization, the zygote develops into a sporophyte and the tiny prothallus structure dies.

4. *Sporophyte enclosing the female gametophyte, and a dispersed male gametophyte (pollen).* The most highly derived life cycle is that in which the female gametophyte remains enclosed within sporophyte tissues. After meiosis of the megaspore, the haploid megagametophyte develops within the spore into a structure of a few cells (see Figure 1–27B). This remains embedded within tissues of the sporophyte. In contrast, the male gametophyte—the pollen—is mobile (airborne or carried by other means such as insects). It lands in the vicinity of a female gametophyte and produces a tube that delivers sperm nuclei through the tissues of the sporophyte (via the micropyle; see Figure 1–27B) to the egg cell of the megagametophyte. This type of life cycle is found in all the seed plants, both extant and extinct.

In summary, then, the first 100 million years of land-plant evolution involved dramatic changes in the life cycle and was accompanied by a relative decrease in the size and complexity of the gametophyte and an increase in complexity of the sporophyte.

Five groups of seed plants live on earth today

The five living groups of seed plants, as noted earlier, are the Cycadales, Ginkgoales, Coniferales, and Gnetales (the gymnosperms), and the angiosperms. To understand their origins—that is, which groups gave rise to which—we need to know the relationships among the five groups. There is as yet no consensus on this, but two sets of hypothetical relationships have been proposed, with radically different implications for the origins of angiosperms. The "anthophyte" hypothesis postulates that the closest living relatives of angiosperms among the living gymnosperms are the Gnetales; this hypothesis is based on comparative morphology. However, comparisons of DNA sequences suggest an alternative set of relationships, dubbed the "gnepine" hypothesis (Figure 1–31). The gnepine hypothesis indicates that living Gnetales are more closely related to conifers than to angiosperms and, by implication, the lineage leading to angiosperms is much more ancient. At present, the gnepine hypothesis seems likely to be accepted.

We now turn our attention to the history of the flowering plants—the angiosperms.

1.5 ANGIOSPERMS

Angiosperms (from the Greek *angeion*, "vessel," and *sperma*, "seed") are the subset of seed plants that develop **flowers**. They constitute one of the major clades of extant seed plants. Angiosperms are the largest group of embryophytes, and they dominate the world's flora in almost every habitat except the highest mountain tops, the poles, and the oceans. They show a tremendous variety of forms. They may be terrestrial plants, rooted in soil; **epiphytes**, growing on other plants; or aquatic plants, rooted or floating in marine or freshwater habitats. Although the exact number of angiosperm species is not known, it has been conservatively estimated as at least 220,000 extant species in more than 450 families.

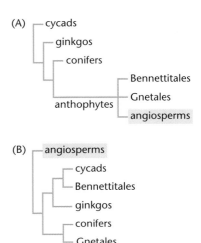

Figure 1–31.
Hypotheses for the relationships among groups of extant seed plants. These relationships are unclear; the two main hypotheses are shown. (A) The anthophyte hypothesis, based on studies of comparative morphology of living and extinct seed plants, proposes that angiosperms are closely related to Gnetales and the extinct Bennettitales, and less closely related to cycads, ginkgos, and conifers. (B) The gnepine hypothesis, based on comparisons of DNA sequences of living seed plants, proposes that the Gnetales are closely related to the conifers and that all of the gymnosperms are more closely related to one another than to the angiosperms.

Angiosperms appear in the fossil record in the Early Cretaceous Period, about 135 million years ago

The oldest angiosperm fossils are from approximately 135 million years ago, with the major lineages evolving between 130 and 90 million years ago. The characteristics of the flower—floral size, organization of floral parts, and structure—varied greatly among the early angiosperms, indicating an early increase in the diversity of floral forms. This initial radiation was followed by a rise to ecological dominance 100 to 70 million years ago. The world flora is now dominated by angiosperms with a huge diversity of floral forms and plant morphologies.

Angiosperms have some shared characteristics that are not present in any other group. These include particular features of the seed (the development of **endosperm**), of the reproductive structures (the presence of flowers in which ovules are surrounded by a **carpel**; **stamens** with two pairs of **pollen sacs**), and of the **phloem**, the transport system for sugars produced in photosynthesis.

Although the angiosperms form a clearly defined monophyletic group, their origin, as indicated above, remains obscure (see Figure 1–31). Determining their origin requires identification of their closest non-angiosperm relative. We might expect the closest relative to lie among the gymnosperms, the other group of seed-bearing plants, with a fossil record older than that of the angiosperms. Unfortunately, as we have noted, it is unclear which of the gymnosperm groups is sister to the angiosperms. More phylogenetic studies are required to resolve the relationships among these groups.

Documentation of the spread of angiosperms from the fossil record highlights the increase in their diversity over time, but tells us nothing about the driving forces behind angiosperm evolution and diversification. Indeed, we may never know what these forces were. Nevertheless, it has been suggested that major environmental changes during the Cretaceous Period may have played an important role, by creating new ecological niches into which the angiosperms radiated. The Cretaceous Period saw the breakup of the supercontinent of Pangea, bringing about volcanic activity, extensive flooding of continental shelf areas, and consequent changes in climate. Another contributory factor may have been an increase in the number and diversity of species of winged insects that could carry pollen from one plant to another (the relationships between plants and pollinating insects are discussed in Section 8.5).

Angiosperms evolved in the tropics and then spread to higher latitudes

The spread of angiosperms from 135 million years ago onward can be documented in detail, because angiosperm pollen, recovered from the fossil record, is easily distinguished on morphological grounds from the pollen of other seed plants (Figure 1–32).

Figure 1–32.
Pollen grains of a gymnosperm and angiosperm. The pollen grains are shown in colored scanning electron micrographs. (A) Pollen from a conifer, *Pinus sylvestris*. The grains have characteristic winglike air sacs, which aid in wind dispersal. (B) Pollen from an angiosperm, the ragweed *Ambrosia*. The patterns on the outer wall of the grain (the exine) differ from one species to another. This is an invaluable aid in reconstructing past vegetation types from the fossil record. It is also important in forensic science: identification of pollen grains on a piece of evidence can allow tracing of its origins and movements. *Ambrosia* pollen grains are also of interest because they are a major cause of hay fever in North America. A particular protein within the pollen grain is an allergen, causing inflammation of the mucosal membrane.

(A)

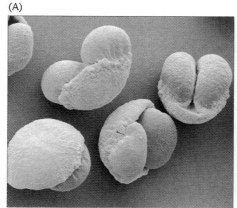

(B)

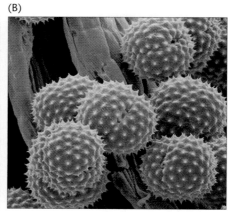

The earliest fossilized angiosperm pollen was found in deposits in Israel and Morocco. Over the next 30 million years, angiosperms spread to most latitudes. Within 60 million years they rose to dominate the equatorial vegetation. By this time they constituted about 30% of the vegetation at higher latitudes.

The impact of angiosperms on the Cretaceous vegetation of Australia illustrates how flowering plants invaded and eventually rose to dominance. In the Early Cretaceous Period, Australia was part of the Gondwana supercontinent. The area that now makes up central Australia had many rivers and lakes and extensive forests. The forest canopy was made up of coniferous trees, including members of the Podocarpaceae and Araucariaceae (including the monkey-puzzle trees), which today are taxa characteristic of the southern hemisphere. The understory included pteridosperms, cycads, Bennettitales, monilophytes (lycophytes, ferns, and horsetails), and bryophytes. When the first angiosperms appeared, around 120 million years ago, they inhabited environments next to rivers and lakes. Over the next 30 million years, they diversified and spread into more varied habitats. By 90 million years ago there is evidence for angiosperm trees, and by 84 million years ago not only had angiosperms displaced the conifers as the dominant forest species, but angiosperms also formed extensive understory communities. The events of Australian angiosperm radiation may mirror the events that occurred across the world during the early radiation of the angiosperms from their sites of origin.

Amborella trichopoda is sister to all living angiosperms

Phylogenetic trees have been constructed to show the hypothetical relationships among the angiosperms. Such trees indicate that angiosperms are divided into the **eudicots** (Eudicotyledons), the **monocots** (Monocotyledons), related groups including the **magnoliids**, the aquatic genus *Ceratophyllum* and the Chloranthaceae, and several other groups that form the lower branches of the tree, which together are called the **basal angiosperms**. Of these basal eudicots, *Amborella trichopoda* is thought to be sister to all other living angiosperms—that is, *A. trichopoda* and all angiosperms are

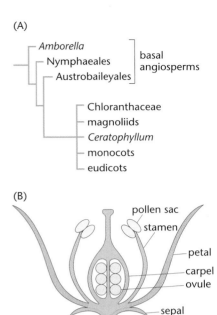

Figure 1–33.
The angiosperms. (A) Phylogenetic tree showing the hypothetical relationships among the extant angiosperms. Ideas about these relationships are being refined as more plant DNA sequences become available and informatics-based methods of deducing the time of divergence of different plant lineages are improved. (B) Section of an angiosperm flower, showing features unique to angiosperms. The ovules (see Figure 1–27) are contained within a carpel. The male gametophytes are produced in pollen sacs on stamens. The reproductive organs are surrounded by petals. It is important to note the huge variation among angiosperms in the size and morphology of flowers; some of this variation is illustrated in the remainder of the chapter.

Figure 1–34.
Members of the basal angiosperms. (A) Flowers and (B) fruit of *Amborella trichopoda*, a shrub native to New Caledonia. (C) Fruit of *Schisandra chinensis*, a woody vine native to China, a member of the Austrobaileyales. (D) The sacred lotus *Nelumbo nucifera*, a member of the Nymphaeales. (A and B, courtesy of Peter Endress; C, courtesy of Tobias Kieser.)

derived from a common ancestor (Figure 1–33A). *A. trichopoda* (Figure 1–34) is a shrub found only in the moist forests of New Caledonia, off the east coast of Australia. It is the sole member of the family Amborellaceae. It is **dioecious**, meaning that male and female floral parts develop on separate plants. The flowers are small with spirally arranged **perianth** segments. These are not clearly differentiated (into **sepals** and **petals**) and are called **tepals**. While there is general agreement that *A. trichopoda* is the most ancestral of living angiosperms, we cannot say whether the proto-angiosperm (the earliest angiosperm) possessed these morphological characteristics.

Among the more recently evolved angiosperms, the magnoliids include the Winterales, Piperales, Laurales, and Magnoliales (Figure 1–35). The monocots are a monophyletic group that includes the grasses, palms, and sedges; these are discussed further below. Altogether, the non-eudicot angiosperms make up about 25% of flowering plant diversity; 97% of these species are monocots.

The remaining 75% of flowering plant diversity is found among the eudicots. The species on the lowest branches of the eudicot tree—the "basal eudicots"—include groups such as the Papaveraceae (poppies) and the Ranunculaceae (buttercups). The more recently evolved (derived) groups, the "core eudicots," are divided into two clades: the Rosid clade, which comprises species with separate petals, and the Asterid clade, which contains species with fused petals. The plants of the core eudicot group show enormous morphological diversity (Figure 1–36). For example, although radial floral symmetry (**actinomorphy**) is considered the ancestral floral form, bilateral symmetry (**zygomorphy**) arose a number of times independently in the core eudicots and

Figure 1–35.
***Ceratophyllum* and the magnoliids.**
(A) *Ceratophyllum demersum*, a plant of freshwater lakes and rivers. (B) *Aristolochia gigantea*, the Dutchman's pipe (Piperales). (C) *Laurus nobilis*, the bay tree (Laurales). (D) *Liriodendron tulipfera*, the tulip tree (Magnoliales). (B-D, courtesy of Tobias Kieser.)

Figure 1–36.
Eudicots. (A) *Papaver somniferum*, the opium poppy (Papaveraceae), a basal eudicot. (B) *Lathyrus odoratus*, the sweet pea (order Fabales), a member of the Rosid clade. This is an example of a strongly zygomorphic (bilaterally symmetric) flower. (C) *Gentiana sino-ornata*, a gentian (order Gentianales), a member of the Asterid clade. (A, courtesy of Tobias Kieser.)

is thought to be the result of **coevolution** with insect pollinators (see Section 8.5). Furthermore, the woody habit (**arborescence**) evolved and was lost multiple times.

Eudicots are distinguished from other flowering plants by the number of pollen apertures

While the phylogeny based on DNA sequences can be used to identify relationships among groups of flowering plants, differences in pollen morphology are used to distinguish between the two main groups of angiosperms: the eudicots and everything else (Figure 1–37). The pollen of angiosperms other than eudicots is **monocolpate**, meaning that it has a single opening. This is the opening through which the **pollen tube** containing the sperm emerges once the pollen has landed on the receptive surface of a stigma. The eudicots, by contrast, produce pollen with three pores, or openings: **tricolpate** pollen. Tricolpate pollen is therefore a derived characteristic shared among the eudicots—a **synapomorphy**. This unique morphological trait reinforces the view that the eudicots constitute a monophyletic group.

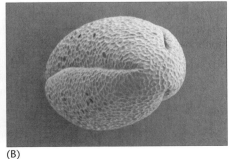

Figure 1–37.
Two distinct types of angiosperm pollen grains. Scanning electron micrographs of (A) a monocolpate pollen grain, showing a single pore through which the pollen tube will emerge, and (B) a tricolpate pollen grain. (A and B, courtesy of Kim Findlay.)

(A)

(B)

(A)

(B)

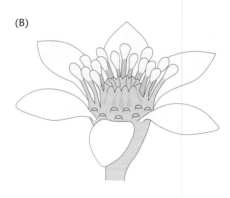

Figure 1–38.
Fossil flower. A fossil flower similar to a modern water lily, a basal angiosperm. This fossil dates from the Early Cretaceous Period, between 125 and 112 million years ago. (A) Fossil viewed from the side. (B) Artist's reconstruction of the flower. (A, from E.M. Friis et al., *Nature* 410:357-360, 2001. With permission from Macmillan Publishers Ltd, courtesy of Else Marie Friis; B, from E.M. Friis et al., *Nature* 410:357-360, 2001. With permission from Macmillan Publishers Ltd, courtesy of Pollyanna von Knorring.)

The earliest angiosperm flowers were small with many parts

Fossilized pollen, as we have noted, has proved useful in revealing when angiosperms first appeared. Pollen morphologies similar to those of basal angiosperm and basal eudicot groups are evident soon after the first appearance of angiosperm pollen in the fossil record, suggesting that the diversification of some of the main groups of flowering plants occurred rapidly during a radiation at about this time.

Given their delicate structure, flowers are not easily preserved in sediments. But exceptional fossils have been found in fine sedimentary deposits that give a picture of the earliest floral morphologies. The oldest fossil flowers (Figure 1–38) are found in deposits between 125 and 112 million years old. These flowers are small, with floral structures arranged in whorls, and radially symmetric. They are similar to the flowers of extant basal angiosperms, and some may be extinct members of these families. Although the basal angiosperms constitute only a small fraction of today's flowering plants, they are the relics of groups that were the dominant angiosperm taxa of the Early Cretaceous Period.

Unfortunately, the floral fossils are found in isolation, unattached to substantial plant remains, so we do not know whether the plants that bore these early flowers were trees, shrubs, or herbs. We can only await the discovery of further fossils that consist of whole branches or herbs that bear flowers.

Monocots are a monophyletic group

The monocots are derived from a single common ancestor. A major shared trait that distinguishes this monophyletic group from other groups of flowering plants is the development of a single **cotyledon**, or seed leaf, rather than two cotyledons (Figure 1–39A, B). Another prominent and characteristic monocot trait is the occurrence of parallel veins in the leaves. Most other angiosperms have netlike patterns of leaf veins: net **venation** is an ancestral character among the flowering plants (Figure 1–39C, D). Some genera of monocots, such as *Dioscorea* (yams), *Trillium*, and *Smilax*, have net venation, thought to result from the loss of parallel venation and independent evolution of net venation.

The oldest fossil monocots are palms and aroids (the arum family), found in rocks approximately 100 million years old. By the end of the Cretaceous Period, 65 million years ago, all the major groups of monocots had evolved. Subsequent speciation and extinction events may have increased the number of species but did not lead to the evolution of dramatically different lineages of monocots.

Most monocots have "non-showy" flowers; their petals and sepals are arranged in whorls of three (that is, the flowers are trimerous) and lack the colorful pigmentation found in most eudicot groups. Notable exceptions are the "petaloid" monocots that comprise the Liliales (lilies), Asparagales (asparagus relatives), and Dioscoreales: these have large petal-like structures (tepals) (Figure 1–40D). There is no evidence, however, that these three monocot orders form a monophyletic group, so they probably developed tepals independently during evolution.

The phylogenetic tree that shows the relationships among the groups of monocots (Figure 1–40A) suggests that *Acorus* is sister to all other living groups. The most derived group of plants is the Commelinid clade, which includes the Arecales (e.g., the palms), the Commelinales (e.g., spider plants and relatives), the Poales (e.g., the grasses), and the Zingiberales (e.g., the gingers and relatives). The Commelinid clade appeared 65 million years ago (the end of the Cretaceous Period), but considerable further diversification has occurred since that time. To highlight the role of more recent evolutionary events in generating diversity, we focus next on the grass family, the Poaceae.

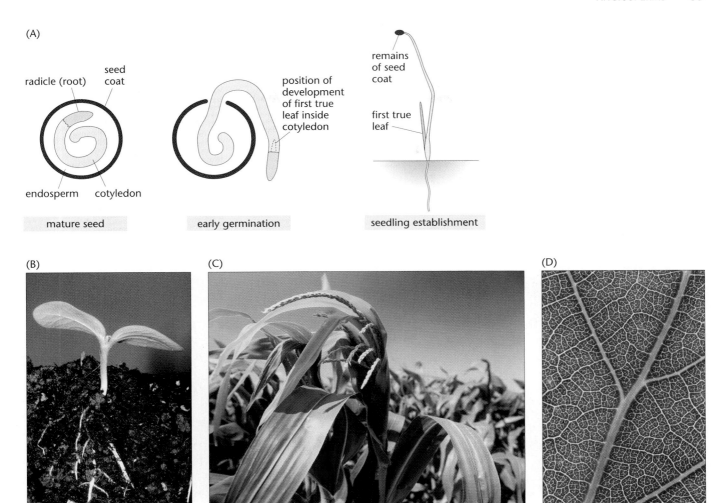

Figure 1–39.
Distinguishing features of monocots and eudicots. (A) Three successive stages in the germination of a monocot seed (an onion, *Allium*). Note the presence of a single cotyledon, from which a single true leaf emerges as the seedling establishes. This feature is characteristic of and unique to the monocots. The endosperm is a nutritive tissue that surrounds the embryo inside the seed (see Chapters 4 and 5 for descriptions of endosperm development and function). (B) Seedling of a eudicot (pumpkin). Note the two large seed leaves. The first pair of true leaves is starting to emerge from between them. (C) Leaves of maize, showing the parallel arrangement of veins (vascular bundles) typical of monocots. (D) The underside of a leaf of *Populus grandidentata*, showing the net arrangement of veins typical of eudicots.

The grass family (Poaceae) evolved about 60 million years ago but diversified more recently

The grass family includes approximately 10,000 species that together cover 20 to 30% of the land surface on the planet. Many of these species are of great economic importance as sources of food, energy (biofuels), and building materials. They include rice, maize, wheat, sorghum, millet, barley, oats, and many others. Grasses have some shared derived traits that distinguish them from other species (Figure 1–41). These traits include the development of distinctive small florets, the formation of a modified cotyledon called a **scutellum** during **embryogenesis**, and a distinctive **fruit** called a **caryopsis**, in which the outer layer of the integument (surrounding the ovule) is fused to the inner wall of the carpel.

Preserved pollen microfossils indicate that grasses evolved between 70 million years ago (late Cretaceous) and 55 million years ago (Paleocene Epoch of the Tertiary Period). These early grasses grew in forest margins and deep shade, where basal members of the family, such as *Anomochloa* and *Streptochaeta*, grow today. A phylogeny of grasses (Figure 1–41A) shows that the subfamily Anomochlooideae is sister to the rest of the Poaceae family. Relatively few species belonging to the ancestral subfamilies have survived. The subfamilies that together form the first three branches of the **cladogram** comprise only 25 extant species. There are two major species-rich clades. The PACCAD clade includes the Panicoideae (including maize, millet, and sorghum), and the BEP clade includes the species-rich subfamily Pooideae (including

Figure 1–40.
Monocots. (A) Phylogenetic tree showing the hypothetical relationships among groups of monocots. Examples of plants within the groups are indicated in parentheses. (B) A flowering spike of *Acorus calamus* (sweet flag), a plant of pond and river margins. The Acorales are at the base of the monocot phylogenetic tree. (C) *Tradescantia occidentalis*, an example of the Commelinales, one of the most recently evolved monocot groups. (D) A lily (*Lilium* sp.), an example of a monocot group that has showy flowers because of the presence of large tepals. (B–D, courtesy of Tobias Kieser.)

wheat) and two less species-rich families, the Bambusoideae (including the bamboos) and the Ehrhartoideae (including rice).

Grass species remained relative rarities in the world's vegetation until around 33 million years ago, when an increase in species diversity started to occur. The increased diversity coincided with continental drying worldwide. The grasses probably expanded into the new drier habitats from the tropical forest edges and the shady habitats typically occupied by ancestral grasses. Evidence for this expansion, and for the first formation of extensive grasslands, is provided by the increased abundance of grass fossils from this time and the first appearance of fossilized soils (paleosols) with the characteristics of grassland soils.

The development of extensive grasslands coincided with the appearance of groups of grazing mammals. Many ungulates (hoofed mammals) developed specialized teeth that allowed the maceration of tough, silicon-containing grass tissues (see Section 4.9). Grass leaves grow through the activity of meristems at the base of the leaf **blade (basal meristems)**. This means that the removal of distal portions of the leaf by grazers does not kill the leaf; it can continue to grow and function. Characteristics such as teeth suitable for a grass diet in mammals and plant growth forms that allow grazing without damage to meristems are likely to be the result of coevolution of grasses and grazers over long periods of time.

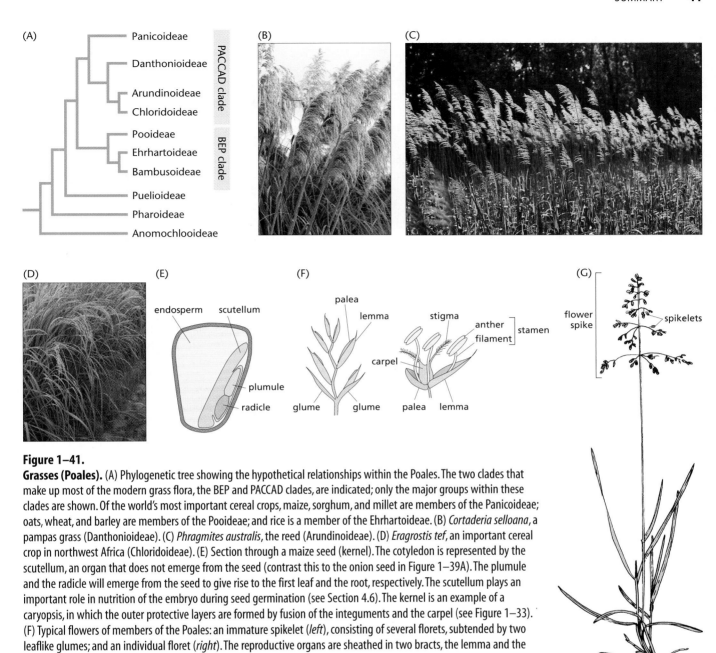

Figure 1–41.
Grasses (Poales). (A) Phylogenetic tree showing the hypothetical relationships within the Poales. The two clades that make up most of the modern grass flora, the BEP and PACCAD clades, are indicated; only the major groups within these clades are shown. Of the world's most important cereal crops, maize, sorghum, and millet are members of the Panicoideae; oats, wheat, and barley are members of the Pooideae; and rice is a member of the Ehrhartoideae. (B) *Cortaderia selloana*, a pampas grass (Danthonioideae). (C) *Phragmites australis*, the reed (Arundinoideae). (D) *Eragrostis tef*, an important cereal crop in northwest Africa (Chloridoideae). (E) Section through a maize seed (kernel). The cotyledon is represented by the scutellum, an organ that does not emerge from the seed (contrast this to the onion seed in Figure 1–39A). The plumule and the radicle will emerge from the seed to give rise to the first leaf and the root, respectively. The scutellum plays an important role in nutrition of the embryo during seed germination (see Section 4.6). The kernel is an example of a caryopsis, in which the outer protective layers are formed by fusion of the integuments and the carpel (see Figure 1–33). (F) Typical flowers of members of the Poales: an immature spikelet (*left*), consisting of several florets, subtended by two leaflike glumes; and an individual floret (*right*). The reproductive organs are sheathed in two bracts, the lemma and the palea. These form the "husk" that surrounds some mature grass seeds. (G) A member of the genus *Poa*, showing the flowering spike. (C, courtesy of Tobias Kieser.)

SUMMARY

The earth formed 4.6 billion years ago from the solar nebula. By the end of the Hadean Era, 3.8 billion years ago, the atmosphere was rich in carbon dioxide and water vapor, but had no oxygen gas and no protective ozone layer. Over the next 2 billion years, the evolution of oxygenic photosynthesis and the gradual oxidation of the oceans resulted in accumulation of oxygen in the atmosphere. Photosynthetic bacterial cells had probably evolved by 3.8 to 3.5 billion years ago, and oxygen-producing photosynthesis was widespread by 2.2 billion years ago.

Eukaryotic organisms had evolved by 2.7 billion years ago. Photosynthetic eukaryotic cells arose from two or more endosymbiotic events: a host cell acquired a bacterium that became the mitochondrion, and this proto-eukaryotic cell type later acquired a cyanobacterium that became a plastid. Three clades were directly derived from this oxygenic eukaryotic ancestor: the glaucophytes, rhodophytes (red algae), and chlorophytes (green algae). For a billion years diversity increased slowly, then the rate accelerated and all major groups of macroscopic algae had appeared by 600 million years ago.

Land plants probably evolved from a small class of predominantly freshwater green algae. Plants (bryophytes) were growing on continental surfaces by 470 million years ago. A gradual increase in plant diversity followed. By the Carboniferous Period (350–300 million years ago), complex forest ecosystems contained plants with a great diversity of size and morphology. A combination of biological and geological events has determined the atmospheric CO_2 and O_2 levels over the past 600 million years, having dramatic effects on the evolution of land plants.

All land-plant life cycles have a multicellular haploid sexual stage (gametophyte) alternating with a multicellular diploid asexual stage (sporophyte). The life cycle went through several transitions, ultimately leading to the situation in seed plants, in which the female gametophyte is diminished and enclosed entirely within a very large sporophyte. The earliest seed plants were probably derived from spore-bearing plants. Seed plants evolved in the Devonian and diversified in the Permian, 290 to 250 million years ago. The angiosperms were the last of the seed plant groups to appear. The oldest angiosperm fossils are found in rocks from the Early Cretaceous Period, 135 to 130 million years ago.

The angiosperms are seed plants that develop flowers; they dominate the world's flora in most land habitats. They form a clearly defined monophyletic group, but their origin remains obscure. The earliest angiosperm flowers were small with many parts arranged in whorls, but angiosperms diversified into species with a tremendous variety of floral forms. Within the angiosperms, two groups dominate in terms of species number and importance in the land flora: the eudicots and monocots (grasses and their relatives).

FURTHER READING

General

Gensel PG & Edwards D (2001) Plants invade the land. New York, NY: Columbia University Press.

Karol KG, McCourt RM, Cimino MT & Delwiche C (2007) The closest living relatives of the land plants. *Science* 294, 2351–2353.

Kellogg EA (2001) Evolutionary history of the grasses. *Plant Physiol.* 125, 1198–1205.

Soltis PS & Soltis DE (2004) The origin and diversification of angiosperms. *Am. J. Bot.* 91, 1614–1626.

Willis KJ & McElwain JC (2002) The evolution of land plants. Oxford: Oxford University Press.

GENOMES

2

When you have read Chapter 2, you should be able to:

- Describe how DNA and protein are organized in plant chromosomes, distinguish between euchromatin and heterochromatin, and describe the functions of centromeres and telomeres.

- Describe the typical structure of a plant gene and the origin of different RNAs produced in the plant nucleus.

- Differentiate between the two classes of transposons, and summarize the role of transposons in genome expansion during evolution.

- Summarize how nuclear gene function is controlled at the transcriptional level, listing the components and activities of the basal transcriptional machinery.

- List the types of covalent modification that can occur in chromatin, and note the functional consequences.

- Summarize the ways in which nuclear gene function is controlled at the RNA level in plant cells.

- Name the types of small regulatory RNAs in plants and their role in nuclear gene function.

- Describe the process of gene annotation, using the Arabidopsis genome as an example.

- Explain how polyploidy arises, and summarize its role in evolution.

- Outline the mechanisms and consequences of gene duplication and divergence.

- Define "synteny" and explain its importance in evolutionary studies and in crop improvement.

- Summarize how genetic markers are used to localize genes in a genome.

- Describe the techniques of transposon tagging, reverse genetics, quantitative trait locus (QTL) analysis, and expression arrays, and give examples of the information they can provide.

- Outline the roles of the plastid genome and the nuclear genome in plastid function, and describe the signaling pathways between plastid and nucleus that affect plant development.

The enormous diversity of plant life on earth is the consequence of differences in genetic content among species, differences that ultimately result from the evolutionary processes described in Chapter 1. For each plant species, its heritable characteristics—those that distinguish the species and can pass from one generation to the next—are determined by its **genome**. For plants and other **eukaryotes**, the genome is the total genetic content of an organism's haploid set of chromosomes; for bacteria, it is the genetic content of a single chromosome.

In this chapter we describe the physical makeup of a plant genome, the type of information encoded in it, and how the genome can change over evolutionary time. Plants are unique in that each cell contains three distinct genomes: the nuclear, mitochondrial, and plastid genomes. By far, the largest genome is the nuclear genome.

The chapter begins with a review of some basic features of the plant **nuclear genome**—those that are shared with all other eukaryotes and those specific to plants. We consider structural sequences, different types of genes, and mobile genetic elements (transposons), then discuss how the activity of the thousands of different genes is regulated coordinately, focusing on the synthesis of RNA, the immediate product of gene activity. We then present an example of what is revealed by the complete DNA sequence of a plant genome, using the much-studied Arabidopsis as an example. As we shall see, many evolutionary insights and biotechnological advances have been gained from understanding the plant nuclear genome.

Reflecting their endosymbiotic origins, as described in Chapter 1, the two other genomes in plant cells—the **plastid genome** and **mitochondrial genome**— have features in common with bacteria. Our discussion of the cytoplasmic genomes will focus mainly on the chloroplast genome, since the chloroplast is the source of many of the distinctive characteristics of plants.

This chapter includes a brief review of gene structure and function, but we assume prior knowledge of basic genetics, including Mendelian inheritance and genetic linkage, and molecular biology, including the structure, function, and biosynthesis of DNA, RNA, and proteins. While reading this chapter, it may also be useful to revisit the mechanisms of mitosis and meiosis, in particular meiotic recombination, which are reviewed in Chapter 3.

2.1 THE NUCLEAR GENOME: CHROMOSOMES

The nuclear genome is subdivided into **chromosomes**, the molecular "vehicles" that enable the sets of genes to be duplicated and the duplicated copies to be accurately segregated to daughter cells. The **haploid** number (n) of chromosomes varies widely among plant species, from 2 (as in *Haplopappus gracilis* of the daisy family) to nearly 500 (in the fern *Ophioglossum petiolatum*). In general, however, related species have similar numbers of chromosomes (for example, all *Pinus* species have 12). The large variation in plant chromosome number is in part due to the fact that many plants are polyploid; that is, their genome has arisen from duplication of an ancestral genome or from the combination of genomes from two ancestral species (see Section 2.4). In these cases, chromosome number generally varies in multiples of the chromosome complement of the basic genome set (also known as the C-value, or "base number," of a species or genus). However, smaller, stepwise changes in chromosome number can also occur, through the loss or gain of individual chromosomes due to defects in segregation during **meiosis** or **mitosis**.

The chromosomes provide a means to package and move large amounts of DNA during cell division. The backbone of each chromosome is a linear, very long piece of DNA, in the order of millions of base pairs (bp), containing thousands of genes. The total amount of DNA in the nuclear genomes of plants is very variable, ranging from 120 million bp in *Arabidopsis thaliana* (a widely studied reference plant species) to 130 billion bp—a thousand times more—in the lily species *Fritillaria assyriaca*. This variation in DNA content is greater in plants than in any other group of eukaryotes and again reflects in part the polyploid nature of plants—although there are additional reasons, described later in this chapter.

The tight packaging of large amounts of DNA in chromosomes occurs through association of the DNA backbones with proteins, forming **chromatin**. The basic packing units of chromatin are the **nucleosomes**, which consist of DNA wrapped around specialized proteins called **histones**. These are small, basic (positively charged) proteins with a high content of arginine and lysine, which allows close interaction with the acidic DNA (Figure 2–1). There are five types of histones, which are largely conserved in structure among the eukaryotic organisms, including plants.

Each nucleosome consists of 146 bp of DNA wrapped in two turns around a core particle containing eight histone molecules (two each of histones H2A, H2B, H3, and H4). This arrangement compacts the length of the DNA about sixfold, with about 80 bp of DNA associated with each full turn of the DNA in the core particle. Between consecutive core particles, a linker region of DNA with 20 to 35 bp is associated with another histone (H1), which is needed for further folding of the nucleosomes into a fiber of about 30 nm in diameter. This results in an approximately fortyfold compaction of the DNA compared with uncondensed DNA.

The 30-nm fibers are organized in loops that are attached to a scaffold of nonhistone proteins. These loops are gene-rich and interact with the **transcription** machinery to produce **RNA transcripts** (see below). The looped fibers are the typical structure of chromatin, called **euchromatin**, in which **gene expression** is actively occurring. In some parts of the chromosomes, however, the chromatin is compacted further in the form of **heterochromatin**. Genes in these regions are less accessible to the transcription machinery and are either silent or expressed at low levels. Heterochromatin is associated with specialized proteins not found in euchromatin, but the exact way in which heterochromatin is organized is not known.

A further level of chromatin condensation is seen during mitosis and meiosis (Figure 2–2), when the looped fibers of euchromatin are coiled together into the compact forms (**metaphase** chromosomes) that are visible with the light microscope (see Chapter 3). This high level of chromatin condensation allows the replicated chromosomes to be sorted to the daughter cells.

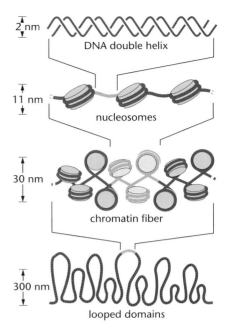

Figure 2–1.
The packaging of DNA in histones.
Increasing levels of compaction of DNA are shown from top to bottom: naked DNA; DNA wrapped around histones to form nucleosomes; nucleosome-wrapped DNA coiled into 30-nm fibers; looped 30-nm fibers forming the typical structure of euchromatin.

2.2 CHROMOSOMAL DNA

Specialized, repetitive DNA sequences are found in the centromeres and telomeres

The **centromeres** of chromosomes associate with a specific set of proteins that provide the attachment points for the **spindle** microtubules that separate chromosomes to opposite poles of the cell during mitosis and meiosis. In plants, as in other higher eukaryotes, a typical centromere spans approximately 1 million bp of DNA. Much of this DNA is made up of repetitive sequences (178 bp long in Arabidopsis). Long arrays of these repeats are found only in the centromeres, implying that they are important for centromere function. The DNA sequence of these repeats, however, is variable even among closely related species, suggesting that the exact sequence itself may not be important. One conserved feature of the centromeric repeats is that they associate with a special type of histone H3 (CenH3), which is found only in the centromeres. In other organisms, such as Drosophila (*Drosophila melanogaster*, the fruit fly), CenH3 is known to be essential for correct segregation of the chromosomes during mitosis.

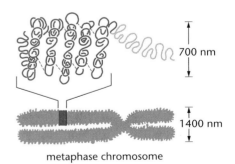

Figure 2–2.
Further compaction of DNA in metaphase chromosomes. The looped domains shown at the bottom of Figure 2–1 are coiled tightly to form the relatively thick (700 nm) chromosome arms seen during metaphase.

In addition to the long arrays of short repeats, the centromeres contain large numbers of DNA elements called **retrotransposons**, described in more detail below. This abundance of retrotransposons is characteristic of heterochromatin, which, as mentioned above, is characterized by a low density of active genes. Nevertheless, the centromeric regions of plant chromosomes do contain actively transcribed genes, albeit at about one-tenth the density found in the euchromatic regions of the chromosomes.

The similarity between centromeres and other heterochromatic regions of the chromosome provides a clue to how chromosome number can change during evolution, resulting in the large variations in chromosome number in plants mentioned earlier. In maize, certain heterochromatic regions known as "knobs," which normally do not have centromeric function, behave as centromeres during meiosis in plants that carry a particular variant of chromosome 10 (Abnormal 10). In these plants, the meiotic spindle attaches not only to the regular centromeres but also to the knobs. The maize example demonstrates that heterochromatic regions have the potential to be transformed into centromeres. The appearance of such "neocentromeres" could possibly be the first step toward the stepwise changes in chromosome number during evolution, because a chromosome fragment can be stably maintained as a separate chromosome only if it acquires its own functional centromere.

The centromeres are not the only regions of the chromosomal DNA that are essential to maintain a stable set of chromosomes through cell division. The ends of the chromosomes, called the **telomeres**, also have an important role in chromosome maintenance. The telomeres are made of hundreds to thousands of copies of a short DNA sequence (TTTAGGG in Arabidopsis and in most other angiosperms). The DNA end in a chromosome is not free, but loops back on itself and associates with telomere-specific proteins (Figure 2–3). This specialized structure is required to prevent the ends of chromosomes from fusing with each other to form joint chromosomes with multiple centromeres, which would disrupt chromosome segregation during mitosis and meiosis. The reason why fusion would occur in the absence of telomeres is that the cell recognizes free DNA ends as broken DNA, and a repair system joins the ends together to fix the damage. The specialized structure of the telomere is believed to conceal the ends of the chromosomal DNA from the cell's damage-surveillance system.

There is another reason for the special structure of the telomeres, which is related to the way DNA is replicated. During replication, one of the strands is synthesized as fragments that are initiated from short RNA molecules (**RNA primers**) hybridized to the template DNA strand (Figure 2–4). The DNA fragments are subsequently extended to replace the RNA primers and then joined into one long molecule. But at the very beginning of a linear DNA template, there is no upstream DNA fragment that can be extended to replace the RNA primer and, at each round of replication, the DNA should become a little shorter. One function of the telomeres is to correct this gradual loss of chromosome sequence. The DNA that makes up the telomeres is added to the ends of the chromosomes by a specialized ribonucleoprotein called **telomerase**; this enzyme uses its RNA subunit as a template to synthesize the short DNA sequences that are attached to the ends of the chromosomal DNA, resulting in the characteristic repetitive sequence of the telomeres.

The importance of the telomeres in maintaining chromosome structure is evident in Arabidopsis plants in which the protein subunit of telomerase has been lost by **mutation**.

Figure 2–3.
Scanning electron micrograph of telomeric DNA from pea. The large loop contains approximately 22,000 base pairs, estimated by comparison with DNA of known size (the plasmid seen as a small circle inside the telomere loop). (Courtesy of Jack Griffith.)

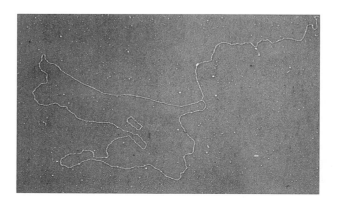

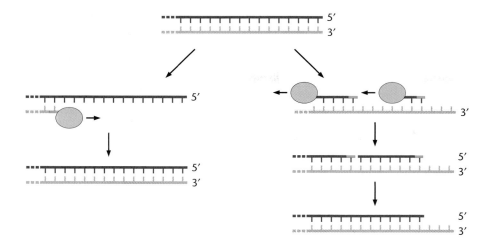

Figure 2–4.
Shortening of linear DNA at each replication round. DNA synthesis is initiated from short RNA molecules (primers, *blue*) hybridized to the template DNA strand. On one replicated strand (*right side*), each new DNA fragment is extended (by the enzyme represented by a *green blob*) until it meets the next primer downstream. The RNA primers are then replaced with DNA, and the DNA fragments are joined to form one long strand and thus a new DNA molecule. However, at the very beginning of this new DNA strand (*red*), a gap remains where there was no primer to initiate DNA synthesis. To correct this progressive shortening of DNA at each replication cycle, a special mechanism is used to extend the end of the DNA molecule without the need for a template strand.

In these plants and in their mutant progeny, the telomeres become progressively shorter (by 250–500 bp per generation). After 6 to 10 generations without telomerase activity the telomeres run out, and the plants become infertile and have serious growth defects due to chromosome fusions that result in chromosome breaks and loss of genetic material.

Just as the absence of telomerase activity causes chromosomal instability, telomerase activity can stabilize abnormal chromosomes such as those that result from joining of a broken fragment to another chromosome. If telomerase establishes new telomeres on the exposed chromosome ends before the DNA repair system is able to join them, the rearranged chromosomes can be stably maintained over subsequent cell divisions. Together with the appearance of neocentromeres, the stabilization of rearranged chromosomes by telomerase has probably been essential for the changes in chromosome number and the reshuffling of chromosome fragments during evolution (we discuss chromosome reshuffling, using the example of cereal genomes, in Section 2.4).

Nuclear genes are transcribed into several types of RNA

The chromosome arms that link the centromeres to the telomeres contain thousands of genes, which provide templates for the production of RNA molecules. These include **ribosomal RNAs (rRNAs)**, **messenger RNAs (mRNAs)**, **transfer RNAs (tRNAs)**, and **small nucleolar RNAs**. In addition to these "classical" RNAs, a more recently described type of RNA transcript is processed into small RNAs that function as negative regulators of gene expression, such as the **microRNAs (miRNAs)**, which are described in more detail in Section 2.3.

In plants, most RNA is produced by the three major types of **RNA polymerase** found in all eukaryotes: RNA polymerase I produces large rRNAs; RNA polymerase II transcribes mRNAs and the RNA precursors that are processed into miRNAs; and RNA polymerase III produces tRNAs, small nucleolar RNAs, and one of the small rRNAs. A small fraction of the cell's RNA is produced by another type of enzyme, using RNA as the template: the **RNA-dependent RNA polymerases**, which are encoded in the nuclear genome. These enzymes convert single-stranded RNA into double-stranded RNA (dsRNA), which then participates in the post-transcriptional regulation of gene expression mediated by **small interfering RNAs (siRNAs)**, as described in Section 2.3.

Three of the larger rRNAs are initially transcribed as a single RNA transcript (45S rRNA), which is subsequently spliced into three separate RNA molecules (in plants, the 26S, 18S, and 5.8S rRNAs). A fourth rRNA (5S rRNA) is transcribed from a separate gene. To sustain the production of the large amounts of rRNA needed in cells, the corresponding genes (45S rDNA and 5S rDNA) are present in hundreds to thousands of copies, which can constitute up to 10% of the nuclear genome (in Arabidopsis, 8%). The rDNA gene copies lie together in one or a few regions of the chromosomes. These regions are contained within the **nucleolus** (Figure 2–5), which is the site where **ribosome** synthesis is initiated (some of the protein components are subsequently assembled in the cytoplasm).

In contrast to the genes encoding rRNA, genes that are transcribed to produce mRNAs are present as single copies or as small families of related genes; they are dispersed throughout the chromosomes, although they are less frequent in the centromeric and other heterochromatic regions. The typical structure of a protein-encoding plant gene is exemplified by the *rbc*S gene, which encodes the small subunit of **Rubisco**, the enzyme responsible for incorporation of most of the carbon dioxide into sugars during **photosynthesis** (see Section 4.2). Plants usually have small families of *rbc*S genes (for example, there are five *rbc*S genes in tomato and five in Arabidopsis). Figure 2–6 shows a comparison of a representative *rbc*S gene from several plant species, highlighting the regulatory region, the transcribed sequence, and the protein-encoding sequence.

In plants, as in all higher eukaryotes, the protein-encoding genes are interrupted by **introns** (recall that the protein-coding sequences are called **exons**); introns are relatively short in plants (usually tens to hundreds of base pairs, compared with thousands in animals). Within a family of related genes, the size, position, and DNA sequence of the introns are more variable than for the protein-coding sequences. What marks certain regions of a transcript to be removed as introns is not completely understood. In part, it depends on short sequences that are characteristic of exon–intron boundaries (usually containing |GU at the start of an intron and AG| at the end, with the bar representing the exon–intron boundary). These short sequences are necessary but not sufficient to mark the intervening sequence to be removed as an intron. Additional features are required; for example, in **eudicots** the introns are also characterized by a higher A and T content than the exons.

The regulatory sequences that determine when and in which cells a gene is transcribed are usually contained within a few hundred to a few thousand base pairs of

Figure 2–5.
Electron micrograph and schematic view of the nucleolus. The chromosomal DNA in the nucleolus contains many copies of the genes that produce ribosomal RNA (rDNA copies in *red*). Each gene copy functions as the template to produce an rRNA precursor (shown as progressively longer *green strands*), and precursors are subsequently spliced to form the final rRNAs. The rRNAs associate with proteins to form ribosomal subunits, which are exported to the cytoplasm to complete the assembly of ribosomes. The electron micrograph shows a nucleolus from pea. (From P.J. Shaw et al., *EMBO J.* 14(12):2896-2906, 1995. With permission from Macmillan Publishers Ltd, courtesy of Peter Shaw.)

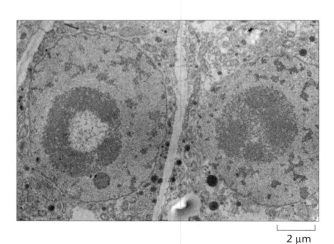

2 µm

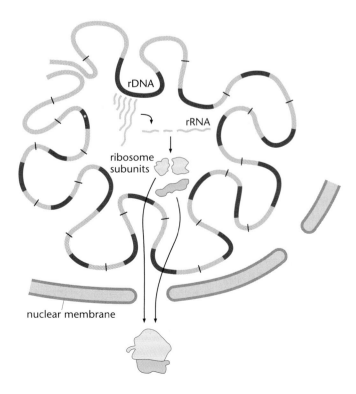

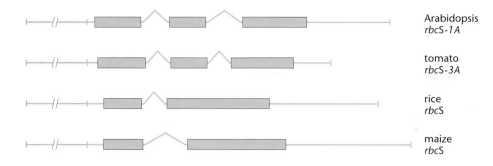

Arabidopsis
*rbc*S-*1A*

tomato
*rbc*S-*3A*

rice
*rbc*S

maize
*rbc*S

Figure 2–6.
**Schematic representation of *rbcS* genes
from several plant species.** The upstream
regulatory sequences are shown in *gray* and
the transcribed sequences in *green*. Protein-
coding sequences (indicated by *green boxes*)
are joined together in the mature mRNA, after
the introns (*angled green lines*) are spliced.
The protein-coding sequences are preceded
and followed by 5′ and 3′ untranslated RNA
sequences (*straight green lines*). The sizes (and
sequences) of regulatory regions, introns, and
untranslated RNA are variable across the gene
family, whereas the coding sequences tend
to be similar.

the transcriptional start site. In many cases, as in the *rbc*S example, the regulatory sequences precede the transcribed region (relative to the direction of transcription). Like introns, the regulatory sequences are much more variable among related genes than are the protein-coding regions, but short sequences that are recognized by regulatory proteins can be conserved (see Section 2.3).

The genes that produce small regulatory RNAs are also dispersed throughout the chromosomes. The best characterized of these genes produce the microRNAs (miRNAs), which are probably encoded by a few hundred genes present as single copies or small gene families. They are transcribed initially as precursor RNA molecules a few hundred bases long, which are subsequently processed into mature miRNAs (as described in Section 2.3). Like other transcripts produced by RNA polymerase II, the miRNA precursors are polyadenylated and can also be interrupted by introns.

Plant chromosomes contain many mobile genetic elements

The rDNA genes, centromeric repeats, and telomere sequences are not the only types of DNA sequence that are present in large numbers of copies in the nuclear genome. Another class of repetitive sequences is the **transposons**, many of which can replicate themselves independent of the rest of the genome and then reinsert in different locations.

Transposons can be separated into two main classes. Class I includes the retrotransposons, so-called because of their similarity to **retroviruses**. They replicate through an RNA intermediate and often encode a **reverse transcriptase**, an enzyme that uses RNA as the template to produce a **complementary DNA (cDNA)** molecule. Thus a reverse transcriptase generates the DNA copies that insert into new locations in the host genome; for this reason, class I transposons are sometimes also called "copy and paste" transposons. Retrotransposons may have evolved from retroviruses, but unlike viruses, they lack the ability to move from cell to cell. Because retrotransposons can move only by replication, further movement causes an increase in the number of transposons inserted in the genome.

Class I transposons can be further divided into two subclasses, according to their sequence organization and content (Figure 2–7). Transposons in the first subclass are flanked by **long terminal repeats (LTRs)**, which have two roles: (1) they control expression of a gene encoding a reverse transcriptase, which uses RNA transcribed from the retrotransposon as a template for cDNA; and (2) they participate in the reactions that integrate the cDNA copy into the host genome. Members of the second subclass of retrotransposons lack LTRs and terminate instead with poly-A sequences. They do not encode a reverse transcriptase and are believed to move passively around the genome, perhaps dependent on reverse transcriptase activity encoded by other retrotransposons.

Class II transposons do not involve an RNA intermediate during transposition. Instead, the DNA sequence is excised from one point of the genome and inserted into another.

Figure 2–7.
Class I ("copy and paste") transposons.
(A) Mechanism of transposition: the retrotransposon (*double red line*) produces an RNA copy of itself (*single blue line*), which encodes, among other proteins, a reverse transcriptase. This enzyme copies the RNA copy back to a DNA molecule, which is inserted elsewhere in the genome.
(B) Structure of a retrotransposon of the subclass containing long terminal repeats (LTRs), which flank coding sequences: *gag* is a gene related to viral structural proteins; *pol* encodes a polypeptide precursor that is cleaved to produce the reverse transcriptase and a protein required for integration of the retrotransposon. (C) Some retrotransposons lack LTRs and instead contain a poly-A sequence.

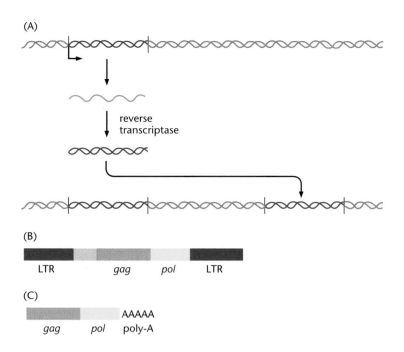

(A)

reverse transcriptase

(B)

LTR   *gag*   *pol*   LTR

(C)

AAAAA

gag   *pol*   poly-A

These are often referred to as "cut and paste" transposons (Figure 2–8). Excision and reintegration are catalyzed by an enzyme known as **transposase**, which is encoded by a gene within the transposon. Because they are excised from one site before reintegration, class II transposons tend not to accumulate to the same degree as retrotransposons, although their preference for moving from replicated DNA to unreplicated DNA does allow for some increase in number of copies. **Transposon insertion** in a gene may give rise to a mutant phenotype. In the case of class II transposons, subsequent excision from the gene may restore the function of the host gene, with a return to the wild-type phenotype. This process is called **reversion**.

Figure 2–8.
Class II ("cut and paste") transposons. The transposon (*red*) is flanked by short repeated sequences (*orange*) and encodes an enzyme called transposase (*green blob*), which catalyzes excision of the transposon from one location in the genome and its insertion in a different location. During this process, a copy of the repeated sequence is left behind where the transposon was originally located ("footprint," *orange*).

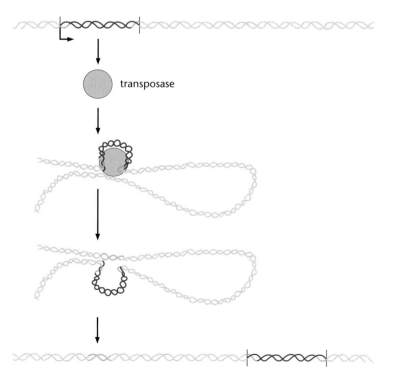

transposase

If reversion occurs at high frequencies, an unstable phenotype may result, comprising a background of mutant cells with sectors made up of cells that have reverted to the wild-type phenotype. Such unstable mutants in maize provided the experimental material used by Barbara McClintock to discover transposons. The most common insertions used in her studies were into genes involved in anthocyanin (a pigment) biosynthesis. When the transposon was inserted, colorless mutant cells were produced. When the transposon was excised, the anthocyanin biosynthetic gene was restored to its function, and this could be observed by the presence of red sectors in the aleurone layer of the maize kernels (Figure 2–9).

Like all random mutations, the genetic changes caused by transposon insertion are occasionally advantageous, but in most cases are expected to be deleterious. This means that during evolution, there has been **selective pressure** in favor of mechanisms that limit transposon activity. One of these mechanisms involves the transcriptional silencing of the chromatin regions where transposons are inserted, through the action of small regulatory RNAs (as described in Section 2.3).

In spite of the mechanisms to keep transposons inactive, over many generations they may still accumulate in large numbers. Transposon insertions in intergenic regions played an important role in genome expansion during the evolution of cereals. Retrotransposons constitute at least 50% of the maize genome, with particularly high

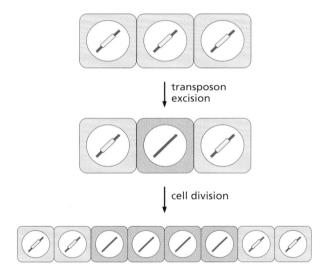

Figure 2–9.
Pigmented sectors formed by random transposon excision during tissue growth. Initially, all cells contain a transposon inserted into a gene that is required for pigment production (the gene, shown in the nucleus, is indicated by a *red line* interrupted by a *yellow box* representing the transposon). In some cells, the transposon is excised and reinserted in a different location (for simplicity, the reinserted transposon is not shown). This reconstitutes the gene that was originally disrupted and restores the cell's ability to produce pigment (cell in *red*). During tissue growth, the descendants of this cell inherit the active gene and form a pigmented region (sector) within the tissue. The photograph shows sectors formed by transposon excision in the aleurone tissue of maize kernels.

density in regions between genes in all cereals. The size of the maize genome has increased two- to fivefold since the divergence of maize and sorghum from a common ancestor about 16 million years ago. The evolution of cereal genomes has been studied by comparing regions that contain the same genes and analyzing DNA differences between the genes. For example, the gene order around the *shrunken1* and *a1* genes is conserved across most cereals, and so the *sh1–a1* region provides a suitable sample region to compare across genomes. Comparison of this region between maize and sorghum reveals many retrotransposons in the maize genome and none in the sorghum genome. Most of the difference in genome size between sorghum and maize can be accounted for by the difference in numbers of retrotransposon copies between these two species. Further analysis of the retrotransposon sequences suggests that most have inserted into the maize genome over the past 6 million years.

2.3 NUCLEAR GENE REGULATION

Differential gene expression underpins most of the phenotypic changes that occur during plant development and in response to environmental signals. Gene expression can be controlled at many different stages from transcription to **translation** of the RNA transcript, and the function of proteins encoded by the genes is often controlled by interaction with other proteins or with small molecules. The initiation of transcription, however, is by far the most widely used point for controlling gene expression in eukaryotes, including plants, so we begin our discussion of regulation there.

Regulatory sequences and transcription factors control where and when a gene is transcribed

As described in Section 2.2, there are three types of eukaryotic RNA polymerase. Most of our understanding of transcriptional initiation has come from studies on RNA polymerase II in yeast and mammalian cell lines. Although having distinct sets of associated proteins, RNA polymerases I and III are believed to initiate transcription in essentially similar ways to RNA polymerase II. **Homologs** of most of the proteins in the **basal transcriptional machinery** of yeast and mammals have also been found in plants, so the principles of transcriptional initiation are assumed to be common to all eukaryotic organisms.

RNA polymerase II cannot initiate gene transcription specifically. It needs to be brought to the start site by the action of other proteins, called **general transcription factors (GTFs)**. The GTFs and RNA polymerase II usually assemble at a short sequence called the **TATA box**. Box 2–1 details the activities of the GTFs and of the regulatory proteins known as **transcription factors**—which recognize short DNA sequences (called **cis-elements**) in the genes they regulate—in the combinatorial control of transcriptional initiation. Some families of transcription factors are conserved across all eukaryotes (for example, the **MYB** and **MADS box** families), whereas other families are found only in single phyla, including several transcription factor families that are unique to plants (Table 2–1).

Combinatorial control is believed to be a feature of all eukaryotic genes transcribed by RNA polymerase II. For example, activation of the *cab* gene (encoding a component of the photosynthetic apparatus) in plants depends on several cis-elements (Figure 2–10). Some of these are bound by the transcription factor CCA1 (of the MYB family), whose activity oscillates with the 24-hour day–night cycle (a **circadian rhythm**). Other cis-elements bind transcription factors that are activated by light of different wavelengths, such as HY5 (of the **basic leucine zipper**, or bZIP, family) and PIF3 (of the **basic helix-loop-helix**, or bHLH, family). In this way, different types of input, such as the time of day or the quality of light reaching the cell, converge to control transcription of the *cab* gene.

Box 2–1 Transcription Factors: Combinatorial Control

The balance between activation and repression of any gene is defined by the combined activity of multiple transcription factors recognizing different regulatory elements.

RNA polymerase II is brought to the start site of transcription by the action of general transcription factors (GTFs). GTFs are also involved in melting (separating) the two strands of the DNA and in switching the polymerase from a form that will initiate transcription to one that will elongate transcripts. The GTFs are the TATA-binding protein (TBP), TFIIB, TFIIE, TFIIF, and TFIIH. Transcription is initiated at defined points in genes, just upstream of the start of the coding sequences. In eukaryotes, a sequence called the "TATA box" (with a consensus sequence of TATAA) usually lies about 25 bp upstream of the transcriptional start site.

The first step in transcriptional initiation is binding of the TBP to the TATA box (Figure B2–1). The TBP protein binds to form a "molecular saddle" astride the TATA box. Binding causes bending of the DNA, which provides the appropriate topology for binding by TFIIB on either side of the TATA box. TFIIB binding specifies the polarity of transcriptional initiation. TFIIB also binds to RNA polymerase II (RP2). The next GTF to bind is TFIIF, which is composed of two protein subunits (three in yeast). TFIIF has high affinity for RNA polymerase II, and RNA polymerase II is recruited to promoters through its association with TFIIF and TFIIB. TFIIE is composed of two peptides and binds next. This protein complex participates in a conformational switch at the active site of RNA polymerase II on its binding to DNA. TFIIE also recruits TFIIH to the transcriptional machinery. TFIIH consists of nine peptide subunits and has three enzymatic functions: a DNA-dependent **ATPase**, an ATP-dependent **helicase**, and a **kinase**. TFIIH acts as a "molecular wrench," rotating the DNA downstream of the basal apparatus through its helicase activity, thus melting the intervening DNA and allowing the start of transcription.

Progression from Transcriptional Initiation to Elongation

Transcriptional initiation progresses into transcriptional elongation via a conformational change in RNA polymerase II. RNA polymerase II contains a repeated amino acid motif (Tyr-Ser-Pro-Thr-Ser-Pro-Ser) in its carboxyl-terminal domain (CTD). When RNA polymerase II enters the pre-initiation complex, the CTD is not phosphorylated. The kinase activity of TFIIH phosphorylates the CTD and moves the polymerase from its initiation conformation to its elongation conformation. When elongation is complete, the action of a specific phosphatase

Figure B2–1.
Stepwise assembly of the pre-initiation complex on a promoter and transition to transcriptional elongation.

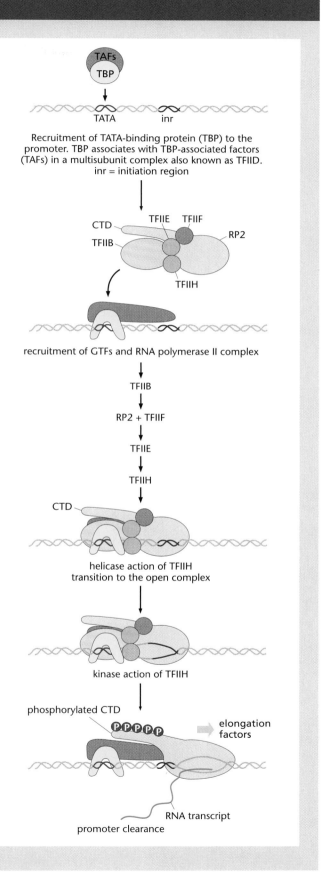

Recruitment of TATA-binding protein (TBP) to the promoter. TBP associates with TBP-associated factors (TAFs) in a multisubunit complex also known as TFIID. inr = initiation region

recruitment of GTFs and RNA polymerase II complex

helicase action of TFIIH
transition to the open complex

kinase action of TFIIH

phosphorylated CTD

elongation factors

RNA transcript

promoter clearance

continued

Box 2–1 Transcription Factors: Combinatorial Control (continued)

dephosphorylates the CTD allowing RNA polymerase II to recycle for further transcriptional initiation.

Activation of the Basal Transcriptional Apparatus

The level of transcription directed by the basal transcriptional apparatus is extremely low. Most genes contain additional regulatory elements recognized by specific DNA binding proteins that can modulate the rate of transcriptional initiation. These proteins are known as transcription factors. The DNA sites recognized by transcription factors are known as cis-elements. Transcription factors can act at a distance from the basal transcriptional machinery and activate transcription from cis-elements located at considerable distances from the regulated gene (up to 25 million bp or more). Cis-elements of this type are often referred to as "enhancers," and may lie downstream of the gene and may operate independent of their orientation. Generally speaking, plant promoters are smaller than those in animals, and most plant regulatory sequences lie within 500 bp of the transcriptional start site.

Transcription factors therefore, by definition, bind specifically to DNA. Generally they are modular in their structure and consist of a **DNA binding domain** and separate domains for influencing transcriptional initiation and for protein-protein interaction. There are relatively few folds that provide for sequence-specific DNA binding, so the DNA binding domains

of transcription factors are highly conserved and allow transcription factors to be divided into families of proteins that contain structurally related DNA binding domains. Some families of transcription factors are conserved across all metazoa (for example, the MYB and bZIP families; see Table 2–1), whereas other families are found only in single phyla. For example, the AP2 family of transcription factors is unique to plants.

Transcriptional activators may also contain **activation domains**. Amino acid sequence requirements for activation domains are fairly flexible but fall into three broad categories: acidic domains, especially where the domain is organized as amphipathic α helix such that the acidic residues align on one side of the helix; glutamine-rich domains; and proline-rich domains. Activation domains work by interacting with the basal transcriptional machinery. They may do this either directly or indirectly through interaction with **co-activators**. Some co-activators are proteins associated with the basal transcriptional machinery (often with TBP) and may transmit the signal from many different transcription factors to the basal transcriptional machinery. These are known as TAFs (**T**BP-**a**ssociated **f**actors). Other co-activators may have more specialized roles related to relaying the signal from one or just a few transcription factors.

The principal way transcriptional activators stimulate transcription is by enhancing recruitment of RNA polymerase II to the promoter of the regulated gene (Figure B2–2). Thus the sequence-specific recognition of target promoters by transcriptional activators enhances recruitment of the basal transcriptional apparatus around the TATA box either through direct protein-protein interaction or indirectly through co-activators.

Transcription factors bound to enhancer elements probably increase transcriptional initiation through DNA looping. Binding to the enhancer element

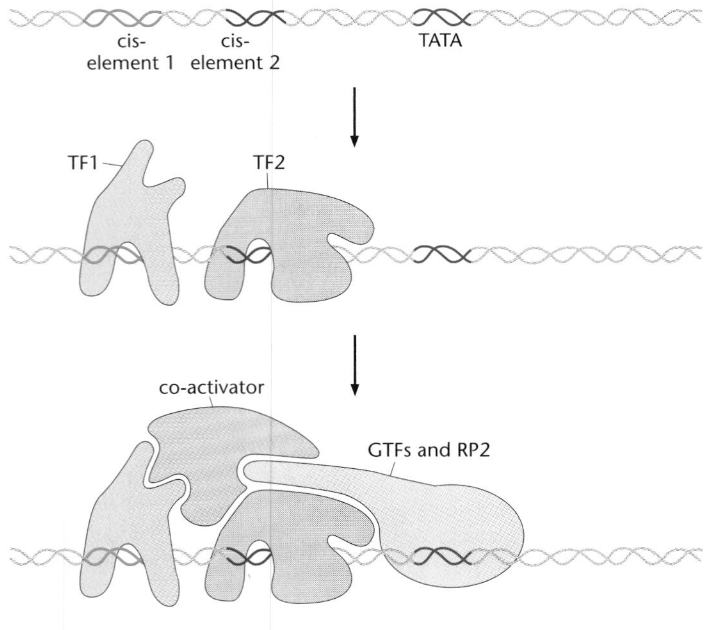

Figure B2–2.
Binding of transcription factors to cis-elements to influence the rate of gene transcription. The cis-elements are short sequences, usually present within a few hundred base pairs of the transcriptional start site. Different cis-elements are recognized by different types of transcription factor (TF). In some cases the TF may interact directly with RNA polymerase II (RP2) or with GTFs to facilitate or hinder assembly of the pre-initiation complex. In other cases the TF may attract one or more intermediary proteins (co-activators), which in turn interact with the pre-initiation complex.

Box 2–1 Transcription Factors: Combinatorial Control (continued)

provides a tether that increases the local concentrations of the interacting proteins, so enhancing the rate of recruitment to the initiation site. DNA binding by transcription factors also often results in bending of DNA, which promotes looping and protein-protein interactions.

Repression of Transcriptional Initiation

Transcriptional repressors work in a variety of ways (Figure B2–3). Some repress the activity of transcriptional activators, so these can be viewed as indirect transcriptional regulators. They may operate by blocking the accessibility of cis-elements to transcriptional activators, through modulation of chromatin structure, or by binding the sequence themselves competitively. They may bind to the transcriptional activator to prevent its DNA binding or to mask the activity of its activation domain. However, there are also transcriptional repressors that can reduce transcriptional initiation in the absence of an activator. Such repressors have defined **repression domains** and may work through interaction with the mediator complex or other TAFs.

Combinatorial Control of Gene Expression

Generally, the specific expression pattern for any gene is defined largely by cell-specific levels of transcriptional initiation, which are determined by the combined activity of multiple transcription factors recognizing different cis-elements within the gene's promoter (and possibly elsewhere). The balance between activation and repression, and the different degrees of activation and repression offered by the different transcription factors recognizing a specific promoter, will define the final level of transcriptional initiation. This is known as **combinatorial control** and is believed to be involved in the control of all eukaryotic genes transcribed by RNA polymerase II.

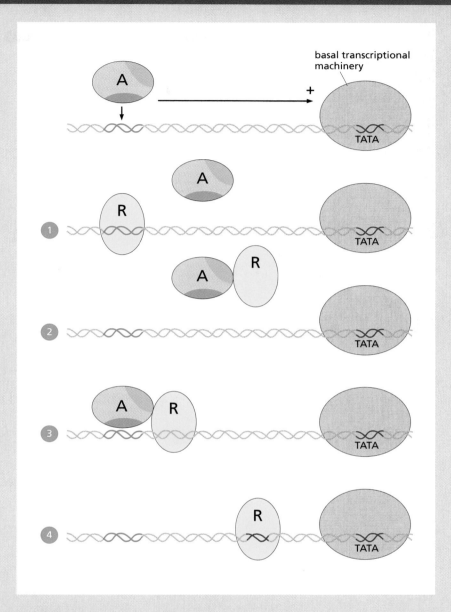

Figure B2–3.
Mode of action of transcriptional repressors. In the top diagram, a transcriptional activator (A) binds to a cis-element (shown in *green* on the DNA strand) and stimulates assembly of the pre-initiation complex (which contains components of the basal transcriptional machinery such as RNA polymerase II and its cofactors). A transcriptional repressor (R) can function in a variety of ways. The repressor may (1) compete for the same cis-element bound by the transcriptional activator, or (2) bind to the activator and prevent its binding to a cis-element on the promoter. (3) Interaction with the repressor may not prevent the activator's binding to its cis-element, but may instead block its ability to stimulate assembly of the pre-initiation complex. (4) The repressor may bind to its own, separate cis-element and have a negative effect on the pre-initiation complex, independent of the activator.

Table 2–1 Transcription factors

Family name	Distribution	Properties of DNA binding domain	Properties of protein-protein interaction domains	Examples in plants
AP2/EREBP	Plant specific	60–amino acid DNA binding domain; resembles bacterial integrases; forms 3-stranded β sheet with parallel α helix. DNA contacted through Arg and Trp residues in β sheet.		Arabidopsis: AP2, ANT homeotic proteins; DREB/CBP stress regulators. Maize: Glossy15. *C. roseus*: ORCA regulators of alkaloid biosynthesis.
ARF/VP1	Plant specific	N-terminal DNA binding domain.	In ARFs, conserved domains III and IV interact with related domains in AUX/IAA proteins. Can act as transcriptional activators or repressors.	Structurally related folds in auxin response factors (ARFs) and Arabidopsis ABI3 and maize VP1 regulators of tissue dormancy.
bHLH	Eukaryotes	Basic domain of conserved sequence.	The helix-loop-helix domain interacts with other HLH domains to form symmetrical dimers composed of bundles of 4 α helices. Association allows DNA binding through the adjacent basic domain.	Arabidopsis: PIF proteins in phytochrome signaling. Maize: R/B proteins controlling anthocyanin biosynthesis.
bZIP	Eukaryotes	Basic domain of conserved sequence.	Leucine zipper consisting of C-terminal α helix with a Leu or hydrophobic residue every 7th amino acid. α helices interact to form dimers.	Maize: Opaque2 regulator of seed storage protein synthesis.
Heat shock factor	Eukaryotes	A 3-helix bundle capped by β sheet. The 3rd α helix (the recognition helix) interacts directly with the DNA of the heat shock element.		HSF
Homeo-domain (HD)	Eukaryotes	A 60–amino acid domain that forms a helix-helix-turn-helix. The 3rd helix is the recognition helix interacting directly with DNA.	May bind DNA as monomers or dimers. If dimerization is involved there may be separate protein-protein interaction domains, as in the HD-ZIP subfamily.	KNOX proteins maintaining stem cells, including STM from Arabidopsis and KN1 from maize; WUS, GL2, PHB, PHV from Arabidopsis.
MADS box	Eukaryotes	A 56–amino acid domain forming a pair of antiparallel coiled-coil α helices packed against an antiparallel 2-stranded β sheet. The α helices make direct contact with DNA, and the N-terminal region of the domain contacts the DNA backbone.	Dimerization is essential for some members to bind DNA. Dimerization domains consist of β sheet and include the I and K domains of MICK-type MADS box proteins. Some MADS box proteins have C-terminal activation domains.	Majority of floral homeotic genes, including AP1, AP3, PI, AG, SEP, and FLC proteins of Arabidopsis.
MYB	Eukaryotes	Proteins containing 1–4 repeats of a 52/53–amino acid motif that forms a helix-helix-turn-helix structure. The 3rd helix is the recognition helix. 2- or 3-repeat MYBs (2R or 3R) bind DNA as monomers. 1-repeat (1R) MYBs bind DNA as dimers and have distinct binding site specificities from 2R/3R MYBs.	2R and 3R MYBs may have C-terminal activation or repression domains. 1R MYBs may contain coiled-coil domains for dimerization/protein-protein interaction.	3R MYBs: regulators of cyclin gene expression. 2R MYBs: GL1, LAF1, AtMYB4, PAP1 of Arabidopsis; C1 of maize. 1R MYBs: PHR1, LHY1, CCA1 of Arabidopsis; Golden2 of maize.
NAC	Plant specific	160–amino acid NAC domain with 5 subdomains (A–E). Subdomains D and E form a 60–amino acid DNA binding domain of twisted antiparallel β sheet with α helix on each side.	160–amino acid NAC domain also contains subdomains for dimerization.	Petunia: NAM1. Arabidopsis: CUC1 and CUV2.
WRKY	Plant specific	60–amino acid region forming a 4-stranded β sheet with a zinc-binding pocket formed by 2 Cys and 2 His residues at one end. The most N-terminal β sheet contains the WRKYGQK motif, which binds to W-boxes with an invariant TGAC core.		Arabidopsis: WRKY1, TTG2.
Zinc finger (C2-H2)	Eukaryotes	DNA binding domain contains 2 Cys and 2 His at fixed points, which together bind zinc. Two antiparallel β sheets with α helix in between bind DNA.		Arabidopsis: SERRATE, DOF proteins involved in seed development.

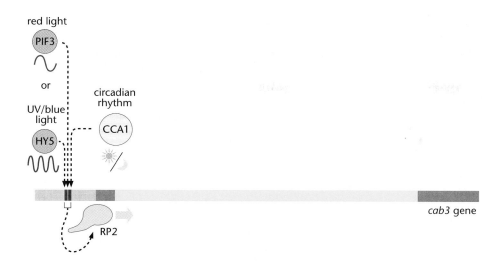

Figure 2–10.
Interaction of multiple transcription factors with cis-elements in the *cab* promoter to combine inputs into gene regulation. PIF3 mediates activation of *cab* by red light; HY5 mediates activation by UV and blue light; and CCA1 provides input from the circadian clock (see Section 2.5).

As noted in Box 2–1, transcription factors control the assembly of the basal transcriptional machinery at the TATA box. In addition, transcription factors can facilitate or repress gene expression by recruiting other proteins that change the way DNA is packaged together with histones and consequently the accessibility of cis-elements and the TATA box. This is described in more detail below.

Gene activity can be regulated by chemical changes in the DNA and proteins of chromatin

The wrapping of DNA around histones and the degree of chromatin condensation affect the accessibility of DNA to proteins, including transcription factors and RNA polymerase, and hence affect gene expression. DNA accessibility is controlled by modifications both to the DNA and to the histone proteins. The addition of acetyl or methyl groups to histones alters the condensation state of the chromatin and hence the transcriptional activity of the associated DNA. Acetylation of histones is associated with transcriptionally active chromatin, whereas removal of acetyl groups from histones and methylation of lysine-9 and lysine-27 of histone H3 are associated with chromatin silencing.

The histone-modifying enzymes (histone acetylases, histone deacetylases, and histone methylases) can be targeted to specific genes or regions of the genome by interaction with transcription factors, which, as described above, recognize specific cis-elements in target genes. The consequent changes in the chromatin of the target genes can lock their expression in an "on" or "off" state, even after the transcription factors that initially targeted the gene are no longer present in the cell.

One example that illustrates the persistence of changes in gene expression long after the initial stimulus is the control of flowering time by temperature. In certain plants, including most lines of Arabidopsis, normal progression to flowering occurs after the plant has been exposed to a long period of cold. In natural conditions, this ensures that the plant does not produce flowers prematurely, before winter has passed. As explained in Section 6.4, the "memory" of a long period of cold is stored in the form of the stable repression of the gene *FLC*, which, when active, prevents flowering. The cold treatment causes *FLC* to shut down permanently, thus allowing the plant to flower (Figure 2–11). This stable repression of *FLC* is caused by methylation of histones in its regulatory regions.

Another example of how changes in chromatin stabilize patterns of gene expression is the regulation of genes that specify each of the different types of organs that make up a flower. As explained in Section 5.5, the arrangement of floral organs is determined by

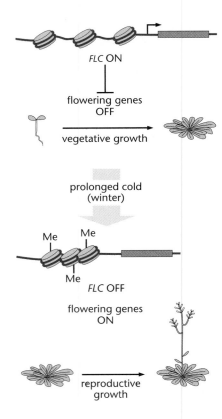

FLC ON

flowering genes
OFF

vegetative growth

prolonged cold
(winter)

Me   Me

Me

FLC OFF

flowering genes
ON

reproductive
growth

Figure 2–11.
Control of flowering by histone modification. Histone modification in response to a temporary cold treatment causes a durable change in expression of a gene that controls flowering in Arabidopsis. If the plants have germinated and have been grown without exposure to winter-like temperatures, the *FLC* gene is constantly on and prevents flowering. When the plants are exposed to a cold period (mimicking winter), methylation of histones (Me = methyl group) associated with the *FLC* promoter causes changes in the chromatin that turn *FLC* off. These changes remain after the cold period has passed, allowing the plant to flower when the temperature becomes favorable. This mechanism minimizes the risk of producing flowers prematurely, before winter has passed.

the expression of **organ identity genes** in specific regions of the developing flower bud. The expression pattern of these genes is set up very early in the developing buds by other transcription factors, but in later stages it is maintained in part by chromatin modification. In Arabidopsis, for example, the organ identity gene *AGAMOUS* is repressed in the regions that develop as petals and sepals, and this repression is maintained by the protein CURLY LEAF; this protein belongs to a family of histone methylases found throughout eukaryotes, the Polycomb group (PcG) proteins (the name originates from the genes originally discovered in *Drosophila*). In *curly leaf* mutants, the absence of this PcG protein causes *AGAMOUS* to be inappropriately turned on, not only in the sepals and petals, but even in leaves (causing the leaf curling that gave the mutant its name).

Histones are not the only components of chromatin that are targeted for covalent modification. Transcriptionally inactive chromatin is also associated with methylation of the DNA itself, and transposons (see Section 2.2) are often maintained in a highly methylated, transcriptionally inactive state. Most DNA methylation in plants involves the transfer of a methyl group from the methyl donor **S-adenosylmethionine** to cytosine residues, at position 5 of the pyrimidine ring (Figure 2–12). This reaction is catalyzed by a methyltransferase. Guanine and thymine are not methylated, and although some adenine methylation occurs in plants, its functional significance is not known. This contrasts with the frequent use of both adenine and cytosine methylation in many prokaryotes.

The pattern of DNA methylation is determined in part by the pattern of histone methylation. Where lysine-9 of histone H3 is methylated, the associated DNA in the nucleosome can become methylated. Evidence for this in plants comes from the *ddm1* mutant of Arabidopsis. DDM1 is a chromatin remodeling factor required for the methylation of lysine-9 in histone H3, because it gives the histone methylation machinery access to the

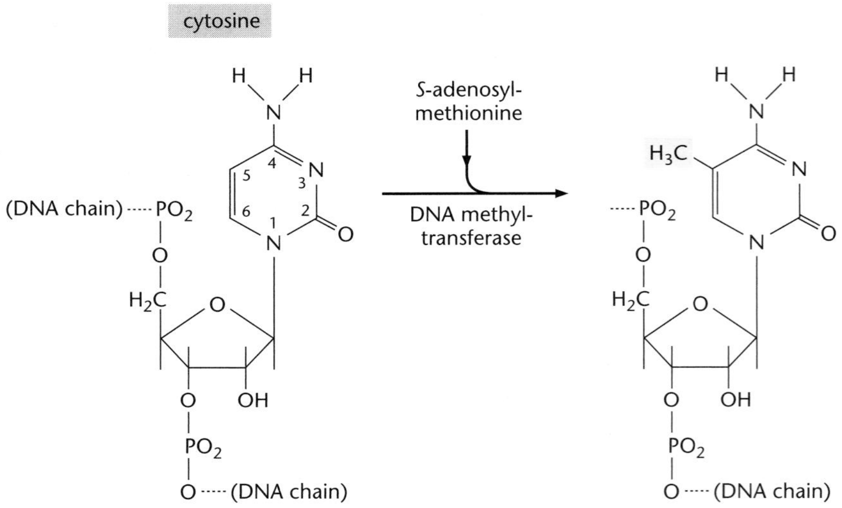

Figure 2–12.
DNA methyltransferase-catalyzed transfer of a methyl group from S-adenosylmethionine to cytosine in DNA. Methyl group shaded *yellow*.

cytosine

(DNA chain) ···· PO$_2$

S-adenosyl-
methionine

DNA methyl-
transferase

H$_3$C

···· PO$_2$

H$_2$C

OH

PO$_2$

O ···· (DNA chain)

proteins of the nucleosome. Even though DDM1 does not itself have methyltransferase activity, levels of DNA methylation are decreased in *ddm1* mutants compared with the wild type. This suggests that the histone methylation pattern laid down by histone methyltransferase (which requires DDM1 activity) dictates where DNA methylation occurs.

Chromatin modification can be inherited through cell division

Part of the reason why chromatin modification is able to maintain stable patterns of gene expression during development is that chromatin changes can be inherited through cell division. In plants, this process is best understood in the case of DNA methylation.

Before replication, both strands of DNA are methylated. After replication, each double helix contains an old strand (which is methylated) and a new strand (which is not methylated). A DNA methyltransferase then uses the methylated DNA strand as the template for the methylation pattern on the newly synthesized strand. This is possible when the methylation sites contain cytosine residues in symmetrical positions on both strands (Figure 2–13). In animals, symmetrical methylation occurs on cytosines that are followed by guanine (CG); in plants, symmetrical methylation can occur in two types of sequence: CG and CNG (where N denotes any base, A, C, T, or G). Specific methyltransferases are associated with the two types of cytosine methylation.

In plants, DNA methylation also occurs at nonsymmetrical sites (such as CNN, where N is any base except G). This type of methylation cannot be copied to the newly synthesized strand after DNA replication, so it is maintained only if methylation is reestablished after each cell division. The mechanism that maintains DNA methylation at nonsymmetrical sites uses small RNAs as guides to identify the DNA sequences to be methylated, as explained later.

Symmetrical DNA methylation and chromatin modification are replicated, through the mitotic divisions that occur during plant growth, in a manner analogous to the replication of genetic information. This form of inheritance is called **epigenetic** (*epi-* meaning "on top of")—that is, it is caused by factors that alter gene function without causing changes in the underlying gene sequence. In most cases, the epigenetic changes accumulated during development are erased during meiosis, but sometimes these changes persist through meiosis. This can result in heritable gene inactivation, with effects similar to those of mutations in the DNA sequence. Gene variants caused by changes in chromatin structure and not by a change in DNA sequence are called **epialleles** (by analogy with **alleles**, multiple versions of a gene differing in DNA sequence).

One example of a naturally occurring epiallele is found in *Linaria* (toadflax), an epiallele that causes loss of floral asymmetry (Figure 2–14). The epimutation is correlated with a decrease in DNA methylation in the gene *CYCLOIDEA*, which is required for the development of bilaterally symmetrical flowers. Plants that lack *CYCLOIDEA* activity develop abnormal, radially symmetrical flowers (see Section 5.5). The *cycloidea* epimutant of *Linaria* is known to have existed in a wild population for more than 250 years, suggesting that this epiallele has been maintained in the population for at least 250 generations.

Figure 2–13.
Maintenance of DNA methylation at symmetrical sites. When methylated DNA is replicated, the cytosines in the newly made strands (*red*) do not have methyl groups (*asterisks*). Maintenance methylases recognize short, symmetrical sequences (*gray boxes*) that contain methylcytosine on one strand and add a methyl group to the corresponding cytosine in the opposite strand, reproducing the pattern of methylated cytosines in the original DNA before replication.

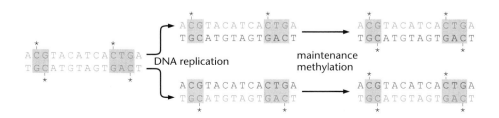

wild type

peloric

Figure 2–14.
***CYCLOIDEA* epimutation in *Linaria*.** The wild-type *Linaria* flower (*top*) is bilaterally symmetrical. The naturally occurring, radially symmetrical (peloric) flower shown below carries a *CYCLOIDEA* epimutation (*cycloidea*). (From P. Cubas et al., *Nature* 401:157–161, 1999. With permission from Macmillan Publishers Ltd, courtesy of Enrico Coen.)

Gene function is also controlled at the RNA level

Transcription factors and chromatin modification control the rate at which RNA transcripts are produced. Production of the RNA transcript, however, does not automatically lead to production of the encoded protein. Accumulation of the final gene product can still be regulated after transcription at several intermediate steps, including RNA splicing, stability, and translation.

As in other eukaryotes, in plants a single transcript may be spliced differently to produce different mRNAs. **Alternative splicing**, however, has been observed less frequently in plants than in animals: it is estimated to occur for approximately 5% of the Arabidopsis genes, while estimates for animal genes range from 10 to 30% or more. The function of alternatively spliced products is known in only a few cases. One striking example occurs in rice: a single gene (*sdhB*) produces two proteins with very different functions, depending on the way the transcript is spliced (Figure 2–15). When exon 1 of this gene is joined directly to exon 3, the encoded product is succinate dehydrogenase subunit B (SDHB), which participates in mitochondrial **respiration**. When exon 1 is joined to exon 2, the encoded product is the mitochondrial ribosomal protein 14 (RPS14). Both proteins have the same N-terminal sequence, which includes a mitochondrial targeting sequence. The probable origin of this unusual gene is that, when the sequence encoding RPS14 was transferred from the mitochondrial genome to the nuclear genome (as happened for many organellar genes; see Sections 2.4 and 2.6), it acquired the sequence encoding the mitochondrial targeting peptide by inserting into the *sdhB* gene. Coexistence of the two splice variants allows both gene functions to be retained.

Another important control point for gene function is mRNA stability. The total amount of an mRNA present in a cell reflects not only the rate at which it is transcribed and processed but also the rate at which it is destroyed by ribonucleases (RNases) in the cytoplasm. Different mRNAs are degraded at different rates, with half-lives that vary from minutes to days. Degradation does not just affect the steady-state mRNA levels, however. The speed of response to changes in gene expression, particularly when genes are repressed, is also affected by mRNA degradation, so regulatory genes that respond quickly to developmental and environmental changes typically encode mRNAs with a short half-life. One example is the mRNAs encoding the ARF transcription factors that mediate cellular responses to **auxin** (auxin and other **phytohormones** are discussed in later chapters).

The differences in mRNA stability are controlled by specific sequences, which vary among the mRNAs. Regulatory regions in mRNAs can also control the rate of degradation in response to external stimuli. In pea, for example, the mRNA that encodes the photosynthetic carrier **ferredoxin I** is stabilized in response to light, and this effect is mediated by sequences close to the 5′ end of the mRNA.

Figure 2–15.
Alternative splicing. In rice, alternative splicing of the *sdhB* gene results in mRNAs that encode proteins with different functions. When exon 1 (*green*) is joined to exon 2 (*blue*), the encoded protein is the mitochondrial ribosomal protein 14 (RPS14). When exon 2 is skipped and exon 1 is joined to exon 3 (*pink*), the mRNA encodes the enzyme succinate dehydrogenase subunit B (SDHB).

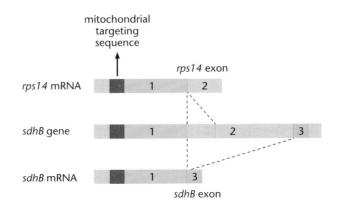

Even during the time it remains intact in the cytoplasm, mRNA is not necessarily free to direct protein synthesis. Another level of gene regulation is mRNA translation, usually at the initial stages of ribosome binding and initiation of protein synthesis. Translation can be inhibited in response to stresses such as desiccation. This response is also regulated differently for specific mRNAs. In the case of desiccation, mRNAs encoding proteins involved in protection from the effects of water loss (such as LEA proteins; see Section 7.3) are still translated under stress. This is clearly seen during seed maturation, when water is lost from the embryo and storage tissues: during seed desiccation, the few ribosomes that remain engaged in protein synthesis are associated with mRNAs encoding dehydrins and LEA proteins.

Translation efficiency and mRNA stability are also linked: when an mRNA contains a defect that affects its ability to make protein (for example, because the mRNA contains a premature **stop codon**), it is targeted for rapid degradation; this process, called "nonsense-mediated decay," is conserved in eukaryotes, including plants.

Small regulatory RNAs control mRNA function

For many mRNAs, stability and translation are also controlled by small RNAs that contain sequences complementary to their target mRNAs. One type of small regulatory RNA are the microRNAs. miRNAs are produced from longer precursor RNAs, which are transcribed by RNA polymerase II from *miR* (*microRNA*) genes. These primary transcripts vary from tens to a few hundred nucleotides in length, depending on the gene, and contain inverted sequence repeats that cause them to fold back on themselves to form double-stranded RNA (dsRNA). The double-stranded regions in this structure are recognized by a specific RNase (a **Dicer** RNase), which cleaves the primary transcript to produce 21- to 24-nucleotide fragments (Figure 2–16).

One strand of the short dsRNA fragment becomes the mature miRNA and associates with the protein ARGONAUTE (AGO), which is part of a multiprotein complex that binds to mRNA with sequences complementary to the miRNA. When a target mRNA is found, either it is cleaved within the region that hybridizes with the miRNA (leading to subsequent degradation of the mRNA fragments) or its translation is inhibited. In plants, the outcome of the interaction depends on the degree of complementarity between the mRNA and the miRNA: perfect or near-perfect complementarity leads to cleavage of the mRNA, whereas imperfect matches can cause translational inhibition (Figure 2–17).

Many of the miRNAs identified in plants target mRNAs that encode transcription factors, and thus miRNAs are an important part of the mechanism that controls where and when these regulatory proteins are expressed. For example, the Arabidopsis miRNAs 165 and 166 cleave mRNAs encoding transcription factors that promote the development of cell types on the **adaxial** side of the leaf (see also the discussion of leaf development in

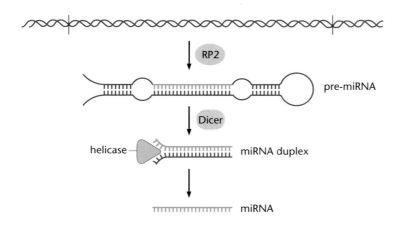

Figure 2–16.
Origin of microRNAs (miRNAs). Specific genes (*miR*) are transcribed by RNA polymerase II (RP2) to produce a pre-miRNA. This precursor transcript has self-complementary sequences and folds back on itself to form regions of double-stranded RNA, which include the miRNA sequence (*blue*). The Dicer RNase cleaves these regions of the pre-miRNA to yield small dsRNAs; the strands are separated by a helicase to yield the mature miRNAs.

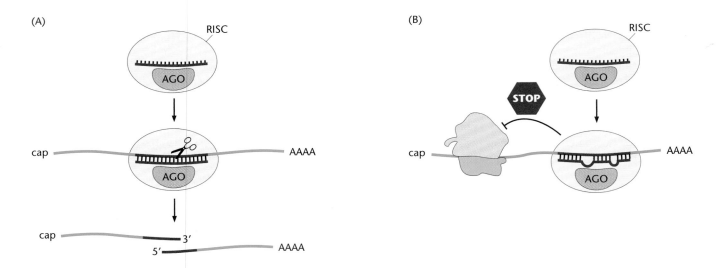

Figure 2–17.
Interaction between miRNA and ARGONAUTE (AGO) protein. Each miRNA is bound by an AGO protein, which functions together with other proteins in a complex known as the **R**NA-**i**nduced **s**ilencing **c**omplex (RISC). The miRNA in RISC serves as a probe to identify matching mRNAs. (A) When the miRNA is perfectly complementary to the mRNA, the mRNA is usually cleaved. (B) A partial match more often leads to translational inhibition.

Figure 2–18.
Control of gene expression in leaf development by miRNA. The *PHABULOSA* (*PHB*) gene controls development of the dorsal (adaxial) side of the leaf, which has specific features such as a higher density of leaf hairs; expression of *PHB* on the lower (abaxial) side of the developing leaf is prevented by miRNA 165 (miR165). The diagram represents a section through a young leaf (*left*), indicating the complementary regions where the *PHB* mRNA (and thus PHB protein) and miR165 accumulate; these regions eventually develop into the two different sides of the mature leaf (shown in cross section on the right). The micrographs show transverse sections through very young leaves arranged around the shoot apex. The sections were hybridized with probes that detect miR165 (*left*) or PHB (*right*) (expression is detected as a darker blue/purple region). Note that miR165 is expressed on the side that faces away from the apex, and in this region accumulation of PHB is decreased. (Courtesy of Catherine Kidner.)

Section 5.4). Expression of these transcription factors on the **abaxial** side of the developing leaf is prevented by miRNA-directed mRNA cleavage (Figure 2–18).

Another mechanism by which small RNAs control mRNA stability is **RNA interference (RNAi)**, which is related to the miRNA mechanism. RNAi is also mediated by 21- to 24-nucleotide RNAs (in this case siRNAs, the small interfering RNAs). Like miRNAs, siRNAs are produced by the Dicer RNase and function together with ARGONAUTE proteins to cleave mRNAs with complementary sequences. The main difference is that unlike miRNAs, siRNAs are not encoded by genes whose specific role is to produce small RNAs. Instead, siRNAs are produced from dsRNA that may not be encoded in the genome—for example, dsRNA formed during viral replication. In addition to a viral origin, dsRNA can be produced from mRNA by cellular RNA-dependent RNA polymerases; this is frequently the cause of RNAi that inhibits expression of genes artificially introduced into a plant (Figure 2–19).

siRNAs are believed to have evolved as a defense mechanism against viruses and against transposable elements. A large fraction of the viruses that infect plants have RNA genomes (see Section 8.3), and their replication mechanism produces dsRNA, which is

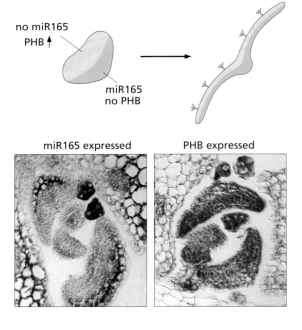

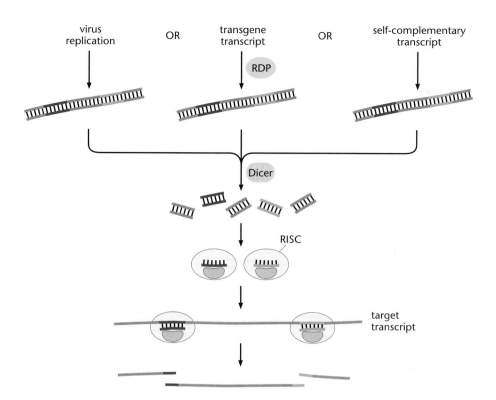

virus replication   OR   transgene transcript   OR   self-complementary transcript

RDP

Dicer

RISC

target transcript

Figure 2–19.
RNA interference (RNAi). RNA interference is triggered by long, double-stranded RNA, which is formed during viral replication, or produced by a cellular RNA-dependent RNA polymerase (RDP; shown here for a transgene transcript), or formed by self-complementary RNA. Dicer cleaves the dsRNA into a mixed population of small RNAs (fragments from different regions of the dsRNA are shown in different colors). Each small RNA is incorporated into RISC (see Figure 2–17) and directs cleavage of the matching sequence in a single-stranded RNA transcript.

the trigger for RNAi. Arabidopsis plants carrying mutations that weaken or eliminate the RNAi response are more susceptible to viral infection. Conversely, many plant viruses have evolved strategies to overcome the defensive RNAi response; for example, the *Tomato bushy stunt virus* produces the protein P19, which binds to small RNAs and prevents their association with ARGONAUTE proteins. (See the discussion of **RNA silencing** in Section 8.4.)

In addition to protecting against viruses, siRNAs inhibit transposon activity. Reflecting this role, a large fraction of the small RNAs detected in plant cells match transposon sequences. While protection against viral infection is based on the ability of siRNAs to target specific RNAs for degradation, the role of siRNAs in preventing overproliferation of transposons probably involves an additional function of siRNAs in chromatin modification, as explained below.

Small RNAs can direct chromatin modification to specific DNA sequences

The role of siRNAs in chromatin modification has been demonstrated in the fission yeast (*Schizosaccharomyces pombe*), which has highly methylated heterochromatin. Regions of heterochromatin are transcribed to produce long RNA molecules, which are converted to dsRNA by an RNA-dependent RNA polymerase. As described above, the dsRNA is cleaved into siRNAs by Dicer. These siRNAs also bind to an ARGONAUTE protein, but instead of targeting mRNAs for degradation or translational inhibition, they direct a histone methyltransferase to the site on the chromosome where the original sequence was transcribed. Methylation of histones, as noted earlier, increases chromatin condensation and promotes DNA methylation in these regions.

In plants, siRNA-directed DNA methylation can sometimes silence foreign genes (known as **transgenes**) that are introduced into plant genomes by technologies described in Section 9.3. When a transgene induces production of siRNAs (for example, because the transcript forms regions of dsRNA), in addition to causing degradation of the mRNA, the siRNAs direct a specific DNA methylase (DRM) to the transgene

(Figure 2–20). DRM initiates DNA methylation—that is, it methylates cytosine residues that were previously unmethylated.

When the RNA-directed DNA methylation happens to include symmetrical sites, these can be copied during DNA replication and maintained in descendant cells (as described above). However, when methylation occurs at nonsymmetrical sites, the methylation of newly synthesized DNA cannot be guided by the methylation pattern of the opposite DNA strand and has to be carried out in new cells by matching siRNAs. Consequently, maintenance of nonsymmetrical methylation (found, for example, in many transposons) probably requires the constant production of siRNAs.

2.4 GENOME SEQUENCES

So far we have described the individual components of the nuclear genome and their functions. Clearly, the functions of different genes and of the structural components of chromosomes are highly interconnected. For example, genes encoding transcription factors regulate many other genes, including other transcription factors. The activity of all these genes can be restricted by chromatin modification, a process controlled by small RNAs, which are also involved in controlling mRNA function. Over long periods of time, genes mutate, sometimes due to transposon activity, changing the interactions among genes. Thus, knowing how each component of the genome behaves individually is not adequate for understanding how an organism functions and evolves. We are a long way from understanding how genomes function as a whole, but the first step toward this goal is to identify the complete set of genes and other components of a nuclear genome. This requires genome sequencing.

Figure 2–20.
Transcriptional gene silencing by siRNAs. In addition to causing mRNA cleavage in the cytoplasm (*left*), siRNAs can direct gene silencing in the nucleus (*right*). Transcription through inverted repeats produces self-complementary RNA (*blue hairpin*), which is recognized and cleaved by different Dicers in the cytoplasm and in the nucleus. Whereas cytoplasmic siRNAs are incorporated into RISC to direct mRNA cleavage, nuclear siRNAs are incorporated into a different protein complex, **R**NA-induced **i**nitiation of **t**ranscriptional gene **s**ilencing (RITS), which also contains an AGO protein. Instead of targeting mRNA, however, RITS targets DNA sequences matching the siRNA. The corresponding gene is then transcriptionally silenced and methylated by the DRM methylase.

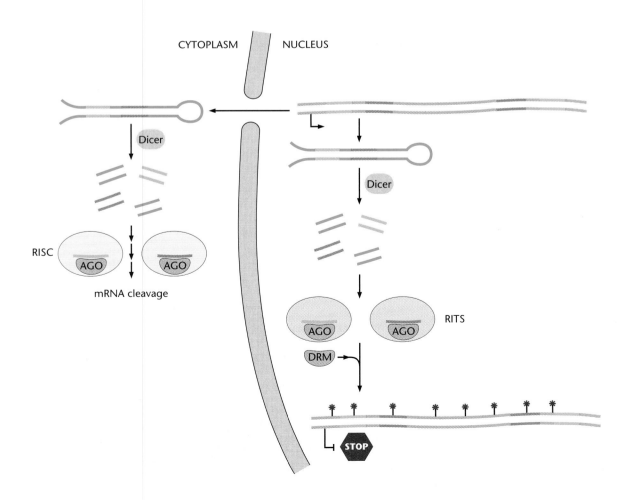

The Arabidopsis genome was the first plant genome to be fully sequenced

A consortium of scientists around the world worked together to sequence the *Arabidopsis thaliana* genome. They selected Arabidopsis for two main reasons: it has one of the smallest nuclear genomes found in plants and, because of its short life cycle, it was already a popular model for genetic studies. The genomes of other, more economically important plants (such as rice and poplar) have now been sequenced, but our discussion here focuses on Arabidopsis as the best-understood plant genome.

Genome sequencing has advanced our understanding of how plant genes function and evolve, in several ways. First, it gives us a better appreciation of the genes necessary to make a plant. Second, the complete genome sequence allows the development and use of methods to monitor the behavior of many genes simultaneously. Third, the complete sequence tells us about the similarities and differences between plants and other organisms with sequenced or partially sequenced genomes. A vast number of plant genes that encode proteins of unknown biochemical function have been unearthed by careful analysis of the Arabidopsis genome sequence. Some of these encoded proteins belong to families that are also found in other organisms, so comparative analysis may help to define their function.

Genome sequences are analyzed to identify individual genes

The final output of the sequencing of an organism's genome is very long sequences of nucleotides corresponding to the DNA backbone of each chromosome (in Arabidopsis, each chromosome contains between 18 and 29 million bp). The next step is to make sense of the sequence and extract information from it, such as how many genes are present, the protein or RNA products they encode, and clues about the biochemical functions of the gene products. This process is called **gene annotation**.

In some cases, individual genes have already been isolated—for example, by map-based cloning or transposon tagging (see Section 2.5)—and their sequences are already known. It is easy to place these within the genome sequence. One of the main reasons for sequencing genomes, however, is to identify previously unknown genes. One clue that allows detection of these genes is that most of them encode proteins and therefore contain sequences that can be translated into relatively long stretches of amino acids. Of the 64 possible triplet **codons**, 3 correspond to stop codons, so translation of a random sequence is expected to terminate in a stop codon every few tens of base pairs. This property can be used to help locate protein-coding regions among long stretches of noncoding DNA. If such an **open reading frame** is found that extends for hundreds of codons without interruption by stop codons, it is likely to be part of a protein-coding gene. One complication with this approach, however, is that the protein-coding sequence in eukaryotic genes is usually split into several exons, which can be small (in some cases as short as 30 bp), and the intervening introns frequently contain stop codons. Although we know that short, specific sequences mark the junctions between exons and introns, and introns tend to have a lower content of G and C nucleotides, these features alone are not sufficient for the accurate prediction of exon and intron structure. Computer programs take into account the length of reading frames and the sequence features that are typical of alternating exons and introns to detect previously unknown genes in a genomic sequence, but reliance on this method alone is prone to error.

Additional clues that a gene has been correctly annotated come from comparing the amino acid sequences encoded by the predicted gene with those of known proteins. If similar sequences are found, this is very unlikely to have arisen by chance and is a good indication that the sequences belong to a protein-encoding gene. Ultimately, however, the most secure way to identify a gene is to show that RNA is transcribed from it. For this reason, genome sequencing is usually complemented by the sequencing of large numbers of cDNA copies of RNAs extracted from numerous tissues and growth stages of the organism (these cDNAs are produced in the test tube by using purified reverse transcriptase; see Section 2.5). These are called **expressed sequence tags (ESTs)**; by matching these sequences with the genomic sequence, transcribed regions and their exon/intron structures can be accurately determined. Databases of ESTs are being developed for many plant species to aid in gene identification and characterization.

Once a potential gene has been identified, both the DNA sequence and the predicted protein sequence it encodes can be compared with all other known or predicted gene and protein sequences in a database of sequences from many organisms. This comparison may reveal similarities between new genes and genes whose functions are already known. Proteins or **protein domains** with similar amino acid sequences often have similar structure and biochemical activity, so sequence similarity gives us important clues about gene function.

Sequencing of the Arabidopsis genome revealed a complexity similar to that of animal genomes and a large proportion of plant-specific genes

The complete Arabidopsis genome has approximately 125 million bp and an estimated 26,000 genes. Thus the number of genes in this plant genome is within the range found in animal genomes (18,000 genes in the nematode *Caenorhabditis elegans*; 13,000 in the fruit fly *Drosophila melanogaster*; 32,000 in humans).

The typical gene in Arabidopsis is 4500 bp long, with average exon and intron lengths of 250 and 170 bp, respectively, and regulatory sequences frequently within a few hundred base pairs upstream of the coding sequence. This average gene structure is similar to that found in *Drosophila* and *C. elegans*, but is more compact than that of mammalian genes (human genes average 30,000 bp, mostly because of large introns that span thousands of base pairs).

Based on sequence similarity to genes found in other eukaryotes, approximately half of the Arabidopsis genes have a clear counterpart in animals or fungi, while approximately 150 of the protein families encoded in the Arabidopsis genome are unique to plants. The latter include about 400 genes that are related to those of **cyanobacteria**. The cyanobacteria-like genes were probably inherited from the plant-specific **endosymbiotic event** that gave rise to the plastid (see Section 1.2) and then transferred from the chloroplast to the nuclear genome during evolution. The transfer of genes from the plastid to the nucleus seems to be an ongoing process, as suggested by the finding that tens of genes that are nuclear in Arabidopsis are still found in the plastids of other plant species. Mitochondrial DNA has also been transferred to the nucleus. This is particularly obvious in Arabidopsis, which contains a nearly complete copy of the mitochondrial genome inserted near the centromere of chromosome 3.

Based on similarity to known proteins (from any organism), 70% of the proteins encoded in the Arabidopsis genome can be placed in functional categories (see Figure 2–21 for the breakdown of gene numbers in these functional categories). The most abundant class (23%) corresponds to metabolic enzymes. This large variety of metabolic enzymes reflects, in part, the complex and extensive **secondary metabolism** found in plants. For example, the biosynthesis of secondary metabolites frequently involves hydroxylation reactions catalyzed by enzymes of the cytochrome P450 family, and approximately 300 of these are encoded in the Arabidopsis genome. The complex cell wall is another feature reflected by the large number of enzymes encoded in the genome: approximately 400 enzymes are probably involved in cell wall metabolism. In addition, the genome sequence suggests a previously unappreciated metabolic complexity. A large number of enzymes encoded by the Arabidopsis genome are related to steps in metabolic pathways that had not been known to function in this plant (for example, alkaloid biosynthesis).

The second most abundant class of functional proteins includes those involved in gene regulation (17%; shown as "transcription" in Figure 2–21). This class includes a large

(A)

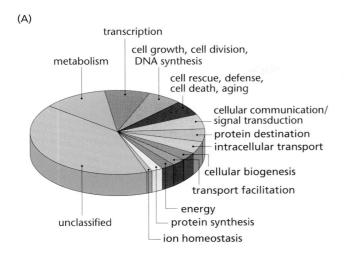

(B)

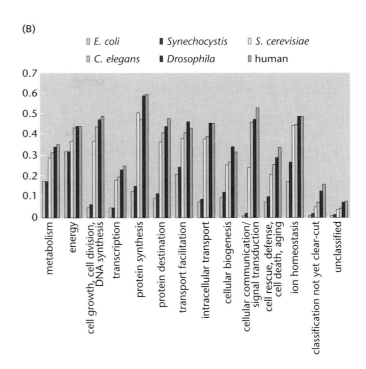

number of transcription factor families, of which 45% are unique to plants (see Table 2–1), indicating that they evolved after separation of the animal, plant, and fungal lineages, approximately 1.5 billion years ago (Precambrian). Conversely, some of the largest families of transcription factors found in animals, such as the nuclear hormone receptor family, are absent in plants. Among the transcription factor families that do have counterparts in animals and fungi, some are greatly expanded in plants. For example, the MADS family contains 4 members in yeast and 2 in *Drosophila* and *C. elegans*, but more than 80 in Arabidopsis, many of them involved in controlling development. It is not clear why specific transcription factor families have expanded to this extent in plants. One possibility is that a rich repertoire of gene expression patterns is required for the continuous adjustment of plant development and metabolism to changes in the environment. (As explained in Chapters 6 and 7, the ability of plants to adjust to environmental changes is essential, given their sedentary lifestyle.)

In addition to certain families of transcription factors, other protein families are over-represented in the Arabidopsis genome compared with other eukaryotes. These include proteins involved in targeted protein degradation and in **RNA processing**. For example, the Arabidopsis genome encodes approximately 400 members of the PPR family of RNA binding proteins, compared with 10 each in yeast, *Drosophila*, and *C. elegans*. Genes involved in intercellular communication are also abundant in Arabidopsis. Receptor kinases of the **leucine-rich repeat (LRR)** family, involved in plant development and defense against pathogens (see Section 8.4), are represented by almost 200 members in the Arabidopsis genome, whereas receptors used in animal cells (such as the **G-protein**–coupled receptors) are rare in plants. This suggests that in both plants and animals, the complex intercellular communication required for development and defense is based on the expansion of different types of receptor gene families.

Apart from the genes described above, the Arabidopsis genome contains a large number of transposons (about 4000), inserted mostly near centromeres and in heterochromatic regions. Class I and II transposons (see Section 2.2) are more or less equally represented (this differs from other plants, in which class I is not so abundant). The majority of these transposons are inactive: of the class I elements, which transpose via an RNA intermediate, only 4% seem to be transcribed.

Figure 2–21.
Predicted functions of Arabidopsis genes. (A) Proportion of genes in different functional categories. (B) Proportion of genes in various organisms that have strong similarity to Arabidopsis genes in each functional category. (Adapted from The Arabidopsis Genome Initiative, *Nature* 408:796–815, 2000. With permission from Macmillan Publishers Ltd.)

Other important nongenic sequences include the telomeres, which contain 2000 to 3000 bp of TTTAGGG repeats, and the centromeres, which consist of more than 3 million bp of mostly repeated sequences (particularly 178-bp repeats). Other regions with large numbers of repeated sequences are the **nucleolar organizing regions (NORs)**, each with approximately 400 copies of the genes encoding rRNAs.

In summary, the Arabidopsis genome sequence provided the first complete view of a plant genome. This revealed a complexity similar to that of animal genomes, with a large number of similar genes, but also a large fraction of plant-specific genes. The accurate and complete data obtained from the sequencing of this genome allowed for the development of techniques to study gene function more efficiently, as described in Section 2.5.

Comparisons among plant genomes reveal conserved and divergent features

Since the Arabidopsis genome sequencing was completed, several other plant genomes have been sequenced, including species of economic interest. Two notable examples are rice and poplar. Rice is particularly important because it serves as a model genome for cereals (see the later discussion of synteny between cereal species). The rice genome is three times bigger (389 million bp) than that of Arabidopsis, but has a lower gene density, with an average of one gene per 9900 bp. The larger intergenic distances are partly due to the presence of numerous transposable elements (35% of the genome). In spite of the lower gene density, there are more predicted genes in rice (37,000, compared with 26,000 in Arabidopsis). The overall representation of gene families is similar in both species, but rice has a great many genes with no homologs in Arabidopsis (89% of Arabidopsis genes have homologs in rice, but only 71% of the predicted rice proteins have homologs in Arabidopsis). These unique genes may underlie many of the developmental and physiological differences between **monocot** and **eudicot** plants.

Poplar was the first tree species to have its genome sequenced. The poplar genome contains approximately 485 million bp, with 45,000 predicted proteins, of which 12% have no homologs in Arabidopsis. The overall representation of known gene families is similar to that in Arabidopsis. Some gene families, however, are over-represented in poplar. For example, as expected for a tree species, families of genes encoding enzymes involved in the synthesis of lignocellulosic cell walls are larger than in Arabidopsis (poplar has 34 genes involved in phenylpropanoid and lignin biosynthesis, compared with 18 in Arabidopsis). Another example is that, like rice, poplar has a larger number of predicted **R proteins**, proteins involved in disease resistance (398 homologs in poplar and 535 in rice, compared with 207 in Arabidopsis). This probably reflects the rapid evolution of this type of gene, which is often subject to intense diversifying selection (see Section 8.4).

Other plant genomes that have been or are being sequenced include maize, tomato, *Medicago truncatula* (a legume species), and the moss *Physcomitrella patens*. While individual genome sequences are invaluable resources for biotechnology (see Section 2.5), comparative genome analysis gives us insights into plant evolution, as we shall see.

Most angiosperms have undergone genome duplication during their evolution

A striking feature of the Arabidopsis, rice, and poplar genomes is that large parts of them seem to have been duplicated—that is, chromosome segments with almost identical stretches of gene sequences are found in multiple locations throughout the genome (Figure 2–22). This suggests that at some point in their evolution, these genomes underwent duplication (entirely or in part), although it is thought that parts of the duplicated sequences were subsequently lost and parts diverged, their genes acquiring different functions.

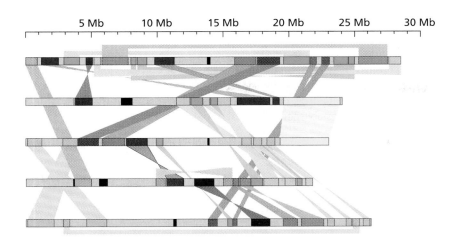

Figure 2–22.
Duplicated regions in the Arabidopsis genome. Each of the five chromosomes is depicted in *gray*, with centromeres shown as *black boxes*. Duplicated segments are connected by colored bands (twisted bands represent inverted duplications). The similarity between rDNA sequences is not shown. (Mb = million base pairs.) (Adapted from The Arabidopsis Genome Initiative, *Nature* 408:796-815, 2000. With permission from Macmillan Publishers Ltd.)

The duplication of large parts of the genome is typical of angiosperms. One way in which these duplications have arisen during evolution is through **polyploidy**—that is, the combination of two or more complete sets of chromosomes, resulting in a larger genome. Polyploids can arise either from multiplication of the set of chromosomes from a single ancestral species (**autopolyploids**; *auto* meaning "self") or as a result of combining two genomes from related species with similar sets of chromosomes (**allopolyploids**; *allo* meaning "other") (Figure 2–23). More than 50% of angiosperm species are believed to be polyploid, a much higher frequency than in animals or in other plant groups, such as gymnosperms.

Polyploidy is a potential source of **genetic variation**; in allopolyploids variation is increased by the contribution of both parental genomes, whereas in autopolyploids variation can be provided by the formation of new allelic combinations. Polyploidy can also lead to the formation of new species, which are reproductively isolated from the parental genome-donor plants because the multiplied genomes generally cannot successfully segregate on crossing to a species with a smaller chromosome number. The formation of polyploids has probably played a major role in the diversification of angiosperms during evolution, and it has been very important in the generation of crop plants, such as wheat and the brassicas (crucifers), as discussed in detail in Chapter 9.

Figure 2–23.
Different origins of polyploidy. (A) In autopolyploidy, an ancestral set of chromosomes is duplicated (in this case, from diploid to tetraploid). (B) In allopolyploidy, the sets of chromosomes from two different species (diploid numbers $2n_A$ and $2n_B$) are combined to form a new species with diploid number $2(n_A + n_B)$.

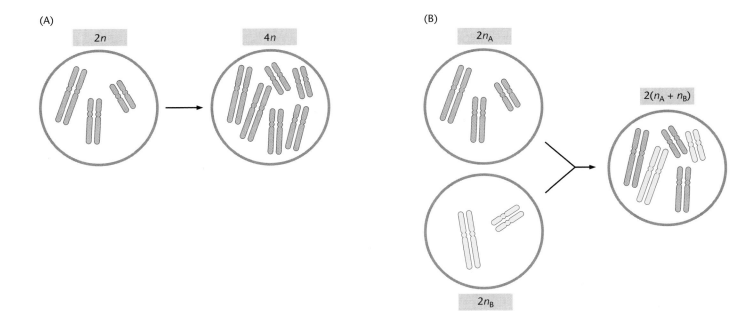

Polyploidy also causes changes in the way genes are expressed. For example, cultivated cotton (*Gossypium hirsutum*) is an allotetraploid derived from *Gossypium herbaceum* and *Gossypium raimondii*. In the tetraploid *G. hirsutum*, many genes are expressed differently from the way they are expressed in the corresponding parental species (i.e., they are expressed at different levels in different organs). When tetraploid cotton is re-created in the laboratory, the same differences in expression patterns found in the cultivated cotton are established shortly after polyploidization.

Changes in gene expression are also seen after polyploidization is induced artificially in Arabidopsis to create an autotetraploid. In this case, inactivation of some of the gene copies is usually not accompanied by changes in DNA sequence. Instead, the genes become hypermethylated and silenced and, once silenced, these alleles remain silenced, even if they segregate into **diploid** offspring. In this example, the modification of chromatin structure that represses gene transcription is stably inherited from one generation to the next. This epigenetic regulation of gene expression is part of the process whereby the genome of a newly formed polyploid is restructured.

Genes can acquire new functions by duplication and divergence

Changes in chromosome number caused by polyploidization or by aberrant chromosome separation are not the only ways to accumulate multiple copies of the same gene within a single genome. Individual genes can be duplicated by imperfect **crossing over** during meiosis. This typically causes the appearance of gene copies arranged side by side at the same chromosomal location.

Whatever the cause of **gene duplication**, one of the gene copies may be inactivated by an epigenetic mechanism (as described above) or by mutation. However, when both copies remain active, they fulfill the function of a single gene in the ancestral plant. Consequently, if one of the genes is lost by mutation, the other member of the pair can compensate for its absence and no phenotypic change occurs. If both genes are lost by mutation, then no compensating protein is available and a mutant phenotype is observed. When the function of two or more genes overlaps in this way, the genes are said to be "functionally redundant." This is exemplified by *SHATTERPROOF1* and *2* (*SHP1* and *SHP2*), which encode highly similar MADS domain transcription factors that control formation of the elongated fruit (silique) of Arabidopsis. The loss of one of these genes by mutation does not result in the formation of defective fruit—one gene can compensate for the absence of another. It is only when both genes are mutated that abnormal fruit development occurs (Figure 2–24).

Figure 2–24.
Gene redundancy. Development of the dehiscence zone in the Arabidopsis fruit is directed by two related genes, *SHP1* and *SHP2*, which occupy different loci but carry out the same function. At the top, the alleles present in each locus are indicated for the wild-type plant, single homozygous mutants, and double homozygous mutant (capital letters are used for wild-type alleles, lowercase for mutants). Arrows indicate that gene function is carried out normally; short lines indicate loss of gene function. Note that normal function at just one of the loci is sufficient for wild-type fruit development. The diagrams and micrographs at the bottom show a normal fruit and a double-mutant fruit lacking the dehiscence zone. (From S. Liljegren et al., *Nature* 404:766–770, 2000. With permission from Macmillan Publishers Ltd, courtesy of Martin Yanofsky.)

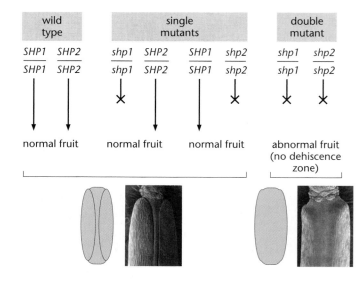

Functional redundancy is more likely among genes that have duplicated relatively recently. Over time, however, each copy of a duplicated gene can acquire a different function. The new functions may correspond to a subset of the functions of the original gene before duplication (in which case the process is called "subfunctionalization"), or they can be different from the functions of the ancestral gene (resulting in "neofunctionalization"). In both cases, the changes in gene function may be caused by changes in the encoded proteins, but they are often caused by mutation of regulatory sequences, so that the two genes are now expressed at different times and places. The latter case is illustrated in Arabidopsis by the two genes encoding the MYB-type transcription factors *WEREWOLF* (*WER*) and *GLABROUS1* (*GL1*). *WEREWOLF* inhibits development of epidermal **root hairs**, whereas *GLABROUS1* promotes epidermal **trichome** (hair) development in shoots (see Chapter 5). Although these genes have very different biological functions, the proteins they encode are interchangeable: expression of the WER protein controlled by GL1 regulatory sequences provides full GL1 function, and vice versa. Thus the difference in their biological roles results primarily from divergence of their regulatory sequences (Figure 2–25).

Another example of divergence of regulatory sequences is seen in genes that control pigmentation in maize. In this plant, which is an ancient tetraploid, two types of transcription

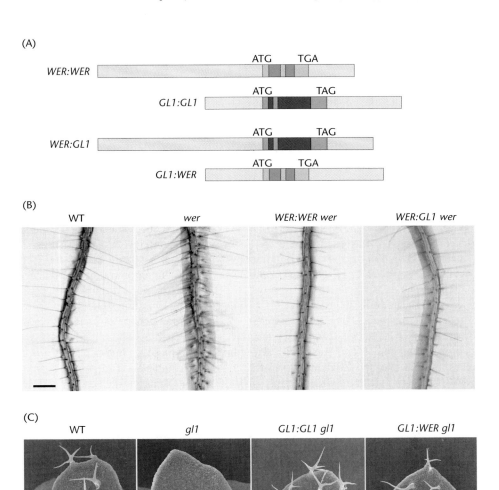

Figure 2–25.
Divergence of gene function through changes in regulatory sequences.
WER and *GL1* genes have different functions in development but encode interchangeable proteins. (A) Structure of *WER*, *GL1*, and artificial genes with the protein-coding sequences exchanged. Regulatory sequences are shown in light color, coding sequences in dark color, and noncoding transcribed sequences (such as introns) in intermediate color. *WER:WER* and *GL1:GL1* correspond to the wild-type genes; *WER:GL1* has the *GL1* coding sequences controlled by *WER* regulatory sequences; *GL1:WER* has the *WER*-coding sequences under *GL1* regulation. (B) The first and second micrographs from the left show the phenotypes of wild-type (WT) and *wer* (mutant) roots; note the increased number of root hairs in the mutant. When *WER:WER* is introduced into the *wer* mutant, *WER* function is recovered and the roots have the wild-type appearance (*3rd from left*). A similar rescue of wild-type development is seen in *WER:GL1 wer* roots (*right*). (C) The first and second micrographs from the left show wild-type and *gl1* leaves; note the absence of trichomes in the mutant. The third and fourth micrographs show how both *GL1:GL1* and *GL1:WER* rescue trichome development in the *gl1* mutant. Scalebar: 0.2 mm. (A–C, with permission of the Company of Biologists.)

(A)

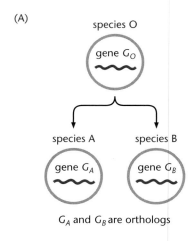

G_A and G_B are orthologs

(B)

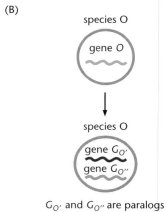

$G_{O'}$ and $G_{O''}$ are paralogs

(C)

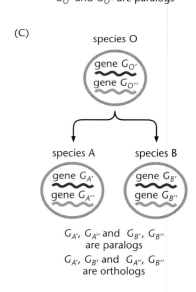

$G_{A'}$, $G_{A''}$ and $G_{B'}$, $G_{B''}$
are paralogs
$G_{A'}$, $G_{B'}$ and $G_{A''}$, $G_{B''}$
are orthologs

Figure 2–26.
Orthologs and paralogs. (A) Orthologs diverge from the same ancestral gene after two species become separate. (B) Paralogs begin to diverge after gene duplication within a single species. (C) When an ancestral species contains paralogs, each paralog gives rise to a set of orthologs in the descendant species.

factors, belonging to the bHLH and MYB families, control anthocyanin biosynthesis. Different copies of these genes control anthocyanin production in different tissues: in the aleurone of the kernel, the MYB-type gene *C1* functions together with the bHLH gene *R*; in the vegetative tissues, *Pl* (a MYB gene) functions along with *B* (another bHLH gene). Although the genes encoding a single type of transcription factor are expressed in different tissues, they still function together with the same type of partner (i.e., a MYB protein with a bHLH protein). This shows that when a set of genes that perform a function together is duplicated, the whole set can diverge together in function; in cases like this, the genes within the set are said to "coevolve."

The process of gene duplication and divergence gives rise to paralogous genes, or **paralogs**—that is, genes that began to diverge after a duplication event within a single species. When similar genes in different species begin to diverge from the same ancestral gene after the species separate, the genes are said to be orthologous, or **orthologs** (Figure 2–26). Frequently, although not invariably, orthologs have the same biological function in different species. For example, the *FLORICAULA* gene, which controls floral development in snapdragon (*Antirrhinum majus*), is an ortholog of the Arabidopsis gene *LEAFY* (see Section 5.5). Mutations in either gene result in developmentally similar phenotypes, indicating that the respective proteins carry out the same function in both organisms.

Gene duplication and divergence have expanded gene families within individual genomes. For example, more than 1500 (5.9%) of the predicted genes in the Arabidopsis genome encode DNA binding proteins that fall into approximately 29 families. The gene families vary greatly in size: in Arabidopsis, the *LEAFY* family comprises only a single member, while the two-repeat MYB family contains 126 members that arose through a series of gene duplication events.

The order of genes is conserved between closely related plant species

The finding that orthologous genes frequently carry out the same function in different species means that identifying orthologs is important not only for evolutionary studies but also for applying our knowledge of model plants to improve crop species. In many cases, however, the presence of multiple paralogs makes it difficult to decide which of the paralogs is functionally equivalent to a gene from another species. Information about the chromosomal context surrounding the genes can be helpful, because the vestiges of a common evolutionary origin are found not just as sequence similarity in the genes but also in the identity and arrangement of other genes in adjacent chromosomal sequences.

Even though the order of genes may have changed during evolution, through chromosomal rearrangements, deletions, and insertions, the arrangement of genes remains broadly similar in related species. For example, grasses such as wheat, barley, maize, rice, sorghum, millet, and sugarcane have different chromosome numbers, but if the chromosomes of each species are dissected into large fragments and realigned, a shared gene order emerges that was probably present in the common ancestor of grasses (Figure 2–27). This conservation of gene order, known as **synteny**, is weaker in more distantly related species, because genome rearrangements have had more time to reshuffle the ancestral gene order.

Synteny helps in the identification of genes responsible for useful phenotypes in crop species. In many important crops, interesting mutations have been located on the **genetic map**, based on **recombination frequencies**. The molecular identification of the genes conferring specific phenotypes, however, requires that the genetic map be translated to a physical map (i.e., based on DNA sequence). For many crops, the physical map is difficult to assemble, because too little DNA sequence is available or the genome is too large and riddled with repetitive sequences, as seen in many grasses such as the cereals (see Section 2.2), but not in rice (or in the model cereal *Brachypodium*). In these cases, it is advantageous initially to identify a similar mutation and isolate the

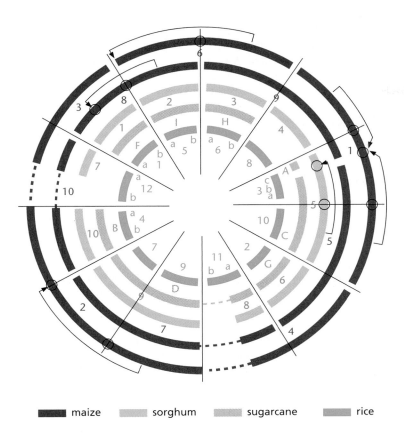

Figure 2–27.
Alignment of the genomes of maize, sorghum, sugarcane, and rice. A set of colored arcs represents the chromosomes of each species, with the chromosome number indicated below each arc. Maize is represented by two sets of arcs because it is an ancient tetraploid. The radiating *black lines* indicate conserved segments that are aligned across the four species. When a chromosome has originated by insertion of one segment into another, a *black arrow* connects the inserted segment to the point of insertion (*circle*). The dashed lines represent duplicated segments.

maize   sorghum   sugarcane   rice

corresponding gene in a related species with a smaller, better characterized genome (in the case of grasses, rice), and then to isolate the gene in the crop of interest using sequence homology. When this approach is taken, it is important to be sure that the gene identified in the model species is the ortholog of the crop gene of interest; this can be confirmed by a combination of similar phenotype and similar order of the surrounding genes.

As more genome sequences become available for plants, such as the genome sequences of rice, poplar, tomato, and the model legumes, more gene comparisons will be possible, and the identification of genes with similar functions should become easier and a lot faster. In this way, information on gene function can be translated from models to crops or even to wild species. The impact of genome sequencing on the technologies used for identifying and ultimately manipulating gene function is described in the next section.

2.5 GENOMES AND BIOTECHNOLOGY

The availability of genome sequences has influenced greatly the manner in which the function of a particular gene can be defined, and the speed at which this can be done. A common aim in plant biology is to understand the function of genes within a particular pathway (either developmental or metabolic), their contribution to how plants work, and what it is that makes one species different from another. Understanding the effects of variation on the function of a gene and its products is key to successful plant breeding (see Chapter 9). An entire genome sequence provides a high density of **genetic markers** for that particular plant species, which can be used to isolate (**clone**) genes through a variety of approaches.

When the role of a gene has been revealed through study of its mutant phenotype, typically the next step is to isolate and sequence the DNA corresponding to the gene. This allows a protein sequence to be predicted, which often gives clues to its biochemical function. With a complete genome sequence, the sequences of thousands of genes and predicted proteins become available, but the challenge then becomes to reveal the role of each of these genes in the organism. This section describes the concepts behind some of the most common techniques used to characterize genes in plant biology and some applications of these techniques (see also Section 9.3 for information on the creation and uses of **transgenic plants**).

Mutated genes can be localized on the genome by co-segregation with known markers

Genes are often first identified by observable changes in phenotype when a gene is mutated. For several plant species, genetic maps have been constructed that provide researchers with a reference of the order of genes along a chromosome. **Map-based cloning** is a technique that uses these maps to navigate the genome and focus on the targeted gene: the most closely associated genetic markers on either side of that gene can be identified, thus restricting the search to a small region of the genome that can be sequenced. The genetic markers used as reference points in the genetic map are often specific DNA sequences. To determine how closely the gene and a particular marker are associated, it is necessary to measure the likelihood that recombination will occur between the marker and the gene during meiosis (see Section 3.2 for a review of meiosis). The likelihood of recombination is related to the length of the chromosome region between the gene and the marker: the shorter the distance between them, the less likely they are to undergo recombination.

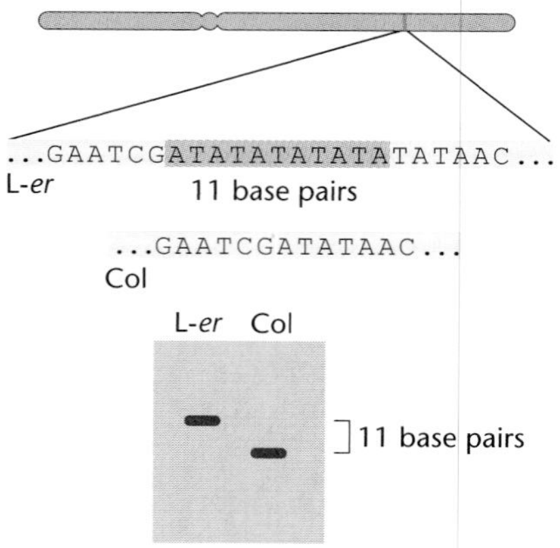

To measure recombination frequencies and thus place a mutation on the genetic map, the mutant plant is crossed with another plant line, called a **mapping line**. A plant from this line is wild-type for the gene being mapped, but its genome sequence is peppered with DNA sequences that differ from the line where the new mutant has been isolated. Because the location of these sequence differences is known, they can be used as reference points in the genetic map (i.e., as genetic markers) (Figure 2–28). When the mutant is crossed with the mapping strain, the genetic markers are mixed with the new mutation and then sorted out in the second (F2) generation (Figure 2–29). Because of recombination and random segregation of the chromosomes, each F2 plant contains a mixture of the genetic markers for the two original parental lines. As mentioned above, however, the closer two DNA sequences are in the chromosome, the less likely is recombination between them. This means that the chromosomal region close to the new mutation will retain the DNA sequence corresponding to the line in which the mutation was originally isolated. In other words, when a particular genetic marker remains associated with the mutation in the F2 plants, this means the new mutation is located in the same region of the genome as the closely linked marker. As researchers look for association with numerous markers, a small part of the genome can be delimited that contains the mutation. This small region can then be sequenced to reveal the new mutation in one of its genes.

Genome sequencing facilitates map-based cloning in two ways. First, it allows the discovery of new genetic markers: when pieces of DNA from a different line are sequenced, the differences from the reference line and their location in the genome are immediately apparent. Second, by providing the wild-type sequence as the reference, genome sequencing simplifies the final steps of finding a mutation. These final steps involve sequencing the genes within the region that has been genetically defined.

Genes that are mutated by insertion of DNA can be isolated by detecting the inserted sequence

Transposon tagging is another commonly used method to clone a gene whose function is first revealed by its mutant phenotype. This method takes advantage of the ability of

Figure 2–28.

Example of a marker used in gene mapping. Two different Arabidopsis strains, Landsberg-*erecta* (L-*er*) and Columbia (Col), contain many small differences in their genome sequences. The figure represents a hypothetical example in which the L-*er* sequence has 11 additional base pairs. When a DNA fragment corresponding to this small region of the genome is amplified by the polymerase chain reaction (PCR; a method for producing *in vitro* many copies of a chosen DNA sequence), the fragment originating from L-*er* is 11 bp longer than that amplified from Col. This size difference can be detected by gel electrophoresis (*bottom*); this method separates DNA fragments according to size. In this way, it is possible to determine whether the DNA sequence for a particular region of the genome originates from the L-*er* or Col strain. (The use of such markers in gene mapping is explained in Figure 2–29.)

...GAATCG**ATATATATATATA**TATAAC...
L-*er*   11 base pairs

...GAATCGATATAAC...
Col

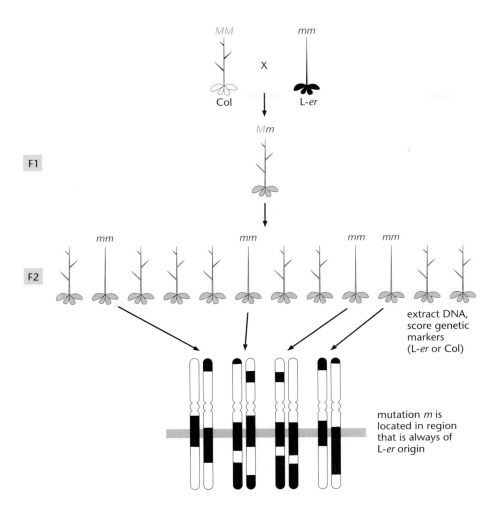

Figure 2–29.
Principle of gene mapping by co-segregation with known markers. The locus *M* has been defined genetically, based on the phenotype caused by the recessive mutation *m* (in this case, the mutant lacks branches). The location of locus *M* in the genome can be determined if the mutant allele *m* segregates in association with known positional markers (as shown in Figure 2–28). To test for co-segregation, a wild-type (*MM*) individual of one Arabidopsis strain (Col; *white*) is crossed with an *mm* mutant of a different strain (L-*er*; *black*). As explained in Figure 2–28, the genome sequences of these two strains contain many small differences that are easily detected. The progeny from this cross has a mixed Col/L-*er* genome (plants shown in *gray*). After self-fertilization of the F1 plants, F2 individuals homozygous for the *m* mutation can be identified by their branchless phenotype. In these individuals, known genetic markers are used to determine whether the DNA in many different regions of the genome is of L-*er* or Col origin. Because the *m* allele was present initially in the L-*er* strain, the DNA sequence in the vicinity of the mutant allele will tend to be of L-*er* origin. Therefore, if the *mm* plants are always homozygous for a particular L-*er* marker, the position of the *M* locus is near that marker.

transposons to insert into new locations in the genome and cause mutations when inserted within a gene. New mutations can be induced at random by stimulating transposons to move (in some cases, by changing the environmental conditions in which the plants are grown). When a mutation is caused by a transposon of known DNA sequence inserting into a gene, the mutated gene is "tagged" with this known sequence. When fragments of genomic DNA containing the transposon sequence are isolated, these fragments also contain sequences of the gene into which the transposon inserted (Figure 2–30). These sequences flanking the transposon can then be used to isolate the complete gene, or to identify it in a database if the genome sequence is already known.

Genes can be screened for mutations at the DNA level independent of phenotype

Reverse genetics is the approach used to determine the function of a gene first identified by its sequence. The technique typically is used to define the functions of the

Figure 2–30.
Procedure to identify a gene that has been disrupted by a transposon. When a transposon (*triangle*) moves to a new location, it may disrupt a gene, indicated by the change from wild-type *M* to mutant *m* allele. The new *m* allele may cause a visible mutant phenotype, but the identity and location of the gene affected are initially unknown. Because the DNA sequence of the transposon is known, methods based on the polymerase chain reaction can be used to amplify a DNA fragment containing transposon sequences, together with flanking sequences corresponding to *m*. By comparing these newly obtained *m* sequences with the known genome sequence, the position and identity of the gene at the *M* locus are revealed.

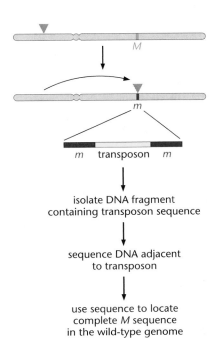

many novel genes (genes with no homology to genes of known function) revealed by genome sequencing.

A range of resources is available—particularly to the Arabidopsis research community—that allows access to plants carrying mutations of almost any gene in the genome. It is relatively easy to generate a population of plant lines with mutations in known genes, with mutations caused by the insertion of DNA with a known sequence, such as transposons or the **T-DNA** that is introduced into plant cells by *Agrobacterium tumefaciens* (see Sections 8.1 and 9.3). In these cases, the known DNA is used as a tag to isolate and amplify the flanking genomic DNA, which allows identification of the gene where the insertion has occurred (see the earlier discussion of transposon tagging). In this way, large collections of seed stocks from populations of plants carrying individual insertions have been generated, along with databases of the genes disrupted by these insertions. By searching these sequence databases for insertions in a gene of interest, researchers can identify the corresponding mutant stock and order it from the population curators. When the line is grown next to a wild-type plant of the same variety of Arabidopsis, the phenotypes of the two plants can be compared. Any observed differences allow the function of the gene to be deduced.

An alternative method for reverse genetics is **TILLING** (**t**argeted **i**nduced **l**ocal **l**esions **in g**enomes), which can be used to identify single nucleotide changes caused by chemical mutagens, such as ethylmethane sulfonate (EMS). With this type of mutagen, it is more difficult to identify the mutated gene because no sequence tag is available. Nevertheless, EMS is a popular mutagen because it induces mutations with a high frequency. Another advantage of EMS is that it can produce more subtle mutant phenotypes such as those caused by amino acid substitutions, which can be very informative about protein function, often complementing information derived from **knockout alleles**. To identify plants with single nucleotide changes in a gene of interest, the gene is amplified by **polymerase chain reaction (PCR)** from a large number of individual lines that have been treated with EMS. The amplified gene fragments are then hybridized with the wild-type sequence. Where a mutation has occurred, there will be a region of unpaired DNA (**heteroduplex**), usually involving a single-base mismatch. This type of DNA can be distinguished from perfectly paired DNA by **high-performance liquid chromatography (HPLC)** or by using an enzyme that cleaves DNA specifically at single-base mismatches. When a particular line with heteroduplex DNA is identified, the mutated gene is sequenced to identify the exact nucleotide change. The mutant line can then be analyzed, along with the wild type, to study the gene's function.

Methods are now available that allow the simultaneous analysis of heteroduplex formation for multiple DNA sequences in the same sample. With these techniques, researchers can identify individual plants carrying a mutant allele in a population of mutagenized plants. TILLING is not dependent on the use of model plants, but instead depends on knowing the sequence of the gene of interest.

RNA interference is an alternative method to knock out gene function

Another way to study the consequences of losing the function of a specific gene is to inactivate the gene by RNAi (see Section 2.3). In this method, a region of the DNA from the gene of interest is expressed in transgenic plants to produce a transcript with a **hairpin structure**. This is normally achieved by constructing an artificial gene that contains two copies of a short DNA sequence (about 500 bp) from the target gene, arranged head-to-head. When this gene is expressed in the host plant, the transcript containing the head-to-head repeat folds into double-stranded RNA, forming a hairpin shape. Processing of the dsRNA through the siRNA pathway results in the production of small RNAs (siRNAs) complementary to the mRNA of the target gene. As described in Section 2.3, the siRNAs are used as guides to locate the target mRNA, which is then cleaved and degraded, effectively preventing expression of the gene. Once the gene has been silenced, its phenotype can be characterized and the gene function deduced.

Multigenic inheritance is analyzed by mapping quantitative trait loci (QTLs)

The methods described above are useful for the analysis of discrete phenotypes that are clearly associated with changes in the function of specific genes. However, in many cases, useful phenotypes such as seed weight or flowering time appear in a population as a continuum of phenotypic variation, rather than as clearly separable classes of phenotype. This **continuous variation** is often seen when the phenotype is controlled by multiple genes, each with a relatively small effect. The absence of clear segregation of phenotypic classes makes it impractical to use classical Mendelian methods for revealing how many genes are involved and the effect of each individual gene on the phenotype. In these cases, a special method is used to map the multiple loci contributing quantitatively to a trait. Each of the genetic loci that contribute collectively to the phenotype is called a **quantitative trait locus (QTL)**, and the mapping method is called **QTL analysis**.

The basic idea behind QTL analysis is that although the presence of a specific allele affecting the trait cannot be revealed by the individual's phenotype, it can be inferred from the presence of genetic markers that are closely linked to the gene and can easily be detected and "scored." Often, the markers used are DNA sequence polymorphisms—that is, variations in the DNA sequence found at the same chromosome position in different individuals in the population. By monitoring many markers covering a large part of the genome, researchers can identify some markers whose presence correlates with the quantitative trait of interest in the population (for example, a particular marker may be found more often in tall individuals). It is unlikely that any of the markers will correspond directly to a gene that causes the phenotypic variation, but correlation with the trait indicates that a locus affecting the trait is nearby on the chromosome. Statistical methods are then used to calculate the likelihood that the correlation between marker and phenotype is not a chance event and to assess how close a QTL may be to the marker. These methods can reveal how many loci are likely to contribute to a quantitative trait, their approximate positions on the chromosomes, and their relative contribution to the final phenotype.

QTLs are the basis of much of the variation seen in natural populations—in contrast to the simple inheritance of mutant alleles selected in the laboratory with the aim of understanding the function of individual genes. This does not mean, however, that individual loci studied in the laboratory are not relevant to natural variation, as exemplified by a study of natural variation in Arabidopsis. There are several natural strains (often referred to as **accessions**) of Arabidopsis with an array of traits that differ from those found in the Columbia accession—whose genome was first fully sequenced. The phenotypic differences among accessions, such as in flowering time, are often determined by QTLs. The Arabidopsis accession Cape Verde originated from islands at the equator, where day and night length are equal, and plants of this accession flower under short-day conditions. Arabidopsis accessions from northern latitudes initiate flowering in response to long days. After crossing of the Cape Verde accession with a northern accession, QTL analysis revealed those regions of the genome that most influence flowering time. One of these genomic regions contained the *CRY2* gene, which encodes a blue-light **photoreceptor** that was already known to promote flowering in long days (more detail on this is provided in Section 6.2). Sequencing of this gene from different accessions revealed that the Cape Verde CRY2 carried an amino acid change that allowed the protein to function also in short days. Thus a naturally occurring, quantitative difference in flowering time was based on a subtle change in a gene for which severe mutations would produce a much more obvious phenotype.

Similar analyses can be carried out in any plant species that exhibits natural variation for a specific trait, and they have been used to good effect to study the changes in crops during domestication (see Chapter 9).

Genome sequencing allows the development of methods to monitor the activity of many genes simultaneously

With the availability of complete genome sequences, tools can be constructed to monitor simultaneously the expression of all genes in an organism. This type of experiment is useful in understanding how genes are expressed coordinately during development or in response to external stimuli.

Genome-wide expression is monitored through the use of **expression arrays**. These consist of large numbers of **DNA probes** (oligonucleotides or longer DNA fragments) attached to a solid substrate (typically, a glass slide). The probes are spotted as tiny dots in a dense matrix, with thousands of genes represented in an area smaller than an ordinary microscope slide. To monitor gene expression, an RNA sample is extracted from a cell type, organ, or whole plant. The whole mRNA population is converted in the test tube into a DNA copy (cDNA), which is tagged with fluorescently labeled nucleotides. The labeled DNA is hybridized to the array, which is then scanned with a laser to measure the amount of fluorescent label captured by each probe. The amount of label reflects the number of DNA copies of each mRNA sequence and thus gives an estimate of the expression level of the gene (Figure 2–31).

One example of what we can learn from this type of experiment is the analysis of gene expression during the 24-hour daily cycle in Arabidopsis. In organisms ranging from bacteria to humans, metabolism, gene expression, and physiological responses oscillate with the day–night cycle. This cycle is not only perceived directly as alternating light and dark periods, but is also mirrored by an internal biological clock, the **circadian clock**, which continues to mark the daily cycle even when there are no external clues as to whether it is night or day (see Section 6.4). Expression arrays have been used to reveal which genes are connected to the circadian clock in Arabidopsis. About 6% of all Arabidopsis genes are regulated by the clock, and their functions give an overview of the metabolic and developmental changes that oscillate daily. For example, expression of genes encoding the proteins involved in photosynthesis is programmed to peak at around midday, whereas genes involved in the production of substances that protect the plant from UV light are expressed before sunrise. Genes encoding enzymes involved in sugar production are programmed to be expressed during the day; enzymes required for the mobilization of starch reserves are preferentially expressed at night. Apart from showing the coordinated expression of hundreds of genes during the daily cycle, these experiments also help to reveal how this coordination is achieved. For

Figure 2–31.
Simultaneous measurement of the expression level of many genes with a cDNA microarray. The image shows part of an Arabidopsis cDNA array. Each spot contained a DNA probe corresponding to a specific gene, immobilized on a glass slide. The spots became visible when the array was hybridized to a mixture of fluorescently labeled cDNA, which was made from RNA extracted from the cells in which gene expression is being monitored. The intensity of each spot reveals the amount of each species of cDNA present in the mixture. The spots have different colors because the array was hybridized at the same time with two different cDNA populations: one prepared with RNA extracted from seedlings grown in white light (cDNA labeled with a green fluorescent dye), the other with RNA from dark-grown seedlings (cDNA labeled with a red fluorescent dye). Thus the green spots on the array show expression of light-activated genes, and the red spots correspond to genes expressed in the dark. Genes expressed in both conditions appear yellow (the mixture of both fluorescent colors).

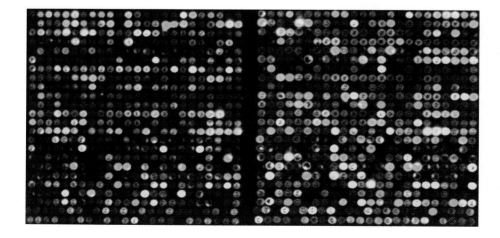

example, examination of the regulatory sequences of genes that peak in the early evening revealed a shared sequence (a cis-element called the "evening element") that mediates the input of signals from the clock into transcriptional control.

The ability to measure mRNA levels for many genes in parallel is mirrored by methods that can identify large numbers of proteins in a biological sample (**proteomics** methods) or can monitor the presence of large numbers of metabolites simultaneously (**metabolomics**). All these techniques are changing the focus of biological research from the activity of single genes, proteins, or biochemical pathways to the ways in which large numbers of genes, proteins, or biochemical pathways function together in a cell. Understanding these networks of interactions is now a major challenge for all biology, including plant biology.

2.6 CYTOPLASMIC GENOMES

The nucleus contains 80 to 90% of the plant cell's DNA; the remainder is found in two other cellular compartments: **mitochondria** and **plastids**. These organellar genomes are usually small and circular, in contrast to the large, linear nuclear chromosomes of higher plants.

The division of plastids and mitochondria is not closely coupled to cell division, and both organelles can multiply in nondividing cells (see Section 3.3). Thus, the number of mitochondria and plastids can vary greatly with cell type and with different metabolic states. Also, depending on the cell type, the plastids develop into different forms, with different functions: **proplastids**, **amyloplasts**, **chromoplasts**, **leucoplasts**, **etioplasts**, and **chloroplasts**. All these types of plastid, however, contain the same genome, which is also referred to as the **plastome**.

Plastids and mitochondria contain their own distinctive machinery for gene expression (Figure 2–32): their genes are transcribed and translated within the organelles themselves. This means that proteins are synthesized at three sites in a plant cell: the cytosol, mitochondria, and plastids. Each site has its own set of ribosomes. Plastid 70S ribosomes and plant mitochondrial 78S ribosomes are different from the 80S ribosomes

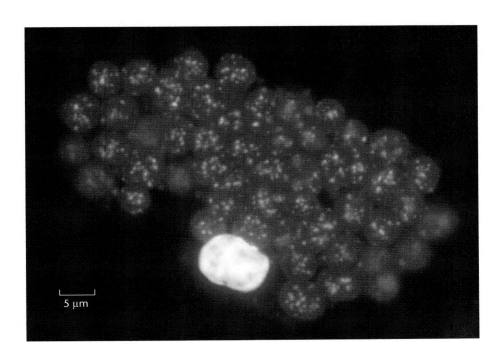

Figure 2–32.
DNA in chloroplasts. The micrograph shows a group of chloroplasts extracted from pea leaves. The chloroplasts are visible as red organelles, due to chlorophyll fluorescence; blue dots inside the chloroplasts correspond to DNA stained with a blue fluorescent dye. (From N. Sato et al., *EMBO J.* 12(2):555–561, 1993. With permission from Macmillan Publishers Ltd, courtesy of Naoki Sato.)

5 µm

found in the cytosol. Organellar ribosomes are more similar to eubacterial ribosomes, reflecting their evolutionary origins.

Plastids and mitochondria evolved from bacteria engulfed by other cells

As discussed in Section 1.2, the genomes of present-day plastids and mitochondria reflect their evolutionary past. Both organelles are probably derived from bacteria that became endosymbionts in proto-eukaryotic cells. Like bacterial DNA, the DNA in organelles is mostly circular and lacks the histones associated with nuclear chromosomes.

Mitochondria are most probably descendants of ancient α-proteobacteria. The closest living relatives of mitochondria are the *Rickettsia* group of α-proteobacteria, and they probably share a common ancestor with mitochondria. *Rickettsia* are obligate intracellular parasites; for example, *Rickettsia prowazekii* is the causative agent of typhus.

Comparisons of organellar genes with bacterial genes suggest that plastids are descended from ancient cyanobacteria, which are a group of photosynthetic bacteria that arose early in evolution. Some cyanobacteria, such as *Synechocystis* sp. PCC6803, are free-living, whereas others can enter symbiotic partnerships with a wide range of eukaryotes, including marine sponges, liverworts, the water fern *Azolla*, and the angiosperm *Gunnera*. It is likely that all plastids in extant algae, land plants, and some protozoa ultimately originate from a single ancestral endosymbiotic event. From this ancestral plastid, two main types of plastids have evolved: those present in red algae and those present in green algae and land plants. These two types of plastids differ from each other in their photosynthetic pigments and in gene content. Subsequently, secondary endosymbiotic events occurred in which nonphotosynthetic eukaryotes engulfed unicellular algae (either red or green algae), giving rise to the plastids found in dinoflagellates, euglenoids, and even some nonphotosynthesizing animal parasites such as *Plasmodium*. Here, we focus on the plastids of the land plants.

Organellar genes do not follow Mendel's laws of inheritance

In contrast to nuclear genes, plastid and mitochondrial genes do not follow a Mendelian pattern of inheritance from one generation to the next. In many angiosperm species, including tobacco (Figure 2–33) and maize, plastids and mitochondria are inherited only from egg cells, not from sperm (pollen) cells. This means not only that the organellar genes are inherited exclusively through the maternal line, but also that any particular set of plastid genes is inherited in association with a specific set of mitochondrial genes.

Less common is the inheritance of plastids from both parents (biparental inheritance). In species exhibiting biparental inheritance of plastids (such as alfalfa and pelargonium), the ratio of male- to female-derived plastid genomes in a zygote can vary, and nuclear genes control the balance. In some gymnosperm species, plastids are predominantly transmitted via sperm cells. This paternal mode of plastid transmission is typical of conifers, including pine. Plastids and mitochondria can show distinct patterns of inheritance. For example, in alfalfa, mitochondria show maternal inheritance and plastids show biparental inheritance.

The genomes of plastids and mitochondria have been reduced during evolution

The original endosymbiotic bacteria that gave rise to mitochondria and plastids probably contained genomes comparable to those of *Rickettsia* and *Synechocystis*, respectively, but during evolution, unnecessary DNA has been lost from both organellar genomes. Consequently, the genomes of plastids and mitochondria are much smaller than those of *Synechocystis* sp. PCC6803 and *R. prowazekii*, respectively. For example, *Synechocystis* has approximately 3000 genes, while the plastids of most land plants have between 100 and 200 genes.

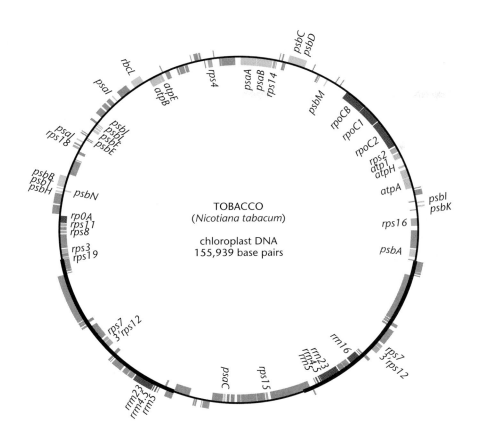

Figure 2–33.
Gene content of tobacco plastid DNA. The
diagram shows a simplified map of the
tobacco plastid genome. The *black circle*
represents the DNA, with genes represented
by boxes or lines on either side of the circle.
Genes involved in similar processes are in the
same color: *green*, genes involved in
photosynthesis; *orange*, genes encoding
subunits of the H+-ATPase (see Chapter 4);
brown, genes encoding ribosomal proteins;
and *blue*, genes involved in transcription.
Genes participating in other processes are
marked in gray.

Some of the reductions in organellar genomes probably occurred because redundant genes (with functions overlapping those of other organellar genes or those of nuclear genes) were lost by chance deletions mediated by DNA replication errors or aberrant DNA recombination events. Other organellar genes, however, have been transferred to the nuclear genome (as described in Section 2.4 and discussed further below). Over evolutionary time, these transferred bacterial genes have developed eukaryotic transcription and translation sequences, allowing efficient transcription in the nucleus and translation on cytosolic ribosomes.

The tendency for genome reduction seems to have diminished for present-day plant mitochondrial genomes, but continues for plastid genomes. For example, the angiosperm *Epifagus virginiana* (beechdrops) is a parasite living on roots. It is unable to photosynthesize and, apart from its flowers, all its aboveground parts, including the reduced leaves, are white. Genes encoding proteins directly involved in photosynthesis that are normally encoded by plastids in green plants have been lost from the small genome in *Epifagus* plastids.

Most polypeptides in organelles are encoded by the nuclear genome and targeted to the organelles

As described above, organellar genomes have only a limited coding capacity. The majority (>95%) of polypeptides located in the organelles are encoded by the nuclear genome. These genes are transcribed in the nucleus, and the RNA transcripts are exported to the cytosol, where they are translated into polypeptides.

Many of the nuclear genes encoding proteins that are targeted to mitochondria and plastids were originally derived from the genomes of the bacterial endosymbionts that gave rise to these organelles. At least 40% of the estimated 2000 to 3000 nuclear-encoded polypeptides found in chloroplasts are derived from the genes of the original cyanobacteria-like endosymbiont that was the ancestor of plastids. However, some nuclear-encoded organellar proteins are apparently derived from the proto-eukaryotic host. Conversion of endosymbionts to organelles meant the complete assimilation of a once free-living bacterium into the metabolic activities of the host cell. During this process of assimilation, many of the host's nuclear-encoded proteins, not in the original endosymbionts, became targeted to the organelles.

Conversely, as described earlier in the chapter, some genes have been transferred from organelles to the nuclear genome. In some cases, it seems that genes originally present in one endosymbiont were transferred to the nucleus and their protein products were subsequently targeted to both types of organelles: mitochondria and plastids. Nuclear-encoded polypeptides destined for organelles contain sequences of amino acids at their N-termini that target them to the organelles. Targeting peptides (often known as **transit peptides**) act as address labels that direct proteins to the correct subcellular location. This is described in more detail in Section 3.3.

Replication and recombination of plastid DNA is not tightly coupled to cell division

The enzymes involved in replication and recombination of the DNA of plastids are encoded by the nuclear genome. During plastid DNA replication, a **displacement loop (D-loop)** is indicative of an **origin of replication**. In a D-loop, only one of the two DNA strands is replicated, and the growing **replication fork** displaces the complementary parental strand as single-stranded DNA. Two origins of replication (oriA and oriB) have been mapped in the large inverted-repeat region of the DNA of tobacco plastids. Replication forks initiated at oriA and oriB give rise to two converging D-loops, which meet to complete the synthesis of both DNA strands (Figure 2–34).

In the nucleus, DNA replication is completed before cell division, when the replicated DNA is partitioned equally between daughter cells. In contrast, the replication of plastid DNA is not tightly linked to organelle division. When plastids divide, they can receive unequal numbers of genome copies, including partially replicated DNA. Because DNA replication and plastid division are uncoupled, the number of genome copies present in plastids can vary during development. For example, in Arabidopsis shoot meristem cells, each plastid contains approximately 40 genome copies; this number increases to 600 in **leaf primordia** and decreases again in mature leaves. The increase in plastid genome copies in the early stages of leaf development may allow the chloroplasts to keep pace with increasing photosynthetic demand.

Gene expression has common features in plastids and eubacteria

Transcription and translation of plastid genes (unlike nuclear genes) take place within the same subcellular compartment. This raises the possibility that transcription, RNA processing, and translation can be coupled as they are in bacteria such as *Escherichia*

Figure 2–34.
Replication of tobacco plastid DNA.
Replication starts from two origins and is initially unidirectional within each origin. When the replication loops (D-loops) meet, they form a single loop with bidirectional DNA replication at each replication fork. The forks progress along the circular DNA until they meet at the opposite side and the two new circular DNA molecules are separated. (With permission from the Company of Biologists.)

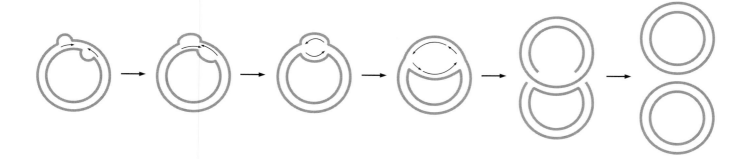

coli: bacterial ribosomes can bind mRNA and start protein synthesis before transcription is complete. This is very unlike nuclear gene transcription, which is physically separated from translation of the resulting mRNA on cytosolic ribosomes. Many bacterial genes are arranged as **operons**, with a single **promoter** that transcribes a set of adjacent genes. Operons allow the coordinated expression of multiple genes (producing a **polycistronic** message) whose products are required in the same metabolic pathway. Many of the genes in the plastid and mitochondrial genomes are also arranged in operons. For example, the adjacent *rps*2, *atp*I, *atp*H, *atp*F, and *atp*A genes of plastids are transcribed together as a polycistronic message. However, other plastid genes, such as the gene encoding the large subunit of Rubisco (*rbc*L), are mainly transcribed as single-gene, or **monocistronic**, mRNAs.

Expression of plastid genes is regulated at several levels, including the steps of transcription, mRNA degradation, translation, and protein turnover, as described below.

Plastids contain two distinct RNA polymerases

At least two types of RNA polymerase are present in plastids. **Plastid-encoded plastidial (PEP) RNA polymerase** is a multisubunit RNA polymerase similar to that of *E. coli*, consisting of five polypeptides in the core polymerase (Figure 2–35). Four types of subunits are encoded by the plastid genes *rpo*A (α subunit), *rpo*B (β subunit), *rpo*C1 (β′ subunit), and *rpo*C2 (β″ subunit), and the fifth (σ) by the nuclear genome. The σ polypeptides bind to the core RNA polymerase to form a complex that specifically recognizes and binds to promoter sequences. Specificity in promoter recognition is thus conferred by the σ factors. PEP RNA polymerase is highly active in leaf chloroplasts and transcribes plastid genes that encode proteins directly involved in photosynthesis; these genes have upstream *E. coli*-like promoters. The photosynthesis-related genes include *rbc*L and *psb*A; the latter encodes **protein D1** of photosystem II (see Sections 4.2 and 7.1).

Evidence for the existence of a **nuclear-encoded plastidial (NEP) RNA polymerase** in land plants first came from studies of cereal plastids lacking functional ribosomes.

nuclear RP2 complex

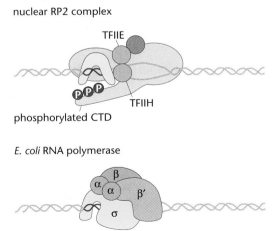

TFIIE

phosphorylated CTD

TFIIH

E. coli RNA polymerase

chloroplast RNA polymerase

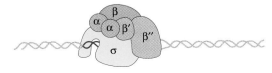

Figure 2–35.
RNA polymerases. Comparison between nuclear RNA polymerase II (RP2) complex (see Box 2-1), a bacterial RNA polymerase (based on *E. coli*), and plastid-encoded plastidial (PEP) RNA polymerase (from a chloroplast). Note that each subunit of the bacterial polymerase has a corresponding homolog in the PEP RNA polymerase (except for the bacterial β′ subunit, which is split into separate β′ and β″ subunits in the plastid RNA polymerase).

The barley *albostrians* and maize *iojap* mutants have recessive nuclear mutations that result in loss of plastid ribosomes. Since ribosomes are required for gene expression in the plastid, including production of ribosomal proteins, loss of the plastid ribosomes is irreversible. Plastids deficient in ribosomes are white because they cannot accumulate chlorophyll, and they are revealed as albino stripes in plants. Plastids deficient in ribosomes cannot translate messages and are unable to express the *rpo*A, B, C, and C1 genes of PEP RNA polymerase. It was the absence of PEP RNA polymerase in these mutants that allowed the detection of messages transcribed by a second polymerase, the NEP RNA polymerase. Similarly, knocking out the synthesis of PEP RNA polymerase by targeted insertional mutagenesis of the plastid *rpo* genes in tobacco has also revealed NEP RNA polymerase activity.

NEP RNA polymerase is a single polypeptide that is structurally similar to bacteriophage T7 RNA polymerase. It recognizes specific promoter motifs in plastid genes. In Arabidopsis, one NEP RNA polymerase is targeted to plastids, another to mitochondria, and a third to both organelles. NEP and PEP RNA polymerases seem to play distinct roles in regulating plastid gene expression. Photosynthetic genes such as *rbc*L contain promoters for PEP RNA polymerase. Genes encoding proteins not directly involved in photosynthesis, such as the plastid rRNA genes, have promoters for both NEP and PEP RNA polymerases. A few nonphotosynthetic genes, such as *clp*P and *rpo*B, contain only promoters recognized by NEP RNA polymerase. It seems that NEP RNA polymerase is generally responsible for low-level transcription of nonphotosynthetic genes with housekeeping functions, because housekeeping functions such as protein synthesis are required in all plastid types. In contrast, PEP RNA polymerase is most active in chloroplasts, where it is responsible for producing the abundant transcripts encoding photosynthetic proteins such as rbcL.

The interplay between the activities of NEP and PEP RNA polymerases allows fine-tuning of plastid gene expression in different plastid types, although the details remain to be fully worked out. One attractive model involves a cascade in which NEP RNA polymerase is replaced by PEP RNA polymerase during the development of chloroplasts from proplastids. The model is supported by the observation that mRNAs (for nonphotosynthetic proteins) generated by NEP RNA polymerase accumulate to their highest levels earlier in development than do the mRNAs generated by PEP RNA polymerase. Because the *rpo* genes encoding PEP RNA polymerase are transcribed by NEP RNA polymerase, the NEP RNA polymerase must be active before PEP RNA polymerase during plastid development.

Post-transcriptional processes are important in regulating plastid gene expression

Plastid genes that encode proteins directly involved in photosynthesis (such as the *psb*A gene) seem to be transcribed at similar rates in organs such as leaves and roots. The finding that the mRNA accumulates to much higher levels in green leaves suggests that these messages are degraded more rapidly in roots than in green leaves. The destruction of plastid mRNAs is initiated by an **endoribonuclease** that hydrolyzes transcripts into smaller pieces by cleaving them internally (Figure 2–36). These decay intermediates are then degraded further by the action of **exoribonucleases**, which remove nucleotides from their ends.

In plastids, **polyadenylation** of the 3′ ends of RNA decay intermediates seems to facilitate their rapid degradation. This is in contrast to nuclear/cytoplasmic mRNAs, which are *stabilized* by polyadenylation; the mature mRNAs have long **poly-A tails** at their 3′ ends. Because polyadenylation of plastid mRNAs marks them for destruction, the presence of poly-A tails at the 3′ ends of plastid messages is transient. Mature plastid mRNAs lack poly-A tails. Short inverted repeats at the 3′ ends, which have the potential to form stem-loop structures, confer stability on plastid mRNAs and are bound by

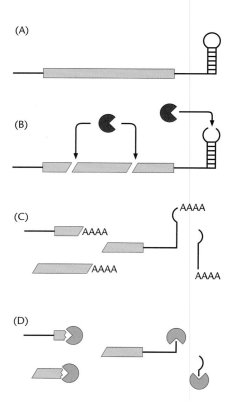

Figure 2–36.
Plastid mRNAs destroyed in a stepwise process. (A) Intact mRNA, with the coding sequence represented by a *green box*; the stem-loop structure at the end of the mRNA is required for stability. (B) The eventual destruction of the mRNA begins with internal cleavage by an endoribonuclease (*red*). (C) The resulting mRNA fragments are modified by addition of poly-A tails. (D) The polyadenylated RNA fragments are degraded by exoribonucleases (*blue*).

several regulatory proteins that seem to control the accessibility of the mRNA to ribonucleases. The 5′ untranslated regions and coding sequences of plastid messages also affect their stability. RNA processing seems to be largely responsible for defining the 5′ and 3′ ends of many plastid mRNAs, but little is known about the ribonucleases involved in message maturation.

Regulation of translation has an important role in determining the relative abundance of plastid proteins. Recruitment of plastid ribosomes to plastid mRNAs is influenced by sequences surrounding the **AUG start codon**. These sequences can include an *E. coli*-like ribosome binding site (**Shine-Dalgarno sequence**) that is complementary to the 3′ end of 16S rRNA. Many plastid messages lack Shine-Dalgarno sequences, and other 5′ sequences that bind nuclear-encoded regulatory proteins seem to facilitate access of plastid ribosomes to the AUG initiation codon. Coding regions might be preferentially translated from a subset of processed transcripts, allowing optimal levels of gene products. In the case of the *rbc*L message, for example, translation of Rubisco protein is only possible following processing of a precursor message.

Organellar transcripts undergo RNA editing

An unusual form of RNA processing that occurs in both plastids and mitochondria is **RNA editing**: controlled alteration of the RNA sequence after it has been transcribed. For example, editing converts an ACG to an AUG in the maize *rpl2* message. This edited AUG is the start codon that initiates translation of the rpl2 protein with tRNA^formyl-Met, the initiator tRNA (carrying formylmethionine) in bacteria, mitochondria, and plastids.

RNA editing has been observed in the plastids of all higher plants examined thus far, including gymnosperms such as blackpine. Since RNA editing has not been observed in eubacteria or in the plastids of algae and lower plants such as the liverwort *Marchantia polymorpha*, it might have originated relatively recently in a common ancestor of higher plants.

Post-translational processes contribute to maintaining the correct ratio of nuclear- and plastid-encoded components of multisubunit complexes

A large number of multisubunit proteins in organelles consist of some polypeptides encoded by nuclear genes and some encoded by organellar genes. In plastids, these polypeptides include the thylakoid membrane complexes: **photosystem I**, **photosystem II**, **cytochrome b_6f**, and **NADH dehydrogenase**; the soluble stromal enzyme Rubisco; and the proteins of plastid ribosomes. Coordinating the expression of nuclear and plastid genes—which are physically separated in the cell—to produce polypeptides at the correct levels for the formation of protein complexes requires a regulatory network that is unique to organelles. Protein degradation plays a role in maintaining the balance of nuclear and plastid gene products. Excess subunits that are not stabilized by incorporation into complexes are rapidly degraded.

The use of **proteolysis** to coordinate the levels of plastid- and nuclear-encoded peptides is exemplified by Rubisco, the central enzyme in the fixation of carbon dioxide (see Section 4.2). Rubisco is composed of eight polypeptides: four identical copies of a nuclear-encoded small subunit, and four identical plastid-encoded large subunits. The synthesis of the large or small subunits can be prevented by mutation or by protein synthesis inhibitors specific for plastid or cytosolic ribosomes, respectively. When synthesis of the large subunits is prevented, the nuclear-encoded small subunits are still synthesized and imported into plastids, but are rapidly degraded there. The reverse is also true: in the absence of small subunits, the large subunits are synthesized in the plastids and then degraded. Assembly of small and large subunits into active Rubisco (the **holoenzyme**) prevents their proteolysis.

Developmental regulation of plastid gene expression includes signaling pathways between plastids and the nucleus

The leaves of cereals and grasses are a particularly good system in which to study the development of plastids. New leaves are formed from a **basal meristem**. This produces a leaf in which the youngest cells are found at the base and the oldest at the tip. From the pale leaf base to the green leaf tip, a gradient of plastid differentiation is seen, starting with proplastids and ending with chloroplasts. Chloroplast development is associated with increases in the number of plastid genome copies and the accumulation of photosynthetic protein complexes.

When higher plants are placed in the dark, they cease to accumulate chlorophyll, and chloroplasts are converted to or replaced by etioplasts, which lack thylakoids. The light-induced conversion of etioplasts to chloroplasts is another well-studied example of chloroplast development. Analysis of this light-regulated developmental pathway in Arabidopsis has allowed the isolation of nuclear mutants defective in various components of the pathway. For example, the *cop1* mutant is able to develop a mature thylakoid system in the dark, even though it does not accumulate chlorophyll. The COP1 protein is a negative regulator of nuclear genes involved in the development of chloroplasts.

As we have seen, proper functioning of organelles requires the coordinated expression of genes located in distinct subcellular compartments. Nuclear control over plastid development is illustrated by the finding that cell type plays a major role in defining the differentiated state of plastids. For example, starch-storing amyloplasts are found in roots and tubers, whereas photosynthetic chloroplasts are found in leaves. Expression of the plastid genome in roots is very different from that in leaves. Unlike amyloplasts, chloroplasts contain abundant ribosomes, are highly active in protein synthesis, and express the proteins directly involved in photosynthesis at high levels.

Research has also suggested that the expression of nuclear genes is sensitive to signals originating in plastids. For example, perturbations of plastid function by **herbicides** such as norflurazon and by inhibitors of electron transport such as dichlorophenyl dimethyl urea (DCMU) can down-regulate the expression of nuclear-encoded photosynthetic polypeptides targeted to plastids. The light-harvesting complex polypeptides (LHCPs) of photosystem II (see Section 4.2) are one example of plastid proteins encoded by nuclear genes that are sensitive to signals indicating that the chloroplasts are functional.

One candidate for a signaling molecule is a **porphyrin**, a precursor of chlorophyll (see Section 4.7). This has been revealed by the *gun* (**g**enome **un**coupled) mutants of Arabidopsis, in which the expression of nuclear genes such as *LHCP* is insensitive to herbicides that affect plastids. *GUN5* encodes a subunit of plastid Mg^{2+} chelatase, which is involved in chlorophyll biosynthesis. In *gun5* mutants, protoporphyrin XI, required for chlorophyll biosynthesis, is not formed, suggesting that this molecule could act as the signal that normally represses *LHCP* gene expression on norflurazon treatment. There is also evidence for other signaling pathways between plastids and the nucleus, involving sugar levels and the oxidation-reduction (redox) status of the electron carrier **plastoquinone**.

SUMMARY

The heritable characteristics of a plant species are determined by its genome, the total genetic content of its haploid set of chromosomes. The nuclear genome is subdivided into chromosomes, each containing DNA tightly packed with histones. Repetitive DNA sequences in the centromeres and telomeres are important in chromosome maintenance and stability. Nuclear genes are transcribed into several types of RNA: rRNAs, mRNAs, tRNAs, small nucleolar RNAs, and RNA transcripts that are processed into small regulatory RNAs (miRNAs, siRNAs). Plant chromosomes contain many mobile genetic elements called transposons, which move by copy-and-paste or cut-and-paste methods. Genetic changes caused by transposon insertion are occasionally advantageous, but mainly deleterious.

Differential gene expression underlies most of the phenotypic changes that occur during plant development and in response to environmental signals. Gene expression can be controlled at many different stages from transcription to translation of the RNA transcript. Regulatory sequences and transcription factors control where and when a gene is transcribed. These act combinatorially to control the initiation of transcription, which requires RNA polymerase II and general transcription factors. Gene activity can be regulated by chemical changes in DNA and histones, including methylation. These changes stabilize patterns of gene expression, and they can persist through cell division, in epigenetic inheritance. Regulation at the RNA level includes alternative transcript splicing, mRNA stability (rate of degradation), and control by small regulatory RNAs, including miRNAs and siRNAs.

Genome sequencing provides an understanding of the genes necessary to make a plant, facilitates the development and use of methods to monitor the behavior of many genes simultaneously, and reveals similarities and differences between plants and other organisms. Gene annotation involves making sense of a genome sequence and extracting information from it. The Arabidopsis genome was the first plant genome to be fully sequenced. Sequencing revealed a complexity similar to that of animal genomes, with a large fraction of plant-specific genes. Several other plant genomes have since been sequenced. A striking feature of the Arabidopsis, rice, and poplar genomes is that large parts seem to have been duplicated; one way that duplication could have arisen is through polyploidy. Genes can acquire new functions by duplication and divergence. The order of genes is conserved between closely related plant species and is called synteny.

An aim in plant biology is to understand the function of genes in a particular pathway, their contribution to how plants work, and what makes one species different from another. The role of a gene can be revealed through study of its mutant phenotype; the next step is to isolate and sequence the gene, then predict a protein sequence, which can give clues to biochemical function. Biotechnology methods developed for achieving these ends include the use of genetic markers, map-based cloning, transposon tagging, reverse genetics (including use of *Agrobacterium tumefaciens* T-DNA and TILLING), polymerase chain reaction, RNA interference, quantitative trait locus analysis, expression arrays, proteomics, and metabolomics methods.

The nucleus contains 80 to 90% of the plant cell's DNA; the remainder is found in mitochondria and plastids. These organellar genomes are derived from bacterial and cyanobacterial endosymbionts, and during evolution some genes were transferred from symbiont to host nuclear genome. Plastid and mitochondrial genes do not follow a Mendelian pattern of inheritance; in many angiosperm species, plastids and mitochondria are inherited exclusively through the maternal line. Most polypeptides of the organelles are encoded by the nuclear genome and targeted to the organelles; many of the nuclear genes encoding these targeted proteins are those derived from the bacterial endosymbionts. Expression of plastid genes is regulated at several levels, including transcription, mRNA degradation, translation, and protein turnover. Developmental regulation of plastid gene expression includes signaling pathways between plastids and the nucleus: the nucleus exerts control over plastid development, and expression of nuclear genes is also sensitive to signals originating in plastids.

FURTHER READING

General

Alberts B, Johnson A, Lewis J et al. (2008) Molecular Biology of the Cell, 5th ed. New York: Garland Science.

Lewin B (2008) Genes IX. Boston: Jones and Bartlett Publishers.

2.1 The Nuclear Genome: Chromosomes

Heslop-Harrison JS (2000) Comparative genome organization in plants: from sequence and markers to chromatin and chromosomes. *Plant Cell* 12, 617–635.

Kornberg RD & Lorch Y (1999) Twenty-five years of the nucleosome, fundamental particle of the eukaryote chromosome. *Cell* 98, 285–294.

2.2 Chromosomal DNA

Copenhaver GP, Nickel K, Kuromori T et al. (1999) Genetic definition and sequence analysis of Arabidopsis centromeres. *Science* 286, 2468–2474.

McKnight TD & Shippen DE (2004) Plant telomere biology. *Plant Cell* 16, 794–803.

Sabot F & Schulman AH (2006) Parasitism and the retrotransposon life cycle in plants: a hitchhiker's guide to the genome. *Heredity* 97, 381–388.

San Miguel P, Gaut BS, Tikhonov A et al. (1998) The paleontology of intergene retrotransposons of maize. *Nat. Genet.* 20, 43–45.

Wilson WA, Harrington SE, Woodman WL et al. (1999) Inferences on the genome structure of progenitor maize through comparative analysis of rice, maize and the domesticated panicoids. *Genetics* 153, 453–473.

2.3 Nuclear Gene Regulation

Baulcombe D (2004) RNA silencing in plants. *Nature* 431, 356–363.

Baurle I & Dean C (2006) The timing of developmental transitions in plants. *Cell* 125, 655–664.

Cao X, Aufsatz W, Zilberman D et al. (2003) Role of the DRM and CMT3 methyltransferases in RNA-directed DNA methylation. *Curr. Biol.* 13, 2212–2217.

Chan SWL, Henderson IR & Jacobsen SE (2005) Gardening the genome: DNA methylation in *Arabidopsis thaliana. Nat. Rev. Genet.* 6, 351–360.

Cubas P, Vincent C & Coen E (1999) An epigenetic mutation responsible for natural variation in floral symmetry. *Nature* 401, 157–161.

Dickey LF, Petracek ME, Nguyen TT et al. (1998) Light regulation of Fed-1 mRNA requires an element in the 5′ untranslated region and correlates with differential polyribosome association. *Plant Cell* 10, 475–484.

Gendrel A-V, Lippman Z, Yordan C et al. (2002) Dependence of heterochromatic histone H3 methylation patterns on the Arabidopsis gene *DDM1. Science* 297, 1871–1873.

Goodrich J, Puangsomlee P, Martin M et al. (1997) A Polycomb-group gene regulates homeotic gene expression in Arabidopsis. *Nature* 386, 44–51.

Kubo N, Harada K, Hirai A & Kadowaki K-I (1999) A single nuclear transcript encoding mitochondrial RPS14 and SDHB of rice is processed by alternative splicing: common use of the same mitochondrial targeting signal for different proteins. *Proc. Natl. Acad. Sci. USA* 96, 9207–9211.

Lippman Z & Martienssen R (2004) The role of RNA interference in heterochromatic silencing. *Nature* 431, 364–370.

Loidl P (2004) A plant dialect of the histone language. *Trends Plant Sci.* 9, 84–90.

Quail PH (2002) Photosensory perception and signalling in plant cells: new paradigms? *Curr. Opin. Cell Biol.* 14, 180–188.

Woychik NA & Hampsey M (2002) The RNA polymerase II machinery: structure illuminates function. *Cell* 108, 453–463.

2.4 Genome Sequences

Adams KL, Cronn R, Percifield R & Wendel JF (2003) Genes duplicated by polyploidy show unequal contributions to the transcriptome and organ-specific reciprocal silencing. *Proc. Natl. Acad. Sci. USA* 100, 4649–4654.

Arabidopsis Genome Initiative (2000) Analysis of the genome sequence of the flowering plant *Arabidopsis thaliana. Nature* 408, 796–815.

International Rice Genome Sequencing Project (2005) The map-based sequence of the rice genome. *Nature* 436, 793–800.

Kellogg EA & Bennetzen JL (2004) The evolution of nuclear genome structure in seed plants. *Am. J. Bot.* 91, 1709–1725.

Liljegren SJ, Ditta GS, Eshed HY et al. (2000) SHATTERPROOF MADS-box genes control seed dispersal in Arabidopsis. *Nature* 404, 766–770.

Min Lee M & Schiefelbein J (2001) Developmentally distinct MYB genes encode functionally equivalent proteins in Arabidopsis. *Development* 128, 1539–1546.

Moore G, Devos KM, Wang Z & Gale MD (1995) Cereal genome evolution—grasses, line up and form a circle. *Curr. Biol.* 5, 737–739.

Riechmann JL, Heard J, Martin G et al. (2000) Arabidopsis transcription factors: genome-wide comparative analysis among eukaryotes. *Science* 290, 2105–2110.

Tuskan GA, DiFazio S, Jansson S et al. (2006) The genome of black cottonwood, *Populus trichocarpa* (Torr. & Gray). *Science* 313, 1596–1604.

Wang J, Tian L, Madlung A et al. (2004) Stochastic and epigenetic changes of gene expression in Arabidopsis polyploids. *Genetics* 167, 1961–1973.

2.5 Genomes and Biotechnology

Alonso JM & Ecker JR (2006) Moving forward in reverse: genetic technologies to enable genome-wide phenomic screens in Arabidopsis. *Nat. Rev. Genet.* 7, 524–536.

Edwards D & Batley J (2004) Plant bioinformatics: from genome to phenome. *Trends Biotechnol.* 22, 232–237.

El-Din El-Assal S, Alonso-Blanco C, Peeters AJM et al. (2001) A QTL for flowering time in Arabidopsis reveals a novel allele of CRY2. *Nat. Genet.* 29, 435–440.

Harmer SL, Hogenesch JB, Straume M et al (2000) Orchestrated transcription of key pathways in Arabidopsis by the circadian clock. *Science* 290, 2110–2113.

Jander G, Norris SR, Rounsley SD et al. (2002) Arabidopsis map-based cloning in the post-genome era. *Plant Physiol.* 129, 440–450.

Tanksley SD (1993) Mapping polygenes. *Annu. Rev. Genet.* 27, 205–233.

Till BJ, Reynolds SH, Greene EA et al. (2003) Large-scale discovery of induced point mutations with high-throughput TILLING. *Genome Res.* 13, 524–530.

2.6 Cytoplasmic Genomes

Deng X-W, Matsui M, Wei N et al. (1992) COP1, an Arabidopsis regulatory gene, encodes a protein with both a zinc-binding motif and a Gβ homologous domain. *Cell* 71, 791–801.

De Pamphilis CW & Palmer JD (1990). Loss of photosynthetic and chlororespiratory genes from the plastid genome of a parasitic flowering plant. *Nature* 348, 337–339.

Douglas SE (1998) Plastid evolution: origins, diversity, trends. *Curr. Opin. Genet. Dev.* 8, 655–661.

Freyer R, Kiefer-Meyer M-C & Kossel H (1997) Occurrence of plastid RNA editing in all major lineages of land plants. *Proc. Natl. Acad. Sci. USA* 94, 6285–6290.

Hoch B, Maier RM, Appel K et al. (1991) Editing of a chloroplast mRNA by creation of an initiation codon. *Nature* 353, 178–180.

Kolodner RD & Tewari KK (1975) Chloroplast DNA from higher plants replicates by both Cairns and rolling circle mechanism. *Nature* 256, 708–711.

Martin W, Rujan T, Richly E et al. (2002) Evolutionary analysis of Arabidopsis, cyanobacterial, and chloroplast genomes reveals plastid phylogeny and thousands of cyanobacterial genes in the nucleus. *Proc. Natl. Acad. Sci. USA* 99, 12246–12251.

Nott A, Jung H-S, Koussevitzky S & Chory J (2006) Plastid-to-nucleus retrograde signaling. *Annu. Rev. Plant Biol.* 57, 739–759.

Rochaix JD (1992) Post-transcriptional steps in the expression of chloroplast genes. *Annu. Rev. Cell Biol.* 8, 1–28.

Stern DB, Higgs DC & Yang JJ (1997) Transcription and translation in chloroplasts. *Trends Plant Sci.* 2, 308–315.

CELLS

When you have read Chapter 3, you should be able to:

- Describe the major structures of plant cells, noting those that distinguish them from animal cells.

- Outline the stages of the cell cycle and the ways in which it is regulated in plants.

- Define "endoreduplication" and explain its relation to ploidy, cell size, and differentiation.

- Describe plant-specific features of cell division, including formation of the preprophase band and the phragmoplast.

- Describe the components of the cytoskeleton and the endomembrane system, and outline their functions.

- Describe the structures and properties of the primary and secondary cell walls.

- Describe how traffic across cellular membranes is regulated, including the roles of pumps, transport proteins, electrical and proton gradients, and aquaporins.

- Summarize the roles of the central plant vacuole.

- Explain how stomata open and close.

- Outline the mechanisms of cell expansion and the factors that determine cell shape at maturity.

- Describe the synthesis of lignin, and distinguish between primary and secondary growth in woody plants.

CHAPTER SUMMARY

Our focus now moves from the plant **genome** to plant cells: their structure, division, expansion, and differentiation. Plant cells have much in common with all other eukaryotic cells, including animal cells. A **plasma membrane** encloses a **cytosol** containing several types of membrane-enclosed subcompartments (**organelles**)—including the nucleus, mitochondria, peroxisomes, endomembrane system, and vacuoles (Figure 3–1)—and a network of tubules and filaments that constitute the **cytoskeleton**. Plant cells, however, have two easily observable features that distinguish them from animal cells. First, the plant cell is surrounded by a **cell wall**; second, it contains **plastids**, which include **chloroplasts**. It was the plant cell wall that allowed the first recognition

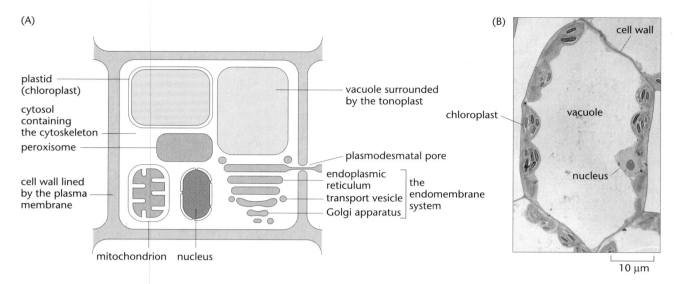

Figure 3–1.
The structure of plant cells. (A) Diagram showing the major structural features of plant cells discussed in this chapter. (B) Transmission electron micrograph of a cell in an Arabidopsis leaf, showing chloroplasts in the cytosol surrounding the large central vacuole. Further pictures of subcellular structures are given elsewhere in this chapter and in Chapter 4. (B, courtesy of Sam Zeeman and Thierry Delatte.)

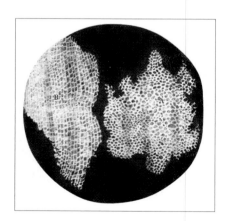

Figure 3–2.
The first illustration of cellular structure. These drawings of cells in cork were made by Robert Hooke, using an early microscope, and were published in 1667 in his book *Micrographia; or, Some Physiological Descriptions of Minute Bodies Made by Magnifying Glasses, With Observations and Inquiries Thereupon.*

of the cellular structure of living organisms. In the mid-seventeenth century, the microscopist Robert Hooke observed that **cork** (from the bark of the cork oak, *Quercus suber*) consists of tiny compartments surrounded by rigid walls, and he named these compartments "cells" (Figure 3–2).

Plant growth and development occur as cells proliferate by division then expand and differentiate to perform functions specific to their position in the plant body. Cell division consists of replication and division of the **nucleus**, followed by partitioning of the two daughter nuclei and the **cytoplasm** (the cytosol and its contents other than the nucleus) into two cells, separated by a new cell wall. Cell expansion and differentiation require the import of water and ions into the cell to increase its volume and the import of substances to provide energy and precursors for the biosynthesis of new structural materials. As we shall see, passage of substances into the cell and cell expansion require special properties of the **plasma membrane** and the cell wall that surround the cell. Within the cell, expansion and differentiation require synthesis of proteins, membranes, new cell wall material, and other cell structural and enzymatic machinery, and the trafficking of the newly synthesized molecules to their correct subcellular locations.

We begin this chapter by describing how the cell divides to form genetically identical daughter cells (the **cell cycle**), and how the organelles replicate and are apportioned to the new cells. We then turn our attention to that unique plant structure, the cell wall, and describe the processes involved in cell expansion: synthesis of the primary cell wall and the import of solutes that drives the increase in cell volume. Finally we describe a critical feature of cell maturation and differentiation: the synthesis of the **secondary cell wall**.

First, a note on another feature, unique to plants, that we refer to frequently in this chapter: the **meristems**. During a plant's vegetative growth, cell division occurs mainly in specific zones—the meristems—and differentiation of the cells produced in these zones gives rise to the enormous range of cell types in the plant (some examples of which are shown in Figure 3–3). These topics are dealt with in detail in Chapter 5.

Figure 3–3.
Some common cell types in plants. (A) Electron micrograph of a cross section of cells in a floral meristem of Arabidopsis (the growing point from which floral structures arise; see Chapter 5). The small, rapidly dividing cells at the tip of the shoot (*center top*) and in the organ primordia arising on its flanks have dense cytoplasm and lack a large central vacuole. (B) Electron micrograph of a section through a rhizome (underground stem) of lily of the valley (*Convallaria majalis*). Most of the organ is made up of parenchyma (*orange*), relatively large, thin-walled cells with very large vacuoles and a thin layer of cytoplasm adjacent to the cell wall. (C) Scanning electron micrograph of the surface of a rice root, showing files of elongated epidermal cells, some of which have long projections called root hairs. Examples of other cell types are shown in other figures in this chapter. (C, courtesy of Paul Linstead.)

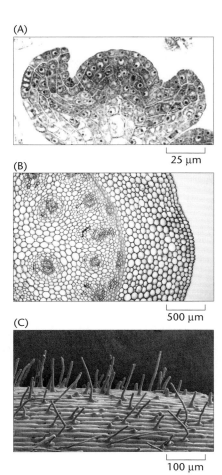

(A)

25 μm

(B)

500 μm

(C)

100 μm

3.1 THE CELL CYCLE

To produce two genetically identical daughter cells, the nuclear **DNA** must be replicated and partitioned accurately into two daughter nuclei. The nucleus of the plant cell is described in Box 3–1. The sequence of events involved in DNA **replication** and partitioning—the cell cycle—is a precisely programmed progression that is similar in all **eukaryotes** (Figure 3–4). In the initial phase of the cell cycle, **G1** (gap 1), cells prepare to replicate their DNA. Replication itself takes place during the **S phase** (synthesis phase). This is followed by another gap period, **G2**, during which cellular changes occur that will permit separation of the two copies of the genome in the subsequent cell division. Breakdown of the nuclear envelope marks the beginning of the **M phase** (**mitosis**), when the replicated copies of the **chromosomes** are separated and pulled to opposite ends of the cell by a structure called the mitotic **spindle**. Two daughter nuclei are then reconstituted from the two sets of chromosomes, and, in plants, the two new cells are separated by formation of a **cell plate** that develops into a dividing wall, as we describe in detail below. The daughter cells then either reenter the cycle at G1 or enter a resting state that is not committed to division.

Transition from one phase of the cell cycle to the next is regulated by a complex set of mechanisms

Progression through the cell cycle is controlled by a set of **protein kinases** called **cyclin-dependent protein kinases (CDKs)**. They act as switches that activate a set of cellular functions during a specific stage of the cell cycle. Protein kinases are **enzymes** that alter the activity of another protein by phosphorylating (adding a phosphate group to) a specific amino acid residue. The CDKs depend for their activity on proteins called **cyclins**, with which they form complexes (Figure 3–5). In the CDK-cyclin complex, the cyclin directs the CDK to act on a specific protein or set of proteins. Plants contain multiple forms of both CDKs and cyclins, each encoded by a different gene. Arabidopsis, for example, has at least 30 genes encoding cyclins and 7 genes encoding CDKs.

Several different CDK-cyclin complexes are active during the cell cycle, acting at different points during the ordered progression from one phase to the next. For example, a complex assembled during G1, containing a member of the G1 family of cyclins, controls the activity of **transcription factors** that allow expression of the genes required for DNA replication during S phase. Another complex, assembled at the end of G2 and containing a member of the mitotic family of cyclins, phosphorylates proteins that control the assembly of the spindle, the structure on which replicated chromosomes are aligned and then separated during mitosis (as described in Section 3.2).

The appearance and removal of different CDK-cyclin complexes during the cell cycle is controlled in a variety of ways. The activity of complexes is regulated in part by the

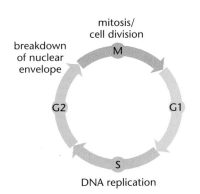

mitosis/
cell division

breakdown
of nuclear
envelope

M

G2   G1

S

DNA replication

Figure 3–4.
Diagrammatic representation of the cell cycle.

Figure 3–5.
Some roles of CDK-cyclin complexes in the cell cycle. DNA replication in S phase and assembly of the mitotic spindle in M phase are both controlled by specific CDK-cyclin complexes that assemble at the appropriate point in the cell cycle.

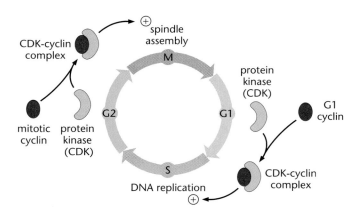

transcriptional control of expression of CDK- and cyclin-encoding genes. One important class of CDKs, CDKA, is **constitutively** expressed (that is, expressed at all times regardless of cell cycle phase), but a second class, CDKB, has a cell-cycle–dependent expression pattern and is active only in G2 and M phase. Genes encoding different cyclins are expressed at different points during the cell cycle. However, much of the control of CDK-cyclin activity occurs by post-translational mechanisms. (Chapter 2 explains more about the transcriptional and post-translational mechanisms for controlling **gene expression**.)

One of the major post-translational mechanisms that controls progression through the cell cycle is **proteolysis** (protein breakdown) of the cyclin subunits of CDK-cyclin complexes. (Specialized proteolytic complexes are important in controlling the abundance of many key proteins in cell growth and differentiation, not just the cyclins.) Two proteolytic complexes important in the control of the cell cycle are SCF and APC. Both complexes attach **ubiquitin** tags to specific target proteins, which are subsequently degraded by a complex of **proteases** (protein-degrading enzymes) called the **proteasome** (see Section 5.4). Cyclins that function during the G1-to-S transition are targeted for breakdown by the SCF complex (the name refers to three of its protein components: **S**kp1, **C**ullin, and **F**-box protein). Exit from mitosis and separation of replicated chromosomes to opposite ends of the dividing cell (in **anaphase**) require breakdown of the mitotic cyclin by APC (**a**naphase-**p**romoting **c**omplex; Figure 3–6). APC also breaks down the proteins, known as cohesins, that attach together each pair of replicated chromosomes during the first stages of mitosis. Removal of cohesins allows chromosome separation at the end of mitosis.

Another mechanism that determines which CDK-cyclin complexes are active is phosphorylation of the CDK subunit. This is best understood in yeast. The *WEE* and *CDC25* genes of yeast control the transition from G2 to M phase (Figure 3–7). *WEE* encodes a protein kinase (WEE) that inhibits the activity of a CDK by phosphorylating two of its amino acids. In wild-type yeast, this inhibition prevents cells from proceeding to M phase—and hence to cell division—until they reach a certain size. In *wee* mutants ("wee" meaning very small), the activity of the mitotic CDK is not inhibited, and this allows cells to divide before they reach the size at which division normally occurs. *CDC25* encodes the **protein phosphatase** (CDC25) that removes the phosphate groups attached by WEE to the CDK, thus reactivating the CDK and promoting progression through M phase. Yeast CDC25 can also promote cell division in higher plants: tobacco plants engineered to produce this protein had larger numbers of cells than normal. However, plants do not seem to have a protein directly equivalent to CDC25. It is possible that control of the G2-to-M transition in plants has a different mechanism, involving the B class of CDKs. This class is not found in organisms other than plants and is specifically expressed in G2 and M phase of the cell cycle.

Box 3–1 The Nucleus

The nucleus is the site of the genomic DNA. Enclosed within the nuclear membrane—the nuclear envelope—are large numbers of proteins and nucleic acids in a complex three-dimensional structure. DNA replication, gene transcription, and **RNA processing** occur within this structure (see Chapter 2). Genes are transcribed by **RNA polymerases**, and the resulting RNA molecules are transported from the nucleus to the cytoplasm through nuclear pores—"molecular gates" in the nuclear envelope. The nuclear pores are also the routes by which proteins are imported from the cytosol to the nucleus. Imported proteins include RNA and DNA polymerases, **histones**, and transcription factors (all of which are described in Chapter 2).

The nucleus comprises several distinct subdomains where the different activities carried out by the imported proteins take place. These subdomains include the nucleoplasm, nucleolus, nuclear speckles, and Cajal bodies (Figure B3–1A).

The **nucleoplasm** is the volume of the nucleus that is occupied by **chromatin** (DNA and its associated proteins) and interchromatin domains (stretches of DNA without associated proteins). Chromatin may be highly condensed **heterochromatin**, in which DNA is packed tightly into a small volume, or less condensed **euchromatin**. In Arabidopsis, the DNA of chromosomal **centromeres** and **telomeres**, and of chromosomal regions rich in repeated DNA sequences rather than genes, tends to occur as heterochromatin. The chromatin of gene-rich regions of the chromosomes, where active transcription is occurring, is in the form of euchromatin.

The **nucleolus** is the site of transcription of the genes that encode **ribosomal RNA (rRNA)** (Figure B3–1B). These genes are arranged in tandem along a region of the chromosome called the **nucleolar organing region** (see Section 2.4). Different regions within the nucleolus have distinct roles in the formation of ribosomal subunits. Transcription of the rRNA genes by RNA polymerase I occurs at numerous foci within regions of decondensed chromatin known as the "dense fibrillar" component. Transcription produces a pre-RNA molecule (the 45S RNA), which is then processed to mature 18S, 5.8S, and 28S components of the ribosome. The first processing steps take place in the dense fibrillar material. The pre-RNA molecule moves to another region of the nucleolus, the "granular" component, in which the remainder of the processing occurs. The mature RNA molecules then associate with ribosomal proteins that have entered the nucleus, and the resulting ribosomal subunits are exported to the cytosol.

Smaller nuclear structures are thought to provide scaffolds for the assembly of complexes and small RNAs. As we describe in Chapter 2, newly made transcripts of most genes contain sequences called **introns**, which must be spliced (removed) to form mature transcripts that can be exported to the cytosol for translation into proteins. Introns are spliced by nucleases located on a particle called a **spliceosome**. The spliceosome contains five different small RNA molecules and at least 200 proteins. It is thought to be assembled in the **nuclear speckle** regions. A similar but much larger structure, the **Cajal body**, is found in all nuclei. Like the speckles, it contains various types of small RNAs and proteins. Cajal bodies are probably involved in the maturation and transport around the nucleus of the small RNAs involved in processes such as RNA splicing and synthesis of mature rRNA.

Figure B3–1.
Nuclear structure. (A) Internal structure of the nucleus.
(B) Structure of the nucleolus, showing the sites of ribosomal RNA synthesis and processing. (Courtesy of Peter Shaw.)

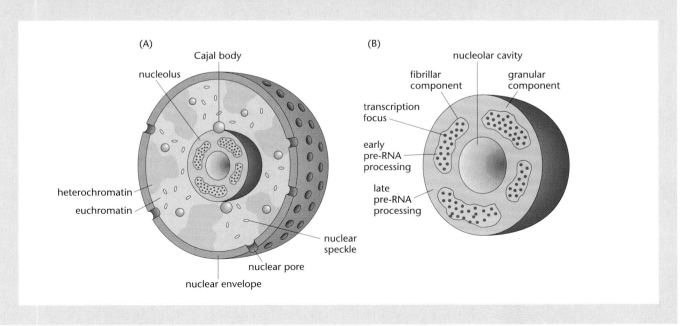

Figure 3–6.

Destruction and assembly of CDK-cyclin complexes in the cell cycle. Specialized proteolytic complexes target specific cyclins active at different points in the cell cycle. The breakdown of these cyclins removes the activity of the specific CDK-cyclin complex, promoting transition of the cell to the next phase of the cycle. The proteolytic complexes consist of several different proteins necessary for the specific recognition of the target proteins. One mechanism controlling the action of the proteolytic complexes is protein phosphorylation. The SCF complex that targets the G1 cyclin does not recognize this cyclin until it has been phosphorylated by a protein kinase. Proteolytic complexes target more than one type of protein. For example, the ACP complex also breaks down proteins essential in an early phase of mitosis, allowing later phases to take place.

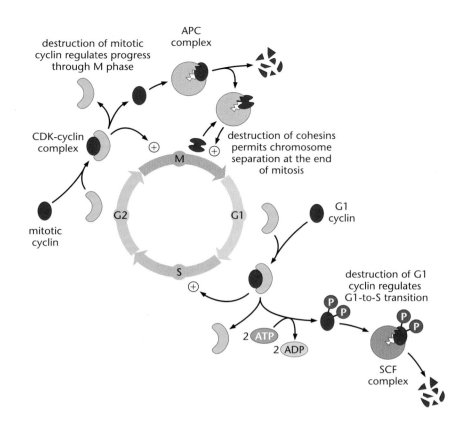

A third mechanism that determines which CDK-cyclin complexes are active also involves inhibition of CDK activity, this time by the action of inhibitory proteins called **cyclin-dependent kinase inhibitors (CKIs)**. Control of the mitosis-promoting CDK of yeast provides a good example (Figure 3–8). SIC1 is a CKI that blocks progression from G2 to M phase by inhibiting the activity of the CDK. This inhibition is released by destruction of SIC1, once again by the proteolytic complex SCF. Arabidopsis has seven genes encoding CKI-like proteins. Evidence that these proteins act to arrest the cell cycle has been provided by genetic manipulation of one of the genes to cause high levels of expression in **flowers**. The development of the flowers was strongly inhibited, indicating that the CKI protein had inhibited passage of cells through the cell cycle.

Figure 3–7.

The WEE/CDC25 pathway in yeast. Phosphorylation of the cyclin-dependent kinase (CDK) that promotes the G2-to-M transition in yeast inhibits the activity of this CDK, preventing the transition. Dephosphorylation, resulting in reactivation of the CDK, allows the transition to proceed. The phosphorylation and dephosphorylation are catalyzed by the WEE protein kinase and the CDC25 protein phosphatase, respectively. Note that this may not be the way in which the G2-to-M transition is controlled in plants; the mechanism in plants is not yet understood.

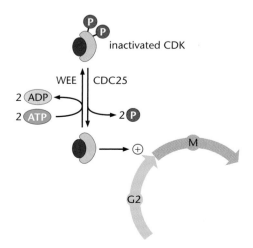

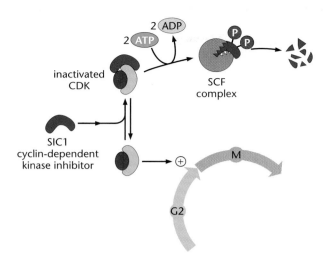

Figure 3–8.
Regulation of the CDK that promotes the G2-to-M transition. This cyclin-dependent protein kinase (CDK) can be inhibited by binding to an inhibitory protein, as well as by phosphorylation of the catalytic subunit (see Figure 3–7). Inhibition by SIC1, a cyclin-dependent kinase inhibitor, is reversed when SIC1 is phosphorylated by a protein kinase. Phosphorylation enables SIC1 to be recognized and broken down by the proteolytic complex SCF, in a manner analogous to the breakdown of the G1 cyclin shown in Figure 3–6.

Clearly, the regulation of the transition from one phase of the cell cycle to another is complex and usually involves several different mechanisms. A good example of this complexity is provided by the retinoblastoma pathway (Figure 3–9), a mechanism that regulates the transition from G1 to S phase. As its name suggests, the **retinoblastoma protein** was found initially in mammals as a growth suppressor of tumors called "retinoblastomas." The protein is a target for inactivation by tumor-inducing **viruses**. This pathway is not ubiquitous in eukaryotes: it is found in animals and plants but not in yeast. The retinoblastoma protein prevents entry into the S phase by inhibiting the activity of the E2F class of transcription factors, which are required for expression of genes necessary for DNA synthesis in S phase. Entry into S phase is promoted by a G1-specific CDK-cyclin complex, which inactivates the retinoblastoma protein by phosphorylation, releasing E2F to activate its target genes and thus allowing DNA replication. The activity of the G1-specific CDK is itself subject to regulation. It is activated by phosphorylation of a specific tyrosine residue by the protein kinase CAK, and is inactivated by binding to a CKI.

There is yet another layer of complexity. Progression through the cell cycle is also controlled by mechanisms to ensure that each phase is initiated only when the previous phase is complete. These mechanisms are often referred to as "checkpoints" (Figure 3–10). One well-studied case is the spindle assembly checkpoint. If cells are treated

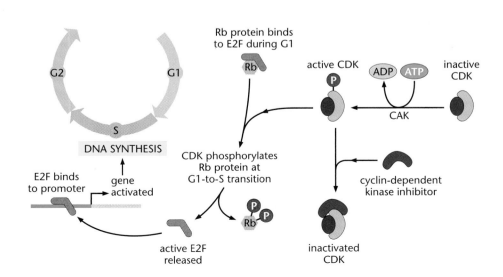

Figure 3–9.
Retinoblastoma pathway.
The retinoblastoma protein (Rb) is part of the mechanism that controls the G1-to-S transition. DNA replication in S phase is activated by the transcription factor E2F. Before S phase, in G1, the action of E2F is prevented by its binding to Rb. The role of the CDK active during the G1-to-S transition is to phosphorylate Rb, which causes the release of Rb from E2F and thus allows E2F to activate DNA synthesis. The CDK itself is also subject to several levels of regulation. As shown here, its activity depends on phosphorylation of the catalytic subunit (by CAK, a protein kinase), and it is inhibited by binding to a cyclin-dependent kinase inhibitor (see also Figure 3–6).

Figure 3–10.

Checkpoints in the cell cycle. Completion of DNA replication is necessary for completion of later stages in the cell cycle. In yeast, if the protein checkpoint kinase 1 (CHK1) detects incomplete DNA replication, it prevents later transition to M phase by acting on proteins that determine the activity of the CDK necessary for this transition. CHK1 activates the protein kinase WEE and inhibits the protein phosphatase CDC25, so the catalytic subunit of the CDK remains in its phosphorylated, inactive form.

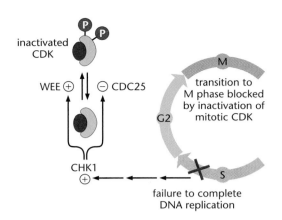

with drugs that delay the assembly of the mitotic spindle, the cell cycle arrests in M phase. The spindle checkpoint genes prevent progression through the cell cycle by blocking the remaining mitotic functions until the spindle is complete. These genes act by preventing degradation of the mitotic cyclin by APC (see Figure 3–6). As long as the mitotic CDK-cyclin complex remains active, the cells cannot leave M phase. Another important checkpoint is the completion of DNA replication. In yeast, incomplete replication of DNA or damage to the DNA promotes the action of another protein kinase, checkpoint kinase 1 (CHK1). CHK1 blocks the activity of the mitotic CDK in two ways: by stimulating the activity of the kinase WEE, which carries out an inhibitory phosphorylation of the catalytic subunit of the CDK, and by inhibiting the CDK phosphatase CDC25 (see Figure 3–10; see also Figure 3–7). The nature of this checkpoint in plants is not yet understood.

The cell cycle in plants is controlled by developmental and environmental inputs

The basic sequence of events in the cell cycle is subject to modulation by many developmental and environmental inputs. The complexity and diversity of mechanisms that activate and destroy CDKs, as described above, allow integration of many different signals from within and outside the plant to control progression through the cell cycle. We discuss here several examples of developmental and environmental controls.

Changes in the rate and pattern of cell division during organ development can be brought about by changes in the temporal pattern of expression of key regulators of the cell cycle. In developing maize **leaves**, for example, cell division occurs only in a meristem at the base of the leaf. This restriction of cell division is thought to be related to the pattern of expression of retinoblastoma protein, which, as noted above, prevents cell division by inhibiting the actions of the E2F transcription factor required for DNA replication (see Figure 3–9). Retinoblastoma protein is expressed at low levels in the **zone of cell division** in the leaf, but at higher levels further up the leaf where cell division is suppressed and cell elongation and differentiation are occurring. (See Sections 5.3 and 5.4 for discussion of growth and differentiation zones in **roots** and **shoots**.)

Events in the cell cycle are also influenced at several different levels by **phytohormones** (plant hormones), small molecules (analogous to the hormones of animals) that act as signals linking growth, development, and environmental conditions. Different classes of phytohormones—including **auxins**, **cytokinins**, **gibberellins**, **brassinosteroids**, and **abscisic acid (ABA)**—influence progression through the cell cycle and thus bring about changes in the growth rate and direction of plant development. Some of these phytohormones regulate the expression of genes encoding CDKs and cyclins; others act less directly by influencing the activity of proteins, such as CKIs, that modulate the activity of CDK-cyclin complexes. In some cases, phytohormones may act by influencing the activity of the proteolytic complexes that degrade negative regulators of the cell cycle. The roles of phytohormones in plant development and responses to environmental conditions are discussed in Chapters 5 and 6.

Many types of environmental stress inhibit plant growth and development (see Chapter 7). The inhibition is at least partly due to effects of stress on events in the cell cycle. For example, transfer of Arabidopsis seedlings from a salt-free medium to a medium containing 0.5% NaCl resulted in a large reduction in the rate of root elongation relative to that of control plants maintained on the salt-free medium. The reduction was found to be due to a decrease in cell division in the meristem, combined with a smaller mature cell length. Treatment with salt resulted in a very rapid reduction in the number of cells undergoing the G2-to-M transition in the meristem, which in turn led to a reduction in the size of the meristem (the population of cells undergoing division) and hence a lower growth rate (Figure 3–11). This pattern of events may be common to many different types of stress. It is possible that cells are more vulnerable to damage by environmental stress during mitosis than during other phases of the cell cycle, so rapid inhibition of the G2-to-M transition following stress may be a protective mechanism that minimizes damage and allows other stress-defense responses to be activated. The consequent reduction in meristem size down-regulates growth to levels appropriate for suboptimal environmental conditions.

Many differentiating cells undergo endoreduplication: DNA replication without nuclear and cell division

Endoreduplication is a modified form of the cell cycle in which cells go through one or several rounds of DNA replication (S phase) without nuclear breakdown and mitosis (M phase). Whereas repetitions of the cell cycle usually produce multiple daughter cells, each with the same DNA content as the initial cell, endoreduplication produces a single **polyploid** cell. **Polyploidy** occurs in almost all plants: at least some cell types undergo endoreduplication after cell division ceases. For example, in Arabidopsis, only about 25% of cells in the leaf have a DNA content of 2C (C is the mass of DNA present in the **haploid** genome, thus 2C is **diploid**). The remaining 75% of cells have

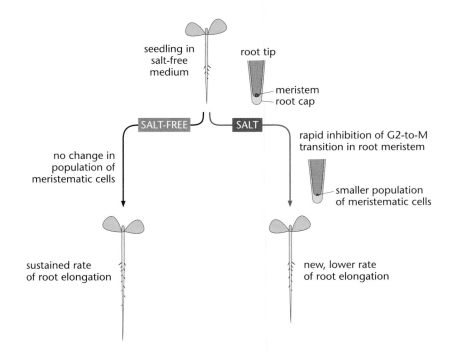

Figure 3–11.
Effect of salt stress on root growth.
Salt stress reduces root elongation by affecting the cell cycle and hence the number of cells in the root meristem. Seedlings were grown without salt, then either transferred to a medium containing salt or maintained on a salt-free medium. In the absence of salt, the population of cells in the root meristem did not change in size and root elongation was maintained. In seedlings in the salt-containing medium, the population of cells in the root meristem declined and a new, lower rate of root elongation was established.

undergone one to four rounds of endoreduplication and as a consequence are polyploid, with DNA contents of 4C to 32C. Very high levels of endoreduplication are found in developing **seeds**, in cells with high rates of synthesis of **storage proteins** and other storage compounds. For example, **endosperm** cells of maize may have more than 200 copies of the haploid genome (the development of endosperm is described in Section 3.2).

There is a good correlation between the ploidy and levels of **messenger RNA** and protein synthesis in plant cells. For example, as noted above, cells with high rates of synthesis of storage products in seeds (in the **embryo** or the endosperm) are usually highly polyploid. There is also a strong correlation between ploidy and cell size: the more genome copies, the larger the cells. This correlation applies to different cells in the same plant, and to different ploidy levels between species: a polyploid plant usually has larger cells than its diploid progenitors. However, ploidy does not directly determine cell size. It is believed to set a maximum size, which may or may not be achieved, depending on other developmental or environmental inputs. A good illustration of the relationship between ploidy and cell size is seen in **chimeric** *Datura stramonium* containing both normal diploid cells and polyploid cells in the same meristem. The polyploid cells are larger than the diploid cells (see Figure 5.34).

The link between endoreduplication and cell size can also be seen in endoreduplication-defective mutants of Arabidopsis (Figure 3–12), such as *roothairless2* (*rhl2*) and *hypocotyl6* (*hyp6*). These mutants lack DNA topoisomerase VI, an enzyme that functions in untangling chromosomes after replication. During replication the DNA double helix unwinds, allowing the **DNA polymerases** to synthesize a new DNA molecule from each of the original template strands. This unwinding of the double-stranded DNA is catalyzed by a group of proteins called DNA **topoisomerases**. In the absence of topoisomerases, the replicated DNA strands are intertwined and cannot be separated at mitosis; endoreduplication is inhibited and plants grow to only a small fraction of the normal size.

Endoreduplication can also be an essential part of cellular differentiation. The trichomes of Arabidopsis are a well-studied example. **Trichomes** are large, branched **epidermal cells**. An increase in DNA content is one of the earliest signs that an epidermal cell will form a trichome (Figure 3–13). As the trichome increases in size by cell expansion, it undergoes more rounds of endoreduplication and forms more branches; mature trichomes with three or more branches can contain 16 times the DNA of a diploid nucleus (i.e., 32C, achieved by four rounds of endoreduplication). The close link between trichome differentiation and endoreduplication is shown by mutants in which both trichome branching and ploidy are affected. Trichomes of *glabra3* mutants have fewer branches and lower ploidy than those of wild-type plants, whereas trichomes of *tryptichon* (*try*) mutants have both increased branching and increased ploidy (see Figure 3–13); *try* mutants also form more trichomes than normal. The finding that a **gene** that normally prevents cells from becoming trichomes also reduces the level of endoreduplication provides extra evidence for the link between differentiation and endoreduplication. (Trichome development is also discussed in Section 5.4.)

Endoreduplication results from a modification of the mitotic cell cycle that promotes S phase and inhibits M phase (Figure 3–14). Regulators that promote S phase in cells undergoing the cell cycle are also active during endoreduplication, whereas regulators that promote M phase in dividing cells are inhibited in endoreduplicating cells. In maize endosperm cells, which become polyploid during seed development, this regulation is achieved by two main mechanisms. First, retinoblastoma protein in the nuclei of endosperm cells is phosphorylated (see Figure 3–9), allowing continuous expression of S-phase–related genes. Second, these cells contain an inhibitor of mitotic CDK that prevents entry into M phase (see Figure 3–14).

Figure 3–12.
Endoreduplication is necessary for normal growth. Comparison of a wild-type Arabidopsis plant (*left*) and a mutant plant lacking DNA topoisomerase VI (*right*). The maximum level of endoreduplication found in the mutant is 8C, compared with endoreduplication in excess of 32C in wild-type plants. (Courtesy of Keiko Sugimoto.)

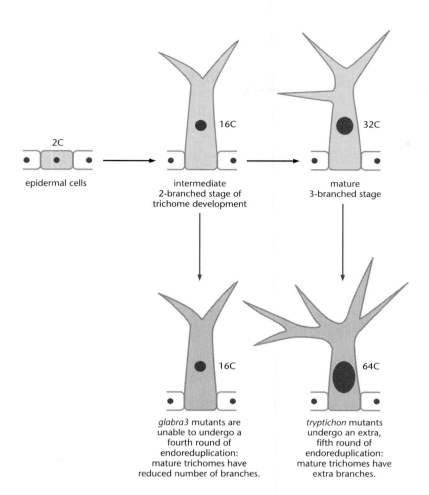

Figure 3–13.
Trichome development in wild-type Arabidopsis and in *glabra3* and *tryptichon* mutants. In the wild type (*top*), endoreduplication leads first to a 16C DNA content, corresponding to an intermediate, two-branched trichome stage, then to a 32C DNA content, corresponding to the mature, three-branched trichome. *glabra3* mutants (*lower left*) are blocked at the final stage of endoreduplication and the trichomes have only two branches. *tryptichon* mutants (*lower right*) achieve DNA contents higher than 32C and have correspondingly multibranched trichomes.

The mechanisms that prevent mitosis during endoreduplication have been studied in detail in **root nodules** of the legume *Medicago truncatula*, in which cells undergo endoreduplication when entering into **symbiotic relationships** with **nitrogen-fixing bacteria** (see Section 8.5). Root nodules have a meristem of small cells where cell division is active, and a region of larger cells colonized by bacteria. DNA levels in the larger cells range from 4C to 32C. During endoreduplication, these larger cells express the *ccs52* gene, which is **homologous** to genes in yeast and animals that promote the destruction of mitotic CDK by proteolysis (see Figure 3–8). This premature proteolysis of mitotic CDK allows M phase to be bypassed.

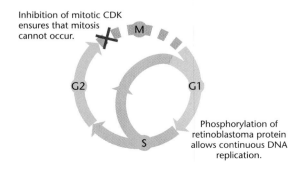

Figure 3–14.
The short-circuited cell cycle that allows endoreduplication in maize endosperm cells.

Figure 3–15.
Mitosis in plants. (A) Before cell division, the cell is in interphase. The first process in division is formation of a band of microtubules, the preprophase band, around the center of the cell. This band determines the position where the new cell wall will form (during cytokinesis). In prophase, the replicated chromosomes in the nucleus condense to a fraction of their interphase length. In metaphase, the nuclear membrane disappears and the chromosomes align at the equator of the newly formed mitotic spindle. In anaphase, the pairs of chromosomes separate and are pulled apart as the spindle contracts away from the equator. In telophase, the two new daughter nuclei form. During cytokinesis, microtubules and actin filaments lying between the daughter nuclei contribute to formation of the phragmoplast, the structure that orchestrates formation of the cell plate that develops into the new cell wall. (B) Micrograph showing microtubules in metaphase (*top*) and anaphase (*bottom*) in tobacco cells. The cells were treated with fluorescent agents so that microtubules appear green and chromosomes blue. (B, courtesy of Sandra McCutcheon and Clive Lloyd.)

3.2 CELL DIVISION

Mitosis consists of two processes: the separation of replicated chromosomes by **karyokinesis** and the formation of two daughter cells by **cytokinesis**. The two processes are coordinated such that each daughter cell has a chromosome complement identical to that of the parent cell—that is, it is genetically identical to the parent cell.

The distinct stages of mitosis are often referred to as **prophase**, **metaphase**, **telophase**, and anaphase. These are described in Figure 3–15.

The cytoskeleton moves cellular components during cell division

Both karyokinesis and cytokinesis are dependent on the cytoskeleton (Box 3–2), a complex of filaments and tubules that also has many functions in nondividing plant cells, including control of cell shape and **cytoplasmic streaming** (see Section 3.3). The cytoskeleton consists of polymers of two main proteins, **tubulin** and **actin**. In combination, these polymers generate molecular tracks and pulleys that can move cellular components to different parts of a cell. They play an important role in karyokinesis, starting with the formation of a **preprophase band**. This is a ring of **microtubules** (consisting of polymerized tubulin) and actin filaments, located around the center of the cell. It is involved in the positioning of the mitotic spindle, which also consists of microtubules. Chromosomes align at the center of the mitotic spindle (the equator) and are separated by the microtubules pulling toward opposite poles. The spindle disassembles and a new cell wall forms between the two daughter cells. Cell wall assembly is orchestrated by a **phragmoplast**, a structure formed by bundles of

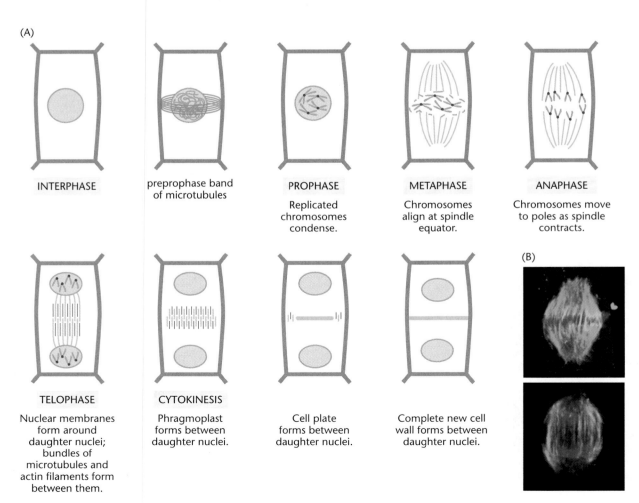

(A)

INTERPHASE

preprophase band of microtubules

PROPHASE
Replicated chromosomes condense.

METAPHASE
Chromosomes align at spindle equator.

ANAPHASE
Chromosomes move to poles as spindle contracts.

TELOPHASE
Nuclear membranes form around daughter nuclei; bundles of microtubules and actin filaments form between them.

CYTOKINESIS
Phragmoplast forms between daughter nuclei.

Cell plate forms between daughter nuclei.

Complete new cell wall forms between daughter nuclei.

(B)

Box 3–2 The Cytoskeleton

The division, growth, and development of plant cells are dependent on the cytoskeleton, a system of filaments and microtubules in the cytosol that is involved in many of the processes described in this chapter. The cytoskeleton forms the spindle on which the chromosomes separate at mitosis, controls the movement of organelles within the cell, guides the synthesis of the cell wall, and determines the direction of cell expansion.

The main components of the plant cell cytoskeleton are actin filaments and microtubules (Figure B3–2A, B3–2B). Actin filaments are linear polymers of the protein actin, which is encoded by a multigene family. The number of genes in the actin family differs greatly from one species to another: Arabidopsis has 10 actin genes, whereas petunia has more than 100. Different members of the gene family are expressed in different parts of a plant, and in response to different environmental and developmental stimuli. Microtubules consist of subunits that are dimers of two proteins, α-tubulin and β-tubulin; the subunits polymerize to form a hollow cylinder. Tubulins are also encoded by a multigene family: there are 19 tubulin genes in the Arabidopsis genome.

The two ends of an actin filament or a microtubule are not identical. Polymerization can occur much faster at the "plus end" than at the "minus end" (Figure B3–2C). Motor proteins, which move along filaments and microtubules, usually move in only one direction. Some move only from plus to minus ends, and others only from minus to plus ends.

Actin filaments and microtubules are highly dynamic structures, undergoing continuous polymerization and depolymerization. Sometimes this involves net loss or gain of these cytoskeletal structures, such as during the formation and disassembly of the spindle during mitosis. However, loss and gain of subunits at either end of filaments and microtubules sometimes occurs without any net change in their length. Actin filaments undergo "treadmilling," in which a net gain of actin proteins at the plus end is balanced by an equal net loss at the minus end (Figure B3–2D). Microtubules can alternate rapidly between growth and shrinkage, a phenomenon known as "dynamic instability" (Figure B3–2E). When subunits are added to a microtubule, they are bound to the energy-rich molecule GTP; after addition, the GTP is hydrolyzed to GDP. The presence of GTP-tubulin at the end of a microtubule favors addition of subunits, whereas the presence of GDP-tubulin favors loss of

Figure B3–2.

(A) Structure of an actin filament. (B) Structure of a microtubule. (C) Polymerization of cytoskeletal components. (D) Treadmilling of an actin filament: net gain at the plus end is matched by net loss at the minus end. (E) Dynamic instability of a microtubule. Subunits are added in the form of GTP-tubulin (*green*). After addition, the GTP is hydrolyzed to GDP-tubulin (*gray*). If the rate of addition of subunits slows down so that newly added subunits are immediately converted to GDP-tubulin, the presence of GDP-tubulin at the end of the microtubule promotes a switch from polymerization to depolymerization (*right*).

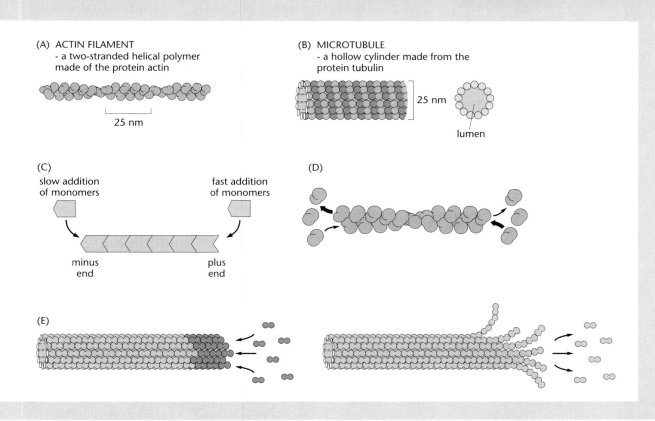

(A) ACTIN FILAMENT
- a two-stranded helical polymer made of the protein actin

25 nm

(B) MICROTUBULE
- a hollow cylinder made from the protein tubulin

25 nm

lumen

(C) slow addition of monomers   fast addition of monomers

minus end   plus end

(D)

(E)

continued

Box 3–2 The Cytoskeleton (continued)

subunits. When polymerization is fast, the rate of hydrolysis of GTP to GDP on the newly added subunits lags behind, and the subunits at the newly forming end are GTP-tubulin rather than GDP-tubulin. This favors further polymerization. If polymerization slows down, hydrolysis of GTP on newly added subunits can catch up, and subunits at the newly formed end are GDP-tubulin. This promotes a switch from polymerization (growth of the microtubule) to depolymerization (shrinkage of the microtubule).

In addition to actin filaments and microtubules, the plant cytoskeleton contains many other protein components. These include motor proteins (e.g., myosin) that drive the movement of subcellular structures such as organelles and paired chromosomes along filaments and microtubules; proteins that bind filaments and microtubules together into bundles; and proteins that regulate the dynamics of polymerization and depolymerization of these cytoskeletal structures. An interesting example in the latter category is profilin, a protein that binds to actin monomers and so determines their availability for polymerization into filaments. Much of the actin in pollen, for example, is bound in complexes with profilin. In fact, profilin is one of the major allergens causing hay fever.

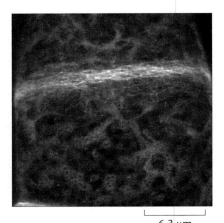

L_____L
6.3 μm

Figure 3–16. Preprophase band in a tobacco cell. The cell was treated with reagents that make microtubules fluorescent. (From C. Lloyd and J. Chan, *Nature Rev. Mol. Cell Biol.* 7:147-152, 2006. With permission from Macmillan Publishers Ltd, courtesy of J. Chan, G. Calder and C. Lloyd, John Innes Centre.)

microtubules and actin filaments that direct vesicles containing precursors into the new cell wall. We describe these processes in more detail below.

A preprophase band forms at the site of the future cell wall

In cells that are about to divide, the preprophase band forms round the equator of the cell (Figure 3–16). It appears late in G2 in the cytoplasm next to the plasma membrane (an area called the "cell cortex"). We have described how different cyclin-dependent kinases function at specific stages of the cell cycle (see Section 3.1); a specific CDK (CDC2) found in the preprophase band probably regulates band formation. The preprophase band narrows to a dense ring and then disappears during prophase, at the time that the nuclear envelope breaks down. This also coincides with the loss of all cortical microtubules from the cell, and the loss of actin filaments specifically from the zone of the cortex that was occupied by the preprophase band. Later, after cytokinesis, the new cell wall that separates the two daughter cells forms in this actin-free zone, exactly where the preprophase band was located (Figure 3–17). The preprophase band thus seems to leave a "molecular footprint" in this region, marking the position where the new cell wall (the cell plate) will attach to the preexisting cell wall.

In most types of cell, the preprophase band regulates the direction of cell division but is not necessary for the division process itself. For example, the *fass* mutant of

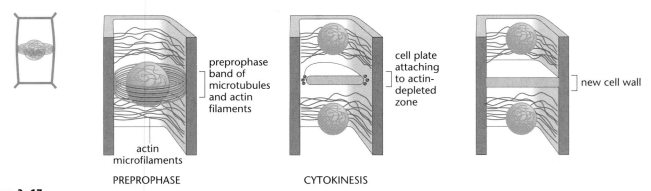

Figure 3–17.
The preprophase band marks the spot where the cell wall will form.
At the end of G2 (interphase), a band of microtubules and actin filaments appears in the cell cortex. This band disappears at the onset of prophase, with the loss of actin filaments from this zone of the cortex. During cytokinesis, the cell plate that gives rise to the new cell wall forms in precisely the position where the preprophase band was located.

Arabidopsis does not form a preprophase band, but cell division is not inhibited. The direction in which cells divide, however, is unregulated in these mutants: in wild-type plants, the majority of divisions in an embryonic **hypocotyl** (the area between root and **cotyledons** in a young seedling) are transverse in orientation, whereas cell division occurs in random directions in the *fass* mutant (Figure 3–18). This suggests that the mitotic spindle can form and function in the absence of a preprophase band, but that control of the direction of cell division in plants requires the presence of a preprophase band.

Replicated pairs of chromosomes are separated on a spindle of microtubules

At the end of G2, the preprophase band disappears, the chromosomes condense, and a spindle forms. The "spindle microtubules" emanate from two, opposite regions of the nuclear envelope (the poles of the spindle). The nucleus becomes surrounded by a dense mass of many thousands of microtubules. This mass becomes organized into the spindle as the nuclear envelope disappears (see Figure 3–15). The two ends of microtubules have different properties (see Box 3–2). One end of a microtubule, the plus end, is capable of rapid growth; the other, minus, end is slower-growing. In the developing spindle, the plus ends of the microtubules grow toward the equator of the spindle and the minus ends are at its poles.

In most organisms other than higher plants, including **algae**, lower plants (such as mosses and ferns), and most animals, the assembly of spindle microtubules involves **centrioles**—proteinaceous structures containing three different forms of tubulin (α-, β-, and γ-tubulin). The centrioles are replicated before mitosis (during S phase) and, just before mitosis, they migrate to opposite sides of the nucleus. The minus ends of the spindle microtubules are associated with these two centrioles (Figure 3–19). Although higher plants lack centrioles, their spindle microtubules are still nucleated at opposite poles of the cell, in diffuse regions called "polar caps."

In plants (and all eukaryotes), before metaphase, the two identical DNA molecules (sister **chromatids**) of a chromosome are attached to each other at the centromeric regions (see Section 2.2). At this paired region, a structure called a **kinetochore** forms on each sister chromatid. Each kinetochore binds to the plus ends of a subset of spindle microtubules—the "kinetochore microtubules." The kinetochores of the two sister chromatids bind to microtubules from opposite poles of the spindle (Figure 3–20). Movement of the bundles of kinetochore microtubules attached to each kinetochore aligns the chromosomes at the spindle equator (at metaphase), then (in anaphase) "pulls" the sister chromatids apart, from the equator toward the poles.

Two types of processes allow the kinetochore microtubules to align the chromosomes at the spindle equator and then pull them to the poles during mitosis: the actions of motor proteins and the depolymerization of microtubules. **Motor proteins** use energy provided by the hydrolysis of **ATP** to generate mechanical energy or force. The spindle microtubules are associated with motor proteins called **kinesins**, which consist of a head domain and a tail domain (the latter of variable structure). Kinesins are often

Figure 3–18.
A mutant that fails to form a preprophase band. Wild-type Arabidopsis seedlings (*right*) and *fass* mutant seedlings (*left*) demonstrate the consequences of loss of the preprophase band for the extent and direction of growth. *fass* mutants are also called *tonneau* (French for "barrel") because of their barrel-shaped embryos.
(Courtesy of Henrik Buschmann.)

Figure 3–19.
Centriole replication and migration in a lower plant. A centriole is present adjacent to the nucleus throughout the cell cycle. It replicates during S phase, and the two centrioles move to opposite sides of the nucleus. The minus ends of the microtubules that form the spindle during mitosis are associated with the centrioles. Each daughter cell formed during cytokinesis receives one centriole.

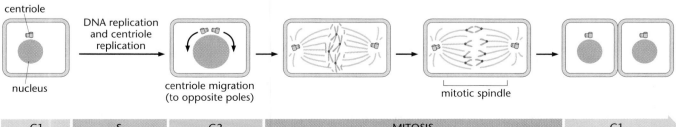

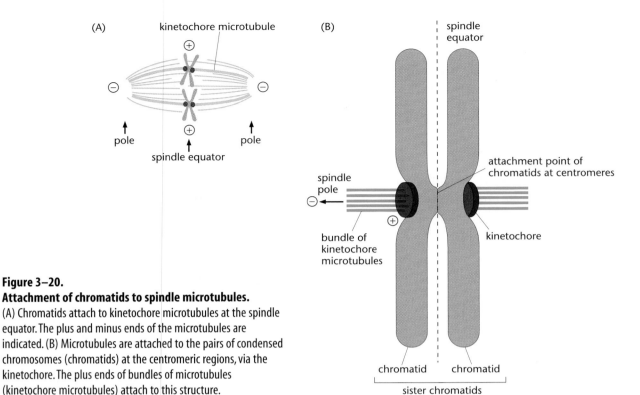

Figure 3–20.
Attachment of chromatids to spindle microtubules.
(A) Chromatids attach to kinetochore microtubules at the spindle equator. The plus and minus ends of the microtubules are indicated. (B) Microtubules are attached to the pairs of condensed chromosomes (chromatids) at the centromeric regions, via the kinetochore. The plus ends of bundles of microtubules (kinetochore microtubules) attach to this structure.

active as dimers, and two molecules dimerize at regions in their tail domains. The head domains bind to microtubules and ATP, and the tail domains associate with the molecules to be moved. When ATP is hydrolyzed, kinesins move along the microtubules carrying their cargo molecules. Kinesins are directional—that is, each type moves in only one direction along the microtubule (Figure 3–21). The best-characterized kinesin in the plant spindle is "kinesin-like calmodulin-binding protein" (KCBP), which moves from the plus to the minus end of the spindle to which it is attached. Exactly how kinesins move chromatids during mitosis is not known, but one possibility is that they attach to kinetochore microtubules and use other, static microtubules in the spindle as tracks along which to move the chromatids. If this model is correct, the kinetochore microtubules must depolymerize at the poles as the distance between the chromatid and the poles decreases (Figure 3–22A). Another possibility is that the head domains of the kinesins bind to the kinetochore microtubules while the tail domains bind to the kinetochore itself, exerting poleward force directly on the kinetochore. If this model is correct, the plus end of the microtubule attached to the kinetochore must depolymerize as movement occurs (Figure 3–22B).

Microtubules direct the formation of the phragmoplast, which orchestrates deposition of the new cell wall

At late anaphase, the new cell wall begins to form between the separated sister chromatids at the spindle equator. The wall is initiated as a cell plate, its deposition directed by the phragmoplast (Figures 3–23 and 3–24). As the chromosomes decondense and new nuclear envelopes form, two bundles of microtubules and actin filaments appear between the daughter nuclei and the position that was occupied by the spindle equator. These bundles are probably formed in part from the remnants of the mitotic spindle and in part from newly polymerized microtubules and actin filaments. They are oriented perpendicular to the plane of division. The microtubules of the two bundles

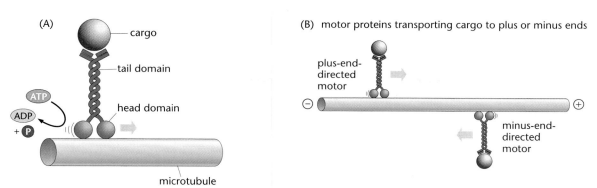

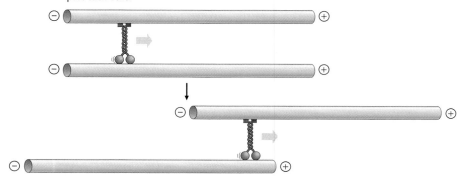

Figure 3–21.
Motor proteins and microtubule motility.
(A) Motor proteins (kinesins) are usually dimers of two identical proteins. The long tail domains bind to a cargo to be transported along a microtubule, and the head domains bind to the microtubule. Hydrolysis of ATP by the head domain provides the energy for movement of the motor protein and its cargo along the microtubule. (B) Some types of kinesin travel only from the plus to the minus end of microtubules; others travel only in the opposite direction. (C) Kinesins cause microtubules to slide past each other by binding (at the tail end) to one microtubule as a cargo and moving (at the head end) along another microtubule.

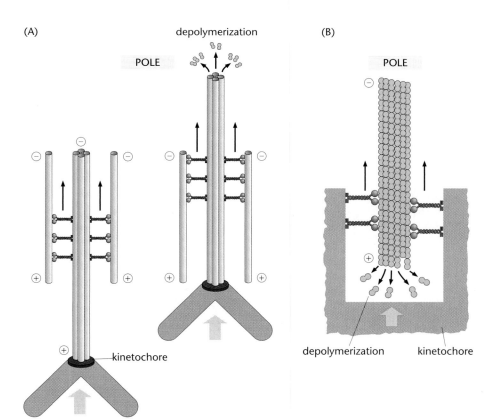

Figure 3–22.
Two models of chromatid movement.
(A) In one model, chromatids (*brown*) are pulled apart by the movement of kinetochore microtubules (*central green rods*) relative to other, static microtubules (*outer green rods*) in the spindle. The kinetochore microtubules are the cargo for kinesins that move along the static microtubules toward the spindle pole, pulling the attached chromatid toward the pole. To allow this movement, the kinetochore microtubules depolymerize at the pole (at their minus ends). (B) In another model, the kinetochore is the cargo for kinesins that move along the kinetochore microtubules (*central green rods*), pulling the kinetochore and its chromatid (*not shown*) toward the pole. To allow this movement, the kinetochore microtubules depolymerize at the kinetochores (at their plus ends).

overlap in a zone formerly occupied by the spindle equator, midway between the daughter nuclei. It is in this "phragmoplast zone" that the cell plate starts to form during telophase. Deposition of the cell plate starts in the center of the cell and, as it proceeds, the bundles of microtubules and actin filaments that constitute the phragmoplast disassemble from the center of the cell and reassemble at the edges of the plate. The phragmoplast thus becomes an annular structure, expanding outward from the center of the cell to the region of the cell cortex that was the site of the preprophase band, eventually meeting the existing cell wall.

Cell wall components are delivered to the site of the new cell boundary in vesicles that fuse to form the cell plate and its bounding membrane; this occurs in three phases. First, the vesicles track along microtubules to the phragmoplast equator. The plus ends of the two sets of phragmoplast microtubules point toward the equator; this allows plus-end–directed motor proteins (such as heavy chain kinesin) to transport vesicles to the growing cell plate (see Figure 3–20). At the equator, the vesicles form a network connected by thin (20 nm) tubes of membrane. In the second phase, these tubes increase

Figure 3–23.
Phragmoplast and cell plate formation.
As the new nuclei form, bundles of actin filaments and microtubules appear on either side of the position formerly occupied by the spindle equator (*top*). The microtubules direct vesicles toward this equatorial region. The vesicles fuse, starting at the center of the cell and spreading out to the periphery (*upper to lower left*). The fused vesicles give rise to a membrane-enclosed compartment in which the new cell plate is formed from components delivered by the vesicles.

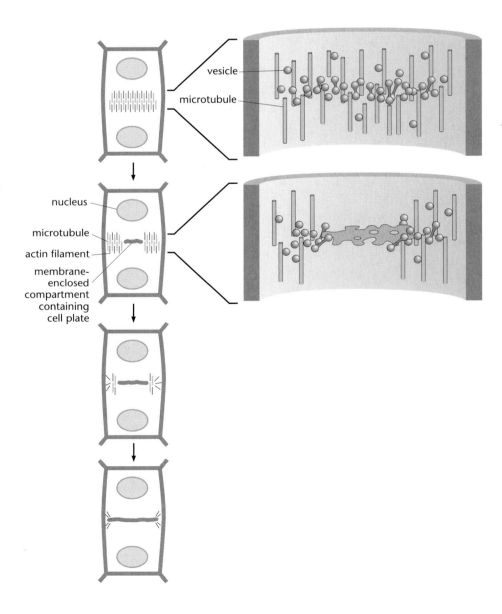

(A)

microtubules

vesicular network

vesicle

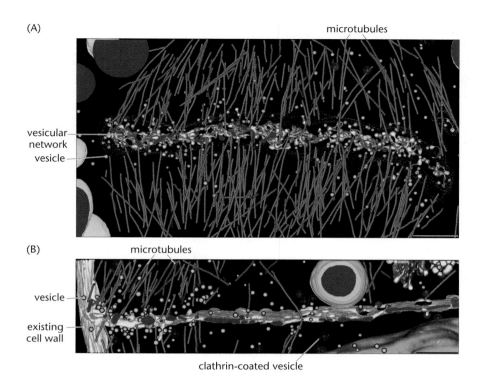

(B)

microtubules

vesicle

existing cell wall

clathrin-coated vesicle

Figure 3–24.
The phragmoplast and cell plate formation in the shoot meristem of Arabidopsis. The images were constructed by electron tomography, a technique that allows three-dimensional visualization of structures observed by transmission electron microscopy. (A) The phragmoplast, viewed side-on, at the center of the cell at the stage when the vesicular network has formed (approximately the first stage shown in Figure 3–23, top). (B) The phragmoplast and cell plate, viewed side-on, at the edge of the cell at a later stage of cell plate formation (approximately the third stage in Figure 3–23, third from top). Note the presence of both smooth-surfaced vesicles and vesicles coated with the protein clathrin (see Figure 3–46). Other vesicle types also contribute to the cell plate (see Figure 3–28). (A and B, courtesy of Jose M. Segui-Simarro.)

in diameter and more vesicles fuse to the expanding network. Third, the microtubules that have delivered the vesicles disappear from the equator, leaving a vesicular network of membranes that fuse to form a membrane-bounded compartment containing the cell plate. The same three phases are repeated at the edge of the phragmoplast as it expands outward, until the cell plate and its bounding membrane make contact with the side walls of the parent cell. The new cell plate then fuses with the existing cell walls and the bounding membrane becomes contiguous with the plasma membrane, thereby separating the two daughter cells (see Figure 3–23).

In many plants, in the early stages of endosperm development in seeds, nuclear division and cell division occur independently rather than together as described above. This process has been extensively studied in the endosperms of cereal seeds, in which large amounts of **starch** are subsequently deposited (see Section 4.2). After **fertilization**, the nucleus that will give rise to the endosperm (see Section 5.2) undergoes several rounds of nuclear division without any cell division. This gives rise to a single, large cell with many nuclei arranged around its periphery (a structure called a **syncytium**). Cell walls then develop perpendicular to the outer edges of this single cell to create compartments, each containing a single nucleus.

The formation of these **anticlinal** walls between the nuclei of the syncytium originates with a phragmoplast formed between bundles of microtubules and actin filaments that radiate from the membrane of each nucleus. Further nuclear division accompanied by cytokinesis then forms the inner, **periclinal**, wall of these cells, and gives rise to nuclei that will form the next layer of cells in the original single cell. Eventually, the entire space in the original cell may become cellularized in this way (Figure 3–25).

Vesicles carry material from the Golgi apparatus to the newly forming cell wall

The **polysaccharide** (glycan) components of the cell plate are synthesized in the **endomembrane system**. The subcellular, membrane-bounded compartments of this

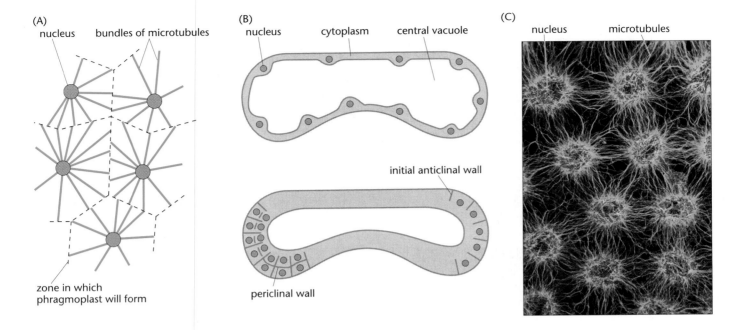

(A) nucleus   bundles of microtubules

zone in which phragmoplast will form

(B) nucleus   cytoplasm   central vacuole

initial anticlinal wall

periclinal wall

(C) nucleus   microtubules

Figure 3–25.
Cellularization of the endosperm.
(A) Diagram showing nuclei and radiating microtubule bundles lining the periphery of the single-celled endosperm in the early stages of development. The view is from the inside of the cell toward the wall.
(B) Development of the endosperm in a cereal seed. The upper diagram shows a cross section of the single-celled, syncytial endosperm at the same stage of development shown in (A). The lower diagram shows cellularization of the endosperm: formation of the first anticlinal wall (*right end*), and formation of periclinal walls and further anticlinal wall deposition (*left end*). (C) Electron micrograph of nuclei and radiating microtubules early in the development of the endosperm of *Coronopus didymus* (Brassicaceae), at the stage shown in (A). (Courtesy of Roy Brown.)

system—comprising the **endoplasmic reticulum** (ER) and the **Golgi apparatus**, the **vacuoles** and vesicles, and the plasma membrane and nuclear envelope—house many metabolic reactions concerned with the synthesis of complex carbohydrates and carbohydrate-linked proteins (**glycoproteins**).

The major carbohydrates that make up the cell plate wall (**pectins** and **hemicelluloses**) are synthesized in Golgi stacks: flattened, membrane-enclosed compartments stacked like coins that lie at the edge of the phragmoplast. The enzymes that synthesize these carbohydrates are delivered to the Golgi stacks from the other main constituent of the endomembrane system: the endoplasmic reticulum. The enzymes are synthesized on **ribosomes** recruited to the ER membrane, then transported into the ER lumen as translation is completed (see Section 3.3). From here, the enzymes are delivered in vesicles to the "*cis* face" of the Golgi (that is, the side of the Golgi stack nearest to the ER; see Figure 3–40). Polysaccharide synthesis continues as the cell wall components move through the Golgi stacks from the *cis* face toward the "*trans* face" adjacent to the cell plate. Vesicles that bud off from the *trans* Golgi deliver the finished polysaccharides directly to the forming cell plate.

Vesicle formation and docking are important in many subcellular processes, in addition to cell plate formation. Vesicles transport materials between the components of the endomembrane system, including the plasma membrane and vacuolar membrane (see Section 3.3) and the cell plate. The details differ depending on the membrane of origin and the target membrane, but the basic process is the same for all vesicles (Figure 3–26).

Vesicle formation is often associated with the acquisition of external proteins that promote the budding-off of the vesicle. For example, the formation of vesicles that traffic between the ER and the Golgi apparatus is initiated by the binding of cytosolic coating proteins to the external surface of the donor membrane, in a process that involves several other specific proteins and the energy-rich compound **GTP**. Once bound, the cytosolic coating proteins polymerize, changing the properties of the donor membrane to form an outgrowing bud. The bud pinches off to form a vesicle entirely coated with these proteins. When the vesicle reaches its target compartment, depolymerization of the coat exposes other proteins on the surface of the vesicle that allow it to dock with the target membrane.

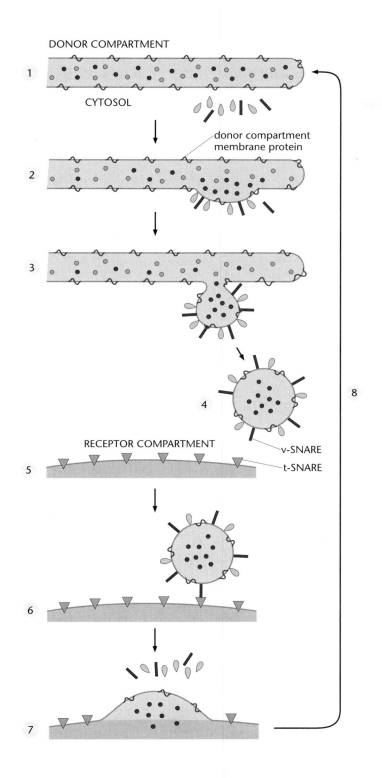

DONOR COMPARTMENT

CYTOSOL

donor compartment membrane protein

RECEPTOR COMPARTMENT

v-SNARE
t-SNARE

Figure 3–26.
Generalized scheme for vesicle transport between subcellular compartments.
(1) Cytosolic proteins (*black rods, green blobs*) are recruited to the membrane of the donor compartment. (2) Some of these proteins form a coat on the membrane that allows it to bud. Specific cargo molecules (*red*) in the lumen of the donor compartment are targeted to the growing bud, leaving other "resident" molecules (*blue*) behind. (3–5) The vesicle buds off and moves through the cytosol to the target compartment. This movement may be guided by motor proteins associated with microtubules (*not shown*). (6) Specific proteins (v-SNAREs) recruited to the surface of the vesicle during its formation recognize specific receptors (t-SNAREs; *blue triangles*) on the surface of the target compartment. This allows the vesicle to dock with the target receptor compartment. (7) The vesicle membrane fuses with the target membrane, releasing the cargo molecules into the compartment. The recruited proteins are released back into the cytosol. (8) Vesicle traffic is a two-way process. Vesicles will move back from the receptor to the original donor organelle to return membrane components.

The delivery of materials from a vesicle to its target compartment requires targeting mechanisms to ensure that the correct compartment is recognized, and mechanisms to promote vesicle fusion with the receptor membrane. Targeting is a function of protein complexes on the vesicle membrane called "tethering complexes." These recognize and dock with specific protein components of the receptor compartment. Membrane fusion is then mediated by a class of proteins called **SNARE proteins**. Generally, a v-SNARE protein on the surface of the vesicle interacts with t-SNARE proteins on the receptor membrane. This interaction brings the vesicle and receptor membranes closer together, resulting in fusion and thus delivery of the vesicular contents (see Figure 3–26).

The importance of these interactions is illustrated by the phenotypes of Arabidopsis mutants that lack components of the membrane-fusion apparatus (Figure 3–27). The *knolle* and *keule* mutants are unable to form normal cell plates. Embryos **homozygous** for these **mutations** have multinucleate cells with incomplete and abnormal cross-walls, and seedlings die soon after **germination**. The KNOLLE protein is a type of t-SNARE called a **syntaxin**. The KEULE protein regulates the ability of the t-SNAREs to form receptor complexes for the v-SNAREs. Thus, loss of either protein reduces the delivery of new material to the cell plate, resulting in incomplete cytokinesis.

At least two distinct classes of vesicle are involved in delivering material to the growing cell plate, as shown in Figure 3–24. One type of vesicle undergoes dramatic shape changes once it enters the cytoplasm surrounding the growing cell plate (known as the "cell plate assembly matrix"). Vesicles fuse in pairs to give hourglass-shaped vesicles, which then undergo severe constriction at the neck region by the recruitment of dynamin proteins. **Dynamins** polymerize to form a spiral around the vesicle, creating a dumbbell shape (Figure 3–28). The squeezing of the vesicle may induce conformational changes in the cell wall polysaccharides it contains, preparing this cargo for incorporation into the cell plate.

Meiosis is a specialized type of cell division that gives rise to haploid cells and genetic variation

Meiosis occurs during **sexual reproduction** and is the process by which a **diploid** cell (a meiocyte) gives rise to **haploid** cells—that is, cells that contain only one copy of each chromosome rather than the two copies found in a diploid cell. In animals meiosis gives rise to **gametes**, whereas in plants meiosis produces **spores**. Unlike gametes, spores do not fuse to form a **zygote**, but instead give rise to a haploid organism that subsequently produces gametes by mitosis. In vascular plants the haploid organism usually consists of only a few cells (in the form of the **pollen** grain and the **embryo sac**), but in nonvascular, lower plants the haploid organism is the dominant form (see Chapter 1 and Chapter 5). During meiosis, chromosomes exchange segments of DNA, thus altering their DNA sequences, in a process called **recombination**. Here we outline the meiotic process, then consider briefly the consequences for the generation of genetic diversity (see also Chapter 2).

Before meiosis, the diploid meiocyte undergoes DNA replication (S phase). Thus, when division begins, each chromosome consists of two identical DNA molecules (sister chromatids) and the nucleus contains pairs of chromosomes similar to each other in size, structure, and DNA sequence (homologous chromosomes, or **homologs**). The production of haploid cells occurs via two consecutive divisions (Figure 3–29). First,

Figure 3–27.
Wild-type and *knolle* mutant embryos in developing Arabidopsis seeds. In the *knolle* mutant (*right*), note the enlarged cells compared with the wild type (*left*). Each cell of the mutant contains multiple nuclei. (Courtesy of Frederic Berger.)

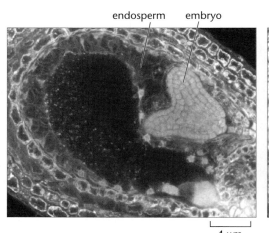

endosperm   embryo

4 μm

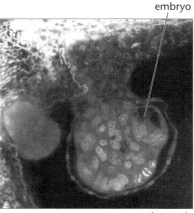

embryo

2 μm

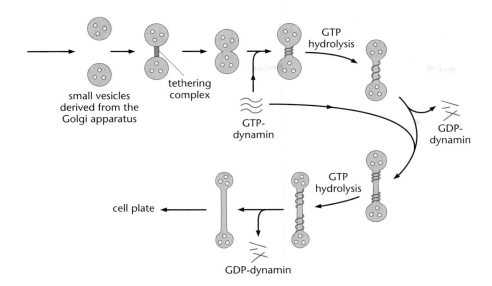

Figure 3–28.
Sequence of events in formation of dumbbell-shaped vesicles during assembly of the cell plate. Small vesicles from the Golgi apparatus become tethered in pairs, then fuse to form an hourglass shape. GTP-dynamin monomers assemble around the neck of the hourglass into "springs," which expand using the energy released by hydrolysis of GTP. GDP-dynamin is released and the neck-stretching process is repeated.

homologous chromosomes are separated from each other on a spindle to give two clusters of chromosomes. Each cluster contains one chromosome from each chromosome pair in the parental cell—in other words, each has a haploid chromosome complement. Note the difference from mitotic division: in mitosis, the sister chromatids of each chromosome are separated on the spindle; in this first meiotic division, the sister chromatids stay together and the pairs of chromosomes separate. In the second meiotic division, the sister chromatids of each cluster are separated on a spindle, giving rise

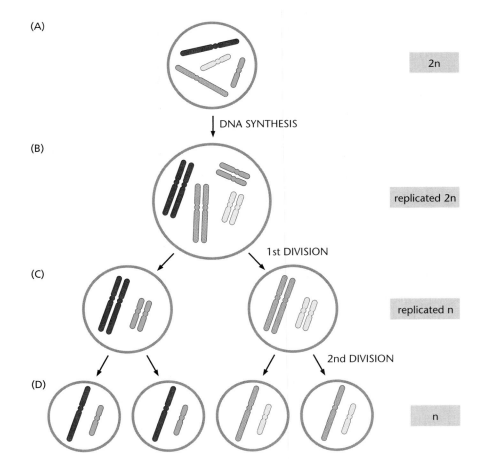

Figure 3–29.
Distribution of chromosomes during meiosis. Each circle represents a cell, containing chromosomes (*colored bars*). (A) Meiosis starts with a diploid (2n) cell containing two homologous copies (a pair of homologs) of each type of chromosome. For simplicity here we show two pairs: a pair of short chromosomes with homologs in light and dark blue, and a pair of longer chromosomes with homologs in light and dark red. (B) At the start of meiosis, each chromosome is replicated, forming two sister chromatids that remain attached to each other (shown here as two aligned colored bars). (C) During the first meiotic division, each daughter cell receives one of the replicated homologous chromosomes, each still consisting of two sister chromatids. (D) In the second meiotic division, the sister chromatids are separated, resulting in four haploid (n) cells, each with a single copy of each chromosome type.

to four clusters of chromosomes, each chromosome consisting of a single DNA molecule (one chromatid). These four clusters become surrounded by nuclear envelopes and undergo cytokinesis to become four haploid cells.

The first meiotic division requires that homologs recognize each other and become paired before separation on the spindle. Pairing ensures that each of the two chromosome clusters formed in this division contains one member of each pair of parental chromosomes. Initial pairing of homologs occurs at localized regions along the chromosomes. The chromosomes align so that regions of equivalent DNA sequence lie adjacent to one another. The two chromosomes then become linked by the formation of a protein complex, the **synaptonemal complex**, between them (Figure 3–30). Recombination events between adjacent regions of the homologous chromosomes occur at this stage. Recombination results in a swapping of segments of DNA between the homologous chromosomes. A double-stranded break occurs in the DNA of one chromatid, and the ends "invade" and become joined to the complementary strand of a chromatid of the other chromosome (Figure 3–30D). The process involves both digestion and synthesis of DNA, and the precise mechanism is not fully understood.

As the first meiotic division proceeds, the synaptonemal complex disappears and the homologous chromosomes remain attached at the points of recombination, called **crossovers** or **chiasmata**. The chiasmata hold the homologous chromosomes together as they align at the spindle equator, ensuring correct assortment of homologs during

Figure 3–30.
Recombination between homologous chromatids during meiosis. (A) The circle represents a cell containing two replicated pairs of homologous chromosomes (the same representation as in Figure 3–29B, just before the first meiotic division). (B) Before the first division is completed, each replicated chromosome (containing two sister chromatids) aligns with its replicated homolog. For simplicity, only the blue pair of homologs is shown here. The position of two genes (M and N) is indicated on the chromosomes, with different alleles in each homolog (alleles M′ and N′ on the light blue homolog; M″ and N″ on the dark blue homolog). The paired homologous chromosomes are held together by the synaptonemal complex (represented by a row of black lines between the light blue and dark blue chromatid pairs). (C) A break occurs at the same position in two adjacent chromatids (light blue and dark blue). (D) Fragments of homologous chromatids are exchanged at the break point (crossover and religation). (E) During the first meiotic division, the synaptonemal complex is dissolved and the homologous chromosomes are separated. One of the light blue chromatids now contains a fragment originally present in a dark blue chromatid, with the alleles M′ and N″; one of the dark blue chromatids contains the new combination of alleles M″ and N′.

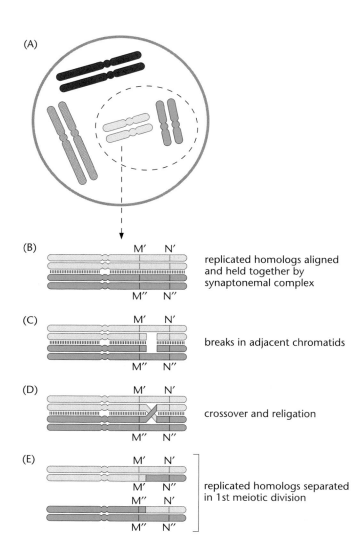

division. The homologs are then pulled apart to opposite poles of the spindle, by a mechanism similar to that which separates sister chromatids during mitotic division. The two groups of chromosomes then undergo a second division (see Figure 3–29), without an intervening S phase, in which the sister chromatids of each chromosome are separated on a spindle. This division is essentially the same as a mitotic division, and it is followed by cytokinesis to form the haploid cells.

The genetic consequences of meiotic division are profoundly different from those of mitotic division. Whereas the cells resulting from mitosis are genetically identical to each other and to the parent cell, the four haploid cells resulting from meiosis are not identical to each other or to the parent cell (the meiocyte). In fact, each meiocyte in an **anther** or **ovule** can potentially produce germ cells that are genetically different from those produced by any other meiocyte. This **genetic variation** arises in several ways. First, in most organisms, homologous chromosomes are similar but not identical in DNA sequence (for example, they may have different allelelic forms—**alleles**—of the same gene; see Chapter 2). Thus the two groups of chromosomes formed in the first meiotic division are different from each other. Second, the direction of separation of one pair of chromosomes in the first meiotic division is independent of that of any other pair of chromosomes. This means that there are several ways of achieving the haploid chromosome complement (Figure 3–31). If the meiocyte has, say, two pairs of homologous chromosomes, there are four possible haploid complements; if it has three homologous pairs, there are eight possible haploid complements, and so on.

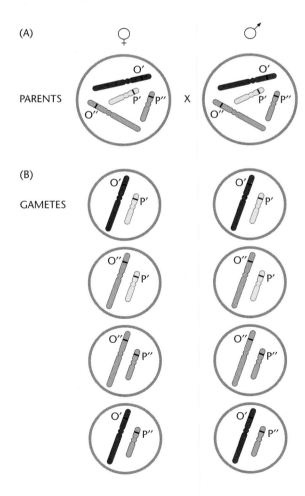

(A) PARENTS

(B) GAMETES

AT FERTILIZATION   Combine any of the left four with any of the right four.

Figure 3–31.
Meiosis creates genetic variability by generating new combinations of alleles.
(A) The circles represent diploid cells that give rise to the male and female gametes, at a stage equivalent to that shown in Figure 3–29A. O and P are genes represented by a different allele in each homologous chromosome (O′ and O″ on the red pair; P′ and P″ on the blue pair). Note that both the female and male parent cells contain the same set of alleles (O′O″P′P″). (B) At the end of meiosis, each diploid parent cell has produced gametes with four possible combinations of alleles: O′P′, O′P″, O″P′, and O″P″. At fertilization, these gametes are combined at random and generate new diploid cells with all possible combinations of O and P alleles, not just those present in the parents.

The third source of genetic variation among germ cells is recombination. As described above, this results in changes in the DNA sequence of chromatids during meiosis, so that the chromatids are no longer identical to those of the parent cell. **Recombination frequency** determines the degree of **genetic linkage**, meaning the linkage between genes on the same chromosome—in other words, the extent to which the genes are inherited together. Genes on different chromosomes are randomly assorted during germ cell formation, so they will segregate independently during sexual reproduction. In the absence of any recombination, genes on the same chromosome would be expected to be inherited together—that is, they would be expected to be linked. Thus, for any parental chromosome pair, an offspring would inherit either all of the genes from one homolog or all of the genes from the other homolog. But with recombination, involving the swapping of DNA segments between homologs (as shown in Figure 3–30), offspring inherit mixtures of genes from the two parental homologs. Genes at opposite ends of a chromosome are likely to be inherited independently, because there is a high probability of recombination between them. Genes that are close together on a chromosome are likely to be inherited together.

Overall, the process of meiosis, together with fertilization (see Section 5.2), ensures that the offspring of most diploid organisms have a genetic makeup unlike that of either parent.

3.3 ORGANELLES

During cell division and cell growth, organelles play a vital role in the manufacture of new cellular components. In many cells, the organelles themselves must multiply to meet the energetic and biosynthetic demands of these processes, and to accommodate the specialized functions of the mature cell. The maintenance and development of organellar function, the biogenesis of new proteins and other organellar materials, and the delivery of these components to the correct location within the cell all require a high level of coordination. The nucleus is described in Box 3–1: in this section we focus on the structure and function of the cytoplasmic organelles. We describe how chloroplasts and **mitochondria** replicate, and how proteins destined for these organelles are synthesized in the cytosol and imported across organellar membranes. We describe the function of the endomembrane system in modifying specific proteins and delivering them to their destinations in the vacuole and cell wall, and the function of the cytoskeleton in moving and positioning organelles in the cell.

Throughout this section, we describe routes and mechanisms by which proteins move to specific destinations in the cell. As we discussed in Chapter 2, the vast majority of proteins in the plant cell are encoded in the nucleus and synthesized on ribosomes, either in the cytosol or on the cytosolic face of the endoplasmic reticulum. Subsequent destinations for these proteins include the nucleus, chloroplasts, mitochondria, **peroxisomes**, vacuoles, components of the endomembrane system, and the cell surface. All proteins, except those that remain in the cytoplasm where they are translated, carry one or more targeting domains (usually amino acid sequence **motifs**). Specific cellular machinery interacts with the information in these domains and "sorts" the proteins according to their destination (Figure 3–32). Each targeting domain is specific to the destination of the protein, and the protein-sorting machinery is specific to the target compartment or membrane system. Thus, for example, a protein carrying a targeting domain for a chloroplast cannot enter any other organelle. These targeting domains are often removed, once they have served their purpose, by proteases (enzymes that cleave peptide bonds at specific points in the amino acid sequence) in the target compartment, thus creating a functional polypeptide.

Protein synthesis occurs throughout the life of a cell, not just during growth and maturation. Almost all proteins in the cell are turned over—synthesized and then degraded (see Figure 3–32)—and must be replenished. The average lifetime of a protein varies

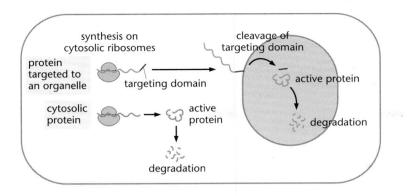

Figure 3–32.
Generalized scheme for protein synthesis, targeting, and turnover.
Proteins synthesized on cytosolic ribosomes are released into the cytosol on completion if they contain no targeting information. Proteins that contain targeting information associate with receptors on the surface of the membrane surrounding their target compartment and are transported through the membrane by transport proteins in the membrane. Inside the target compartment, the targeting sequence may be cleaved to form the active protein. Regardless of subcellular location, proteins have a finite lifetime. Degradation is not a random process, but is under tight and protein-specific control. Examples of the importance of regulated protein degradation are given elsewhere in this book. The amount of a particular protein in a cell is a function of both its rate of synthesis and its rate of degradation.

enormously from one type to another, ranging from a few minutes to many days. Protein degradation serves to remove damaged or incorrectly folded proteins, and is also a means of altering the levels of some proteins in response to developmental and environmental cues. Thus, for almost all proteins, some level of synthesis and transport to the appropriate cellular compartment is necessary simply to maintain levels of functional protein in the cell: the level is a function of the balance between the rate of synthesis and the rate of degradation.

Plastids and mitochondria replicate independent of cell division

Plastids and mitochondria replicate by **binary fission**. In dividing cells, this replication occurs during the cell cycle, and plastids and mitochondria are segregated into daughter cells at cell division. Organelle replication often continues during cell growth and differentiation, after cell division has ceased. For example, the number of plastids in a leaf **mesophyll** cell typically increases from about 20 to 50 as the cell expands and matures.

During binary fission of plastids, the parent plastid initially becomes dumbbell-shaped (Figure 3–33A). The constriction becomes progressively narrower until the two daughter plastids are separated. Constriction is brought about by contraction of two concentric ring structures associated with the constriction zone: an outer ring located on the cytosolic face of the outer membrane of the chloroplast envelope and an inner ring on the stromal face of the inner membrane (the contents of the chloroplast outside the photosynthetic membranes are called the **stroma**). **FtsZ proteins**, related to the cytoskeletal component tubulin, are crucial components of the inner ring structure (Figure 3–33B). Arabidopsis mutants lacking these FtsZ proteins have impaired plastid development and have fewer, larger chloroplasts than wild-type plants. The positioning of the contractile rings is determined by **MIN proteins**, which are synthesized in the cytosol and imported into the plastid (see below). Mutant plants that lack MIN gene function, such as the *minD* mutant, develop rings in asymmetrical locations along the chloroplast, resulting in the formation of a large and a small daughter plastid rather than identically sized daughter plastids (Figure 3–33C). Further proteins required for chloroplast division are being discovered through characterization of the *arc* (**a**ccumulation and **r**eplication of **c**hloroplasts) mutants of Arabidopsis, which have fewer chloroplasts per leaf mesophyll cell than wild-type plants. The chloroplasts of *arc5* mutants, for example, are large and frequently dumbbell-shaped, implying that division is blocked at a late stage (Figure 3–33D). The ARC5 protein, which belongs to the dynamin family of self-assembling proteins, is thought to play an essential role in the outer ring structure at the constriction zone in the late stages of chloroplast division.

The process of plastid division is similar in many respects to bacterial cell division, reflecting the endosymbiotic evolutionary origin of plastids (as discussed in Section 1.2;

Figure 3–33.
Electron micrographs of dividing plastids. (A) Three successive stages in the division of *Pelargonium* chloroplasts. The position of the FtsZ ring is indicated by arrows. (B) Position of the FtsZ ring during sequential stages of chloroplast division in a cell of the red alga *Cyanidioschyzon merolae*. The FtsZ protein was labeled with a fluorescent molecule and appears yellow. Note the formation and contraction of a ring of FtsZ protein as division proceeds. (C) Chloroplasts in leaf cells of wild-type Arabidopsis (*left*), of mutant plants lacking one of the FtsZ proteins (*center*), and of mutant plants lacking the protein MIN D (*right*). Note the single giant chloroplast of the *ftsZ* mutant (*center*) and the irregular sizes of chloroplasts in the *minD* mutant (*right*). The irregular sizes arise from asymmetrical chloroplast division (Scalebar: 5μm.). (D) Leaf cell of wild-type Arabidopsis (*left*) and the *arc5* mutant (*right*). Note the large, dumbbell-shaped chloroplasts in the *arc5* mutant. (A, courtesy of Haruko Kuroiwa; B, courtesy of Shin-ya Miyagishima; C, courtesy of Katherine Osteryoung; D, courtesy of Joanne Marrison.)

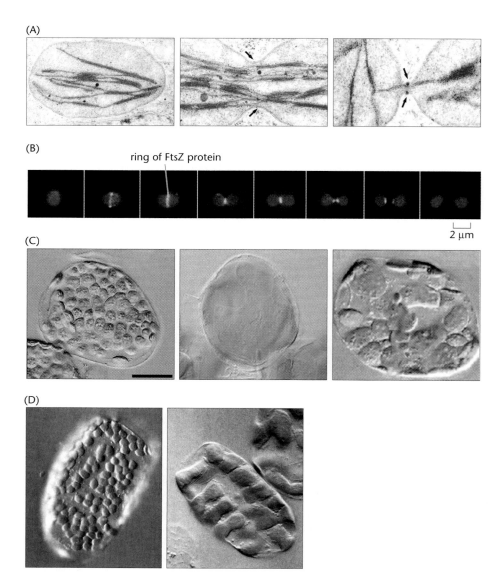

see also Section 2.6). In bacteria, constriction is brought about by a contractile ring composed of a FtsZ protein, and the position of the ring—around the center of the long axis of the cell—is determined by Min proteins very similar to those in chloroplasts. Mutant bacteria lacking Min proteins develop rings in asymmetrical positions along the cell during fission, resulting in the formation of nonviable mini-cells that lack DNA.

As immature plant cells differentiate, plastids also differentiate according to the particular cell type. For example, chloroplasts develop in green photosynthetic tissues; **amyloplasts** (starch-storing plastids) develop in storage organs such as tubers and the endosperms of cereal seeds; and **chromoplasts** (plastids with a high content of the yellow pigment β-carotene or the red pigment **xanthophyll**) develop in some **petals**, ripening **fruits**, and storage roots. All of these types of plastids are derived ultimately from **proplastids**—the small plastids present in undifferentiated cells (Figure 3–34).

Mitochondrial division is less well understood than that of plastids. Like plastids, mitochondria constrict before division, and ring structures form at the point of constriction. Mitochondria of higher plants lack FtsZ-like proteins, and the rings contain a dynamin-like protein.

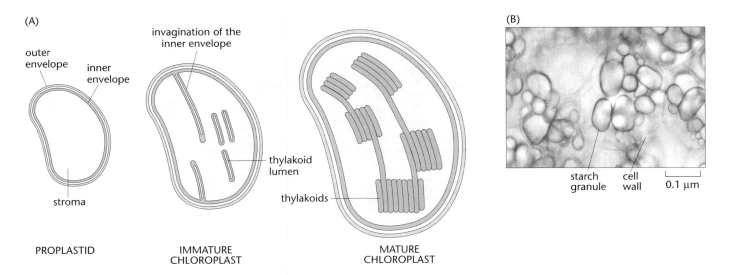

(A)

outer envelope

inner envelope

invagination of the inner envelope

stroma

thylakoid lumen

thylakoids

PROPLASTID

IMMATURE CHLOROPLAST

MATURE CHLOROPLAST

(B)

starch granule

cell wall

0.1 μm

Figure 3–34.
Plastid development. (A) Development of a chloroplast from a proplastid. Thylakoids form from invaginations of the inner envelope of the proplastid. For simplicity, the thylakoid lumens are not shown for the mature chloroplast. (B) Light micrograph of living cells in a slice of potato tuber. The large oval structures are starch granules; each is contained in an individual amyloplast.

Plastid and mitochondrial biogenesis involves post-translational import of many proteins

Most **plastid genomes** encode only a small proportion of plastid proteins: about 100 of the thousands found in the organelle (see Section 2.6). All other plastid proteins are encoded in the nucleus, synthesized on ribosomes in the cytosol, and imported into the plastid via **protein-import complexes** (also called "translocation apparatus" or "translocases") on the plastid envelope. Many of these proteins function in the stroma or in the inner envelope membrane. However, in chloroplasts, a specific subset of imported proteins is further transported into the photosynthetic membranes (the **thylakoids**), or across these membranes and into the space they enclose (the thylakoid lumen) (Figure 3–35).

The import of proteins into plastids requires specific N-terminal signal sequences—usually about 50 amino acid residues—on the proteins. These signals, or **transit peptides**, interact with the protein-import apparatus on the target membrane and are then cleaved as part of the import process. Transit peptides are recognized by the import apparatus not by their specific amino acid sequences but by their secondary structures. We describe here the way in which proteins destined for the inner envelope membrane, stroma, thylakoid membrane, or thylakoid lumen cross the chloroplast envelope (Figure 3–36) and the mechanisms by which some of these proteins are further translocated into the thylakoids.

In addition to the protein-import apparatus, movement of proteins into plastids requires interactions with proteins called **chaperones** (Figure 3–36A). Chaperones are important in many cellular functions involving the folding and unfolding of proteins. Almost all proteins are functional only when they assume a particular three-dimensional structure, acquired by folding of the amino acid chain after synthesis. Chaperones bind to unfolded proteins, stabilizing their structures so that correct folding can occur. They thus prevent incorrect folding and aggregation with other proteins. As well as functioning in normal cellular development, they are also important in protecting cells against the effects of exposure to high temperature, a condition that causes protein **denaturation** (unfolding). High temperature induces rapid synthesis of chaperones: they bind to

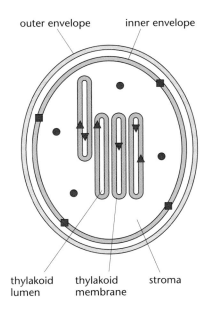

outer envelope

inner envelope

thylakoid lumen

thylakoid membrane

stroma

Figure 3–35.
Destinations for nuclear-encoded proteins in the chloroplast. Proteins that are synthesized on cytosolic ribosomes and destined for plastids (i.e., have plastid transit peptides) include components of the inner envelope (*red squares*), components of the thylakoid membranes and soluble proteins of the thylakoid lumen (*red triangles*), and soluble proteins of the stroma (*red circles*).

Figure 3–36.
Translocation into the plastid of a protein destined for the stroma. (A) The newly synthesized protein is chaperoned to the plastid, and (B) is translocated into the stroma via an import apparatus with components in the outer and inner envelopes. The transit peptide associates with a receptor on the outer envelope, and the protein crosses the envelope. (C) In the stroma, the transit peptide is cleaved by a protease, and a stromal chaperone binds to the imported protein to ensure correct folding to its fully active state.

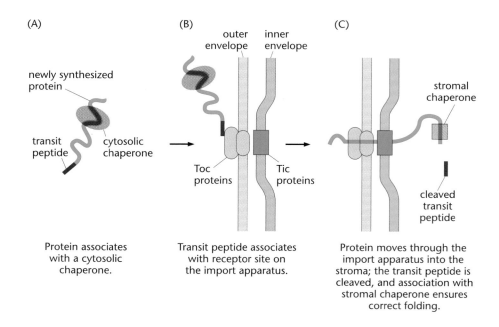

denaturing proteins to allow refolding rather than aggregation and loss of function. For this reason, chaperones are often known collectively as **heat shock proteins (HSPs)** (see Section 7.2).

Immediately after synthesis on ribosomes in the cytosol, a protein destined for the plastid is bound by a type of chaperone called HSP70. This chaperone maintains the protein in the unfolded state necessary for its passage through the protein-import apparatus. The transit peptide is recognized by a receptor on the outer surface of the plastid envelope, allowing the unfolded protein to associate with the protein-import apparatus (Figure 3–36B). The part of the apparatus in the outer envelope membrane consists of four Toc proteins (**t**ranslocase of the **o**uter **c**hloroplast envelope). One of these, Toc75, forms the translocation pore through which the protein passes. Another Toc protein is a membrane-bound chaperone of the HSP70 class. Translocation of the protein across the inner chloroplast membrane is mediated by a second part of the import apparatus, consisting of several Tic proteins (**t**ranslocase of the **i**nner **c**hloroplast envelope). The energy for translocation is provided by hydrolysis of ATP. As the protein enters the stroma, the transit peptide is cleaved by a stromal protease (Figure 3–36C). Proteins that will function in the stroma then fold to assume their functional forms, in association with a complex consisting of two chaperones, HSP60 and HSP10.

Proteins destined for the thylakoids contain further targeting information. For proteins that function in the thylakoid lumen or on the lumenal face of the thylakoid membrane, this information is often in the form of a lumenal targeting sequence (lumenal transit peptide) immediately behind the original transit peptide. The lumenal targeting sequence is exposed by removal of the transit peptide as the protein enters the stroma (Figure 3–37). For some proteins destined for the thylakoid membranes, the nature of the targeting is less clear. In some cases, hydrophobic parts of the proteins may allow insertion into these membranes without specific translocation mechanisms.

Proteins may be translocated from the stroma into the thylakoid lumen by one of three distinct pathways: the SEC pathway, the ΔpH pathway, and the signal recognition particle (SRP) pathway. Two proteins are required for transport by the **SEC pathway**: SECA and SECY. SECA is an ATP-hydrolyzing enzyme (an **ATPase**) that moves in and out of the thylakoid membrane, driving proteins through a channel formed by the SECY protein as it does so (Figure 3–38). An example of a protein transported in this manner is **plastocyanin**, an **electron carrier** that is a component of the **electron transport**

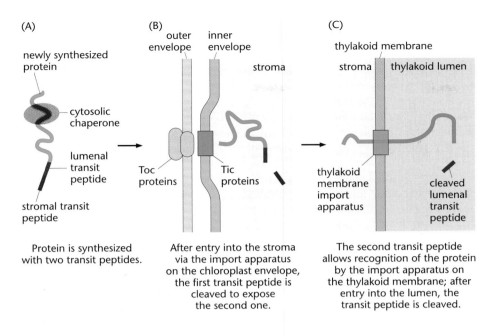

(A)

newly synthesized
protein

cytosolic
chaperone

lumenal
transit
peptide

stromal transit
peptide

Protein is synthesized
with two transit peptides.

(B)

outer
envelope

inner
envelope

stroma

Toc
proteins

Tic
proteins

After entry into the stroma
via the import apparatus
on the chloroplast envelope,
the first transit peptide is
cleaved to expose
the second one.

(C)

thylakoid membrane

stroma   thylakoid lumen

thylakoid
membrane
import
apparatus

cleaved
lumenal
transit
peptide

The second transit peptide
allows recognition of the protein
by the import apparatus on
the thylakoid membrane; after
entry into the lumen, the
transit peptide is cleaved.

Figure 3–37.
**Translocation across the chloroplast
envelope and thylakoid membrane of
proteins destined for the thylakoid
lumen.** (A, B) The newly synthesized protein
has two transit peptides: a terminal peptide
targeting it to the stroma (see Figure 3–36)
and a second transit peptide that will be
exposed after removal of the stromal transit
peptide in the stroma. (C) The stromal transit
peptide associates with the import apparatus
of the thylakoid membrane (as described in
the text), and is removed after the protein
enters the thylakoid lumen.

chain that links the light-capturing photosystems in the thylakoid membrane (the components of the photosystems and electron transport chain are described in Section 4.2). Mutants of maize in which the SEC pathway is defective (*tha1* mutants, lacking SECA proteins) are unable to form functional thylakoid membranes. Seedlings of these plants have severely impaired **photosynthesis** and die soon after germination.

The **ΔpH pathway** uses energy provided by the pH gradient between the stroma and the thylakoid lumen, rather than energy from ATP hydrolysis as in the SEC pathway (see Figure 3–38). Proteins dependent on this transport pathway include components of the **water-splitting complex** of **photosystem II**, located on the lumenal face of the thylakoid membrane. Proteins transported by the SRP pathway associate with a stromal factor called a **signal recognition particle** before translocation (see Figure 3–38). Energy for translocation is provided by hydrolysis of GTP, and the process is also stimulated by the pH gradient across the thylakoid membrane. As for the SEC pathway,

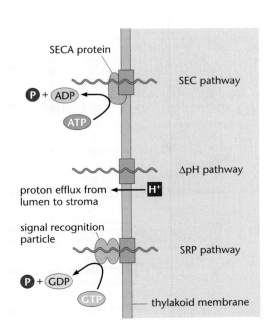

SECA protein

P + ADP

ATP

SEC pathway

ΔpH pathway

proton efflux from
lumen to stroma

H+

signal recognition
particle

SRP pathway

P + GDP

GTP

thylakoid membrane

Figure 3–38.
**Pathways of translocation of proteins
across thylakoid membranes.** In the SEC
pathway, translocation is driven by ATP
hydrolysis by a stromal protein, SECA. In the
ΔpH pathway, translocation is driven by the
proton concentration gradient between the
lumen and the stroma. During photosynthesis,
energy from light capture is used to pump
protons across the membrane from the stroma
to the lumen (see Section 4.2). The proton
concentration in the lumen becomes much
higher than that in the stroma. Movement of
protons back into the stroma down this
concentration gradient is harnessed primarily
to provide energy for ATP synthesis (see
Section 4.2), but also for other
energy-requiring processes such as the ΔpH
pathway. In the SRP pathway, translocation
involves binding of the protein to the stromal
signal recognition particle, which then
associates with the import apparatus. The
energy is provided by GTP hydrolysis.

mutant maize plants lacking components of either the ΔpH pathway or the SRP pathway have defective thylakoids and hence impaired photosynthesis.

Like the plastid genome, the **mitochondrial genome** encodes only a very small proportion of the proteins found inside the organelle. At least 95% of mitochondrial proteins are synthesized in the cytosol and enter via an import apparatus in the mitochondrial membrane. The import mechanism is similar in several respects to that of chloroplasts, although the two sorts of organelle have different types of translocation apparatus and the targeted proteins have different types of transit peptides.

Proteins synthesized in the cytosol and destined for the mitochondrial matrix must cross the outer and inner mitochondrial membranes (Figure 3–39). As with the plastid, this process is aided by the action of cytosolic HSP70 chaperones. The protein-import apparatus of the mitochondrion consists of a Tom complex (**t**ranslocase of the **o**uter **m**itochondrial membrane) of at least eight different proteins and a Tim complex (**t**ranslocase of the **i**nner **m**itochondrial membrane). Translocation requires energy from ATP and also uses energy provided by the **electrochemical gradient** across the inner mitochondrial membrane (see Section 4.5). The transit peptide—often known as a "presequence"—is cleaved by a protease in the matrix. The matrix proteins then fold into their active conformations with the aid of a chaperone complex consisting of HSP60 and HSP10 proteins. Some proteins destined for the intermembrane space or the inner membrane are inserted directly into these locations after import by the Tom complex. Others follow a pathway resembling that of proteins destined for the thylakoid lumen of chloroplasts. An initial presequence allows import into the matrix, then cleavage of this presequence exposes a second presequence that targets the protein back into the inner membrane or across this membrane and into the intermembrane space.

The endomembrane system delivers proteins to the cell surface and to vacuoles

Material is transferred between the endoplasmic reticulum and the Golgi apparatus, and between these compartments, the vacuoles, and the cell surface, by vesicles that bud off from one part of the endomembrane system and fuse with another (Figure 3–40), as we described in Section 3.2 for the transfer of cell-wall–synthesizing enzymes from the ER to the Golgi apparatus to the cell plate. The endomembrane system is the site of synthesis of many types of molecules: for example, **lipids** are synthesized in the

Figure 3–39.
The protein-import apparatus of mitochondria. As for the chloroplast, proteins (*not shown*) destined for the mitochondrial matrix associate with a cytosolic chaperone, cross the inner and outer membranes via an import apparatus with components in both membranes (as described in the text), and undergo folding mediated by a chaperone in the matrix.

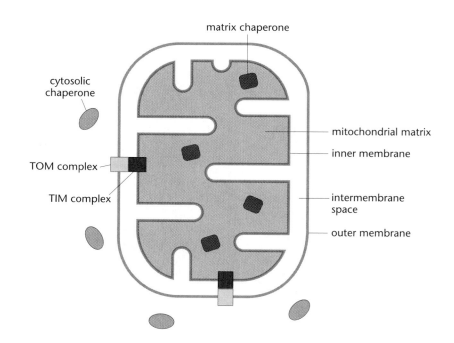

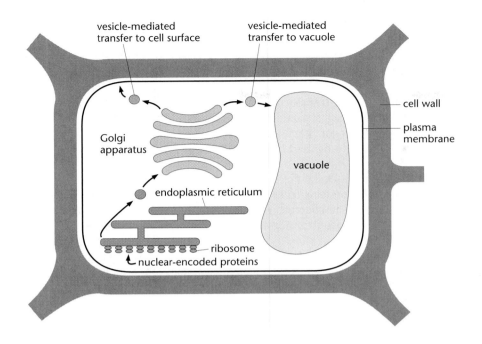

Figure 3–40.
Overview of the endomembrane system. Proteins destined for the vacuole or the cell surface are synthesized on ribosomes associated with the endoplasmic reticulum (ER) membrane and imported into the ER lumen during this process. (Endoplasmic reticulum with associated ribosomes is known as rough ER; smooth ER—lacking ribosomes—is the site of other biochemical processes, including lipid synthesis; see Chapter 4.) After post-translational modifications in the lumen, the proteins are exported in vesicles to the Golgi apparatus. After further modification there, the proteins are exported either to the membrane surrounding a vacuole (the tonoplast) for entry into the vacuole or to the plasma membrane for secretion to the outside of the cell (formation of the cell plate, described in Section 3.2, is a special case of this latter transport route).

ER (see Section 4.6); complex cell wall polysaccharides are assembled in the Golgi apparatus; and proteins destined for cell walls, vacuoles, and various types of storage organelles are modified by glycosylation (addition of carbohydrate) and cleavage during their passage through the endomembrane system.

Proteins that enter the endomembrane system, either in transit or as enzymes and chaperones destined to function within the system, are translated on cytoplasmic ribosomes that become associated with the cytosolic face of the ER. Regions of endoplasmic reticulum with associated ribosomes are known as "rough" ER, and those with no ribosomes as "smooth" ER (Figure 3–41). The ratio of rough to smooth ER varies with cell type and with developmental phase in a single cell. Rough ER is particularly abundant in cells secreting large amounts of proteins or synthesizing storage proteins destined for vacuoles (see Section 4.8).

Proteins destined for the endoplasmic reticulum have a **signal peptide** at the N-terminus. As the newly synthesized N-terminus protrudes from the ribosome, it is recognized by

Figure 3–41.
Rough and smooth endoplasmic reticulum. The electron micrographs show (A) rough ER in a tapetal cell, the cell type that provides nutrition for developing pollen grains in the anther, and (B) smooth ER in a cell of a farina gland on the surface of sepals of *Auricula* (Primulaceae). Farina glands secrete protective compounds including flavonoids onto the surface of the sepal, where they crystallize to form a flour-like coating called farina. (A, courtesy of M.W. Steer; B, courtesy of B.E.S. Gunning.)

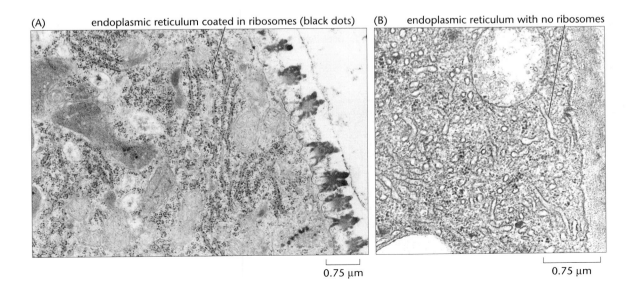

(A) endoplasmic reticulum coated in ribosomes (black dots)

(B) endoplasmic reticulum with no ribosomes

0.75 μm

0.75 μm

and binds to a signal recognition particle (a complex of RNA and protein analogous to the SRP of the chloroplast, described above) (Figure 3–42). The SRP stops further synthesis of the protein, and targets the ribosome and the nascent protein to the ER membrane by binding to a membrane surface receptor. Synthesis of the protein resumes, and as it proceeds, nascent protein is translocated across the membrane into the ER lumen via a hydrophilic pore, or **translocon**. A protease on the inner face of the ER membrane cleaves the N-terminal signal peptide of the translocated protein. This cleavage is a necessary step for translocation; mutations that affect the cleavage site of the signal peptide so that it is no longer recognized by the protease disrupt protein transport.

Within the ER lumen, several mechanisms are responsible for the correct folding and assembly of imported proteins. These mechanisms are essential for further transport of the proteins through the endomembrane system: incorrectly folded proteins are not transported and are targeted for degradation.

First, partially folded proteins are bound by chaperones that stabilize their structure and allow full folding to occur. One of the major chaperones of the ER is BiP, a chaperone of the HSP70 class. Second, newly imported proteins are glycosylated by the addition of branched oligosaccharides (small sugar polymers) to specific asparagine residues of the protein, in a process called **N-glycosylation**. The oligosaccharides contain N-acetyl-glucosamine, mannose, and glucose residues (Figure 3–43). Removal of the glucose residues (de-glucosylation) allows the protein to bind the chaperones calnexin and calreticulin, which promote correct folding. If correct folding does not occur, the attached oligosaccharide may be re-glucosylated and de-glucosylated again, allowing a further opportunity for chaperone binding and correct folding.

Finally, in the ER, some classes of protein require the formation of specific intramolecular disulfide bonds between cysteine residues for correct folding. These include many of the important storage proteins of seeds such as the **glutenins** of wheat and the **vicilins** and **legumins** of peas and beans (see Section 4.8). Storage proteins accumulate in large amounts during seed development and are degraded during germination to provide amino acids to the seedling during the early stages of growth. The ER lumen is a relatively oxidizing environment that favors the **oxidation** of cysteine sulfhydryl groups to form disulfide bonds. Formation of the appropriate disulfide bonds for correct protein folding is promoted by protein disulfide isomerase, an enzyme that breaks and re-forms disulfide bonds until the correct conformation is achieved.

Figure 3–42.
Import of proteins into the lumen of the endoplasmic reticulum. (1) Synthesis begins on a cytosolic ribosome. (2) The signal peptide of the nascent protein is recognized by the signal recognition particle (SRP), which forms a complex with the ribosome and nascent protein and halts protein synthesis. (3) The SRP recognizes and binds to a receptor that is part of the translocation apparatus on the ER membrane. (4) The SRP dissociates; protein synthesis now resumes, with the nascent protein directed through the pore (translocon) into the ER lumen. (5) The signal peptide is cleaved by a protease and the mature protein is released into the lumen.

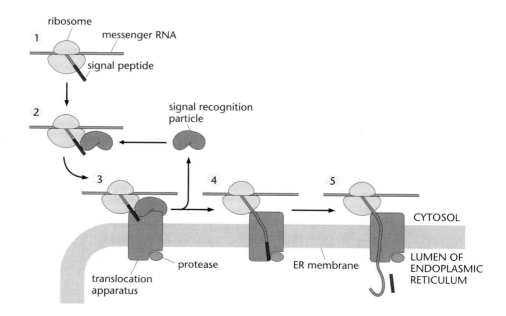

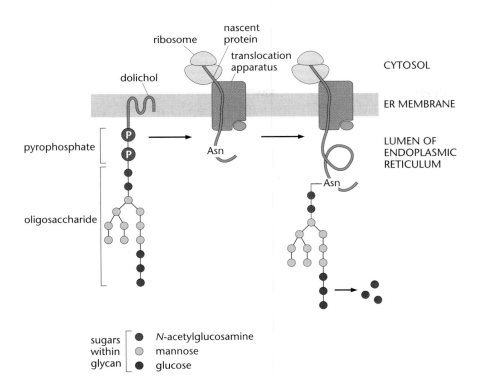

sugars within glycan
- N-acetylglucosamine
- mannose
- glucose

Figure 3–43.
Glycosylation of proteins as they enter the lumen of the endoplasmic reticulum.
The oligosaccharide (glycan) is synthesized by transfer of sugars from sugar nucleotides (UDP-acetylglucosamine, UDP-mannose, and UDP-glucose) to a chain constructed on the pyrophosphate group of dolichol pyrophosphate. Dolichol is a hydrophobic, isoprenoid (terpenoid) compound found in the membranes of the endomembrane system (see Section 4.7). When complete, the oligosaccharide is transferred to specific asparagine (Asn) residues in proteins entering the ER lumen. This process is called *N*-glycosylation because the glycan is linked to the protein via the amide nitrogen of an asparagine residue. Removal of the terminal glucose residues of the oligosaccharide allows the protein to associate with chaperones in the ER lumen.

Proteins that are subsequently secreted at the cell surface or targeted to the vacuole first pass from the ER to the Golgi apparatus (Figure 3–44). We described in Section 3.2 how polysaccharides destined for the cell plate move from the ER to the Golgi apparatus as they are elaborated, and then move in vesicles from the Golgi to the membrane

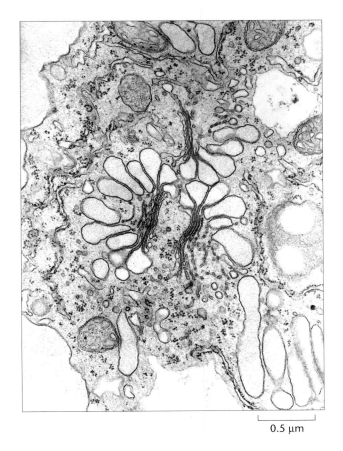

0.5 µm

Figure 3–44.
Golgi apparatus. Electron micrograph showing two Golgi stacks in a root-cap cell of maize. These large Golgi stacks and the associated secretory vesicles are involved in producing the mucilage secreted from the root cap as it penetrates the soil.
(Courtesy of Chris Hawes.)

surrounding the cell plate. Many of the proteins destined for the vacuole and the cell surface follow a similar path. Once the protein has entered the Golgi apparatus, sugar groups can be added to or removed from the *N*-linked glycan side chains that were attached to asparagine residues in the ER (Figure 3–45). A further glycosylation process, **O-glycosylation**, may also take place: in the *cis* Golgi, glycans (usually consisting of arabinose and galactose residues) are linked to hydroxyl groups on some amino acids, including serine, threonine, and hydroxyproline.

The final destination of a protein is determined by specific secondary structural motifs that act as targeting signals, as noted above. Proteins with these motifs are transported to vacuoles; proteins without the motifs are secreted across the plasma membrane to the outside of the cell (for example, the cell wall proteins described in Section 3.4).

The route taken by proteins destined for vacuoles varies from one type of protein to another (Figure 3–46). We describe below the development and functions of the large central vacuole that develops in most cells as they mature (see also Section 3.5). Proteins destined for this vacuole (known as the "lytic vacuole") bud off from the *trans* Golgi network in vesicles that are coated with the protein **clathrin**, and fuse to form a prevacuolar compartment that later fuses with the vacuole. These proteins include proteases, **lipases**, and **nucleases**. In some storage organs, particularly developing seeds, storage proteins accumulate in a second type of vacuole known as a "protein-storage vacuole." These vacuoles differ from the lytic vacuole not only in the protein composition of the vacuolar contents but also in having a neutral rather than an acidic internal environment, and in having different types of proteins in their vacuolar membrane (**tonoplast**). Proteins destined for protein-storage vacuoles bud off from the Golgi apparatus in dense vesicles that are not coated with clathrin. These fuse directly with the tonoplast of the protein-storage vacuole. Other proteins are stored in **protein bodies** that bud off directly from the ER, bypassing the Golgi apparatus. The synthesis of storage proteins is discussed further in Section 4.8.

Figure 3–45.
Protein glycosylation in the Golgi apparatus. (A) Some examples of modifications to the *N*-glycan added in the ER lumen (see Figure 3–43). (B) An example of *O*-glycosylation in a portion of the protein extensin. This cell wall protein contains many repeats of the amino acid sequence Hyp-Hyp-Hyp-Hyp-Ser. In the Golgi apparatus, single sugars or short oligosaccharides are added to the hydroxyl groups of hydroxyproline (Hyp) and serine (Ser) residues—hence the term *O*-glycosylation. Typically, a single galactose residue is added to serine, and one to four arabinose residues are added to hydroxyproline.

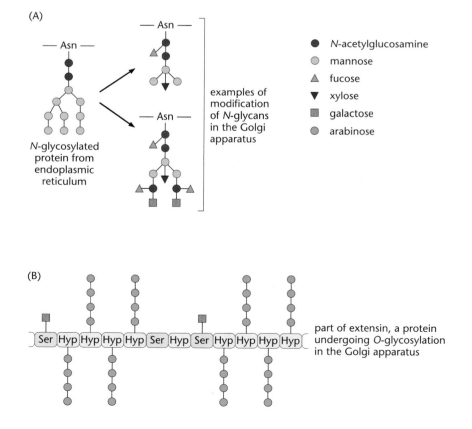

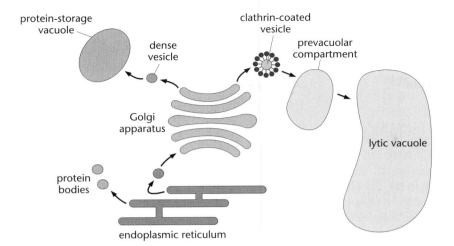

Figure 3–46.
Routes of protein trafficking to vacuoles.
Proteins are moved by various routes from the endoplasmic reticulum and the Golgi apparatus to lytic and protein-storage vacuoles and protein bodies. Proteins destined for a lytic vacuole pass through the ER and Golgi apparatus to a prevacuolar compartment via clathrin-coated vesicles. Prevacuolar compartments fuse to form the vacuole. Storage proteins either pass through the ER and Golgi apparatus to protein-storage vacuoles via dense vesicles, or leave the ER in protein bodies that bud off directly from the ER membrane.

Organelles move around in the cell on actin filaments

Organelles are not randomly distributed within a plant cell, but are held in fixed positions or moved around in a coordinated manner through association with the cytoskeleton. In many plant cells, organelles and cytoplasm undergo bulk movement around the cell known as cytoplasmic streaming. Specific organellar movements also occur during development and in response to environmental changes. Positioning and movement are brought about by **myosin** motor proteins on the surfaces of the organelles, which interact with a network of actin cables composed of bundles of actin filaments. The interaction of myosin motors with actin filaments is also responsible for the movement of vesicles delivering components from the Golgi apparatus to the cell plate during cell division (see Section 3.2) and to the growing points of tip-growing cells such as **pollen tubes** and **root hairs** (see Section 3.5). The direction of movement of organelles and vesicles along actin cables is determined by the polarity of the filaments. Like microtubules, actin filaments have plus and minus ends. Myosin proteins generally move toward the minus end of an actin filament.

Cytoplasmic streaming is easily observed in the large cells of some green algae and in root hairs. In the giant internodal cells of algae such as *Chara* and *Nitella*, the chloroplasts are contained in cortical cytoplasm (located at the edge of the cell near the plasma membrane), which surrounds a zone of cytoplasm rich in actin cables along which organelles in the central cytoplasm migrate (Figure 3–47).

In leaf cells, chloroplasts move in response to changes in light intensity and quality (see Section 7.1). In low light, chloroplasts are held by the cytoskeleton along the sides of

Figure 3–47.
Cytoplasmic streaming in giant algal cells. (A) *Nitella hyalina.* Each branch of this freshwater alga is a single cell. The terminal cells are about 1 to 2 mm long.
(B) Longitudinal view of part of a giant algal cell. Chloroplasts are anchored in the stationary, cortical cytoplasm adjacent to the plasma membrane. Inside this layer, the remainder of the cytoplasm, containing mitochondria and the nucleus, streams around in the cell. The movement is brought about by interaction of cables of actin filaments with myosin motors located on the surfaces of the organelles. (Courtesy of Charles Delwiche.)

(A)

(B)

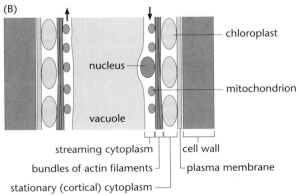

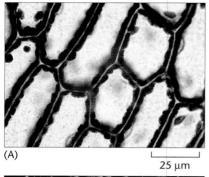

(A)

25 µm

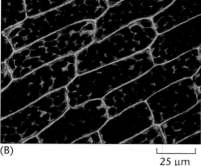

(B)

25 µm

Figure 3–48.
Chloroplast reorientation in leaf cells in response to changing light levels.
Electron micrographs showing surface views of a leaf of the moss *Physcomitrella patens* in two different light conditions. (A) Chloroplasts are located down the sides of the cell under high light levels. (B) Chloroplasts are at the upper surface under low light. Chloroplasts appear brown because they were stained with iodine, which binds to starch granules in the stroma. (Courtesy of Matilda Crumpton-Taylor.)

the cell that are perpendicular to incident light, maximizing light absorption. High light levels cause chloroplasts to move along the actin cables to the sides of the cell parallel with the incident light, reducing the percentage of incident light absorbed and thus reducing the likelihood of damage to the photosynthetic apparatus by excess energy (Figure 3–48).

3.4 PRIMARY CELL WALL

The primary cell wall consists of microfibrils of **cellulose** embedded in a matrix that consists mainly of pectins and cross-linking glycans (Figure 3–49). The matrix is laid down in the cell plate (see Section 3.2), and the cellulose microfibrils are synthesized after the plate has reached the sides of the cell. Cross-linking glycans form hydrogen bonds with the cellulose microfibrils, and together the glycans and microfibrils form the main structural framework of the wall, which is embedded in a gel-like network composed of pectins. The cell wall is not a static structure. After cell division, the wall continues to grow as the cell expands. This process is made possible by continuing introduction of new polysaccharide materials and enzymes, secreted into the wall by fusion of vesicles with the plasma membrane (the process of **exocytosis**). The enzymes break and synthesize bonds between sugars or **methylate** sugar groups. These changes can profoundly affect the physical properties of the wall.

When cells reach their final size, many lay down a secondary cell wall within the primary wall as part of the differentiation process. The nature and synthesis of the secondary cell wall is considered in Section 3.6. Here we describe the structure and properties of the primary wall and the way in which it grows as the cell expands.

Figure 3–49.
The primary cell wall. (A) Electron micrograph showing the middle lamella separating the primary cell walls of adjacent leaf cells. (B) Scanning electron micrograph of deep-etched cell wall preparation from elongating carrot cells, showing the network of cellulose microfibrils. (C) Diagram showing arrangement of the three main polysaccharide components of the wall. Cellulose microfibrils are linked by hydrogen bonding of the cross-linking glycans, within a network of pectins. Junction zones in the pectin matrix are described later in the text. (B, with permission from the Company of Biologists.)

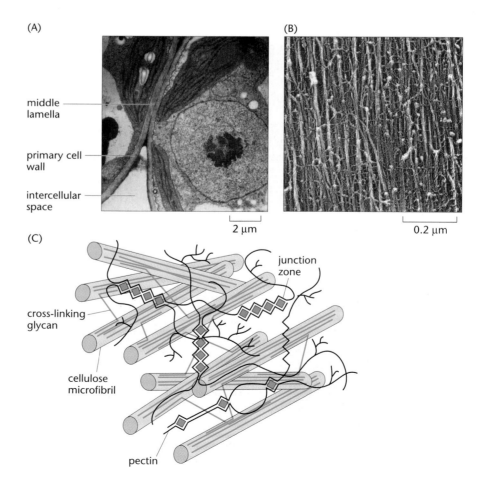

(A)

middle lamella

primary cell wall

intercellular space

2 µm

(B)

0.2 µm

(C)

cross-linking glycan

cellulose microfibril

junction zone

pectin

The matrix of the cell wall consists of pectins and hemicelluloses

As described earlier in the chapter, the major components of the cell wall matrix—pectins and cross-linking glycans—are synthesized in the Golgi apparatus and transported to the cell plate in vesicles. Both classes of polysaccharides are complex and show considerable variation among species and cell types.

Pectins are a heterogeneous group of linear and branched sugar polymers, rich in galacturonic acid, but frequently containing several other sugars (Figure 3–50). They are

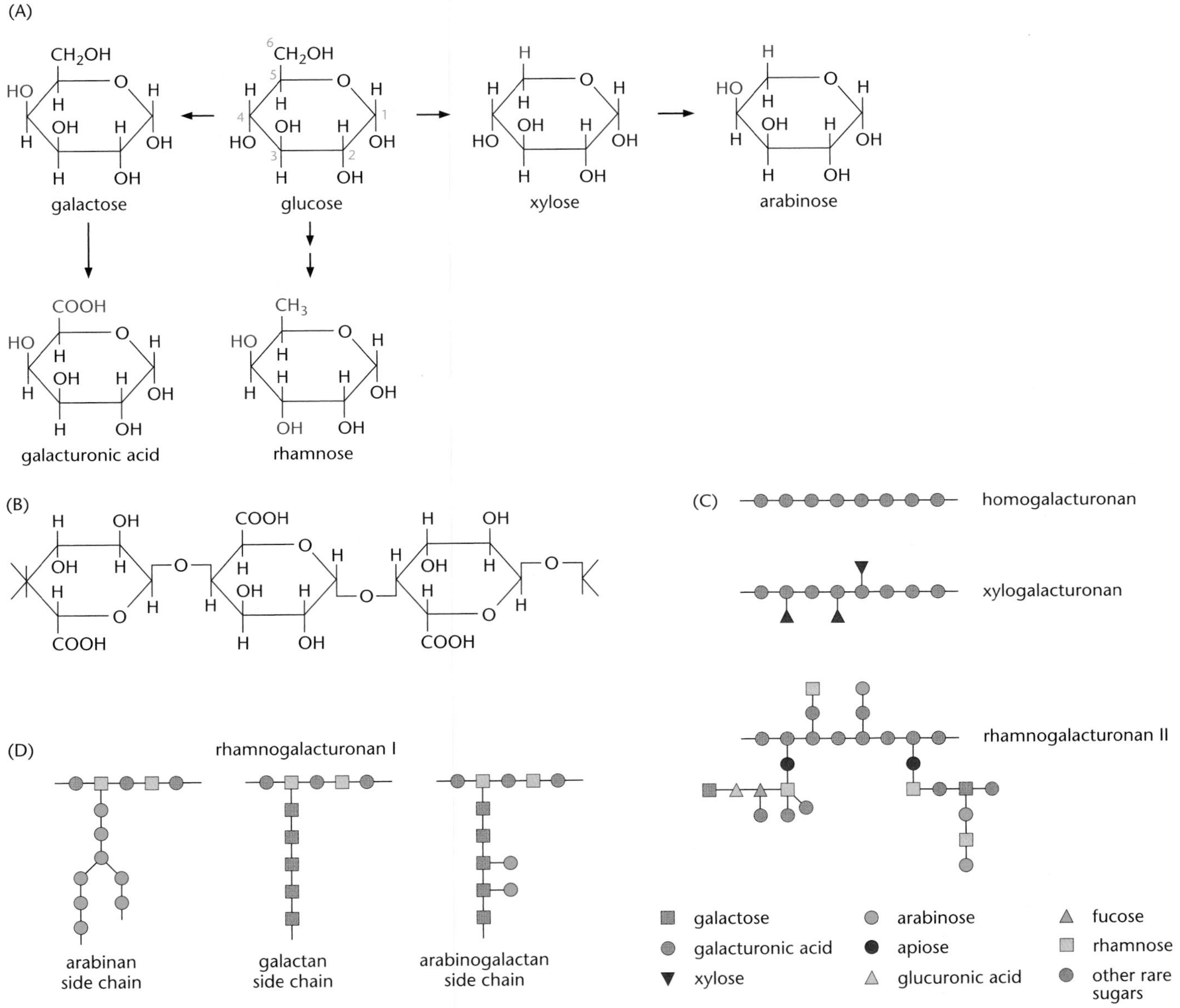

Figure 3–50.

Components and structure of pectins. (A) Some of the major sugars found in pectins and cross-linking glycans. All are derived from glucose, for which the numbering system of the carbon atoms is shown. The positions at which the sugars differ from glucose are indicated (*red*). The enzymes that interconvert the sugars use sugar nucleotides, not free sugars. For example, UDP-glucose is acted on by the enzyme UDP-glucose epimerase to produce UDP-galactose. This is further acted on by UDP-galactose dehydrogenase to

produce UDP-glucuronic acid. Formation of the various sugar nucleotides from UDP-glucose is thought to take place in the endomembrane system. Polymerization of the sugars to form pectins and cross-linking glycans is catalyzed by glycosyltransferase enzymes in the Golgi apparatus, which transfer the sugar from the sugar nucleotide to the growing polymer. (B) The polygalacturonic acid backbone of pectins. (C) The homogalacturonan class of pectins. (D) The rhamnogalacturonan I class of pectins.

classified into two main groups according to the composition of their backbones, and further classified by the structure of their side chains. Pectins of the **homogalacturonan** class have a backbone of α1,4-linked galacturonic acid. Homogalacturonan itself has no side chains. Xylogalacturonan has single xylose side chains on about half of the backbone residues. **Rhamnogalacturonan** II is a complex, branched structure in which the galacturonic acid backbone is substituted with four distinct types of side chains. Pectins of the rhamnogalacturonan I class have a backbone of alternating galacturonic acid and rhamnose residues. Side chains attached to some of the rhamnose residues consist of linear chains of galactose, branched chains of arabinose, or linear chains of galactose with arabinose side branches.

Cross-linking glycans are also introduced into the primary wall in the cell plate. There are three major classes of these polymers, with different distributions among higher plants (Figure 3–51). **Xyloglucans** consist of a β1,4-linked glucose backbone with single xylose side chains. The precise structure varies from one species to another; for example, the number of side chains varies, and xylose may be substituted with another sugar such as fucose or arabinose. Xyloglucans are found in most **eudicots**—where they are the main form of cross-linking glycan—and in about half of the **monocots**. **Glucuronoarabinoxylans** consist of a xylose backbone with side branches containing glucuronic acid and arabinose. They are found in all higher plants, and in monocots— including grasses, bromeliads (e.g., pineapple), gingers, and palms—they are often the main form of cross-linking glycan. **Mixed-linkage glucans**, found in the Poales (all grasses and related families), are composed of β1,3- and β1,4-linked glucose residues.

In addition to the pectins and cross-linking glycans, the matrix of the cell plate also contains a β1,3-linked glucose polymer called **callose**. This is synthesized within the cell plate rather than imported in vesicles. It accumulates during the second stage of cell plate formation, when the vesicles form a tubular network (see Figure 3–23). At the end of cytokinesis the callose gradually disappears, and most mature cell walls lack this polymer altogether. A notable exception is the pollen tube, which contains callose in its cell walls and forms plugs of callose along its length (see Sections 5.2 and 5.6).

Cellulose is synthesized at the cell surface after the cell plate has formed

Cellulose begins to accumulate in the primary cell wall after the cell plate has reached the sides of the cell and cytokinesis is complete. By the time the primary wall is mature (i.e., cell expansion has ceased), cellulose accounts for 15 to 30% of wall mass in most cell types. It is organized in microfibrils that wrap around the cell, like thread around a spool.

Cellulose is synthesized and assembled at the surface of the plasma membrane by the **cellulose synthase complex**, a multimeric (multisubunit) enzyme known as a **terminal complex**. Terminal complexes are organized into rosettes, consisting of six hexagonally arranged complexes (Figure 3–52A); cellulose microfibrils are extruded from the cellulose synthase complexes into the cell wall. Cellulose is the most abundant biopolymer on earth, but the structure and function of the cellulose synthase complex is not yet understood. It has proved extremely difficult to study, because the process cannot

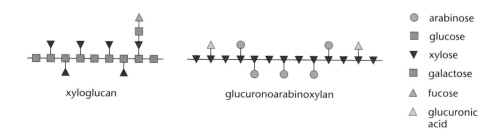

Figure 3–51.
Cross-linking glycans. Some common repeat units of xyloglucans and glucuronoarabinoxylans.

xyloglucan   glucuronoarabinoxylan

- arabinose
- glucose
- xylose
- galactose
- fucose
- glucuronic acid

be completely reconstituted outside a living plant. Disruption of the plant cells results in loss of cellulose-synthesizing capacity, probably because essential associations between proteins within the enzyme complex are lost. According to current models, each terminal complex in **angiosperms** synthesizes 3 or 6 cellulose polymers in parallel; thus each rosette synthesizes 18 or 36 cellulose polymers (Figure 3–52C). The polymers are glucans consisting of β1,4-linked glucose residues. The glucans are synthesized from a sugar **nucleotide, UDP-glucose,** which is derived from **sucrose** by the action of the enzyme **sucrose synthase**. It is thought that sucrose synthase is associated with the terminal complexes on the inner face of the plasma membrane (Figure 3–52D; see also Section 4.5). The parallel glucan chains emerging from each rosette become organized into a crystalline rod or microfibril as they enter the wall matrix.

The cellulose synthases of higher plants are encoded by small **multigene families**. In Arabidopsis, three different cellulose synthases seem to be necessary for normal primary

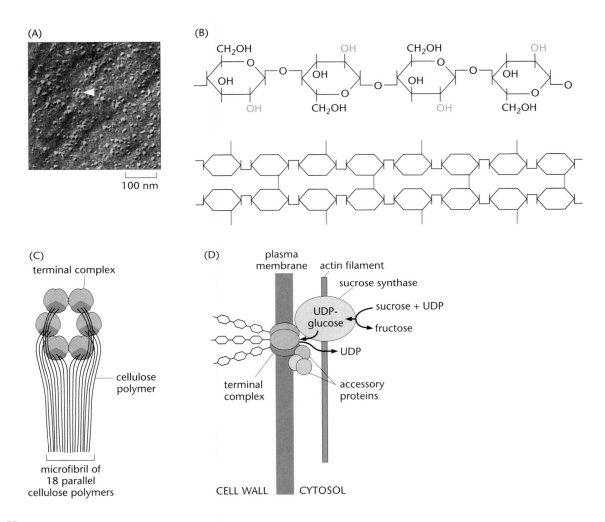

Figure 3–52.
Cellulose synthesis. (A) Electron micrograph showing rosettes (example arrowed) consisting of cellulose synthase terminal complexes on the outer face of the plasma membrane of a maize root cell. (B) Structure of cellulose. Given the nature of the bonding between glucose residues, adjacent chains can form large numbers of interchain hydrogen bonds between hydroxyl groups (*blue*) that project from both sides of the chains. (C) Hypothetical structure of a rosette, seen from outside the cell. Each terminal complex may contain three different types of cellulose synthase proteins. Eighteen cellulose molecules are shown; some research indicates that each rosette may produce 36 molecules. (D) Hypothetical structure of part of the cellulose synthase complex. Sucrose synthase, anchored to actin filaments adjacent to the plasma membrane, provides UDP-glucose from sucrose to the terminal complex, which consists of three different types of cellulose synthase and unknown accessory proteins. Cellulose polymers emerge on the outer face of the plasma membrane. This scheme is highly speculative. (A, from T. Arioli et al., *Science* 279(5351):717-720, 1998. With permission from AAAS.)

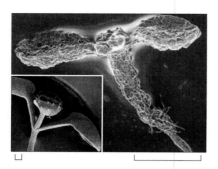

Figure 3–53.
An Arabidopsis *root swelling* (*rsw*) mutant. The mutant reveals the impact of the loss of a β1,4-glucanase necessary for cellulose synthesis; the inset shows a wild-type plant. The seedling of the *rsw* mutant, which fails to form cellulose microfibrils correctly, shows swelling and bulging of cells. The scale bars represent the same distance on the two seedlings. (From T. Arioli et al., *Science* 279(5351):717-720, 1998. With permission from AAAS.)

cell wall synthesis. Plants carrying mutations in any of these three genes exhibit stunted growth, such as *root swelling1* mutants. In addition to cellulose synthase, other enzymes are thought to be necessary for normal cellulose synthesis. For example, Arabidopsis plants that lack a plasma-membrane–associated **endoglucanase** (an enzyme that cleaves β1,4 linkages) have reduced cellulose synthesis and stunted growth; these include *korrigan* and *root swelling2* mutants (Figure 3–53). The function of this enzyme in cellulose synthesis is unclear. It might release completed chains from the terminal rosettes or degrade chains that fail to associate correctly to form microfibrils.

Carbohydrate components of the cell wall interact to form a strong and flexible structure

The carbohydrate components of the cell wall are assembled to form a structure that is sufficiently strong to withstand the outward pressure of the cell (**turgor** pressure; see Section 3.5) yet sufficiently flexible to allow the cell to expand. As we described above (see Figure 3–49), the wall consists of a framework of cellulose and cross-linking glycans embedded in a network of pectins. Covalent and noncovalent interactions between these components, plus the actions of enzymes and other proteins in the wall (also described above), provide flexibility and the potential for expansion. Here we describe major features of these interactions and their effects on the properties of the cell wall.

Cellulose microfibrils have a tensile strength equivalent to that of steel. They provide much of the mechanical strength of the cell wall, resisting turgor pressure and supporting the plant. The orientation of microfibrils in the primary wall determines the direction in which the cell can expand (see Section 3.5). Where microfibrils are laid down in a crisscrossing network, cells tend to expand equally in all directions. Where microfibrils are laid down parallel to each other and arranged in hoops around the cell, the cell expands along the axis perpendicular to the hoops and becomes elongated or cylindrical (Figure 3–54). Cells of mutant plants with decreased amounts of cellulose often fail to expand appropriately; for example, *rsw* mutants have short, swollen cells in their roots and hypocotyls (see Figure 3–53).

Figure 3–54.
Expansion of cell walls. Where microfibrils crisscross (*top*), cell expansion occurs equally in all directions. Where microfibrils are arranged in hoops (*bottom*), expansion is restricted to elongation along the axis allowed by this microfibril arrangement.

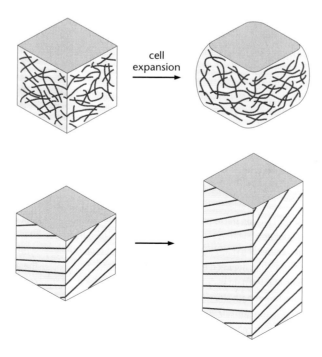

The arrangement of cellulose synthase rosettes in the plasma membrane determines the positioning of cellulose microfibrils in the cell wall, and hence the direction and extent of cell expansion. The arrangement of rosettes is in turn closely linked to the orientation of the cytoskeleton in the cell cortex, immediately inside the plasma membrane. In many types of expanding cells, the cortical microtubules are co-aligned with the cellulose microfibrils on the outside of the plasma membrane. This suggests that microtubules are involved in the orientation of new cellulose microfibrils deposited in the wall. The relationship between the orientation of cortical microtubules and the direction of cell expansion is discussed in Section 3.5.

The mechanism by which cellulose microfibrils and cortical microtubules are aligned is not yet clear. It has been proposed that the rosettes are connected via transmembrane bridges to the microtubules in the cell cortex. According to this model, these bridging proteins move along microtubules carrying the rosettes in the direction determined by the microtubules (Figure 3–55).

The cross-linking glycans provide flexible links between cellulose microfibrils. The glycan chains either span the distance between microfibrils, hydrogen-bonding to a microfibril at either end, or form a link between a microfibril and a framework of other glycan chains that link to further microfibrils. This cross-linking of the long, strong cellulose microfibrils adds to the strength of the wall.

Pectins interact to form an amorphous network in which the other components of the wall are embedded. One of the main ways in which the network is formed is via cross-linking with calcium ions (Ca^{2+}). Homogalacturonan chains are negatively charged and bind strongly to each other through Ca^{2+} cross-bridges (Figure 3–56A). The frequency of formation of these strengthening bridges, called "junction zones," changes during cell growth. Many of the sugar residues of homogalacturonans are modified by the addition of methyl groups when the polymer is secreted into the wall. This reduces their negative charge and prevents formation of Ca^{2+} bridges, rendering the pectin network extensible. When cell expansion has stopped, the methyl groups are enzymically cleaved from the homogalacturonan by pectin methyl esterases, enzymes in the cell wall; this allows Ca^{2+}-mediated cross-linking and thus strengthening and loss of flexibility in the wall. The structure of the complex pectin rhamnogalacturonan II (see Figure 3–50C) is very highly conserved across both monocot and eudicot species, hence it probably plays a specific and important role in determining cell wall structure. Rhamnogalacturonan II molecules form dimers cross-linked by boron (as borate esters), and these are believed to add strength to the wall matrix. This idea is supported by studies of an Arabidopsis mutant, *mur1*, that cannot synthesize fucose in most root cells.

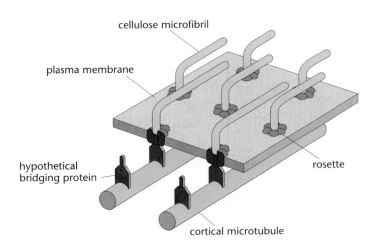

Figure 3–55.
Hypothetical model for mechanism of alignment of cellulose microfibrils and cortical microtubules. Cut-away section of the cell surface shows proposed attachment of cellulose-synthesizing rosettes in the plasma membrane to microtubules in the underlying cortex, via hypothetical bridging proteins. Parallel microfibrils are typically about 30 nm apart.

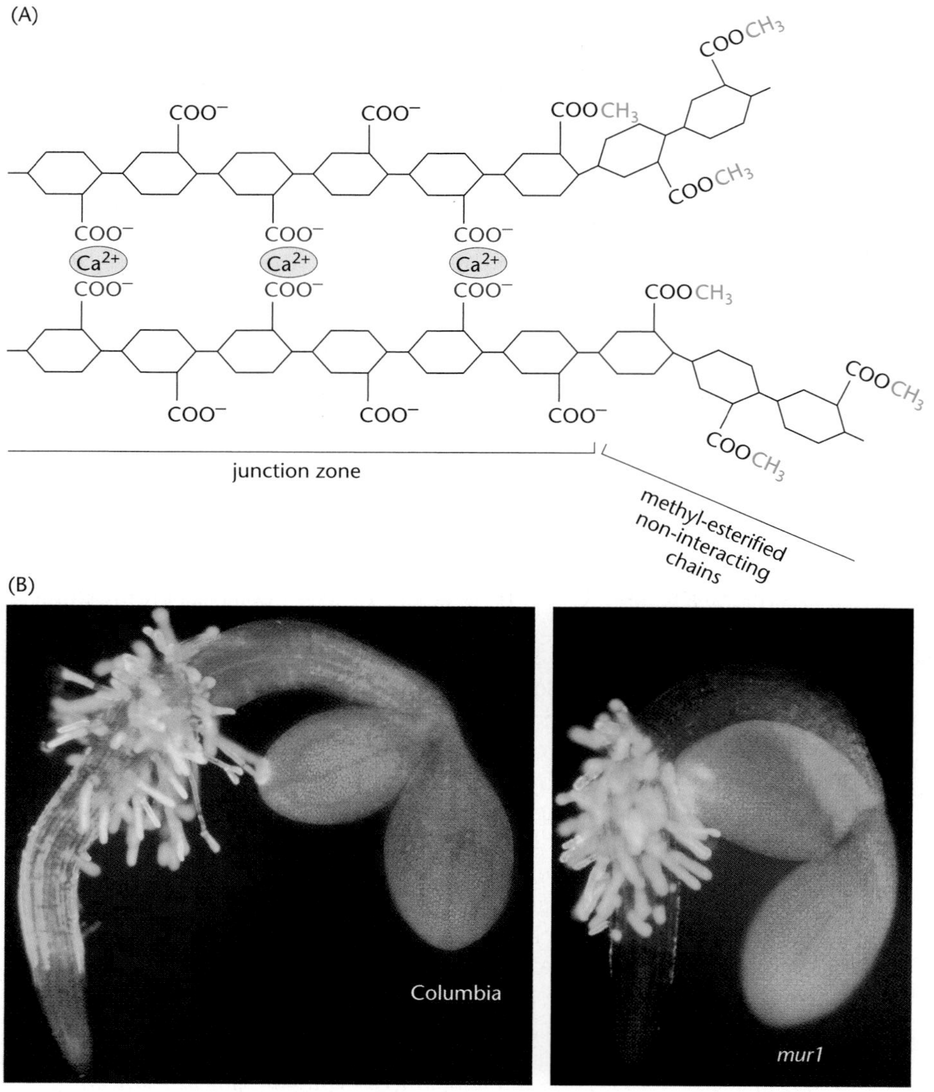

Figure 3–56.
Formation of pectin networks.
(A) Formation of junction zones between stretches of homogalacturonan molecules that are not methylated; methylated regions cannot interact in this way. (B) Seedlings of wild-type (Columbia variety) Arabidopsis (*left*) and the *mur1* mutant (*right*). The seedlings were treated so that the fucose of cell wall polymers is fluorescent. In the wild-type seedling, fucose is present in cell walls of both the root hairs and the body of the root. In the *mur1* seedling, fucose is present in the cell walls of root hairs but absent from the body of the root. As a consequence, the root cells of the *mur1* mutant are smaller than those of the wild type and root extension is severely restricted. (Courtesy of Michael Hahn and Glenn Freshour.)

(A)

junction zone

methyl-esterified non-interacting chains

(B)

Columbia

mur1

The consequent disruption of the structure of rhamnogalacturonan II reduces the degree of boron cross-linking in the wall, and this drastically affects cell expansion (Figure 3–56B).

The nature of the pectin polysaccharides regulates the porosity of the cell wall. A uniform gel of homogalacturonan contains few pores because of the regular cross-linking of the polymers. Branched pectin polymers, such as rhamnogalacturonan I and II, obstruct the formation of junction zones, resulting in the formation of pores. Typical pores in the pectin matrix are about 5 to 7 nm in diameter. Differences in wall porosity among species and changes in porosity during cell expansion are attributable to different relative amounts of homogalacturonan, rhamnogalacturonan, and Ca^{2+} cross-bridging in the pectin network.

Glycoproteins and enzymes have important functions in the cell wall

In addition to its polysaccharide components, the plant cell wall also contains proteins that play structural roles and enzymes that metabolize the carbohydrate polymers to allow cell expansion.

Most of the structural proteins have large polysaccharide side chains, and the protein may be only a small component (as little as 5%) of the whole molecule. Like the polysaccharide components of the wall matrix, these glycoproteins are synthesized in the endomembrane system and transferred to the wall via vesicles. Both the proteins and their carbohydrate side chains are extremely variable: the nature of the cell wall glycoproteins differs from one type of tissue to another within the plant. The functions of these glycoproteins are not fully understood, but they may be involved in the cross-linking of carbohydrates in the wall. The structure of one glycoprotein, **extensin**, is shown in Figure 3–45B. Other classes of structural protein include the arabinogalactan proteins, in which large, highly branched galactan chains carrying arabinose residues are attached to the protein, and glycine-rich proteins, in which up to 25% of the protein consists of glycine. Glycine-rich proteins have "pleated sheet" structures and are thought to form plates of protein at the plasma membrane–cell wall interface.

To accommodate cell expansion and elongation, cell walls must be dynamic structures capable of extension. This is achieved in part by the extensibility of the matrix itself, and in part by the actions of enzymes in the cell wall that loosen the matrix by cleaving specific bonds in the cell wall polysaccharides, allowing the wall to extend and new polysaccharides to be inserted as the cell grows. The xyloglucan endotransglycosylases, for example, break and re-form xyloglucan chains in the wall during cell elongation. Their action may allow transient slippage of cellulose microfibrils as the cell expands. Another important enzyme family is the **expansins**, which break the hydrogen bonds between cellulose microfibrils and xyloglucans, thus allowing the cell wall to expand. We mentioned above the importance of the pectin methyl esterases in influencing wall strength and extensibility by allowing formation of junction zones.

Plasmodesmata form channels between cells

The cell plate is not a continuous barrier between the two daughter cells. As noted in Section 3.2, membrane-lined channels are formed in the developing wall as it is laid down around strands of endoplasmic reticulum running between the daughter cells. When the cell plate fuses with the existing cell wall, the membrane lining the channels becomes contiguous with the plasma membrane, providing a passage through which molecules can move directly from the cytoplasm of one cell to the cytoplasm of another. The strand of ER and the gap between it and the membrane lining the channel together make up a **plasmodesma** (plural, **plasmodesmata**; Figure 3–57). The central portion of the plasmodesma—the strand of ER—is called the **desmotubule**, and the surrounding cylinder of cytoplasm is the **cytoplasmic sleeve**. These connections between cells

Figure 3–57.
Plasmodesmata. The diagram shows the structure of a single plasmodesma. The nature of the particles lining the cytoplasmic sleeve, the "spokes" that connect them, and the rodlike structure inside the desmotubule is not known. The electron micrographs show a cross section of plasmodesmata in a cell wall of a leaf of sugarcane (*left*) and a face view of a cell wall showing plasmodesmatal pores (*right*).

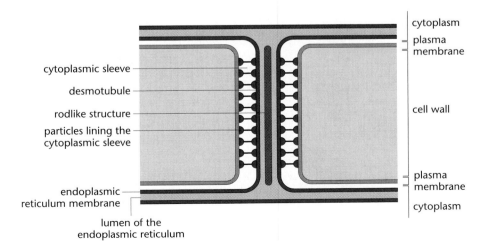

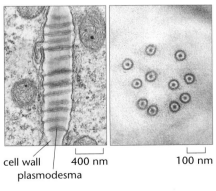

are vitally important to cell and tissue function, but they are extremely difficult to isolate and study. The nature of the proteins that make up the structures inside plasmodesmata is not known, and the way in which molecules move through plasmodesmata is also poorly understood.

Most of the plasmodesmatal movements that are understood involve movement of molecules through the cytoplasmic sleeve. For example, sucrose produced during photosynthesis in leaf mesophyll cells passes from cell to cell toward the **phloem** through plasmodesmatal cytoplasmic sleeves. After transport in the phloem to nonphotosynthetic parts of the plant, sucrose leaves the phloem and enters surrounding cells, again through plasmodesmatal cytoplasmic sleeves. (These movements are described in Section 4.4.)

The cytoplasmic sleeve is not freely permeable to all molecules in the cytoplasm. By injecting individual cells with fluorescently tagged molecules of known size, researchers can ascertain the maximum mass of molecules that can move freely through plasmodesmata. For most plasmodesmata, this mass is about 1 kDa (kilodalton): this is called the **size exclusion limit** of the plasmodesmata.

Proteins exceed the size exclusion limit of normal plasmodesmata, so cannot move between cells. However, some transcription factors (proteins that coordinate the expression of genes involved in plant development) are able to move from cell to cell through the cytoplasmic sleeve. The mechanism that allows these proteins to pass when most are excluded is not known. An example is the transcription factor KNOTTED, which initiates and maintains **shoot meristems** (see Section 5.4). In the maize shoot meristem, KNOTTED protein is found in all cells, but no corresponding mRNA is found in the outermost layer of meristem cells (Figure 3–58). This is because the gene is transcribed and translated in the inner meristem cells, producing a protein (the transcription factor) that then moves to the outer cells through plasmodesmata.

For those proteins that can move through plasmodesmata, both the distance (number of cells crossed) and the direction of movement are regulated. Another transcription factor, SHORT ROOT, regulates cell differentiation in Arabidopsis roots and can pass through plasmodesmata in just one direction. The *SHORT ROOT* gene is transcribed and translated in **endodermal cells** of the root (see Section 5.2), and the protein then moves radially outward to cells in the neighboring root **cortex** layer. SHORT ROOT moves only to cells immediately adjacent to the cell in which it is made (that is, it

Figure 3–58.
Movement of KNOTTED1 (KN1) protein through plasmodesmata in developing shoot meristems of maize. Sections through three spikelet meristems on the flank of a maize ear inflorescence. The mRNA for the transcription factor KN1 is present in most cells (*left*), but completely absent from the outer layer of cells (*arrow*). The KN1 protein is present in the nuclei of all cells (*right*), including those in the outermost layer (*arrow*). (A and B, with permission of the Company of Biologists.)

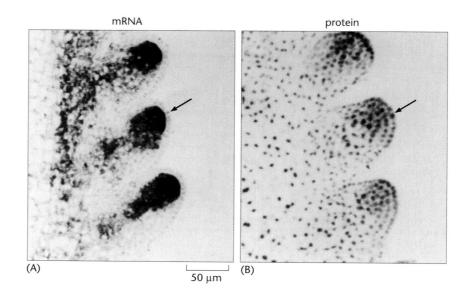

mRNA   protein

(A)   (B)
50 µm

moves a distance of only a single cell), and moves only in an outward direction across the root. This has been established by experiments with **transgenic** Arabidopsis plants in which the *SHORT ROOT* gene is expressed in the cortex rather than the endodermal cells. In these plants, too, the protein moves outward by one cell layer from the cell in which it is made.

When a plant is infected with viruses, the size exclusion limit of the plasmodesmata may increase. The increase is caused by virus-encoded **movement proteins** that facilitate the movement through plasmodesmata of viral **nucleic acid** or entire viral particles, thus spreading the infection from cell to cell (see Section 8.3). Some movement proteins interact with viral nucleic acid molecules, enabling them to take up a conformation in which they can pass through plasmodesmata; these movement proteins do not modify the plasmodesmata themselves. Other movement proteins do directly modify the host plasmodesmata, replacing the normal structure with tubules that extend through the cell wall and permit the movement of viral particles and other large molecules from cell to cell. The exact means by which these tubules are formed is not understood, and it is very likely that the mechanism differs from one type of virus to another.

There are two classes of plasmodesmata in plants. "Primary plasmodesmata" form in the cell plate during cytokinesis, as we have already described. "Secondary plasmodesmata" form between cells during cell expansion and at maturity, rather than during cytokinesis. Unlike primary plasmodesmata, they are often branched. In the early stages of formation of secondary plasmodesmata, the ER associates with the plasma membrane in the region of the cell where a plasmodesma will form. The cell wall disassembles at this location, and ER on either side fuses to form the desmotubule. The cell wall then re-forms around the plasmodesma as it matures.

Some cell types do not contain plasmodesmata, so they are cytoplasmically isolated from all other cells in the plant. A striking example is the **guard cells** of the stomata. **Stomata** are the pores on the leaf surface that control the exchange of water and carbon dioxide between the internal spaces of the leaf and the outside **atmosphere**. The pore is formed by two flanking guard cells, which change in volume to bring about the opening and closure of the pore. These volume changes occur rapidly in response to environmental conditions. When conditions are favorable for photosynthesis, the guard cells take up solutes and water from the surrounding intercellular space; as a result, their turgor increases (see Section 3.5) and the cells swell. Because of differential thickening of the cell walls, swelling opens the stomatal pore (Figure 3–59). The volume changes of guard cells, and hence the opening of the pore, can occur rapidly (within a few minutes). These large, rapid changes in cell volume would not be possible if guard cells were connected to their neighbors by plasmodesmata: solutes and water taken up by the cells could move out via the plasmodesmata, rather than bringing about a rapid increase in internal pressure. (The crucial role of stomata in gas exchange between leaves and the atmosphere is described in later chapters.)

(A) radially arranged cellulose microfibrils

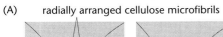
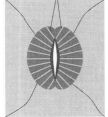
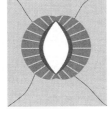

closed stoma   open stoma

(B)

Figure 3–59.
Role of changes in guard cell volume.
(A) The radial arrangement of microfibrils in the walls of guard cells allows the opening of the pore as the cells expand. The guard cells lack plasmodesmata through which solutes and water could escape. (B) Surface of a leaf of Arabidopsis, showing two stomata.

3.5 CELL EXPANSION AND CELL SHAPE

After a plant cell has divided and the daughter cells have laid down their primary cell walls, cells undergo expansion to acquire their mature size and shape. Cell expansion involves the movement of solutes and water into the cell, accompanied by loosening and insertion of new material in specific regions of the cell wall. The movement of solutes and water depends on the properties of the plasma membrane, and we begin by considering the role of the plasma membrane in this process. We then describe how the internal pressure—turgor pressure—generated by entry of solutes and water into the cell, and in particular into the vacuole, acts to extend the cell wall, and how the direction of cell expansion is determined by orientation of the cortical cytoskeleton. Specialized cases of cell expansion—the stomatal guard cell and the tip-growing root hair—further illustrate these important concepts.

The properties of the plasma membrane determine the composition of the cell and mediate its interactions with the environment

The plasma membrane is the defining boundary of the cell: it maintains an internal environment in which the cell's metabolic activities can take place (Figure 3–60). It is selectively permeable and thus controls the movement of water, inorganic solutes, and **metabolites** in and out of the cell, allowing large differences to be maintained between the inside and outside of the cell in chemical composition, pH, **osmotic potential**, and electrical potential. The plasma membrane is also the site of detection of external signals that coordinate the activities of different cells and enable the plant to respond to its changing environment. These signals are often detected at the membrane surface by protein **receptors** located in the **lipid bilayer** of the membrane. The synthesis of the cell wall is also coordinated at the plasma membrane and, as described above (Section 3.4), complexes that synthesize cellulose microfibrils are embedded in the membrane.

Proton transport across the plasma membrane generates electrical and proton gradients that drive other transport processes

The maintenance of differences between the inside and outside of a cell usually requires transport of molecules and ions across the plasma membrane *against* concentration and electrical potential gradients. This sort of transport—**active transport**—requires

Figure 3–60.
The plasma membrane controls the movement of molecules into and out of the cell. Differences between the inside and outside of the cell—that is, gradients across the membrane—in osmotic potential, electrical potential, and pH are maintained by movement of specific solutes across the membrane. Molecules secreted from the cell across the plasma membrane include components of the cell wall, defense compounds that deter or disable potential pathogens (see Chapter 8), and molecules that facilitate a plant's uptake of mineral nutrients (chelators; see Section 4.9). External signals perceived at the plasma membrane include phytohormones and molecules secreted by pathogens (see Chapters 6 and 8). Signals are perceived by receptor proteins (*green circle*) embedded in the lipid bilayer. Small molecules cross the membrane in part by diffusion through the lipid bilayer but primarily via specific channel and transporter proteins embedded in the bilayer. Large molecules (such as proteins and carbohydrate polymers) usually cross the membrane by fusion of vesicles with the membrane and release of their cargo to the outside of the cell (see Section 3.4).

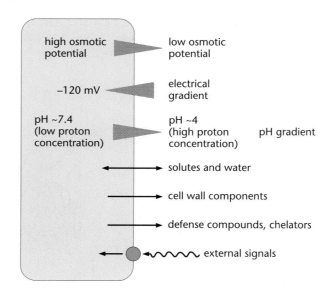

energy, either chemical or electrical energy, generated by specific types of **transport proteins** located in the membrane.

Some small, uncharged molecules can permeate the lipid bilayer of the plasma membrane relatively freely by simple diffusion (e.g., oxygen, carbon dioxide, water). In contrast, larger substances and ions can enter and leave the cell only via highly specific transport proteins. Some substances use only one type of transport protein, whereas others may have several different routes across the plasma membrane (e.g., potassium, as described below).

Movement of many kinds of substances across the plasma membrane is directly or indirectly dependent on a class of transport proteins termed **ion pumps**, which use energy from the hydrolysis of ATP. **Proton pumps**, also called proton transporters, are by far the most important pumps in plant plasma membranes. They are proteins that span the membrane. They hydrolyze the energy-rich terminal phosphate bond of ATP to produce ADP and phosphate on the inner face of the membrane, and use the energy released to transport H^+ ions from the inside to the outside of the cell (Figure 3–61). The action of proton pumps maintains a higher pH (i.e., lower H^+ concentration) inside the cell than outside and—most important—it generates an electrical potential across the membrane. Because positively charged ions are being moved out, the inside of the membrane has a more negative charge than the outside. The **membrane potential** of a plant cell is typically −60 to −240 mV. The activity of proton pumps is controlled by factors outside and inside the cell. For example, cellular **metabolism** usually produces net amounts of protons, causing acidification of the cytosol. This activates the proton pumps on the plasma membrane, which export the excess protons and thus maintain a stable intracellular pH and membrane potential. Auxin (a phytohormone) promotes cell elongation. One of the ways it does this is by activation of the plasma membrane proton pumps. This causes acidification of the cell wall, and this is thought to activate cell-wall–loosening enzymes (see Section 3.4), allowing the cell to expand.

The proton gradient and the membrane potential provide the driving force for the movement of many other charged and uncharged substances across the plasma membrane (Figure 3–62). Two sorts of transport proteins are involved in these movements: transporters that utilize the H^+ gradient to move another substance, and channels that allow passive flux of ions across the membrane down the electrical potential gradient.

Proton flux down the concentration gradient (from outside to inside) through membrane-spanning transporters provides the energy to drive the inward or outward movement of another substance. For example, potassium, phosphate, and nitrate ions all move into cells with the proton flux, catalyzed by specific transporters (known as **symporters**). Sodium ions move out of the cell against the proton flux on a specific transporter (known

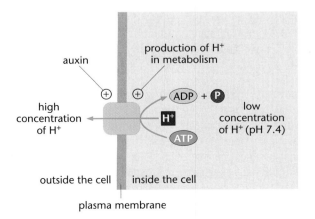

Figure 3–61.
Action and activation of the plasma membrane proton pump. The pump exports protons from the cell, using energy provided by hydrolysis of ATP. Removal of protons from the cytosol maintains a relatively high cytosolic pH and a negative charge on the inside of the membrane; it also maintains an acidic environment in the cell wall. The pump is activated by accumulation of protons (i.e., a fall in pH) in the cytosol and by the phytohormone auxin.

Figure 3–62.
Transporters and channels allowing movement of inorganic ions across the plasma membrane. The proton pump (*top*) generates a gradient of proton concentration and an electrical potential across the membrane (see also Figure 3–61). Specific transporter proteins provide a route for the passive flux of protons back into the cell down the concentration gradient, and couple this proton flux to transport of ions into or out of the cell. Channel proteins form pores that allow flux of ions into the cell in response to the electrical potential across the membrane.

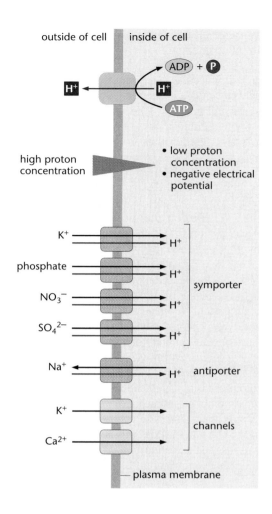

as an **antiporter**). In plant cells, potassium is maintained at a high level and sodium at a low level; this contrasts with animal cells, in which sodium level is kept high and potassium level low.

Channels are transmembrane proteins that act as selective pores for particular ions: in particular, potassium, calcium, and chloride (see Figure 3–62). Most channels open only in response to changes in electrical potential across the membrane—the so-called **voltage-gated channels**. For example, AKT1, a member of the Arabidopsis potassium channel family, is activated by a high membrane potential (hyperpolarization) and is the major route by which potassium from the soil enters root cells. Although plasma membranes also have potassium transporters, as noted above, these cannot completely compensate for loss of AKT1 in Arabidopsis mutants. Plants that lack the AKT1 protein are stunted because cell expansion is restricted by low concentrations of intracellular solutes (as discussed below).

The types of transporters and channels in the plasma membrane vary considerably from one cell type to another, and with environmental conditions. For example, the presence of nitrate in the soil induces the expression of genes encoding **nitrate transporters** in the plasma membranes of root cells (see Section 4.8). In contrast, lack of potassium in the soil induces the expression of a gene encoding one particular potassium transporter. Functionally specialized cells such as stomatal guard cells (see below) and **salt glands** (see Section 7.4) have specific transporters and channels associated with their unique functions.

Movement of water across the plasma membrane is facilitated by aquaporins

In plants, movement of water into cells is driven by the difference in solute concentration between the inside and outside of the cell. Water moves by **osmosis** from the solute-poor environment outside the cell to the solute-rich interior (as discussed below). Water crosses the plasma membrane both by diffusion through the lipid bilayer and through water-selective pores formed by membrane proteins called **aquaporins** (Figure 3–63). Both types of route are "passive," meaning that neither requires energy.

The rate of water movement across the membrane is determined in part by physical differences between conditions inside and outside the cell (**water potential** and **hydrostatic pressure**; see Figure 3–65). However, large and rapid changes in the permeability of the plasma membrane to water can also be brought about by changes in the number or activity of aquaporins. Expression of some aquaporin genes changes in response to external conditions that affect the water status of the cell, such as high levels of salt and low water availability in the environment (see Section 7.3). These conditions increase the abundance of specific aquaporins, increasing the permeability of the plasma membrane to water and thus facilitating the inward movement of any water available outside the cell. Some phytohormones that promote cell expansion also increase the activity of aquaporins and thus the rate at which water enters the cell.

The aquaporins are encoded by a large gene family, members of which have very different patterns of expression throughout the plant and during plant development. For example, some aquaporin genes are expressed specifically when cells are expanding and rates of water uptake are high. Others are expressed during rehydration of dry seeds before germination.

Aquaporins allow the very high fluxes of water across the membranes of specialized cell types that swell and contract rapidly in response to environmental stimuli. These include stomatal guard cells and cells associated with the rapid movements of whole plant organs. For example, closure of the trap of the Venus flytrap (Figure 3–64), closure and drooping of leaves of the sensitive plant *Mimosa pudica*, and the springing apart of **stamens** of *Sparmannia* in response to touch are all brought about by sudden changes in turgor in "motor cells" at the base of these organs, involving very rapid efflux of water via aquaporins. Touch stimuli to the leaves or stamens trigger collapse of membrane potential in the motor cells (depolarization); potassium and chloride ions rapidly leave the cell, and loss of water follows. Aquaporins facilitate a rapid outward

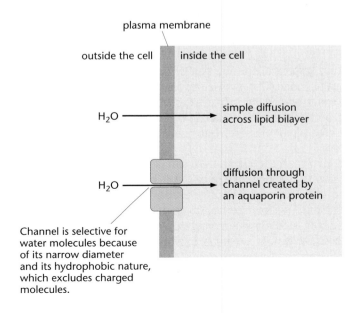

Figure 3–63.
Movement of water across the plasma membrane. There are two ways in which water crosses the plasma membrane: simple diffusion, and diffusion through aquaporin channels.

plasma membrane

outside the cell   inside the cell

H_2O ——→ simple diffusion across lipid bilayer

H_2O ——→ diffusion through channel created by an aquaporin protein

Channel is selective for water molecules because of its narrow diameter and its hydrophobic nature, which excludes charged molecules.

Figure 3–64.
Closure of a Venus flytrap (*Dionaea muscipula*). When hairs on the inner surface of the trap (a modified leaf) are stimulated, such as by contact with an insect, the trap snaps shut. The precise mechanism is not understood, but it is likely to involve a dramatic change in the turgor of mesophyll cells that underlie the inner epidermis of the trap. The snapping shut takes less than 0.5 seconds. A slower, continued change in turgor then brings the sides of the trap closer together so that the prey is brought in contact with glands that secrete enzymes capable of digesting it. Flytraps are found in soils with very low nitrogen content. The plant obtains nitrogen from amino acids released by digestion of the proteins of its insect prey.

movement of water, so that cell turgor drops dramatically within seconds. The thin-walled motor cells collapse, resulting in rapid movement of the whole organ.

Movement of solutes into the cell vacuole drives cell expansion

The increase in volume of a plant cell during cell expansion is largely due to the uptake of water into its central vacuole. Before describing the coordinated events that promote water uptake into the vacuole and allow expansion, we need to explain more fully the concept of cell turgor, and this is most easily done at the whole-cell level.

As described above, the proton gradient and electrical potential across the plasma membrane drive the accumulation of solutes such as potassium inside the cell, so that the concentration of solutes inside the cell is greater than that in the extracellular space (Figure 3–65). Water moves into the cell by osmosis, down this gradient of solute concentration. Osmosis results in a net transfer of water from the low-solute (high water potential) environment to the high-solute (low water potential) environment. The contents of the cell tend to swell as the amount of water inside increases. The tendency to swell is counterbalanced by the resistance of the cell wall, leading to a buildup of pressure in the cell. It is this hydrostatic pressure that is referred to as "turgor" pressure. As turgor pressure builds up, the net movement of water into the cell slows and eventually stops.

Turgor pressure is responsible for the rigidity of nonwoody plant parts. The cell walls alone cannot maintain the stiffness of leaves and stems: this stiffness is a function of cell turgor. When water availability in the external environment drops (in other words, water potential outside the cell is reduced), cells can lose water across the plasma membrane and thus lose turgor pressure. This results in wilting or drooping of plant organs.

Cell expansion involves the integration of osmotically driven water uptake (resulting in turgor pressure) with the loosening of cell wall structure and addition of new matrix material to the wall (Figure 3–66). The actions of the cell-wall–loosening enzymes described in Section 3.4 allow the wall to stretch under the influence of turgor pressure. Stretching reduces the pressure, and this in turn allows more water to flow into the cell by osmosis. New matrix material must be delivered to the cell wall to maintain its thickness as stretching and expansion occur. This material is delivered in vesicles from the endomembrane system where it is synthesized, as described earlier for the cell plate (see Section 3.2).

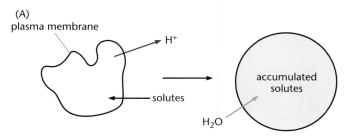

Import of solutes into the protoplast decreases internal water potential.

Water enters down the gradient of water potential; the protoplast swells and becomes spherical.

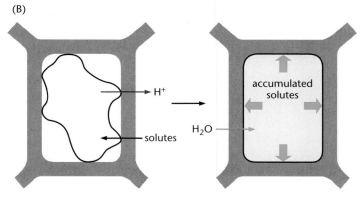

Protoplast is not exerting pressure on the cell wall, so there is no turgor pressure; the rigidity of the cell is determined by the properties of the cell wall alone.

Presence of the cell wall resists the swelling of the protoplast, leading to buildup of turgor pressure and increased rigidity of the cell.

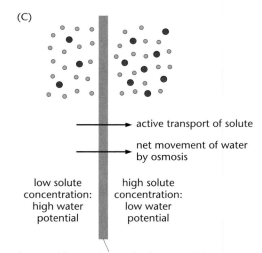

active transport of solute

net movement of water by osmosis

low solute concentration: high water potential

high solute concentration: low water potential

semipermeable membrane, freely permeable to water but not to solute

Figure 3–65.
Turgor pressure, water potential, and osmosis. Diagrams illustrating the same sequence of events (A) in a protoplast (a cell without a cell wall) and (B) in an intact cell (protoplast surrounded by a cell wall). (A) Solutes enter the protoplast via transporters and channels, driven primarily by the export of protons by the proton pump. The accumulation of solutes inside the protoplast reduces its water potential, causing the inward movement of water by osmosis. In the absence of a cell wall, the protoplast swells like a balloon, forming a spherical shape. (B) In the intact cell, when the protoplast is not exerting pressure on the wall (*left*), it is said to be "plasmolyzed." In these conditions the turgor pressure of the cell is zero. When solutes enter the cell, reducing its water potential and causing the inward movement of water by osmosis, the protoplast starts to swell (*right*). Its expansion is resisted by the cell wall, leading to buildup of turgor pressure. (C) The principle of water movement by osmosis. A net movement of water occurs across a selectively permeable membrane from a solution of high water potential on one side to a solution of low water potential on the other.

Returning now to the role of the vacuole: most of the water in a plant cell is contained in the central vacuole. During cell expansion it is the vacuole rather than the surrounding cytoplasm that increases in volume. In many mature plant cells, the central vacuole occupies most of the volume of the cell. Water movement across the tonoplast (vacuolar membrane) occurs by the same routes as water movement across the plasma membrane: diffusion across the lipid bilayer and flow through pores formed by aquaporins. These pores are extremely abundant in the tonoplast. Indeed, one of the most abundant **mRNA transcripts** in Arabidopsis leaf cells is that encoding the tonoplast aquaporin α-TIP.

Figure 3–66.
Cell expansion driven by loosening of the cell wall and turgor pressure. In a nonexpanding cell (*upper left*), turgor pressure is high and net inward movement of water by osmosis is prevented. As cell expansion starts (*upper right*), cell-wall–loosening enzymes break cross-linking bonds in the cell wall, increasing its plasticity. The turgor pressure stretches the loosened cell wall (*lower left*), causing thinning of the wall and cell expansion. This reduces the turgor pressure. More water now flows into the cell (*lower right*), and turgor pressure increases again. The thickness of the wall is regained by the addition of new material.

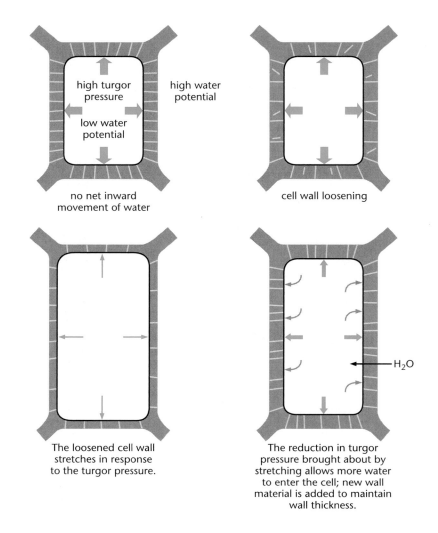

Water movement across the tonoplast into the vacuole is driven by the accumulation of solutes in the vacuole, which provides a difference in water potential between the cytosol and the vacuole. Solutes are actively transported from the cytosol into the vacuole, driven by a proton gradient and an electrical potential set up by the actions of two types of proton pumps (Figure 3–67). One of these uses energy from the hydrolysis of ATP, and the other uses energy from the hydrolysis of **inorganic pyrophosphate**. As with the plasma membrane, movement of solutes occurs in two ways: anions enter via channels in response to the electrical potential across the tonoplast, and cations and neutral compounds such as sugars enter via transporters, coupled to the proton gradient. During cell expansion, the activity of tonoplast proton pumps causes a large net movement of solutes into the vacuole, thus driving water movement from the cytoplasm into the vacuole.

The importance of solute transport into the vacuole for cell expansion is illustrated by the *de-etiolated3* (*det3*) mutant of Arabidopsis, which has reduced amounts of one protein subunit of a major proton pump in the tonoplast. The hypocotyl of the mutant plant fails to elongate because reduced proton transport across the tonoplast restricts solute, and hence water, movement into the vacuole and thus restricts movement of water into the cell as a whole, inhibiting cell expansion (Figure 3–68).

The vacuole acts as a storage and sequestration compartment

In addition to its role in maintaining cell turgor, the vacuole sequesters substances that could potentially inhibit metabolic processes elsewhere in the cell. Included among

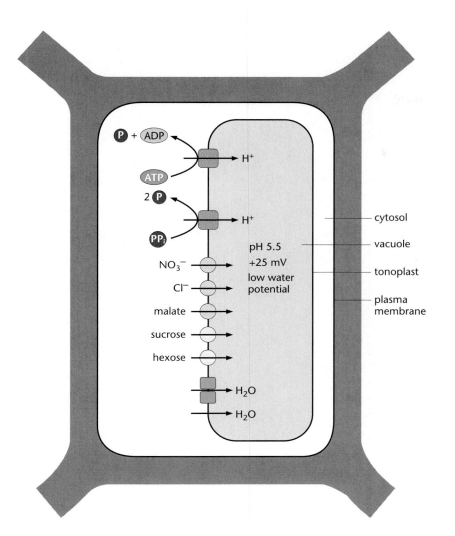

Figure 3–67.
Movement of solutes into the vacuole.
Solute movement is driven by tonoplast proton pumps, which use the energy of hydrolysis of ATP or inorganic pyrophosphate (PP_i) to move protons against a concentration gradient from the cytosol into the vacuole. This sets up a proton gradient and a membrane potential across the tonoplast. Thus the contents of the vacuole are typically more acidic than the cytosol and have a more positive charge. Anions such as nitrate and chloride (both imported across the plasma membrane) and malate (synthesized in the cytosol) move from the cytosol into the vacuole in response to the electrochemical gradient. Uncharged molecules such as sucrose and hexoses (glucose and fructose) enter via transporters (antiporters) that use the passive flux of protons from the vacuole into the cytosol down the proton concentration gradient. The accumulation of solutes in the vacuole lowers its water potential, promoting the entry of water via aquaporins and by passive diffusion through the lipid bilayer.

these are "waste products" of cellular metabolism (e.g., breakdown products of **chlorophyll** degradation), toxic compounds that have entered the cell from the environment (e.g., herbicide-derived compounds in herbicide-tolerant plants; see Section 9.3), and defensive substances that deter or poison herbivores when released as the plant is eaten (e.g., **cyanogenic glycosides** and **glucosinolates**; see Section 8.4). The vacuole also contains enzymes capable of degrading macromolecules, including proteases, lipases, and nucleases, which degrade proteins, lipids, and nucleic acids, respectively (hence the name "lytic vacuole," as noted above). These enzymes probably break down cellular components during **senescence** (see Section 5.4), but it is unclear whether they degrade cellular components at other times in the life of the cell. The vacuolar enzymes may also form part of the plant's defense mechanism, rendering tissues less palatable to herbivores.

Colored compounds (pigments) found in petals as attractants for pollinators also accumulate in the central vacuole, particularly the **anthocyanins**. The palatability of fruits is often determined by sugars and acids contained in vacuoles. For example, the huge vacuoles of the thin-walled juice sac cells of citrus fruits contain citric acid and sugar (Figure 3–69), and vacuoles in the **parenchyma** of apple fruits contain high concentrations of malic acid. Some of these vacuoles can accumulate particularly high concentrations of protons, reducing the pH to as low as 2 and driving the accumulation of high concentrations of organic anions (such as citrate and malate) from the cytosol, where they are synthesized.

ABC transporters (**A**TP **b**inding **c**assette transporters) are pumps involved in the transport of many of these compounds across the tonoplast. They use energy from the

Figure 3–68.
The *det3* mutant of Arabidopsis.
(A) Wild-type and (B) *det3* mutant seedlings grown in the dark. The elongation of the hypocotyl (etiolation) that occurs in the wild type is severely restricted in the mutant. (C) Adult *det3* mutant and (D) adult wild-type plants. The reduced cell size in the *det3* mutant leads to a dwarf phenotype. (Courtesy of Karin Schumacher.)

(A)

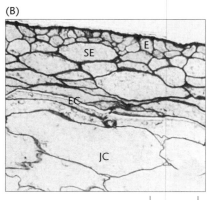

(B)

10 μm

(C)

hydrolysis of ATP to pump organic molecules across the membrane. ABC transporters are encoded by a multigene family, and it is likely that each member of the ABC family transports a different compound or class of compounds. They are best studied in the context of herbicide tolerance and some defense compounds. For example, in plants tolerant to thiocarbamate herbicides, the herbicide is detoxified by conjugation to the small peptide **glutathione**, catalyzed by **glutathione-*S*-transferases**. This takes place in the cytosol, and the conjugate is transported into the vacuole via an ABC transporter.

Coordinated ion transport and water movement drive stomatal opening

The stomatal guard cell provides a good example of the way in which the transport of solutes and water across membranes is coordinated. As described in Section 3.4, the opening and closure of stomatal pores on the leaf surface is controlled by volume changes of the two guard cells that flank the pore. When their turgor pressure is high, the guard cells swell and the pore between them opens; when pressure decreases, the guard cells shrink and the pore closes (Figure 3–70). The size of the guard cells is regulated by the

Figure 3–69.
Some storage functions of vacuoles. (A) Juice sacs of the citrus fruit pummelo (*Citrus grandis*). The sacs are about 2 cm long. (B) Cross section of part of a juice sac, showing the outer layers of epidermal (E), subepidermal (SE), and elongated (EC) cells surrounding an inner tissue of large, thin-walled juice cells (JC). The vacuoles of juice cells occupy most of the cell volume and contain high concentrations of sugars and citric acid. (C) Flowers of *Antirrhinum*. The magenta color of the petals (*left*) is due to flavonoid pigments (in this case, cyanidin 3-rutinoside) stored in the vacuoles of epidermal cells The white *nivea* mutant (*right*) cannot make flavonoids because it lacks the enzyme chalcone synthase, which catalyzes the first step in the biosynthetic pathway.

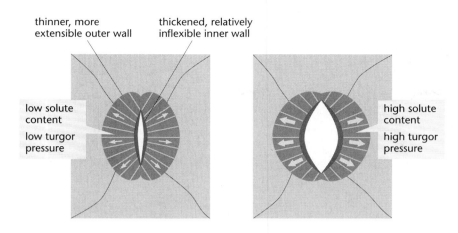

thinner, more
extensible outer wall

thickened, relatively
inflexible inner wall

low solute
content

low turgor
pressure

high solute
content

high turgor
pressure

Figure 3–70.
Opening and closing of a stoma.
The stoma is closed when the guard cells have a low turgor pressure. When turgor pressure increases, due to the net influx of solutes followed by water, the guard cells swell. The positioning of cellulose microfibrils and the differential thickening of the walls cause the cells to bulge outward, opening the stoma.

movement of water and solutes into and out of the cell. In most plants, stomata open in the morning in response to blue light (see Section 6.2) and close at night.

Opening is dependent on a blue-light–activated proton pump that creates a large electrical potential across the plasma membrane (Figure 3–71A). This negative potential activates a specific class of potassium channels, known as "voltage-gated inward rectifying K^+ channels," that allow potassium ions to move into the cell. The entry of potassium is accompanied by the movement of water, leading to an increase in turgor pressure in the guard cell. This causes guard cells to change shape so that the stoma opens. In the dark, deactivation of the proton pump leads to membrane depolarization (a reduction in electrical potential across the membrane), which inactivates the potassium channel. The influx of potassium and water into the cell stops, turgor pressure drops, the shape of the guard cells changes, and the stoma closes.

Control of guard cell turgor by light is overridden during conditions when water is scarce, enabling stomata to close in the light to restrict water loss from the leaf (see Section 7.3). Closure is triggered by increased levels of the phytohormone abscisic acid (ABA), which is synthesized in response to low availability of water. The responses of the guard cell to ABA are mediated by several **signaling cascades**: chains of events that link the arrival of a signaling molecule (such as a phytohormone) to a downstream response. The downstream effect of ABA is a depolarization of the guard cell plasma membrane, which allows loss of solutes, followed by loss of water, and hence a drop in guard cell turgor and stomatal closure (Figure 3–71B).

A key step in the signaling cascade that leads to depolarization is an increase in Ca^{2+} ion level in the cytosol of the guard cell. ABA triggers this increase by at least two mechanisms. One of these involves the activation of a **G protein** on the inner face of the plasma membrane. G proteins are involved in many signaling cascades in plant and animal cells. The activated G protein in turn activates **phospholipase C**, an enzyme associated with the membrane, and this cleaves a specific type of lipid in the membrane to release the small, soluble molecule inositol 1,4,5-trisphosphate (IP3). IP3 diffuses through the cytosol to interact with IP3 receptors associated with calcium channels on the ER membrane and the tonoplast. The ER and the vacuole contain much higher Ca^{2+} ion concentrations than the cytosol (millimolar as opposed to nanomolar concentrations). IP3 activates the calcium channels, allowing an efflux of Ca^{2+} into the cytosol and an increase of up to twentyfold in the cytosolic Ca^{2+} concentration (Figure 3–72). A second type of signaling cascade triggered by ABA involves the production of **reactive oxygen species (ROS)** (molecules such as hydrogen peroxide) at the cell surface, which trigger the opening of plasma membrane calcium channels and allow Ca^{2+} ions to enter the cell. Another signaling cascade involves the production of nitric oxide (NO), a molecule that also promotes release of Ca^{2+} ions from organelles into the cytosol.

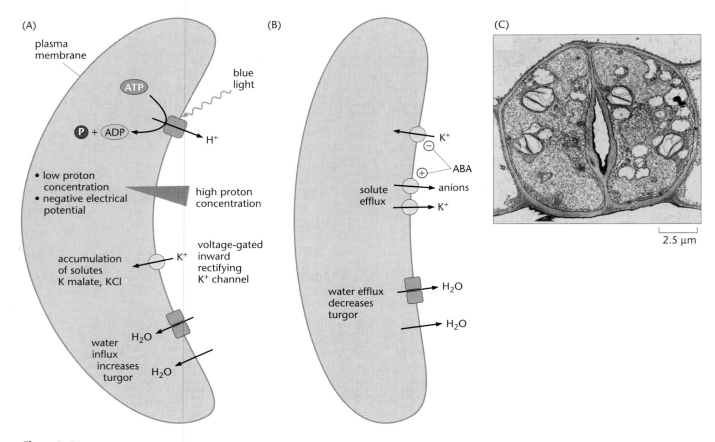

Figure 3–71.
Outline of events in the guard cell leading to light-induced stomatal opening and ABA-induced stomatal closure. (A) Blue light activates a proton pump, and a large membrane potential develops. This activates a potassium channel, which allows an influx of potassium into the cell. The K^+ influx is accompanied by an influx of chloride (Cl^-) ions and the synthesis of malate, an organic anion, in the cytosol. The net increase in solute concentration in the cell drives an influx of water, increasing the turgor pressure of the cell. (B) Abscisic acid (ABA) activates anion channels that allow efflux of anions from the cell. This reduces the membrane potential, causing inhibition of the inward K^+ channel; an outward K^+ channel is activated. The net efflux of solutes drives the outward movement of water, leading to loss of turgor. (C) Electron micrograph of a pair of guard cells in an Arabidopsis leaf. As described earlier in the text, much of the accumulation of solutes in a guard cell occurs in the large vacuole—not shown in (A) and (B). (C, courtesy of Liming Zhao.)

Figure 3–72.
Abscisic acid begins a signaling cascade in guard cells. ABA binds to a receptor on the plasma membrane, which results in activation of a G protein on the inner face of the membrane. This in turn activates phospholipase C, which metabolizes a membrane lipid, phosphatidylinositol bisphosphate (PIP2), to release inositol 1,4,5-trisphosphate (IP3) into the cytosol. IP3 activates calcium channels in the tonoplast (and in the ER membrane, not shown here), and Ca^{2+} ions are released into the cytosol from the vacuole (and from the ER lumen).

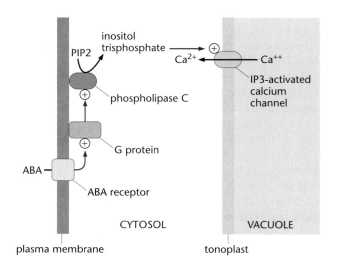

The increase in calcium in the cytosol in response to ABA causes depolarization of the plasma membrane by activating anion channels in the membrane (Figure 3–73). These channels mediate the loss of anions such as chloride and malate from the guard cell. The consequent decrease in membrane potential inactivates the voltage-gated inward rectifying potassium channel and activates an outward potassium channel. There is thus a large net efflux of solutes from the guard cell, increasing its water potential and leading to outward movement of water. The consequent loss of turgor changes the shape of the guard cells so that the stoma closes.

The signaling cascades that promote stomatal opening and closure are similar to those that mediate the effects of signaling molecules in other plant cells. Examples of signaling molecules that trigger effects on development and response to **pathogens** are given in Chapters 6 and 8, respectively. The guard cell is a convenient system in which to study these phenomena because it can be isolated in an active form and its responses—turgor and shape changes—are easily monitored. Several methods have been developed to study intracellular processes in isolated guard cells. An example is the introduction into the cell, by microinjection or by transgenic means, of molecules that fluoresce when bound to calcium, so that the researcher can use fluorescence microscopy to monitor changes in Ca^{2+} ion levels in subcellular compartments in real time.

The direction of cell expansion is determined by microtubules in the cell cortex

Most cells do not expand in all directions at once. The more or less equal growth on all faces of pith cells (cells in the central column of **ground tissue** in stems) leads to the development of isodiametric cell shapes, whereas growth primarily along one axis in root epidermal cells leads to the elongated shape characteristic of these cells (see Figure 3–54). (See Chapter 5 for descriptions of the various cell and tissue types of plants.) Extreme examples of directional growth are root hairs and pollen tubes, where growth is localized to the tips of the cells, resulting in unidirectional growth and the formation of pin-shaped cells (see the discussion of this below). We described above how the positioning of cellulose microfibrils in the cell wall controls the direction in which cells expand, and how the alignment of microfibrils is determined by the alignment of cytoskeletal microtubules in the cell cortex. Here we focus on the relationship between the organization of the cytoskeleton and the direction of cell expansion.

The microtubules in the cell cortex are derived from a population that grows out from the surface of the nucleus after cytokinesis. At the beginning of cell expansion, the cortical microtubules assume a transverse orientation, perpendicular to the direction of cell expansion. The growth phenotypes of Arabidopsis mutants with disoriented microtubules demonstrate the importance of microtubule orientation in determining the direction of cell expansion during plant growth (Figure 3–74). Cells of *katanin* and *mor1* mutants, for example, have cortical microtubules that are randomly rather than transversely oriented, leading to isodiametric rather than longitudinal growth and hence stunted organ development. The katanin protein severs microtubules, thereby generating free ends that rapidly depolymerize. In mutants that lack katanin activity, the microtubules depolymerize less and cannot reorient correctly. The MOR1 (microtubule organization 1) protein stabilizes the microtubule array. In mutants that lack MOR1, microtubules are short and disoriented.

Environmental stimuli affect the orientation of cortical microtubules and hence the direction of cell elongation (Figure 3–75). For example, the direction of cell expansion in the hypocotyls (stems) of seedlings is affected by light. Seedlings grown in the dark grow much taller than when light is supplied. They are said to be "etiolated." **Etiolation** involves elongation of cells in the hypocotyls: under these conditions, microtubules in the cortical arrays are oriented at right angles to the long axis of the stem (transverse organization). If the etiolated seedlings are exposed to blue light, the

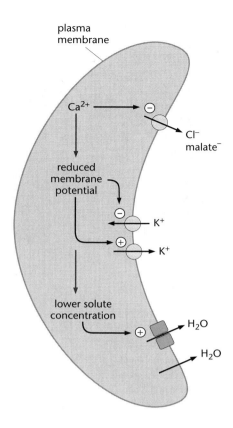

Figure 3–73.
Events triggered by increased calcium in the guard cell cytosol. Calcium activates an anion channel in the plasma membrane, allowing an efflux of anions (such as chloride and malate) from the cell. This leads to a reduction in membrane potential, which inhibits the voltage-gated inward rectifying K^+ channel and activates an outward K^+ channel. The efflux of anions and K^+ from the cell drives a net outward movement of water, resulting in loss of turgor and stomatal closure. (See also Figure 3–71B for a simplified view of the ion and water movements.)

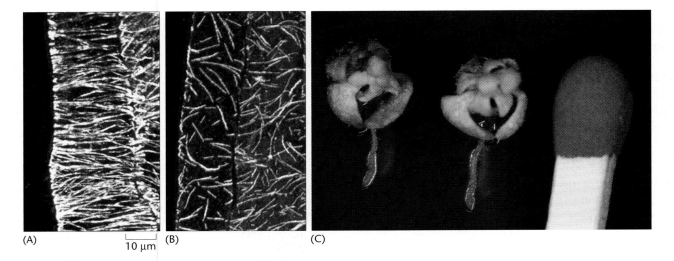

(A)   10 µm   (B)   (C)

Figure 3-74.
Role of microtubules in the direction of cell expansion. (A) Cortical array of microtubules in an epidermal cell of a cotyledon of wild-type Arabidopsis, treated to make the microtubules fluorescent. Note the transverse orientation of the microtubules, perpendicular to the direction of growth. (B) The same cell type in a *mor1* mutant, which lacks a protein essential for the organization of microtubules; the microtubules are short and disoriented. (C) *mor1* mutant plants are severely stunted. Two plants are shown, with the head of a match as an indicator of size. (A and B, with permission of the Company of Biologists; C, from Whittington et al., *Nature*, 411:610–613, 2001. With permission from Macmillan Publishers Ltd, courtesy of Geoffrey Wasteneys.)

transverse microtubules depolymerize and re-form in a longitudinally aligned array. Cells now expand radially and not longitudinally.

The phytohormones **gibberellic acid** and **ethylene** regulate the direction of plant growth by regulating microtubule orientation. Treatment of seedlings of many species of plants with gibberellic acid induces a transverse orientation of microtubules, which promotes longitudinal stem growth: treated seedlings grow taller than untreated seedlings. Ethylene, in contrast, promotes the longitudinal alignment of microtubules and hence the lateral expansion of cells. We provide more information on the importance of these phytohormones in plant development and responses to the environment in Chapters 5 and 6.

Microtubules of the cortical array in expanding cells are bundled together in parallel to form clusters. The spacing between adjacent tubules is constant, like the distance between the two rails of a train track. This bundling is mediated by cross-bridging proteins, the microtubule associated proteins (MAPs). MAPs isolated from plants can

Figure 3–75.
Orientation of microtubules in cell expansion in the hypocotyl of an etiolated seedling. (A) In darkness, the seedling grows tall because of elongation of hypocotyl cells. Microtubules in the cells are oriented transversely, promoting longitudinal expansion. (B) When the etiolated seedling is given light, the microtubules reorganize into a longitudinal array, promoting lateral rather than longitudinal expansion. Upward growth of the seedling stops and the hypocotyl thickens. (C) Reorientation of microtubules in cells of a gametophyte of the fern *Ceratopteris richardii* in response to blue light. After a period of darkness (*top*), microtubules (treated to make them fluorescent) are transversely oriented in the cells. After four hours of blue light (*bottom*), the microtubules are obliquely or longitudinally oriented. (C, courtesy of Takashi Muarata.)

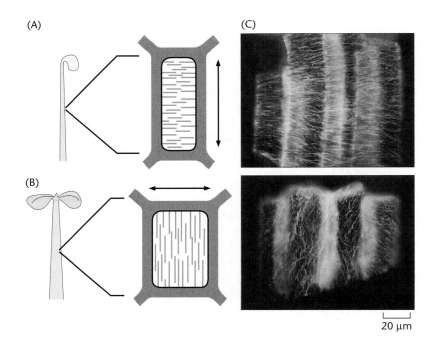

(A)

(B)

(C)

20 µm

mediate the bundling and fixed spacing of microtubules in a test tube (Figure 3–76). This bundling and spacing is important to the function of the cortical microtubules. For example, the MAP protein MOR1 stabilizes the microtubule array by promoting the correct microtubule bundling. In the *mor1* mutant, cells expand isodiametrically rather than longitudinally (see Figure 3–74).

Actin filaments direct new material to the cell surface during cell expansion

As cells expand, new membrane and cell wall material must be deposited at the cell surface specifically in expanding regions. The direction of the material to the appropriate regions of the surface is brought about by actin filaments. The final shape of the mature cell, then, is determined by the coordination of the activity of actin filaments and of the microtubules that control the direction of cell expansion.

Trichomes have been used to study how this coordination of elements of the cytoskeleton determines final cell shape. As described in Section 3.1, trichomes are large epidermal cells with a distinctive branched morphology. Because of their surface location, their development can be studied relatively easily. Epidermal cells that develop into trichomes become larger than the surrounding epidermal cells, eventually forming a spike that elongates away from the surface of the epidermis. Branches grow out from this spike by expansion of localized regions of its cell wall. During this process, microtubules control the initiation and formation of branches, while actin filaments ensure deposition of new material at appropriate places on the cell surface.

The involvement of microtubules in the initiation and formation of trichome branches is revealed by study of mutant plants with abnormal trichomes. Although some such mutants have defects in endoreduplication (see Section 3.1), others are defective in proteins that regulate microtubule organization (Figure 3–77). For example, *zwichel*

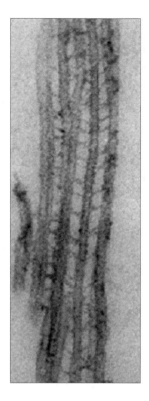

Figure 3–76.
Bundling of microtubules by a MAP protein. When purified MAP protein (isolated from carrots) is added to purified microtubules in a test tube, it creates cross-links that bind the microtubules together at a fixed spacing apart. The spacing (approximately 25–30 nm) is the same as that seen in bundles of microtubules in plant cells.
(Courtesy of Jordi Chan and Clive Lloyd.)

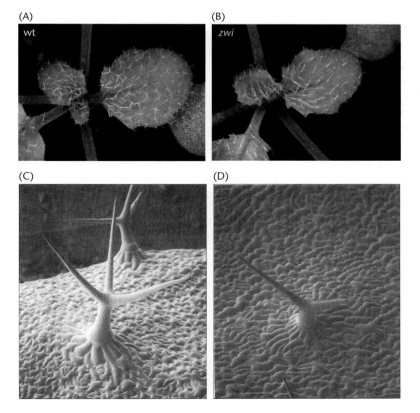

Figure 3–77.
Abnormal trichome development caused by microtubule defects. (A) Photograph and (C) scanning electron micrograph of the leaf surface of wild-type (wt) Arabidopsis. The trichome cells have three branches. (B) Photograph and (D) scanning electron micrograph of the leaf surface of a *zwichel* (*zwi*) mutant. Trichome development is abnormal because of microtubule defects that lead to abnormal deposition of the cell wall. (A and B, courtesy of David Oppenheimer; C and D, courtesy of Jordi Chan and Clive Lloyd.)

mutants of Arabidopsis lack a kinesin-like microtubule motor protein (also involved in spindle and phragmoplast formation; see Section 3.2). The trichomes of these mutants have reduced branching (Figure 3–77), suggesting that microtubule reorganization mediated by the motor protein is required for the formation of branches. As well as a general role in determining the direction of cell expansion (as described above), the protein katanin is important in branch formation in trichomes. In *katanin* mutants, epidermal cells undergo the initial stages of trichome formation (swelling) but rarely elongate to form a spike and do not form branches. Katanin is necessary for normal depolymerization of microtubules, part of the depolymerization–repolymerization process by which microtubule reorientation occurs.

Actin filaments deliver vesicles containing new material to the surface of growing parts of the cell during trichome growth. Filaments are present at the growing regions of the cell in the early stages of growth and accumulate at branch sites where new growth is initiated. Treatment of expanding trichomes with inhibitors of actin polymerization distorts trichome formation, demonstrating the importance of the regulated deposition of new cell surface material for determining cell shape. However, the same inhibitors have no effect on trichome branching, suggesting that actin is not required for branch initiation.

In root hair cells and pollen tubes, cell expansion is localized to the cell tips

As most plant cells grow, expansion is distributed over large areas of their surface. In some specialized cell types, however, expansion is highly localized so that only the tips of the cells grow. These cells include root hairs and pollen tubes of angiosperms (Figure 3–78), **rhizoids** of mosses and liverworts, and filaments produced by moss spores as they germinate.

Initiation of tip growth is brought about by a high concentration of calcium in a localized region of the cytosol. In all tip-growing cells, the concentration of Ca^{2+} ions in the cytosol at the tip is higher than elsewhere in the cell (see Figure 3–79A, B). This calcium is transported into the tip from outside the cell, via channels in the plasma membrane. The import of Ca^{2+} and the presence of a higher cytosolic concentration at the tip than further back in the cell is necessary and sufficient to instigate tip growth. Treatment of tip-growing cells with drugs that block Ca^{2+} transport into the cell abolishes growth. Conversely, if an artificial calcium gradient is created, tip growth initiates near the site of the new gradient (Figure 3–79C). A mutant of Arabidopsis that fails to form root hairs, known as *root hair defective2* (*rhd2*), is defective in the ability to form a calcium gradient (Figure 3–79D). Swellings appear on the epidermal cells that would normally become root hairs, but these are not associated with a calcium gradient and no further development occurs.

The mechanism by which Ca^{2+} ions promote tip growth probably involves the regulation of cytoskeletal organization. For example, calcium regulates the activity of the protein profilin, which is necessary for the correct formation of actin filaments. As described below, changes in the dynamics of actin filaments affect the delivery of vesicles containing materials for cell wall growth to the cell tip.

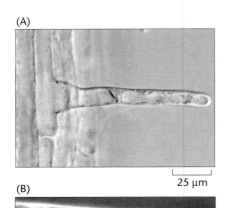

(A)

⊢—————⊣
25 μm

(B)

⊢—⊣
4 μm

Figure 3–78.
Tip-growing cells in plants. (A) Root hair of Arabidopsis. The width of the root hair just behind the tip is 7 μm. (B) Pollen tube of the lily *Lilium longiflorum*. This structure is formed when a pollen grain germinates on the stigma of a flower (see Chapter 5). The transmitted light image shows a clear zone at the tip and organelles behind the tip. (A, courtesy of Seiji Takeda; B, courtesy of N. Moreno and J. Feijo.)

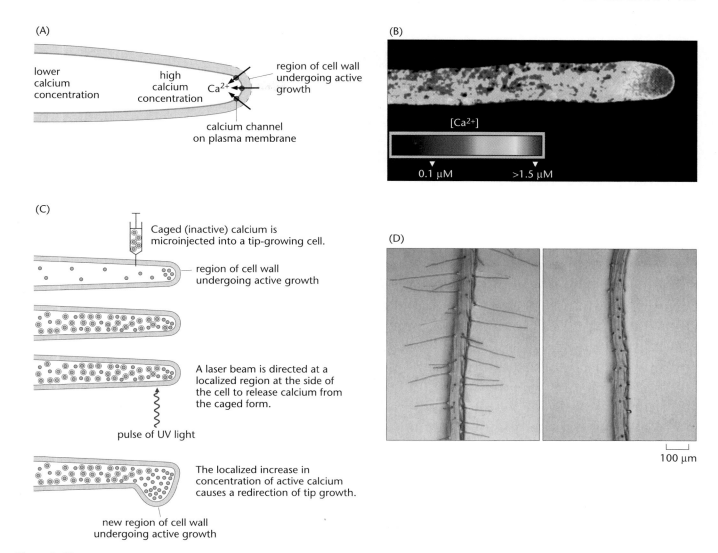

Figure 3–79.
The role of calcium in tip growth. (A) Tip growth is associated with the import of Ca^{2+} ions and establishment of a Ca^{2+} gradient. (B) Demonstration of the association between formation of a Ca^{2+} gradient and growth of a pollen tube. The pollen tube was injected with a calcium-sensitive dye. The Ca^{2+} ion concentration increases from blue through green to yellow and orange. (C) Diagram showing an experiment in which an artificial Ca^{2+} gradient is created in a tip-growing cell. Calcium is injected into the cell in a "caged" form, as a complex with the compound diphenyl EDTA; calcium cannot promote tip growth in this form. Free Ca^{2+} ions can be released from the caged form inside the cell by treatment with UV light, which causes photolysis of the diphenyl EDTA–calcium complex. If a microbeam pulse of UV light is applied to the side of the cell, behind the growing tip, the resulting localized increase in free Ca^{2+} inside the cell can redirect tip growth to that point on the cell wall. (D) Roots of wild-type Arabidopsis (*left*) and *rhd2* mutant (*right*). Swellings develop on epidermal cells of the *rhd2* root but do not grow out to form root hairs. (B, courtesy of N. Moreno and J. Feijo.)

The highly polar and often rapid growth of tip-growing cells (pollen tubes of *Tradescantia* grow at a rate of up to 0.24 mm/sec) is reflected in the distribution of cellular organelles. The region immediately behind the tip is rich in vesicles containing cell wall and membrane material. These bud off from Golgi apparatus located behind the tip and fuse with the plasma membrane to deposit their contents into the growing wall at the cell tip. As in the formation of the cell plate (see Section 3.2), the motor protein myosin binds to the outside of the vesicles and transports them along actin filaments to the tip of the cell (Figure 3–80A). Tip growth is prevented by inhibitors and mutations that disrupt actin filaments. For example, the *deformed root hairs1* (*der1*) mutant lacks a major form of actin present in wild-type plants and develops short, bulging root hairs (Figure 3–80B).

Figure 3–80.
The role of actin in tip growth. (A) Vesicles containing cell wall matrix material synthesized in the endomembrane system move along actin filaments to the growing tip of the cell, where they deliver their contents to the wall by fusion with the plasma membrane. (B) The *der1* mutant of Arabidopsis (*right*), defective in the formation of actin filaments in root hairs, develops much shorter, thicker root hairs than wild-type plants (*left*). (Courtesy of Christoph Ringli.)

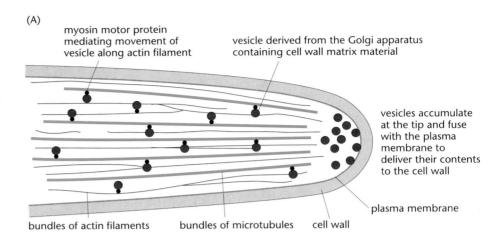

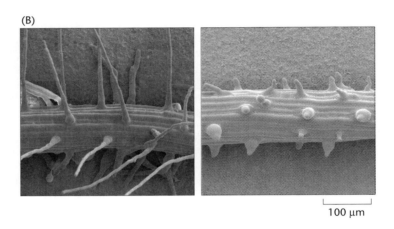

3.6 SECONDARY CELL WALL AND CUTICLE

As we have seen, the primary cell wall is formed at the cell plate during cytokinesis and is added to during cell expansion. Once cell expansion is complete, many cells lay down a secondary wall inside the primary wall (Figure 3–81). The formation of the secondary cell wall is a crucial feature in the differentiation of many cell types in higher plants. The nature of this wall differs profoundly from one cell type to another and confers unique properties on each cell type. Secondary cell walls are often thick and multilayered. They may contain a variety of polymers that serve as storage compounds or waterproofing for the wall, or cross-link the cellulose microfibrils to give the wall extra mechanical strength and durability.

In describing the development of secondary cell walls, we focus in particular on the synthesis of the cross-linking polymer **lignin**; the importance of the secondary wall in the differentiation of two classes of cells: the water-conducting cells of **xylem** and the cells of **wood**; and the synthesis of **cuticle**, the outer covering of epidermal cell walls in the aerial parts of plants.

The structure and components of the secondary cell wall vary from one cell type to another

After cell expansion ceases, the extensibility of the primary cell wall is reduced by covalent cross-linking of some of its glycoprotein components, including extensin, and the formation of junction zones (see Section 3.4). The secondary cell wall is then laid down inside the primary wall, restricting the volume of the cell within the plasma membrane. Most secondary walls are rich in cellulose, and in some cases—for example, the fibers of cotton—the wall may be almost pure cellulose. The low concentration or absence of matrix components such as pectins means that the secondary wall is less extensible than the primary wall. In some cells, cellulose microfibrils in the secondary wall are deposited in a series of layers of differing orientation (see Figure 3–81), adding mechanical strength to the wall.

In Arabidopsis, the cellulose synthase complexes responsible for the synthesis of secondary cell walls are different from those that synthesize primary walls, and this may well be the case in other species too. Arabidopsis *rsw1* mutants lack a cellulose synthase necessary for depositing normal amounts of cellulose in the primary walls of all cell types: mutant seedlings have short swollen roots and hypocotyls (see Figure 3–53). Mutations affecting a different cellulose synthase gene, *IRREGULAR XYLEM 3* (*IRX3*), have no effect on the primary wall, but reduce the secondary thickening of xylem **vessels** (as discussed below). Stems of the mutant plants contain only about 25% of the amount of cellulose found in stems of wild-type plants, and the xylem vessels collapse under the negative pressure generated during water transport (Figure 3–82).

In addition to cellulose, many secondary cell walls contain other types of polymers that confer specialized properties on the wall. For example, the secondary walls in some seeds contain large deposits of polysaccharides that serve as a reserve material. **Mannans**—polymers of the sugar mannose, which may also contain glucose and galactose residues—are deposited in the secondary cell walls of many species, then degraded to provide materials for growth during seed germination. The cell walls of the endosperm of lettuce seeds are 70% mannans. The galactomannans of some legume seeds have valuable viscosity properties when extracted, and these are used in the food industry (Figure 3–83A, B). The secondary walls of some cell types contain a hydrophobic (water-repellant) polymer called **suberin** (Figure 3–83D, E), which restricts the movement of water through the wall. A suberin-containing zone, called the **Casparian strip**, is found in the walls of the **endodermis**, a cylinder of cells important in controlling the movement of water and solutes across the root (Figure 3–83C; see also Section 4.10). Suberin is also a principal component of the cell walls of cork, which forms the outer bark of woody plants.

Lignin is a major component of many secondary cell walls

After cellulose, lignin is by far the most common component of secondary cell walls (Figure 3–84), and it is the second most abundant organic substance on earth. About 30% of the dry weight of wood is lignin. Lignin cross-links cellulose microfibrils in the secondary wall, providing a rigid and impermeable structure that confers mechanical strength, chemical stability, durability, and resistance to attack by **pests** and pathogens. The development of the ability to synthesize lignin was arguably a critical step in the

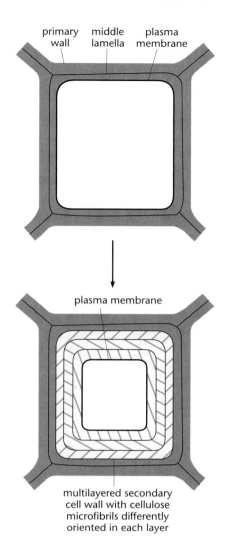

Figure 3–81.
Deposition of the secondary cell wall. The secondary, often multilayered, cell wall is deposited inside the primary wall.

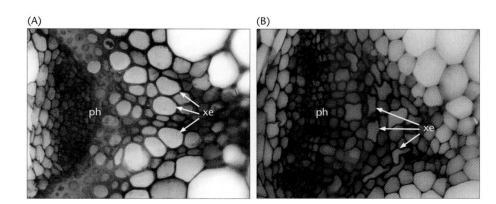

Figure 3–82.
Mutation affecting secondary thickening of xylem vessels. (A) Wild-type Arabidopsis and (B) the *irx3* mutant. The phloem (*small pink cells to left of* ph) is toward the outside and xylem toward the inside of the stem. Note the collapsed appearance of the xylem vessels (xe) in the *irx3* mutant. (A and B, courtesy of Simon Turner.)

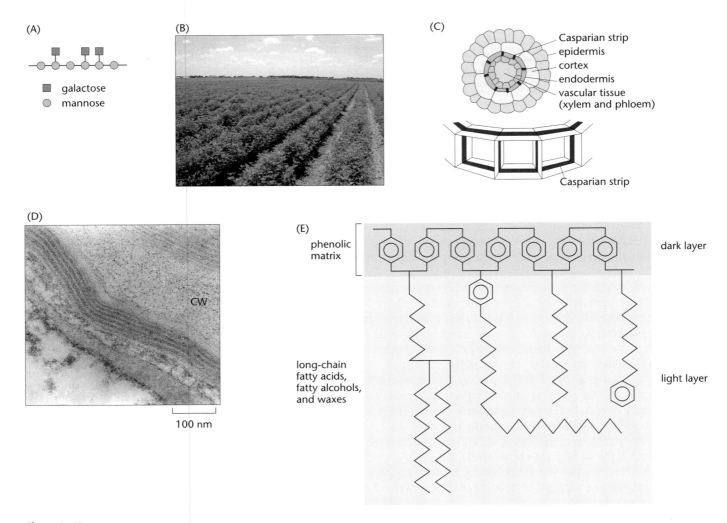

Figure 3–83.
Some secondary cell wall polymers. (A) Structure of a galactomannan. (B) A field of guar (*Cyamopsis tetragonoloba*), a legume grown for the valuable galactomannan gum that can be extracted from the cell walls of the seed endosperm. The gum has a wide range of uses, including in cosmetics, as a thickener in ice cream and puddings, and as a sizing agent for paper. (C) Position of the suberized Casparian strip in the endodermal cell walls of a root. The endodermis separates the cortex from the stele, which contains the xylem and phloem. Water and minerals taken up from the soil and destined for water-transporting cells (the xylem), and sugars in the phloem destined for the cortex and the epidermis, must pass through the endodermis. Almost all of the traffic through the endodermis must occur via the interior of cells, because the walls are rendered highly impermeable by the deposition of suberin. (See Section 4.10 for a discussion of transport across the endodermis.) (D) Electron micrograph showing the layered appearance (lamellae) of suberin in the cell wall between two root cells in Arabidopsis (CW marks the nonsuberized part of the cell wall). (E) Much-simplified model for the structure of suberin in the cell wall. The dark layers evident in (D) contain a matrix of cross-linked phenolic compounds, like those found in lignin (discussed later in the text). Linked to this matrix are long-chain hydrophobic fatty acids, fatty alcohols, and waxes, which make up the light layers. (B, courtesy of John Sij; D, courtesy of Christiane Nawrath.)

evolution of higher plants. Lower plants—mosses and liverworts (bryophytes)—lack the capacity to synthesize lignin. As a consequence, they do not have cell walls with the great mechanical strength and rigidity of the **lignified** walls of higher plants, or efficient vessels that can move water over significant vertical distances to aerial parts of the plant, or strong, rigid tissues that can support tall, load-bearing stems.

Lignin is a complex, irregular polymer composed of aromatic alcohol monomers (subunits) called **monolignols** (Figure 3–85). Monolignols are synthesized from the aromatic amino acid phenylalanine (see Section 4.5). The first steps in synthesis occur in the cytosol and the later steps in the endomembrane system. The monolignols are

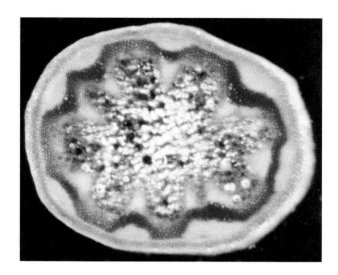

Figure 3–84.
Lignin in cells of a stem. Section through the base of an Arabidopsis inflorescence stem, stained for the presence of lignin (*pink*). Note the location of lignin in the thickened walls of a ring of fiber (sclerenchyma), xylem cells that give the stem strength and rigidity. (Courtesy of Zheng-Hua Ye.)

exported from the endomembrane system to the outside of the cell. Once in the wall, they are polymerized by **oxidases** to form the lignin polymer.

Three structural units are found in lignin polymers: *p*-hydroxyphenyl (H), guaiacyl (G), and sinapyl (S) moieties. These are derived from three precursor monolignols: *p*-coumaryl alcohol, coniferyl alcohol, and sinapyl alcohol, respectively, which differ in the degree of **methylation** of their phenolic ring (Figure 3–85B). The composition of lignin varies considerably among species. In ferns and **gymnosperms** only *p*-coumaryl

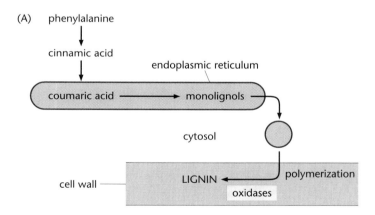

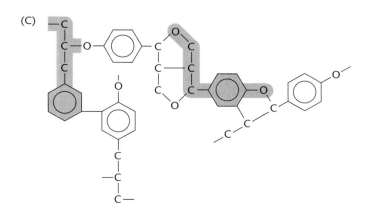

Figure 3–85.
The structure of lignin. (A) Lignin monomers, the monolignols, are synthesized from the amino acid phenylalanine in a pathway that starts in the cytosol and continues in the endoplasmic reticulum. The monolignols are transferred in vesicles to the cell wall, where they are polymerized to lignin in reactions catalyzed by oxidase enzymes in the wall.
(B) Generalized structure of the three types of monolignol (see also Figure 3–86). All have a phenolic ring and a three-carbon side chain with a double bond; this bond is oxidized during polymerization. (C) Generalized structure of a portion of lignin. Note that lignin is a complex, heterogeneous substance; this illustrates the kind of structures found. The contributions of two individual monolignols are highlighted (*red*).

and coniferyl alcohols are synthesized, hence lignin is composed of H and G units only. In angiosperms, sinapyl alcohol is also synthesized, but most angiosperms do not use *p*-coumaryl alcohol for lignin synthesis. Thus the wood of angiosperm trees is composed of S and G units only. However, grasses use all three alcohols in lignin synthesis, so lignin in these species contains S, H, and G units. Lignin composition also varies among cell types and in different regions of the cell wall in the trunks of trees. For example, in black spruce (*Picea mariana*, a gymnosperm), lignin in the secondary wall of water-conducting cells (**tracheids**) comprises predominantly G units. However, lignin in the **middle lamella**—the zone separating the cell walls of adjacent cells—comprises predominantly H units. In birch (*Betula papyrifera*, an angiosperm), the lignin of fiber cells in the wood contains both G and S units, but the lignin of the water-conducting cells in the wood contains predominantly G units.

Lignin composition has very significant effects on the properties of cell walls, with important consequences for the value of the plant organ to agriculture and industry. For example, the composition of lignin in the wood of different tree species renders some trees far more useful for papermaking than others; lignin must be removed from wood by mechanical and chemical means during conversion to pulp for papermaking. The composition of lignin also affects the digestibility of plants by farm animals and hence their value as forage crops.

The first step in the conversion of phenylalanine to monolignols is catalyzed by the enzyme phenylalanine ammonia lyase (Figure 3–86A). The product, cinnamic acid, is a precursor not only of monolignols but also of a whole family of other phenolic molecules known as **phenylpropanoids**. Conversion of cinnamic acid to *p*-coumaric acid by the enzyme cinnamate 4-hydroxylase occurs on the ER membrane, and this is generally regarded as a critical step that strongly influences the overall rate of monolignol synthesis. Changes in expression of the gene encoding cinnamate 4-hydroxylase bring about changes in the amount of enzyme protein and hence the activity of the enzyme in the cell, and this in turn changes the rate at which cinnamic acid is converted to monolignols.

The pathways by which cinnamic acid is converted to monolignols are complex, but involve three basic types of modification of the molecule: hydroxylation (addition of a hydroxyl group) of the aromatic ring, methylation (addition of a methyl group) at a position on the aromatic ring, and **reduction** of the carboxyl group to a hydroxyl group (Figure 3–86B).

The pathways to coniferyl alcohol and sinapyl alcohol are the same as far as the intermediate coniferaldehyde. At this point the pathways diverge. The path to coniferyl alcohol involves reduction of the aldehyde group by the enzyme cinnamyl alcohol dehydrogenase (CAD; this enzyme can act on several aldehydes in the pathways of phenylpropanoid metabolism). The path to sinapyl alcohol involves hydroxylation of coniferaldehyde by ferulate 5-hydroxylase, followed by methylation of the new hydroxyl group, and finally reduction of the aldehyde group.

Much important information has been gained from reducing or enhancing the expression of genes that encode enzymes believed to participate in these pathways. For example, Arabidopsis plants with mutations that eliminate expression of the gene encoding ferulate 5-hydroxylase have no S units in their lignin. Overexpression of this enzyme in tobacco, poplar, and Arabidopsis results in lignin composed almost entirely of S units. These observations confirm that the reaction catalyzed by ferulate 5-hydroxylase is necessary for the synthesis of sinapyl alcohol. Similarly, examination of plants with reduced activity of cinnamyl alcohol dehydrogenase has confirmed the broad importance of this enzyme in monolignol synthesis. In alfalfa plants with reduced cinnamyl alcohol dehydrogenase activity, amounts of both S and G units in the cell wall are

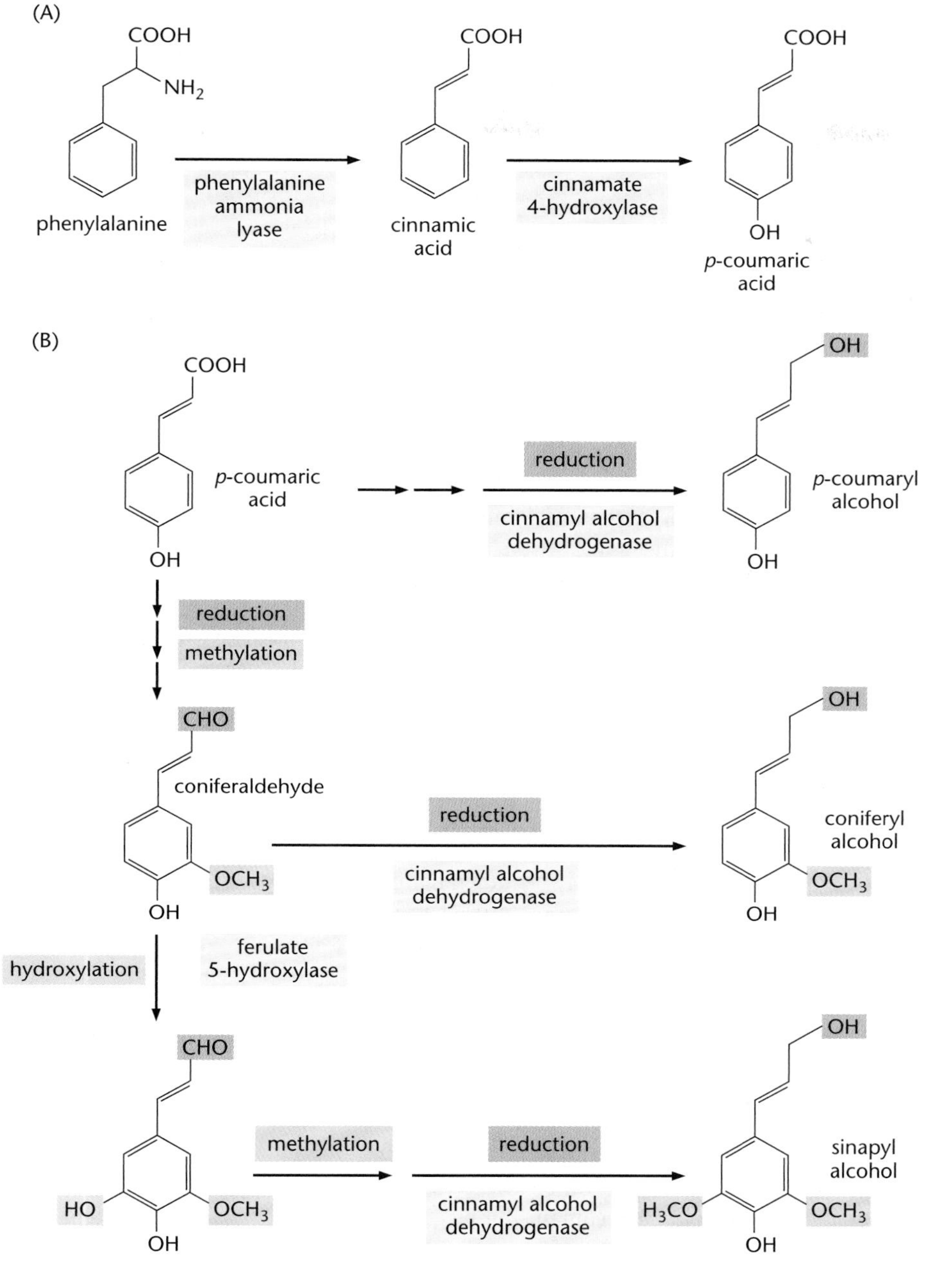

(A)

(B)

Figure 3–86.
Synthesis of the monolignols.
(A) Conversion of phenylalanine to coumaric acid, the starting point for synthesis of the monolignols. (B) Synthesis of the monolignols from coumaric acid.

reduced. This suggests that the enzyme is involved in the synthesis of both coniferyl and sinapyl alcohol. It is also responsible for the reduction of p-coumaraldehyde to form the third monolignol, p-coumaryl alcohol (see Figure 3–86B).

Polymerization of monolignols occurs in specific regions of cell walls. Polymerization involves the formation of oxidized forms of the monolignols (free-radical intermediates), probably involving cell wall enzymes that use oxygen (such as laccase) and hydrogen peroxide (peroxidases). The free-radical intermediates then couple together to form chemically stable polymeric networks. The coupling of free-radical intermediates can happen without the participation of enzymes or other proteins; indeed, lignin-like polymers can be made simply by oxidizing monolignols in a test tube. However, the linkages that form between monolignols during lignin synthesis in the cell wall are

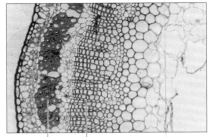

sclerenchyma   xylem

Figure 3–87.
Fiber cells (sclerenchyma) of xylem. Cross section of a stem of flax (*Linum usitatissimum*), showing the thick-walled cells forming the fibers (*pink*) that are extracted for use in the textile industry. (Courtesy of Isabelle His-Mauger.)

not the same as those that form in a test tube. This observation, plus the finding that lignin is deposited at specific sites in the cell wall rather than uniformly throughout the wall, implies that cell wall proteins determine the distribution of exported monolignols in the wall and promote the formation of specific types of bond between the free-radical intermediates. The nature of the cell wall proteins having this function is not yet fully understood.

Lignification is a defining characteristic of xylem vessels and tracheids

The xylem is the principal water-conducting tissue of vascular plants, transporting water and dissolved ions from the roots to the aerial parts of the plant. The specialized functions of the two main cell types in the xylem—the fiber cells and the water-conducting cells —are dependent on specific types of secondary cell walls.

The fiber cells (**sclerenchyma cells**) are elongated cells in which the lignified secondary wall occupies much of the volume of the cell. Continuous files of fiber cells run throughout the vascular system, providing rigid support for the water-conducting cells and for the phloem—the transport system for sugars and other nutrients that is the other component of the **vascular bundles**. (See Sections 4.4 and 4.10 on phloem transport and xylem transport.) The strong files of fiber cells in the vascular bundles of some plants provide the raw materials for the textile industry. The cellulose-rich fibers of flax and hemp stems, for example, are used in the manufacture of cloth (linen in the case of flax fiber) and rope (Figure 3–87).

Water-conducting cells also occur in continuous files through the vascular system. They are of two types in the wood of eudicot species: tracheids and xylem vessels (Figure 3–88). As they mature, both cell types undergo massive secondary wall development in very specific patterns. Depending on the species and the position within the xylem tissue, water-conducting cells may lay down lignified secondary wall in rings or helices around the cell, or in a reticulate pattern, or in more continuous plates perforated by **pits** in which no secondary wall develops. After the secondary wall is deposited the water-conducting cells die, leaving the walls as a strong, rigid "pipe" through which water flows relatively unimpeded. Wall strength is essential for the functioning of the xylem, as the system often operates with a high negative internal pressure (see Section 4.10): the rings and plates of secondary cell wall prevent the collapse of the pipes under these conditions.

Tracheids are found in most vascular plants and are the only type of water-conducting cell in gymnosperms and other lower vascular plants. The tapering end walls of the elongated tracheid cells contain the pits, where adjacent tracheids are separated only by a thin primary wall (Figure 3–88B). Xylem vessels are found only in angiosperms, and may occur in xylem tissue alongside tracheids. The cells of these vessels are usually less elongated than tracheids. They also form files, but in this case the end walls contain perforations in which no secondary or primary cell wall is present (Figure 3–88B, C).

Wood is formed by secondary growth of vascular tissues

The vascular bundles of the stems of eudicots and gymnosperms develop from groups of cells arranged in a cylinder in the stem, called the **vascular cambium**. This is a meristematic tissue in which cell division gives rise to phloem on the outside of the cylinder and xylem on the inside (see Section 5.4). The initial formation of discrete bundles is called **primary growth**. Stems of **annual plants** and **herbaceous** plants (plants with stems that die during the winter) have only this type of growth. In woody plants, a continuous cylinder of cambial cells develops within the stem, producing new xylem cells each year. This increases the girth of the stem and gives rise to extensive lignified tissue—wood—in the center of the stem. This is **secondary growth** (Figure 3–89).

(A)

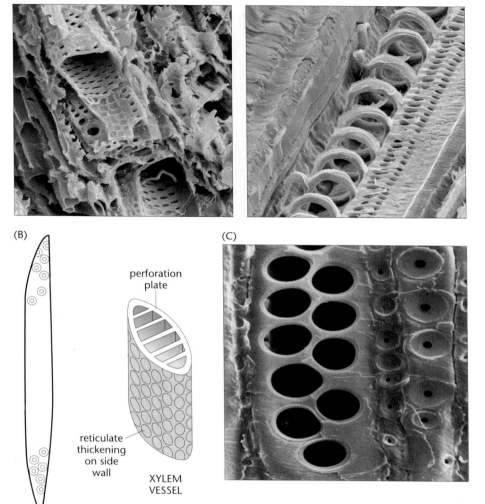

(B)

perforation
plate

reticulate
thickening
on side
wall

XYLEM
VESSEL

(C)

Figure 3–88.
Water-conducting cells of xylem.
(A) False-color scanning electron micrographs of xylem vessels in a bamboo stem, showing annular wall thickenings (*left*) and spiral wall thickenings (*right*). (B) Diagram of a tracheid and a xylem vessel. Note the pits in the tapered end walls of the tracheid and the large perforations in the end wall of the xylem vessel. (C) Scanning electron micrograph (face-on view) of a perforation plate from a xylem vessel.

Figure 3–89.
Stages of secondary growth. (A) Mature primary stem. (B) Development of a ring of cambium. (C) Stem after a year of secondary growth, showing development of secondary phloem and secondary xylem. (D) After further development and lignification of the secondary xylem in successive years, a large proportion of the stem is occupied by woody secondary xylem wood. The outer epidermal layer of cells is replaced by a protective periderm, created by extra cell divisions around the stem in the cork cambium. This gives rise to the bark of woody stems. (E) Cross section of a stem of the Virginia creeper (*Parthenocissus inserta*), at the stage shown in (D).

Secondary growth starts with activation of cell division in the vascular cambium of existing bundles. Division is coordinated with divisions initiated in the adjacent parenchymatous region—the **interfascicular region** ("between bundles"), also called **pith rays**—resulting in a continuous cylinder of vascular cambium around the stem (see Figure 3–89). Within the vascular and interfascicular cambium, two types of **initial cell** can be identified (Figure 3–90A). The **fusiform initials** divide longitudinally to produce cells with long vertical axes. Differentiation of these cells on the outside of

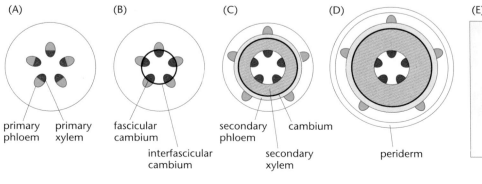

(A)

primary
phloem

primary
xylem

(B)

fascicular
cambium

interfascicular
cambium

(C)

secondary
phloem

cambium

secondary
xylem

(D)

periderm

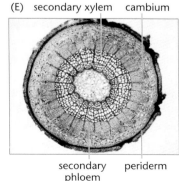

(E) secondary xylem   cambium

secondary
phloem

periderm

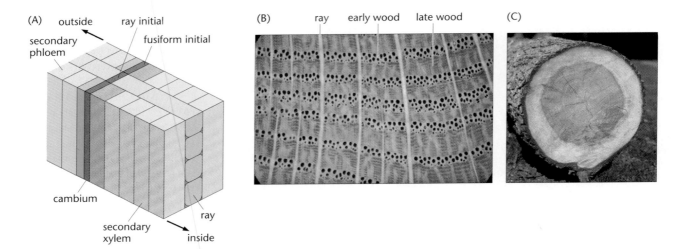

Figure 3–90.
Formation of a woody stem. (A) Diagram (cut-away section) of a woody stem, showing cambial cells giving rise to secondary phloem and xylem and to ray cells. (B) Cross section of the wood (secondary xylem) of an oak tree, showing about six years of growth. Each year's growth (a growth ring) consists of large vessels laid down early in the growing season (early wood), followed by much smaller vessels formed later in the year (late wood). (C) Section through the trunk of an oak tree, showing the distinction between heartwood (inner dark wood) and sapwood (outer light wood).

the cambium gives rise to secondary phloem, while on the inside it gives rise to cell types of the xylem. Xylem tissue makes up the bulk of secondary growth in stems; in other words, wood is mostly xylem. The **ray initials** give rise to **ray cells** with long horizontal axes. These cell files form rays that run radially across the secondary vascular tissue. They transport water, dissolved gases, and organic nutrients between the secondary phloem and the secondary xylem, and also store reserve compounds such as starch (see Section 4.6)

The cambium in woody plants usually shows periodic activity coinciding with growth and **dormancy** periods. In spring and summer, the cambium produces wide **tracheary elements** with relatively thin cell walls, which is called **early wood**. Later in the year, in the late summer and fall, narrow tracheary elements with thick walls are formed; this is **late wood**. Together, the rings formed from early and late wood comprise an **annual growth ring**. Growth rings are visible in cross sections of stems and trunks, and can be used to measure the age of the plant (Figure 3–90B).

Wood does not have an unlimited lifetime in conducting water. The outer, younger parts of a woody stem (the **sapwood**) fulfill this function, while in the inner, older regions (the **heartwood**), tracheids and vessels cease to function and the living cells of the xylem die (Figure 3–90C). The heartwood becomes filled with resinous material. This is often composed of **polyphenols** (polymers of compounds based on aromatic rings), consisting of tannins, dyes, resins, and gums. Heartwood is more resistant to penetration by moisture and to rot by microorganisms than sapwood, and is particularly prized for use in the outdoor construction industry.

Monocots differ from eudicots and gymnosperms in the distribution of vascular bundles in the stem and the ways in which stem thickening is achieved. Vascular bundles are distributed throughout the parenchymatous ground tissue. They lack meristematic cells. Stem thickening in most monocots, including palms, occurs from the top (crown) of the stem. An intensive zone of cell division, the **primary thickening meristem**, lies immediately beneath the **apical meristem** and the **leaf primordia**. Division in this zone gives rise to vascular bundles and parenchymatous ground tissue, and can produce a very wide crown. No further thickening occurs below this zone, resulting in trunks that differ little in diameter over their length. Stem thickening in a few tree-like monocots (e.g., *Cordyline*) results from true secondary growth, occurring along the length of the trunk. In these species, the primary thickening meristem extends downward around the flanks of the stem to its base. This flanking meristem (**secondary thickening meristem**) gives rise to new parenchyma cells and vascular bundles from its inner face, resulting in thickening along the entire length of the trunk (Figure 3–91).

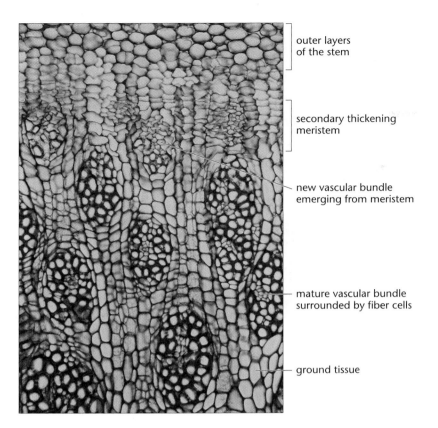

outer layers
of the stem

secondary thickening
meristem

new vascular bundle
emerging from meristem

mature vascular bundle
surrounded by fiber cells

ground tissue

Figure 3–91.
Stem of *Cordyline*, a tree-like woody monocot. At the outer edge of this cross section (*top*), a secondary thickening meristem gives rise to new vascular bundles and ground tissue, resulting in lateral thickening of the stem.
(Courtesy of David T. Webb.)

The cuticle forms a hydrophobic barrier on the aerial parts of the plant

All plant organs exposed directly to the atmosphere—the aboveground parts of land plants—are covered in a layer of polymeric material called the cuticle, which serves to reduce water loss to the atmosphere, reduce penetration by chemicals and potentially pathogenic organisms on the plant surface, and perhaps prevent adhesion of organ surfaces during development. The precise nature and the thickness of the cuticle vary considerably from one plant and one organ to another. In Arabidopsis the cuticle is 80 nm or less in thickness, but in drought-resistant plants it may be many times thicker (Figure 3–92) (see Section 7.3).

Two main types of polymer make up the cuticle. **Cutin** is a three-dimensional network of long, cross-linked **fatty acid** molecules (fatty acids are described in Section 4.6). It is laid down within the outer regions of the cell wall and in a layer covering the outer surface of the wall (Figure 3–93). The second component, **cuticular wax**, is laid down within the cutin network and also in a layer on the outer surface above this network. The **waxes** are esters of fatty acids and long-chain alcohols. Within the cutin network they exist in an amorphous state, but at the outer surface they crystallize to form plates and protrusions that may be very elaborate. The waxes render the cuticle highly impermeable to water. Some plant waxes are commercially important. Carnuba wax, found on the surfaces of a Brazilian palm tree, is harvested for use as a floor and car wax.

Both cutin and the waxes are very complex molecules, and their synthesis is not fully understood. The fatty acid monomers of cutin are synthesized in the endoplasmic reticulum of epidermal cells and exported to the wall, where cross-linking occurs. The mechanisms of export and cross-linking are not known. The waxes are a very diverse group of molecules, and their composition varies from one plant organ to another. Like the cutin monomers, they are synthesized in the ER and exported to the wall. Progress in understanding cuticle synthesis and assembly comes from the discovery of mutants

Figure 3–92.
Cuticle of a desert plant. False-color scanning electron micrograph of a section through a leaf of *Haworthia pumila*, a desert plant. The cuticle (*green*) is exceptionally thick, preventing loss of water from the plant surface. In many other plants the cuticle is much thinner. For example, in Arabidopsis it is only about 5% of the thickness of the epidermal cell wall.

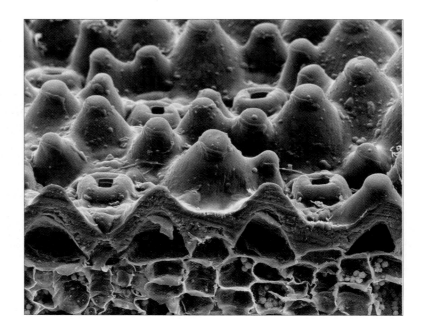

with abnormal cuticles. Mutants with altered amounts and types of cuticular waxes (such as *eceriferum*, "without wax," mutants) are relatively easy to identify because the reflective properties of the plant surface are changed (Figure 3–94). Large numbers of gene loci influencing wax synthesis have been identified in some species. For example, there are more than 80 *Eceriferum* genes in barley and more than 20 discovered thus far in Arabidopsis. This implies that very large numbers of different gene products are necessary for the synthesis and secretion of cuticular wax.

Several Arabidopsis mutants with defects in cutin synthesis and assembly have severely abnormal growth patterns and reduced fertility, implying that cuticle deposition is essential for normal growth and development. In the *fiddlehead* mutant, for example, leaves and floral organs tend to fuse together at an early developmental stage (Figure 3–95). Although *fiddlehead* plants have a cuticle, the fatty acid composition of the cutin is abnormal and this seems to result in "stickiness" of the surfaces of developing organs. The gene affected in this mutant encodes an enzyme involved in fatty acid elongation. The *lacerata* mutant, which also has organ fusions and stunted growth, is defective in an enzyme that modifies fatty acids on the pathway to cutin synthesis. In these plants the cuticle is missing in the regions where organs are fused.

Figure 3–93.
Surface layers of the aerial parts of a plant. The diagram shows the multiple layers of an epidermal cell surface. The scanning electron micrograph shows ridges of wax protruding from the cuticle on the leaves of a rubber tree (*Hevea braziliensis*). This pattern of wax deposition is very different from that of the Arabidopsis stem surface (see Figure 3–94B).

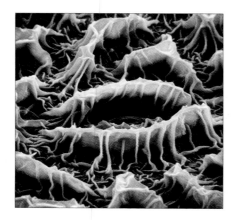

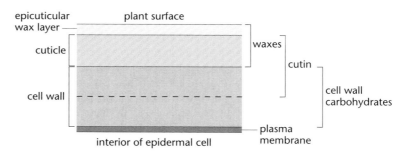

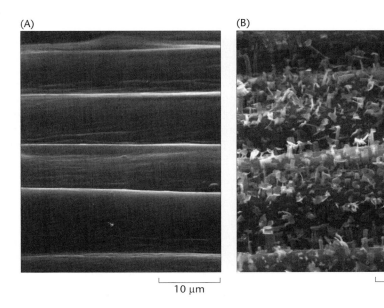

Figure 3–94.
Surfaces of *eceriferum* mutant and wild-type Arabidopsis flower stems. Scanning electron micrographs of (A) the waxless (*eceriferum*) mutant *cut1* and (B) wild-type Arabidopsis. The flakes on the surface of the wild-type stem are wax crystals. These are missing from *cut1*, which lacks an enzyme necessary for the synthesis of the long-chain fatty acids found in wax. (Courtesy of Ljerka Kunst.)

SUMMARY

In a plant cell, a plasma membrane encloses the cytosol, which contains organelles (including nucleus, mitochondria, plastids, peroxisomes, endomembrane system, and vacuoles) and a network of tubules and filaments (the cytoskeleton); the membrane is surrounded by a cell wall. Plasmodesmata lined with plasma membrane form channels between adjacent cells.

The cell cycle is a programmed progression of events, including mitosis, that results in replication and separation of chromosomes into two daughter nuclei, which are separated by a wall to form two daughter cells. Transition from one phase of the cycle to the next is regulated by complex mechanisms, with developmental and environmental inputs. Cell wall assembly between daughter cells is orchestrated by a phragmoplast, which directs vesicles containing wall materials to the new cell wall. Many differentiating plant cells undergo endoreduplication, resulting in polyploidy and increased cell size. Meiosis is a specialized type of cell division that produces haploid cells (spores) and genetic variation; the spores give rise to a haploid organism that subsequently produces gametes. Meiosis and fertilization ensure that offspring have a genetic makeup unlike that of either parent.

Plastids and mitochondria replicate by binary fission, both during the cell cycle and during cell growth and differentiation. Most plastid and mitochondrial proteins are nuclear-encoded and imported into the organelle via protein-import complexes and chaperones. The endomembrane system synthesizes lipids and polysaccharides and modifies proteins for delivery to the cell wall and vacuole.

The primary cell wall consists of cellulose microfibrils embedded in a matrix of pectins and hemicelluloses. The cellulose microfibrils are synthesized by the cellulose synthase complex. Cells acquire their mature size and shape by controlled expansion, through the movement of solutes and water into the cell, thus increasing turgor pressure, and the loosening of and insertion of new material into the wall. Turgor also maintains rigidity in nonwoody plants. Movement of solutes and water into plant cells is determined by properties and components of the plasma membrane, including transporters, channels, and aquaporins. Proton transport across the membrane generates electrical and proton gradients that drive other transport processes.

Most of the water that maintains plant cell turgor is contained in the central vacuole. Water movement into the vacuole is driven by the accumulation of solutes in the vacuole,

(A)

(B)

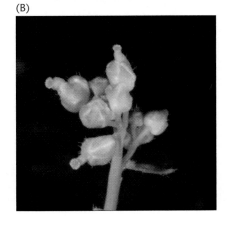

Figure 3–95.
Inflorescences of wild-type Arabidopsis and the *fiddlehead* mutant. (A) Wild type and (B) *fiddlehead* mutant. The *fiddlehead* mutant lacks an enzyme necessary for synthesis of the long-chain fatty acids found in the cuticle. The floral organs stick together and fail to develop properly.

which enter via transporters in the tonoplast. The vacuole also acts as a storage and sequestration compartment, containing pigments, defensive substances, and breakdown products. Coordinated ion transport and water movement in guard cells controls stomatal opening and closure.

When cell expansion is complete, many plant cells lay down a secondary wall inside the primary wall. The main components of this secondary wall are cellulose and lignin. Formation of the secondary wall is important in the differentiation of tracheary elements of xylem and in the cells of wood. The cuticle forms a hydrophobic barrier on the aerial parts of plants.

FURTHER READING

General references

Alberts B, Johnson A, Lewis J, et al. (2002) *Molecular Biology of the Cell.* New York, NY: Garland Science. 1463 pp.

Jurgens G. (2004) Membrane trafficking in plants. *Annual Review of Cell and Developmental Biology* 20, 481–504.

3.1 The Cell Cycle

Dewitte W, Murray JAH (2003) The plant cell cycle. *Annual Review of Plant Biology* 54, 235–264.

Cnudde F, Gerats T (2005) Meiosis: inducing variation by reduction. *Plant Biology*, 321.

Sugimoto-Shirasu K, Roberts K (2003) "Big it up": endoreduplication and cell-size control in plants. *Current Opinion in Plant Biology* 6, 544.

Stals H and Inzé D (2001) When plant cells decide to divide. *Trends Plant Sci* 6, 359-364.

3.2 Cell Division

Verma, D.P.S., Hong, Z (Eds.) (2008) Cell Division Control in Plants. *Plant Cell Monographs Vol. 9.*

Jurgens, G (2005) Cytokinesis in higher plants. *Annual Review of Plant Biology* 56, 281–299.

3.3 Organelles

Kenneth Cline and Carole Dabney-Smith (2008) Plastid protein import and sorting: different paths to the same compartments. *Current Opinion in Plant Biology* 11:585–592.

Cassie Aldridge , Jodi Maple and Simon G. Møller (2005) The molecular biology of plastid division in higher plants. *Journal of Experimental Botany* 56(414):1061–1077.

3.4 Primary Cell Wall

Clive Lloyd and Jordi Chan (2008) The parallel lives of microtubules and cellulose microfibrils. *Current Opinion in Plant Biology* 11:641–646.

Olivier Lerouxel, David M Cavalier, Aaron H Liepman and Kenneth Keegstra (2006) Biosynthesis of plant cell wall polysaccharides—a complex process. *Current Opinion in Plant Biology* 9: 621–630.

3.5 Cell Expansion and Cell Shape

Ken-ichiro Shimazaki, Michio Doi, Sarah M. Assmann, and Toshinori Kinoshita (2007) Light Regulation of Stomatal Movement. *Annu. Rev. Plant Biol.* 58: 219–47.

Dolan L and Davies J (2004) Cell expansion in roots. *Curr Opin Plant Biol* 7, 33–39.

Martin C, Bhatt K and Baumann K (2001) Shaping in plant cells. *Curr Opin Plant Biol* 4, 540–549.

3.6 Secondary Cell Wall and Cuticle

Lacey Samuels, Ljerka Kunst, and Reinhard Jetter (2008) Sealing Plant Surfaces: Cuticular Wax Formation by Epidermal Cells. *Annual Review of Plant Biology* Vol. 59: 683–707.

Laigeng Li; Shanfa Lu; Vincent Chiang (2006) A Genomic and Molecular View of Wood Formation. *Critical Reviews in Plant Sciences* 25, 215–233.

METABOLISM

4

When you have read Chapter 4, you should be able to:

- Summarize the importance of compartmentation in the control of metabolic pathways, and give examples of processes occurring in each plant cell compartment.

- Outline the role of transport proteins in the plasma membrane and in plastidial, mitochondrial, and vacuolar membranes in regulating metabolism.

- Summarize the ways in which metabolism is controlled at the level of enzyme activity, distinguishing between coarse and fine control.

- Describe the capture of light energy, the synthesis of reducing power and ATP, and the assimilation of carbon in the Calvin cycle, noting how carbon assimilation and energy supply are coordinated.

- Summarize the role of photorespiration in carbon and nitrogen metabolism.

- Outline how sucrose moves from leaves to nonphotosynthetic parts of the plant, noting the alternative routes used at each end.

- Describe how sucrose is metabolized in nonphotosynthetic cells to provide reducing power, ATP, and biosynthetic precursors.

- Summarize the central role of the Krebs cycle in metabolism.

- Describe the main storage carbohydrates in plants, how they are synthesized, where they are stored, and how they are degraded to release energy and carbon.

- Describe the synthesis and storage of lipids and how they are converted to sugars.

- Outline how inorganic nitrogen is assimilated into amino acids, and describe the forms in which nitrogen is stored in plants.

- Summarize how phosphorus, sulfur, and iron are taken up from the soil and their roles in metabolism.

- Describe how water and minerals move from the soil to the leaves and the role of this movement in plant metabolism.

CHAPTER SUMMARY

4.1 CONTROL OF METABOLIC PATHWAYS

4.2 CARBON ASSIMILATION: PHOTOSYNTHESIS

4.3 PHOTORESPIRATION

4.4 SUCROSE TRANSPORT

4.5 NONPHOTOSYNTHETIC GENERATION OF ENERGY AND PRECURSORS

4.6 CARBON STORAGE

4.7 PLASTID METABOLISM

4.8 NITROGEN ASSIMILATION

4.9 PHOSPHORUS, SULFUR, AND IRON ASSIMILATION

4.10 MOVEMENT OF WATER AND MINERALS

Most of the organic matter on earth is manufactured by plants through the assimilation of inorganic carbon and nitrogen from the environment into organic molecules, in processes driven by energy from sunlight. These processes, **photosynthesis** and **nitrogen assimilation**, are essential not only for the plants themselves but for almost all other forms of life—animals, fungi, and most bacteria, which obtain their carbon and nitrogen only from organic compounds and thus are completely dependent on plants for their nutrition.

Plants obtain the major elements that make up the plant body—carbon, oxygen, hydrogen, and nitrogen—mainly as carbon dioxide, water, and nitrate. They also take up and use many other minerals and elements, albeit in much smaller quantities. This means of nutrition, from inorganic compounds, is known as **autotrophy** ("self-feeding"). Organisms that obtain their carbon and nitrogen only from organic compounds—that is, ultimately, from plants—have a form of nutrition known as **heterotrophy** ("other-feeding").

Photosynthesis is not only a basic process of plant and animal nutrition, it is also a crucial determinant of the composition of the atmosphere. Each year, plants incorporate approximately 100 billion metric tons of carbon, about 15% of the total carbon dioxide in the atmosphere. Respiration by plants and heterotrophic organisms converts the same amount of carbon in organic compounds back into carbon dioxide. The effects of carbon assimilation and respiration on the atmosphere during the earth's history, and the evolutionary relationships between autotrophic and heterotrophic organisms, are described in Chapter 1. In this chapter our focus is on photosynthesis as a primary process of plant metabolism.

Plants show huge genetic variation in **metabolism**, the multitude of interrelated biochemical reactions that maintain plant life. Tens of thousands of different organic compounds have been discovered in plants. Some of these are ubiquitous. These include compounds involved in the **metabolic pathways** that assimilate nutrients from the environment, in energy metabolism, and in biosynthetic pathways that produce the basic components of the plant cell: proteins, membranes, and cell walls. Many of the metabolic processes that are ubiquitous in plant cells also occur in the cells of animals, fungi, and bacteria. For example, the process by which sugars are oxidized to produce the energy-rich compound **ATP (adenosine triphosphate)** is broadly similar in plants and animals, and many of the **metabolites** (compounds involved in metabolic pathways) are identical. Some ubiquitous plant compounds and metabolic processes—such as those involved in photosynthesis—are also found in some bacteria, while others are unique to plants. However, the vast majority of plant compounds discovered thus far occur only in particular families or genera of plants, and some in only one species. Many of these metabolites are synthesized only in specialized organs or particular cell types, and only at certain developmental stages or under particular conditions of growth. These substances play roles in a plant's adaptation to changing environmental conditions and in defense against disease and predation.

The metabolic networks responsible for the assimilation and utilization of carbon, nitrogen, and other elements are highly regulated. Regulation allows the integration and coordination of processes, as well as rapid responses to environmental changes that directly influence plant metabolism, such as temperature, light intensity, and water availability. For example, enzymes of the metabolic pathway that assimilates carbon dioxide during photosynthesis (the **Calvin cycle**) are regulated by a host of factors that allow the rate of assimilation and the fate of the immediate products to respond rapidly and sensitively to change—in the availability of carbon dioxide inside the leaf, in the supply of energy from light-driven processes, and in the demand for organic compounds by the nonphotosynthetic cells of the plant.

We begin the chapter with an overview of the ways in which plant metabolism is regulated and integrated. The principles of this discussion provide an important background for understanding all of the metabolic processes described in the chapter. The discussion of metabolism starts with the photosynthetic assimilation of carbon in leaves, the movement of assimilated carbon in the form of sucrose from the leaves to nonphotosynthetic parts of the plant, and the use of sucrose as a source of carbon for biosynthesis, energy production, and storage processes. We provide examples of biosynthetic pathways that generate important structural, regulatory, and defense compounds, and emphasize the central role of the plastid in metabolism in both photosynthetic and nonphotosynthetic cells. We then consider the assimilation of nitrogen, phosphorus, sulfur, and iron, and the response of these processes to availability in the soil and demand from plant tissues. All of the metabolic and developmental processes of the plant are critically dependent on the uptake of water from the soil and its movement through the plant body; this is considered in the final section of the chapter.

4.1 CONTROL OF METABOLIC PATHWAYS

Broadly speaking, metabolic processes in plant cells are controlled at two levels: by compartmentation of metabolic pathways into different **organelles** and by mechanisms that modulate the activities of individual **enzymes** or blocks of enzymes that catalyze the reactions of these pathways. We consider first the importance of compartmentation in increasing the potential for metabolic diversity in the cell, and then the ways in which metabolic processes are integrated and controlled by modulation of enzyme activities.

Compartmentation increases the potential for metabolic diversity

Compartmentation of the plant cell into discrete organelles with distinct sets of internal conditions increases the potential for metabolic flexibility and diversity. First, it allows metabolic reactions requiring very different conditions to occur simultaneously in the same cell. Second, it allows pathways and reactions that take place in more than one compartment to proceed with opposite net fluxes and at different rates in the same cell at the same time. Third, it provides a mechanism by which one metabolite can be used in two different processes without those processes being in direct competition. The **cytosol, plastid, mitochondrion, vacuole**, and **endomembrane system** are all metabolically distinct compartments (see Chapter 3). Within **chloroplasts** (a type of plastid) and mitochondria, internal membranes create further compartments (Figure 4–1).

Metabolic processes are also compartmented between soluble and membrane phases in subcellular compartments. For example, the **electron transport chains** of the mitochondrion and chloroplast operate exclusively within membranes, whereas the Calvin cycle (which utilizes the products of the chloroplast electron transport chain) and **Krebs cycle** (which provides electrons for the mitochondrial electron transport chain) are in the soluble phase of the organelles.

Many major metabolic processes in plant cells occur exclusively in the plastid, an organelle unique to plants. Others occur in both plastids and cytosol: these processes may be completely duplicated in the two compartments or may involve some exclusively plastidial and some exclusively cytosolic reactions. Photosynthesis is the most obvious exclusively plastidial process; others include the synthesis of fatty acids, chlorophyll and related pigments, and starch, and the conversion of nitrite to amino acids. Processes involving both plastids and cytosol include the synthesis of several amino acids (see Section 4.8) and of lipids and terpenoids (see Section 4.7).

Plastids evolved from free-living photosynthetic organisms that formed an endosymbiotic relationship with a nonphotosynthetic host (see Section 1.2). Although the genes encoding most of the functions inherited from the photosynthetic organism moved to the nucleus during the evolution of modern plants, the functions themselves have largely been retained by the plastid. The maintenance of the separate metabolic identity of the plastid during plant evolution is testament to the importance of intracellular compartmentation in plant metabolism.

The intracellular compartmentation of metabolism requires that different compartments exchange metabolites while retaining distinct sets of conditions (of pH, ionic

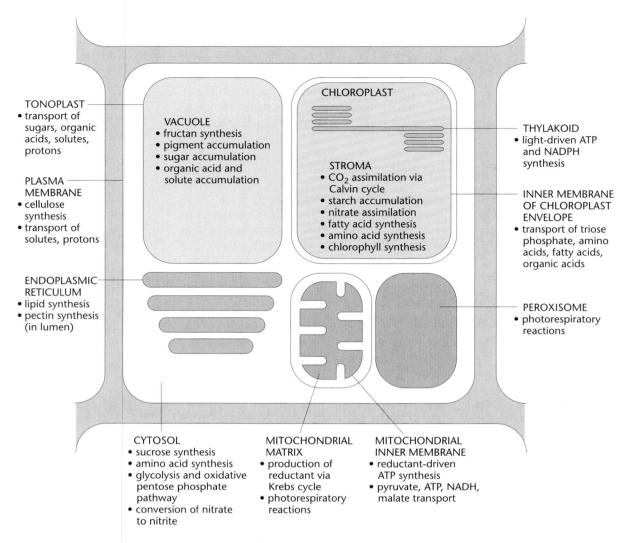

TONOPLAST
• transport of sugars, organic acids, solutes, protons

PLASMA MEMBRANE
• cellulose synthesis
• transport of solutes, protons

ENDOPLASMIC RETICULUM
• lipid synthesis
• pectin synthesis (in lumen)

VACUOLE
• fructan synthesis
• pigment accumulation
• sugar accumulation
• organic acid and solute accumulation

CHLOROPLAST

STROMA
• CO_2 assimilation via Calvin cycle
• starch accumulation
• nitrate assimilation
• fatty acid synthesis
• amino acid synthesis
• chlorophyll synthesis

THYLAKOID
• light-driven ATP and NADPH synthesis

INNER MEMBRANE OF CHLOROPLAST ENVELOPE
• transport of triose phosphate, amino acids, fatty acids, organic acids

PEROXISOME
• photorespiratory reactions

CYTOSOL
• sucrose synthesis
• amino acid synthesis
• glycolysis and oxidative pentose phosphate pathway
• conversion of nitrate to nitrite

MITOCHONDRIAL MATRIX
• production of reductant via Krebs cycle
• photorespiratory reactions

MITOCHONDRIAL INNER MEMBRANE
• reductant-driven ATP synthesis
• pyruvate, ATP, NADH, malate transport

Figure 4–1.
Compartmentation of metabolic processes in a leaf cell. For each subcellular compartment, some of the major metabolic processes are shown. Many processes occur exclusively in a single compartment but may obtain their substrates from, and export their products to, other compartments. Others take place in parallel in more than one compartment, or start in one compartment and are completed in another. Many processes take place in the soluble phase of subcellular compartments, others within or associated with specific membranes. The transporter proteins that allow metabolites to cross the membranes separating compartments play a vital role in coordinating metabolic processes in the cell.

composition, and so forth). The membranes separating the compartments contain multiple types of **transport proteins**, each allowing the movement of particular metabolites or ions (see Section 3.5). These transport proteins are as important in the control of cellular metabolism as are the enzymes in each compartment. Because of practical difficulties in studying the structure and function of proteins embedded in membranes, the transport proteins are generally less well understood than soluble enzymes.

Metabolic processes are coordinated and controlled by regulation of enzyme activities

The integration and coordination of metabolic processes in the cell, and their modulation in response to environmental and developmental changes, is brought about by regulation of the activities of enzymes. Two types of mechanism act to regulate enzyme activity (Figure 4–2). The first type, known as "coarse control," determines the amount of enzyme protein present in a cell. Coarse control operates through changes in the relative rates of synthesis and degradation (turnover) of the enzyme protein. As with any protein, the rate of synthesis depends on the rate of **transcription** of the gene encoding the enzyme (transcriptional control), the turnover of the **messenger RNA (mRNA)**, and the rate of **translation** of the mRNA (translational control). These regulatory mechanisms are discussed in Chapter 2. Coarse control can be regarded as setting the capacity of the cell to catalyze a particular reaction or metabolic pathway. Generally speaking, it brings about changes in amounts of enzymes in hours or days rather than minutes.

The other type of regulation, known as "fine control," works on a much shorter timescale to modulate the activities of enzymes within the framework set by coarse control. Thus, coarse control determines how many molecules of enzyme protein are present in the cell, while fine control determines the activity of these molecules. Developmentally programmed changes in metabolism, and changes in response to major and sustained environmental change, usually involve coarse control. Coarse-control mechanisms can act on a whole pathway or on single enzymes or blocks of enzymes in the pathway. For example, amounts of all Calvin cycle enzymes increase due to the synthesis of new enzyme protein as a young leaf develops and acquires photosynthetic capacity, and the amounts of all decrease during leaf **senescence**. Alterations in light intensity and CO_2 concentration also bring about changes in the amounts of Calvin cycle enzymes, but much greater changes for some enzymes than for others.

Fine control operates in a highly enzyme-specific manner, and by no means all enzymes are subject to this type of regulation. There is often great variation within a metabolic pathway in the extent to which its individual enzymes are subject to fine control, and in the nature of the fine control. A single enzyme may be subject to several kinds of fine-control mechanisms that modulate its actions in a variety of ways. This provides a very sensitive and rapid means of altering enzyme activity in response to perturbations in related areas of the metabolic network. Fine control operates either through **covalent modification** of specific amino acid residues of the protein, or through noncovalent interaction between the enzyme protein and specific metabolites.

The two most important types of covalent modification that modulate enzyme activity are **phosphorylation/dephosphorylation** of specific amino acid residues (often serine residues) and the **reduction** and **oxidation** of the sulfhydryl (–SH) groups of cysteine residues (Figure 4–2). Phosphorylation/dephosphorylation is brought about by specific **protein kinases** and **protein phosphatases**. (A kinase is an enzyme that adds a phosphate group to a specific protein or metabolite; a phosphatase is an enzyme that removes a phosphate group.) The phosphorylated enzyme has a different level of activity from the unphosphorylated enzyme. For example, phosphorylation of the

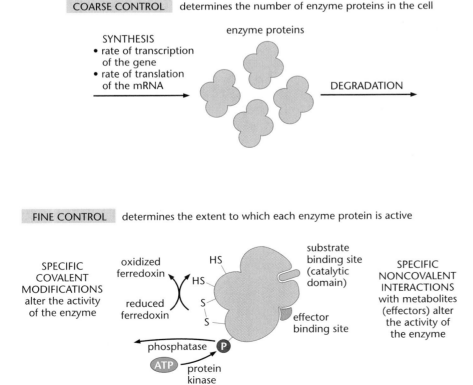

COARSE CONTROL determines the number of enzyme proteins in the cell

SYNTHESIS
• rate of transcription of the gene
• rate of translation of the mRNA

enzyme proteins

DEGRADATION

FINE CONTROL determines the extent to which each enzyme protein is active

SPECIFIC COVALENT MODIFICATIONS alter the activity of the enzyme

oxidized ferredoxin

reduced ferredoxin

HS
HS
S
S

substrate binding site (catalytic domain)

effector binding site

SPECIFIC NONCOVALENT INTERACTIONS with metabolites (effectors) alter the activity of the enzyme

phosphatase P
ATP protein kinase

Figure 4–2.
Regulation of enzyme activity by coarse and fine control. The intracellular levels of all enzymes, like those of all proteins, depend on a balance between synthesis and degradation; this is known as coarse control. In contrast, fine-control mechanisms are highly specific for particular enzymes. Some enzymes are not subject to fine control, some are subject to one particular type of fine control, and others may be subject to several different types—for example, oxidation/reduction of sulfhydryl (–SH) groups plus interaction with one or more effector metabolites. Effectors may interact with the enzyme protein at the substrate binding site, thus directly influencing the binding of the substrates, or may bind at different sites (as shown here). When they bind at different sites (allosteric effectors), they bring about conformational changes in the protein that affect the binding of substrates and/or other effectors. (Note that the *red shaded* "P" bonded to the protein represents a phosphate group; this symbol is used to denote phosphate in many figures in this chapter.)

enzymes **sucrose phosphate synthase** (see Section 4.2) and **nitrate reductase** (see Section 4.8) substantially inhibits the activity of these enzymes. Oxidation of sulfhydryl groups to form disulfide (S–S) bonds between cysteine residues in different regions of the protein can result in conformational changes that alter the activity of the enzyme. Oxidation and reduction of sulfhydryl groups are usually mediated by the redox (oxidation-reduction) proteins **thioredoxin** and **ferredoxin**. Several enzymes of the Calvin cycle are regulated in this way (see Section 4.2): reduced and active in the light, oxidized and inactive in the dark.

Changes in the activities of many enzymes are brought about by specific, noncovalent interactions of the enzyme protein with metabolites. These metabolites—often products of other steps in the same or related metabolic pathways—can affect enzyme activity in a variety of ways, and their effects may be complex. For example, a metabolite may alter the relationship between the activity of an enzyme and the concentration of its **substrate**, and may influence the extent to which other metabolites interact with the enzyme to alter its activity. The chloroplast enzyme **ADP-glucose pyrophosphorylase** catalyzes the conversion of glucose 1-phosphate and ATP to ADP-glucose, the substrate for synthesis of starch (see Section 4.6). ADP-glucose pyrophosphorylase is activated by reduction of S–S bonds, as described above. Once activated, its activity can be further enhanced or inhibited by its interaction with metabolites. Interaction with 3-phosphoglycerate increases the maximum possible activity of the enzyme and reduces the concentrations of the substrates (glucose 1-phosphate and ATP) required for maximum activity. 3-Phosphoglycerate also reduces the extent to which activity of the enzyme is inhibited by another interacting metabolite, **inorganic phosphate**. Thus changes in the relative concentrations of 3-phosphoglycerate and inorganic phosphate in the chloroplast have large and complex effects on the activity of ADP-glucose pyrophosphorylase. Metabolites such as 3-phosphoglycerate and inorganic phosphate that interact with enzymes to affect their activity are referred to as **allosteric effectors** (activators or inhibitors). Further examples of enzymes undergoing complex and important fine control are phosphofructokinase and pyrophosphate–fructose 6-phosphate 1-phosphotransferase, both of which can catalyze the first reaction of **glycolysis** (see Section 4.5).

The regulation of enzyme activity serves two purposes. First, changes in enzyme activity can change the rate at which a pathway operates—that is, increase or decrease the flux through the pathway (as discussed below). Second, regulation serves to coordinate the activities of different enzymes in a pathway in response to changes in flux and thus prevent large perturbations in metabolite levels that might disrupt the cell's metabolic network. For example, **phosphoribulokinase**, an enzyme of the Calvin cycle, is highly regulated by oxidation and reduction of sulfhydryl groups and by noncovalent interactions involving various Calvin cycle and related metabolites. In this way, the activity of this enzyme can change rapidly in response to changes in flux through the Calvin cycle resulting from, for example, changes in the availability of carbon dioxide or the supply of energy for the pathway. The coordination of phosphoribulokinase activity with flux ensures that its substrates and products do not undergo large perturbations in concentration when the flux changes.

Control of the flux through a metabolic pathway is a function of interactions between the individual steps in the pathway. In some pathways, the amount of a single enzyme is of overriding importance in controlling the overall flux. Relatively small changes in the activity of this enzyme translate directly into changes of comparable magnitude in the flux through the pathway; similar changes in the activities of other enzymes in the pathway have much smaller effects (Figure 4–3). In many pathways, however, no single enzyme is of overriding importance in controlling flux. Instead, many enzymes each make a small contribution. For any given enzyme, a large change in activity is required to bring about a small change in flux. The importance of an individual enzyme in the control of flux through a pathway can be calculated using a theoretical framework called **metabolic control analysis**. Some important principles of this theory are shown in Figure 4–3.

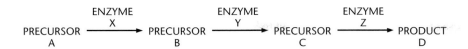

Figure 4–3.
Control of flux through a metabolic pathway. The analysis here considers flux through a simple, linear pathway—conversion of precursor A to product D via enzymes X, Y, and Z—in the soluble phase of a subcellular compartment. The application of metabolic control analysis in other situations is more complex, such as in branched pathways and in pathways for which the enzymes are associated together or are associated with a membrane.

1. For all of the enzymes in the pathway there is a hyperbolic relationship between the amout of enzyme present and the flux through the pathway.

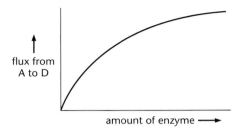

2. For most enzymes, the normal amount in the cell lies on the flat part of the curve. Large changes in amount are needed to bring about significant changes in flux.

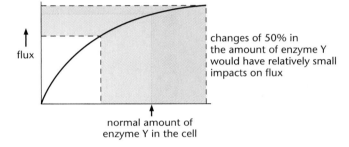

3. For a few enzymes, the normal amount in the cell lies on the steep part of the curve. Changes in amount cause comparable changes in flux.

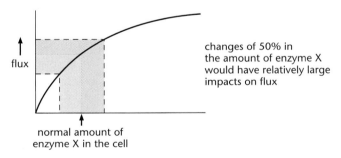

4. The slope of the curve can be used to calculate the FLUX CONTROL COEFFICIENT (FCC) for an enzyme. This is a measure of its importance in controlling flux through the pathway.

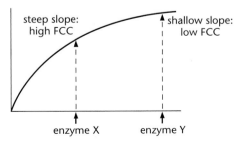

5. For any given pathway, the sum of the flux control coefficients for each enzyme is 1.

$$FCC_X + FCC_Y + FCC_Z = 1$$

If the FCC for X is 0.8 and that for Y is 0.08, the FCC for Z is 0.12.
In the pathway under these conditions, most of the control of the flux from A to D lies with enzyme X. Enzymes Y and Z are relatively unimportant in controlling flux.

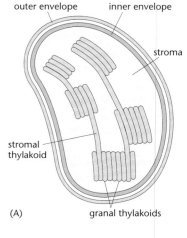

(A)

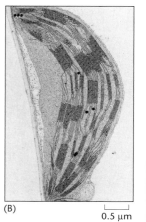

(B)

0.5 µm

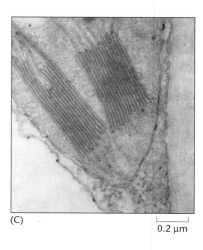

(C)

0.2 µm

The contribution of a particular enzyme to the control of flux through a pathway is not fixed. It is likely to change during cell development and with changing environmental conditions. Thus an enzyme that plays an important role in controlling flux through a pathway under one set of environmental conditions may be of minor importance under a different set of conditions.

4.2 CARBON ASSIMILATION: PHOTOSYNTHESIS

Photosynthesis converts atmospheric carbon dioxide into organic compounds that are either used in biosynthesis by the photosynthetic tissues themselves or converted to low-molecular-weight sugars (usually sucrose) and transported to nonphotosynthetic cells. Carbon dioxide is assimilated via a cycle of reactions that takes place in the **stroma** of the chloroplast. This process uses energy and reducing power supplied from light-driven reactions occurring in a specialized set of chloroplast membranes—the **thylakoids** (Figure 4–4). In this section we describe how physical energy (light) and inorganic carbon (CO_2) are captured and used by plants.

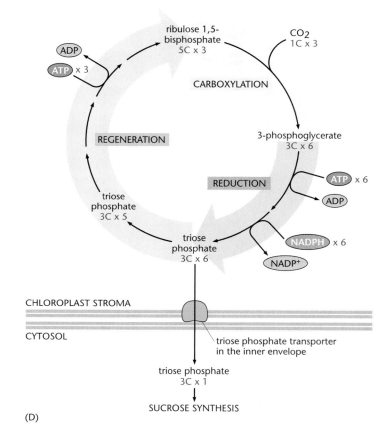

(D)

Figure 4–4.
The chloroplast and the Calvin cycle. (A) The chloroplast is surrounded by a double membrane, the envelope. The internal membranes, the thylakoids, are the location of light-driven electron transport chains that provide the energy (as ATP) and reducing power (as NADPH) for CO_2 assimilation. Assimilation occurs in reactions catalyzed by enzymes located in the stroma. (B) Electron micrograph of a chloroplast lying against a cell wall. (C) Closer view of one end of a chloroplast, showing the structure of the stacks of thylakoids (the grana). (D) Overview of the assimilation of CO_2 and its conversion to sucrose. Assimilation occurs in the Calvin cycle, taking place in the chloroplast stroma. Initially (the carboxylation stage), carbon from CO_2 is brought into organic combination by reaction with ribulose 1,5-bisphosphate (a 5-carbon compound). The resulting 3-phosphoglycerate (a 3-carbon compound) is reduced to triose phosphate. Some of this intermediate is exported from the chloroplast via a specific transporter protein, and is converted to sucrose in the cytosol. The remainder of the triose phosphate enters the regeneration stage of the Calvin cycle, in which it is converted to ribulose 1,5-bisphosphate. The diagram indicates the number of molecules of CO_2, Calvin cycle intermediates, NADPH (the reductant), and ATP involved in the net synthesis of one molecule of triose phosphate for sucrose synthesis (see also Figure 4–7). (C, courtesy of K. Plaskitt.)

Net carbon assimilation occurs in the Calvin cycle

Carbon dioxide is assimilated through a cycle of reactions known as the **reductive pentose phosphate pathway** or the Calvin cycle (named after Melvin Calvin, who discovered it). This is the only pathway by which plants can convert a net amount of inorganic carbon into sugars. Its reactions are common to all autotrophic organisms and are the source of most of the organic compounds on earth.

The cycle reactions are catalyzed by soluble enzymes in the chloroplast stroma, and can be divided into three stages (Figure 4–4D): (1) carboxylation of a five-carbon compound (the CO_2 "acceptor") to give two molecules of a three-carbon compound; (2) reduction of the three-carbon compound to give triose phosphates; and (3) regeneration of the five-carbon acceptor from the triose phosphates.

The initial, **carboxylation** step is catalyzed by the enzyme **Rubisco** (Figure 4–5): carbon dioxide is combined with water and the five-carbon compound ribulose 1,5-bisphosphate to form two molecules of 3-phosphoglycerate (we'll have more to say about the naming of this enzyme in Section 4.3). 3-Phosphoglycerate is then reduced to three-carbon sugars, the triose phosphates glyceraldehyde 3-phosphate and dihydroxyacetone phosphate (Figure 4–6). The assimilation of three molecules of carbon dioxide in this cycle produces one more molecule of triose phosphate than is required for the regeneration of three molecules of ribulose 1,5-bisphosphate. Thus, for every five triose phosphate molecules used to maintain the supply of ribulose 1,5-bisphosphate for carboxylation, one can be exported from the chloroplast for the synthesis of sucrose (Figure 4–7). The cycle is "autocatalytic," in that it allows for removal of an intermediate for synthesis of net amounts of a product (sucrose) without compromising the rate of regeneration of the acceptor compound.

The reactions of the Calvin cycle require energy in the form of ATP and reductant in the form of **NADPH (reduced nicotinamide adenine dinucleotide phosphate)** (Figure 4–7). The assimilation of 3 molecules of carbon dioxide—enabling the net synthesis of 1 molecule of triose phosphate for sucrose synthesis—uses 9 molecules of ATP and 6 molecules of NADPH. ATP and NADPH are synthesized in the thylakoid membranes, where light energy is captured and harnessed by electron transport chains.

Energy for carbon assimilation is generated by light-harvesting processes in the chloroplast thylakoids

The capture of energy from light and its use in the synthesis of ATP and NADPH for the Calvin cycle involves a complex series of physical and chemical reactions. Figure 4–8 illustrates the overall process.

Photons of light excite (energize) molecules of **chlorophyll**, a specialized pigment embedded in the thylakoid membranes. The energy of **photoexcitation** is then transferred by excited electrons to a **reaction center**, where this physical form of energy is

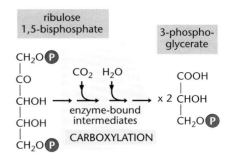

Figure 4–5.
The carboxylation reaction catalyzed by Rubisco. One molecule of ribulose 1,5-bisphosphate reacts with one CO_2 molecule to produce two molecules of 3-phosphoglycerate, in a complex reaction involving several intermediate steps that take place on the surface of the enzyme protein.

Figure 4–6.
The reduction stage of the Calvin cycle. The enzymes 3-phosphoglycerate kinase and glyceraldehyde 3-phosphate dehydrogenase catalyze the conversion of 3-phosphoglycerate to glyceraldehyde 3-phosphate, using ATP and NADPH. Glyceraldehyde 3-phosphate is isomerized to dihydroxyacetone phosphate by the enzyme triose phosphate isomerase; this reaction is close to equilibrium *in vivo*. One-sixth of the triose phosphate (glyceraldehyde 3-phosphate and dihydroxyacetone phosphate) produced is exported from the chloroplast for sucrose synthesis; the remaining five-sixths enters the regenerative phase of the Calvin cycle.

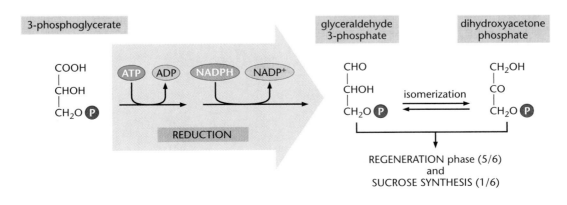

Figure 4–7.
Requirements for the net synthesis of one molecule of triose phosphate. The balance sheet shows the number of molecules of Calvin cycle intermediates, ATP, and NADPH involved at each stage of the cycle.

CARBON AND ENERGY BALANCE SHEET FOR THE CALVIN CYCLE

at the end of the REGENERATION phase	3 x 5C ribulose 1,5-bis Ⓟ	= 15C
in the CARBOXYLATION phase	3 x 5C + 3 x CO_2	= 18C
in the REDUCTION phase	6 x 3C 3-phosphoglycerate triose phosphate	= 18C
removed for SUCROSE SYNTHESIS	1 x 3C triose phosphate	= 3C
entering the REGENERATION phase	5 x 3C	= 15C

required for

REGENERATION	3 ATP	—
CARBOXYLATION	—	—
REDUCTION	6 ATP	6 NADPH
total requirement for net gain of 3C	9 ATP +	6 NADPH

converted to chemical energy by driving the transfer of one electron from a specific chlorophyll molecule to an acceptor molecule, thus leaving the chlorophyll molecule in an oxidized state ("oxidation" is loss of one or more electrons; "reduction" is the gain of electrons). This conversion of energy from a physical to a chemical state is known as the **primary charge-separation** event.

Figure 4–8.
Overview of the conversion of physical energy from light to chemical energy in the form of ATP and reductant (NADPH) for the Calvin cycle. Light energy excites chlorophyll (Chl) molecules in the thylakoid membrane. The excitation energy passes from chlorophyll to chlorophyll to a reaction center, where excitation of a specific chlorophyll molecule results in the loss of an electron (the primary charge-separation event). The electron passes through electron transport chains to NADP⁺ to form NADPH. The energy generated by the transfer is used to create an electrochemical gradient across the thylakoid membrane, and this gradient drives the synthesis of ATP from ADP and inorganic phosphate. (Note that a *red shaded* "P" not bonded to another molecule represents inorganic phosphate, P_i.) NADPH and ATP provide the reductant and energy for the operation of the Calvin cycle, allowing the net synthesis of triose phosphate and hence sucrose.

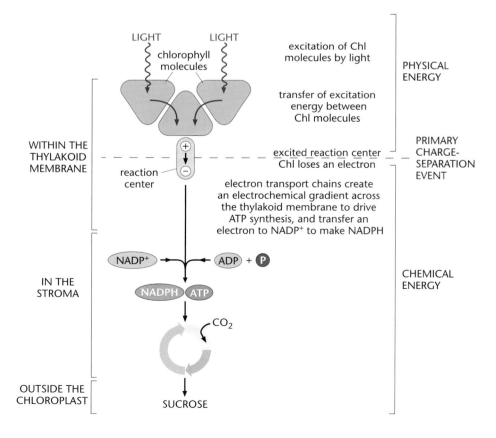

The reduced acceptor molecule in turn donates its newly received electron to an **electron carrier**, a molecule that accepts electrons then donates them to another molecule. Several different electron carriers exist in the thylakoid membrane, and together these make up electron transport chains that transfer electrons from the acceptor molecules in the reaction centers to $NADP^+$, reducing it to NADPH. The transfer of electrons along the chain drives the transport of protons across the thylakoid membrane from the stroma to the lumen (the internal space contained within the thylakoid membranes), setting up an **electrochemical gradient** across the membrane. The flux of protons back across the membrane from the lumen to the stroma, down the electrochemical gradient, drives the activity of membrane-bound **ATP synthases**, which phosphorylate ADP to ATP (Figure 4–9).

We will describe the stages of this process in detail: the initial capture of light energy and transfer of energy to the reaction centers; the transfer of electrons via electron transport chains to reduce $NADP^+$ and set up a proton (pH) gradient; and the powering of ATP synthesis by the proton gradient. (See Box 4–1 on some general properties of light and light absorption by plants.)

Light energy is captured by chlorophyll molecules and transferred to reaction centers

Light energy is captured by large, discrete complexes of chlorophyll molecules bound to proteins and embedded in the thylakoid membranes. Each of these discrete complexes constitutes a photosystem, each containing about 250 chlorophyll molecules. The majority of the chlorophyll molecules in a photosystem make up the **antenna.** The antenna molecules channel excitation energy toward the remaining chlorophyll molecules, which make up the **core complex** (Figure 4–10A). The reaction center is contained within the core complex, and it is here that light energy is transduced into chemical energy. In the antenna, chlorophyll is bound to proteins known as "light-harvesting complex polypeptides" (LHCPs); these pigment–protein complexes span the thylakoid membrane, with chlorophyll molecules positioned to maximize the efficiency of energy transfer among them. The **light-harvesting complexes** contain two types of chlorophyll molecule, chlorophyll *a* and chlorophyll *b*, in an average ratio of about 3 to 1. The complexes also contain related pigment molecules known collectively as **carotenoids** (Figure 4–10B).

When antennal chlorophyll absorbs a photon, energy is transferred to adjacent chlorophyll molecules and then channeled toward the reaction center in the photosystem. Chlorophylls *a* and *b* have different **absorption spectra**: chlorophyll *a* is excited by

Figure 4–9.
Overview of ATP synthesis in the chloroplast. The energy produced by transfer of an electron from an excited chlorophyll molecule along the electron transport chains of the thylakoid membrane is used to drive the transfer of protons from the stroma into the thylakoid lumen. This sets up both a proton (pH) gradient and an electrical charge across the thylakoid membrane (an electrochemical gradient). Because of this transfer of protons, the pH in the lumen may be as low as 4, while that of the stroma may be as high as 8 (compared with a pH of about 7 in the cytosol). Movement of protons down the electrochemical gradient from the lumen to the stroma is coupled, via ATP synthase proteins embedded in the membrane, to the phosphorylation of ADP to produce ATP in the stroma.

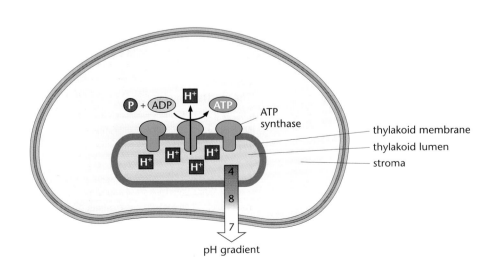

ATP synthase
thylakoid membrane
thylakoid lumen
stroma

pH gradient

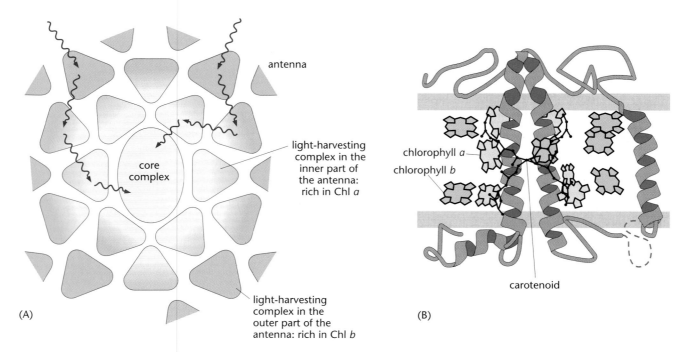

(A)

(B)

Figure 4–10.
Structures of a photosystem and a light-harvesting complex.
(A) View of a photosystem from above the thylakoid membrane. The core complex, containing the reaction center where the primary charge-separation event takes place, is surrounded by the antenna. The antenna contains light-harvesting complexes, each consisting of three light-harvesting complex polypeptides (LHCPs) and associated pigments embedded in the membrane. Light-harvesting complexes in the outer parts of the antenna contain a higher ratio of chlorophyll b to chlorophyll a than those closer to the reaction center: this distribution channels the excitation energy captured by chlorophyll molecules toward the reaction center.
(B) Structure of an LHCP and its associated pigments, seen from the side. The protein spans the thylakoid membrane, and the chlorophyll and carotenoid pigments lie within the membrane. LHCPs are encoded by a multigene family: each type of LHCP is associated with particular ratios of chlorophyll a to chlorophyll b and other pigment molecules.

Box 4–1 Light

Plants use light in two distinct ways: as the source of energy for photosynthesis (discussed in this chapter) and as a signal that enables many plant functions to be adjusted to changes in the light environment. These light-responsive functions—for example, chloroplast development, stem elongation, germination, and directional growth—are considered in Chapter 6.

Light is part of the spectrum of **electromagnetic radiation** (Figure B4–1). The part of the spectrum visible to the human eye largely overlaps the part used by plants for photosynthesis and as a regulatory signal. The human eye perceives radiation with wavelengths between about 380 nm (the violet side of the rainbow) and 720 nm (the red side of the rainbow). The light that provides the energy for photosynthesis also falls in this range. Electromagnetic radiation perceived as signals by plants includes wavelengths on either side of this visible range—in the near-ultraviolet and far-red range.

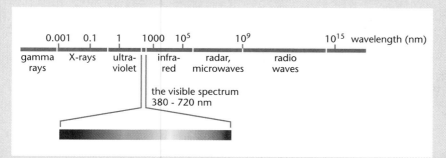

Figure B4–1.
The spectrum of electromagnetic radiation.
Energy from the sun that reaches the earth's surface has wavelengths in the range of 350 nm (in the ultraviolet) to 1500 nm (in the infrared, perceived as heat). The visible part of this energy—white light—is in the range 380 to 720 nm. White light consists of a spectrum of colors, seen when sunlight passes through a prism or through raindrops (in the form of a rainbow).

Box 4–1 Light (continued)

Light is perceived by plants via absorption by pigments. Photosynthetically active light is absorbed by chlorophyll *a* and chlorophyll *b*. Both pigments have two peaks of absorption in the visible spectrum, between about 400 and 480 nm (the blue end) and between 550 and 700 nm (the red end). Neither chlorophyll absorbs light in the middle part of the visible range—the green part of the spectrum (Figure B4–2A). This is why leaves appear green: the red and blue components of the visible spectrum are absorbed by chlorophyll, but the green wavelengths are reflected or pass through the leaf.

Several types of pigment are responsible for the perception of light as a signal. The best known of these light sensors—known as **photoreceptors**—is phytochrome, which absorbs light in the red and far-red parts of the spectrum. Phytochrome exists in two forms, with different biological activities and different absorption spectra (Figure B4–2B). The two forms are interconvertible: when one form absorbs light, it is converted to the other form. Because the two forms have different signaling functions, the plant can respond to differences in the ratio of red to far-red light. This ratio varies with environmental conditions: for example, full sunlight is rich in red light and poor in far-red light, whereas the

shade under a leaf canopy is enriched in far-red relative to red light, because the red component of sunlight is absorbed by chlorophyll in the canopy. The importance of the phytochrome system in plant responses to light is discussed in Chapter 6.

In considering how much light is needed for a particular plant process, it is convenient to consider light in terms of particles, known as "photons," rather than as a wave form. For most plants, the minimum amount of light required for net photosynthesis is in the range 1 to 20 μmol of photons m^{-2} s^{-1}. Below this value (called the **light compensation point**), there is a net loss of carbon dioxide from the plant through respiration (Figure B4–3). Plants adapted to growth in low light conditions (e.g., plants of woodland floors) often have low light compensation points, in the range 1 to 5 μmol of photons m^{-2} s^{-1}, whereas plants adapted to full sunlight have higher light compensation points. For the vast majority of plants (C3 plants; see Section 4.3), photosynthesis is saturated (i.e., reaches a maximal value) with respect to light at values of about 500 μmol of photons m^{-2} s^{-1} (Figure B4–3), well below the value for full sunlight, which is about 2000 μmol of photons m^{-2} s^{-1}. For one group of plants, the C4 plants, which have a higher energy requirement for photosynthesis (see Section 4.3), photosynthesis is not saturated even in full sunlight.

The amount of light required to trigger light-responsive functions in plants varies widely from one type of photoreceptor to another. In some cases, the amount required is tiny in comparison to the amount required for photosynthesis. For example, Arabidopsis seeds, which require light for maximum rates of germination, can be induced to germinate by light intensities as low as 1 nmol of photons m^{-2} s^{-1}.

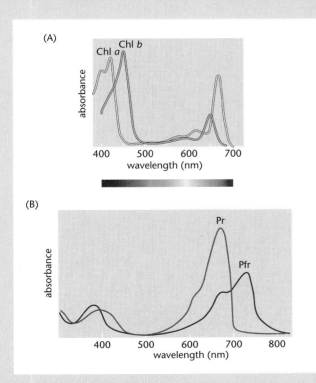

(A)

(B)

Figure B4–2.
Light absorption by chlorophylls and phytochromes. (A) Light absorption by chlorophyll *a* and chlorophyll *b*, plotted as a function of wavelength. This type of plot is known as an "absorption spectrum." Note that both pigments absorb blue and red light, but not green light. (B) The absorption spectra of the two forms of phytochrome, Pr and Pfr.

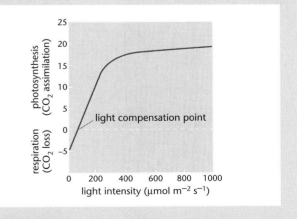

Figure B4–3.
A typical plot of CO$_2$ assimilation and CO$_2$ loss by a plant against light intensity. At light intensities below the light compensation point, the rate of CO$_2$ assimilation by photosynthesis is less than the rate of CO$_2$ loss by respiration. Light intensities of about 500 μmol of photons m^{-2} s^{-1} are sufficient to give maximal rates of photosynthesis in most plants. The values on the y axis are a commonly used way of expressing the rate of CO$_2$ loss or assimilation—as μmol of CO$_2$ m^{-2} (of leaf surface) s^{-1}.

photons of lower energy (longer wavelengths of light) than chlorophyll *b*, and thus achieves a lower energy state than *b*. This means that excitation energy passes readily from chlorophyll *b* to chlorophyll *a*, but not in the other direction. The ratio of chlorophyll *a* to *b* increases toward the reaction center (Figure 4–10A), thus channeling excitation energy toward the reaction center. In addition, chlorophyll *a* molecules associated with the reaction centers are excited by photons of lower energy (i.e., longer wavelength) than antennal chlorophyll *a*—similarly promoting the transfer of excitation energy toward the reaction center. The absorption spectra of antennal and reaction-center chlorophyll *a* differ because the chlorophylls in the two locations are bound to different types of protein.

Electron transfer between two reaction centers via an electron transport chain reduces NADP⁺ and generates a proton gradient across the thylakoid membrane

Higher plants contain two types of photosystem: **photosystem I (PSI)** and **photosystem II (PSII)**; the two work in series, connected by an electron transport chain (Figure 4–11). In essence, primary charge-separation events in the PSI reaction center cause the reaction-center chlorophyll *a* molecule to donate electrons to NADP⁺ via an electron transport chain, thus making the NADPH required by the Calvin cycle. The oxidized chlorophyll *a* molecule in the PSI reaction center is then re-reduced by electrons generated by primary charge-separation events in PSII and transferred to PSI by an electron transport chain. In turn, the oxidized chlorophyll *a* in the PSII reaction center is re-reduced by electrons donated from water, via a **water-splitting complex** that is an integral part of the PSII reaction center. The photosystems and the components of the electron transport chain are asymmetrical complexes, organized across the thylakoid membrane in such a way that reductive reactions (in which protons are taken up) occur on the stromal side of the membrane, and oxidative reactions (in which protons are released) occur on the lumenal side of the membrane. This leads to a net transfer of protons from the stroma to the lumen (Figure 4–11), setting up both a proton (pH) gradient and an electrical potential across the membrane. This electrochemical gradient drives the synthesis of ATP by the ATP synthase complex embedded in the membrane. We consider first the structure and operation of the photosystems and the electron transport chain that links them.

Figure 4–11.
Overview of the flow of electrons between the photosystems in the thylakoid membrane. At the reaction center of photosystem I (PSI), excitation of the chlorophyll *a* molecule P700 by light energy results in an excited state and the consequent loss of electrons (e⁻). The electrons are transferred to NADP⁺ to form NADPH, required in the Calvin cycle. P700 is re-reduced by electrons produced by excitation of the chlorophyll *a* molecule P680, in the reaction center of photosystem II (PSII). The electrons are transferred to PSI via a series of electron carriers in the thylakoid membrane. P680 is re-reduced by transfer of electrons from water, producing oxygen. The energy generated by the electron transport chain drives the transfer of protons from the stroma into the thylakoid lumen, building up an electrochemical gradient. Protons can move back down this gradient via the ATP synthase complex in the membrane; this movement drives synthesis of the ATP required for the Calvin cycle. The antennae of the photosystems are not shown.
(PQ = plastoquinone; PC = plastocyanin)

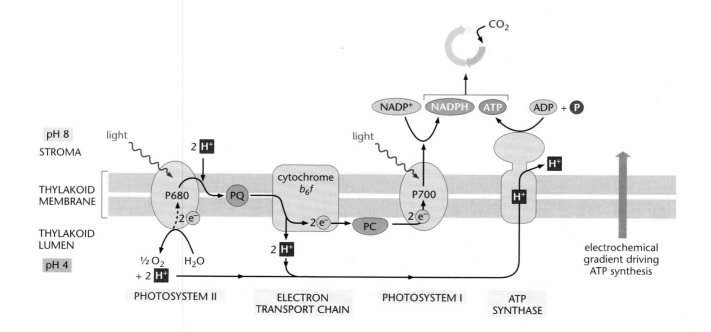

The chlorophylls of the PSI and PSII reaction centers are termed **P700** and **P680**, respectively. The numbers correspond to the wavelengths of light (in nanometers) at which the chlorophyll in each reaction center is maximally excited. The antennal chlorophylls associated with each reaction center also absorb maximally at different wavelengths; these absorbance maxima are geared to match the energy required to excite their reaction-center chlorophylls. Overall, this means that light energy is efficiently used over a wider range of wavelengths than would be possible with only one type of photosystem. The wavelengths of light used in photosynthesis are described in Box 4–1.

When the sequential operation of the photosystems is displayed so as to show the reduction potential of the components, as in Figure 4–12A, the scheme takes on the appearance of a "Z," and thus is called the "Z scheme." The excited P700 chlorophyll (P700*) formed by charge-separation events in PSI is a very strong reductant, able to donate electrons for the reduction of NADP$^+$. The excited P680 chlorophyll (P680*) formed by charge-separation events in PSII is a weaker reductant, which can donate electrons to re-reduce the oxidized P700 chlorophyll of PSI. The oxidized P680 chlorophyll is a strong oxidant that can accept electrons from water, resulting in the production of oxygen.

Not all photosynthetic organisms have two photosystems that operate sequentially. The Z scheme is found in higher plants, algae, and cyanobacteria, but other photosynthetic bacteria have only one photosystem. In purple bacteria, the electrons lost from the reaction-center chlorophyll cycle back via an electron transport chain (Figure 4–12B). The energy released is used to generate a transmembrane electrochemical gradient that drives the synthesis of ATP and reductant.

The structure and hence the function of the core parts of the photosystems and their reaction centers are not yet completely understood. Much of our present understanding comes from studies of the single photosystem of purple bacteria, which is similar to PSII (Figure 4–12B). Researchers have been able to put together a three-dimensional model of the core part of PSII by analyzing the extracted and crystallized photosystem from purple bacteria (e.g., *Rhodobacter*).

The core complex of PSII contains about 20 proteins. Some of these are part of the reaction center where the charge-separation event takes place; others are involved in binding the antennal complexes and in the oxidation of water and electron transfer. The reaction center and antennal binding proteins are highly hydrophobic and so are deeply embedded in the nonpolar membrane; they are encoded by genes in the chloroplast genome (see Section 2.6). The proteins involved in the oxidation of water are exposed on the lumenal side of the thylakoid membrane and are nuclear-encoded.

Two reaction-center proteins, **protein D1** and **protein D2**, mediate the transfer of electrons released from P680 during primary charge separation to **plastoquinone**, a molecule that can move around in the membrane and bind reversibly to proteins (Figure 4–13). Electrons pass from P680 to a tightly bound plastoquinone molecule (Q_A) on D1, then to a loosely bound plastoquinone molecule (Q_B) on D1. After accepting two electrons, this plastoquinone acquires two protons from the stromal side of the membrane, thus becoming reduced. It then dissociates from D1 and diffuses into the membrane, where its electrons are transferred to further components of the electron transport chain (see Figure 4–11 and below). It is replaced on D1 by an oxidized plastoquinone from the mobile pool of plastoquinone in the membrane. The D1 protein is particularly unstable and has a very short half-life. It probably becomes damaged by the highly reactive states of oxygen (**reactive oxygen species (ROS)**) generated during the oxidation of water (see below). The plastoquinone-binding site of D1 is the site of action of several important herbicides, in particular atrazine, which inhibit binding and thus the whole photosynthetic process.

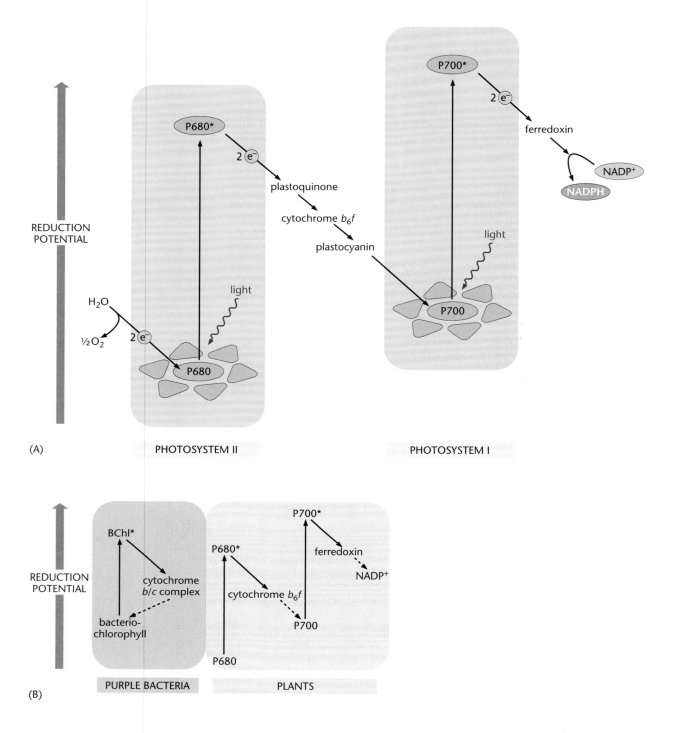

Figure 4–12.
The Z scheme, and a comparison of the photosystems of purple bacteria and plants. (A) The plant photosystems and their interconnecting electron transport chain are shown as a function of the relative strength of the various components as reductants (their reduction, or redox, potential). Strong reductants (with the most negative redox potentials) are at the top of the diagram, and strong oxidants (with the most positive redox potentials) are at the bottom. (B) In contrast to higher plants and cyanobacteria, the photosynthetic apparatus of purple bacteria consists of a single type of photosystem. Electrons lost from excited bacteriochlorophyll (BChl) are transferred via an electron transport chain back to the chlorophyll, re-reducing it. This "cyclic electron transport" generates an electrochemical gradient across the bacterial membrane, which drives ATP synthesis via an ATP synthase complex, as in plants. It also provides energy to drive an additional electron transport chain (*not shown*) that transfers electrons from a reduced substrate (often hydrogen sulfide, H_2S) to NAD^+, generating NADH as the reductant for carbon assimilation via the Calvin cycle.

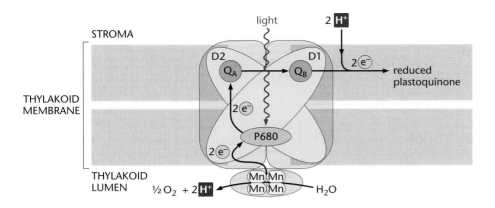

Figure 4–13.
Structure of the PSII core complex.
The electron-transfer components of the PSII reaction center are associated with two main proteins, D1 and D2. Excitation of P680 by energy transferred from antennal chlorophylls results in loss of electrons, which are transferred to the mobile plastoquinone pool in the thylakoid membrane via two bound plastoquinones, Q_A and Q_B. P680 is re-reduced by transfer of electrons from a cluster of manganese ions, located on proteins (the "water-splitting complex") on the lumenal side of the reaction center. The cluster accepts electrons from water to yield oxygen.

The transfer of electrons from water to P680 after the charge-separation event involves a cluster of four manganese ions associated with proteins in the water-splitting complex at the lumenal face of PSII. These ions act as a "charge buffer" between the conversion of water to oxygen and the donation of electrons to P680. The complete oxidation of water to oxygen requires the loss of four electrons (Figure 4–14). Successive passage of individual electrons from water to replace individual electrons lost from P680 would generate ROS—oxygen radicals—that are potentially highly damaging to cell components. This problem is largely prevented by the manganese cluster, which can contain from zero to four electrons at a time (that is, it can exist in five different oxidation states). The manganese cluster donates the electrons one at a time to re-reduce P680 after each of four successive charge-separation events. The fully oxidized state of the cluster (4+) then takes four electrons simultaneously from water, converting two water molecules to oxygen with minimal production of ROS. Note that the oxidation of each water molecule results in the release of its two protons on the lumenal side of the membrane, contributing to the electrochemical gradient.

Electrons generated in the primary charge-separation event in PSII are transferred along an electron transport chain in the thylakoid membrane to PSI. The second component of the chain, the **cytochrome b_6f** complex, is a large, membrane-spanning association of chloroplast- and nuclear-encoded proteins. In contrast, the first and last components, plastoquinone and **plastocyanin** (a single, small protein), are small molecules that are mobile in the membrane and can diffuse between cytochrome b_6f and the photosystems. Electrons from reduced plastoquinone are transferred to plastocyanin via electron-transfer components of the cytochrome b_6f complex on the lumenal side of the membrane. This reoxidation of plastoquinone releases into the lumen the protons that it acquired from the stromal side of the membrane during its reduction by electrons

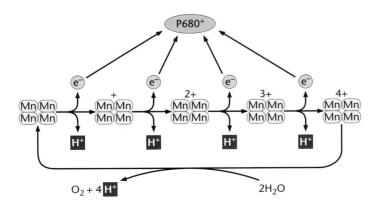

Figure 4–14.
Transfer of electrons from water to oxidized P680. A cluster of four manganese ions in the water-splitting complex feeds electrons to oxidized P680 (P680$^+$) until all four ions are oxidized. The manganese ions are re-reduced by electrons from two molecules of water, yielding four protons and molecular oxygen (O_2).

from P680 of PSII. Thus two protons are transferred from the stroma to the lumen for every two electrons transferred from PSII to plastocyanin, contributing to the electrochemical gradient across the membrane. Reduced plastocyanin moves from the cytochrome b_6f complex and donates its electrons to oxidized P700 (Figure 4–11).

The core complex of PSI contains about 15 proteins, some nuclear- and some chloroplast-encoded. By far the largest proteins, PsaA and PsaB, are chloroplast-encoded and make up the reaction center. Other proteins link the photosystem to its antenna and are involved in the transfer of electrons to a soluble electron-carrier protein, ferredoxin, on the stromal side of the membrane. Ferredoxin forms a complex with the enzyme ferredoxin-NADP reductase, which mediates the transfer of electrons to NADP$^+$ (Figure 4–15). The herbicide paraquat (methyl viologen) acts as an alternative electron acceptor to ferredoxin, thus disrupting the operation of the electron transport chain and hence the whole process of photosynthesis.

The proton gradient drives the synthesis of ATP by an ATP synthase complex

The way in which electron transport chains in the membranes of chloroplasts and mitochondria drive ATP synthesis was first explained by Peter Mitchell in the 1960s. In his **chemiosmotic model**, Mitchell proposed that electron-transfer processes in the membrane result in a net movement of protons from one side of the membrane to the other, generating both a chemical gradient (a gradient of proton concentration) and an electrical potential across the membrane. Mitchell called this electrochemical gradient the **proton-motive force**. The energy released by dissipation of the proton-motive force—by passage of protons down the electrochemical gradient through a channel in a membrane-spanning complex—is used to drive the synthesis of ATP. In chloroplasts this process is called **photophosphorylation**.

In chloroplasts, a net transfer of protons across the thylakoid membrane, from the stroma to the lumen, takes place at two points in the transfer of electrons from water to NADP$^+$ (Figure 4–16). First, oxidation of water on the lumenal side of the membrane frees protons into the lumen. Second, electron transfer to and from plastoquinone results in the transfer of protons from the stroma to the lumen. The release of protons

Figure 4–15.
Structure of the PSI core complex. P700 is associated with the proteins PsaA and PsaB at the reaction center. Excitation of P700 by energy transferred from antennal chlorophylls results in loss of an electron. This electron is transferred by electron carriers (iron-sulfur proteins) on the Psa proteins and on associated proteins on the stromal side of the complex to a soluble protein, ferredoxin. Electrons from reduced ferredoxin are then transferred to NADP$^+$ by the stromal enzyme ferredoxin-NADP reductase. P700 is rereduced by transfer of electrons from the mobile electron carrier plastocyanin.

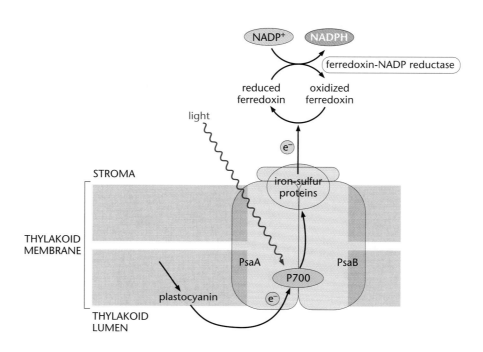

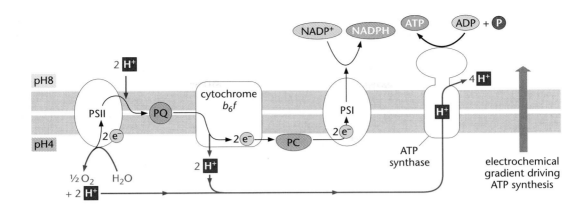

at these two points produces a large gradient of protons and a small electrical potential across the thylakoid membrane. In conditions favorable for photosynthesis, the difference in proton concentration between the lumen and the stroma may be as much as four orders of magnitude (4 pH units): the pH values of the lumen and the stroma are about 4 and 8, respectively.

The ATP synthase complex is located in the stromal thylakoid membrane. Its three-dimensional structure is well understood (Figure 4–17). The complex consists of two major components: the transmembrane portion (the CF_0 segment), which contains the channel for proton movement, and the portion that projects into the stroma (the CF_1 segment), which synthesizes ATP. The CF_0 segment consists of four types of protein, three chloroplast-encoded and one nuclear-encoded. The CF_1 segment consists of three chloroplast-encoded and two nuclear-encoded proteins.

Figure 4–16.
Transfer of protons across the thylakoid membrane. Net transfer of protons from the stroma into the thylakoid lumen is brought about by the oxidation of water in PSII and by the transfer of electrons from plastoquinone (PQ) to cytochrome b_6f. This sets up an electrochemical gradient across the membrane. The proton-motive force thus generated is dissipated as protons move back down the gradient, into the stroma, via a channel in the ATP synthase complex that spans the thylakoid membrane. This flux of protons drives the synthesis of ATP on the stromal side of the ATP synthase.

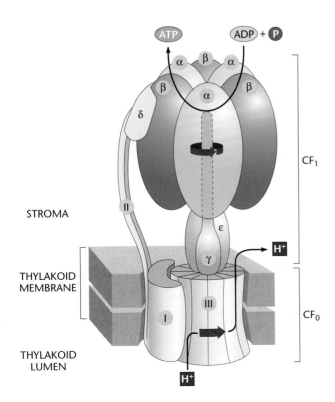

Figure 4–17.
The ATP synthase complex. Within the thylakoid membrane, protein III of the CF_0 segment of the complex forms the channel through which the proton-motive force is dissipated by flux of protons from the lumen to the stroma. This drives conformational changes in the CF_1 segment (indicated by the *red arrow*; see Figure 4–18) on the stromal side of the membrane, and these changes allow the synthesis of ATP. Components I and III of CF_0 are encoded by the chloroplast genome, whereas component II is encoded in the nucleus. Components I and II are responsible for binding CF_0 to CF_1. In CF_1, the α, β, and ε components are chloroplast-encoded, whereas the γ and δ components are nuclear-encoded.

How the passage of protons through CF_0 is coupled to the synthesis of ATP by CF_1 is not clear. It is generally agreed, though, that the flux of protons brings about a cycle of conformational changes in the catalytic sites of CF_1, and these changes drive ATP synthesis. At a given time, the three catalytic sites of CF_1 exist in three different states: loose, tight, and open (Figure 4–18). Conformational changes brought about by proton flux through CF_0 drive the three sites through a cycle of all three states. In the loose state, a site binds the substrates for ATP synthesis, ADP and inorganic phosphate; conversion to the tight state promotes the synthesis of ATP; and finally, conversion to the open state releases the ATP.

The ATP synthase complex is strictly regulated by light and by the pH gradient across the thylakoid, such that it is inactive in the dark and when rates of electron transport are low. This regulation is important because the ATP synthase activity is potentially reversible—if active in the absence of a pH gradient sufficient to drive ATP synthesis, it could hydrolyze stromal ATP and drive proton movement from the stroma to the lumen (i.e., acting as an **ATPase**). The catalytic sites of CF_1 are inactivated when the pH gradient drops below a certain level. Their activity also depends on the reduction of a disulfide bond on one of the CF_1 subunits. The reduction is brought about by reduced thioredoxin, an important sensor of the rate of electron transport: thioredoxin is reduced in the light and oxidized in the dark, hence CF_1 is active only in the light (thioredoxin is further discussed below).

Light-harvesting processes are regulated to maximize the dissipation of excess excitation energy

Under many circumstances, the light energy absorbed by the light-harvesting complexes greatly exceeds that needed to meet the energy requirements of the Calvin cycle. This occurs at high light intensities and when the rate or capacity of the Calvin cycle is limited—for example, by low temperature (which slows chemical reactions but not physical processes such as absorption of light energy, transfer of energy between chlorophylls, and the primary charge-separation events), by stresses such as nutrient deficiency and disease, and by lack of carbon dioxide as **stomata** close to conserve water (see Chapter 7). In these circumstances, the electron carriers of the electron transport chains can potentially become highly reduced, as NADPH is consumed only slowly and little $NADP^+$ is available to receive electrons.

When the electron acceptors of PSI and PSII are highly reduced, they cannot accept electrons lost from P680 and P700 in the primary charge-separation events. This causes excitation energy to build up in the photosystem and antennal chlorophylls, which can cause serious damage to the light-harvesting apparatus, known as **photoinhibition**. Mechanisms that prevent damage caused by photoinhibition are described in Section 7.1.

Figure 4–18.
Synthesis of ATP in CF_1 of the ATP synthase complex. This scheme depicts the complex from the stromal side. The three α and three β subunits of CF_1 make up three nucleotide-binding sites. These exist in three different states and are driven from one state to the next by rotation of the γ subunit, which forms part of the "stalk" linking CF_0 to CF_1 (see Figure 4–17). Rotation of the γ subunit (shown by the *red arrow*) is driven by proton flux through CF_0. In the loose state (L), the nucleotide-binding site binds ADP and inorganic phosphate from the stroma. In the tight state (T), ATP and P_i are converted to ATP. In the open state (O), ATP is released from the complex.

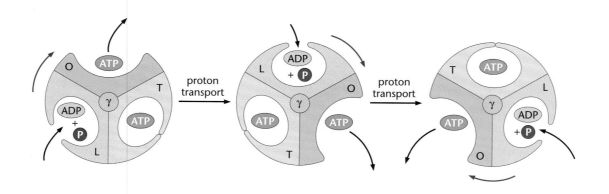

In addition to mechanisms for dissipating excess energy, the light-harvesting processes also require mechanisms to regulate the distribution of excitation energy between the two photosystems. For effective flow of electrons through the two photosystems to $NADP^+$, excitation energy should be evenly distributed between PSI and PSII. As described above, the two photosystems have different wavelengths of maximum light absorption, so the distribution of excitation energy between them could change if light conditions—intensity and wavelength distribution—changed. This would result in suboptimal and variable efficiency in the use of light energy. The mechanism that maintains a balance between the two systems relies on the different spatial locations of PSI and PSII in the thylakoids.

As illustrated in Figure 4–4, thylakoids exist in both stacked (**granal thylakoids**) and unstacked (**stromal thylakoids**) forms, which has consequences for the functioning of complexes in the membrane. Complexes on the appressed surfaces of granal thylakoids, which include the PSII complexes, exist in a hydrophobic environment with relatively little access to the stroma. Complexes in the stromal thylakoids are more exposed to the hydrophilic stromal environment; these include the PSI complexes, which can donate electrons to the stromally located ferredoxin (Figure 4–19), and the ATP synthase complex. The cytochrome b_6f complex, which forms part of the electron transport chain linking the two photosystems, is evenly distributed between granal and stromal thylakoids. As we described earlier, transfer of electrons between PSII and PSI is made possible by the mobility within the membrane of plastoquinone, which can migrate between granal and stromal regions. Thus there is no requirement for a strict stoichiometry between PSI and PSII: electrons move from one to the other via a mobile pool of electron carrier. In fact, the ratio of PSII to PSI complexes in many plants is about 1.5:1.

The balancing of excitation energy between the photosystems is mediated by changes in the size of the antennae of PSII, brought about by the movement of some of the antennal light-harvesting complexes between granal and stromal thylakoids (Figure 4–20). When excitation of PSII exceeds that of PSI, this triggers phosphorylation of LHCII, a light-harvesting complex associated with PSII, giving the complex a negative charge. This results in electrostatic repulsion between the LHCII complexes, which tends to unstack the granal thylakoids. A part of the LHCII population then migrates from the granal to the stromal thylakoids, reducing the size of the PSII antennae.

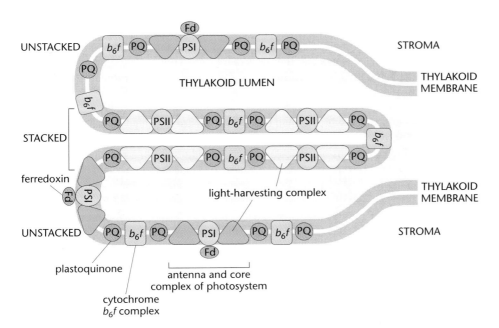

Figure 4–19.
Distribution of protein complexes between stacked and unstacked thylakoid membranes. PSII is found in granal stacks in which the membranes are closely appressed, whereas PSI is found in unstacked membranes. Cytochrome b_6f, and the mobile pools of plastoquinone (PQ) and plastocyanin (*not shown*), are found in both regions. The *green* and *yellow triangles* are light-harvesting complexes.

Figure 4–20.
Control of the distribution of excitation energy between photosystems I and II by phosphorylation of LHCII. In its unphosphorylated state (*top*), LHCII (*yellow triangles*) is exclusively associated with PSII in appressed thylakoid membranes. Excess excitation of PSII leads to phosphorylation of LHCII. This confers a negative charge on the surface of the membranes, leading to partial unstacking (*middle*) and migration of some of the LHCII to PSI in the stromal thylakoids (*bottom*). The resulting change in the relative size of the antennae of PSI and PSII redresses the imbalance in excitation level between the two photosystems.

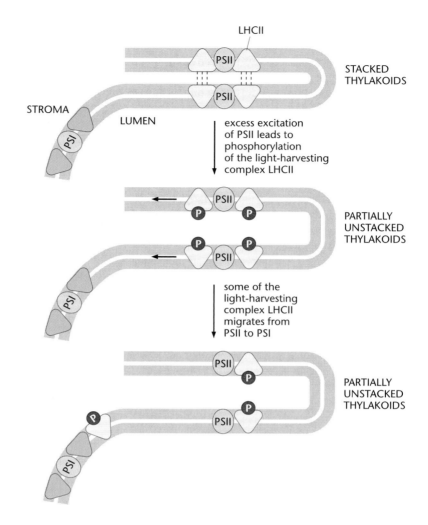

Carbon assimilation and energy supply are coordinated by complex regulation of Calvin cycle enzymes

Integration of Calvin cycle activity with the supply of ATP and NADPH is essential for a plant's efficient and sustained assimilation of carbon dioxide. When energy supply changes due to changing environmental conditions, coordinated regulation of certain Calvin cycle enzymes ensures that the pools of intermediates are maintained at optimal levels, thus allowing sensitive modulation of the balance between energy supply and carbon assimilation.

The main mechanism by which Calvin cycle enzymes are regulated by the cell's energy supply is **reductive activation** (Figure 4–21). Some enzymes of the cycle are active when the sulfhydryl groups of particular cysteine residues are in a reduced state (–SH) and inactive when these groups are oxidized as disulfide bonds (S–S). The reduction state, and hence the level of activation, is controlled by a system that responds directly to the state of reduction of ferredoxin, the electron-carrier protein of PSI that donates electrons to NADP+. Besides reducing NADP+, reduced ferredoxin also reduces thioredoxin, in a reaction catalyzed by ferredoxin-thioredoxin reductase. Reduced thioredoxin in turn reduces the disulfide bonds of some Calvin cycle enzymes, thus activating them (Figure 4–21B). Reduced thioredoxin can also be oxidized by oxygen, and the balance between the oxidized and reduced states depends on the state of reduction of ferredoxin, and hence on the flux of electrons through the photosystems. In its oxidized state, thioredoxin oxidizes the sulfhydryl groups and thus deactivates the Calvin cycle

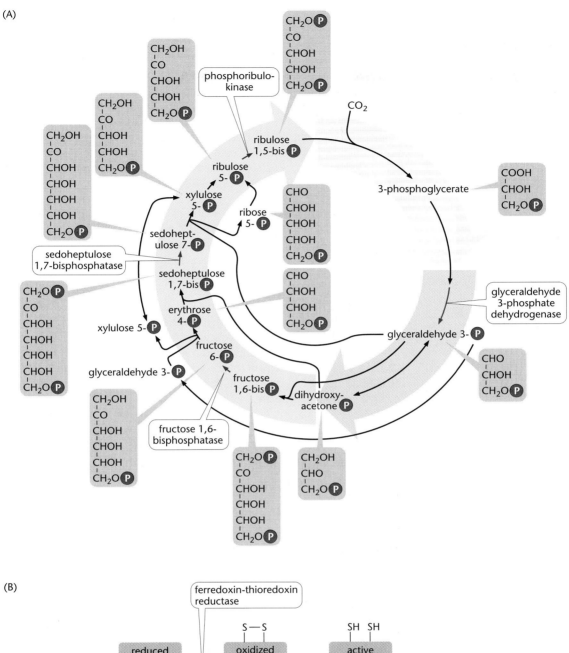

Figure 4–21.
Regulation of Calvin cycle enzymes by electron flow through the photosystems. (A) Four enzymes of the Calvin cycle are activated when particular cysteine residues of the enzyme proteins are reduced, and inactivated when these residues are oxidized. Coordinated regulation of enzymes at four different points in the cycle in response to changes in energy supply ensures that changes in flux through the cycle are not accompanied by large changes in relative amounts of its metabolic intermediates. (B) Activation of a Calvin cycle enzyme by transfer of electrons from PSI. When ferredoxin is in a reduced state, indicating availability of reductant for the Calvin cycle, it reduces the small, soluble stromal protein thioredoxin, breaking a disulfide bond between two cysteine residues. Thioredoxin in turn reduces a disulfide bond of the Calvin cycle enzyme, thus activating the enzyme. When ferredoxin is in an oxidized state—as, for example, in the dark—the other components of this ferredoxin-thioredoxin pathway are also oxidized and the Calvin cycle enzyme is inactive.

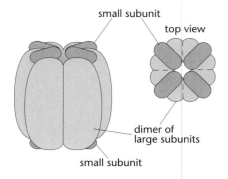

Figure 4–22.
Structure of Rubisco. The enzyme is a large complex consisting of eight large and eight small subunits. It constitutes about half of the total protein in the chloroplast and is believed to be the most abundant protein on earth. Like all other Calvin cycle enzymes—and the vast majority of other proteins in the chloroplast—the small subunit is encoded in the nucleus, synthesized in the cytosol, and imported into the chloroplast. However, the large subunit is encoded in the plastid genome and synthesized inside the chloroplast (see Chapter 2). Complex mechanisms act to coordinate the synthesis and assembly of the two sorts of subunit to form the mature protein.

Figure 4–23.
Activation of Rubisco by Rubisco activase. Activation of Rubisco (E) requires the carbamylation of a specific lysine residue (binding of a CO_2 molecule—a separate event from the carboxylation reaction itself) in the presence of magnesium. For carbamylation to occur, bound molecules of the substrate ribulose 1,5-bisphosphate (RuBP) must be removed from the enzyme. This is achieved by the activase.

enzymes. In this way, the level of activation of several enzymes is directly linked to the flux of electrons through the photosystems and hence to the availability of NADPH and ATP for Calvin cycle reactions.

The level of activation of Rubisco, the carbon dioxide–assimilating enzyme (Figure 4–22), is also linked to light-driven reactions, but by a different mechanism that involves the enzyme Rubisco activase (Figure 4–23). Rubisco activase is activated by the reduction state of PSI, probably via the ferredoxin-thioredoxin system in the same manner as the Calvin cycle enzymes described above (see Figure 4–21B). When active, Rubisco activase promotes the binding of carbon dioxide to a specific lysine residue in the active site of Rubisco, located on the large subunits of the enzyme. This **carbamylation** reaction promotes Rubisco activity.

The light-driven reactions of photosynthesis also regulate the activity of Calvin cycle enzymes in more generalized ways. The pH of the stroma (where the Calvin cycle reactions occur) is strongly influenced by the electron-transfer processes in the thylakoid membranes. As described above, the generation of the proton gradient that drives ATP synthesis removes protons from (increases the pH of) the stroma. This proton movement results in the movement of other cations—notably Mg^{2+}—from the thylakoid lumen into the stroma. Thus the stroma is at about pH 7 with 1 to 3 mM Mg^{2+} at night, but about pH 8 with 3 to 6 mM Mg^{2+} in the light, when electron transport is occurring. The higher pH and Mg^{2+} concentration increase the activation state of several Calvin cycle enzymes, including Rubisco (these conditions promote activation by favoring carbamylation). Activity of several Calvin cycle enzymes is also modulated by intermediates of the Calvin cycle itself. For example, Rubisco is inhibited by a buildup of its product, 3-phosphoglycerate.

Sucrose synthesis is tightly controlled by the rate of photosynthesis and the demand for carbon by nonphotosynthetic parts of the plant

The main fate of triose phosphate (the net gain) from the Calvin cycle is the synthesis of **sucrose** in the cytosol (Figure 4–24). Triose phosphate leaves the chloroplast via a **triose phosphate transporter**, which exchanges dihydroxyacetone phosphate for inorganic phosphate. In the cytosol, triose phosphate is converted to fructose 1,6-bisphosphate, and then to fructose 6-phosphate via **fructose 1,6-bisphosphatase**. Interconversion of hexose phosphates yields glucose 1-phosphate. Reaction of glucose 1-phosphate with uridine triphosphate (UTP; a high-energy compound similar to ATP), yields the sugar nucleotide **UDP-glucose**. In a reaction catalyzed by **sucrose phosphate synthase**, UDP-glucose and fructose 6-phosphate are converted to sucrose phosphate, which is converted to sucrose by **sucrose phosphate phosphatase**.

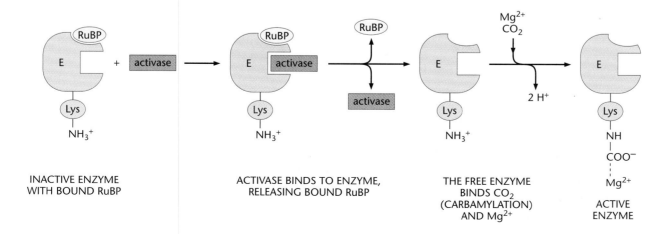

INACTIVE ENZYME WITH BOUND RuBP

ACTIVASE BINDS TO ENZYME, RELEASING BOUND RuBP

THE FREE ENZYME BINDS CO_2 (CARBAMYLATION) AND Mg^{2+}

ACTIVE ENZYME

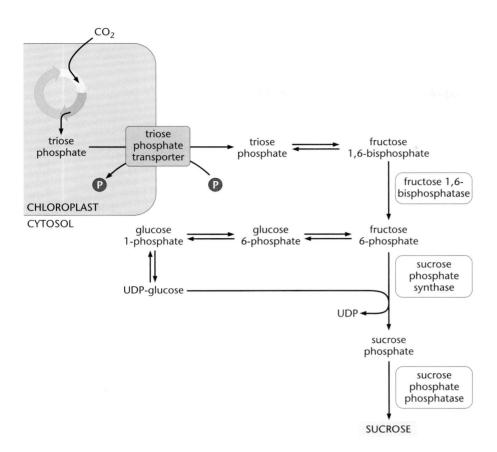

Figure 4–24.
Synthesis of sucrose in the cytosol.
Triose phosphate generated in the Calvin cycle is transported to the cytosol, where it undergoes a series of conversions involving six-carbon sugars—as sugar phosphates or sugar nucleotides—followed by joining of a glucose and a fructose residue to yield the disaccharide sucrose. Four triose phosphate molecules yield one sucrose molecule.

If optimal rates of CO_2 assimilation are to be maintained, the rate of sucrose synthesis in the cytosol must be coordinated with the operation of the Calvin cycle in the chloroplast (Figure 4–25). We described earlier how five-sixths of the triose phosphate synthesized in the Calvin cycle is needed for regeneration of ribulose 1,5-bisphosphate (the CO_2 acceptor), and the remaining one-sixth is available for sucrose synthesis. If a higher proportion of triose phosphate is withdrawn, ribulose 1,5-bisphosphate levels fall progressively and the Calvin cycle collapses. If triose phosphate is *not* withdrawn for sucrose synthesis, Calvin cycle intermediates accumulate and the level of inorganic phosphate in the chloroplast falls (because it is no longer returned to the chloroplast from the cytosol in exchange for triose phosphate; see Figure 4–26). Decreased inorganic phosphate availability in the chloroplast restricts ATP synthesis and hence reduces the supply of ATP to the Calvin cycle.

The rate of sucrose synthesis in the cytosol must also be coordinated with the demand for carbon by the nonphotosynthetic parts of the plant. A high rate of sucrose synthesis is needed when demand is high, and a low rate when demand is low. Sucrose synthesis must be regulated both by **feed-forward** mechanisms that respond to the availability of carbon from the Calvin cycle and by **feedback** mechanisms that respond to demand for sucrose from the rest of the plant (Figure 4–25).

These complex and potentially conflicting requirements for control of the flux from triose phosphate to sucrose are met by a battery of integrated mechanisms, operating at several points in the chloroplast and the cytosol. Here we consider three major mechanisms: (1) the central role of the triose phosphate transporter in regulating exchange of metabolites between the chloroplast and the cytosol; (2) the complex regulation of two enzymes of sucrose synthesis in the cytosol (fructose 1,6-bisphosphatase and sucrose phosphate synthase); and (3) the diversion of carbon from the Calvin cycle into starch synthesis in the chloroplast rather than sucrose synthesis in the cytosol.

Figure 4–25.

Control of the rate of sucrose synthesis in a photosynthetic cell. Feed-forward mechanisms coordinate the rate of sucrose synthesis in the cytosol with the rate of CO_2 assimilation, and hence the rate of triose phosphate production, in the chloroplast. The triose phosphate transporter in the chloroplast envelope, which exchanges triose phosphate for inorganic phosphate released during sucrose synthesis, plays a central role in these mechanisms. Feedback mechanisms coordinate the rate of sucrose synthesis with the demand for sucrose in the nonphotosynthetic parts of the plant to which it is exported.

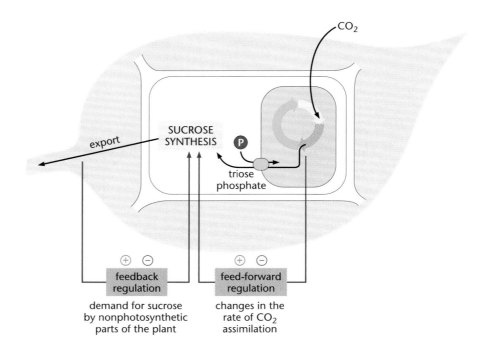

The triose phosphate transporter in the inner membrane of the chloroplast envelope plays a vital role in coordinating sucrose synthesis with the operation of the Calvin cycle. The release of inorganic phosphate during conversion of triose phosphate to sucrose means that the availability of inorganic phosphate in the cytosol closely reflects the rate of sucrose synthesis (Figure 4–26). Thus the rate of withdrawal of triose phosphate from the chloroplasts in exchange for inorganic phosphate is closely matched to

Figure 4–26.

Export of triose phosphate from the chloroplast in exchange for inorganic phosphate released in sucrose synthesis. The triose phosphate transporter catalyzes a one-for-one exchange of triose phosphate for inorganic phosphate, P_i. Phosphate groups from three of the four triose phosphates required to make a molecule of sucrose are released as P_i in direct reactions in the sucrose-synthesizing pathway. The fourth P_i is derived from metabolism of the inorganic pyrophosphate (PP$_i$; represented by *red shaded* "PP") generated during conversion of glucose 1-phosphate to UDP-glucose. Pyrophosphate metabolism is discussed in Section 4.5.

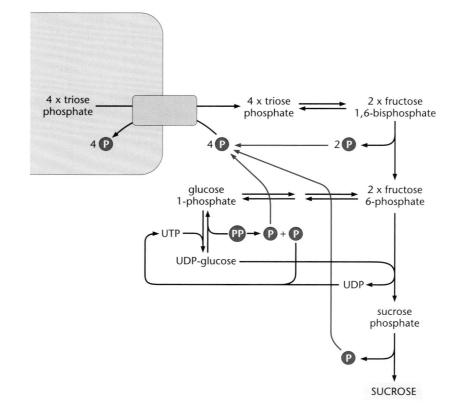

the rate of sucrose synthesis. Conversely, changes in the levels of triose phosphate and inorganic phosphate in the chloroplast, resulting from changes in the rate of CO_2 assimilation, are reflected in changes in the levels of these metabolites in the cytosol. Regulatory mechanisms in the cytosol respond to these changes by modulating the activities of fructose 1,6-bisphosphatase and sucrose phosphate synthase and thus changing the rate of sucrose synthesis.

Cytosolic fructose 1,6-bisphosphatase catalyzes the first essentially irreversible step of sucrose synthesis. Its activity is regulated primarily by a powerful inhibitor, fructose 2,6-bisphosphate (Figure 4–27). Fructose 2,6-bisphosphate is synthesized from, and hydrolyzed to, fructose 6-phosphate by two enzymes found exclusively in the cytosol: fructose 6-phosphate 2-kinase and fructose 2,6-bisphosphate phosphatase. These enzymes are regulated by cytosolic metabolites, including triose phosphate, inorganic phosphate, and fructose 6-phosphate. The level of fructose 2,6-bisphosphate and hence the activation state of fructose 1,6-bisphosphatase are thus exquisitely sensitive to factors that signal both the availability of triose phosphate (feed-forward regulation) and the downstream conversion of sugar phosphates to sucrose (feedback regulation).

To illustrate feed-forward regulation of cytosolic fructose 1,6-bisphosphatase, we consider what happens when the rate of CO_2 assimilation in the chloroplast rises due to an increase in light intensity, potentially making more carbon available for sucrose synthesis. The level of triose phosphate in the chloroplast tends to rise, and the level of inorganic phosphate falls as the rate of ATP synthesis increases. These changes are transmitted to the cytosol via the triose phosphate transporter: more triose phosphate is exported to the cytosol and more inorganic phosphate is imported to the chloroplast. Triose phosphate inhibits fructose 6-phosphate 2-kinase, whereas inorganic phosphate activates this enzyme and inhibits fructose 2,6-bisphosphate phosphatase. Thus the rise in the ratio of triose phosphate to inorganic phosphate in the cytosol inhibits fructose 2,6-bisphosphate synthesis and promotes its hydrolysis. Fructose 2,6-bisphosphate levels fall and fructose 1,6-bisphosphatase activity rises, allowing an increased flux of carbon toward sucrose.

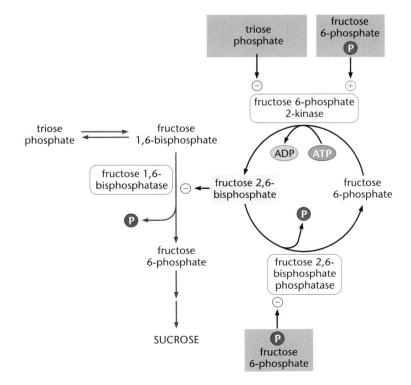

Figure 4–27.
The role of fructose 2,6-bisphosphate in the regulation of sucrose synthesis.
The enzyme fructose 1,6-bisphosphatase is extremely sensitive to inhibition by fructose 2,6-bisphosphate, the levels of which are determined by a highly regulated cycle of synthesis and degradation. The kinase that catalyzes the synthesis of fructose 2,6-bisphosphate is inhibited by triose phosphate and activated by fructose 6-phosphate and P_i. The phosphatase that catalyzes the degradation of fructose 2,6-bisphosphate to fructose 6-phosphate is inhibited by fructose 6-phosphate and P_i.

Feedback regulation of cytosolic fructose 1,6-bisphosphatase is illustrated by what happens when the demand for sucrose in nonphotosynthetic parts of the plant falls, so that the rate of sucrose synthesis exceeds the rate of export. Sucrose levels in the leaf rise, and this inhibits the latter stages of sucrose synthesis. Hexose phosphates accumulate in the cytosol. Fructose 6-phosphate, like inorganic phosphate, activates fructose 6-phosphate 2-kinase and inhibits fructose 2,6-bisphosphate phosphatase. Thus the increase in fructose 6-phosphate promotes the synthesis and inhibits the hydrolysis of fructose 2,6-bisphosphate, fructose 2,6-bisphosphate levels rise, and fructose 1,6-bisphosphatase activity is inhibited, reducing the flux of carbon to sucrose.

Sucrose phosphate synthase, too, is regulated by feed-forward and feedback mechanisms (Figure 4–28). Its regulation is linked to that of fructose 1,6-bisphosphatase, ensuring sensitive, modulated responses of flux through the pathway to changes in photosynthesis and the demand for sucrose. Two mechanisms render sucrose phosphate synthase highly sensitive to the ratio of glucose 6-phosphate to inorganic phosphate in the cytosol. First, these two metabolites directly modulate enzyme activity: a rise in glucose 6-phosphate levels relative to inorganic phosphate levels activates the enzyme. Second, sucrose phosphate synthase is phosphorylated and dephosphorylated by enzymes that are themselves modulated by the ratio of glucose 6-phosphate to inorganic phosphate in the cytosol. Phosphorylation of the sucrose phosphate synthase protein by a protein kinase reduces the activity of the enzyme, and removal of the phosphate by a protein phosphatase activates it. The kinase is inhibited by glucose 6-phosphate, and the phosphatase is inhibited by inorganic phosphate. Thus a rise in the ratio of glucose 6-phosphate to inorganic phosphate decreases the level of phosphorylation of sucrose phosphate synthase, activating the enzyme and reinforcing the direct stimulation of activity brought about by this change in metabolite levels.

The importance of feed-forward regulation of sucrose phosphate synthase by the ratio of glucose 6-phosphate to inorganic phosphate is illustrated when the rate of photosynthesis rises. We have described how an increased ratio of triose phosphate to inorganic phosphate in the cytosol activates fructose 1,6-bisphosphatase. The resulting increase in fructose 6-phosphate synthesis leads to an increase in glucose 6-phosphate levels, and thus an increase in the ratio of glucose 6-phosphate to inorganic phosphate in the cytosol. This activates sucrose phosphate synthase, so that its activity rises in concert with the rise in fructose 1,6-bisphosphatase activity and allows an increase in flux to

Figure 4–28.
Regulation of sucrose phosphate synthase by phosphorylation. Levels of glucose 6-phosphate and inorganic phosphate modulate enzyme activity both directly and through their effects on a protein kinase and a protein phosphatase. These two enzymes mediate the phosphorylation and dephosphorylation of a serine residue of the sucrose phosphate synthase protein: phosphorylation inactivates the enzyme.

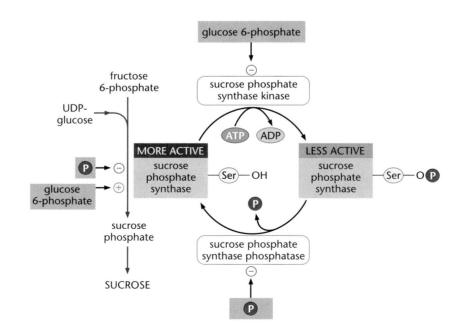

sucrose. Feedback regulation of sucrose phosphate synthase is not fully understood; changes in sugar levels in the leaf are reflected in changes in the degree of activation of sucrose phosphate synthase, but the mechanism by which this occurs is not known.

In our discussion of feedback regulation of sucrose synthesis so far, we have shown how changes in demand for sucrose can alter the rate of its synthesis from triose phosphate in the cytosol, through alterations in the activity of fructose 1,6-bisphosphatase and sucrose phosphate synthase. This matches the rate of synthesis to demand, but creates a further potential problem. If the rate of sucrose synthesis falls because demand is low, but the rate of CO_2 assimilation remains high, triose phosphate export from the chloroplast is restricted because insufficient inorganic phosphate is being released in the cytosol for exchange on the transporter. As we discussed earlier, this potentially leads to a decrease in inorganic phosphate and an accumulation of phosphorylated intermediates in the chloroplast. This situation restricts ATP synthesis and thus reduces the rate of CO_2 assimilation. In many plants, this problem is overcome by the synthesis of starch from phosphorylated intermediates in the chloroplast.

Synthesis of starch allows the photosynthetic rate to remain high when sucrose synthesis is restricted

Starch synthesis in the chloroplast provides an alternative fate for Calvin cycle intermediates when triose phosphate export is restricted because of low rates of sucrose synthesis. Diversion of carbon from the stromal fructose 6-phosphate pool into starch prevents the buildup of phosphorylated intermediates and allows high rates of CO_2 assimilation to continue (Figure 4–29). At night, carbon stored as starch is released by starch degradation, exported from the chloroplast, and used to synthesize sucrose (see Section 4.6). Sucrose is then exported from the leaf to maintain a supply of carbon to the nonphotosynthetic parts of the plant during the period of darkness.

The regulation of starch synthesis is dominated by mechanisms that reflect the rate of sucrose synthesis in the cytosol, such that low rates of sucrose synthesis potentially

Figure 4–29.
Conversion of Calvin cycle intermediates to sucrose or starch. (A) If export of triose phosphate is restricted because of a low rate of sucrose synthesis, phosphorylated intermediates of the Calvin cycle can instead be diverted into starch synthesis. Some of the fructose 6-phosphate formed in the regeneration stage of the cycle is converted to glucose 1-phosphate, which is converted via an ATP-consuming reaction (catalyzed by ADP-glucose pyrophosphorylase) to ADP-glucose, the substrate for starch synthesis. (B) Electron micrograph of a starch-containing chloroplast in an Arabidopsis leaf. (The spaces around the starch granules are an artifact of fixation of the specimen for electron microscopy.)

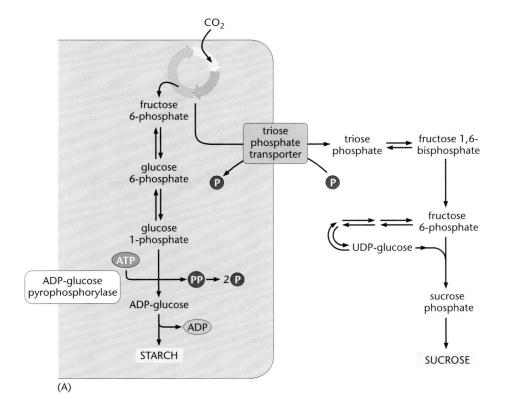

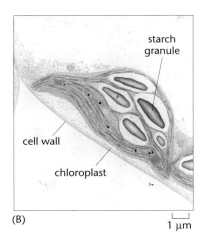

result in high rates of starch synthesis, and vice versa. In other words, sucrose synthesis exerts negative control over starch synthesis. The main point at which flux of sugar phosphates into starch is regulated is the conversion of glucose 1-phosphate to the sugar nucleotide ADP-glucose, catalyzed by ADP-glucose pyrophosphorylase (Figure 4–29). This reaction is effectively irreversible because the pyrophosphate produced is rapidly hydrolyzed to inorganic phosphate. Like some Calvin cycle enzymes discussed above, ADP-glucose pyrophosphorylase is activated by reduction. The activated enzyme is regulated by 3-phosphoglycerate and inorganic phosphate: activity is high when the ratio of 3-phosphoglycerate to inorganic phosphate is high, and low when the ratio is low. These properties allow activity to respond sensitively to changes in sucrose synthesis in the cytosol.

As we discussed above, if the rate of sucrose synthesis decreases, the level of inorganic phosphate in the chloroplast falls and the level of phosphorylated intermediates rises. The ratio of 3-phosphoglycerate to inorganic phosphate in the chloroplast therefore rises, activating ADP-glucose pyrophosphorylase and diverting fructose 6-phosphate from the Calvin cycle into starch synthesis. Conversely, if the rate of sucrose synthesis rises, the ratio of 3-phosphoglycerate to inorganic phosphate in the chloroplast falls, lowering the activity of ADP-glucose pyrophosphorylase and ensuring that carbon from the Calvin cycle is exported for sucrose synthesis rather than being converted to starch (Figure 4–30). The state of reduction of the ADP-glucose pyrophosphorylase protein, and hence its general level of activity, is also linked to the level of sucrose in the cell, but the mechanism by which this occurs is not yet understood.

The control of starch synthesis in the chloroplast by cytosolic sucrose synthesis is illustrated by studies of mutant and **transgenic plants** with a reduced capacity for sucrose

Figure 4–30.
Negative control of starch synthesis by sucrose synthesis. The diagram illustrates how a reduced rate of sucrose synthesis leads to increased starch synthesis, and an increased rate of sucrose synthesis leads to reduced starch synthesis, through modulation of the activity of ADP-glucose pyrophosphorylase.

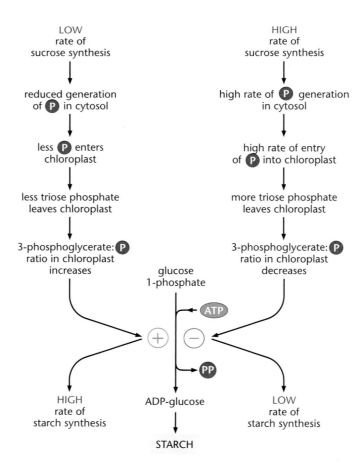

synthesis. (A description of the creation and uses of transgenic plants is given in Section 9.3.) In mutant plants of *Clarkia xantiana* that have lowered activity of the cytosolic phosphoglucose isomerase (the enzyme that interconverts fructose 6-phosphate and glucose 6-phosphate in the sucrose synthesis pathway), and in transgenic potato plants with lowered activity of either the triose phosphate transporter or cytosolic fructose 1,6-bisphosphatase, rates of sucrose synthesis during photosynthesis are much lower than in wild-type plants. In all of these mutant plants there is a reciprocal increase in the rate of starch synthesis during photosynthesis so that it is much higher than normal. For example, transgenic potato plants with 10% of the normal activity of cytosolic fructose 1,6-bisphosphatase had three times more starch in their leaves at the end of the day than wild-type plants.

Although the rate of sucrose synthesis is a major factor in determining the rate of starch synthesis, it is by no means the only one. In many plants, some starch synthesis occurs in the chloroplast regardless of the rate of sucrose synthesis. This basal rate is related to the light regimen (light intensity and length of dark period) under which the plant is grown. The rate of starch synthesis during the day seems to be programmed in a complex way—probably involving both fine and coarse control—to meet the plant's requirement for a continued supply of carbon in the dark. It is important to note that species vary enormously in the extent of starch synthesis in chloroplasts. At the extremes, some species (e.g., soybean and cotton) sequester up to half of the carbon assimilated by the Calvin cycle as starch, whereas others (e.g., spinach and many grasses) synthesize very little starch and rely largely on sucrose synthesis.

4.3 PHOTORESPIRATION

Rubisco can use oxygen instead of carbon dioxide as substrate

As well as catalyzing the carboxylation of ribulose 1,5-bisphosphate, Rubisco can also catalyze its oxygenation. (The full name of Rubisco is **ribu**lose 1,5-**bis**phosphate **c**arboxylase/**o**xygenase.) Instead of adding one carbon from carbon dioxide to the five-carbon ribulose 1,5-bisphosphate to yield two three-carbon 3-phosphoglycerate molecules, the enzyme adds oxygen to yield one molecule of 3-phosphoglycerate and one of the two-carbon compound 2-phosphoglycolate (Figure 4–31). The affinity of the enzyme for oxygen is very much lower than its affinity for carbon dioxide; in other words, the O_2 concentration required for a given rate of the oxygenation reaction is much higher than the CO_2 concentration required for the carboxylation reaction to proceed at the

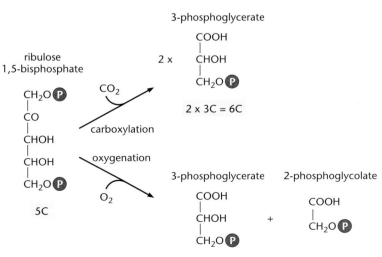

Figure 4–31.
The carboxylase and oxygenase reactions of Rubisco. Carboxylation of (5-carbon) ribulose 1,5-bisphosphate yields two (3-carbon) molecules of 3-phosphoglycerate, which enter the Calvin cycle. Oxygenation of ribulose bisphosphate yields one molecule of (3-carbon) 3-phosphoglycerate and one molecule of (2-carbon) 2-phosphoglycolate.

Figure 4–32.
Overview of the photorespiratory cycle.
(A) Electron micrograph of part of a leaf cell, showing the three subcellular compartments involved in metabolism of 2-phosphoglycolate: the chloroplast (where 2-phosphoglycolate is generated), the peroxisome, and the mitochondrion. Peroxisomes and mitochondria are often closely associated with chloroplasts in photosynthetic cells. The large rectangular inclusion in the peroxisome is a crystal of the abundant enzyme catalase. (B) The photorespiratory cycle, in which 2-phosphoglycolate is converted to 3-phosphoglycerate, involves export of glycolate (derived from 2-phosphoglycolate) from the chloroplast to the peroxisome, followed by a series of reactions in the peroxisome, the mitochondrion, and then the peroxisome again, followed by generation of 3-phosphoglycerate in the chloroplast. Overall, two molecules of 2-phosphoglycolate are converted to one molecule of 3-phosphoglycerate, which returns to the Calvin cycle, and one molecule of CO_2 (in the mitochondrion). Thus two oxygenation reactions catalyzed by Rubisco result in the loss of one molecule of CO_2. (RuBP = ribulose 1,5-bisphosphate.) (A, courtesy of Eldon Newcomb and Department of Botany, University of Wisconsin Madison. Originally published in *J. Cell Biol.* 43:343–353, 1969. With permission from Rockefeller University Press.)

same rate. The O_2 concentration required for the enzyme to proceed at half its maximal oxygenation rate is 535 μmol/L, whereas the CO_2 concentration required to achieve half the maximal carboxylation rate is one-sixtieth of this: 9 μmol/L.

In the chloroplast, both carbon dioxide and oxygen are present at much lower concentrations than those required to achieve the maximal rate of either reaction of Rubisco (O_2 concentration is 250 μmol/L; CO_2 concentration is about 8 μmol/L), so these two substrates compete for the active site of the enzyme. Although the enzyme has a very much lower affinity for oxygen than for carbon dioxide, the fact that the O_2 concentration in the chloroplast is very much higher than CO_2 concentration means that the oxygenation reaction usually proceeds at about 25% of the rate of carboxylation; that is, one oxygenation reaction occurs for every three carboxylation reactions.

The oxygenation rather than carboxylation of ribulose 1,5-bisphosphate has profound consequences for the metabolism of the photosynthesizing leaf. Carbon from the 2-phosphoglycolate produced by oxygenation is converted to 3-phosphoglycerate and returned to the Calvin cycle (Figure 4–32). This requires a complex cycle of reactions that consumes ATP, involves at least 10 enzymes in three subcellular compartments, and—most important—results in the loss of one molecule of carbon dioxide for every 3-phosphoglycerate molecule recovered. This loss of carbon dioxide during the conversion of 2-phosphoglycolate to 3-phosphoglycerate is called **photorespiration**, and the cycle of reactions by which this conversion occurs is the **photorespiratory cycle**.

The evolutionary origins of the oxygenase function of Rubisco and of the photorespiratory cycle have been the subject of much speculation and debate among plant physiologists. There is general agreement from comparisons of Rubisco sequence and structure in bacteria and plants that a Rubisco-like protein was responsible for CO_2 assimilation in the very earliest organisms capable of using inorganic carbon for growth, at least 3.5 billion (3.5×10^9) years ago. These organisms existed in an essentially anoxic (oxygen-free) environment (see Chapter 1). Over about 1.5 billion years, the O_2 concentration in the earth's atmosphere gradually increased due to production of oxygen

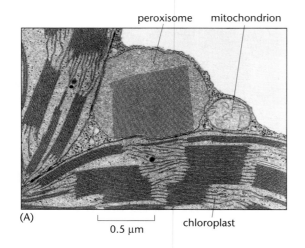

(A)   0.5 μm

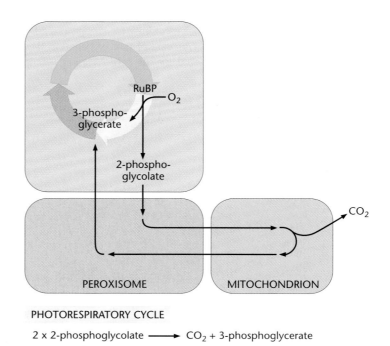

PHOTORESPIRATORY CYCLE

2 x 2-phosphoglycolate ⟶ CO_2 + 3-phosphoglycerate

2 x 2C   1C   3C

(B)

by organisms using water as an electron donor in photosynthesis. As the O_2 concentration increased, so did the oxygenase reaction of Rubisco.

The deleterious impact of phosphoglycolate production on photosynthetic CO_2 assimilation generated selection pressure for mechanisms that either recovered the carbon from phosphoglycolate (as in the photorespiratory cycle) or prevented the oxygenation reaction. The ratio of oxygenase to carboxylase activity of Rubisco probably fell due to selection for modifications to the active site; evidence for this evolutionary trend comes from the study of modern photosynthetic bacteria occupying anoxic environments. The Rubisco of these species has a higher ratio of oxygenase to carboxylase activity than that found in the higher plants. This evolutionary change in properties of Rubisco is relatively small, however, and the enzyme of modern higher plants retains a very substantial oxygenase activity. Recovery of the 2-phosphoglycolate produced by the oxygenase is essential for normal plant growth.

Why has the apparently expensive and wasteful oxygenase reaction of Rubisco been retained? First, it may be that the active site of Rubisco cannot be modified to reduce the oxygenase activity without adversely affecting the carboxylase activity. The Rubisco of modern higher plants may represent the most favorable ratio of oxygenase to carboxylase activity achievable by natural selection from the ancestral Rubisco—in other words, the oxygenase reaction is an inevitable consequence of the structural requirements for the carboxylase reaction. Some support for this idea comes from extensive attempts to engineer an "improved" Rubisco, carried out from the 1970s onward. Determination of the three-dimensional structure of the enzyme enabled researchers to alter the structure of the active site by substituting amino acids that they hoped would decrease the oxygenase reaction without adversely affecting the carboxylase reaction. Although this work resulted in much valuable information about the reaction mechanism of the enzyme, no significant improvements in its properties were achieved.

Another possible reason why the oxygenase reaction has been retained is that, under some conditions, it can help prevent photoinhibition, a damaging side effect of excess excitation energy in the photosystems that can occur in high light conditions (see Section 4.2; see also Section 7.1). After explaining the theory behind this idea, we will describe some experimental findings that support it.

When the stomata of illuminated leaves shut to reduce water loss, the concentration of carbon dioxide inside the leaf falls as it is used up in photosynthesis. The ratio of oxygenase to carboxylase activity of Rubisco rises as the $CO_2:O_2$ ratio falls. This means that the amount of carbon dioxide released in photorespiration rises as the rate of assimilation of carbon dioxide falls. Eventually a point is reached at which the rate of CO_2 loss in photorespiration equals the rate of CO_2 assimilation, and there is no further net gain of carbon by the leaf (Figure 4–33). This is called the carbon dioxide **compensation point**.

If the oxygenase reaction, and hence photorespiratory CO_2 loss, did not occur, the leaf could in theory continue to gain net amounts of carbon at lower and lower internal CO_2 concentrations, potentially increasing the plant's productivity in conditions of water stress. But, although photorespiration reduces productivity under these conditions, it also helps to prevent photoinhibition. As the CO_2 concentration inside the leaf falls, so does the rate of consumption of ATP and NADPH by the Calvin cycle. The electron carriers of the electron transport chains could potentially become highly reduced, because $NADP^+$ is not available to receive electrons and the electrochemical gradient is not dissipated by ATP synthesis, thus causing serious damage—photoinhibition—to the light-harvesting apparatus. The consumption of ATP and NADPH by the photorespiratory cycle makes a major contribution toward dissipating excitation energy and hence preventing photoinhibition. Approximately 2 ATP and 1 NADPH are consumed for each CO_2 released in photorespiration.

Figure 4–33.
Stomatal closure and the carbon dioxide compensation point. The three diagrams illustrate the impact of stomatal closure on assimilation of CO_2 and O_2 by Rubisco and loss of CO_2 in photorespiration. The CO_2 concentration in the atmosphere outside the leaf is 0.038% (380 ppm). Inside the leaf, the concentration is about 0.025% (250 ppm) when the stomata are open (*left*). When the stomata close, the concentration inside the leaf falls as CO_2 is used up in photosynthesis (*middle*). Eventually the rate of CO_2 assimilation by Rubisco equals the rate of CO_2 loss in photorespiration (*right*). At this point, the carbon dioxide compensation point, the CO_2 concentration inside the leaf is about 0.005% (50 ppm).

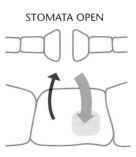

STOMATA OPEN

- unrestricted entry of CO_2 into leaf
- carboxylase reaction is 3 or more times oxygenase reaction rate
- CO_2 assimilation is 6 or more times rate of photorespiratory CO_2 loss

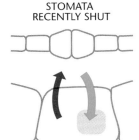

STOMATA RECENTLY SHUT

- no entry of CO_2 into leaf
- CO_2 assimilation rate falls
- photorespiratory CO_2 loss rises

  much-reduced rate of net carbon assimilation

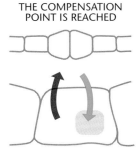

THE COMPENSATION POINT IS REACHED

- no entry of CO_2 into leaf
- carboxylase reaction is half oxygenase reaction rate
- CO_2 assimilation rate equals photorespiratory CO_2 loss

  no net carbon assimilation

The importance of photorespiration in preventing photoinhibition under these circumstances is illustrated by experiments in which both the carboxylation and the oxygenation reactions of Rubisco are prevented in leaves in the light. At normal CO_2 concentrations, no photoinhibition is observed when the oxygenase reaction, and hence the photorespiratory cycle, is prevented by placing illuminated leaves in low (1–2%) O_2 concentrations. However, photoinhibition rapidly occurs if CO_2 concentrations are also reduced, preventing operation of both the Calvin and the photorespiratory cycles.

Photorespiratory metabolism has implications for both the carbon and the nitrogen economy of the leaf

The photorespiratory cycle is complex and interacts closely with other metabolic processes—notably, nitrogen assimilation. The sequence of reactions was worked out by classical biochemical techniques (Figure 4–34) and has been confirmed and refined by the isolation and characterization of mutant plants lacking components of the cycle (Figure 4–35).

Barley and Arabidopsis plants with impaired photorespiratory capacity are either unable to grow at all or develop very slowly in normal air, because 2-phosphogycolate cannot be converted to 3-phosphoglycerate and thus its carbon cannot be returned to the Calvin cycle. In addition, the buildup of photorespiratory cycle intermediates due to loss of one of the cycle enzymes can inhibit other aspects of leaf metabolism. To select these photorespiratory mutants, mutagenized seeds are germinated and grown in high CO_2 concentrations, saturating the carboxylase reaction of Rubisco and inhibiting the oxygenase reaction. Under these conditions no 2-phosphoglycolate is produced and the mutant plants grow normally. When the plants are then transferred to normal air, those that fail to grow and can be "rescued" by transfer back to high CO_2 conditions are the photorespiratory mutants.

In addition to its importance in the carbon economy of the photosynthesizing leaf, the photorespiratory cycle also has major implications for nitrogen economy. The cycle involves metabolism of the amino acid glycine in a reaction that releases ammonia. Escape of this ammonia from the leaf would potentially result in a large loss of assimilated nitrogen; instead, the ammonia is reassimilated into amino acids in the chloroplasts. Thus,

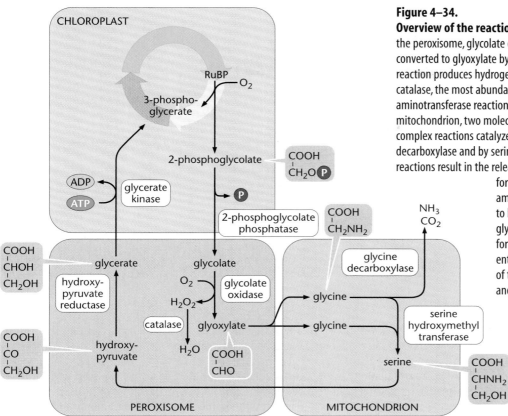

Figure 4–34.
Overview of the reactions of the photorespiratory cycle. In the peroxisome, glycolate derived from 2-phosphoglycolate is converted to glyoxylate by the enzyme glycolate oxidase. This reaction produces hydrogen peroxide, which is the substrate for catalase, the most abundant enzyme in peroxisomes. An aminotransferase reaction converts glyoxylate to glycine. In the mitochondrion, two molecules of glycine are metabolized via complex reactions catalyzed by the multisubunit enzyme glycine decarboxylase and by serine hydroxymethyl transferase. These reactions result in the release of CO_2 and ammonia, and formation of serine. In the peroxisome, an aminotransferase reaction converts serine to hydroxypyruvate, which is reduced to glycerate. This returns to the chloroplast for conversion to 3-phosphoglycerate and entry into the Calvin cycle. Further details of this pathway are shown in Figures 4–36 and 4–37.

although the photorespiratory cycle involves rapid metabolism of nitrogen-containing compounds, it is an essentially closed system with little or no net loss or gain of nitrogen (Figures 4–36 and 4–37).

The synthesis of glycine in the photorespiratory cycle takes place in the **peroxisome**, where glyoxylate derived from 2-phosphoglycolate receives amino groups from serine and glutamate (Figure 4–36). Glycine then moves to the mitochondrion, where it is metabolized via two enzymes, **glycine decarboxylase** and **serine hydroxymethyl transferase**, which convert 2 glycine to 1 serine, 1 carbon dioxide, and 1 ammonia, with the generation of 1 NADH (the other major form of reducing power in the cell; see Section 4.5). Glycine decarboxylase is a very abundant protein complex (comprising half of the soluble protein in leaf mitochondria) consisting of four enzymes. These

normal air   high carbon dioxide

wild type   mutant   wild type   mutant

Figure 4–35.
A photorespiratory mutant of Arabidopsis. When grown in normal air a mutant lacking glycine decarboxylase grows much more slowly than wild-type plants and eventually dies.

Figure 4–36.
Implications of the photorespiratory cycle for nitrogen metabolism. The cycle involves two aminotransferase reactions, and release of ammonia in the reaction catalyzed by glycine decarboxylase (GDC). One of the aminotransferases uses glutamate as an amino group donor for the formation of glycine; the other uses serine as an amino group donor for the formation of glycine. (SHMT = serine hydroxymethyl transferase)

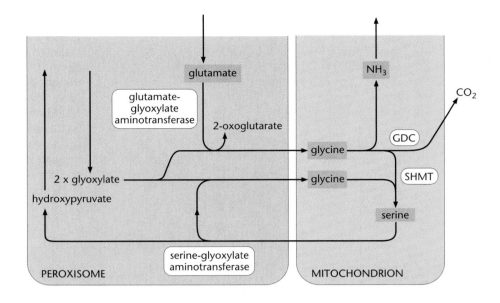

mitochondrial reactions are a key step in the photorespiratory cycle: they release the "photorespiratory" carbon dioxide and release ammonia.

The reassimilation of ammonia occurs in the chloroplast, using ATP and reducing power (in the form of reduced ferredoxin) produced in photosynthesis, in reactions catalyzed by glutamine synthase and glutamine–2-oxoglutarate aminotransferase (the **glutamine synthetase–GOGAT system**; see Figure 4–37). As described in Section

Figure 4–37.
Recapture of the ammonia released in photorespiration. Ammonia released in the glycine decarboxylase (GDC) reaction in the mitochondrion is reassimilated in the chloroplast, via reactions requiring 2-oxoglutarate and generating glutamate. Conversion of glutamate to 2-oxoglutarate is brought about by one of the aminotransferase reactions that synthesizes glycine as the substrate for glycine decarboxylase (GDC), thereby completing a closed system in which no nitrogen is lost or gained. Glutamate is exported from the chloroplast via a transporter that exchanges it for malate, and 2-oxoglutarate is imported into the chloroplast via another transporter, also in exchange for malate (see Figure 4–93A).

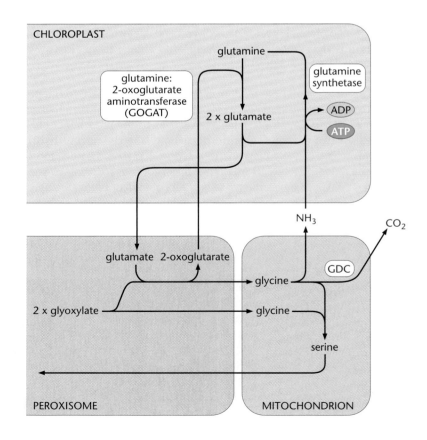

4.8, these enzymes are also involved in the assimilation of ammonium taken up from the soil or produced by other enzymes from nitrate taken up from the soil—collectively known as **primary nitrogen assimilation.** Small multigene families encode both glutamine synthetase and GOGAT. The forms of these enzymes present in the chloroplast are involved in ammonia reassimilation during photorespiratory metabolism and in primary nitrogen assimilation, whereas the forms located elsewhere are involved only in primary nitrogen assimilation.

The glutamine synthetase and GOGAT reactions convert ammonia to glutamate, which can move from the mitochondrion to the peroxisome to provide the amino group for glycine synthesis. Serine also moves back to the peroxisome and contributes amino groups to the synthesis of glycine via an aminotransferase reaction. The glycerate produced from serine then moves to the chloroplast and is converted to 3-phosphoglycerate, completing the photorespiratory cycle.

C4 plants eliminate photorespiration by a mechanism that concentrates carbon dioxide

Some species of plants, known as **C4 plants**, have a mechanism that prevents photorespiration by providing Rubisco with saturating concentrations of carbon dioxide (these plants are said to have **C4 photosynthesis**). The Rubisco of these plants is essentially the same as that of other plants—the capacity for the oxygenase reaction is still present—but the high CO_2 concentrations in the chloroplast allow the carboxylase reaction to proceed at its maximal rate, effectively eliminating the oxygenase reaction. No 2-phosphoglycolate is produced, and so there is no photorespiratory cycle and no loss of carbon dioxide.

The CO_2-concentrating mechanism involves both biochemical and anatomical adaptation in the C4 leaf. In overview (Figure 4–38), carbon dioxide is assimilated in the outer layer of cells adjacent to the leaf air space by a highly efficient, oxygen-insensitive carboxylase enzyme to produce the four-carbon compound oxaloacetate (hence the name "C4"; other plants are known as **C3 plants**, with **C3 photosynthesis**, because the first product of assimilation is the three-carbon compound 3-phosphoglycerate). A four-carbon compound synthesized from oxaloacetate (malate or aspartate) moves to inner cells where the Rubisco-containing chloroplasts are located. Here, decarboxylation of the four-carbon compound releases carbon dioxide, creating high concentrations in the chloroplast. The remaining three-carbon fragment moves back to the outer cells for regeneration of the CO_2 acceptor in the next round of the cycle. The cycle thus acts as a CO_2 pump, taking up carbon dioxide from a low concentration in the leaf air space and releasing it at a high concentration in the chloroplast.

In a C3 leaf, the chloroplast-containing cells are arranged in layers (the upper **palisade** and the lower **spongy mesophyll**); in a C4 leaf, chloroplast-containing cells are arranged in two ring-shaped layers around the **vascular bundles**. This arrangement is known as **Kranz anatomy,** from the German word meaning "wreath" (Figure 4–38B, C). Although both layers of cells—the outer **mesophyll** and the inner **bundle sheath**—contain chloroplasts, Rubisco is present only in the bundle-sheath cells.

In the mesophyll cells, a cytosolic enzyme, **phosphoenolpyruvate (PEP) carboxylase**, catalyzes the addition of carbon dioxide to the three-carbon compound phosphoenolpyruvate to produce the four-carbon oxaloacetate. PEP carboxylase has a very high affinity for carbon dioxide (in the form of bicarbonate, HCO_3^-, rather than CO_2) and achieves its maximal rate at CO_2 concentrations well below those in the mesophyll cell. Unlike Rubisco, PEP carboxylase cannot use oxygen as a substrate. Oxaloacetate is converted to malate or aspartate, which moves down a concentration gradient through the plasmodesmata into the bundle-sheath cells (Figure 4–38D).

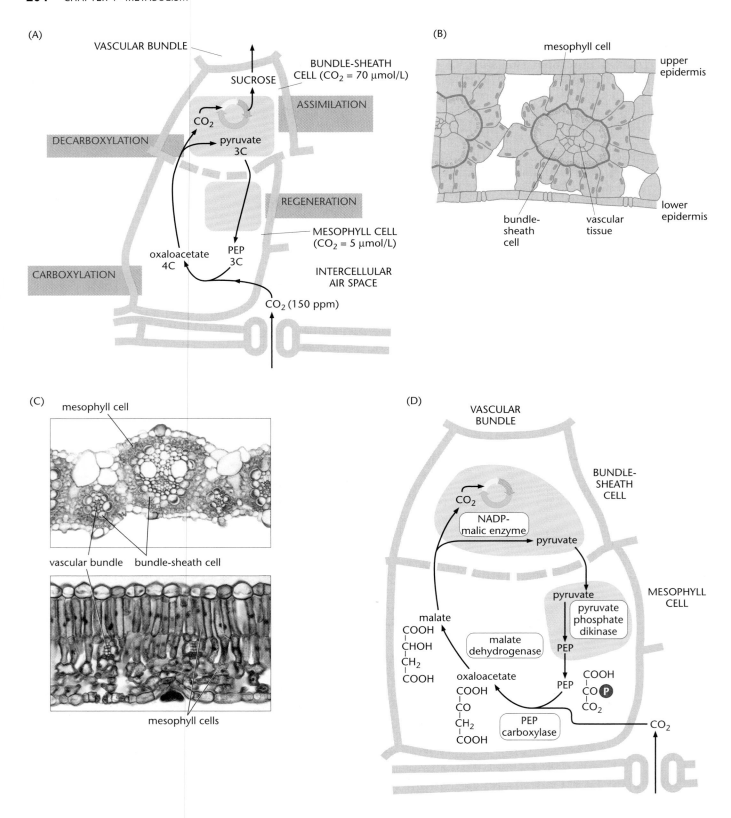

Release of carbon dioxide in the bundle-sheath cells can proceed by three different routes, depending on the plant species: via **NADP-malic enzyme** in the chloroplast, **NAD-malic enzyme** in the mitochondrion, or **PEP carboxykinase** in the cytosol (Figure 4–39). Species that use NADP-malic enzyme (e.g., maize) are the best studied, and this is the route we describe here.

Figure 4–38.
The C4 cycle and C4 leaf anatomy. (A) In overview, C4 photosynthesis can be divided into four phases. Carbon dioxide entering the leaf is taken up in a carboxylation reaction in the outer, mesophyll layer. The (4-carbon) product moves to the inner, bundle-sheath cells, where a decarboxylation reaction releases CO_2. The CO_2 is assimilated via the Calvin cycle, which is present in bundle-sheath but not in mesophyll cells. The (3-carbon) fragment remaining after decarboxylation returns to the mesophyll, where a regeneration reaction produces the (3-carbon) CO_2 acceptor for another turn of the cycle. Because of the high affinity for CO_2 of the carboxylation reaction in the mesophyll cell, this cycle acts as a CO_2 "pump" that achieves high CO_2 concentrations in bundle-sheath cells. Typically, the CO_2 concentration in a bundle-sheath cell may be 70 μmol/L (compared with 8 μmol/L in the chloroplast of a C3 leaf). (B) Anatomy of a C4 leaf. Bundle-sheath cells are arranged in a ring around the vascular bundle. The cells are thick-walled and have no direct contact with the air spaces of the leaf. They are surrounded by a ring of thinner-walled mesophyll cells. Both types of cell have chloroplasts, but Rubisco is present only in bundle-sheath cells. (C) Light micrographs of sections through a leaf of sugarcane, a C4 plant (*top*), and a leaf of a C3 plant (*bottom*). (D) The C4 cycle. The carboxylation reaction is carried out in the mesophyll by phosphoenolpyruvate (PEP) carboxylase, which converts CO_2 (as bicarbonate) and PEP to oxaloacetate. The oxaloacetate is reduced to malate, which moves to the bundle sheath for decarboxylation. In species that use NADP-malic enzyme (e.g., maize), as illustrated here, decarboxylation produces CO_2 and pyruvate. The pyruvate moves to the chloroplast of the mesophyll cell, where it is converted to PEP by pyruvate phosphate dikinase, in a reaction that consumes ATP. (C, courtesy of Ray F. Evert.)

Malate entering the bundle sheath is transported into the chloroplast, where it is decarboxylated by NADP-malic enzyme. The carbon dioxide thus released is assimilated by Rubisco. The pyruvate produced by the decarboxylation diffuses back into the mesophyll cell and enters the chloroplast, where it is converted to PEP by the enzyme **pyruvate phosphate dikinase**. Export of PEP to the cytosol as the substrate for PEP carboxylase completes the cycle (Figure 4–38D).

Other anatomical features have developed in C4 plants, in addition to biochemically distinct mesophyll and bundle-sheath cells, that ensure the efficient concentration of carbon dioxide. For example, in some species the bundle-sheath walls are made impermeable to carbon dioxide by the deposition of layers of the waxlike substance suberin (see Section 3.6). The bundle-sheath cells are completely surrounded by mesophyll cells, so they have no direct contact with leaf air spaces; any carbon dioxide that escapes through the bundle-sheath walls must pass through a mesophyll cell, where it can be recaptured by PEP carboxylase. The cell walls between the mesophyll and bundle-sheath cells have many plasmodesmata, which allow high rates of diffusion of C4 cycle intermediates between the two cell types.

The division of labor between the mesophyll and bundle-sheath cells extends beyond the operation of the C4 cycle to the Calvin cycle itself. Although Rubisco is present only in the chloroplasts of bundle-sheath cells, conversion of its product, 3-phosphoglycerate, to triose phosphate (the reduction stage of the Calvin cycle) takes place in both bundle-sheath and mesophyll chloroplasts. In maize leaves, about half of the

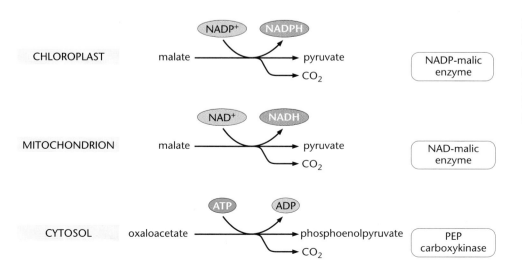

Figure 4–39.
Decarboxylating enzymes. The diagram shows the intracellular locations of and the reactions catalyzed by the three types of decarboxylating enzyme found in the bundle-sheath cells of C4 species.

3-phosphoglycerate generated by Rubisco moves out of the bundle-sheath chloroplast, into the mesophyll cell, and into the mesophyll chloroplast, where it is converted to triose phosphate. This triose phosphate is exported to the cytosol, where about one-third is used for sucrose synthesis and the remainder diffuses back to the bundle-sheath chloroplast to reenter the Calvin cycle (Figure 4–40A).

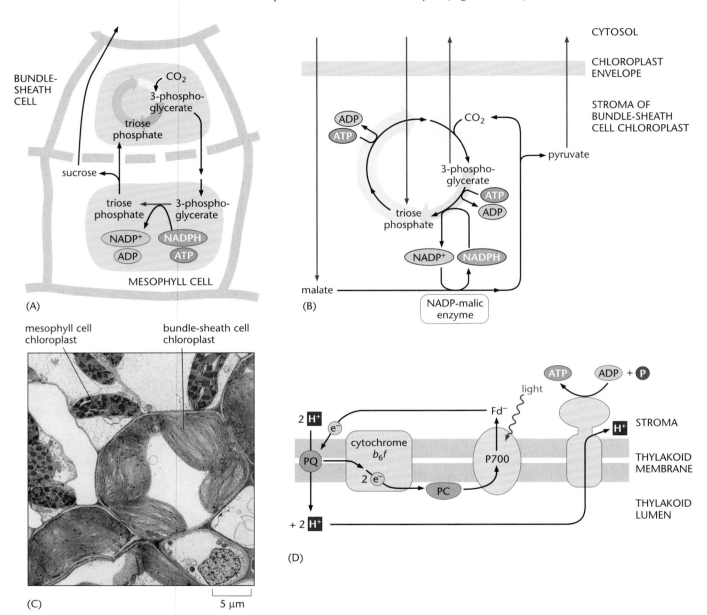

Figure 4–40.
Production of reducing power for the Calvin cycle in C4 plants with NADP-malic enzyme. (A) Half of the 3-phosphoglycerate formed by Rubisco-catalyzed CO_2 assimilation in the bundle sheath moves to the mesophyll chloroplasts for conversion to triose phosphate. Some of the triose phosphate is converted to sucrose for export from the leaf; some returns to the bundle sheath and reenters the Calvin cycle. (B) In the bundle-sheath cell chloroplast, the remaining 3-phosphoglycerate is reduced to triose phosphate, using NADPH generated by the chloroplast decarboxylating enzyme, NADP-malic enzyme. Thus the Calvin cycle requires no NADPH from photosynthetic electron transport, but does require ATP from that process. (C) Electron micrograph of a section through a maize leaf, showing adjacent bundle-sheath and mesophyll cells. The bundle-sheath chloroplasts of C4 plants with NADP-malic enzyme do not contain photosystem II, hence their thylakoid membranes are not stacked to form grana (see Section 4.2). (D) Cyclic electron transport around photosystem I allows formation of a proton gradient, and hence ATP synthesis, without $NADP^+$ reduction. Excitation of the P700 chlorophyll results in the reduction of ferredoxin (Fd). Electrons pass from reduced ferredoxin to plastoquinone (PQ), then through the cytochrome b_6f complex, and via plastocyanin (PC) back to PSI to re-reduce P700 (compare this with noncyclic electron transport; see Figure 4–11). Passage of electrons through plastoquinone is coupled to transfer of protons across the thylakoid membrane, forming the electrochemical gradient that drives ATP synthesis. (C, courtesy of Ray F. Evert.)

The lower flux through the reduction stage of the Calvin cycle in bundle-sheath chloroplasts in C4 plants means that the ratio of NADPH to ATP required is much lower than in the C3 chloroplast. Almost all of the requirement for NADPH can be met by the NADPH produced by the decarboxylating enzyme NADP-malic enzyme, as it converts malate to carbon dioxide and pyruvate (Figure 4–40B). This means that the demand for NADPH from the light-driven electron transport chains is negligible. The demand for ATP, however, remains high. This unusual ratio of NADPH and ATP requirements is matched by unusual light-harvesting and electron-transfer processes in the bundle-sheath chloroplasts. These chloroplasts have only very low levels of photosystem II, and the granal thylakoids in which PSII is primarily located are almost absent (Figure 4–40C). The proton gradient that drives ATP production is generated by cycling of electrons around photosystem I, in a process known as **cyclic electron transport** (Figure 4–40D).

Why has this complex partitioning of the Calvin cycle arisen? One possibility is that its operation augments the effectiveness of the C4 cycle in suppressing the oxygenase reaction of Rubisco. Noncyclic photosynthetic electron transport results in the production of oxygen from water, whereas cyclic electron transport in the bundle-sheath cell does not. Thus cyclic electron transport maximizes the ratio of carbon dioxide to oxygen and helps suppress the oxygenation reaction.

C4 photosynthesis seems to have obvious selective advantages over C3 photosynthesis. C4 plants are highly productive—in fact, the world's most productive crops are C4 plants: maize, sugarcane, millet, and sorghum (Figure 4–41). C4 plants are theoretically more productive than C3 plants because there is no loss of carbon via photorespiration and Rubisco's carboxylase reaction proceeds at its maximum rate because it is saturated with carbon dioxide. The elimination of photorespiration reduces the amount of ATP needed for a given net assimilation of carbon dioxide. There is a drawback, however. Assimilation of carbon dioxide via a C4 mechanism requires a greater input of energy than the operation of the Calvin cycle alone: ATP is required for the regeneration step in which pyruvate is converted to PEP. In addition, the C4 mechanism requires the development of an ancillary cell type—the mesophyll cell—containing functions that are not needed in C3 photosynthesis. The higher energy requirement of C4 photosynthesis is reflected in the light intensity required to achieve maximum photosynthetic rates. For C3 plants, CO_2 assimilation reaches a maximum rate at light intensities well below the intensity of normal sunlight, whereas the rate in a C4 plant continues to increase up to very high light intensities.

We would expect the elimination of photorespiration to place C4 plants at a particular advantage in high temperatures and in conditions of water stress. When the stomata

(A)   (B)   (C)

Figure 4–41.
Some C4 crop plants. (A) Maize (*Zea mays*), showing female flowers ("silks.")
(B) Sugarcane (*Saccharum officinarum*), showing a detail of the stem in which sucrose is stored. (C) Sorghum (*Sorghum vulgare*) with ripening seed heads.
(B, courtesy of Tobias Kieser.)

close to conserve water, the efficient mechanism for CO_2 uptake and concentration in a C4 plant means that almost all of the carbon dioxide in the leaf air spaces can be assimilated. The carbon dioxide compensation point of a C4 leaf—the point at which no further net CO_2 assimilation is possible—is less than 5 ppm. This contrasts sharply with the situation in a C3 leaf, in which this point is reached at 50 ppm (see Figure 4–33). In this way, C4 plants can assimilate more carbon dioxide per unit of water lost from the leaf than C3 plants; that is, they have a higher water-use efficiency than C3 plants.

The particular advantages of C4 over C3 photosynthesis at high temperature and in water stress suggest that the C4 mechanism evolved primarily as an adaptation to hot, arid climates. This idea is supported by analyses of the distribution of C4 species. For example, a survey found that 60 to 80% of the grass species in three natural habitats on the U.S.–Mexico border were C4, whereas 0 and 12% of the grass species were C4 in two natural habitats in central Canada. However, this is not a simple picture. Many C3 species are also confined to hot, arid habitats. In fact, although C4 species are abundant in hot, dry areas of North America, in some other parts of the world there are very few C4 species in habitats with the same climate. This means that C4 metabolism is only one of many adaptations that allow survival in these climates, and it does not represent an overwhelming selective advantage under these conditions. Note that a related type of photosynthetic metabolism, **crassulacean acid metabolism (CAM)**, has obvious selective advantages in hot, dry conditions. (C4 and CAM plants are further discussed in relation to drought stress in Section 7.3.)

Some scientists believe that the **selective pressure** that gave rise to C4 photosynthesis was not hot, dry climates but rather episodes of very low CO_2 concentration in the atmosphere. Until the end of the Cretaceous Period (about 100 million years ago), the CO_2 concentration in the atmosphere was probably higher than it is now (see Chapter 1). Levels then fell drastically, and at times during the past 100 million years have been as low as 200 ppm. Such low levels would have resulted in very high rates of photorespiratory CO_2 loss and hence poorer growth rates than under present atmospheric conditions. Selective pressure for CO_2-concentrating mechanisms would have been very strong, and it seems possible that C4 photosynthesis arose during one of these episodes of low carbon dioxide.

Whatever the selective pressure that led to the evolution of C4 photosynthesis, the predicted doubling of atmospheric CO_2 concentrations during the past century will have important consequences for the growth of C4 relative to C3 plants. At atmospheric CO_2 concentrations of 500 to 600 ppm, photorespiration will be strongly suppressed in C3 plants and any advantage held by C4 over C3 plants at the present time may well diminish.

Analysis of the distribution of C4 photosynthesis among higher plants shows that the mechanism has arisen independently many times, and it appears in about 18 families of higher plants. All of these families contain both C3 and C4 species (Figure 4–42). The complexity of the C4 mechanism—involving a host of changes to both the anatomy and the biochemistry of the leaf—raises interesting questions about the way in which it has evolved. Some clues have come from the study of a few species of "C3-C4 intermediate" plants. These were first discovered because their carbon dioxide compensation points are lower than those of C3 plants but not nearly as low as those of C4 plants. Like C4 plants, they have arisen several times during the evolution of higher plants. The genus *Moricandia* (Brassicaceae) contains C3 and C3-C4 members, and the genera *Panicum* (Poaceae) and *Flaveria* (Asteraceae, or Compositae) contain C3, C4, and C3-C4 members. Like C4 plants, C3-C4 plants have a mechanism to reduce the release of photorespiratory carbon dioxide. However, most of these species do not have true Kranz anatomy, a C4 cycle, or differentiation of chloroplast function. Instead, the reduction in compensation point is achieved by a differentiation of mitochondrial

(A)   (B)

function (Figure 4–43), which increases the efficiency with which photorespiratory carbon dioxide is recaptured by the Calvin cycle before it can escape from the leaf.

In photosynthetic cells of the C3-C4 intermediate leaf, as in the C3 leaf, 2-phosphoglycolate produced in the oxygenase reaction is processed via the photorespiratory cycle to glycine. Only the cells immediately surrounding the vascular bundle can carry out the next step of the cycle: conversion of glycine to serine via the mitochondrial glycine decarboxylase. Other photosynthetic cells lack glycine decarboxylase activity. Glycine diffuses to the cells surrounding the vascular bundle for decarboxylation, and serine diffuses back so that the photorespiratory cycle can be completed. In C3-C4 intermediate plants, therefore, photorespiratory carbon dioxide is released only from the cells surrounding the vascular bundle. The mitochondria in these cells lie on the inner wall, beneath a layer of chloroplasts, and all of the photorespiratory carbon dioxide must pass through the layer of chloroplasts (Figure 4–43). This means

Figure 4–42.
C3 and C4 members of the genus *Suaeda*.
(A) *Suaeda maritima*, a C3 species found on seashores. (B) *Suaeda moquinii*, a C4 species of deserts in the southwestern United States. *Suaeda* is in the family Chenopodiaceae, in which C4 photosynthesis has arisen independently several times. (B, courtesy of Keir Morse.)

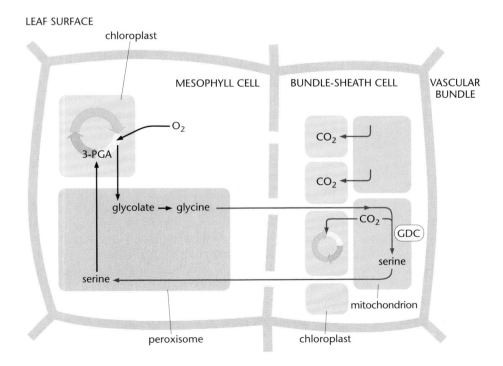

LEAF SURFACE

Figure 4–43.
Metabolism and leaf anatomy of C3-C4 intermediate species. Glycine produced in mesophyll cells during photorespiration diffuses to the bundle-sheath cells, the only leaf cells that contain glycine decarboxylase (GDC). In the bundle sheath, the CO_2 released via glycine decarboxylase must pass through a layer of chloroplasts, increasing the chance that it will be re-assimilated before it can escape from the leaf.
(3-PGA = 3-phosphoglycerate.)

that recapture of photorespiratory carbon dioxide via the Calvin cycle is much more likely in a C3-C4 leaf than in a C3 leaf.

4.4 SUCROSE TRANSPORT

Sucrose moves to nonphotosynthetic parts of the plant via the phloem

Sucrose synthesized in photosynthetic cells is the source of carbon for all other cells in the plant. It must be transported from the leaves (**source organs**) to nonphotosynthetic parts of the plant (**sink organs**). Sink organs include roots, **meristems**, young leaves that do not yet export sucrose, flowers, seeds and fruits, and vegetative storage organs such as **tubers** and **rhizomes**.

Sucrose is transported in part of the vascular tissue—the **phloem**—that consists of two types of cell: **sieve elements** and their **companion cells** (Figure 4–44). These two types of cell are formed when a common mother cell divides asymmetrically. As the sieve element matures, the plasmodesmata in the end walls adjacent to other sieve elements widen to form **sieve plates**, through which movement of solutes and proteins is relatively unimpeded. At the same time, the sieve element loses many of its subcellular structures, including the nucleus, vacuole, much of the endomembrane system, and many of the ribosomes. Large amounts of sieve element–specific proteins called **P-proteins** accumulate in tubular or fibrillar aggregations. These are thought to be important in plugging the sieve plates to prevent loss of sugars in the event of damage to the phloem.

Mature sieve elements stack on top of each other to form long conducting vessels for the translocation of sugars (**assimilate**) and other solutes throughout the plant. Companion cells are connected to sieve elements by numerous plasmodesmata (which allow passage of much larger molecules than plasmodesmata connecting most other cell types). The companion cells provide a route of entry of sugars into the sieve elements, and also provide some of the metabolic functions lost or reduced during differentiation of the sieve elements. For example, many of the proteins of the sieve elements, including the P-proteins, are synthesized in the companion cells and pass into the sieve elements via the large plasmodesmata.

The movement of sugars in the sieve elements probably occurs by a process called **pressure flow** (Figure 4–44C). The contents of the sieve elements in leaves, where sugars are loaded, are under high pressure, whereas the contents in sink organs, where sugars are withdrawn, are under lower pressure. This pressure gradient in the phloem causes the movement of phloem contents from the source to the sink organs. The pressure gradient is generated by the process of **phloem loading** in leaves. Sugars are loaded into the phloem from the leaf mesophyll cells by an energy-consuming process that results in a large difference in sucrose concentration between the cytosol of the mesophyll cells and the phloem. Typically, the concentration of sucrose in the mesophyll cytosol is 10 to 50 mM, whereas that in the phloem is about 1 M. The osmotic potential in the sieve element is thus much higher than that in surrounding tissues. Water is drawn into the phloem from the adjacent **xylem** by this high osmotic potential, creating a high pressure within the phloem. In the sink organs, sugars move out of the phloem into the surrounding tissue, and water follows. This means that the pressure in the phloem in sink organs is much lower than in the leaf.

Phloem loading may be apoplastic or symplastic

Sucrose moves from the cytosol of the mesophyll cells in which it is synthesized to mesophyll cells adjacent to the phloem by passive diffusion through plasmodesmata. The energy-consuming step required to load the phloem against a concentration gradient

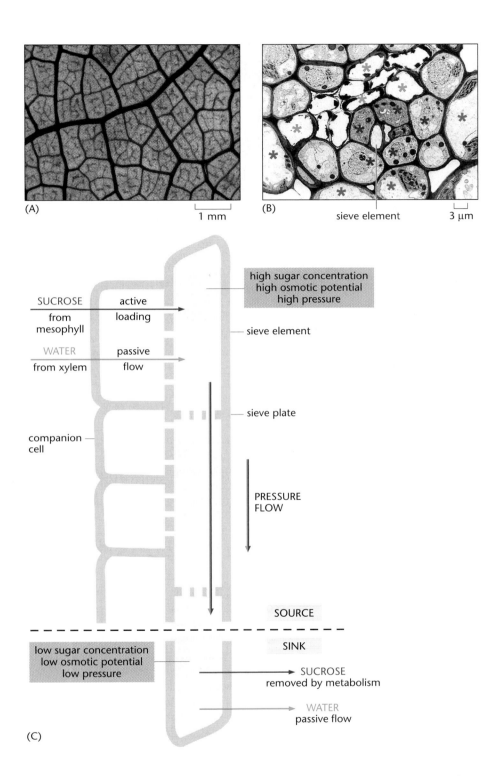

(A) 1 mm

(B) sieve element 3 μm

(C)

Figure 4–44.
Transport in the phloem. (A) Part of a leaf of poplar or cottonwood (*Populus deltoides*), stained to reveal the network of vascular bundles (veins) containing phloem and xylem elements. The small veins are the sites of loading of sucrose from the photosynthetic mesophyll cells into the phloem. (B) Electron micrograph of a cross section of a small vein in a leaf of ivy (*Hedera helix*); labels show some companion cells (*red stars*), xylem elements (*blue stars*), and mesophyll cells (*green stars*). (C) Mechanism of phloem transport. Sucrose is loaded into the phloem by an active, energy-consuming process that creates a high sucrose concentration, and hence a high osmotic potential, in the sieve element; water is drawn in, leading to a high pressure. Pressure in the sieve element of sink organs is much lower, because sucrose moves out as it is consumed in metabolism, and water follows. Movement of sugars in the phloem is driven by the difference in pressure between the source end and the sink end, a process called pressure flow. (A, courtesy of William A. Russin; B, courtesy of Gudrun Hoffmann-Thoma and Katrin Ehlers.)

occurs within the phloem itself. Two different mechanisms of phloem loading have been identified (Figure 4–45): apoplastic (that is, through the **apoplast,** the region out-side the plasma membrane, consisting mainly of cell walls) and symplastic (through the **symplast,** the cellular volume bounded by the plasma membranes and connected by plasmodesmata). In most species, one of the two mechanisms predominates. However, both may operate in the same plant, usually at different times or locations.

In apoplastic phloem loading, sucrose moves out of mesophyll cells adjacent to the phloem and into the apoplast. It is then taken into companion cells or sieve elements

Figure 4–45.
Apoplastic and symplastic phloem loading. In apoplastic loading (*left*), sucrose passes from mesophyll cells to the apoplast, then is transported across the plasma membrane of companion cells by an energy-requiring process that allows accumulation of sucrose to a high concentration in the sieve element. The inner wall of the companion cell is often deeply convoluted, increasing the area across which sucrose transport can occur. In symplastic phloem loading (*right*), sucrose passes from mesophyll cells into companion cells via plasmodesmata. In the companion cells sucrose is converted to larger sugars (such as raffinose; see Figure 4–49), which move into the sieve element. It is thought that raffinose and related sugars are too large to diffuse back through the plasmodesmata, so they can accumulate to high concentrations in the phloem.

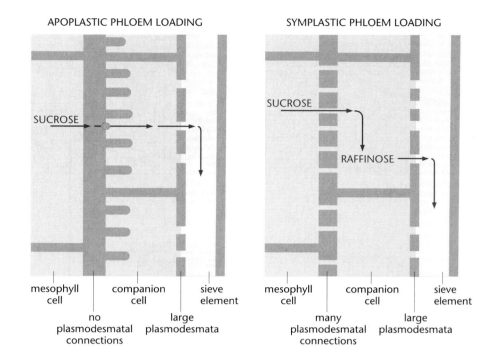

against a concentration gradient by an energy-requiring **sucrose transporter** protein in the plasma membrane of these cells. In symplastic phloem loading, sucrose moves from mesophyll cells into companion cells via the plasmodesmata, and in the companion cells it is converted in energy-requiring reactions to higher-molecular-weight sugars—typically the oligosaccharides raffinose, stachyose, or verbascose (Figure 4–45). The precise mechanism of symplastic loading of the sieve elements is not understood, but one suggestion is that the high-molecular-weight sugars produced in the companion cells are too large to pass back through the plasmodesmata into the mesophyll cells, and thus can only move into the phloem.

It is technically very difficult to study which kind of loading mechanism is being used at a given time, because the loading occurs in only a small number of cells in the leaf tissues. Circumstantial evidence, which is relatively easily gathered, includes the type of phloem anatomy in the minor leaf veins where loading occurs. Striking differences in phloem anatomy between species are thought to reflect whether phloem loading is apoplastic or symplastic. There are few, if any, plasmodesmatal connections between mesophyll cells and companion cells in tissues believed to use apoplastic loading. Instead, the inner face of the phloem cell wall next to the mesophyll cells is deeply invaginated, creating a very large surface area of plasma membrane—a hallmark of cells involved in major transfer processes. In contrast, vascular tissue thought to use symplastic loading has many plasmodesmatal connections and no invaginations of the cell wall at the phloem–mesophyll interface (Figure 4–46).

Direct evidence for apoplastic phloem loading in tobacco leaves has been provided by experiments with transgenic tobacco plants containing a yeast gene encoding **invertase,** an enzyme that hydrolyzes sucrose to the six-carbon sugars (hexoses) glucose and fructose. Neither glucose nor fructose can be loaded into the phloem. The yeast invertase carried signals that prevented its accumulation inside leaf cells, and instead targeted it to the apoplast. If phloem loading were symplastic, the presence of the invertase in the apoplast would be expected to have no effect on the process. If phloem loading were apoplastic, the invertase would be expected to disrupt the process because it would hydrolyze sucrose entering the apoplast from the mesophyll cells. This

(A)

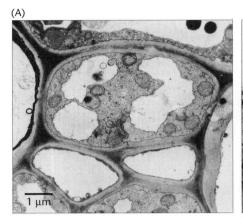 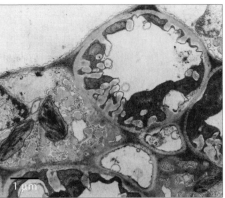

(B)

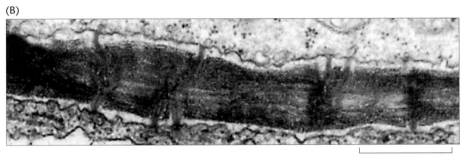

0.5 µm

Figure 4–46.
Differences in anatomy of companion cells, which may reflect different types of phloem loading. (A) Electron micrographs of companion cells in leaves of *Lythrum salicaria* (*left*), a species with symplastic loading, and *Zinnia elegans* (*right*), a species with apoplastic loading. Note the many cell-wall ingrowths in the *Zinnia* cell, increasing the surface area over which sucrose can be taken up from the apoplast. No ingrowths are present in the *Lythrum* companion cell. (B) The cell wall between a photosynthetic cell (*top*) and a phloem companion cell (*bottom*) in *Ajuga reptans*, a species with symplastic loading. Note the large number of plasmodesmata, allowing movement of sucrose into the companion cell through the symplast. (B, courtesy of Gudrun Hoffman-Thoma and Katrin Ehlers.)

would lead to an accumulation of hexoses and a drastic reduction in the amount of sucrose entering the sieve element (Figure 4–47). The transgenic plants were found to show severe reductions in growth of roots and shoots and reduced photosynthetic function in source leaves—consistent with apoplastic phloem loading. The reduced growth rate suggested that sink organs received reduced amounts of sucrose from the leaves. The effects on source leaves were probably caused by the very high levels of glucose and fructose in the apoplast. These sugars would be taken up by the surrounding mesophyll cells, causing osmotic stress and reduced expression of genes encoding components of the photosynthetic apparatus.

The proteins responsible for the movement of sucrose from the apoplast to the phloem in apoplastic loaders belong to a class of transporters known as "sucrose–proton cotransporters," which operate in conjunction with an ATP-driven **proton pump**, or **H$^+$-ATPase** (Figure 4–48). The proton pump exports protons from the phloem into the apoplast, using energy from the hydrolysis of ATP, thus creating a proton gradient across the plasma membrane. The sucrose transporter protein allows the diffusion of protons down the gradient from the apoplast into the phloem cell, and uses this proton flux to drive the uptake of sucrose. Transgenic plants with greatly reduced numbers of sucrose transporter proteins show the symptoms of a deficiency in phloem loading: reduced growth and loss of photosynthetic function in source leaves.

Figure 4–47.
Experiment in transgenic tobacco to determine type of phloem loading. Expression of yeast invertase in the apoplast of tobacco leaves reveals whether phloem loading is symplastic or apoplastic. If phloem loading were symplastic (*top*), the presence of invertase in the apoplast would not affect the process. If phloem loading were apoplastic (*bottom*), the invertase would hydrolyze sucrose leaving the mesophyll cells. This would prevent phloem loading and lead to accumulation of hexoses in the apoplast and in the mesophyll cells.

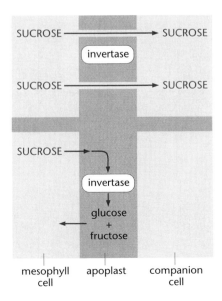

Figure 4–48.
Sucrose transport into the phloem.
An ATP-consuming proton pump in the plasma membrane of the companion cell sets up a proton gradient, which is used by the sucrose–proton co-transporter to drive sucrose uptake from the apoplast into the companion cell.

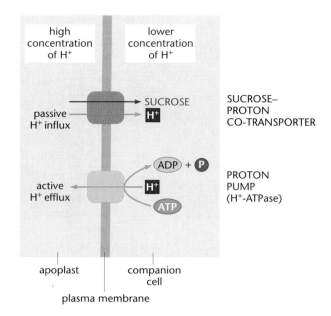

Apoplastic phloem loading is more widespread among plant species than symplastic loading. The phloem transport of raffinose, stachyose, verbascose, and related sugars rather than sucrose is confined to species in relatively few families of plants. Transport of these sugars is common in, for example, members of the Lamiaceae (Labiatae), Scrophulariaceae, and Cucurbitaceae (Figure 4–49). However, by no means all species in these families transport sugars other than sucrose, and some species may use both apoplastic and symplastic loading at different developmental stages or in different locations in the leaf.

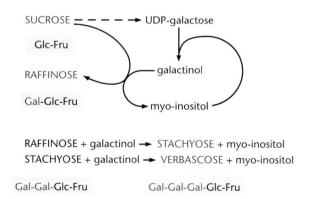

Figure 4–49.
Symplastic phloem loading. The photographs show species typical of families in which symplastic loading is common: *Stachys macrantha* (Labiatae), of the genus for which stachyose is named (*upper left*); mullein, *Verbascum thapsus* (Scrophulariaceae), of the genus for which verbascose is named (*upper right*); and zucchini (courgette), *Cucurbita pepo* (Cucurbitaceae) (*bottom*). The diagram shows synthesis of raffinose and larger sugars, stachyose and verbascose, in companion cells (see Figure 4–45). The process starts with the energy-requiring synthesis of UDP-galactose from sucrose. UDP-galactose and myo-inositol are substrates for the synthesis of galactinol. Galactinol donates a galactosyl unit (Gal) to sucrose to form raffinose, with the regeneration of myo-inositol. Similarly, donation of a galactosyl group to raffinose yields stachyose, and donation of a galactosyl group to stachyose yields verbascose. (Glc = glucose; Fru = fructose) (A and B, courtesy of Tobias Kieser.)

The path of sucrose unloading from the phloem depends on the type of plant organ

After transport through the phloem to the sink organs of the plant, sucrose must be unloaded. This process may also occur via the symplast or the apoplast or both, depending on the type of organ and its developmental stage. Unloading in vegetative sink organs, such as roots, young developing leaves, and meristems, is primarily symplastic. Sucrose moves through plasmodesmata from the sieve elements into surrounding cells by passive diffusion, down a concentration gradient created by consumption of sucrose in these cells.

Symplastic phloem unloading may also be accompanied by apoplastic unloading—for example, in the supply of sugars to developing sugar-beet roots and potato tubers. During apoplastic unloading, sucrose moves across the plasma membrane of the sieve elements, or of adjacent cells into which symplastic unloading has occurred, into the apoplast. This apoplastic sucrose may then be taken up by neighboring cells across their plasma membranes by the same type of sucrose transporter that operates in phloem loading, or it may first be hydrolyzed to hexoses by invertases in the cell walls; the hexoses are then transported into neighboring cells by specific hexose transporters. Analysis of the Arabidopsis genome has revealed several families of hexose transporter genes and a family of at least six sucrose transporter genes. Different members of these families are expressed in different sink organs.

In a few specialized situations, the movement of assimilate from the phloem to a sink organ proceeds entirely via the apoplast. This occurs in developing **seeds**. The maternal plant and the developing **endosperm** and **embryo** are genetically distinct and there is no direct symplastic connection between them. Apoplastic unloading is the only route for passage of sugars from the maternal plant into these organs. In maturing seeds of many eudicot plants, sucrose is unloaded from the phloem in the **testa** (seed coat) by symplastic movement into adjacent testa cells. From these cells it moves across the plasma membrane into the apoplast, either by passive diffusion or via some form of transporter. Sugars are then taken up from the apoplast into the outer cells of the embryo by plasma membrane transporters (Figure 4–50A). In some cases, this uptake is facilitated by a layer of specialized **transfer cells** at the periphery of the organ. For example, cells at the periphery of the developing bean embryo have deep invaginations of the inner face of cell wall adjacent to the testa, and thus a very large surface area of plasma membrane through which sugars can be taken up (Figure 4–50B).

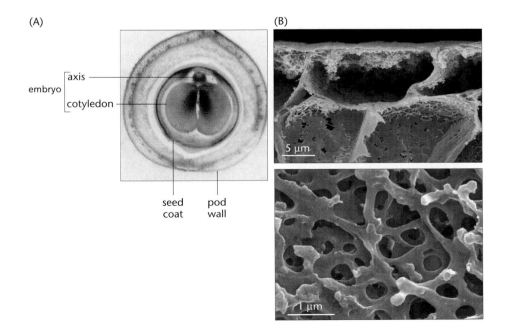

(A)

embryo — axis
— cotyledon

seed coat | pod wall

(B)

5 μm

1 μm

Figure 4–50.
Developing embryos of bean (*Vicia faba*).
(A) Section through a developing bean pod. The pod and the seed coat are part of the mother plant—in other words, they are maternal organs. The cells in these organs are supplied with sucrose by symplastic unloading from the phloem. The embryo is surrounded by the seed coat but not connected to it; it takes up sucrose released by inner cells of the seed coat into the apoplastic space.
(B) Transfer cells at the periphery of the embryo. Scanning electron micrograph of a cut-away section of the embryo (*top*) shows massive ingrowths of the outer cell wall of the surface (epidermal) cells (the upper layer of cells in the photograph) and some ingrowths of the walls in the next layers of the cell. A higher magnification view (*bottom*) shows the inner face of the labyrinthine cell wall.
(A, courtesy of Hans Weber.)

The supply of assimilate from the leaf is coordinated with demand elsewhere in the plant

The fate of assimilate that enters the phloem is determined largely by the demand for carbon in the sink organs. The sink organs can be regarded as competing with each other for assimilate, giving rise to the concept of "sink strength." A rapidly growing organ with a high demand for carbon will remove sugars rapidly from the phloem, and a relatively large proportion of the assimilate entering the phloem will thus be directed toward it. Such an organ is said to have a "high sink strength." Conversely, a slow-growing organ with a low demand for carbon will remove only small amounts of assimilate from the phloem and will thus command a relatively small proportion of the assimilate entering the phloem. The relative sink strengths of different organs change during the life of the plant and with changing environmental conditions. For example, during the early stages of growth of an **annual plant**, the main sinks are regions of vegetative growth: the shoot and root **apical meristems** and the young leaves. When flowering commences, the **floral meristem**, followed by the flowers themselves and ultimately the seeds, become the major sinks. In **perennial** and **biennial plants**, the main sinks are not only regions of vegetative growth and reproductive structures but also vegetative storage organs and storage structures such as tubers, roots, rhizomes, and specialized regions of cells in the wood of deciduous trees. The reserves in these storage organs and structures enable regrowth of the plant after a period of dormancy.

The importance of sink strength in determining how much assimilate is received by a sink organ is demonstrated by experiments in which the ability of potato tubers to metabolize sucrose was altered by genetic manipulation (Figure 4–51). The capacity to metabolize sucrose was increased by expression of a yeast invertase in the tuber apoplast (the same enzyme used in the experiments with transgenic tobacco plants, described above). The presence of invertase increased tuber sink strength by rapidly hydrolyzing sucrose and thus increasing the sucrose concentration gradient between the phloem and the tuber tissue. In these invertase-expressing potatoes, fewer potato tubers were produced than in the wild type, but they were significantly larger. In other experiments, the capacity to metabolize sucrose was decreased by creating transgenic plants with reduced activity of ADP-glucose pyrophosphorylase, an enzyme necessary for starch synthesis (see Section 4.2). Tubers of these transgenic plants had lower rates of conversion of sucrose to starch than wild-type plants, leading to the accumulation of sucrose rather than starch. This decreased the overall sink strength of the tuber because it reduced the sucrose concentration gradient between the phloem and the tuber tissue, and thus decreased the rate of phloem unloading. These plants had many more, smaller tubers than wild-type controls.

Taken together, the results from these experiments suggest that individual tubers compete with each other for sucrose from the phloem. When sink strength is abnormally high, the first few tubers formed consume all of the available sucrose, and no further tubers develop. Conversely, when sink strength is abnormally low, sucrose availability remains high and many tubers develop.

Although the fate of sucrose entering the phloem is determined by the relative strengths of the sink organs, the total flux of carbon from source to sink may be controlled primarily by factors in the source organs. This is illustrated by the transgenic potato plants with altered tuber sink strengths (Figure 4–51). Although changes in sink strength clearly altered the level of competition between individual sinks (i.e., between tubers), the total weight of tuber material was not greatly affected by the changes. In other words, increasing or decreasing sink strength had relatively little effect on the total flux of carbon from source to sink. The rate of CO_2 assimilation and sucrose synthesis in the leaves was thus more important in controlling source-to-sink carbon flux than was the rate of consumption of sucrose in the tubers.

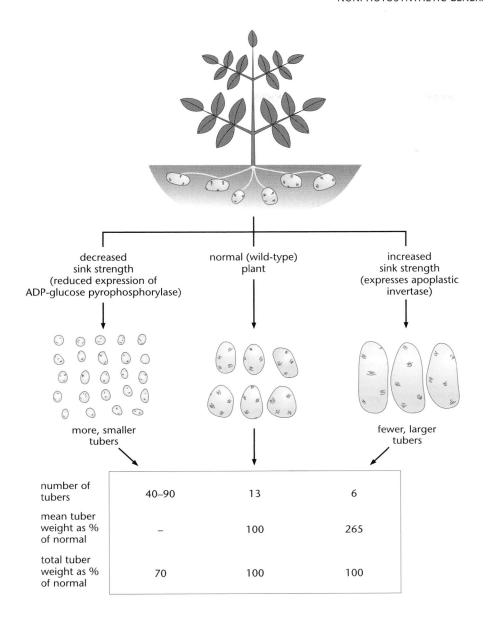

Figure 4–51.
Experiments demonstrating the importance of sink strength. Altering the sink strength changed the partitioning of sucrose among tubers, but not the total allocation to tubers; sink strength was very important in determining how many tubers could develop. At low sink strength (*left*), many, small tubers developed. At high sink strength (*right*), high competition among tubers for available sucrose meant that fewer, larger tubers developed. The total tuber weight, however, was not strongly affected by the changes in sink strength: it was reduced only 30% (compared with wild type) when sink strength was low, and was unaltered when sink strength was high. Thus, in this case, the source organs were more important than the sink organs in determining how much sucrose was allocated to tubers.

	decreased sink strength	normal (wild-type)	increased sink strength
number of tubers	40–90	13	6
mean tuber weight as % of normal	–	100	265
total tuber weight as % of normal	70	100	100

The control of source-to-sink carbon flux is a very important consideration in breeding crop plants for increased productivity. The harvested parts of most crop plants are sink rather than source organs (e.g., tubers and seeds). In order to increase the productivity of the crop, the carbon flux into these organs must be increased. Whether this is best achieved by altering metabolism in the sink organs themselves or in the source leaves is likely to depend on the species. The relative importance of sources and sinks in determining crop yield is also likely to be influenced by a host of interacting developmental and environmental factors.

4.5 NONPHOTOSYNTHETIC GENERATION OF ENERGY AND PRECURSORS

Three basic metabolic ingredients are essential for the biosynthesis of the host of molecules necessary for cell maintenance and growth (Figure 4–52): reducing power, usually in the form of NAD(P)H; energy, in the form of ATP; and precursor molecules.

Figure 4–52.
Sucrose as the source of carbon for the growth and maintenance of nonphotosynthetic cells. Metabolism of imported sucrose via glycolysis, the oxidative pentose phosphate pathway, and the Krebs cycle provides the reducing power, ATP, and precursor molecules for all of the processes involved in cell maintenance and growth. Some of the major purposes for which reducing power, ATP, and precursors are required are illustrated.

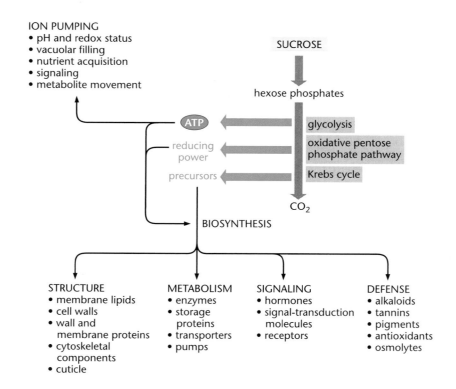

ION PUMPING
• pH and redox status
• vacuolar filling
• nutrient acquisition
• signaling
• metabolite movement

SUCROSE

hexose phosphates

ATP

reducing power

precursors

glycolysis

oxidative pentose phosphate pathway

Krebs cycle

CO_2

BIOSYNTHESIS

STRUCTURE
• membrane lipids
• cell walls
• wall and membrane proteins
• cytoskeletal components
• cuticle

METABOLISM
• enzymes
• storage proteins
• transporters
• pumps

SIGNALING
• hormones
• signal-transduction molecules
• receptors

DEFENSE
• alkaloids
• tannins
• pigments
• antioxidants
• osmolytes

In a photosynthetic cell in the light, the requirements for reducing power, ATP, and precursor molecules are met by photosynthesis. In a nonphotosynthetic cell, these requirements are met almost entirely by carbon compounds derived from sucrose imported from the leaves via the phloem. Sucrose entering a nonphotosynthetic cell is initially metabolized to hexose phosphates in the cytosol. Hexose phosphates are then further metabolized via three interrelated metabolic pathways, which can be regarded as the "metabolic backbone" of the cell: glycolysis, the **oxidative pentose phosphate pathway**, and the Krebs cycle (named after Hans Krebs, who discovered it; also known as the **tricarboxylic acid cycle** or **citric acid cycle**). These pathways provide all of the reducing power, ATP, and precursor molecules for the cell (Figure 4–52). In this section we describe the initial metabolism of sucrose and the three "metabolic backbone" pathways, then emphasize the way in which the partitioning of sucrose between these pathways responds to the changing demands of the cell.

Interconversion of sucrose and hexose phosphates allows sensitive regulation of sucrose metabolism

The cytosol of nonphotosynthetic cells has two routes for the interconversion of sucrose and hexose phosphates. These routes require different amounts of energy, are subject to different forms of regulation, and have different consequences for the subsequent metabolism of the hexose phosphates. This flexibility potentially allows the rate of sucrose metabolism to respond to the availability of sucrose, requirements for biosynthetic precursors in the cell, and the energy status of the cell. The complexity of sucrose metabolism reflects the central role of sucrose as the supply of carbon for all metabolic processes in the cell.

Conversion of sucrose to hexose phosphates is catalyzed either by invertase or by **sucrose synthase** (Figure 4–53). Invertase hydrolyzes sucrose to its component hexoses, which are then phosphorylated by **hexokinases** with the consumption of ATP. Sucrose synthase converts sucrose to UDP-glucose and fructose. UDP-glucose is

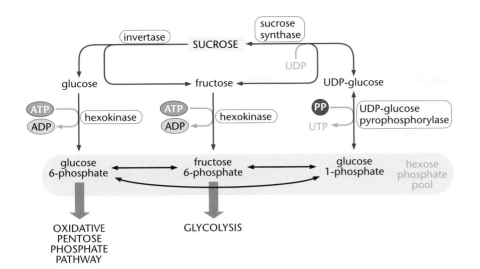

Figure 4–53.
Sucrose as a source of hexose phosphates. Conversion of sucrose to hexose phosphates via invertase involves effectively irreversible reactions and the consumption of two molecules of ATP for each sucrose metabolized. Conversion via sucrose synthase has lower energy costs and involves reversible reactions.

converted to glucose 1-phosphate by **UDP-glucose pyrophosphorylase**, and fructose is converted to fructose 6-phosphate via a hexokinase.

The relative importance of invertase and sucrose synthase in catalyzing the conversion of sucrose to hexose phosphates has been assessed using mutant pea plants and transgenic potato plants in which the activity of sucrose synthase is strongly reduced but the activity of cytosolic invertase is not. Starch synthesis is a major fate for hexose phosphates in developing pea seeds and in potato tubers. The altered plants showed substantial reductions in the rate at which sucrose was converted to starch in seeds (pea) and tubers (potato), demonstrating the importance of sucrose synthase in the conversion of sucrose to hexose phosphates in starch-synthesizing organs. Cytosolic invertase cannot compensate for large reductions in sucrose synthase activity.

Hexose phosphates can be converted back to sucrose in two ways (Figure 4–54). First, both sucrose synthase and UDP-glucose pyrophosphorylase catalyze effectively reversible reactions. Although net flux is usually in the direction of sucrose degradation, both enzymes can work in the opposite direction to maintain an equilibrium between cytosolic pools of sucrose and hexose phosphates. Depletion of hexose phosphates because of demand for carbon substrates in the cell shifts the equilibrium toward increased sucrose degradation. Conversely, reduced demand for carbon substrates, leading to accumulation of hexose phosphates, leads to reduced degradation of sucrose. Second, in addition to sucrose synthase, the cytosol of many types of nonphotosynthetic

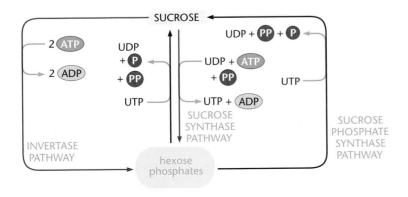

Figure 4–54.
Interconversion of sucrose and hexose phosphates. Hexose phosphates can be converted back to sucrose via the sucrose synthase pathway (*center*) and via the sucrose phosphate synthase pathway (*right*), involving sucrose phosphate synthase and sucrose phosphate phosphatase. The latter pathway is the same as that operating in sucrose synthesis in the cytosol of photosynthetic cells (see Section 4.2). The invertase pathway (*left*) is irreversible and so is not a route for conversion of hexose phosphates to sucrose.

cell contains the enzymes sucrose phosphate synthase and sucrose phosphate phosphatase (see Section 4.2), which catalyze the effectively irreversible synthesis of sucrose from hexose phosphates.

Altogether, these enzymes of sucrose synthesis and degradation create a complex cycle of reactions between sucrose and hexose phosphates. This cycle allows net flux between sucrose and hexose phosphates to respond in a very sensitive and flexible way to changes in demand for and availability of sucrose in the cell.

Glycolysis and the oxidative pentose phosphate pathway generate reducing power, ATP, and precursors for biosynthetic pathways

Glycolysis and the oxidative pentose phosphate pathway catalyze the partial oxidation of hexose phosphates to pyruvate with the generation of reducing power, ATP, and many important precursors of biosynthetic pathways (Figure 4–55). Glycolysis produces reducing power in the form of **NADH (reduced nictotinamide adenine dinucleotide)** in the reaction catalyzed by **glyceraldehyde 3-phosphate dehydrogenase**. ATP is produced in the glycolytic reactions catalyzed by **3-phosphoglycerate kinase** and **pyruvate kinase**. Note that in most cells these glycolytic steps are only a minor source of ATP: most of the ATP required by the cell is generated by mitochondrial metabolism. The first two steps of the oxidative pentose phosphate pathway produce reducing power in the form of NADPH, the primary reductant for biosynthetic reactions (see Section 4.2).

Many of the intermediates of glycolysis and the oxidative pentose phosphate pathway are precursors for biosynthetic pathways. For example, pyruvate is the precursor for the amino acid alanine, phosphoenolpyruvate for aspartate and several other amino acids derived from it, and ribose 5-phosphate for histidine. Phosphoenolpyruvate and erythrose 4-phosphate are together the precursors for the amino acids phenylalanine, tyrosine, and tryptophan, and these are in turn the precursors for the synthesis of a host of compounds including **lignin** and **suberin**, the phytohormone **indoleacetic acid**, and **flavonoids**. Five-carbon compounds from the oxidative pentose phosphate pathway are precursors for **nucleotides**, and dihydroxyacetone phosphate provides the glycerol backbone for **lipid** synthesis.

Flux through glycolysis and the oxidative pentose phosphate pathway is closely regulated in response to a cell's demand for particular biosynthetic precursors, ATP, NADH, and NADPH, and to the availability of hexose phosphates. The control of flux is shared by many steps in each pathway, and the relative importance of particular steps in controlling flux depends on many developmental and environmental factors. Consideration of the control of flux is further complicated because, for each of these pathways, most of the reactions are catalyzed by two separate sets of enzymes, one in the cytosol and one in the plastid, which in most circumstances are likely to differ in the supply of substrate, the demand for products, and hence the way in which flux is controlled. This division of labor between the cytosol and the plastid is discussed further in Section 4.7.

Here we consider the first step of glycolysis—the conversion of fructose 6-phosphate to fructose 1,6-bisphosphate—as an illustration of the complexity of regulation of "metabolic backbone" pathways. **Phosphofructokinase** and **pyrophosphate–fructose 6-phosphate 1-phosphotransferase (PFP)** are both capable of catalyzing this conversion, and both are found in the cytosol of most plant cells (Figure 4–56). The reaction catalyzed by phosphofructokinase is effectively irreversible. The enzyme is inhibited by the glycolytic intermediate phosphoenolpyruvate, and the strength of this inhibition is reduced by inorganic phosphate. The activity of phosphofructokinase is thus strongly dependent on the ratio of phosphoenolpyruvate to inorganic phosphate in the cytosol, and this ratio is in turn sensitive to the demand for ATP in the cell. The demand for ATP influences the flux through the Krebs cycle, and hence the rate of consumption of the

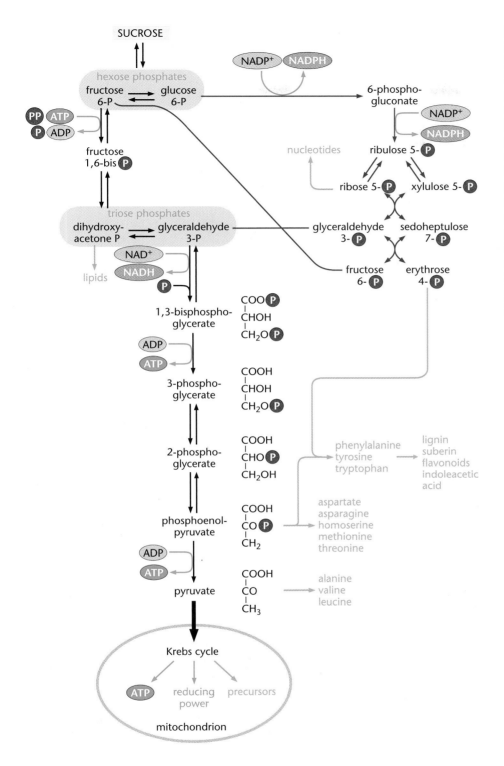

Figure 4–55.
Summary of pathways that generate ATP, reducing power, and precursors for biosynthetic processes. The hexose phosphate pool derived from sucrose is metabolized (partially oxidized) via glycolysis (*black arrows*) and the oxidative pentose phosphate pathway (*red arrows*) to produce ATP, reducing power in the form of NADH and NADPH, and precursors for biosynthetic pathways. Some examples of important compounds derived from intermediates of glycolysis and the oxidative pentose phosphate pathway are shown (*green*). Pyruvate, the product of glycolysis, enters the mitochondrion, where it is oxidized via the Krebs cycle to yield large amounts of ATP, reducing power, and further precursors for biosynthetic pathways.

pyruvate produced from phosphoenolpyruvate in glycolysis. The demand for ATP also influences the inorganic phosphate concentration in the cytosol, since ATP consumption produces ADP and phosphate.

The reaction catalyzed by PFP is readily reversible, thus allowing the operation of a cycle between fructose 6-phosphate and fructose 1,6-bisphosphate that is sensitive to and controlled by other cytosolic metabolites. PFP uses inorganic pyrophosphate (PP$_i$) rather than ATP to drive the reaction. It is strongly activated by the regulatory

Figure 4–56.
Interconversion of fructose 6-phosphate and fructose 1,6-bisphosphate.
Phosphofructokinase converts fructose 6-phosphate to fructose 1,6-bisphosphate via an effectively irreversible, ATP-consuming reaction sensitive to the ratio of phosphoenolpyruvate to inorganic phosphate in the cytosol. Pyrophosphate–fructose 6-phosphate 1-phosphotransferase catalyzes a reversible reaction, and is strongly activated by fructose 2,6-bisphosphate. Further details of modulation of this reaction are given in the text.

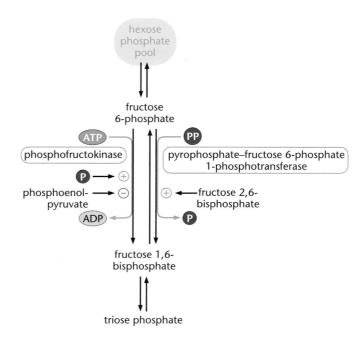

metabolite fructose 2,6-bisphosphate (described above as a regulator of sucrose synthesis; see Section 4.2) and is also subject to modulation by inorganic phosphate and pyrophosphate.

The regulation of the pyrophosphate concentration in the cytosol is of great importance in primary metabolism. Inorganic pyrophosphate is involved in many cytosolic metabolic reactions and also provides the energy for a class of proton pumps located on the **tonoplast** (see Section 3.5). In many respects it can be viewed as an alternative to ATP as an energy donor for cytosolic reactions. The cytosol has alternative routes using either ATP or pyrophosphate for conversion of sucrose to hexose phosphates (the invertase versus the sucrose synthase route), conversion of fructose 6-phosphate to fructose 1,6-bisphosphate (phosphofructokinase versus PFP), and maintenance of an appropriate proton gradient across the tonoplast, between the vacuole and cytosol (ATP- versus pyrophosphate-driven proton pumps). Reactions using pyrophosphate rather than ATP may be particularly favored when ATP availability is limited, such as when the O_2 concentration in plant tissues falls, potentially restricting the capacity for mitochondrial ATP synthesis (see below and Section 7.6). The central role of pyrophosphate is revealed in transgenic tobacco and potato plants with reduced pyrophosphate concentrations in the cytosol caused by the expression of a bacterial pyrophosphatase (an enzyme that converts pyrophosphate to phosphate). These plants display large and complex changes in their metabolism, many of which result from a reduced rate of conversion of sucrose to hexose phosphates.

The Krebs cycle and mitochondrial electron transport chains provide the main source of ATP in nonphotosynthesizing cells

Most of the ATP requirements of a nonphotosynthesizing cell (including leaf cells at night) are met by **respiration** in mitochondria (Figure 4–57). This process comprises the oxidation of pyruvate to generate reducing power, which drives electron transport chains in the inner mitochondrial membrane. Electron transport sets up an electrochemical gradient across the membrane that drives the synthesis of ATP. In theory, the complete oxidation of 1 pyruvate molecule to carbon dioxide can potentially drive the synthesis of about 12 ATP molecules. However, the actual yield is usually less than this because, as we describe below, biosynthetic precursors are removed from the Krebs cycle, and the yield of ATP is modulated according to the requirements of the cell.

Pyruvate produced by glycolysis in the cytosol enters the mitochondrion via a specific **pyruvate transporter** in the inner mitochondrial membrane. Once in the matrix, pyruvate is converted to **acetyl-CoA** with the production of reducing power (NADH).

Figure 4–57.
The Krebs cycle. In brief, a two-carbon unit derived from pyruvate (*top*) is condensed with a four-carbon unit; oxidative decarboxylation of the resulting six-carbon unit then generates reducing power in the form of NADH and $FADH_2$, produces carbon dioxide, and regenerates the four-carbon unit to allow a further turn of the cycle. The summary of the Krebs cycle products (*bottom*) shows that each turn of the cycle produces 1 ATP, 3 NADH, and 1 $FADH_2$. The NADH and the $FADH_2$ supply electrons to electron transport chains that drive further ATP synthesis. In theory, one turn of the cycle provides enough reducing power for the synthesis of 12 ATP, and thus a total of 13 ATP per turn.

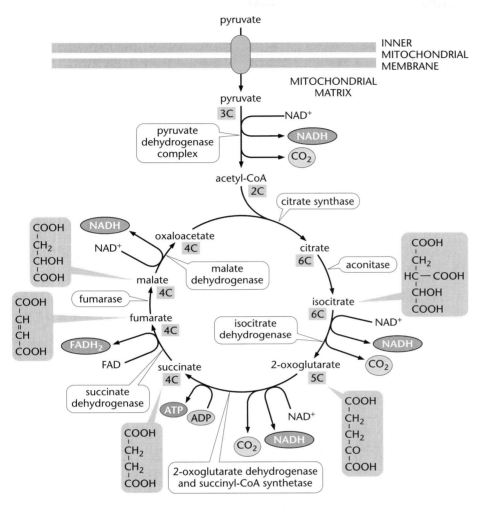

IN THE KREBS CYCLE:

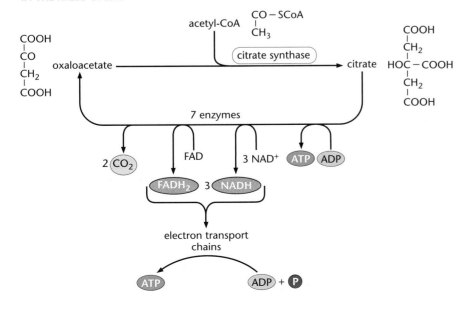

Acetyl-CoA enters the Krebs cycle and condenses with oxaloacetate to form citric acid, a reaction catalyzed by **citrate synthase**. In a series of oxidative reactions, which result in the loss of carbon dioxide and the production of large amounts of reducing power, citric acid is converted back to oxaloacetate, which reenters the cycle (Figure 4–57).

The reducing power—mainly in the form of NADH—drives electron transport chains in the inner mitochondrial membrane, transferring electrons from NADH to oxygen to form water. The electron carriers are arranged within the membrane so that electron transport drives a net transfer of protons from one side of the membrane to the other, setting up a proton and electrical gradient (an electrochemical gradient) across the membrane (Figure 4–58). The flux of protons back across the membrane, down the electrochemical gradient, drives membrane-bound ATP synthases, which convert ADP and phosphate to ATP. This process, called **oxidative phosphorylation**, is very similar to the synthesis of ATP by photophosphorylation in the thylakoid membranes of chloroplasts (see Section 4.2).

The basic electron transport chain—known as the **cytochrome pathway**—in the inner mitochondrial membrane consists of three electron-transfer complexes, I, III, and IV (Figure 4–59). All three complexes contain several different proteins, some embedded in the membrane and some exposed to either the mitochondrial matrix or the intermembrane space. Each complex contains some proteins encoded by the nuclear genome and some encoded by the mitochondrial genome.

In **complex I**, an **NADH dehydrogenase** accepts electrons from NADH in the matrix, regenerating NAD^+. The electrons are transferred from complex I to **complex III** via the electron carrier **ubiquinone**, and from complex III to **complex IV** via the electron carrier **cytochrome c**. In complex IV, which is known as **cytochrome c oxidase**, the electrons are transferred to oxygen on the matrix side of the membrane,

Figure 4–58.
Overview of ATP synthesis in the mitochondrion. NADH from the Krebs cycle feeds electrons into electron transport chains in the inner mitochondrial membrane. The transfer of electrons to oxygen, generating water, drives the transfer of protons across the membrane from the matrix to the intermembrane space, setting up an electrochemical gradient across this membrane. The flux of protons back into the matrix via the ATP synthase complexes in the membrane drives the synthesis of ATP.

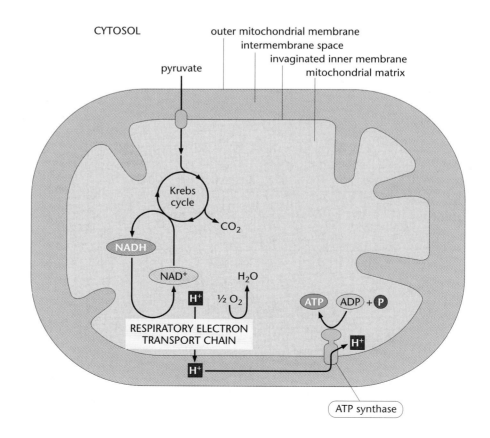

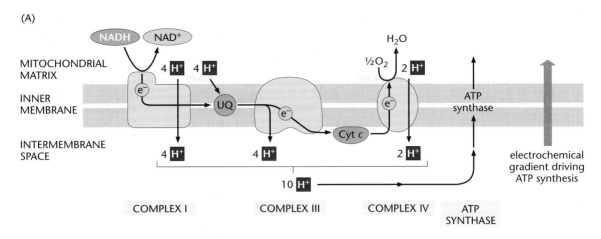

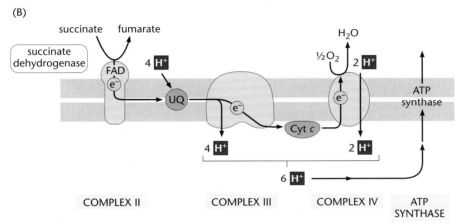

forming water. The passage of electrons through each of the complexes provides energy that drives the movement of protons from the matrix side of the membrane to the intermembrane space. Approximately 10 protons are transferred per NADH oxidized (Figure 4–59A).

The oxidation of succinate to fumarate in the Krebs cycle does not produce NADH. Instead, electrons are transferred to a **flavin adenine dinucleotide (FAD)** cofactor to produce $FADH_2$. (A **cofactor** is a molecule attached to an enzyme that participates in the catalytic reaction.) This cofactor is covalently bound to **succinate dehydrogenase**, which forms part of a membrane-bound complex known as **complex II**. Ubiquinone carries electrons directly from complex II to complex III, omitting the complex I step of the basic electron transport chain. Transfer of electrons through complex II does not result in a sufficient gain of energy to move protons across the membrane, hence electron transport from $FADH_2$ to oxygen results in the transfer of fewer protons than transport from NADH to oxygen (Figure 4–59B).

The main component of the electrochemical gradient across the inner mitochondrial membrane is a large electrical potential (~200 mV). The difference in proton concentration across the membrane is small: only 0.2 to 0.5 pH units—compared with a difference of about 4 pH units across the thylakoid membrane in chloroplasts (see Section 4.2). What is the reason for this difference between the mitochondrial membrane and thylakoid membrane gradients? In the mitochondrion, protons are transferred from the matrix to the intermembrane space (between the inner and outer membranes); because the outer membrane is freely permeable to protons, only a small proton concentration gradient can be maintained across the inner membrane. In the

Figure 4–59.
The electron transport chains of the inner mitochondrial membrane.
(A) Electrons from NADH are transferred via three complexes in the membrane to oxygen, generating water. The energy made available during this transfer is used to drive passage of protons across the membrane: a total of 10 protons for each electron. The electrochemical gradient thus generated promotes a flux of protons back across the membrane via the transmembrane ATP synthase complexes. (B) Conversion of succinate to fumarate via succinate dehydrogenase in the Krebs cycle reduces FAD rather than NAD^+. The energy generated by transfer of electrons from $FADH_2$ is less than that generated from NADH: 6 rather than 10 protons are transferred across the membrane per electron. (UQ = ubiquinone; Cyt c = cytochrome c)

chloroplast, electron transport chains transfer protons from the stroma into the restricted space of the thylakoid lumen, so a much larger pH gradient can be established.

The electrochemical gradient across the inner mitochondrial membrane drives the synthesis of ATP by an F_1-F_0 ATP synthase embedded in the membrane and extending into the matrix (Figure 4–60). The mechanism of ATP synthesis is essentially the same as in the thylakoid membrane. Energy for ATP synthesis is provided by the movement of protons down the electrochemical gradient from the intermembrane space to the matrix through a channel in the ATP synthase complex. The synthesis of one molecule of ATP requires the movement of about three protons through the ATP synthase.

Most of the ATP generated in the mitochondrial matrix by the F_1-F_0 ATP synthase is required for reactions that take place in other compartments of the cell. ATP must therefore be exported from the matrix, in exchange for ADP and phosphate generated as ATP is consumed in cellular processes (Figure 4–60). This exchange is achieved by two transporters in the inner membrane, both of which are driven by the electrochemical gradient. The **ATP–ADP transporter** catalyzes a one-to-one exchange of ATP for ADP. Since ATP has one more negative charge than ADP, the electrical potential across the membrane favors the movement of ATP outward and ADP inward. Inorganic phosphate is imported via a transporter that is driven by the efflux of hydroxyl groups, again favored by the electrochemical gradient across the membrane.

In addition to its role in ATP synthesis, mitochondrial metabolism also contributes to the maintenance of appropriate levels of reducing power in the cytosol and provides precursors for biosynthetic pathways. Mitochondria can both import and export reducing power, in two main ways. First, the inner membrane contains NAD(P)H dehydrogenases directed toward the intermembrane space. When the ratio of NAD(P)H to $NAD(P)^+$ in the cytosol is very high, the dehydrogenases accept electrons from cytosolic NAD(P)H and transfer them to ubiquinone, which transfers them to oxygen via the electron transport chains (Figure 4–61A). Second, reducing power can be imported or exported via metabolite **shuttles**. These shuttles consist of the exchange across the inner membrane of compounds in different oxidation states. An example is the exchange of malate or citrate for oxaloacetate, via a specific transporter. Interconversion of these metabolites by **malate dehydrogenase**, present in both the cytosol and the mitochondrion, generates or consumes reducing power (Figure 4–61B). The exchange of these metabolites thus effectively shuttles reducing power across the membrane.

Figure 4–60.
ATP synthesis and export from the mitochondrion. Movement of protons through a channel in the ATP synthase complex, down the electrochemical gradient across the inner mitochondrial membrane, drives the synthesis of ATP. ATP is exported from the mitochondrion via a transporter that exchanges it for ADP generated during consumption of ATP in cellular processes. Inorganic phosphate required for ATP synthesis is imported into the mitochondrion via a phosphate–hydroxyl exchange transporter.

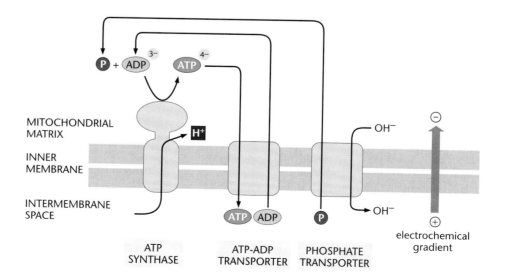

(A)

(B)

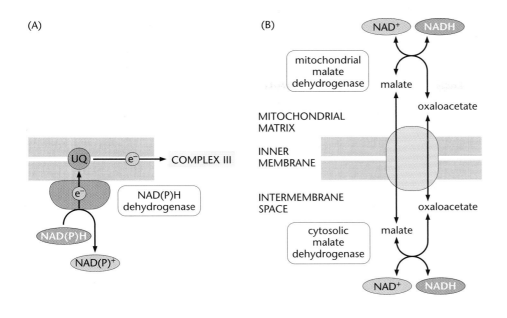

Figure 4–61.
Movement of reducing power into and out of the mitochondrion. (A) NAD(P)H dehydrogenases on the outer face of the inner mitochondrial membrane can feed excess reducing power from the cytosol into the mitochondrial electron transport chains for ATP synthesis. The dehydrogenases transfer electrons from NAD(P)H to the electron carrier ubiquinone (UQ), which transfers them to complex III of the electron transport chain (see Figure 4–59).
(B) A malate–oxaloacetate exchange transporter exchanges these two compounds across the inner mitochondrial membrane. This effectively transfers reducing power across the membrane, because interconversion of malate and oxaloacetate catalyzed by malate dehydrogenase, an enzyme found in both the cytosol and the mitochondrial matrix, consumes or generates NADH. Thus, transfer of malate into the mitochondrion in exchange for oxaloacetate drives oxaloacetate synthesis in the matrix and malate synthesis in the cytosol, resulting in the simultaneous generation of NADH in the matrix and consumption of NADH in the cytosol.

Intermediates of the Krebs cycle are precursors for several important biosynthetic pathways; for example, 2-oxoglutarate is a precursor for the amino acid glutamate. As Krebs cycle intermediates are diverted into other pathways, an equivalent amount of another intermediate must be added to the cycle, because the Krebs cycle—unlike the Calvin cycle (see Section 4.2)—cannot maintain a constant flux if intermediates are removed. If no intermediates were removed during a single turn of the Krebs cycle, the amount of oxaloacetate regenerated (molecule for molecule) would be the same as the initial amount of citrate formed. However, if intermediates were removed, the amount of oxaloacetate regenerated would be less than the original amount of citrate, so less citrate could be synthesized at the start of the next turn. In other words, the flux through the cycle would fall. This potential problem is prevented by a "topping-up" or **anaplerotic pathway**, which adds malate to the Krebs cycle to compensate for the removal of other intermediates for biosynthesis (Figure 4–62). Malate is synthesized in the cytosol from the glycolytic intermediate phosphoenolpyruvate, via the enzymes PEP carboxylase and malate dehydrogenase, and imported into the mitochondrion.

Mitochondrial metabolism must be tightly and sensitively controlled in response to demand within the cell as a whole for ATP, reducing power, and biosynthetic precursors, and the availability of substrates for oxidation in the Krebs cycle. The relative importance of these factors changes continuously, in response to a host of environmental and developmental factors. The basic pattern of mitochondrial metabolism described above is regulated at several levels, allowing an integrated response to these changes. The **pyruvate dehydrogenase complex**, which catalyzes the conversion of pyruvate to acetyl-CoA in the mitochondrion, is regulated so that its activity responds both to substrate supply and to the operation of the Krebs cycle. For example, it is inhibited by phosphorylation by a protein kinase that is itself inhibited by pyruvate, thus ensuring that the enzyme complex is active when pyruvate concentrations are high. The complex is also inhibited by NADH and acetyl-CoA, reducing its activity when demand for its products is low.

Some Krebs cycle enzymes are inhibited by a high $NADH:NAD^+$ ratio in the matrix, which signals that ATP demand is low. When the cell's need for ATP is low, the ATP:ADP ratio is high, restricting the activity of the ATP synthase. This means that the electrochemical gradient across the inner membrane is not dissipated by movement of protons through the channel in the ATP synthase complex. This in turn restricts proton transfer across the membrane via the electron transport chains, and hence the overall

Figure 4–62.
An anaplerotic pathway that maintains flux through the Krebs cycle when intermediates are removed as precursors for biosynthesis. Removal of a Krebs cycle intermediate for biosynthesis—for example, removal of 2-oxoglutarate for amino acid synthesis—necessitates the addition of another intermediate so that flux through the cycle is maintained. This is achieved by the synthesis of malate via phosphoenolpyruvate (PEP) carboxylase and malate dehydrogenase in the cytosol, and import of the malate into the mitochondrion.

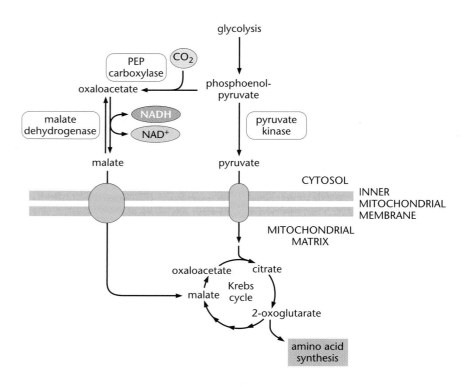

flow of electrons: electrons cannot be accepted from NADH because the electron carriers remain in a reduced state, and the NADH:NAD$^+$ ratio in the matrix increases.

The levels of regulation described above are found in both plants and animals, but plant mitochondria have two additional regulatory features that greatly increase the flexibility and potential for regulation of mitochondrial metabolism: an alternative route for providing pyruvate to the Krebs cycle, and an alternative to the cytochrome pathway for reoxidizing NADH.

NAD-malic enzyme catalyzes the conversion of malate to pyruvate in the mitochondrial matrix, so pyruvate can be synthesized in the matrix from malate imported from the cytosol. This pathway thus provides pyruvate for mitochondrial metabolism without the involvement of either cytosolic pyruvate kinase or the pyruvate transporter, and allows the entire operation of the Krebs cycle to proceed from imported malate (Figure 4–63). This route of pyruvate synthesis operates in parallel with the pyruvate kinase route in many plant cells, although in most cases the flux of carbon via malic enzyme seems to be considerably lower than that via pyruvate kinase. In bypassing pyruvate kinase, the route via malic enzyme reduces the yield of ATP from the conversion of sucrose to pyruvate. This route may be of particular importance when demand for ATP is low.

In addition to the cytochrome pathway illustrated in Figure 4–59, plant mitochondria also contain a pathway for transfer of electrons that bypasses complexes III and IV. This is known as the **alternative oxidase pathway** (Figure 4–64). Electrons transferred to ubiquinone from complex I are passed to the alternative oxidase (a nuclear-encoded protein located in the inner mitochondrial membrane) rather than to complex III. The alternative oxidase passes these electrons to oxygen, forming water, and protons are transferred across the membrane only via complex I.

This alternative oxidase provides a pathway for electron flow from NADH that is much less tightly coupled to the synthesis of ATP than is the cytochrome pathway. Only 4 protons are transferred across the inner mitochondrial membrane for each NADH molecule

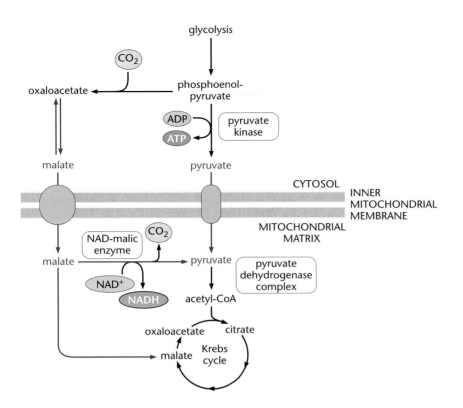

Figure 4–63.
An alternative route for providing pyruvate to the Krebs cycle. Malate imported from the cytosol into the mitochondrion can provide both a "top up" for the Krebs cycle in response to removal of intermediates for biosynthesis (see Figure 4–62) and, via NAD-malic enzyme, a source of pyruvate. This route of pyruvate synthesis bypasses pyruvate kinase and the pyruvate transporter.

oxidized, compared with 10 protons in the cytochrome pathway. Thus the contribution to the electrochemical gradient, and hence to ATP synthesis, is much smaller. The alternative oxidase probably functions to dissipate reducing power as heat, with minimal ATP synthesis, when demand for ATP is low. This is supported by the finding that the alternative oxidase is activated by a high $NADH:NAD^+$ ratio and high concentrations of pyruvate. Both of these conditions are likely to occur when electron flow through the electron transport chains is restricted by a low demand for ATP.

An extreme example of the use of these alternative routes is provided by thermogenic plant organs of a small number of flowering plants: floral parts that are capable of generating sufficient heat to raise their surface temperature several degrees above the temperature of the surrounding air. In the best-known examples of this phenomenon, the

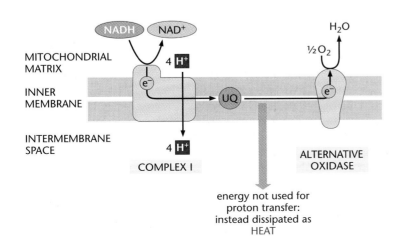

Figure 4–64.
The alternative oxidase pathway. Electrons from NADH pass via ubiquinone (UQ) to the alternative oxidase, bypassing complexes III and IV. This results in the transfer of only four protons across the membrane for each electron transferred, and hence a smaller electrochemical gradient and less ATP formed per NADH reoxidized. The energy generated in electron transport is instead dissipated as heat.

spadix of some species of *Arum* (Figure 4–65), the increased temperature is part of a mechanism for attracting pollinating insects.

In the *Arum* spadix, the generation of heat is triggered by an enormous increase (up to 150-fold) in the rate of glycolysis, fueled by the degradation of large reserves of starch. Almost all of the pyruvate synthesis proceeds via PEP carboxylase, malate dehydrogenase, and NAD-malic enzyme, rather than via pyruvate kinase and the pyruvate transporter. And almost all of the NADH generated in the Krebs cycle is oxidized via the alternative oxidase rather than the cytochrome pathway. The heat generated by the extremely high flux through the alternative oxidase pathway is sufficient to warm the spadix to about 10°C above ambient temperature. The use of the malic enzyme and alternative oxidase pathways rather than the pyruvate kinase and cytochrome pathways means that ATP generation is minimal; the spadix at this stage has no requirement for ATP, and it dies when the short period of thermogenesis (just a few hours) is finished.

Partitioning of sucrose among "metabolic backbone" pathways is extremely flexible and is related to cell function

So far we have discussed features of the regulation of "metabolic backbone" pathways that allow their outputs to be modulated in response to a host of metabolic factors that reflect the cell's requirements for ATP, reducing power, and biosynthetic precursors. These requirements vary enormously from one cell type to another, and from one developmental stage to another. As a consequence, cells vary greatly in the way that incoming sucrose is partitioned among metabolic backbone pathways. Here we illustrate the flexibility of the metabolic backbone by considering its relationship to cell function in four cell types in the tip of a root (Figure 4–66): meristematic cells of the root apex; elongating cells of differentiating **cortical parenchyma**; mature cortical parenchyma; and cells differentiating to form xylem.

Each cell type in the root arises from a meristem close to the apex (see Section 5.3). Cells in this apical meristem divide and grow rapidly; there is a high demand for precursors for the biosynthesis of a wide range of cell constituents, especially cell walls, membranes, and proteins, and for ATP and reducing power to drive these biosynthetic reactions. The flux through glycolysis and the Krebs cycle is high, and fine control of enzymes at many steps maintains a balance between the removal of intermediates as precursors and the oxidation of sufficient substrate to produce the ATP required to meet the biosynthetic demand. About half of the carbon entering glycolysis is used as metabolic precursors, and

Figure 4–65.
Thermogenesis in the *Arum* spadix. (A) During development of the inflorescence—a period of several weeks—the spadix is enclosed in a leaflike sheath. Sucrose is imported into the spadix from the leaves and used in the synthesis of large amounts of starch. At the same time, the capacity for glycolysis in the spadix increases to a high level, but the flux through this pathway is low. (B) When the inflorescence is mature, the sheath opens to reveal the spadix. Within just a few hours, all of the starch is degraded to hexoses and metabolized via glycolysis and the Krebs cycle. The very high rate of metabolism, and the engagement of the alternative oxidase pathway rather than the cytochrome pathway, results in the generation of heat (thermogenesis). The spadix temperature can rise to several degrees above ambient (the spadix feels very warm to the touch). The heat volatilizes amines produced in the spadix, and the smell attracts small insects. These crawl down into the base of the sheath, where they are trapped by a ring of downward-pointing hairs. Female flowers below these hairs are pollinated by insects that have visited other *Arum* flowers. (C) On the day following thermogenesis, the downward-pointing hairs shrivel and the insects can escape. The male flowers now shed pollen, which is carried by the escaping insects to other *Arum* flowers. (D) Mature inflorescence of *Arum maculatum* (*left*) cut away to show the downward-pointing hairs and male and female flowers enclosed in the sheath base, and (*right*) in its natural state in shady woodland. (E) Metabolism in a spadix cell during thermogenesis. PEP generated from starch via glycolysis is converted to pyruvate by the enzymes PEP carboxylase, malate dehydrogenase, and malic enzyme. Pyruvate is oxidized via the Krebs cycle, and the NADH generated donates electrons to the alternative oxidase pathway, resulting in the production of heat. The spadix at this stage has no requirement for ATP or precursor molecules—it senesces when its starch supply is exhausted—and essentially all of the carbon derived from starch is converted to carbon dioxide and reducing power for heat production in the mitochondria. (D, courtesy of Tobias Kieser.)

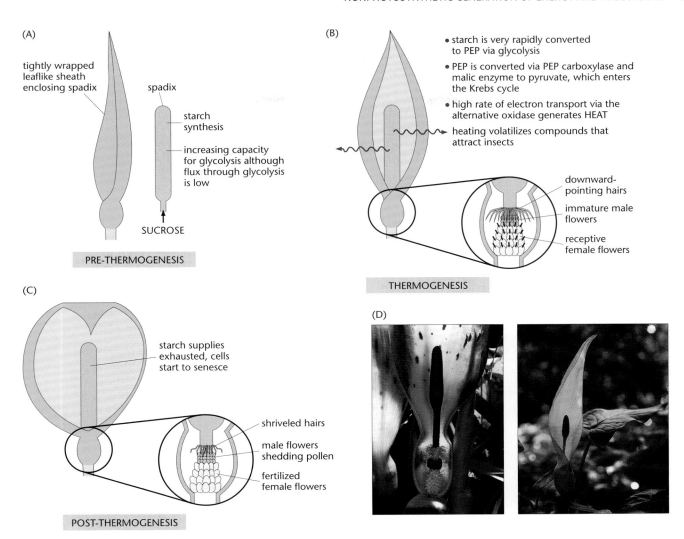

(A)

tightly wrapped
leaflike sheath
enclosing spadix

spadix

starch
synthesis

increasing capacity
for glycolysis although
flux through glycolysis
is low

SUCROSE

PRE-THERMOGENESIS

(B)

- starch is very rapidly converted
  to PEP via glycolysis
- PEP is converted via PEP carboxylase and
  malic enzyme to pyruvate, which enters
  the Krebs cycle
- high rate of electron transport via the
  alternative oxidase generates HEAT

heating volatilizes compounds that
attract insects

downward-
pointing hairs

immature male
flowers

receptive
female flowers

THERMOGENESIS

(C)

starch supplies
exhausted, cells
start to senesce

shriveled hairs

male flowers
shedding pollen

fertilized
female flowers

POST-THERMOGENESIS

(D)

(E)

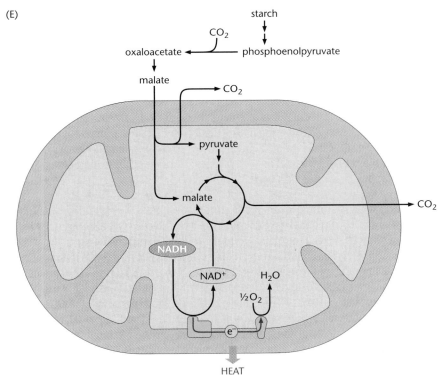

starch

CO_2

oxaloacetate ← phosphoenolpyruvate

malate

CO_2

pyruvate

malate

CO_2

NADH

NAD^+ H_2O

$\frac{1}{2}O_2$

e⁻

HEAT

Figure 4–66.
Cell types in a root tip. The four cell types referred to in the text are indicated on the right. For further details of root development see Section 5.3.

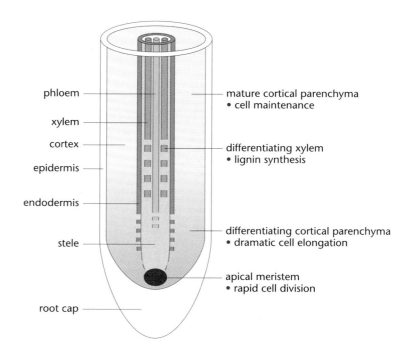

phloem

xylem

cortex

epidermis

endodermis

stele

root cap

mature cortical parenchyma
• cell maintenance

differentiating xylem
• lignin synthesis

differentiating cortical parenchyma
• dramatic cell elongation

apical meristem
• rapid cell division

is ultimately converted to structural and protein components of the cell. The other half is oxidized to carbon dioxide in the Krebs cycle to drive ATP synthesis.

After its final division, a cell destined to become part of the cortical parenchyma undergoes dramatic elongation. The huge increase in cell volume is dependent on the development of a central vacuole, which fills most of the internal volume of the cell, and on the synthesis of new membrane and cell wall material (see Sections 3.4 and 3.5). These elongating cells have a very high requirement for ATP for the ion pumps in the plasma membrane and tonoplast that drive the filling of the vacuole. Much of the carbon entering glycolysis is ultimately oxidized to carbon dioxide to fuel the ATP demand, and the yield of ATP per sucrose molecule oxidized is closer to the theoretical maximum (about 60–64 ATP per sucrose) than in the meristematic stage. At the same time, the demand for precursors for the synthesis of membrane lipids, **cellulose**, and **pectins** for wall and membrane extension must be met.

Having reached maturity, the parenchyma cell is much less metabolically active. Cellular structures and functions are maintained: lipids and proteins are replaced as they are turned over, and ATP is supplied to the ion and proton pumps that maintain the ionic, pH, and turgor balance of the cell. However, overall demand for ATP, reducing power, and biosynthetic precursors—and hence the rate of utilization of sucrose—is far lower than in the earlier stages of the cell's life.

A cell differentiating to form a xylem vessel undergoes massive thickening of the cell wall, brought about by the deposition of cellulose, and the cell wall also becomes lignified. Cellulose is synthesized from UDP-glucose, derived from sucrose via sucrose synthase (see Section 3.4). Lignin is synthesized from monolignol monomers derived from the aromatic amino acid phenylalanine (see Section 3.6), for which the precursors are PEP (a glycolytic intermediate) and erythrose 4-phosphate (an intermediate of the oxidative pentose phosphate pathway) (see Figure 4–55). Monolignol synthesis requires reducing power in the form of NADPH, which is generated in the oxidative pentose phosphate pathway. Because of these specific demands on the oxidative pentose phosphate pathway, much more of the carbon from sucrose enters this pathway in the differentiating cell than in the meristematic cell.

The changes in flux through metabolic backbone pathways during differentiation to form xylem are facilitated both by fine control of enzyme activities and by changes in **gene expression**. For example, the increased proportion of carbon from sucrose that enters the oxidative pentose phosphate pathway is accompanied by an increased capacity for the pathway as a whole (that is, increased amounts of all pathway enzymes due to increased expression of the genes that encode them) and by changes in fine control of the enzyme steps.

Other important aspects of the flexibility of the metabolic backbone are discussed in the remainder of this chapter. These include the capacity to store carbon as sucrose, and to remobilize this carbon (Section 4.6); duplication of pathways in the cytosol and the plastid (Section 4.7); and coordination of carbon metabolism with the requirements of nitrogen assimilation (Section 4.8) and the assimilation of other essential nutrients (Section 4.9).

4.6 CARBON STORAGE

Besides using imported sucrose for biosynthesis and energy production, nonphotosynthetic cells can also convert the sucrose to storage compounds. These stores of carbon can be mobilized at a later time, when the requirements of biosynthesis and energy production cannot be met by sucrose import. Imported sucrose itself can also serve as a short-term reserve of carbon in many cells, but longer-term storage is usually in the form of **starch**, **fructan**, or lipid (Figure 4–67).

Carbon storage is most obvious in the specialized organs that allow the plant to survive periods when no photosynthesis is possible—seeds, tubers, rhizomes, and other sorts of vegetative **propagule**. With very few exceptions, these organs contain large reserves of stored carbon synthesized during conditions favorable for photosynthesis, when sucrose is readily available to nonphotosynthetic organs. The organs persist through periods of dormancy or unfavorable conditions when no photosynthesis is possible, or

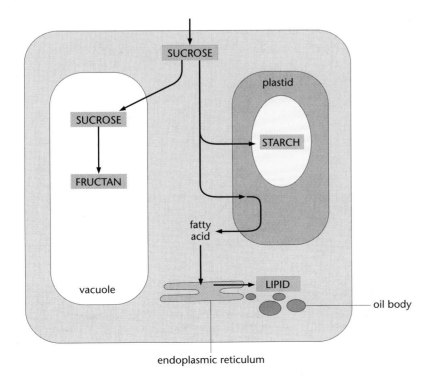

Figure 4–67.
The main forms in which carbon is stored in plant cells. Sucrose entering the cell can be stored in the vacuole or can be used in the synthesis of fructan in the vacuole, starch granules in the plastid, and/or storage lipids in oil bodies derived from the endoplasmic reticulum. Formation of storage lipids involves the synthesis of fatty acids in the plastid followed by export to the endoplasmic reticulum for lipid synthesis. Note that cells do not simultaneously accumulate all of these storage compounds: the main form in which carbon is stored varies from one cell type to another, and from one developmental stage to another.

after the organs are detached from the plant on which they developed. They can then germinate or sprout to produce new leaves, using the carbon released by degradation of the stored compounds for growth and respiration until photosynthetic capacity is regained.

For example, potato tubers develop from underground stems (**stolons**) during active growth of the potato plant and synthesize large quantities of starch from imported sucrose (Figure 4–68). The aboveground parts of the plant die at the end of the growing season, and the tuber persists in the ground through the winter. When conditions become favorable in the spring, degradation of the starch reserves to produce hexoses and hexose phosphates is triggered. These sugars provide substrates for respiration and growth that allow the development of **axillary buds** and eventually new photosynthetic shoots.

Storage of carbon is not confined to these specialized organs. Many sink organs accumulate storage compounds as a relatively short-term reserve at some stage of their growth and development. For example, starch is stored transiently (for a few days or hours) in cells close to apical meristems undergoing a developmental switch from division to expansion. In such cases, storage compounds can be viewed as a "carbon buffer," ensuring the maintenance of an adequate supply of carbon to the cell through a developmental period that may involve large changes in metabolic requirements. During cell division, the rate of sucrose import into the cell exceeds the immediate requirement for growth and respiration, and the excess is converted to starch. As the cells begin to expand, demand for carbon exceeds the rate of sucrose import, and degradation of the stored starch makes up the deficit.

Sucrose is stored in the vacuole

In many nonphotosynthetic cells, a degree of flexibility in the timing of use of imported sucrose is provided by a pool of sucrose in the vacuole, at equilibrium with the sucrose pool in the cytosol. Since the volume of the vacuole normally greatly exceeds that of the cytosol, most of the sucrose in the cell is contained in the vacuole. Short-term differences between the rate of entry of sucrose into the cell and the rate of its utilization for growth and respiration are thus buffered by the large vacuolar pool. For example, a rapid increase in the rate of metabolism—perhaps in response to an external stress—can be accommodated immediately by the movement of sucrose from the vacuole into the cytosol, as cytosolic sucrose is used up faster than the rate of import.

Figure 4–68.
A year in the life of a potato plant. During the summer, carbon is stored as starch in specialized vegetative storage organs, the tubers. After a period of dormancy during the winter, this starch provides the carbon supply for regrowth of the plant in the spring.

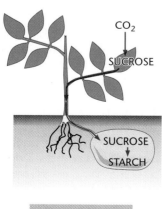

SUMMER
sucrose produced via photosynthesis in the leaves is exported to developing tubers where it is converted to starch

WINTER
aboveground parts die, tuber remains dormant in the soil

EARLY SPRING
initial growth of sprouts is supported by sucrose synthesized from starch in the tuber

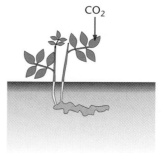

LATE SPRING
the young plant gains carbon from photosynthesis and is no longer dependent on the tuber

In addition to this short-term carbon buffering, a few plants also use sucrose as a long-term store of carbon. The best-known examples are crops that have been selected during domestication for the storage of very large amounts of sucrose in stems or roots—for example, in the internodal cells of sugarcane stems and the roots of sugar beet (Figure 4–69). In the wild relatives of these species, smaller stores of sucrose persist during unfavorable growing seasons and are then mobilized to provide carbon for growth and respiration when conditions become favorable. A vacuolar invertase is synthesized at the onset of mobilization of vacuolar sucrose, and hydrolysis of sucrose to hexoses inside the vacuole is followed by rapid movement of carbon to the cytosol via hexose transporters in the tonoplast.

The starch granule is a semi-crystalline structure synthesized by small families of starch synthases and starch-branching enzymes

Starch is by far the most common form of stored carbon in plants. Most plant cells contain starch at some point during their development. The harvested parts of the world's main crops are starch-storing organs: the grains of cereals such as maize, rice, and wheat; tubers such as potato; roots such as cassava and sweet potato; and pulses such as peas and many beans (Figure 4–70). In these organs, starch makes up 50 to 80% of the dry weight.

Starch synthesis occurs exclusively inside plastids, but the location of the synthetic pathway to the precursor for starch synthesis—the sugar nucleotide ADP-glucose—differs from one type of starch-storing organ to another. In most nonphotosynthetic organs (Figure 4–71A), including roots, tubers, and the seeds of peas and beans, ADP-glucose is synthesized in the plastid from a hexose phosphate, usually glucose 6-phosphate, that enters the plastid via a specific transporter in exchange for inorganic phosphate. In the starch-storing endosperm of cereal and other grass seeds, however, ADP-glucose can also be synthesized in the cytosol and taken up by a specific transporter into the plastid (Figure 4–71B). The endosperm has two **isoforms** (proteins encoded by different members of a gene family) of the enzyme ADP-glucose pyrophosphorylase, one in the plastid and the other in the cytosol.

Figure 4–69.
Swollen taproot of sugar beet. Sucrose may contribute as much as 20% of the weight of the root. A stem of sugarcane, in which sucrose is stored in the internodes, is shown in Figure 4–41B.

(A)

(B)

Figure 4–70.
Important tropical crops in which starch is the main storage product. (A) Women in Ghana preparing cassava (*Manihot esculentum*) roots. (B) A crop of the leguminous plant chickpea (*Cicer arietinum*). (A, courtesy of FAO/18293/P. Cenini; B, courtesy of ICARDA Photo.)

The relative importance of the cytosolic and plastidial pathways of ADP-glucose synthesis in developing maize endosperm is illustrated in mutants that lack active cytosolic ADP-glucose pyrophosphorylase. *shrunken2* mutants and *brittle2* mutants have **mutations** in the genes encoding subunits that make up the cytosolic enzyme. Both types of mutation considerably reduce the starch content of the seed (Figure 4–72). This finding, together with the fact that the cytosolic isoform accounts for more than

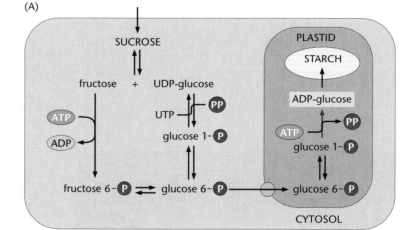

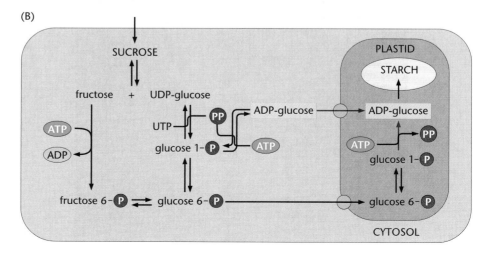

Figure 4–71.
Synthesis of ADP-glucose and starch.
(A) In most species, ADP-glucose is synthesized in plastids. Sucrose is metabolized to hexose phosphates in the cytosol, and glucose 6-phosphate is transported into the plastid (in exchange for inorganic phosphate; not shown) and converted to glucose 1-phosphate. ADP-glucose pyrophosphorylase converts glucose 1-phosphate to ADP-glucose. (B) In cereal endosperm, ADP-glucose is synthesized in both the plastid and the cytosol. A cytosolic isoform of ADP-glucose pyrophosphorylase, probably operating close to equilibrium, synthesizes ADP-glucose, which is transported into the plastid. ADP-glucose is also synthesized inside the plastid from imported hexose phosphate, as in other plants (A).

90% of the ADP-glucose pyrophosphorylase activity in the endosperm, suggests that most of the ADP-glucose for starch synthesis comes from the cytosolic rather than the plastidial pathway.

A starch granule is composed of two sorts of glucose polymer: **amylose** and **amylopectin** (Figure 4–73). Amylose consists of long, linear chains of glucose residues with few branch points. The overall structure of the granule is conferred by the amylopectin component, which makes up at least 70 to 80% of most starches. Amylopectin is a branched polymer of glucose residues consisting of α1,4-linked chains joined by α1,6 linkages. The branch points occur at regular intervals along the axis of the amylopectin molecule, so that the molecule consists of clusters of short chains (around 12–20 glucose residues) linked together by longer chains spanning two or more clusters (Figure 4–73B). This arrangement allows the organization of amylopectin molecules to form the semi-crystalline matrix of the starch granule. Adjacent chains within the clusters form double helices, which pack together in ordered arrays to form crystalline lamellae; these lamellae alternate with amorphous lamellae in which the branch points occur. The semi-crystalline "sandwiches" thus formed account for only part of the granule matrix. They alternate with amorphous zones in which the amylopectin chains are not so highly organized, giving rise to "growth rings" in the matrix (Figure 4–73C). Amylose may be located primarily in the amorphous zones of the starch granule.

The main factors controlling the rate of starch synthesis differ from one type of organ to another, and during the development of an organ. If starch is the main product of an organ's sucrose metabolism, there is a dramatic increase in transcription of genes encoding the enzymes of starch synthesis at the onset of starch accumulation. Flux through the pathway in potato tubers is largely controlled at the ADP-glucose pyrophosphorylase step, but in pea embryos control is more evenly distributed among different steps of the pathway.

The starch polymers are synthesized by starch synthases and **starch-branching enzymes** (Figure 4–74). Five classes of starch synthase and two of starch-branching enzyme have been identified in higher plants, and these are highly conserved across the plant kingdom. The sizes and branching patterns of the starch polymers are probably controlled by interactions of different isoforms of the two enzymes, but the way in which the polymers become packed together to form the granule matrix is still not fully understood.

Figure 4–72.
A maize cob with kernels segregating for the _brittle2_ mutation. The plump kernels have normal activity of ADP-glucose pyrophosphorylase and contain large amounts of starch (>70% of the dry weight). The shrunken kernels lack one subunit of the enzyme; the starch content of these kernels is about one-fourth that of the plump kernels, and the sucrose content is about 10 times higher.

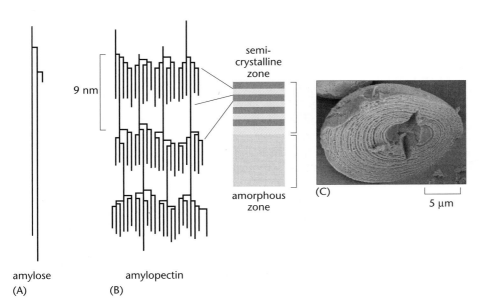

Figure 4–73.
Starch polymers and starch granule structure. (A) Amylose is a relatively linear polymer. (B) Amylopectin is highly branched. Because of the positioning of the branch points (α1,6 linkages), branches of 12 to 20 glucose units cluster at regular intervals along the axis of the molecule. Chains in these clusters become organized into crystalline lamellae (_blue bars_) alternating with amorphous lamellae (_gray bars_) that contain the branch points. These semi-crystalline zones themselves alternate with amorphous zones (_pale gray_) in which the amylopectin is less organized. (C) The alternating semi-crystalline and amorphous zones are visible as "growth rings" when the inner face of a starch granule is digested with amylase, an enzyme that preferentially attacks the amorphous zones. The scanning electron micrograph shows a starch granule from a potato tuber, cracked open and treated in this way.

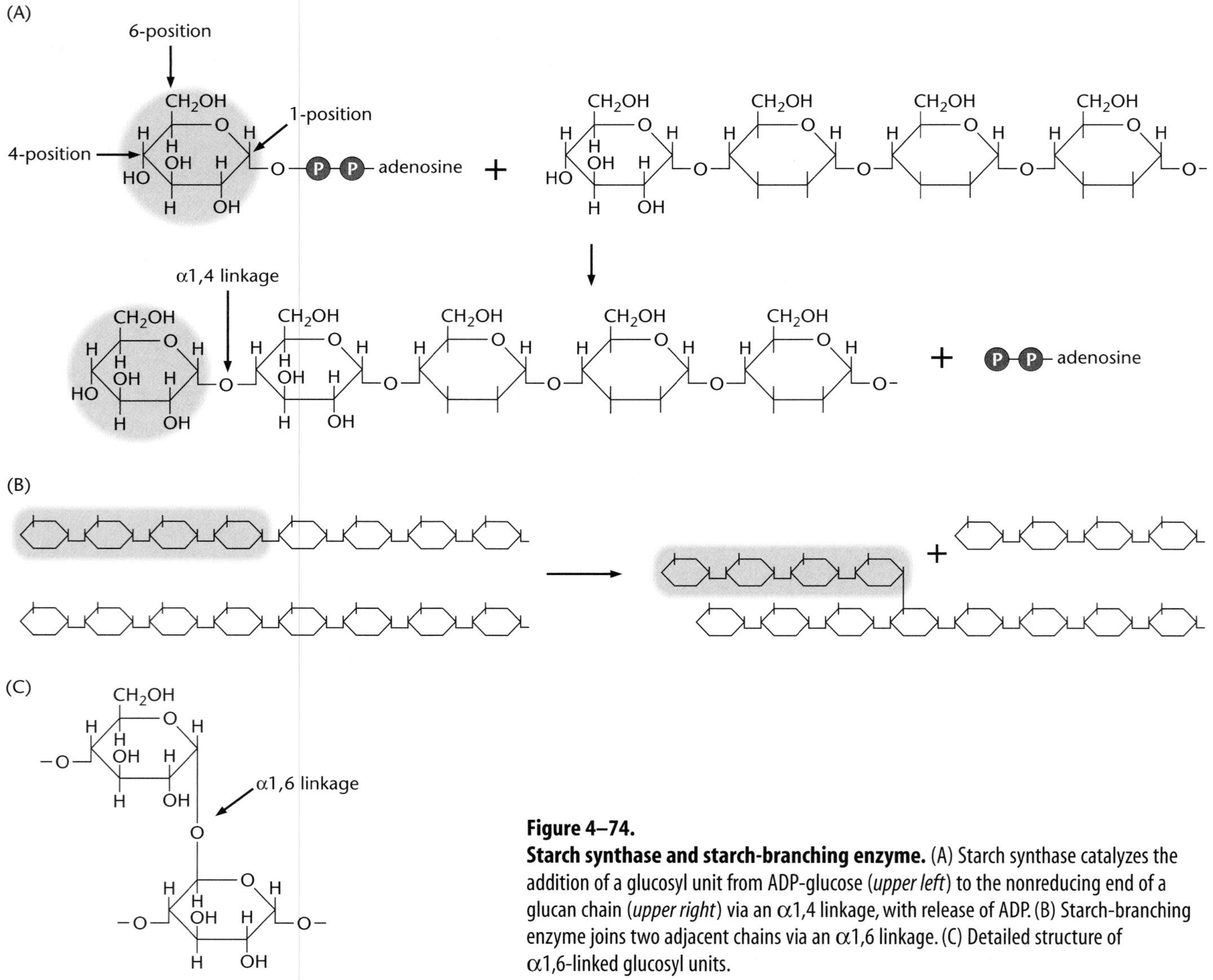

(A)

(B)

(C)

Figure 4–74.
Starch synthase and starch-branching enzyme. (A) Starch synthase catalyzes the addition of a glucosyl unit from ADP-glucose (*upper left*) to the nonreducing end of a glucan chain (*upper right*) via an α1,4 linkage, with release of ADP. (B) Starch-branching enzyme joins two adjacent chains via an α1,6 linkage. (C) Detailed structure of α1,6-linked glucosyl units.

Evidence that isoforms of starch synthase and starch-branching enzyme have distinct roles in the synthesis of the starch polymers comes from studies of mutant and transgenic plants in which the activity of individual isoforms has been reduced or eliminated. Elimination of different isoforms has radically different effects on the structure and composition of starch (Figure 4–75). One of the most striking revelations is that amylopectin is the product of complex synergistic interactions between the activities of four of the starch synthases and the two starch-branching enzymes, whereas amylose is exclusively synthesized by the fifth isoform of starch synthase. This latter isoform—known as "granule-bound starch synthase"—differs from the other four in several respects, including many of its catalytic properties and its location. Granule-bound starch synthase is exclusively located in the amylopectin matrix of the granule. Mutants lacking granule-bound starch synthase, such as the *waxy* mutant of maize, cannot synthesize amylose, and their starch consists entirely of amylopectin.

A third type of enzyme, a **debranching enzyme** called **isoamylase**, also plays a role in starch synthesis. Debranching enzymes cleave the α1,6 linkages of starch, releasing linear chains. Mutant plants that lack isoamylase contain both starch and another

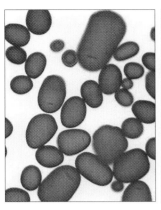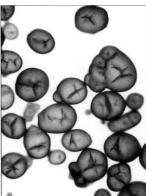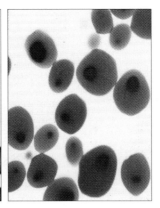

Figure 4–75.
Alterations in starch granules of potato tubers caused by reductions in specific starch synthases. Starch granules from the tubers of three potato plants, stained with iodine solution that interacts with amylose to give a blue color. In a normal (wild-type) tuber (*left*), the largest granule is about 60 μm long. In one transgenic plant (*middle*), the activity of an isoform of starch synthase (starch synthase III) is reduced. The structure of amylopectin is altered, resulting in deep cracking of the granules. In a second transgenic plant (*right*), the activity of granule-bound starch synthase is reduced. The amylose content is very low and most is confined to the central core of the granule, giving rise to blue staining only in the center and reddish staining of the amylopectin component in the outer regions.

glucose polymer, **phytoglycogen**. The best known of these mutants is the *sugary-1* mutant of maize, which is the basis of many of the low-starch, high-sugar sweet corns grown for human consumption. Phytoglycogen is, like amylopectin, an α1,4-, α1,6-linked glucose polymer, but it is more highly branched. It is not organized to form a granule and is found in plastids in a soluble form (as shown for another isoamylase-deficient mutant, *isa2*, in Figure 4–76). How the presence of isoamylase prevents the accumulation of phytoglycogen during normal starch synthesis is not yet clear.

The pathway of starch degradation depends on the type of plant organ

The degradation of starch to provide carbon for growth and respiration probably proceeds in two main stages (Figure 4–77). First, the granule is attacked by **endoamylases**, enzymes (such as α-amylase) that hydrolyze α1,4 linkages and release soluble **glucans** from the granule. In the second stage, the soluble glucans are broken down to glucose or hexose phosphate. Four types of enzyme are involved in this second stage. Debranching enzymes hydrolyze α1,6 linkages to release linear chains; an **exoamylase** (such as β-amylase) cleaves two-glucose units (maltose) from the end of these chains; and **maltases** hydrolyze the maltose to glucose. In addition to an exoamylase, **starch phosphorylase** can act on the released linear chains, adding phosphate to the terminal glucosyl residue and releasing glucose 1-phosphate.

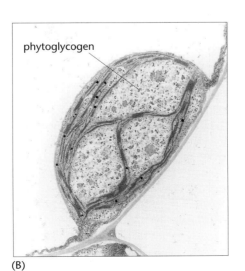

starch granule

phytoglycogen

(A) (B)

Figure 4–76.
Accumulation of phytoglycogen in chloroplasts of an Arabidopsis mutant lacking isoamylase. (A) Chloroplast from a wild-type plant, containing starch granules. Chloroplast starch granules are typically 1 to 2 μm in diameter. (B) Chloroplast from an *isa2* mutant, containing residual starch and a large quantity of soluble and semi-soluble phytoglycogen.

Figure 4–77.
Generalized scheme for starch degradation. The granule is attacked by an endoamylase (α-amylase), which releases soluble, branched glucans, which in turn are acted on by debranching enzymes to give linear chains. These can be attacked progressively from their nonreducing end, either by an exoamylase (β-amylase) to produce maltose or by starch phosphorylase to produce glucose 1-phosphate.

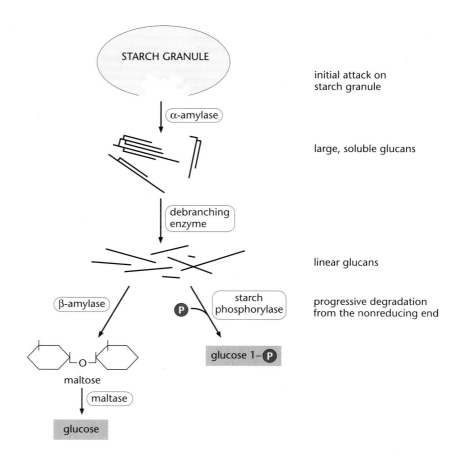

The relative importance of the exoamylase and starch phosphorylase (both of which act on the linear chains released in earlier stages of starch degradation) is not understood for most plant organs, and there is also little understanding of how the overall process of starch degradation is controlled. It is clear, however, that both the pathway and the control of starch breakdown differ from one type of organ to another. Examples of very different types of starch degradation are provided by the endosperm of germinating cereals and by leaves at night.

In the endosperm, as in all starch-synthesizing tissues, starch synthesis and storage occur inside plastids. However, at seed maturity, the membranes within and around the endosperm cells disappear; when the seed imbibes water prior to germination, the starch is contained in a nonliving tissue. The process of starch degradation is triggered by the embryo, which lies at one end of the seed. The embryo secretes the phytohormone **gibberellin**, which causes the **aleurone** (the layer of cells surrounding the endosperm) to release an endoamylase (such as α-amylase) into the endosperm (Figure 4–78). Gibberellin is detected by receptors on the plasma membrane of the aleurone cells, and transduction of this signal inside the cell leads to up-regulation of the expression of a **transcription factor** of the **MYB family**. This transcription factor binds to elements in the **promoter** of the gene encoding the endoamylase and activates its expression. Unlike the endoamylase, the exoamylase (such as β-amylase) is synthesized in the endosperm during seed development and is sequestered in protein-storage bodies in an inactive form as the seed matures. Imbibition by the mature seed triggers release and activation of the enzyme. The hydrolytic degradation of starch within the endosperm by enzymes secreted from surrounding cells (an endoamylase) or activated within the endosperm itself (an exoamylase) results in the production of glucose, which moves via transporters across the **scutellum** layer and into the embryo, where it is used as the substrate for respiration and growth of the new plant.

(A)

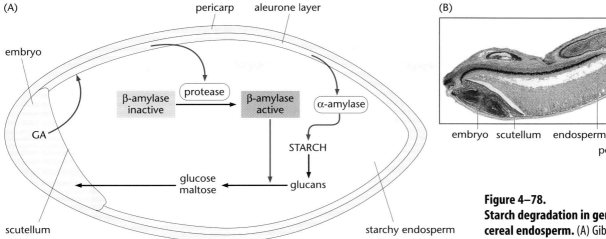

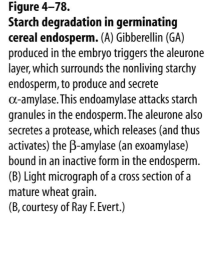

Figure 4–78.
Starch degradation in germinating cereal endosperm. (A) Gibberellin (GA) produced in the embryo triggers the aleurone layer, which surrounds the nonliving starchy endosperm, to produce and secrete α-amylase. This endoamylase attacks starch granules in the endosperm. The aleurone also secretes a protease, which releases (and thus activates) the β-amylase (an exoamylase) bound in an inactive form in the endosperm. (B) Light micrograph of a cross section of a mature wheat grain.
(B, courtesy of Ray F. Evert.)

In leaves, starch degradation takes place in chloroplasts and is very tightly regulated. Although the enzymes of the degradative pathway are present in the chloroplast during the day, when starch is being synthesized (see Section 4.2), little starch breakdown occurs in the light. The process is switched on in darkness, by unknown mechanisms. Many chloroplasts contain both an exoamylase and starch phosphorylase, so can potentially degrade starch either to free maltose or to glucose 1-phosphate (Figure 4–79). In Arabidopsis leaves, the main product is maltose. Maltose is exported from the chloroplast and converted to hexose phosphate in the cytosol; the hexose phosphate is then converted to sucrose for export to nonphotosynthetic parts of the plant. Glucose 1-phosphate can be converted inside the chloroplast to triose phosphates, which are exported by the triose phosphate transporter for use in respiration and biosynthesis in the leaf cell at night.

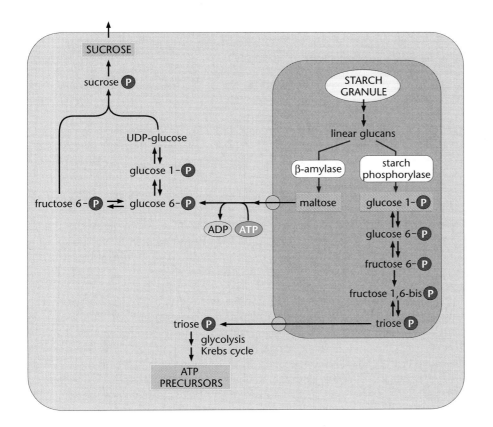

Figure 4–79.
Proposed pathways of starch degradation in a leaf cell at night. The carbon in starch accumulated as a product of photosynthesis during the day is needed at night for sucrose synthesis to maintain export from the leaf, and for the growth and maintenance of the cell. In this proposed pathway, carbon for sucrose synthesis is provided by hydrolytic degradation of starch via β-amylase, with export of glucose to the cytosol; carbon for cell growth and maintenance may be provided by phosphorolytic degradation via starch phosphorylase, with conversion of glucose 1-phosphate to triose phosphate for export to the cytosol. This scheme is speculative: the precise pathways of starch degradation in chloroplasts are not yet known.

Some plants store soluble fructose polymers rather than starch

In some plants, carbon is stored, in the long or short term, as fructans rather than starch. Fructans are synthesized from sucrose in the cell vacuoles of the leaves and stems of many grasses and of members of the Liliales, including onions and asparagus, and in the underground storage organs of some members of the Asterales, such as *Inula*, *Cichorium* (*Chichorium intybus* is chicory), and *Helianthus* (*Helianthus tuberosus* is the Jerusalem artichoke) (Figure 4–80).

The fructans are a complex series of soluble fructose polymers. The precise structure of the polymer is species-dependent. In the simplest case, polymers are linear chains of either α1,2-linked or α2,6-linked fructose residues. Other types of fructans contain a mixture of linkage types and may be branched. Synthesis proceeds by transfer of the fructosyl residue of sucrose onto a fructan chain (Figure 4–80B), and several different types of **fructosyl transferase** enzyme may be required, depending on the type of fructan. The glucose residue of the sucrose is exported from the vacuole to the cytosol, phosphorylated, and recycled into sucrose.

Fructan synthesis in grass leaves takes place when the rate of sucrose synthesis exceeds the rate of export; high sucrose levels induce expression of genes encoding fructan-synthesizing enzymes. When sucrose levels fall, fructans are enzymically degraded to release fructose from the ends of chains. The fructose is exported from the vacuole to the cytosol, where it is phosphorylated and converted to sucrose. Fructans in nonphotosynthetic storage tissues—such as the stems of perennial grasses, bulbs of onions, and roots of chicory and Jerusalem artichoke—fulfill the same role as starch in similar organs of other species: they are synthesized during active growth of the plant, persist during periods when photosynthesis is reduced or absent, and are degraded to provide carbon for regrowth when conditions become favorable.

Storage lipids are synthesized from fatty acids in the endoplasmic reticulum

The fruits and seeds of many plant families store lipids rather than starch or fructans as their carbon reserve. Seeds of peanut and soybean (legumes) store lipid, as do seeds of oilseed rape and Arabidopsis (crucifers), marrow and squash (cucurbits), and sunflower (Asterales family). In some species, lipid is the major storage compound in the seed. In other species, some parts of the seed accumulate lipid and others starch—for example, the endosperm of cereal seeds accumulates starch and the embryo may contain some storage lipid.

Lipid is stored in the form of **triacylglycerols**, highly hydrophobic molecules consisting of three **fatty acid** chains attached to a glycerol backbone (Figure 4–81). This form of stored carbon has the potential to yield more energy than starch because lipids are a more reduced form of carbon (i.e., they contain less oxygen per unit of carbon) than carbohydrates such as sucrose and starch; lipid metabolism potentially yields more ATP per unit mass. This difference in energy yield per unit mass is enhanced by the hydrophobic nature of storage lipids. Starch granules contain considerable amounts of water (about 40% of their volume), whereas the **oil bodies** in which lipids are stored contain no water. However, the synthesis of storage lipid is a more complex and energetically expensive process than the synthesis of starch.

Fatty acid molecules consist of a hydrocarbon chain with a terminal carboxyl group. Fatty acids containing 16 or 18 carbon atoms are synthesized in plastids (see Section 4.7). On export to the cytosol as substrates for triacylglycerol synthesis, the fatty acids are converted to esters of **coenzyme A (CoA)**, or acyl-CoAs. The acyl-CoAs then move to the endoplasmic reticulum, where membrane and storage lipids are synthesized. In steps common to both membrane and storage lipid synthesis, the fatty acid is transferred from an acyl-CoA to position 1 of glycerol 3-phosphate, and a second fatty

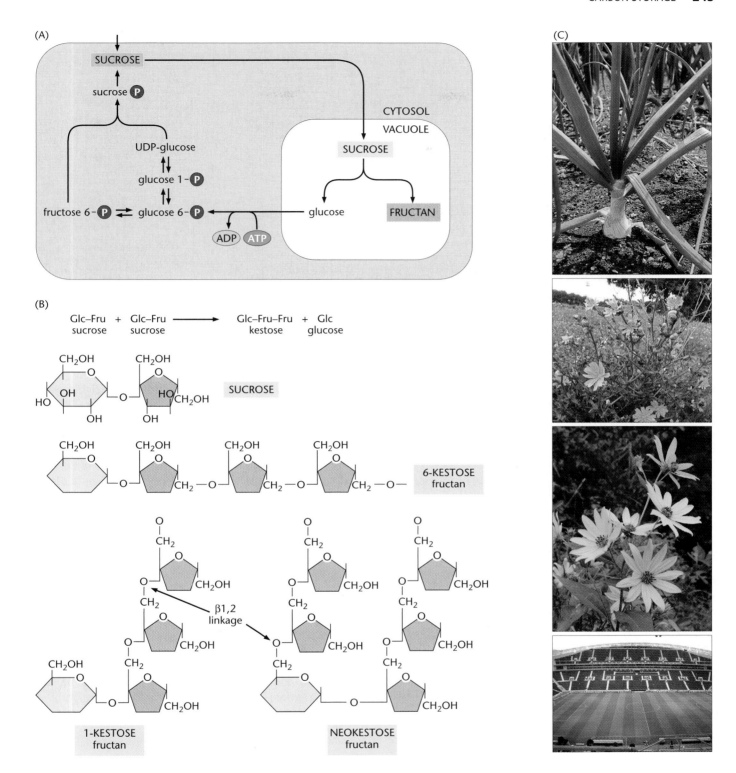

Figure 4–80.
Synthesis of fructans. (A) Use of sucrose for fructan synthesis in the vacuole releases glucose, which is exported to the cytosol and can be recycled into sucrose. (B) Fructan synthesis starts with the transfer of a fructosyl (Fru) unit from one sucrose molecule to another, yielding glucose and kestose. Fructans retain a glucosyl (Glc) unit from the original sucrose acceptor molecule. The number of fructosyl units and the types of linkage in the fructan molecule vary with species and developmental stages. Three classes of fructans are shown here. (C) Some fructan-accumulating species. From top to bottom: onion, flowers of chicory, and flowers of Jerusalem artichoke (all of which accumulate fructans in swollen roots), and rye grass of a stadium field (*Lolium* sp.; it accumulates fructan in its leaves during photosynthesis). (Ci and ii, courtesy of Tobias Kieser; Ciii, courtesy of David G. Smith; Civ, courtesy of Thomas Kramer/Stadionwelt.)

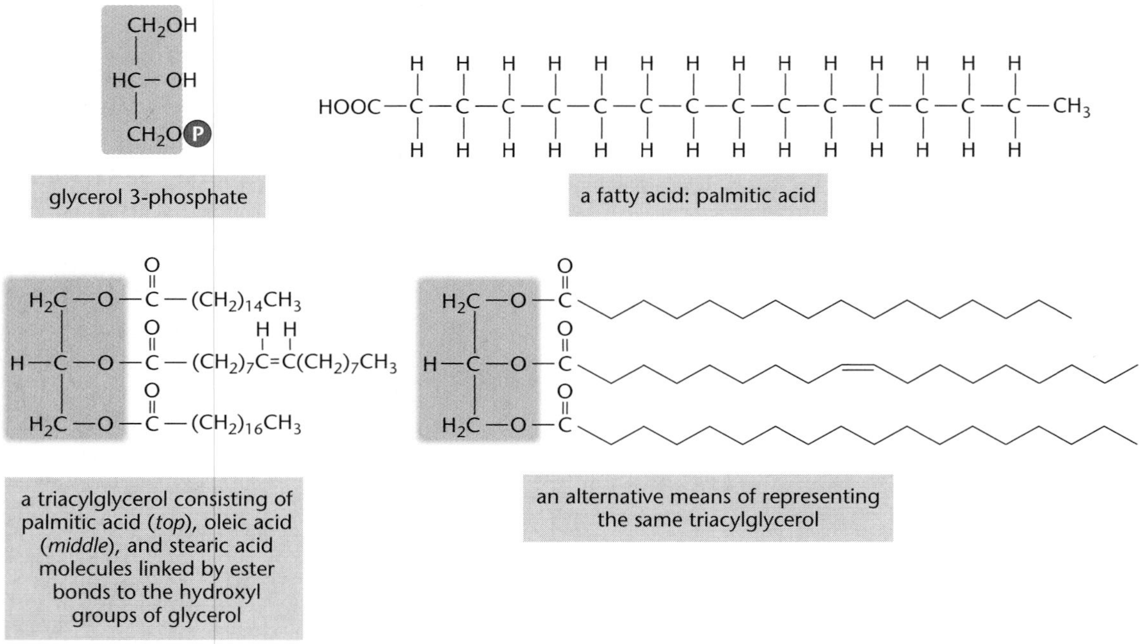

glycerol 3-phosphate

a fatty acid: palmitic acid

a triacylglycerol consisting of palmitic acid (*top*), oleic acid (*middle*), and stearic acid molecules linked by ester bonds to the hydroxyl groups of glycerol

an alternative means of representing the same triacylglycerol

Figure 4–81.
Structure of triacylglycerols. A glycerol backbone is linked to three fatty acid units—in this case, palmitic acid, oleic acid, and stearic acid.

acid is similarly transferred to position 2, forming a **phosphatidic acid** (Figure 4–82, top). At this point, the membrane and storage lipid pathways branch. Here we describe storage lipid synthesis; membrane lipid synthesis is described in Section 4.7.

The conversion of the phosphatidic acid to a triacylglycerol can proceed in two ways (Figure 4–82). In the first, the phosphate group is removed from position 3 of the glycerol backbone (forming a diacylglycerol), and a third fatty acid is added by a **diacyl glycerol acyltransferase**. In the second pathway, the diacylglycerol reacts with CDP-choline to form **phosphatidylcholine** (CDP is cytidine diphosphate). This lipid is a major component of membranes, but it can also undergo a reaction catalyzed by a **phospholipid:diacylglycerol acyltransferase** in which the fatty acid on position 2 of phosphatidylcholine is transferred to position 3 of a diacylglycerol to form a triacylglycerol.

The synthesis of triacylglycerols occurs in specialized regions of the endoplasmic reticulum. The newly synthesized material is believed to accumulate within the lipid of the endoplasmic reticulum membrane, which then buds off to form oil bodies in the cytosol. The oil body is a droplet of triacylglycerol surrounded by a monolayer of membrane lipid (Figure 4–83). In most oil bodies this membrane also contains **oleosins**, proteins that stabilize the oil body during desiccation (e.g., during seed maturation). Oleosins are unusual proteins, with central hydrophobic regions and terminal regions that are more hydrophilic; this means that the central region of an oleosin is embedded inside the oil body and the termini are on the outer surface.

The fatty acid composition of storage lipids varies among species

The nature of the three fatty acids attached to the glycerol backbone of a triacylglycerol varies from one plant species to another. There are two main reasons for this variation: first, the starting materials vary, as different species synthesize different types of fatty acid; second, the enzymes that attach fatty acids to the glycerol backbone have different specificities. We describe first how the different types of fatty acid are produced.

Fatty acid synthase complexes usually release, as their end products, fatty acids that are 16 or 18 carbons in length (Figure 4–84; see also Figure 4–86), with the carbon

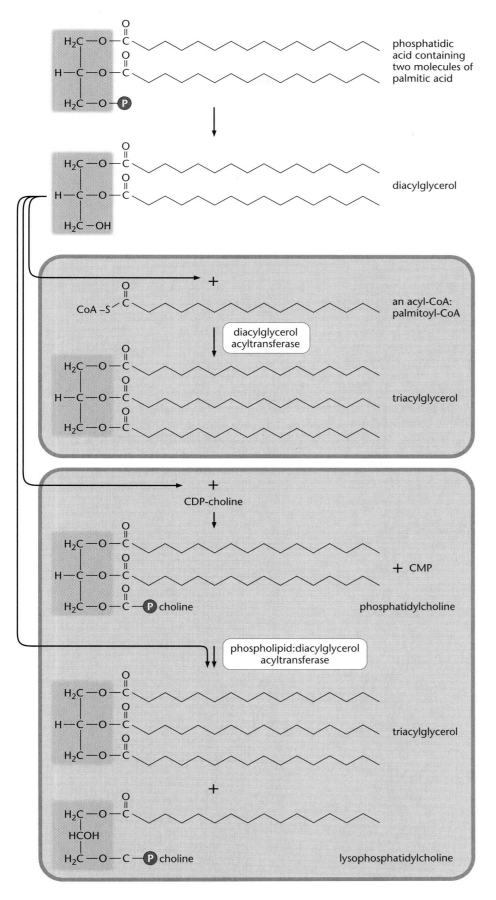

Figure 4–82.
Two routes of triacylglycerol synthesis from phosphatidic acid. Diacylglycerol is derived from phosphatidic acid (*top*). In one route (*upper gray box*), diacylglycerol is joined to a third fatty acid in the form of an acyl-CoA (here shown as palmitoyl-CoA) by the enzyme diacylglycerol acyltransferase. In the other route (*lower gray box*), diacylglycerol and CDP-choline form phosphatidylcholine, with release of CMP. A second diacylglycerol molecule then undergoes a transferase reaction with phosphatidylcholine, catalyzed by phospholipid:diacylglycerol acyltransferase, to give triacylglycerol and lysophosphatidylcholine.

phosphatidic acid containing two molecules of palmitic acid

diacylglycerol

an acyl-CoA: palmitoyl-CoA

diacylglycerol acyltransferase

triacylglycerol

CDP-choline

+ CMP

phosphatidylcholine

phospholipid:diacylglycerol acyltransferase

triacylglycerol

lysophosphatidylcholine

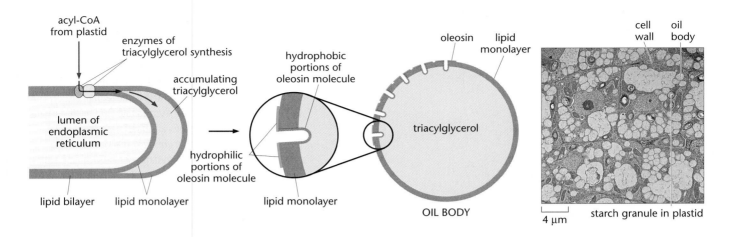

Figure 4–83.
Triacylglycerol accumulation in oil bodies. It has been proposed that triacylglycerol synthesis takes place within membranes of the endoplasmic reticulum. In this model, newly formed triacylglycerol accumulates between the lipid monolayers of the endoplasmic reticulum membrane, and the triacylglycerol-filled region buds off to form an oil body surrounded by a lipid monolayer. The oil body is stabilized by the insertion of oleosins, proteins with both hydrophobic and hydrophilic regions, which coat the surface of the oil body. The electron micrograph shows cells in an Arabidopsis cotyledon from a developing seed. The cell volume is starting to fill with oil bodies. (Courtesy of Sue Bunnewell and Vasiolos Andriotis.)

atoms linked by single bonds. Palmitic acid is 16 carbons long, with no double bonds, and is designated 16:0. Likewise, the 18-carbon stearic acid is designated 18:0. Fatty acids with no double bonds are referred to as **saturated fatty acids**. Palmitic and stearic acids can undergo two main types of modification to generate the range of fatty acids found in membrane, cuticular, and storage lipids: elongation and desaturation—the introduction of double bonds.

The introduction of double bonds into fatty acids is catalyzed by a family of **fatty acid desaturases**. A soluble desaturase inside the plastid (see Figure 4–86) converts stearic acid to oleic acid (18 carbons with one double bond, designated $18{:}1^{\Delta 9}$; Figure 4–84). Other desaturases are bound to the endoplasmic reticulum membrane. Arabidopsis has at least 15 of these enzymes, each with a different effect on the complement of fatty acids present in leaves and seeds. All desaturases catalyze the introduction of a double bond at a specific position in a defined fatty acid substrate, with each catalyzing a different desaturation reaction.

We consider two fatty acid desaturases—FAD2 and FAD3—as examples of the different specificities of these enzymes. Both are located in the endoplasmic reticulum and both catalyze the desaturation of the fatty acids of phosphatidylcholine molecules (Figure 4–85). FAD2 catalyzes the introduction of a double bond between carbon-11 and carbon-12 of the 18:1 fatty acid in which the existing double bond is between carbon-9 and carbon-10, thus converting an $18{:}1^{\Delta 9}$ fatty acid to an $18{:}2^{\Delta 9,12}$ fatty acid (the Δ indicating the position of the double bond). FAD3 then converts an $18{:}2^{\Delta 9,12}$ fatty acid to an $18{:}3^{\Delta 9,12,15}$ fatty acid. These 18:1, 18:2, and 18:3 fatty acids (oleic, linoleic,

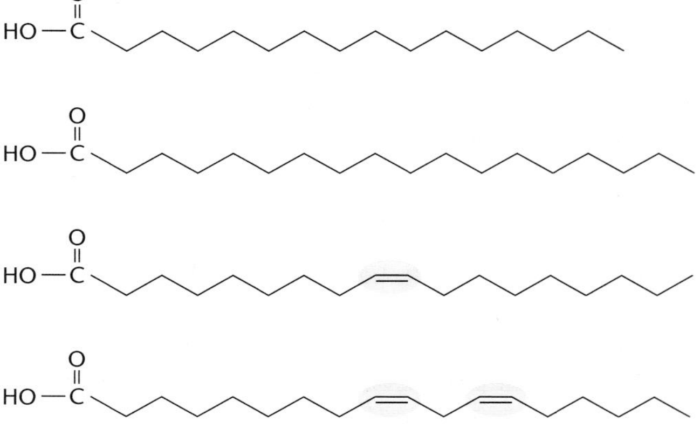

Figure 4–84.
Some fatty acids. Palmitic, stearic, and oleic acids are synthesized in plastids. The action of desaturase enzymes in the endoplasmic reticulum introduces further double bonds at specific positions in the fatty acid chain. Oleic and linoleic acids are very widely distributed in plants. The Δ designation indicates the position of double bonds.

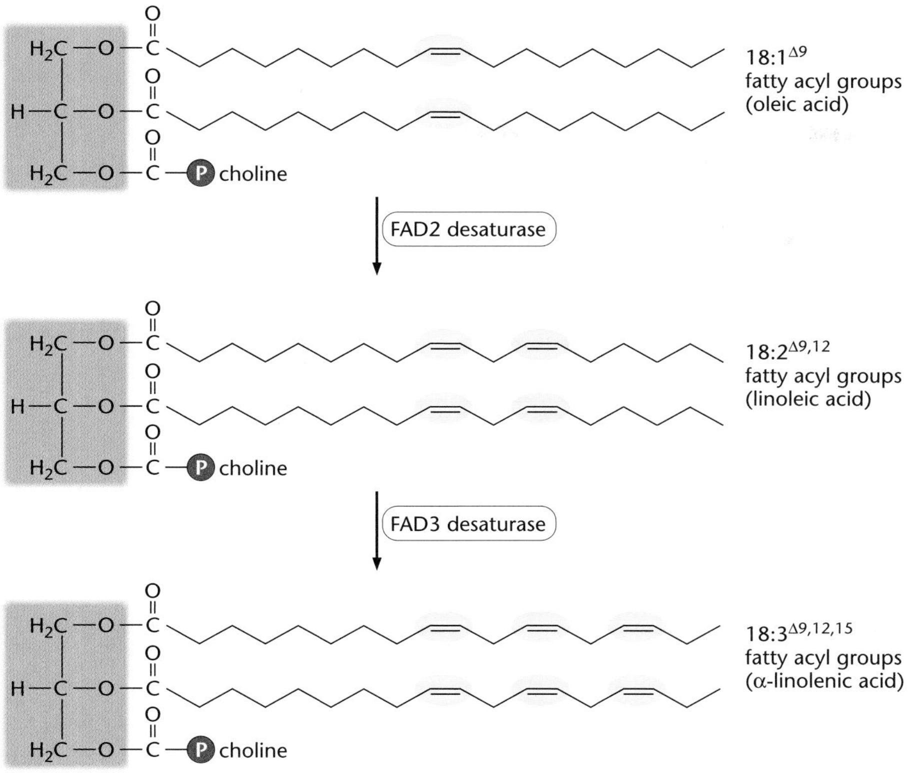

Figure 4–85.
Desaturation of fatty acyl groups in phosphatidylcholine. The FAD2 desaturase acts on the oleic acid component of phosphatidylcholine to introduce a second double bond, and FAD3 desaturase introduces a third double bond. The resulting linoleic or α-linolenic fatty acyl groups can then be introduced into triacylglycerols as shown in Figure 4–82.

and α-linolenic acids, respectively) are of almost universal occurrence in plants as major constituents of membrane lipids, and are also found in triacylglycerols. Linoleic and α-linolenic acids are so-called "essential fatty acids." They are not synthesized in the human body yet are essential precursors for some membrane lipids and cell-signaling molecules in our bodies. All of the linoleic and α-linolenic acid required by humans comes from the diet, largely from seeds and leafy vegetables. Other plant fatty acids described below are not essential, but some are thought to have therapeutic and other health benefits.

Some plants synthesize highly unusual, species-characteristic, desaturated fatty acids, which are formed by the activity of additional desaturases with specificities for fatty acid chain length and double-bond position that differ from those of the common desaturation reactions. For example, the oils of borage and evening primrose are rich in γ-linolenic acid ($18:3^{\Delta 6,9,12}$), a fatty acid believed to have important benefits for human health. These species contain a Δ6 desaturase with only a low level of similarity to the common fatty acid desaturases; this enzyme introduces a double bond into an $18:2^{\Delta 9,12}$ fatty acid of a phosphatidylcholine molecule.

The triacylglycerols of some species, such as coconut and oil palm, are rich in fatty acids that are less than 16 carbon atoms in length, formed by the early release of fatty acid (acyl) chains from the fatty acid synthase complex in the plastid. Fatty acid synthase catalyzes the sequential addition of two-carbon units to the growing acyl chain (Figure 4–86). Throughout elongation, the chain is attached to an **acyl carrier protein (ACP)**. Release of the free fatty acid from ACP by an **acyl-ACP thioesterase** terminates the process, and the fatty acid can then be exported from the plastid. (Further details of these reactions of fatty acid synthesis are discussed in Section 4.7.) The acyl-ACP thioesterases are encoded by the *FAT* gene family. In most species these enzymes release fatty acids of 16 or 18 carbon atoms. The plastids of seeds and fruits that contain triacylglycerols with shorter fatty acids have unusual enzymes of the FATB subclass of acyl-ACP thioesterases. Different FATB enzymes may release fatty acids of

Figure 4–86.
Release of fatty acids of various chain lengths from fatty acid synthase. The fatty acid synthase complex sequentially adds two-carbon units to acyl chains attached to acyl carrier protein (ACP). In most species, as shown here, 16:0, 18:0, and 18:1 fatty acids are released from the complex by acyl-ACP thioesterases, which remove the ACP. The free fatty acids are converted to acyl-CoAs as they leave the plastid. This is the form in which they are used for triacylglycerol synthesis (see Figure 4–82). Different types of acyl-ACP thioesterase have different chain-length preferences. Seeds of some species contain unusual acyl-ACP thioesterases (the FATB class) that act on acyl-ACPs with acyl chains shorter than 16 carbons; these give rise to triacylglycerols with short-chain fatty acids.

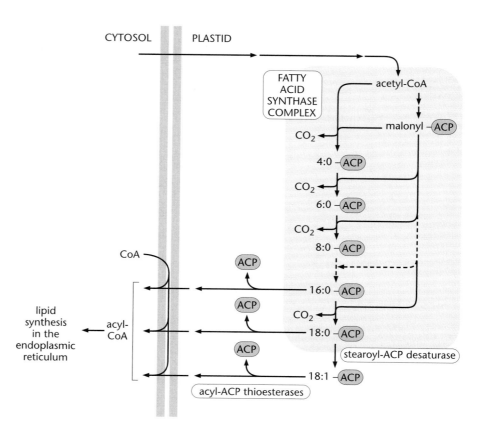

8, 10, 12, and 14 carbons. For example, coconut and oil palm contain an acyl-ACP thioesterase that specifically releases 12-carbon chains (lauric acid). Different species of the genus *Cuphea* have FATB enzymes with different specificities: the triacylglycerol of *Cuphea hookeriana* is very rich in 8:0 fatty acids, while that of *Cuphea wrightii* is 30% 10:0 and 50% 12:0 fatty acids. Short-chain fatty acids from plants have many industrial uses, including the manufacture of detergents, soaps, paints, and plastics.

Fatty acids with more than 18 carbons are formed by **elongase complexes** bound to the endoplasmic reticulum membrane (Figure 4–87). In a series of reactions involving at least four different proteins, one or more two-carbon units are added to an 18-carbon substrate. The triacylglycerols of crucifers such as Arabidopsis contain a high proportion of these longer fatty acids. Long-chain fatty acids are also found in the waxes that form the **cuticle** on the aerial surface of plants (see Section 3.6) and in **sphingolipids**, minor components of membranes that may have signaling functions. Different elongase complexes encoded by separate genes provide the fatty acids for these various purposes.

We have described several examples of how the fatty acid composition of a triacylglycerol is strongly influenced by the availability of different fatty acids. Also important in determining the fatty acid content of triacylglycerols are the substrate preferences of the acyltransferases that add the fatty acids to the glycerol backbone. The acyltransferase that adds the fatty acid chain to position 1 of glycerol has a low selectivity for the type of fatty acid, and many different sorts of fatty acid appear in this position in triacylglycerols. The transferase acting at position 2 of glycerol preferentially uses unsaturated fatty acids. The chain-length preferences of this enzyme differ among species. For example, in *Brassica napus* (oilseed rape) the position-2 transferase will not accept erucic acid (22:1), but the acyltransferases acting at positions 1 and 3 will. In the seed oil of *Limnanthes*, long-chain fatty acids appear at all three positions on the glycerol backbone.

Although the triacylglycerol-synthesizing enzymes exercise some control over the fatty acid composition of triacylglycerols, most will accept a relatively wide range of substrates.

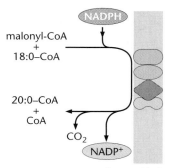

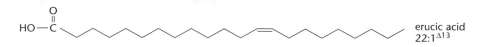

erucic acid
$22:1^{\Delta 13}$

fatty acid elongase
complex in the membrane
of the endoplasmic reticulum

Figure 4–87.
Fatty acid elongation by an elongase. Fatty acids of 20 or more carbons are produced by membrane-bound elongase complexes (*left*). These enzymes catalyze addition of two-carbon units to a fatty acyl-CoA (here, stearic acid, 18:0), using malonyl-CoA as substrate. This reaction is analogous to elongation of acyl-ACP by fatty acid synthase in the plastid, using malonyl-ACP as substrate (see Figure 4–86). The 22-carbon erucic acid (*right*) is found in the triacylglycerols of some crucifers, including some cultivars of oilseed rape.

This is dramatically illustrated by experiments in which genes encoding enzymes that synthesize unusual fatty acids are introduced into species that do not normally make these fatty acids, or fatty acid composition is altered by down-regulation of the expression of a particular synthetic enzyme. Many such experiments have been conducted in oilseed rape and soybean, in attempts to improve the quality of the oil of these commercial oil crops for both health benefits and industrial purposes. Genes for these experiments are obtained from species that accumulate triacylglycerols with unusual fatty acids (Figure 4–88).

Altering the profile of fatty acids synthesized in a plant has in most cases altered the fatty acid composition of triacylglycerols. For example, down-regulation of a desaturase acting on oleic acid increased the oleic acid content of triacylglycerols in soybean seed from less than 10% to more than 85%. Introduction of an acyl-ACP thioesterase specific for a 12-carbon fatty acid (lauric acid) in oilseed rape resulted in triacylglycerol that was up to 50% lauric acid, even though oilseed rape oil naturally contains no short-chain fatty acids.

Figure 4–88.
Some species that produce high proportions of unusual fatty acids.
Left: oil palm (*Elaeis guineensis*)—oil from the outer part of the fruit contains about 50% palmitic acid, whereas oil from the inner part (kernel oil) contains 50% lauric acid; middle: meadow foam (*Limnanthes douglasii*)—oil from seeds contains a high proportion of 20:1, 22:1, and 22:2 fatty acids; right: coriander (*Coriandrum sativum*)—oil from seeds contains a high proportion of petroselenic acid, $18:1^{\Delta 6}$. (middle image, courtesy of Tobias Kieser; right-hand image courtesy of A. Fiedler.)

Triacylglycerols are converted to sugars by β oxidation and gluconeogenesis

During the germination of lipid-storing seeds the triacylglycerols are converted to sucrose (Figure 4–89), which is exported from the storage cells to rapidly growing tissues for use as a source of energy and precursors for biosynthetic pathways. Triacylglycerol breakdown starts with hydrolysis by **lipases**, which yields fatty acids. The fatty acids are broken down into two-carbon units by the process of **β oxidation**, which takes place in **glyoxysomes**, specialized organelles related to peroxisomes.

Figure 4–89.

Conversion of triacylglycerol to sucrose.
Storage lipids are mobilized to acyl-CoAs by lipases. The process of β oxidation converts acyl-CoAs to acetyl-CoA, which is converted to oxaloacetate via the glyoxylate cycle in the glyoxysome and two reactions of the Krebs cycle in the mitochondrion. Oxaloacetate is converted to sucrose by gluconeogenesis in the cytosol, in a series of reactions that is, effectively (though not using all the same enzymes), a reverse of glycolysis.

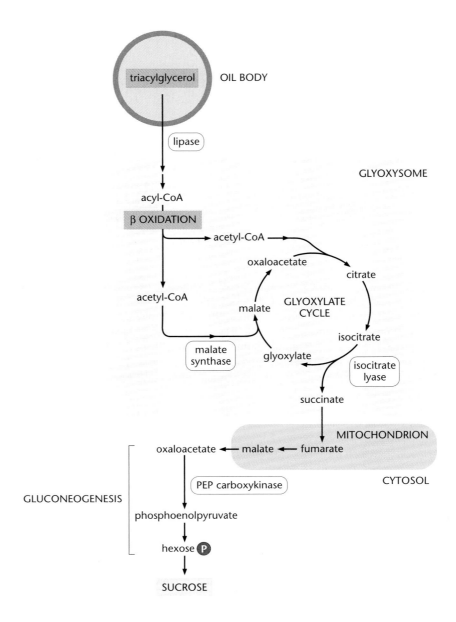

The β oxidation of fatty acids to the two-carbon acetyl-CoA is catalyzed by a four-enzyme complex. Two molecules of acetyl-CoA are then converted in a specialized cycle, called the **glyoxylate cycle**, to the four-carbon succinate molecule (Figure 4–89). Note an important difference between the glyoxylate cycle and the Krebs cycle (see Figure 4–57). Both convert acetyl-CoA to succinate, but whereas the Krebs cycle includes two decarboxylation reactions, with loss of carbon dioxide, there is no loss of carbon in the glyoxylate cycle. Succinate from the glyoxylate cycle moves to the mitochondrion, where Krebs cycle reactions convert it to oxaloacetate. The oxaloacetate is exported to the cytosol and converted to sucrose, via reactions that are effectively the reverse of those occurring in glycolysis. This process is called **gluconeogenesis**.

Many of the reactions of glycolysis are reversible, and for these reactions gluconeogenesis can use the same enzymes as glycolysis. However, conversion of oxaloacetate to phosphoenolpyruvate (Figure 4–89) cannot use PEP carboxylase, because this enzyme catalyzes an effectively irreversible reaction in the direction of oxaloacetate synthesis (see Figure 4–62). Gluconeogenesis instead employs the decarboxylating enzyme PEP carboxykinase, which plays a key role in the control of flux from lipid to sucrose.

The glyoxylate cycle is involved not only in mobilizing carbon reserves from storage lipids but also in recovering the carbon from membrane lipids in senescing leaves. The genes encoding two enzymes unique to gluconeogenesis—**isocitrate lyase** and **malate synthase**—are expressed both during the germination of lipid-storing seeds and during the programmed degradation of structural components in senescing leaves (see Section 5.4).

Sugars may act as signals that determine the extent of carbon storage

The processes of storage and subsequent remobilization of carbon are controlled at the transcriptional level (see Section 2.3) by sugars as well as by developmental signals. In essence, research results suggest that high levels of sucrose activate the expression of genes encoding the enzymes that synthesize storage compounds, whereas glucose promotes the expression of genes involved in cell growth and division and the mobilization of stored carbon. The evidence falls into two categories: (1) direct demonstration that the expression of particular genes is influenced by sugars, and (2) correlation between sugar levels and the patterns of synthesis and mobilization of storage compounds in storage organs.

Genes known to be activated by sucrose include those encoding many of the enzymes of starch synthesis and fructan synthesis. In potato, for example, expression of the genes encoding starch synthases, starch-branching enzymes, and ADP-glucose pyrophosphorylase is strongly up-regulated in leaves when exogenous sucrose is supplied. In leaves of rye grass, expression of genes encoding the enzymes of fructan synthesis is up-regulated in response to increases in sucrose. Conversely, expression of genes encoding some enzymes of starch degradation is down-regulated by sucrose and up-regulated by glucose.

Detailed information on the effect of sucrose on gene expression has been gained from transgenic plants that express different parts of the promoter regions of sugar-responsive genes that are fused to reporter genes. **Reporter genes** are research tools that enable discovery of the location and conditions under which a particular promoter is active. They encode proteins that can be visualized (for example, green fluorescent protein, see Chapter 8), or that produce easily visualized products within the plant. A commonly used reporter gene encodes the bacterial enzyme **β-glucuronidase (GUS)**. This enzyme will convert a synthetic colorless substrate to a blue product. To study the promoters of sugar-responsive genes, the *GUS* gene is fused to a specific part of the promoter region of a gene of interest. In the transgenic plant, the involvement of that part of the promoter region in the sugar response of the gene can then be tested by supplying the plant with different amounts of sugars, then testing for GUS activity by adding the colorless substrate of GUS and measuring the production of blue color (Figure 4–90). Experiments of this type have revealed short DNA sequences (known as **DNA motifs**) that are common to the promoter regions of many sugar-responsive genes and are necessary for switching on gene expression in response to sugars. It is likely

(A)

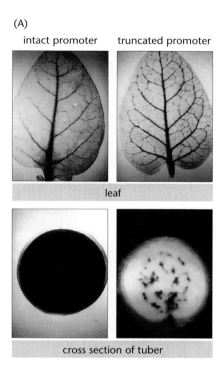

intact promoter   truncated promoter

leaf

cross section of tuber

(B)

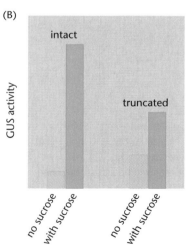

intact

truncated

GUS activity

no sucrose   with sucrose   no sucrose   with sucrose

Figure 4–90.
Use of the *GUS* reporter gene to study promoter function.
This experiment was designed to study the promoter region of a gene encoding the sucrose synthase of potato. The *GUS* reporter gene was fused to the intact promoter region of the sucrose synthase gene, or to a truncated promoter. The fused gene was then introduced into potato plants. (A) *GUS* gene activity, and hence the activity of the promoter region, was detected by supplying the colorless substrate for the GUS enzyme to parts of the plant and observing how much blue product was produced. This was done for leaves and for tubers. (B) To test for sugar-responsiveness, detached leaves were incubated with or without sucrose, then extracts were supplied with the substrate and the blue product was quantified as a measure of GUS activity. As shown in (A), the intact promoter of the sucrose synthase gene is very active in the tuber but not in the leaves (*upper and lower left*). The truncated version of the promoter—about half of the promoter region—is active in the leaves but much less active in the tubers. As shown in (B), the intact promoter is highly sucrose-responsive (*left*). The truncated promoter responds only weakly to sucrose (*right*). Taken together with results for other truncated versions of the promoter, this experiment pinpoints the parts of the promoter responsible for the spatial expression pattern of the gene and its responsiveness to sucrose.

that these motifs are targets for specific transcription factors that are expressed or activated in response to sugars. (See Section 2.3 and Box 2–1 on the role of transcription factors in gene expression.)

Relatively little is known about how changes in sugar levels are sensed in the plant, or how this leads to the expression or activation of specific transcription factors. However, recent evidence strongly implicates a particular isoform of hexokinase as a sensor of glucose levels. In yeast, sugar sensors are linked to the control of expression of sugar-responsive genes by mechanisms that involve specific protein kinases of the SNF (**s**ucrose **n**on-**f**ermenting) family. Plants have SnRK1 protein kinases, which are very closely related to the yeast SNF kinases involved in sugar responses, and it seems likely that plant sugar-response pathways involve SnRK1 kinases.

Some evidence for a role for sugar signaling in development of storage organs comes from studies of correlations between sugars and development in potato tubers. Potato tubers develop by lateral expansion of the tip of a stolon (underground shoot). Before tuber development, the sucrose concentration in the stolon tip is low and the hexose concentration is relatively high. The stolon tip has high levels of invertase and low levels of sucrose synthase. Lateral expansion of the tip and the onset of starch accumulation to form a tuber coincide with a radical change in sugar metabolism: invertase activity and hexose levels fall, and sucrose synthase activity and sucrose levels rise (Figure 4–91). These results, and similar data from other developing storage organs, indicate that the sucrose:hexose ratio influences gene expression to determine whether incoming carbon is used immediately for growth or is stored for later use.

Other important metabolites also influence the balance between growth and storage by affecting gene expression. For example, the expression of genes encoding the enzymes of storage-product synthesis is affected by levels of nitrogen-containing compounds as well as by sugars. The mechanisms by which sugars activate and repress gene expression probably interact with the mechanisms by which major phytohormones such as **auxins** bring about changes in gene expression. Thus the switch from active growth to carbon storage in plants is controlled in a complex way by many interacting factors.

Figure 4–91.
Sugar signaling. Proposed roles for sugar signaling in the transition from growth to storage functions during the transition from actively growing stolon to developing, starch-storing tuber in the potato plant. (A) During active growth and cell division, sucrose is converted to hexoses by invertase. No starch is accumulated. The developmental shift toward cell expansion and accumulation of starch is accompanied by loss of invertase, increased sucrose:hexose ratio, and increased sucrose synthase activity. (B) Stolons during the early stage of tuberization, stained to reveal the amount and location of invertase activity (*left*) and sucrose synthase activity (*right*). Note that invertase activity is high in the stolon before tuberization but low in the developing tuber. Conversely, sucrose synthase activity is low in the stolon but high in the developing tuber.

(A)

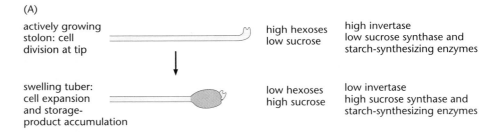

(B)

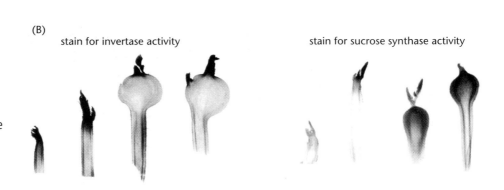

4.7 PLASTID METABOLISM

The metabolic processes occurring in plastids vary enormously with cell type and developmental stage. The most obvious example is photosynthesis. This is by far the dominant process in leaf plastids (chloroplasts), yet is absent altogether from the plastids of, for example, roots. During its lifetime, a photosynthetic cell undergoes considerable changes in the dominant form of metabolism in its plastids, from a massive synthesis of membranes, pigments, and Calvin cycle enzymes in the young leaf, through a phase when light-harvesting and CO_2 assimilation predominate, to a final stage in which chlorophyll is degraded as the leaf senesces (Figure 4–92). Profound changes may also occur in plastid metabolism during the life of a nonphotosynthetic cell. For example, the plastids of oilseed rape and Arabidopsis embryos go through an early phase in which the major metabolic process is starch synthesis and degradation, then switch to high rates of fatty acid synthesis to provide substrates for lipid synthesis in the endoplasmic reticulum.

In this section we consider first how plastid metabolism is coordinated with overall cellular metabolism via metabolite transporters in the plastid envelope, and the ways in which plastids obtain energy and biosynthetic precursors for their metabolism. We then describe some metabolic processes that occur exclusively in plastids (synthesis of fatty acids and chlorophyll) and some that occur both inside and outside plastids (synthesis of membrane lipids and terpenoids).

Plastids exchange specific metabolites with the cytosol via metabolite transporters

The partition of metabolic processes between the cytosol and the plastid requires mechanisms that connect and coordinate metabolism in the two compartments. As noted earlier in our discussion of metabolic processes, this role is filled by transporters in the inner envelope of the plastid that allow selective movement of metabolites (Figure 4–93).

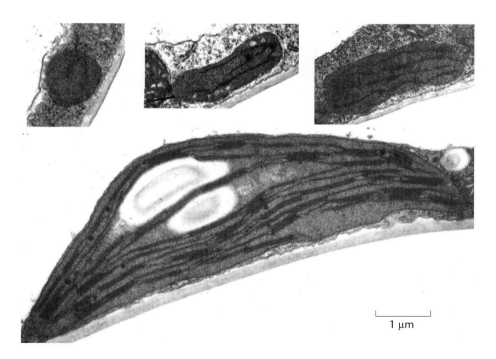

Figure 4–92.
Stages in the development of chloroplasts in an Arabidopsis leaf. These electron micrographs show development from proplastid to mature chloroplast. Meristematic cells in a very early stage of leaf development (*upper left*) contain undifferentiated proplastids. After cell division ends and leaf expansion starts, plastids rapidly increase in size, elongate, and acquire thylakoids.

1 μm

Figure 4–93.
Metabolite transporters connecting cytosolic and plastidial metabolism.
(A) In the chloroplast, most of the carbon assimilated via the Calvin cycle is exported as triose phosphate during the day or as maltose and glucose (hexose) at night (see Section 4.2). (Note that hexose and maltose are shown here on a single transporter for simplicity; in fact, maltose and hexoses have distinct transporters.) Phosphoenolpyruvate is imported for biosynthetic reactions. Exchange of 2-oxoglutarate, glutamate, and malate between the chloroplast and the cytosol occurs during photorespiration (see Section 4.3) and assimilation of nitrogen into amino acids (see Section 4.8). (B) In nonphotosynthetic plastids, carbon from sucrose in the cytosol is imported as hexose phosphate, PEP, and pyruvate. Carbon produced from starch degradation inside the plastid (see Section 4.6) is exported as maltose and glucose. Reducing power is imported by the action of a malate–oxaloacetate shuttle, similar to the shuttle operating in mitochondria (see Figure 4–61), and ATP is imported in exchange for ADP generated inside the plastid as ATP is used in plastidial processes.

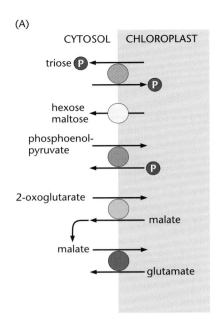

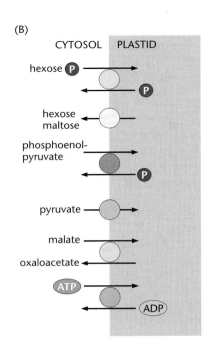

Most metabolite transporters that researchers have identified thus far share a common structure: a pore formed by membrane-spanning helices. Most transporters show preference for one particular substrate but will also transport a range of related compounds. For example, the triose phosphate transporter will also transport 3-phosphoglycerate, and the hexose phosphate transporter will also transport triose phosphate. Thus the movement of metabolites in and out of the plastid is regulated by competition for transporters among several metabolites present in the plastid and the cytosol. Movement of metabolites across the plastid envelope thus responds to changes in plastidial and cytosolic metabolism that affect the concentrations of the relevant metabolites in the two compartments.

Nonphotosynthetic plastids ultimately obtain all of their biosynthetic precursors and energy from sucrose in the cytosol. This is achieved largely by the import of a limited number of metabolites derived from sucrose in the cytosol (Figure 4–93B). Which metabolites are imported varies from one part of the plant to another, and depends largely on which metabolic processes are occurring in the plastid. All nonphotosynthetic plastids seem to be able to take up and metabolize glucose 6-phosphate, and in many cases this is the main form in which carbon enters the plastid. Energy, reductant, and biosynthetic precursors required for plastidial processes are then provided by metabolism of glucose 6-phosphate via glycolysis and the oxidative pentose phosphate pathway inside the plastid. These two central pathways of primary metabolism are duplicated in the cytosol and the plastid. (Both pathways are discussed in Section 4.5.)

The plastidial and cytosolic versions of each of the enzymes involved in glycolysis and the oxidative pentose phosphate pathway are encoded by different genes. Generally speaking, the plastidial version more closely resembles the bacterial enzyme and the cytosolic version resembles that found in other eukaryotic organisms, reflecting the different evolutionary origins of the plastid and the rest of the cell (see Chapter 1). Some of the plastidial enzymes of glycolysis and the oxidative pentose phosphate pathway are very similar to their cytosolic counterparts, whereas others differ considerably.

The level of duplication of metabolic pathways in plastids and the cytosol varies with the stage of organ development. Specialized metabolite transporters can compensate for this variation. For example, the capacity of plastids in the embryo of oilseed rape to carry out glycolysis diminishes as the embryo switches from starch to lipid accumulation (Figure 4–94). During the starch-accumulation phase, the plastids have a high capacity for importing glucose 6-phosphate. Most of this hexose phosphate is converted to starch, and some is metabolized via plastidial glycolysis to provide precursors for fatty acid synthesis. During the later lipid-accumulation phase, the plastid imports most of its carbon as phosphoenolpyruvate (PEP), supplying precursors directly to fatty acid synthesis.

Nonphotosynthetic plastids can synthesize ATP by glycolysis and can also import ATP from the cytosol via a transporter that exchanges it for ADP (Figure 4–93B). Research has shown that import of ATP from the cytosol is essential for normal rates of starch synthesis in the plastids of potato tubers: ATP is required for the ADP-glucose pyrophosphorylase reaction, which synthesizes ADP-glucose as the substrate for starch synthases (see Section 4.6). Transgenic plants with a reduced ATP–ADP transporter activity have a lower rate of starch synthesis than normal plants. Conversely, overexpression of the ATP–ADP transporter in transgenic plants increases the rate of starch synthesis.

Reducing power can be generated in plastids by glycolysis and the oxidative pentose phosphate pathway, and can also be exchanged across the plastid envelope via a metabolite shuttle. This shuttle exchanges malate for oxaloacetate, as also occurs across the inner mitochondrial membrane (Figure 4–93B; see also Section 4.5).

The main movement of metabolites between chloroplasts and the cytosol is the export of triose phosphate in exchange for inorganic phosphate via the triose phosphate transporter (Figure 4–93A; this is described in more detail in Section 4.2). Three other major classes of transporters linking the metabolic processes of chloroplast and cytosol are the **hexose transporters**, the **maltose transporter**, and **dicarboxylate transporters**. The hexose and maltose transporters allow the export of sugars formed by the

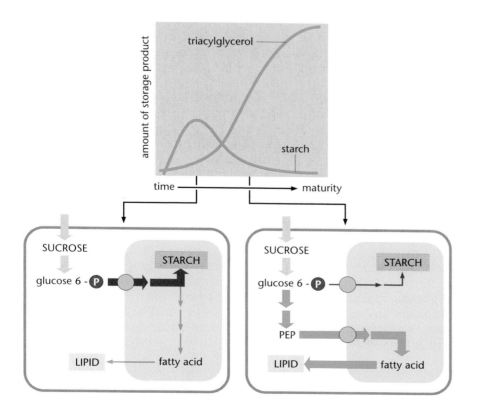

Figure 4–94.
Changes in plastid metabolism during development of oilseed rape (canola) embryos. Metabolism in the plastid changes radically during development of the embryo. In the early stages, starch is the main storage compound. It is synthesized in the plastid from hexose phosphate imported from the cytosol (*lower left*). Some fatty acids are also synthesized in the plastid: hexose phosphate is converted via glycolysis to substrates for fatty acid synthesis. At later stages of development (*lower right*), the rate of starch synthesis declines and starch stores are degraded. The main fate for sucrose entering the embryo is now the synthesis of storage lipids (triacylglycerols). Carbon for fatty acid synthesis is thought to enter the plastid via a PEP transporter. The importance of hexose phosphate import and plastidial glycolysis declines.

breakdown of starch at night, providing the substrate for sucrose synthesis in the cytosol (see Section 4.6). The main function of the dicarboxylate transporters is in the recapture of ammonia released during photorespiration (see Section 4.3). One dicarboxylate transporter exchanges malate in the plastid for 2-oxoglutarate in the cytosol. Inside the plastid, the 2-oxoglutarate is the substrate for the glutamine synthetase–GOGAT system that converts ammonium to glutamate (see Section 4.8). A second dicarboxylate transporter then exports the glutamate in exchange for cytosolic malate. The net outcome of the actions of the two transporters is the exchange of 2-oxoglutarate for glutamate (Figure 4–93A).

Although chloroplasts can obtain many of the precursors for biosynthesis from the Calvin cycle, they are not completely self-sufficient in this respect. Like nonphotosynthetic plastids, chloroplasts may have a very limited capacity for glycolysis and may lack some glycolytic enzymes altogether. Some glycolytic intermediates required for biosynthetic pathways in the chloroplast are imported from the cytosol rather than generated within the chloroplast. In Arabidopsis, phosphoenolpyruvate is imported into both nonphotosynthetic plastids and chloroplasts as a precursor for the aromatic amino acids tyrosine, tryptophan, and phenylalanine (see Section 4.8); this import occurs via a transporter (CUE1) that exchanges PEP for inorganic phosphate (Figure 4–93). Plants with mutant *cue1* genes are slow-growing and have **chlorotic** leaves, and they have reduced levels of many aromatic compounds derived from the aromatic amino acids, including simple phenolic compounds, flavonoids, and **anthocyanins**. The severity of the mutant phenotype can be greatly reduced by supplying these plants with phenylalanine, tyrosine, and tryptophan.

Fatty acids are synthesized by an enzyme complex in plastids

We described in Section 4.6 how some plant species store carbon predominantly in the form of lipids, synthesized from fatty acids, and how the early stages of synthesis of storage lipids and membrane lipids occur exclusively in plastids and share a common pathway. As we noted, fatty acids are assembled from two-carbon (acetyl) units by a fatty acid synthase complex (Figure 4–95; see also Figure 4–86). In most plastids, the source of acetyl units is acetyl-CoA, which is synthesized by a pyruvate dehydrogenase complex similar to that in mitochondria (see Section 4.5). The other substrate for fatty acid synthase is malonyl-CoA, which is synthesized from acetyl-CoA by the enzyme **acetyl-CoA carboxylase**.

Higher plants have plastidial and cytosolic forms of acetyl-CoA carboxylase. The plastidial enzyme is a complex of four different proteins (Figure 4–96), which together catalyze the carboxylation of acetyl-CoA to produce malonyl-CoA. It is similar in structure to bacterial forms of the enzyme, reflecting again the evolutionary origin of plastids. The cytosolic form of acetyl-CoA carboxylase consists of a single protein with four different regions (domains) that catalyze the separate reactions carried out by the four proteins in the plastidial complex. This type of monomeric enzyme is also found in animals and fungi, reflecting the eukaryotic origins of the nonplastidial parts of the plant cell. The cytosolic enzyme in plants provides malonyl-CoA for the synthesis of long-chain fatty acids, flavonoids, and **stilbenes**.

Grasses are an important exception to the picture presented above. In the grass family, both the plastidial and the cytosolic forms of acetyl-CoA carboxylase are monomeric. The monomeric plastidial enzyme is inhibited by synthetic compounds that have no effect on the multimeric plastidial enzyme of other plants. This has allowed the development of grass-specific herbicides, which kill grasses by preventing fatty acid synthesis but have no effect on other plants.

The fatty acid synthase is a complex consisting of several proteins, each catalyzing a separate reaction in the stepwise addition of acetyl groups to an acyl chain (Figure 4–97).

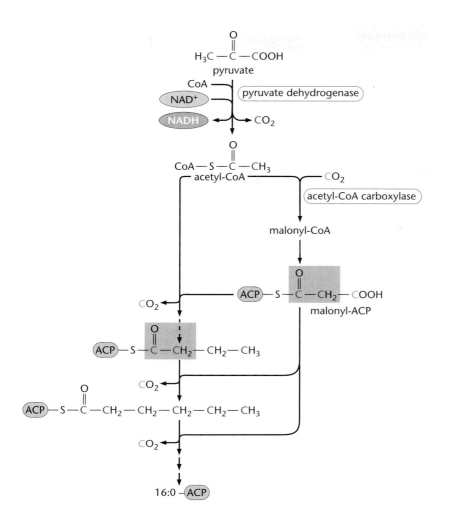

Figure 4–95.
Overview of fatty acid synthesis.
Acetyl-CoA, formed from pyruvate by the pyruvate dehydrogenase complex, is the substrate for the fatty acid synthase complex. Initially, acetyl-CoA is carboxylated and attached to acyl carrier protein (ACP) to form malonyl-ACP. Malonyl-ACP is condensed with a further molecule of acetyl-CoA through a series of reactions that produces a four-carbon chain linked to ACP: two of the carbons are derived from malonyl-ACP and two from acetyl-CoA (*pink boxes*). The ACP-linked four-carbon chain is the substrate for addition of further two-carbon units, derived from malonyl-ACP—eventually reaching a chain length of 16 or 18 carbons. These chain-lengthening reactions are shown in more detail in Figure 4–97.

The plastidial fatty acid synthases are similar to those of bacteria, whereas the animal and fungal synthases consist of one or two very large, multifunctional proteins.

Fatty acid synthesis starts when a malonyl unit is transferred from malonyl-CoA to the acyl carrier protein (ACP), to which it becomes covalently linked by a thioester bond.

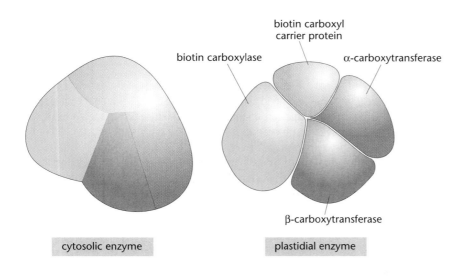

cytosolic enzyme

plastidial enzyme

Figure 4–96.
Structure of cytosolic and plastidial acetyl-CoA carboxylases. The carboxylation of acetyl-CoA begins with carboxylation (with bicarbonate as reactant) of a biotin cofactor on the enzyme, in an ATP-requiring reaction. The carboxyl group is then transferred to acetyl-CoA to form malonyl-CoA. The plastidial enzyme in most plants consists of four separate proteins (*right*): biotin carboxyl carrier protein, to which biotin is attached; biotin carboxylase; and two carboxytransferase subunits, which transfer the carboxyl group from biotin to acetyl-CoA. The cytosolic enzyme (*left*), and the plastidial enzyme in grasses, is a single protein with different domains carrying out these functions. The cytosolic enzyme is not involved in fatty acid synthesis; instead it provides malonyl-CoA for related two-carbon molecular elongation reactions outside the plastid.

Figure 4–97.
One round of elongation of a fatty acid chain by fatty acid synthase. The set of reactions shown here results in elongation of a four-carbon to a six-carbon fatty acid. Condensation of a molecule of three-carbon malonyl-ACP (*blue carbons*) with the four-carbon acyl-ACP (*red carbons*) with loss of one molecule of CO_2, catalyzed by ketoacyl-ACP synthase, gives a six-carbon chain. Three reactions, catalyzed by three distinct components of the fatty acid synthase complex, result in reduction, dehydration, and further reduction of the third carbon in this chain (*purple boxes*). The resulting acyl-ACP can undergo a further round of these four reactions to produce an eight-carbon acyl-ACP, and so on, up to 16 or 18 carbons.

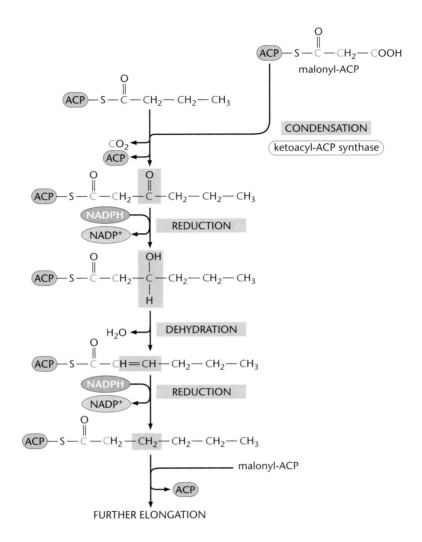

Four reactions are involved in the addition of each successive acetyl unit to the acyl chain (Figure 4–97). The first reaction, a condensation reaction, is carried out by one of three different isoforms of **ketoacyl-ACP synthase**, depending on the length of the fatty acid chain already attached to the ACP. In the condensation reaction, two of the three carbons of the malonyl unit of malonyl-ACP are transferred to an acyl chain, with loss of the third carbon as carbon dioxide. One isoform of ketoacyl-ACP synthase catalyzes the condensation of malonyl-ACP with the acetyl unit of acetyl-CoA (Figure 4–95), forming a four-carbon acyl chain attached to ACP. A second isoform of the enzyme is involved in the condensation reactions that elongate the acyl chain from 4 to 14 carbons. The third isoform carries out the final condensation reactions to produce a chain of 16 or 18 carbons, depending on the plant species. Following the condensation reaction, the fatty acid synthase complex carries out successive reduction, dehydration, and further reduction of the newly introduced two-carbon unit, creating the substrate for the next condensation reaction.

Release of the completed 16- or 18-carbon acyl chain from the fatty acid synthase complex occurs by breakage of the thioester bond, catalyzed by acyl-ACP thioesterases (see Figure 4–86). The fatty acid thus released may be used either for the synthesis of membrane lipids inside the plastid or, after export, for synthesis of membrane lipids, storage lipids, and lipid derivatives such as waxes elsewhere in the cell, as described below (see also Section 4.6)

Fatty acids are highly reduced molecules, and their synthesis from acetyl units requires very large amounts of reducing power. Two NAD(P)H molecules are required for each two-carbon unit added to the acyl chain. In chloroplasts, this requirement for NADPH can be met by the light-driven reactions of photosynthesis (see Section 4.2). In non-photosynthetic plastids, the requirement may be met by the import of reducing power via the malate–oxaloacetate shuttle and by the import of glucose 6-phosphate and its subsequent metabolism in the plastidial oxidative pentose phosphate pathway (see Section 4.5). The enzyme catalyzing the first step of the oxidative pentose phosphate pathway (**glucose 6-phosphate dehydrogenase**) is inhibited by NADPH. This means that a high demand for NADPH for fatty acid synthesis, and hence a low NADPH:NADP$^+$ ratio in the stroma, will activate glucose 6-phosphate dehydrogenase and thus increase the rate of NADPH regeneration via the oxidative pentose phosphate pathway.

Membrane lipid synthesis in plastids proceeds via a "prokaryotic" pathway different from the "eukaryotic" pathway elsewhere in the cell

The major lipid constituents of plant membranes are **glycerolipids** (Figure 4–98). Like the storage lipids described in Section 4.6, all glycerolipids have a glycerol backbone, but they differ in the chemical groups attached to the backbone. They have a hydrophilic "head" (a polar group) and a hydrophobic "tail." Molecules with both hydrophilic and hydrophobic regions are referred to as **amphipathic**. The hydrophobic portion consists of two fatty acid chains esterified to positions 1 and 2 of the glycerol backbone. The polar head group is attached to position 3. The amphipathic nature of these lipids enables them to form **lipid bilayers**, in which the hydrophilic heads face outward and the hydrophobic tails are aligned together in an internal double layer. In all cell membranes apart from those of plastids, the main glycerolipids are **phospholipids**. They contain any one of a variety of chemical groups linked via a

Figure 4–98.
Glycerolipids. These are the main lipid constituents of membranes. Glycerolipids consist of two fatty acid chains (R$_1$ and R$_2$) attached to positions 1 and 2 of a glycerol backbone (*pink box*), with a polar group at position 3 of the backbone (*blue box*). (A) In nonplastidial membranes, the most abundant glycerolipids are phospholipids. The polar group on position 3 of glycerol is attached via a phosphate group. Three of the most common types of phospholipids are shown. In plastidial membranes, the most common glycerolipids are two galactolipids: monogalactosyl-diacylglycerol and digalactosyl-diacylglycerol, with one or two galactose groups, respectively, attached to position 3 of glycerol. (B) In membranes, the hydrophobic fatty acid chains ("tails") are directed inward as a bilayer, and the hydrophilic groups ("heads") face outward.

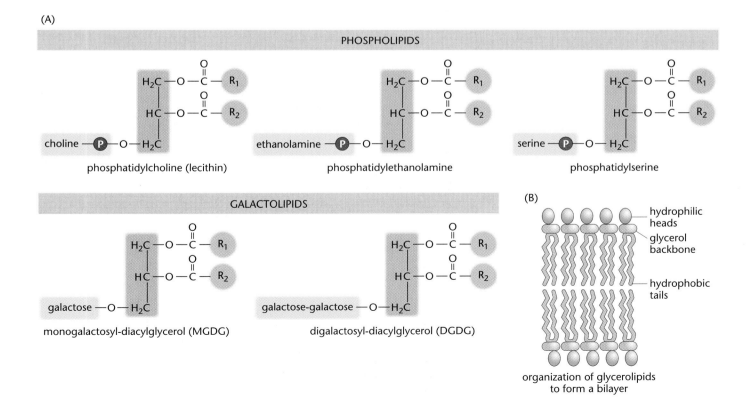

(A)

PHOSPHOLIPIDS

phosphatidylcholine (lecithin)

phosphatidylethanolamine

phosphatidylserine

GALACTOLIPIDS

monogalactosyl-diacylglycerol (MGDG)

digalactosyl-diacylglycerol (DGDG)

(B)

hydrophilic heads
glycerol backbone
hydrophobic tails

organization of glycerolipids to form a bilayer

phosphate group to position 3 of the glycerol backbone. **Galactolipids** make up most of the thylakoid and other plastid membranes. In these compounds, galactose is linked to position 3 of the glycerol backbone. Thylakoids also contain **sulfolipids**, in which the hydrophilic head is a glucose residue carrying a sulfite group (see Figure 4–131).

The fatty acids attached to the glycerol backbone differ between plastid and nonplastid membrane lipids (Figure 4–99). This is because phosphatidic acid, the precursor of glycerolipids, is synthesized by different pathways in the plastid and in the rest of the cell. Phosphatidic acid is a glycerol backbone with fatty acids attached at positions 1 and 2 and a phosphate group at position 3 (see Figure 4–82). Different fatty acids are attached in different compartments of the cell, depending on the differing substrate specificities of the acyltransferase enzymes responsible for their attachment. Acyltransferases in the endoplasmic reticulum attach either 16- or 18-carbon fatty acids to position 1, but mainly 18-carbon fatty acids to position 2. Acyltransferases in the plastid use acyl-ACPs as substrates. They attach 18-carbon chains to position 1 of the glycerol backbone, but

Figure 4–99.
Origin and fatty acid composition of the phosphatidic acid used in plastidial and nonplastidial glycerolipid synthesis. The 16:0 and 18:1 fatty acid chains released from the fatty acid synthase complex in the plastid are used for phosphatidic acid synthesis both inside and outside the plastid. In the plastid (the "prokaryotic pathway"), the fatty acid chains are in the form of acyl-ACPs. Acyltransferases first transfer an 18:1 chain to position 1 of a glycerol phosphate backbone, then transfer a 16:0 chain to position 2. Outside the plastid (the "eukaryotic pathway"), fatty acid chains are first released from ACP by thioesterases and converted to acyl-CoAs as they leave the plastid. In the endoplasmic reticulum, acyltransferases transfer an 18:1 or a 16:0 chain to position 1 of a glycerol phosphate backbone, then transfer an 18:1 chain to position 2.

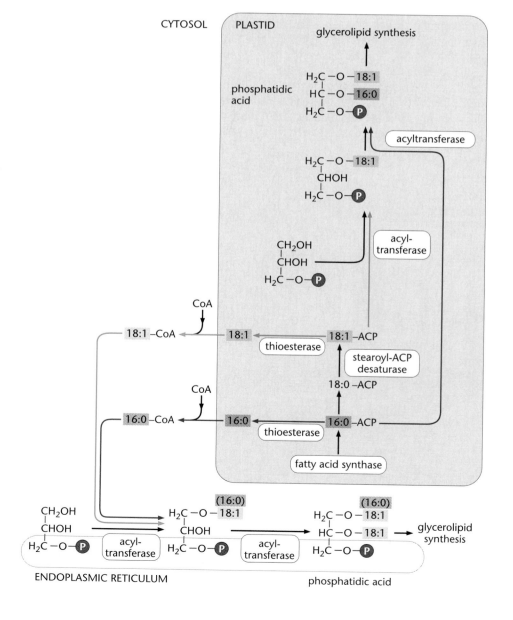

mainly 16-carbon saturated chains to position 2. The pathway of glycerolipid synthesis in the endoplasmic reticulum is sometimes known as the "eukaryotic pathway," and the pathway in the plastid as the "prokaryotic pathway."

Although the prokaryotic and eukaryotic pathways are confined to the plastid and endoplasmic reticulum, respectively, the use of their products is not confined to the compartment in which they are synthesized. Products of the eukaryotic pathway may be imported into the plastid for membrane lipid synthesis, and products of the prokaryotic pathway are sometimes exported for use elsewhere in the cell. In higher plants, the synthesis of membrane lipids in the plastid always employs at least some lipid made in the eukaryotic pathway. In pea and barley leaves, for example, the galactolipids of the chloroplast membranes are derived from the eukaryotic pathway. The only plastid glycerolipid derived from the prokaryotic pathway is **phosphatidylglycerol**, a minor component of thylakoids. In contrast, in Arabidopsis and spinach leaves, a substantial proportion of the membrane lipid of the plastid is derived from the prokaryotic pathway.

The transfer of lipid between the plastid and the rest of the cell is closely controlled to coordinate the lipid requirements of different subcellular compartments. Evidence for this comes from analysis of Arabidopsis mutants with altered membrane lipid compositions. For example, the wild-type *ACT1* gene encodes the first enzyme of the prokaryotic pathway of phosphatidic acid synthesis (an acyltransferase called acyl-ACP–glycerol 3-phosphate acyltransferase; Figure 4–100). Mutations in this gene result in drastically reduced flux through the prokaryotic pathway, which supplies most of the phosphatidic acid for galactolipid synthesis in chloroplasts in wild-type plants. In *act1* mutants, normal rates of glycerolipid synthesis in chloroplasts are maintained because flux through the eukaryotic pathway is increased and some of its product is diverted into the chloroplast to support galactolipid synthesis.

Synthesis of phospholipids from the precursor phosphatidic acid takes place in several cellular membranes, including the mitochondrial membranes and chloroplast envelope. A cytidine diphosphate (CDP) group is added either directly to the phosphatidic acid or to the compound that will be added to the phosphatidic acid to become the polar head group. When CDP is added directly to phosphatidic acid, the resulting CDP-diacylglycerol reacts with a polar molecule to form a phospholipid, with the release of CMP. When CDP is added to the polar compound itself, the product then reacts with diacylglycerol—formed by the action of a phosphatase on phosphatidic acid—to form a phospholipid, with the release of cytidine monophosphate (CMP) (Figure 4–101). Synthesis of galactolipids in the chloroplast also uses diacylglycerol formed from phosphatidic acid. The galactosyl group is transferred from UDP-galactose to diacylglycerol to form a monogalactosyl-diacylglycerol (MGDG). A galactosyltransferase then transfers the galactosyl group from one MGDG molecule to another, forming digalactosyl-diacylglycerol (DGDG) and diacylglycerol.

Figure 4–100.
Coordination of the requirements for membrane lipids in different cellular compartments. This coordination is illustrated by the *act1* mutant of Arabidopsis. (A) *ACT1* (the wild-type gene) encodes the chloroplast acyltransferase that transfers a fatty acid chain to position 1 of glycerol phosphate in the pathway of galactolipid synthesis (MGDG; see Figure 4–98). Most of the chloroplast phospholipid is synthesized by this prokaryotic pathway (*green arrow*), with only a small contribution from the eukaryotic pathway (*blue arrows*). (B) In the *act1* mutant, synthesis of MGDG via the prokaryotic pathway is no longer possible, due to loss of the acyltransferase. However, this loss is compensated for by much higher import of the products of the eukaryotic pathway for chloroplast galactolipid synthesis (MGDG, DGDG).

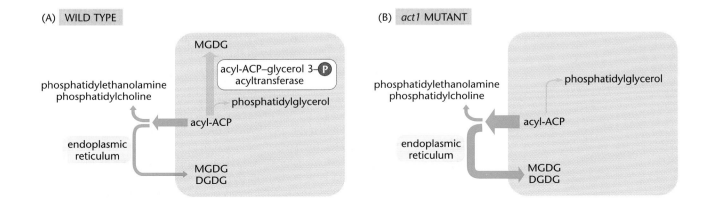

Figure 4–101.

Examples of the two routes of phospholipid synthesis from phosphatidic acid. Synthesis of phosphatidylserine (*left*) proceeds by replacement of the phosphate group of phosphatidic acid by CDP. CDP is then replaced by phosphoserine, and CMP is released. In the synthesis of phosphatidylethanolamine (*right*), CDP replaces the phosphate group of phosphoethanolamine, which will become the polar head group of the phospholipid. CDP-ethanolamine then reacts with a diacylglycerol formed from phosphatidic acid to give phosphatidylethanolamine, with release of CMP.

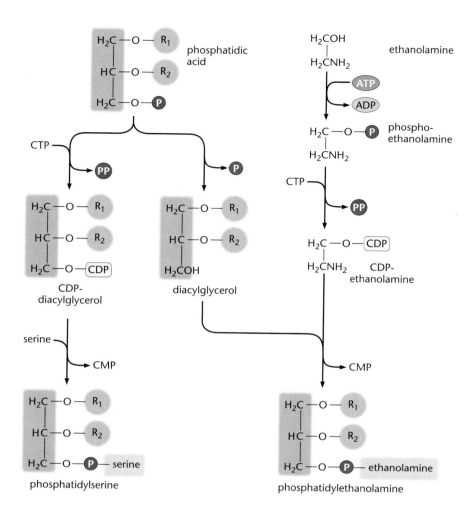

Different pathways of terpenoid synthesis in the plastid and the cytosol give rise to different products

Terpenoids are another group of compounds that can be synthesized both inside and outside the plastid, from the five-carbon precursor isopentenyl diphosphate (Figure 4–102). Examples of terpenoids include plastid components such as β-carotene, parts of the electron carriers plastoquinone and ubiquinone, and the phytol side chain of chlorophyll, and other compounds with a variety of subcellular locations, including the phytohormones gibberellin and abscisic acid, **phytoalexins** (produced in defense responses), resins, surface waxes, rubber, and many scent and flavor compounds (known as "essential oils") (Figure 4–102). The plastidial and nonplastidial pathways of terpenoid synthesis are radically different, both in the way the isopentenyl diphosphate precursor is synthesized and in the types of terpenoid produced. As we have seen for other aspects of metabolism, the pathway in the plastid resembles that in bacteria more closely than does the nonplastidial pathway.

Outside the plastid, isopentenyl diphosphate synthesis starts with acetyl-CoA (Figure 4–103). The condensation of three acetyl-CoA molecules produces the intermediate 3-hydroxy-3-methyl-glutaryl-CoA (HMG-CoA). This is converted to mevalonic acid, which is phosphorylated then decarboxylated to produce isopentenyl diphosphate. The enzyme that synthesizes mevalonic acid, **HMG-CoA reductase**, is located in the endoplasmic reticulum and is important in the control of nonplastidial terpenoid synthesis. This reductase is subject to several levels of regulation: expression of the genes

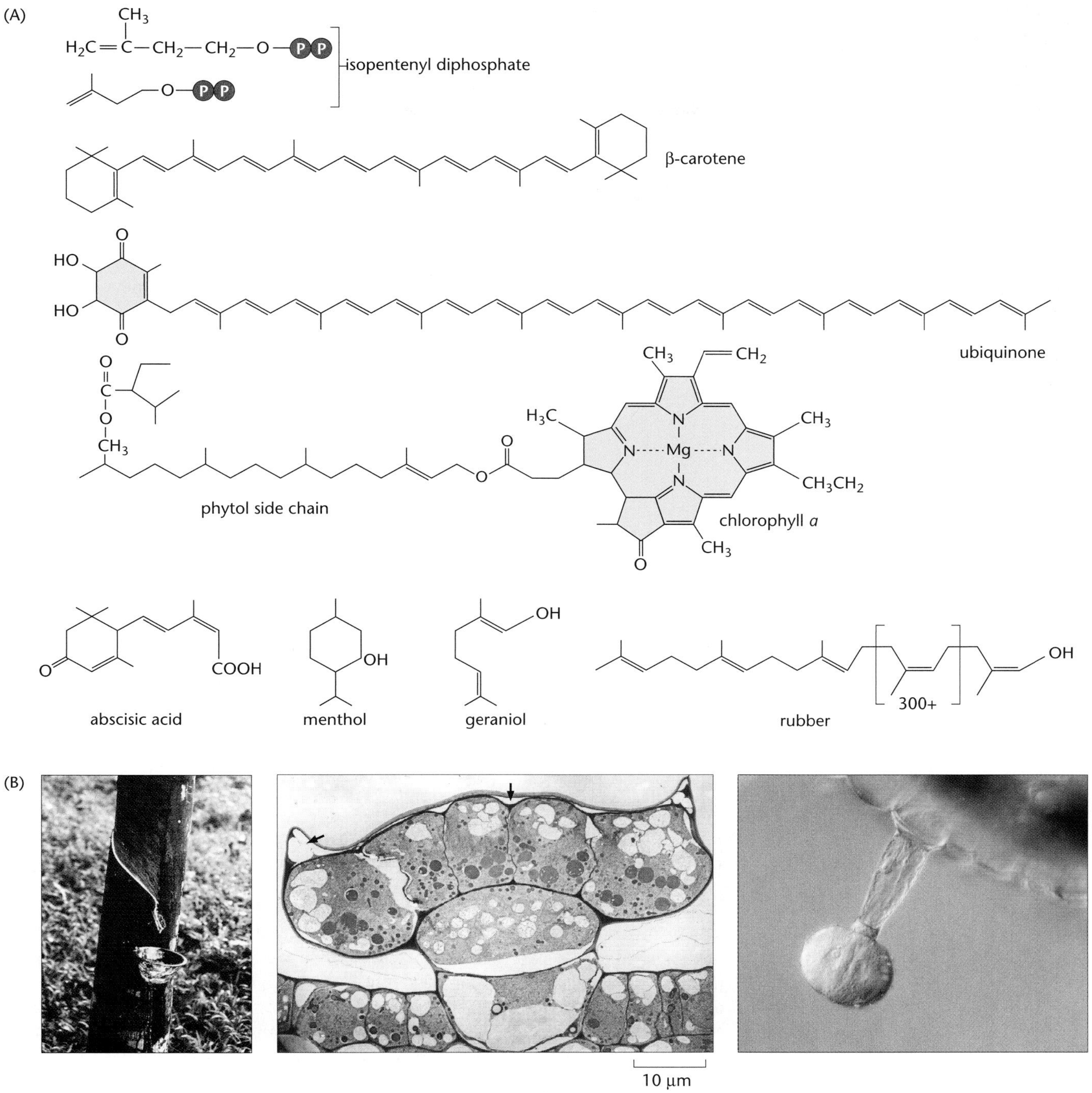

Figure 4–102.

Isopentenyl diphosphate and some terpenoid compounds derived from it. (A) The full structure of isopentenyl diphosphate (*top*), together with a representation that shows only the carbon–carbon bonds. Terpenoids consist of multiple repeats of this basic unit, or of a derivative of this unit. In some terpenoids the repeated structure is linear, and in others cyclization creates ring structures. In ubiquinone and chlorophyll, only part of the molecule is derived from the terpenoid pathway; the rest (*shaded blue*) is synthesized in a different pathway (see Figure 4–107 for the chlorophyll pathway). Carotene, ubiquinone, and chlorophyll have roles in energy-transfer reactions (see Sections 4.2 and 4.5). Abscisic acid is an important phytohormone (see Chapters 6 and 7). Menthol, geraniol, and rubber are so-called secondary metabolites that accumulate in large amounts in a small number of plant species, and may act as defense compounds (see Chapter 8). (B) The photograph (*left*) shows collection of rubber-containing latex from a rubber tree (*Hevea brasiliensis*). The electron micrograph (*middle*) shows a gland on the surface of a mint leaf; the gland is formed from outgrowth and division of an epidermal cell. The upper cells (apical disc cells) synthesize menthol, which accumulates in a space beneath the waxy cuticle that overlies the outer cell walls (*arrow*). The light micrograph (*right*) shows a gland on the surface of a tomato leaf. The cells that form the "head" of the gland produce a range of chemicals (including terpenoids) that may deter pathogens. (Bi, courtesy of Dennis W. Woodland.)

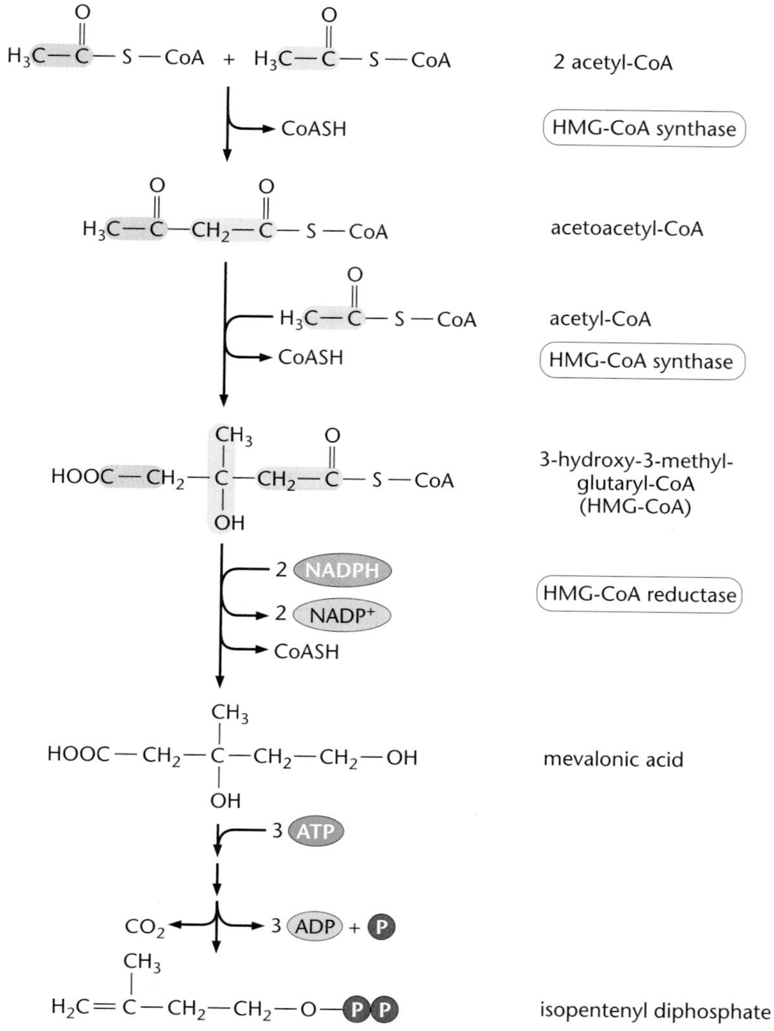

Figure 4–103.
Synthesis of isopentenyl diphosphate in the cytosol. Three molecules of acetyl-CoA are condensed by the enzyme 3-hydroxy-3-methyl-glutaryl-CoA (HMG-CoA) synthase to form the six-carbon HMG-CoA. Colors indicate the fate of each two-carbon unit of the HMG molecule. HMG-CoA reductase produces mevalonic acid, with NADPH as reductant. Three further, ATP-requiring reactions add phosphate groups and decarboxylate the product to form the five-carbon isopentenyl diphosphate.

encoding the enzyme is affected by developmental stage, hormonal signals, attack by pathogens, and wounding; and the enzyme is inactivated by phosphorylation by a protein kinase that also phosphorylates sucrose phosphate synthase and nitrate reductase (the latter enzyme is discussed in Section 4.8). The rate of terpenoid synthesis can thus be adjusted to match both the demand for terpenoids and for signaling and defense compounds and the availability of sugars and nitrogen-containing compounds in the plant. In the plastid, isopentenyl diphosphate synthesis begins with the condensation of pyruvate and glyceraldehyde 3-phosphate, followed by an intramolecular rearrangement to form 2-methylerythritol 4-phosphate (Figure 4–104). This pathway was discovered relatively recently, and its regulation is not yet understood.

Different classes of terpenoids are synthesized inside and outside the plastid (Figure 4–105). Terpenoids are produced by the linking of two or more isopentenyl diphosphate units by **prenyltransferase** enzymes, followed by modification (often ring formation) to form the terpenoid. Linkage of 2, 3, 4, 6, and 8 isopentenyl diphosphate units produces the monoterpene, sesquiterpene, diterpene, triterpene, and tetraterpene classes of terpenoids, respectively. The prenyltransferases of the plastid produce the mono-, di-, and tetraterpenes, whereas those of the cytosol and endoplasmic reticulum produce sesqui- and triterpenes. Isopentenyl diphosphate and terpenoids can move between cellular compartments. Terpenoid synthesis in the plastid may use isopentenyl diphosphate produced by the cytosolic pathway as well as that produced by the

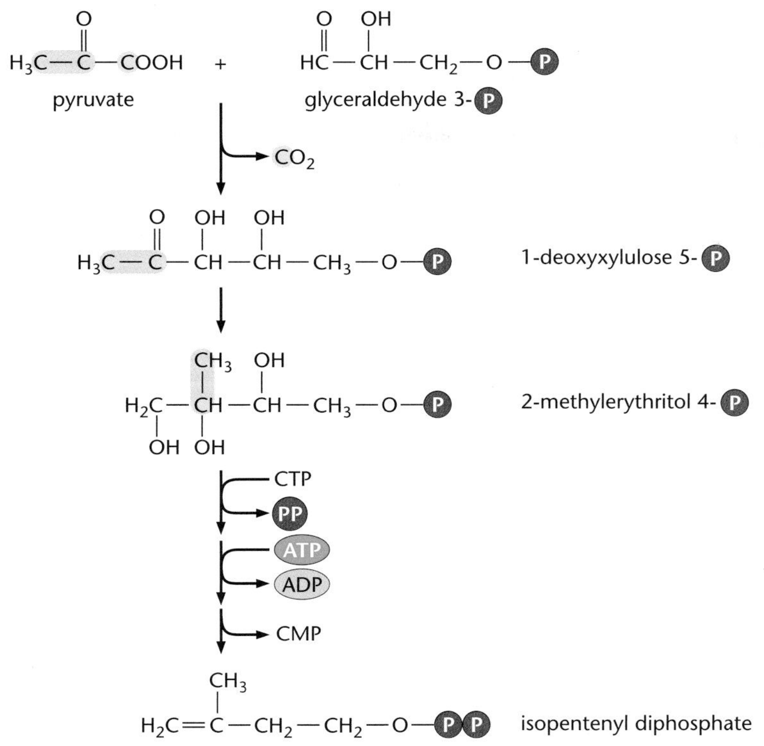

Figure 4–104.
Synthesis of isopentenyl diphosphate in the plastid. Condensation of a molecule of pyruvate with a molecule of glyceraldehyde 3-phosphate, followed by a decarboxylation and a complex intramolecular rearrangement, yields the 2-methylerythritol 4-phosphate. Colors trace the fate of the three carbons from pyruvate. Further ATP- and CTP-requiring reactions produce isopentenyl diphosphate.

plastidial pathway, and terpenoid synthesis in the mitochondrion uses isopentenyl diphosphate from the cytosol. Furthermore, terpenoids are not necessarily synthesized where they are used; for example, gibberellin is a diterpene made in the plastid and exported to the cytosol.

Terpenoids also serve an important role as membrane anchors for other types of molecule. Prenyltransferases add terpenoid units to pigments and components of electron transport chains, such as chlorophyll, plastoquinone, and ubiquinone, allowing such molecules to associate reversibly with membranes. **Prenylation** by specific protein prenyltransferases also allows proteins to become anchored to membranes (Figure 4–106). Examination of the Arabidopsis genome reveals many proteins with amino acid sequences appropriate for prenylation, including transcription factors, signaling proteins, and cell cycle regulators. The importance of prenylation is evident from studies of an enzyme that catalyzes the first step in the synthesis of the cytokinin class of phytohormones, adenosine phosphate isopentenyltransferase 3 (IPT3). Prenylation changes the location of IPT3 from the chloroplast to the nucleus, and influences the types of cytokinins made in the cell. This, in turn, influences the growth and development of the plant.

Tetrapyrroles, the precursors of chlorophyll and heme, are synthesized in plastids

Tetrapyrrole rings, or **porphyrin rings**, are complex molecular structures capable of binding a metal ion at their center. These rings have many different functions in the plant cell. Chlorophyll has a tetrapyrrole ring with bound magnesium. **Heme**, a tetrapyrrole ring binding iron, is the prosthetic group that mediates electron transfers in cytochromes, nitrate reductase, peroxidases (see Chapter 7), leghemoglobin (see Section 8.5), and other "heme proteins." **Siroheme**, another iron-containing tetrapyrrole ring, is the electron-transferring group of sulfite and nitrite reductases (see Section 4.8). Linear tetrapyrroles also have important functions, as in the light

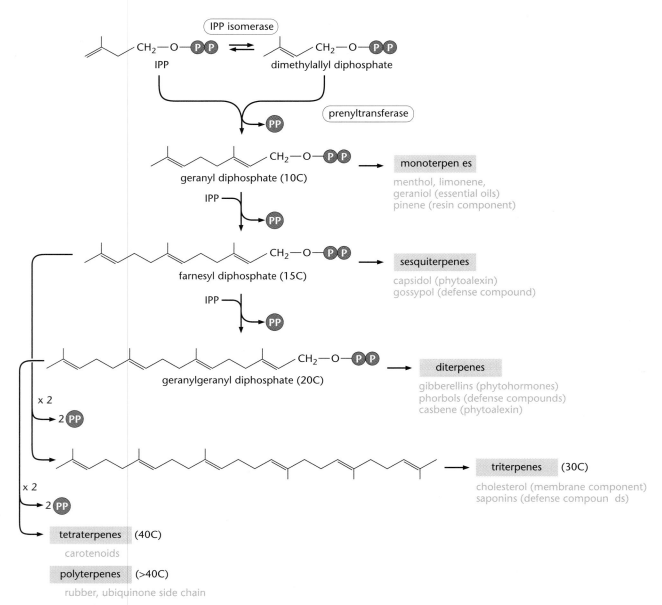

Figure 4–105.
Synthesis of terpenes from isopentenyl diphosphate. Isopentenyl diphosphate (IPP) is isomerized to form dimethylallyl diphosphate. A prenyltransferase creates the 10-carbon geranyl diphosphate, from which monoterpenes are derived. Addition of further IPP units to geranyl diphosphate via prenyltransferases produces 15- and 20-carbon compounds, which give rise to the sesquiterpenes and diterpenes, respectively. Terpenes of 30, 40, or more carbons are produced from addition of 15-carbon and 20-carbon units. Examples of the multitude of terpenoid compounds are listed (*green*). Some, such as gibberellins, cholesterol, and carotenoids, are found in essentially all plants. Others, such as the essential oil limonene (found in lemon fruits), occur in only a small number of species. The roles of terpenoids as defense compounds and phytoalexins are discussed in Chapter 8.

receptor **phytochrome** (see Sections 6.2 and 7.1). Although tetrapyrrole-containing molecules are found in most cellular compartments, the synthesis and initial elaboration of the ring structure occurs exclusively in the plastid (Figure 4–107).

The first step in tetrapyrrole synthesis is the conversion of the amino acid glutamate to 5-aminolevulinic acid. This is a highly unusual reaction because it proceeds via a glutamyl-tRNA (transfer RNA) intermediate (Figure 4–107A); the usual function of tRNAs is in protein synthesis. Once again, this mechanism reflects the bacterial origins of the plastid: the same reaction is found in cyanobacteria but not in animals. The

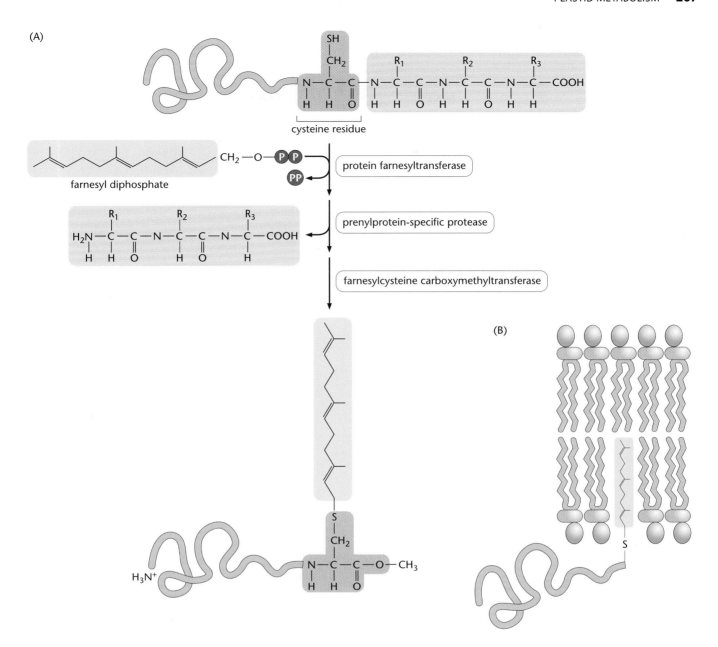

Figure 4–106.
Prenylation provides proteins with membrane anchors. (A) Proteins that become prenylated have a cysteine residue as the fourth amino acid from the C-terminus. A protein farnesyltransferase enzyme transfers a 15-carbon terpenoid chain from farnesyl diphosphate to the sulfur of the cysteine residue. A prenylprotein-specific protease then cleaves the three amino acids from the C-terminus, and a methyl group is added by a carboxymethyltransferase. (B) The terpenoid chain at the C-terminus acts as a hydrophobic "anchor," and is inserted into a membrane so that the protein is associated with the membrane surface.

5-aminolevulinic acid is cyclized (Figure 4–107B), and then undergoes a series of four condensations to form uroporphyrinogen, followed by further modifications to produce protoporphyrin IX (Figure 4–107C, D). The protoporphyrin IX may be retained in the chloroplast for chlorophyll and heme synthesis, or is exported to the cytosol and into mitochondria for heme synthesis. Heme synthesis is necessary in both chloroplasts and mitochondria for the manufacture of components of the electron transport chains. Heme is also exported from plastids for synthesis of heme proteins elsewhere in the cell.

Given that tetrapyrroles are usually not end products in themselves but are assembled into other types of molecules, there are particular regulatory demands on the pathway of their synthesis. Synthesis must be coordinated with that of the proteins and other molecules into which the tetrapyrroles are incorporated, and many of these molecules are synthesized in different cellular compartments and at rates that change differentially in response to developmental and environmental cues. Failure to match the synthesis of different types of tetrapyrroles with their assembly into appropriate enzyme or electron-transfer complexes would prevent the functioning of these complexes.

(A)

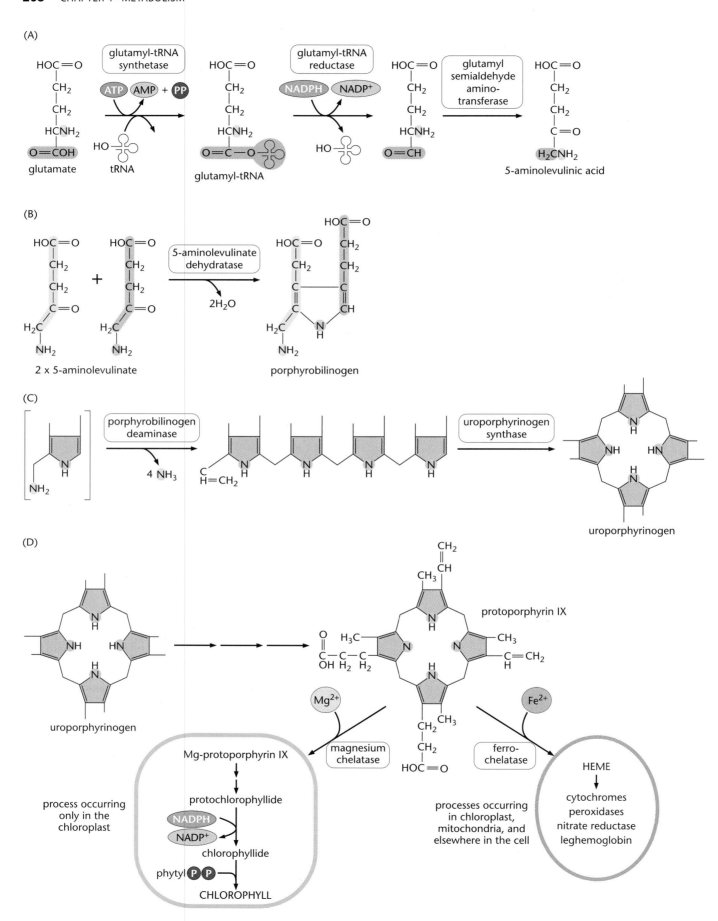

glutamate

tRNA

glutamyl-tRNA

5-aminolevulinic acid

(B)

2 x 5-aminolevulinate

porphyrobilinogen

(C)

uroporphyrinogen

(D)

uroporphyrinogen

protoporphyrin IX

Mg-protoporphyrin IX

protochlorophyllide

chlorophyllide

CHLOROPHYLL

process occurring
only in the
chloroplast

processes occurring
in chloroplast,
mitochondria, and
elsewhere in the cell

HEME

cytochromes
peroxidases
nitrate reductase
leghemoglobin

Figure 4–107.
Synthesis of chlorophyll and heme. (A) In the first stage, the five-carbon compound 5-aminolevulinic acid is synthesized from glutamate in the chloroplast, in a highly unusual pathway in which glutamyl-tRNA is an intermediate. (B) Condensation of two molecules of 5-aminolevulinate produces porphyrobilinogen, the "building block" of the tetrapyrrole ring. (C) Four molecules of porphyrobilinogen are joined together and cyclized to form the tetrapyrrole uroporphyrinogen. (D) Modification of the tetrapyrrole ring produces protoporphyrin IX, the branch point in the synthesis of chlorophyll and heme. Insertion of magnesium into this compound leads to chlorophyll; insertion of iron leads to heme. The last step in chlorophyll synthesis is addition of a phytyl unit to form the phytol (terpenoid) side chain (see Figure 4–102).

Accumulation of "free" tetrapyrroles is also potentially damaging to the cell, because free chlorophyll (i.e., not incorporated into a reaction center) and chlorophyll precursor molecules can absorb light energy. In the absence of regulated energy transfer, such as occurs when the chlorophyll is incorporated in a reaction center, damaging reactive oxygen species may be produced.

Complex mechanisms regulate the flux to different tetrapyrrole end products and, as noted above, coordinate this with synthesis of the proteins into which the tetrapyrroles are to be inserted (Figure 4–108). First, the pathway is strongly regulated by light. Synthesis of 5-aminolevulinic acid is stimulated by light and suppressed in the dark. The conversion of protochlorophyllide to chlorophyllide, catalyzed by protochlorophyllide oxidoreductases, is completely dependent on light. Second, 5-aminolevulinic acid synthesis is also regulated by a host of factors, internal and external to the pathway, that act on the expression or the activity of glutamyl-tRNA synthetase. These factors include phytohormones, circadian signals (discussed in Chapter 6), and free heme. Third, flux through the two main branches of tetrapyrrole synthesis, the chlorophyll and heme pathways, is controlled at the level of the two **chelatase** enzymes that compete for the

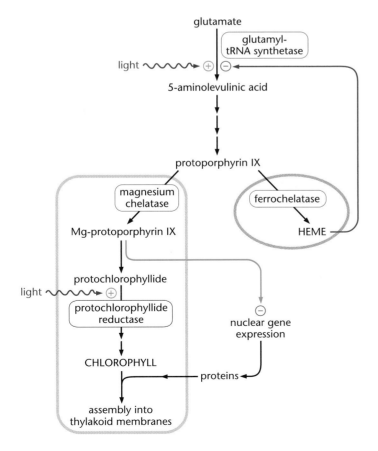

Figure 4–108.
Complex regulation of tetrapyrrole synthesis. Important regulation is believed to occur at the steps catalyzed by glutamyl-tRNA synthetase, ferrochelatase, magnesium chelatase, and protochlorophyllide reductase, and at the assembly of chlorophyll molecules into the thylakoid membranes. Chlorophyll synthesis is regulated by light and by signals affecting nuclear gene expression—probably mediated by Mg-protoporphyrin IX—that coordinate the rate of synthesis of chlorophyll with that of the proteins to which it is attached in the thylakoid membranes. Reduced rates of chlorophyll synthesis result in an increase in heme, which feeds back to inhibit the initial stages of tetrapyrrole synthesis.

common precursor, protoporphyrin IX. These two enzymes have very different kinetic and regulatory properties, and may have different locations in the chloroplast. Finally, coordination of the synthesis of tetrapyrroles and of the proteins into which they are inserted seems to be mediated by signals from the chloroplast to the nucleus that influence nuclear gene expression. This is best understood for chlorophyll synthesis.

Blocks in the pathway of chlorophyll synthesis lead to down-regulation of expression of the nuclear genes encoding chlorophyll binding proteins and other proteins involved in photosynthesis, including the small subunit of Rubisco. In Arabidopsis, *gun* (**g**enomes **un**coupled) mutants lack the signaling pathway from chloroplasts to the nucleus. When chlorophyll synthesis is blocked, there is no down-regulation of the nuclear genes encoding the photosynthetic proteins. From comparisons of *gun* mutants with wild-type plants, it has been suggested that an intermediate of chlorophyll synthesis—Mg-protoporphyrin IX—is the primary chloroplast signal that normally promotes down-regulation of the nuclear genes. The *gun* mutations prevent synthesis of this intermediate. Thus tetrapyrroles seem to play an important role in the overall coordination between nuclear gene expression and chloroplast assembly and development.

4.8 NITROGEN ASSIMILATION

Many classes of molecules in plants contain nitrogen: amino acids and proteins, nucleotides and nucleic acids, and a host of other macromolecules including tetrapyrroles such as chlorophyll and **phenylpropanoids** such as flavonoids. Nitrogen is the most abundant element in plants after carbon, hydrogen, and oxygen. Its assimilation from inorganic sources into organic compounds is one of the major metabolic activities of many plant cells.

Nitrogen assimilation is regulated at many levels and at many stages in the process. Environmental signals are integrated with signals from the plant's carbon metabolism so as to couple assimilation with the availability of inorganic nitrogen in the soil, the demand for synthesis of various nitrogen-containing compounds, and the availability of biosynthetic precursors, energy, and reductant for the assimilatory pathway. For most plants, the main source of nitrogen is nitrate (NO_3^-) in the soil. Other plants—mainly legumes—form tight symbiotic relationships with **nitrogen-fixing bacteria**; we describe these processes in Section 8.5. Here we focus on the processes by which nitrate is taken up from the soil, reduced to ammonium, and assimilated into the amino acids glutamate and glutamine (Figure 4–109). The latter part of this pathway is also involved in reassimilation of the ammonia released in photorespiration (see Section 4.3). Glutamate and glutamine are, directly or indirectly, the sources of nitrogen for almost all other nitrogen-containing compounds in the plant. We consider the synthesis of some of these, including other amino acids and the nitrogen-containing compounds derived from them.

Plants contain several types of nitrate transporter, regulated in response to different signals

Nitrate from the soil enters epidermal and cortical cells of the root via **nitrate transporters**. Its subsequent fate depends on its availability and on the species of plant. In many species, some of the nitrate is converted to glutamate and glutamine in the root cells, and some is transported via the xylem to the leaves, where it is metabolized to glutamate and glutamine in mesophyll cells (Figure 4–110). In general terms, herbaceous plants assimilate nitrate primarily in the leaves, whereas many trees and shrubs assimilate nitrate primarily in the roots. In a single species, the location of nitrate assimilation is often determined by nitrate availability: leaves are the primary location for assimilation when nitrate is plentiful, but assimilation may be mainly in the roots when the nitrate supply is limited.

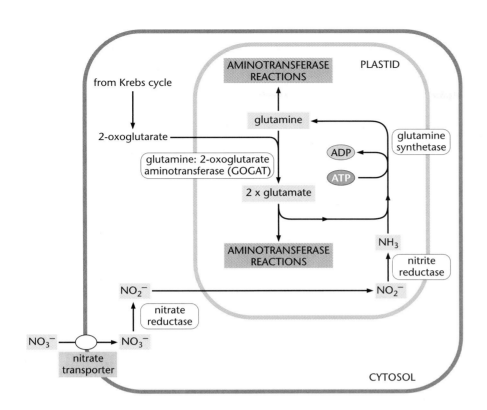

Figure 4–109.
Overview of nitrogen assimilation.
Nitrate is imported into the cell via a nitrate transporter, reduced to nitrite in the cytosol, and then reduced to ammonium (shown here as NH_3) in the plastid. The nitrogen is introduced into amino acids (as the $-NH_2$ group) via glutamine synthetase and glutamine: 2-oxoglutarate aminotransferase (GOGAT), using 2-oxoglutarate from the Krebs cycle as the carbon backbone. Glutamate and glutamine produced by these reactions donate nitrogen-containing groups to other biosynthetic processes in the cell, mainly via aminotransferase reactions.

The nitrate transporters in the plasma membrane of root cells, and of leaf cells, are of several different types. All are single proteins with 12 membrane-spanning domains. The transporters are nitrate–proton co-transporters, acting in conjunction with the plasma membrane ATP-driven proton pump (Figure 4–111A), in the same manner as

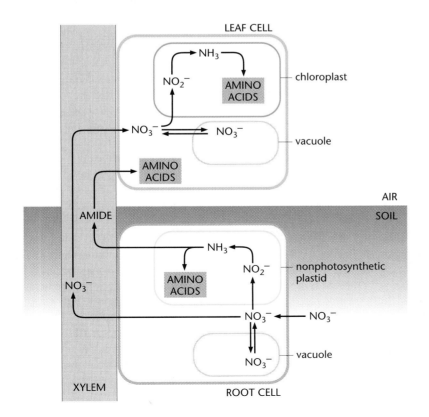

Figure 4–110.
Location of nitrogen assimilation in the plant. Nitrogen enters root cells as nitrate from the soil and is either reduced to ammonium for amino acid synthesis in the root cell and exported to the shoot as amino acids, or transported directly to the shoot as nitrate. Export from root to shoot occurs in the xylem stream. In most plants the main transported amino acids are glutamine and asparagine. Nitrate transported to the shoot is converted to ammonium and then amino acids in chloroplasts. In both root and shoot cells, cytosolic nitrate is in equilibrium with nitrate in the vacuole (as described later in the text).

Figure 4–111.
Transport of nitrate into plant cells.
(A) A proton pump exports protons from the cell, a process driven by ATP hydrolysis. Influx of protons back into the cell via the nitrate transporter is coupled to the import of nitrate. (B) Two major families of nitrate transporters are encoded in the Arabidopsis genome. The families differ in the location in the plant, affinity for nitrate, and the nature and extent of regulation of their gene expression by nitrate, light, sucrose, and immediate products of nitrogen assimilation.

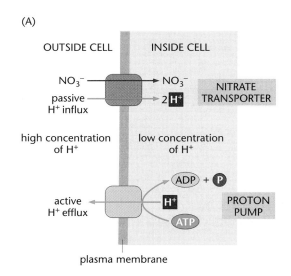

the sucrose–proton co-transporter (see Figure 4–48). There are two families of nitrate transporters, NRT1 and NRT2, with different affinities for nitrate: all NRT2 transporters have a very high affinity for nitrate, whereas NRT1 transporters have high or low affinities (Figure 4–111B).

The transporters differ not only in their affinity for nitrate but also in their location in the plant and their response to external signals. This diversity allows sensitive regulation of nitrate uptake in a variety of conditions. Nitrate is efficiently taken up over a very wide range of extracellular concentrations, from about 5 μM to 50 mM. Expression of some of the genes encoding nitrate transporters is induced by nitrate and repressed by ammonium. Other nitrate transporter genes are expressed constitutively—that is, even when no nitrate is present. The availability of precursors for amino acid synthesis is another factor influencing the expression of some nitrate transporters. For example, some transporter-encoding genes show a diurnal pattern of expression, and expression patterns are altered by supplying plants with sucrose or glutamine, as discussed below.

Although nitrate is usually by far the major form of nitrogen in the soil, most soils also contain nitrogen in the form of ammonium ion (NH_4^+). This may become the main form of nitrogen under some conditions, such as when soils are waterlogged. Almost all plants have the ability to use soil ammonium as well as nitrate. Ammonium transporters, like the nitrate transporters, are encoded by a multigene family; the transporter types differ in their affinity for ammonium and their pattern of expression. Expression of some root ammonium transporters is induced when nitrogen levels in the soil are low (nitrogen starvation), whereas expression of others is induced by the presence of ammonium.

Nitrate reductase activity is regulated at many different levels

Nitrate reductase is a cytosolic enzyme that catalyzes the reduction of nitrate to nitrite (NO_2^-). It consists of two identical subunits, each with three covalently bound cofactors—FAD, a heme, and **molybdopterin** (a molybdenum-containing cofactor; see Figure 4–128B)—which participate in the transfer of electrons from NADH or NADPH to nitrate. Each cofactor is associated with a specific domain of the enzyme (Figure 4–112). FAD is bound to a domain resembling the photosynthetic ferredoxin-NADP reductase (see Section 4.2). The FAD receives electrons from NADPH and passes them to the heme cofactor, bound to a domain that resembles a family of cytochromes. This in turn passes an electron to molybdopterin, bound to a third enzyme domain; this cofactor passes the electron to nitrate, reducing it to nitrite.

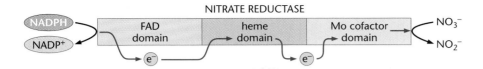

Figure 4–112.
Domains of a subunit of nitrate reductase. Electrons from NADPH are transferred to nitrate via three cofactors, each associated with a specific domain of the enzyme.

Levels of nitrate reductase protein and activity are controlled by many factors that link the capacity for nitrate reduction to the availability of nitrate, the availability of biosynthetic precursors for nitrogen assimilation, and the plant's need for the products of nitrogen assimilation. Thus the expression of the nitrate reductase protein is affected by factors as diverse as nitrate abundance, light levels, diurnal and circadian cycles, and the concentrations of sucrose, glutamine, and the phytohormone **cytokinin** (Figure 4–113). Expression of nitrate reductase genes is usually strongly and rapidly induced by nitrate, and is also induced by light. In the dark, sucrose induces gene expression. These mechanisms link the capacity for nitrate assimilation with the availability of sugars to provide precursor molecules. Conversely, expression of nitrate reductase is repressed by glutamine, a primary product of nitrogen assimilation.

Synthesis of the functional nitrate reductase requires not only expression of the nitrate reductase gene but also insertion of the three cofactors. The synthesis of these cofactors must be regulated so that their availability is coordinated with that of nitrate reductase protein. In Arabidopsis, expression of the genes encoding enzymes involved in the last stages of molybdopterin synthesis are induced by nitrate, in parallel with induction of nitrate reductase gene expression.

Nitrate reductase activity is also regulated by several post-translational mechanisms. The best understood of these is phosphorylation (Figure 4–113). A protein kinase phosphorylates the nitrate reductase on a serine residue between the FAD and heme binding sites. Phosphorylation enables this part of the protein to bind to a 14-3-3 protein. The **14-3-3 proteins** are a class of proteins involved in mediating changes in the activity of crucial metabolic enzymes in response to phosphorylation. The binding of the phosphorylated nitrate reductase to the 14-3-3 protein inactivates the enzyme. The enzyme phosphorylation that allows this inactivation by 14-3-3 protein binding is inhibited by triose and hexose phosphates; hence nitrate reductase is in the active form when there is a plentiful supply of precursors for amino acid synthesis. The protein kinase that phosphorylates nitrate reductase also phosphorylates and thus inactivates sucrose phosphate synthase (see Figure 4–28), providing a direct link between control of carbon assimilation and control of nitrogen assimilation. Dephosphorylation

Figure 4–113.
Complex regulation of expression and activity of nitrate reductase. Expression of the nitrate reductase gene is regulated by light, the circadian clock (see Chapter 6), sucrose, nitrate, and immediate products of nitrogen assimilation. The enzyme is regulated by phosphorylation/dephosphorylation by protein kinase/phosphatase activities. (The protein kinase, shown here as nitrate reductase kinase, is the same kinase that phosphorylates and thus regulates sucrose phosphate synthase; see Figure 4–28.) Phosphorylation enables the binding of a 14-3-3 protein, which inactivates the enzyme. The inactivation process is inhibited when triose phosphate and hexose phosphate are abundant, so the availability of biosynthetic precursors promotes nitrate reduction.

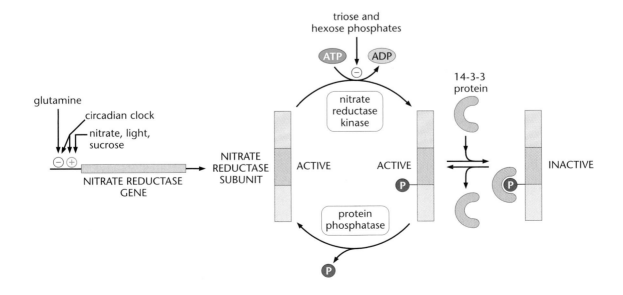

Figure 4–114.
Nitrite reductase. Electrons from ferredoxin, reduced in photosystem I (see Section 4.2), are transferred to nitrite via three cofactors bound to the enzyme.

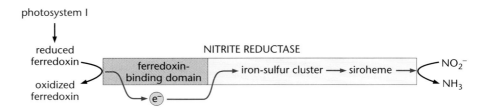

of nitrate reductase by a protein phosphatase prevents 14-3-3 binding and restores enzyme activity.

Nitrite formed by nitrate reductase is converted to ammonium by the plastidial enzyme **nitrite reductase**. Like nitrate reductase, the enzyme contains three cofactors; these transfer electrons to nitrite, converting it to ammonium (Figure 4–114). Nitrite reductase generally uses reduced ferredoxin as the source of reductant, but can also use NADPH. In chloroplasts in the light, reduced ferredoxin is a product of photosynthetic electron transport (see Section 4.2). It can also be generated in nonphotosynthetic plastids by ferredoxin-NADP reductase. Electrons from the reduced ferredoxin are passed to nitrite via a ferredoxin-binding domain, a structure known as an **iron-sulfur cluster** (see Section 4.9), and a siroheme cofactor, reducing the nitrite to ammonium. Nitrite is highly toxic to plant cells, and there are several mechanisms that maintain nitrite reductase activity at a sufficiently high level to prevent nitrite accumulation. Indeed, the activity of nitrite reductase is generally far higher than that of nitrate reductase, although expression of both enzymes is induced by many of the same factors.

Assimilation of nitrogen into amino acids is coupled to demand, nitrate availability, and availability of biosynthetic precursors

The ammonium generated by nitrate and nitrite reductases is assimilated by the activity of the enzymes glutamine synthetase and glutamine–2-oxoglutarate aminotransferase (GOGAT; Figure 4–115)—the same enzymes responsible for the reassimilation of ammonium released in photorespiration (see Section 4.3) and the assimilation of ammonium produced by symbiotic nitrogen fixation in legumes (see Section 8.5). Note that in most plants in the light, the rate of reassimilation of ammonium produced in photorespiration is much greater than the rate of "primary" assimilation of ammonium derived from nitrate uptake.

Plants have both cytosolic and plastidial forms of glutamine synthetase. The plastidial isoform predominates in leaves, where it assimilates ammonium generated from nitrate and by the photorespiratory cycle. The ammonium produced from nitrate can also be assimilated via cytosolic glutamine synthetase. In fact, mutant plants lacking the plastidial glutamine synthetase grow normally in conditions that suppress photorespiration, suggesting that the cytosolic enzyme alone has sufficient capacity to assimilate all of the ammonium generated from nitrate.

The conversion of glutamine to glutamate takes place in plastids, which have two isoforms of GOGAT. One form accepts electrons from NADPH, and the other from reduced ferredoxin. The ferredoxin-linked GOGAT predominates in leaves, and the NADPH-linked GOGAT in roots. Both forms contain an iron-sulfur cluster that transfers electrons during the reductive synthesis of two glutamate molecules from one 2-oxoglutarate and one glutamine molecule.

Assimilation of nitrate into glutamate and glutamine has a complex regulatory network that links many aspects of the plant's nitrogen and carbon metabolism. When nitrate is plentiful, there are profound diurnal changes in the activities of many enzymes necessary for the conversion of nitrate to amino acids, brought about mostly by diurnal

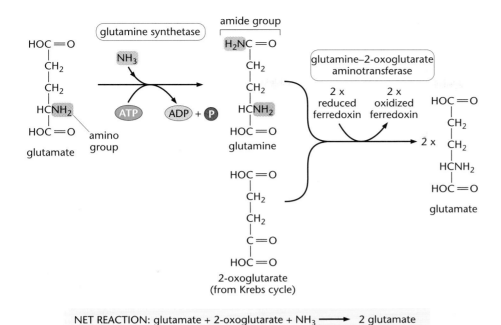

Figure 4–115.
Assimilation of nitrogen into amino acids via glutamine synthetase and GOGAT. Ammonium released in the nitrite reductase reaction is transferred to glutamate via glutamine synthetase, forming the amide, glutamine. An aminotransferase reaction between glutamine and 2-oxoglutarate, catalyzed by GOGAT, then generates two molecules of glutamate.

changes in gene expression. Diurnal changes occur in the activities of some nitrate transporters, nitrate and nitrite reductases, and glutamine synthetase and GOGAT. Enzymes necessary for the generation of 2-oxoglutarate (as substrate for GOGAT) also show diurnal changes, including cytosolic pyruvate kinase, mitochondrial citrate synthase, and the NADP-linked isocitrate dehydrogenase. The diurnal changes in enzyme activities are not in phase with each other: conversion of nitrate to ammonium peaks at one time of day, nitrate uptake at another time, and conversion of assimilated nitrogen to amino acids at another (Figure 4–116). These differences serve to coordinate nitrate assimilation with diurnal patterns of carbon assimilation and export in leaves, and with photorespiration.

Nitrate availability greatly influences the activities of many enzymes directly and indirectly involved in nitrate assimilation and amino acid synthesis. As noted above, it stimulates the expression of genes encoding transporters and the enzymes of the pathway from nitrate to glutamate and glutamine. Nitrate also stimulates expression of genes encoding the enzymes that generate reducing power for conversion of nitrate to ammonium in the root (ferredoxin-NADP reductase, glucose 6-phosphate dehydrogenase, and

Figure 4–116.
Diurnal patterns of nitrate assimilation. At night, the capacity for reduction of imported nitrate is low, so nitrate levels in the plant increase. Early in the day, the capacity for reduction of nitrate and its conversion to glutamine increases dramatically, so plant nitrate levels fall and ammonium and glutamine levels rise. Later in the day, the capacity for reduction of nitrate falls while the use of glutamate and glutamine for amino acid synthesis remains high, so ammonium and glutamine levels also fall. This scheme is an example only: the diurnal pattern of nitrate uptake and assimilation varies with species and with growth stage and is subject to substantial modification by factors such as nitrate availability.

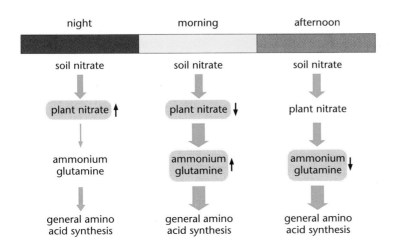

6-phosphogluconate dehydrogenase) and enzymes that synthesize 2-oxoglutarate. Other genes involved in carbon metabolism are also affected by nitrate levels. For example, expression of genes encoding ADP-glucose pyrophosphorylase is repressed by nitrate; this reduces the capacity for starch synthesis and directs carbon toward sucrose synthesis. Overall, the effect of nitrate is to increase the capacity for its own import, reduction, and conversion to amino acids, and to increase the availability of carbon for that purpose. Nitrate is also an important signal in root growth. High levels of nitrate inhibit root growth and decrease proliferation of lateral roots. On the other hand, local application of nitrate to part of a nitrogen-starved root system causes proliferation of lateral roots in that region (see Section 6.5).

Other intermediates of carbon and nitrogen metabolism also act as signals that coordinate growth and metabolism. For example, glutamine and other amino acids have effects opposite to those of nitrate on the capacity for nitrate assimilation: glutamine inhibits both nitrate uptake and its subsequent assimilation, thereby linking the rate of nitrogen assimilation to the plant's need for amino acids. High levels of sucrose stimulate both glutamine and 2-oxoglutarate synthesis, thus increasing the availability of precursors for synthesis of nitrogenous compounds when precursor molecules are plentiful. Low levels of sugars strongly repress the expression of nitrate reductase, reducing nitrogen assimilation when the appropriate precursors for amino acid synthesis are in short supply.

Amino acid biosynthesis is partly controlled by feedback regulation

Cells throughout the plant body require 20 different amino acids for protein synthesis, and also use many amino acids as precursors for other nitrogen-containing compounds. Most plant cells are highly autonomous with respect to amino acid synthesis: they can synthesize all the necessary amino acids from a few imported amino acids (as sources of amino groups) and sucrose (as source of precursor molecules). In many plants, the main amino acid transported from sites of nitrogen assimilation is glutamine, the product of primary ammonium assimilation. The high nitrogen content of glutamine (Figure 4–115; note that it has both an amino and an amide group) makes it a particularly effective nitrogen-transport compound. Glutamine synthesized in the leaves is exported via the phloem to other organs of the plant, and glutamine synthesized in the roots is exported via the xylem. It is the single most abundant nitrogen-containing organic compound in both the xylem and the phloem of many plant species. However, depending on the species and the growth stage, a variety of other amino acids can also be transported in both the xylem and the phloem.

The pathways of amino acid synthesis from glutamate and glutamine are grouped into families, according to the precursors involved (Figure 4–117). Some amino acids are synthesized directly from glutamate, but most are synthesized via pathways in which the amino or amide group of glutamate or glutamine is transferred to a precursor molecule by a family of enzymes called **aminotransferases**. Some aminotransferases preferentially use aspartate or alanine, which are themselves formed from glutamate by aminotransferase reactions. Aspartate can serve both as an amino group donor and as a direct precursor of other amino acids.

Amino acid synthesis must be coordinated both with the availability of nitrogen donors and precursor molecules and with a wide range of demands for amino acids in the synthesis of proteins and many "secondary" products. The control of most pathways of amino acid synthesis is not fully understood, but two mechanisms have been studied in some detail: feedback regulation of early steps in the pathway by amino acid products, and the breakdown of products when synthesis exceeds demand. We consider here two examples of these synthetic routes and their regulation: first, the pathway leading to the aspartate family (Figure 4–117) and other products derived from aspartate (such as the purines and pyrimidines); and second, the pathway leading to the aromatic amino acids—tyrosine, tryptophan, and phenylalanine—and some of their many derivatives.

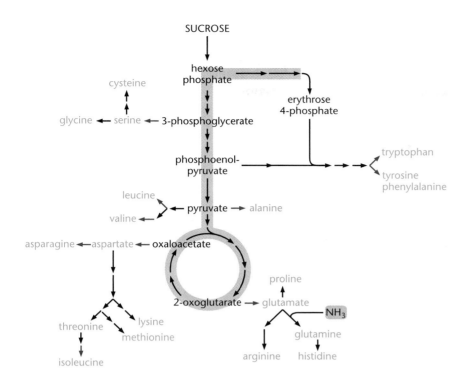

Figure 4–117.
Derivation of carbon backbones for
amino acid synthesis. Intermediates of
glycolysis, the oxidative pentose phosphate
pathway, and the Krebs cycle (*shaded purple*)
are the starting points for the synthesis of
amino acids. The amino groups are introduced
directly or indirectly from glutamate and
glutamine, the products of primary
ammonium assimilation. The introduction of
amino groups is usually catalyzed by
aminotransferases (positions of important
aminotransferase reactions are indicated by
red arrows).

Aspartate is the starting point for the synthesis of five other amino acids: threonine, isoleucine, lysine, methionine, and asparagine (Figure 4–118). Asparagine synthesis takes place in the cytosol, but most of the steps to the other amino acids (except the final stages of methionine synthesis) occur in the plastid. The first two steps in threonine, isoleucine, lysine, and methionine synthesis are the same. Aspartate is phosphorylated then reduced to aspartate semialdehyde. This compound is the substrate for two enzymes: dihydrodipicolinate synthase, which feeds into lysine synthesis, and homoserine dehydrogenase, which feeds into threonine, isoleucine, and methionine synthesis. Both the first enzyme of aspartate metabolism, aspartate kinase, and these two branchpoint enzymes are inhibited by the amino acid products of the pathway. Aspartate kinase is inhibited by products of both branches of the pathway, lysine and threonine, and also by an important product of methionine metabolism, **S-adenosylmethionine** (important in methylation reactions; e.g., see Section 2.3). Homoserine dehydrogenase is inhibited by threonine, and dihydrodipicolinate synthase is inhibited by lysine. These feedback mechanisms serve to regulate flux through the pathway in response to demand for its products: if the concentration of products increases, flux through the pathway is reduced.

The control of lysine synthesis has received particular attention because this is one of the nutritionally "essential" amino acids for humans and other mammals: lysine cannot be synthesized by mammals and hence must be obtained from plant foods. We describe here some of the efforts by plant scientists to produce plants with increased levels of lysine, because this work provides a good illustration of factors that control the levels of some amino acids.

Most of these efforts to produce high-lysine plants have centered on abolishing feedback regulation, so that accumulation of lysine no longer restricts the rate of its synthesis. The most successful strategy has been to manipulate dihydrodipicolinate synthase to make it feedback-insensitive. This was achieved in two ways. First, mutant populations of plants were screened for plants resistant to a toxic analog of lysine (toxic to wild-type plants because it competes with lysine for incorporation into proteins, which

Figure 4–118.
Synthesis of the aspartate family of amino acids. Aspartate is formed in the plastid from oxaloacetate, by transfer of the amino group from glutamate. The starting point for synthesis of methionine, threonine, isoleucine, and lysine is the addition of a phosphate group, catalyzed by aspartate kinase. This is an important regulatory point in amino acid synthesis: aspartate kinase is subject to feedback inhibition by lysine and threonine and by *S*-adenosylmethionine, a product of methionine that is of great importance in the introduction of methyl groups in biosynthetic reactions and is the precursor of the phytohormone ethylene (see Figure 6–21). Lysine, threonine, isoleucine, and methionine also inhibit enzymes farther down their synthetic pathways. Aspartate is also the starting point for synthesis of asparagine, by transfer of the amino group from glutamine to one of the two aspartate carboxyl groups (forming the amide), catalyzed by the cytosolic enzyme asparagine synthase.

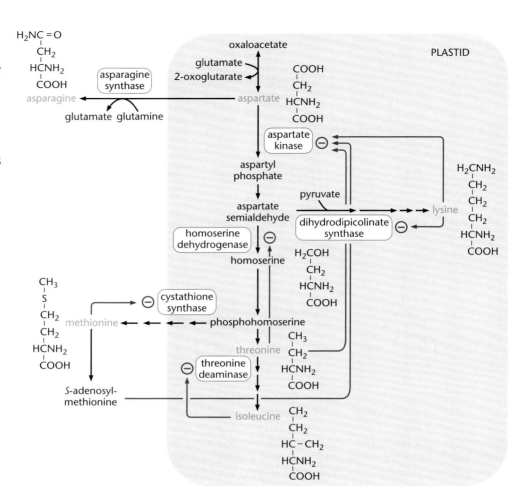

are then dysfunctional). Plants resistant to this analog survive because they have a mutant dihydrodipicolinate synthase that is not subject to feedback inhibition. The resulting high lysine levels reduce the frequency of incorporation of the lysine analog into protein, so the plant has sufficient normal protein to survive. The second strategy was to make transgenic plants that express bacterial genes encoding dihydrodipicolinate synthase. The bacterial enzyme is much less sensitive to lysine inhibition than the plant enzyme, so this strategy resulted in elevated lysine levels in the transgenic plants (Figure 4–119).

Unfortunately, increasing lysine synthesis by reducing feedback inhibition did not immediately have the expected nutritional benefits for mammals. Little of the extra lysine was incorporated into stable storage proteins, and much of it was degraded to glutamate by a pathway in which expression of the first enzyme is induced by lysine (Figure 4–119). Overcoming this problem required a way of preventing lysine degradation. This strategy has been tested in Arabidopsis. Introduction of both a feedback-insensitive dihydrodipicolinate synthase and a mutation that eliminates activity of the first enzyme of the lysine-degradation pathway resulted in very much higher levels of lysine in seeds than did introduction of the feedback-insensitive dihydrodipicolinate reductase alone. These experiments illustrate that lysine levels in plant cells are determined not only by the rate of lysine synthesis and incorporation into protein, but also by the rate of lysine degradation. The levels of many other amino acids may also be determined by similar interactions between synthesis, utilization, and degradation.

Besides providing the precursor molecules and the amino group for synthesis of the five amino acids described above, aspartate is involved in the synthesis of **purines** and

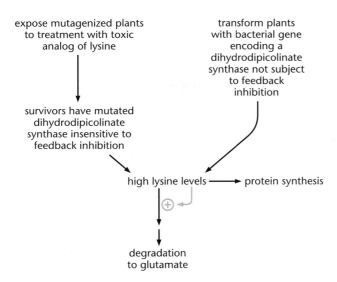

Figure 4–119.
Two strategies for increasing lysine levels in plants. Both strategies targeted dihydrodipicolinate synthase—the first enzyme in the aspartate-to-lysine pathway after the branch point to lysine (see Figure 4–118)—with the aim of introducing a dihydrodipicolinate synthase insensitive to feedback inhibition by lysine. Both strategies were successful in that the feedback insensitivity of the introduced dihydrodipicolinate synthase did allow higher rates of lysine synthesis. However, the increases in level of free lysine (and of lysine-containing storage proteins) were lower than expected, because high lysine levels induced the activity of an enzyme of lysine degradation. In other words, lysine was involved in feed-forward promotion of its own degradation.

pyrimidines, precursors in the synthesis of **nucleic acids** and nucleotides (such as ATP). There is a close and complex relationship between amino acid metabolism and the synthesis of purines and pyrimidines: both synthetic pathways derive precursor molecules from ribose 5-phosphate and aspartate and nitrogen from the amide group of glutamine via aminotransferase reactions.

The purines guanine and adenosine are synthesized from aspartate and glycine through a series of aminotransferase reactions. The pyrimidine orotate is synthesized from carbon dioxide, aspartate, and the amide group of glutamine (Figure 4–120). Other pyrimidines, including uracil and cytidine, are made by further modifications of orotate. All of the enzymes of pyrimidine synthesis are found in the plastid; purine synthesis is thought to take place in the cytosol.

Synthesis of the aromatic amino acids tryptophan, tyrosine, and phenylalanine has been the subject of intensive research, for two main reasons: first, they are precursors for thousands of other compounds (essentially every compound containing a phenolic ring); second, one of the enzymes involved in production of chorismate, the common precursor of the aromatic amino acids, is the target of the commercially important **herbicide** glyphosate (see Section 9.3). Production of chorismate takes place in the plastids via the **shikimate pathway** (Figure 4–121). This pathway begins with the condensation of phosphoenolpyruvate and erythrose 4-phosphate, followed by cyclization and conversion via shikimate (in a series of reactions that utilize another PEP molecule) to chorismate.

Chorismate is the substrate for two enzymes: chorismate mutase, which provides the substrate for the short pathways to phenylalanine and tyrosine, and anthranilate synthase, which provides the substrate for tryptophan synthesis (Figure 4–122). Chorismate mutase is inhibited by both phenylalanine and tyrosine, and the enzymes catalyzing the unique final steps in the synthesis of these two amino acids are inhibited by their respective amino acid products. The final enzyme of phenylalanine synthesis is also activated by tyrosine. Some isoforms of anthranilate synthase are feedback-inhibited by tryptophan (Figure 4–123). As with other pathways of amino acid synthesis, these feedback loops serve to prevent accumulation of the amino acids when demand for them is low, and to coordinate their synthesis.

Many of the metabolites derived from the aromatic amino acids are discussed elsewhere in this book, particularly in the context of biotic and abiotic stress (Chapters 7 and 8). They include the monomers of lignin, the flavonoids (including anthocyanin pigments and defense and signaling molecules), and numerous other defense, signaling, taste, and odor compounds such as stilbenes, **coumarins**, and many **alkaloids**. There is enormous variation among species, and among organs, cell types, growth conditions, and developmental stages in a single plant, in the extent and nature of these pathways. Thousands of metabolites derived from aromatic amino acids have been discovered, many found in only one organ of one or a handful of species. In many cases,

Figure 4–120.
Synthesis of pyrimidines and purines.
The involvement of aspartate in these pathways is highlighted (*pink shading*). In pyrimidine synthesis (*left*), the ring compound orotate is formed from aspartate and the one-carbon nitrogen-containing compound carbamoyl phosphate. Cytidine 5′-triphosphate is synthesized by addition of a ribosyl group and further modification of the ring structure. In purine synthesis (*right*), a ring structure is formed on ribose 5-phosphate, from glycine, carbon dioxide, a one-carbon unit, and nitrogen derived from glutamine. Further complex modifications—involving addition of two aspartate molecules and a one-carbon unit, and elimination of two fumarate molecules—form a second ring to give the purine adenosine 5′-monophosphate (AMP).

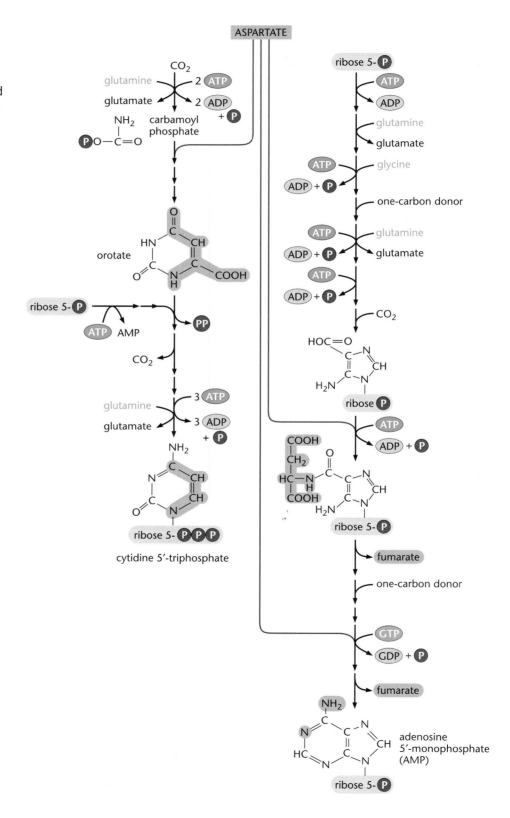

Figure 4–121.
Synthesis of the aromatic amino acids: the shikimate pathway. The glycolytic intermediate phosphoenolpyruvate is condensed with erythrose 4-phosphate, an intermediate of the oxidative pentose phosphate pathway and the Calvin cycle, to give the ring compound shikimate. Modifications of the ring and addition of a further molecule of PEP yield 5-enolpyruvylshikimate 3-phosphate (EPSP). (The reaction in which PEP is added is catalyzed by EPSP synthase, the enzyme that is the target of the commercially important herbicide glyphosate.) Removal of the phosphate group from EPSP yields chorismate, the precursor for pathways to tyrosine and phenylalanine and to tryptophan (see Figure 4–122).

their function and importance in the plant is not known. Collectively, metabolic pathways thought not to be essential for basic growth and development—including most of the pathways for which aromatic amino acids are precursors—are often referred to as **secondary metabolism**. However, the distinction between secondary metabolism and those metabolic pathways regarded as essential ("primary metabolism") is not clear-cut. Deeper understanding of the function and importance of the "secondary" pathways may well reveal that many are necessary for plant survival under natural conditions.

Flux through pathways leading from the aromatic amino acids, and hence demand for the aromatic amino acids as precursors, changes in response to a multitude of developmental and environmental signals. The flux through the shikimate pathway must be coordinated with these multiple and changing demands. Conditions that place a large demand for aromatic amino acids as precursors of secondary metabolites—for example, wounding or fungal invasion—often induce expression of the enzymes of the shikimate pathway, and also of various enzymes of primary metabolism that supply biosynthetic precursors and reducing power for that pathway. The importance of coordination between primary and secondary metabolism in the case of the shikimate pathway is strikingly illustrated by detailed analysis of transgenic tobacco plants with reduced activities of **transketolase**, the Calvin cycle enzyme that catalyzes the production of erythrose 4-phosphate (a precursor for the shikimate pathway; see Figure 4–121). A 50% reduction of transketolase activity in the leaf has large and complex effects on the accumulation of metabolites derived from the shikimate pathway, including a reduction of 50% or more in the levels of aromatic amino acids and lignin.

Nitrogen is stored as amino acids and specific storage proteins

Plants store nitrogen both as a short-term buffer against imbalances between demand and uptake and as a longer-term reserve (in storage organs such as seeds, roots, and tubers, for example) to provide amino acids for biosynthesis when new growth occurs (Figure 4–124). In the short term, nitrogen is stored as nitrate in the vacuole. Here, nitrate can accumulate to high concentrations when in plentiful supply, and can be exported to the cytosol for metabolism when circumstances change.

Nitrogen is also stored in the short to medium term in the form of amino acids. One of the most commonly stored amino acids is asparagine, which like glutamine has a high nitrogen content. Asparagine is also a common form in which nitrogen moves in the xylem and phloem: in many legumes it is the major form in which assimilated nitrogen is exported from the root nodules, following fixation of atmospheric nitrogen by *Rhizobium* bacteria (see Section 8.5). Asparagine is synthesized from aspartate in the cytosol by **asparagine synthase**. Some isoforms of asparagine synthase are specifically induced by darkness and repressed by sucrose, so that asparagine is synthesized as a nitrogen reserve when the supply of precursor molecules and energy for biosynthesis of amino acids and other molecules is minimal.

Many plant species accumulate one or a few unusual, "nonprotein" amino acids. These may accumulate to high levels, and tend to be characteristic of an individual species.

Figure 4–122.
Synthesis of the aromatic amino acids from chorismate. Chorismate is the precursor of two pathways of amino acid synthesis. Metabolism via anthranilate synthase is the first step in a complex pathway in which a ribosyl group and a serine are added and the product is modified to yield tryptophan. Metabolism of chorismate via chorismate mutase followed by an aminotransferase reaction yields arogenate. Two different modifications of arogenate produce the amino acids tyrosine or phenylalanine.

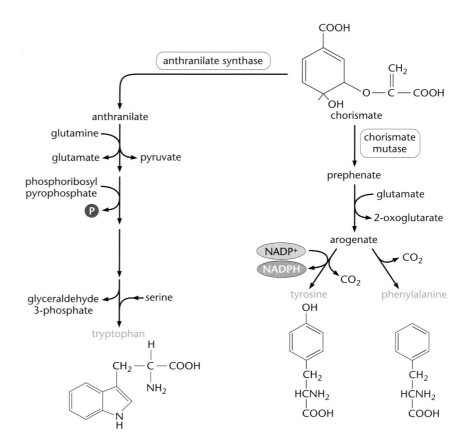

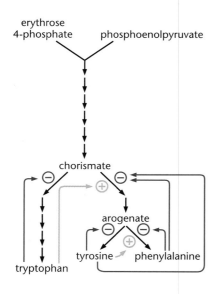

Figure 4–123.
Feedback regulation of the pathways of aromatic amino acid synthesis. Inhibition of the enzymes that metabolize chorismate and arogenate by the end products of the pathways they initiate links the rate of synthesis of the aromatic amino acids to the rate at which they are consumed in protein synthesis and other biosynthetic pathways. Activation of tyrosine and phenylalanine synthesis by tryptophan, and of phenylalanine synthesis by tyrosine, controls the distribution of carbon from chorismate among the amino acids according to the rates at which they are being consumed.

Examples include the accumulation of canavanine (similar to arginine) and homoserine (an intermediate in threonine, isoleucine, and methionine synthesis) in the seeds of legumes, and pyrrolidone 5′-carboxylic acid (similar to proline) in lily-of-the-valley (*Convallaria majalis*). These unusual amino acids may serve as nitrogen reserves, but may equally act as deterrents against predation (see Section 8.4). As close analogs of the amino acids of proteins, many are toxic to animals because tRNAs "recognize" and incorporate them into proteins in place of the correct amino acid, thus disrupting protein synthesis.

Long-term reserves of nitrogen in storage organs are held in the form of **storage proteins**. These proteins have been studied in great detail in the seeds of legumes (where they may make up more than 40% of the dry weight) and cereals (up to 10–15% of the dry weight), because of their nutritional importance. The storage proteins of legumes and cereals supply a large portion of the amino acids in the diets of humans and farmed animals. They also give the characteristic elastic properties to dough made from flour produced from these seeds, and hence determine the texture and baking or cooking properties of bread, noodles, pasta, and so forth. Storage proteins are also found at lower levels in vegetative storage organs such as potato tubers and cassava roots.

The most widespread storage proteins are known as **globulins**, which include the major storage proteins of legume seeds (**legumins** and **vicilins**) and potato tubers (**patatin**). The primary storage proteins of cereal seeds are the **prolamins**. Prolamins of high molecular weight, called "HMW-glutenins," are critical in determining the suitability of flour for bread making. These forms are abundant in some varieties of wheat, but absent from barley, oats, and maize. Prolamins contain regions of tandemly repeated amino acid sequences about 20 amino acids in length, and the combination of long repeated sections and the formation of disulfide bonds between adjacent protein molecules allows the HMW-glutenins to form flexible networks in hydrated flour

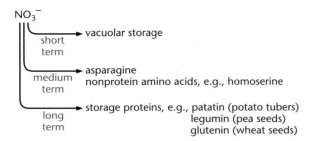

**Figure 4–124.
Short-, medium-, and long-term storage forms of nitrogen.**

(Figure 4–125). These networks trap gases during bread making, so that the dough will "rise" to form a good-quality bread. (In Section 9.1 we discuss the domestication of wheat for these particular properties.)

Storage proteins usually accumulate in specific organelles, called **protein bodies**. Most of the storage proteins are synthesized on the rough endoplasmic reticulum and secreted into the endoplasmic reticulum lumen (Figure 4–126). Globulins are then transported via the Golgi apparatus to the vacuole, and the vacuole subdivides to form many individual protein-containing vesicles. Vesicles containing prolamins also bud off directly from the endoplasmic reticulum. During passage from the site of synthesis to the protein body, many storage proteins undergo considerable post-translational modification. For example, the original translation products of a legumin form trimers in the endoplasmic reticulum lumen. On reaching the vacuole, each of these products is cleaved by a specific protease, and the final legumin molecule is a hexamer (Figure 4–127). Other storage proteins, such as vicilins, are modified by the covalent addition of carbohydrate side chains, often rich in mannose and *N*-acetylglucosamine.

Storage proteins are expressed in highly specific temporal and spatial patterns. In developing pea seeds, for example, members of the multigene family encoding vicilins are mostly expressed before members of the family encoding legumins, but the developmental timing of expression also differs among individual members of each family. The transcription factors that control these patterns of expression have a highly conserved function. This is revealed by experiments in which genes for legume storage proteins, with their own promoters, are introduced into tobacco plants. Gene expression follows the same spatial and developmental pattern in the transgenic tobacco seeds as in the legume seed. Consistent with these observations, the promoter regions of storage-protein genes from a wide range of species contain near-identical DNA motifs to which transcription factors bind. Synthesis of seed storage proteins is also influenced by the availability of sulfur for the synthesis of sulfur-containing amino

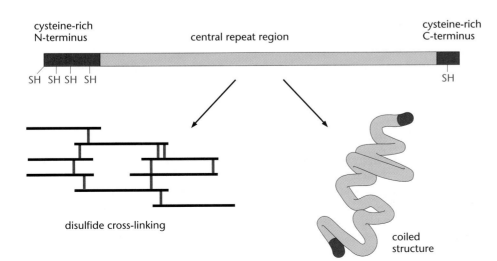

**Figure 4–125.
Structure of HMW-glutenins.** The N- and C-terminal regions are rich in cysteine residues, and the central region consists of large numbers of tandem repeats. Both of these features contribute to the ability of the protein to form a viscoelastic network in dough. The tandem repeats allow hydrogen bonding and other types of bonding between amino acid residues in the protein so that it forms an elastic, coiled structure. The cysteine residues allow the formation of disulfide bonds between molecules to form a glutenin network.

Figure 4–126.
Synthesis of prolamins and globulins.
(A) Both types of storage protein are initially synthesized on ribosomes attached to the cytosolic face of the endoplasmic reticulum (rough ER) and then released into the endoplasmic reticulum lumen. Vesicles containing prolamins bud off from the endoplasmic reticulum as protein bodies. Globulins are transferred by vesicle traffic from the endoplasmic reticulum to the Golgi apparatus and then to the vacuole, usually undergoing post-translational modification en route. Regions of the vacuole in which globulins have accumulated fragment into protein bodies. (B) Scanning electron micrograph of a cell in the embryo of a mature jack bean seed, containing many protein bodies and larger starch granules.

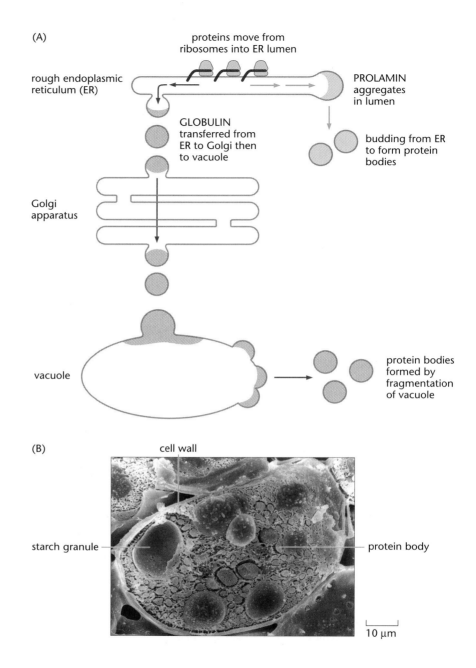

acids. In a sulfur-deficient environment, legume and cereal seeds contain more of the types of storage protein poor in sulfur-containing amino acids (see Section 4.9) and less of the types rich in sulfur-containing amino acids than when sulfur is in abundant supply. Both transcriptional and post-transcriptional mechanisms mediate this relationship between sulfur supply and synthesis of particular storage proteins.

4.9 PHOSPHORUS, SULFUR, AND IRON ASSIMILATION

In addition to carbon, hydrogen, oxygen, and nitrogen, plants require 13 other elements for growth. The required elements are conventionally divided into **macronutrients**, needed in relatively large amounts for normal growth, and **micronutrients**, needed in only very small amounts (Table 4–1). All of these are obtained mainly by uptake from the soil across the plasma membrane of root cells, but the routes by which they are assimilated and their fates within the plant are highly diverse. Some, like the macronutrient potassium (as K^+ ions) and the micronutrient chlorine (as chloride, Cl^- ions), are

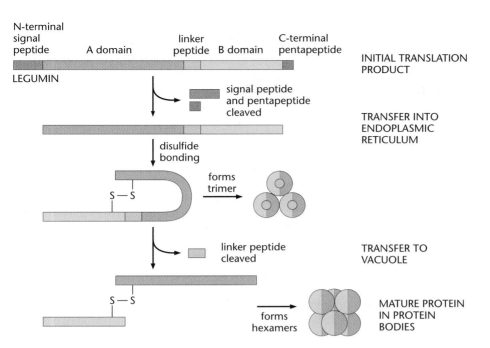

Figure 4–127.
Post-translational modification of legumin (a globulin). The initial translation product has N- and C-terminal regions that allow entry into the endoplasmic reticulum lumen, and these are cleaved from the protein in the process. In the lumen, the protein undergoes internal disulfide bonding and association to form trimers. An internal peptide is then removed, leaving a protein with two separate peptides linked by a disulfide bond. In the vacuole, these proteins form hexamers.

Table 4–1 The elements essential for plant growth and their major functions

	Typical amount in a healthy plant (mg g^{-1} dry weight)	Functions in the plant
MACRONUTRIENTS		
Nitrogen	15,000	See Section 4.8
Potassium	10,000	• Osmotic and ionic balance in cells (see Chapter 3) • Activator of many enzymes especially in respiration and photosynthesis
Calcium	5000	• Pivotal position in intracellular signal transduction • Complex formation with pectin in the cell wall (see Chapter 3)
Magnesium	2000	• Component of the chlorophyll molecule (see Section 4.7) • Activator of many enzymes
Phosphorus	2000	See Section 4.9
Sulfur	1000	See Section 4.9
MICRONUTRIENTS		
Chlorine	100	• Osmotic and ionic balance in cells (see Chapter 3)
Iron	100	• Component of enzymes involved in electron transfer; for example, the heme groups in cytochromes and nitrate reductase
Manganese	50	• Electron transfer during O_2 production in PSII (see Section 4.2) • Activator of many enzymes
Boron	20	• Required for integrity of plasma membrane and cell wall matrix
Zinc	20	• Activator of many enzymes • Component of zinc finger transcription factors
Copper	6	• Component of proteins involved in electron transfer; for example, plastocyanin
Molybdenum	0.1	• Component of enzymes, including nitrate reductase (see Section 4.8)
Nickel	0.005	• Component of urease, an enzyme involved in degradation of arginine

Figure 4–128.
Some roles of micronutrients. (A) Many transcription factors that interact with the DNA helix are proteins with zinc finger motifs (see Table 2–1). The zinc is bound (coordinated) by cysteine and histidine residues. This results in a protein conformation with projections (fingers) that fit into the major groove of the DNA molecule. **(B)** One of the cofactors of nitrate reductase is molybdopterin, a ring compound with chelated molybdenum. Changes in the oxidation state of molybdenum (between 4+ and 6+) are involved in the transfer of electrons from NAD(P)H during reduction of nitrate to nitrite (see Figure 4–112). **(C)** Interaction between copper and amino acid residues of the electron carrier plastocyanin. The copper is coordinated by a cysteine, a methionine, and two histidine residues. Electron transfer from PSII to PSI (see Section 4.2) involves a change in the oxidation state of copper ion between Cu$^+$ and Cu^{2+}. **(D)** The copper-sulfur cluster of cytochrome *c* oxidase, the final protein complex (complex IV) in the electron transport chain of mitochondria (see Section 4.5). The copper atoms are coordinated by two histidines, two cysteines, a methionine, and the carboxyl group of a glutamate, and linked through the sulfur atoms of the cysteines. This structure transfers an electron from cytochrome *c* to further electron carriers in complex IV.

Silicon is not an essential nutrient for higher plants, but is found in grasses and horsetails and in diatoms. **(E)** Horsetails (*Equisetum* spp.) have high concentrations of silicon dioxide in their cell walls. **(F)** Scanning electron micrograph of the silicon-dioxide–based outer casing (frustule) of a marine diatom, a single-celled alga. Diatoms are massively abundant in the oceans: there are estimated to be approximately 7 billion beneath one square meter of sea surface in the Gulf of Maine, off the northeast coast of the United States. Diatoms are responsible for much of the photosynthesis in the oceans (about 2.5–3 times more carbon is assimilated by photosynthesis in the oceans than on land). Diatom frustules, deposited on the ocean floor when the cells die, form the basis of many silicon-containing rocks and mineral deposits. (E, courtesy of Tobias Kieser.)

freely soluble in the soil and are taken up via transporters in root cell plasma membranes. Others, such as phosphorus and iron, are present in the soil mainly in insoluble chemical compounds. Several different mechanisms increase the availability of these elements to plants, including the formation of symbiotic relationships with soil fungi and the secretion from roots of substances that increase the solubility of the element in the soil.

Once inside the plant, two of the nutrients—sulfur and phosphorus—are assimilated into organic compounds; they are components of many essential molecules in the cell. Other nutrients have a wide variety of roles. Metal ions act as activators for enzymes and serve as cofactors for macromolecules involved in oxidation-reduction reactions (Figure 4–128); potassium and chloride are essential to the ionic and osmotic balance of the cell (see Section 3.5); calcium is a component of many signal-transduction pathways (see Chapter 7); boron is necessary for the integrity of the cell wall matrix (see Section 3.4). Although not essential for the growth of higher plants, silicon is found in strengthening deposits in the cell walls of grasses and horsetails, and is the main component of the cell walls of diatoms (Figure 4–128E, F).

In this section we focus on the uptake and fate of three nutrients, selected for their interest from a metabolic perspective: the macronutrients phosphorus and sulfur, and the micronutrient iron. Although the mechanisms involved are superficially very different, in all three cases the capacity for uptake and metabolism of the nutrient is controlled

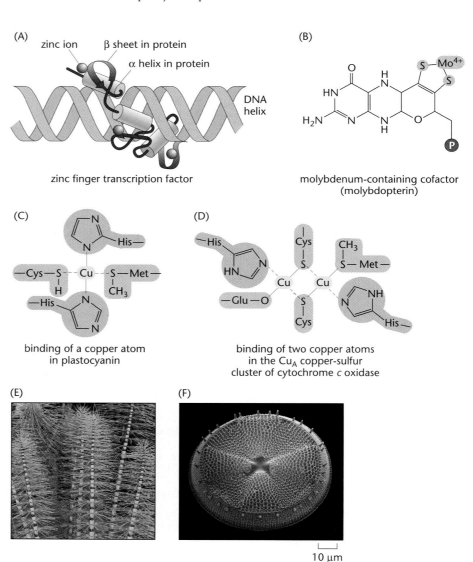

both by the local availability of the nutrient in the soil and by the general nutrient status of the plant. As we described for nitrate (Section 4.8), extensive regulatory networks throughout the plant body serve to link the availability of a particular nutrient with the plant's requirement for that nutrient for growth and development. Thus the extent to which a nutrient is taken up from the soil is regulated not only by its availability in the soil but also, in a complex manner, by the availability in the plant of sugars, nitrogen, and other macro- and micronutrients necessary for growth.

The availability of phosphorus is a major limitation on plant growth

Phosphorus in the form of phosphate is essential for an array of functions in the plant cell, some of which we have already described in this chapter. The synthesis of ATP from ADP and inorganic phosphate in photophosphorylation and oxidative phosphorylation is the major energy-consuming function of plant cells (using light energy in photosynthetic cells, and energy from the oxidation of sugars in nonphotosynthetic cells), and the hydrolysis of ATP is the major driving force for energetically unfavorable reactions in the cell. Phosphorylated intermediates occur in many metabolic pathways, and phosphorylation and dephosphorylation of specific proteins modulates their activity in response to a wide range of signals. This central role of phosphate means that the precise control of phosphate levels in different cellular compartments is essential for the coordinated regulation of cellular processes. Phosphate can be stored in cells in the form of phytate (see below), and phosphate in the vacuole acts as a buffer for phosphate levels in other cellular compartments. Phosphate is exchanged for key phosphorylated intermediates across the plastid envelope (e.g., via triose phosphate and hexose phosphate transporters; Figure 4–93). Phosphate levels modulate the activities of important enzymes and signaling molecules, promoting sensitive coupling between phosphate availability and metabolism (e.g., sucrose phosphate synthase activity; see Section 4.2).

After carbon, oxygen, hydrogen, and nitrogen, phosphorus is the most abundant element in plants, making up approximately 0.2% of plant dry weight. Plants achieve this relatively high level of phosphorus despite the extremely limited availability of phosphate in the soil; most soil phosphate is strongly adsorbed to soil particles and is thus insoluble. The free phosphate concentration is often as low as 1 μM. Most plants

Figure 4–129.
Acquisition of phosphate by plant roots.
In the first route (*lower part of diagram*) phosphate is acquired directly from the soil, in which most phosphate is bound to clay particles and organic matter. A high-affinity phosphate transporter in the plasma membrane of root epidermal cells takes up free phosphate from the soil. Free phosphate diffuses very slowly in the soil, and a zone of 1 to 5 mm beyond the root surface is usually depleted of phosphate. Phosphate availability is increased by secretion of phosphatases and of organic anions (e.g., malate and citrate) and protons—which serve to acidify the zone immediately surrounding the root—by the root epidermal cells. Acidification solubilizes bound phosphate from soil particles. A further means of phosphate acquisition (*top*) by plants growing in natural conditions is a symbiotic association with a fungus (a mycorrhizal association). The fungus invades the root cortex, forming branched arbuscules that invaginate the plasma membrane to create a large membrane surface between the fungal hyphae and the cytosol of the cortical cell. Fungal hyphae, growing out into the soil from the root, take up phosphate and transfer it to the arbuscules. Specific phosphate transporters in the cortical cell plasma membrane transfer the phosphate into the root cell. The fungus can exploit a much larger soil volume than the plant roots alone, enhancing phosphate availability to the plant. The fungus takes up carbon nutrients—probably sugars—from the host plant's cortex.

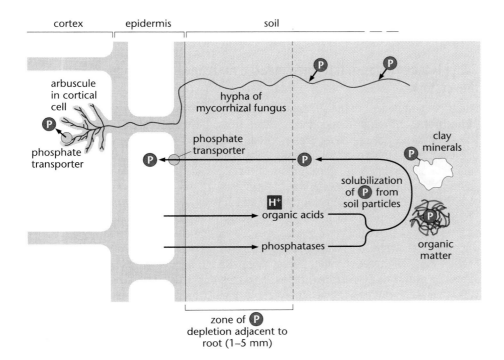

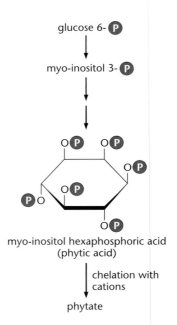

glucose 6-**P**

↓

myo-inositol 3-**P**

↓

↓

myo-inositol hexaphosphoric acid
(phytic acid)

↓ chelation with
cations

phytate

Figure 4–130.
Storage of phosphate as phytate.
Phytate is synthesized from glucose
6-phosphate and consists of large
aggregations of molecules of a six-carbon,
six-phosphate compound chelated with
cations. About 60 to 80% of the phosphate in
the seeds of cereals and soybeans exists as
phytate.

acquire free phosphate in two ways (Figure 4–129). First, the plasma membrane of root cells has high-affinity phosphate transporters that import phosphate in exchange for protons, the uptake driven by the proton gradient generated by H$^+$-ATPases. When free phosphate levels in the soil are low, root cells can secrete organic acids, protons, or phosphatases—all of which increase the solubility of soil phosphates. Second, many plant species form symbiotic associations with mycorrhizal fungi, forming **mycorrhizae** (described in more detail in Section 8.5). This greatly enhances nutrient transfer from the soil to the root, as the fungal **hyphae** take up phosphate from a large volume of soil surrounding the roots, transferring it to fungal structures (**arbuscules**) inside the cortical cells of the root. Here, a specific phosphate transporter in the plant cell plasma membrane (different from the transporters that take up free phosphate directly from the soil) transfers the phosphate from the arbuscule into the cytosol of the cortical cell.

The mechanisms for acquiring phosphate respond to the availability of phosphate in the soil in several ways. First, in most plant species the balance between direct uptake by root cells and reliance on mycorrhizal fungi is affected by the concentration of soluble phosphate in the soil. At low levels of phosphate, roots are heavily infected with mycorrhizal fungi and levels of expression of the plasma membrane transporters that take up phosphate directly from the soil are low. At high levels of phosphate, the mycorrhizal association is suppressed and the expression of the direct transporters is increased. (An exception to this is plants of the Brassicaceae family—including Arabidopsis—which do not form mycorrhizal associations.) Second, the secretion of compounds that increase phosphate solubility is increased at low external phosphate levels. Finally, root growth is affected by free phosphate levels: low levels promote root elongation and proliferation and the proliferation of **root hairs**, thus increasing the volume of soil from which phosphate can be acquired. These effects of phosphate on root development are analogous to those of nitrate (see Section 6.5), but the mechanism of phosphate signaling is not yet known.

Many seeds and some vegetative tissues contain stores of phosphate in the form of insoluble, crystalline **phytates** (Figure 4–130). Phytic acid (myo-inositol hexaphosphoric acid) is thought to be synthesized from glucose 6-phosphate (via myo-inositol 3-phosphate) in the endoplasmic reticulum during seed development The phosphate groups of the phytic acid become associated with a variety of cations, especially Mg^{2+} and K$^+$, and the resulting crystalline phytate is exported in vesicles from the endoplasmic reticulum into protein bodies (see Section 4.8). During seed germination, phosphate is released by stepwise hydrolysis of the phosphate groups of the stored phytate by **phytases**.

Sulfur is taken up as sulfate, then reduced to sulfide and assimilated into cysteine

Sulfur is a component of the amino acids cysteine and methionine and of the tripeptide **glutathione** and other compounds derived from these amino acids. It is also a constituent of a host of metabolites primarily involved in defense (Figure 4–131A shows one example). Sulfur-containing groups play some critical roles in plant metabolism: sulfur-containing lipids are constituents of thylakoid membranes (Figure 4–131B); the sulfhydryl groups of cysteine residues in proteins are oxidized to form disulfide bonds essential in determining the conformation and activity of many enzymes (for examples see Section 4.2); and iron-sulfur clusters are involved in many electron-transfer reactions (Figure 4–131C). Redox reactions involving the sulfhydryl group of glutathione are an essential part of the mechanisms that protect cells from oxidative stress (see Section 7.7). The reactive sulfur group of glutathione (Figure 4–131D) is also the substrate for **glutathione-S-transferases**, enzymes that link glutathione to other molecules that are then transferred to the vacuole as glutathione conjugates. This is a mechanism for transfer of flower pigments and other molecules into the vacuole, and also for the detoxification and vacuolar deposition of toxic substances such as herbicides (Figure 4–131E).

Sulfur occurs in the soil primarily as sulfate (SO_4^{2-}), and this is the form in which it is taken up into plants. Uptake into root cells occurs via sulfate transporters that exchange sulfate for protons (as with phosphate uptake, this is driven by the proton gradient generated by H^+-ATPases in the plasma membrane). Arabidopsis has at least seven types of

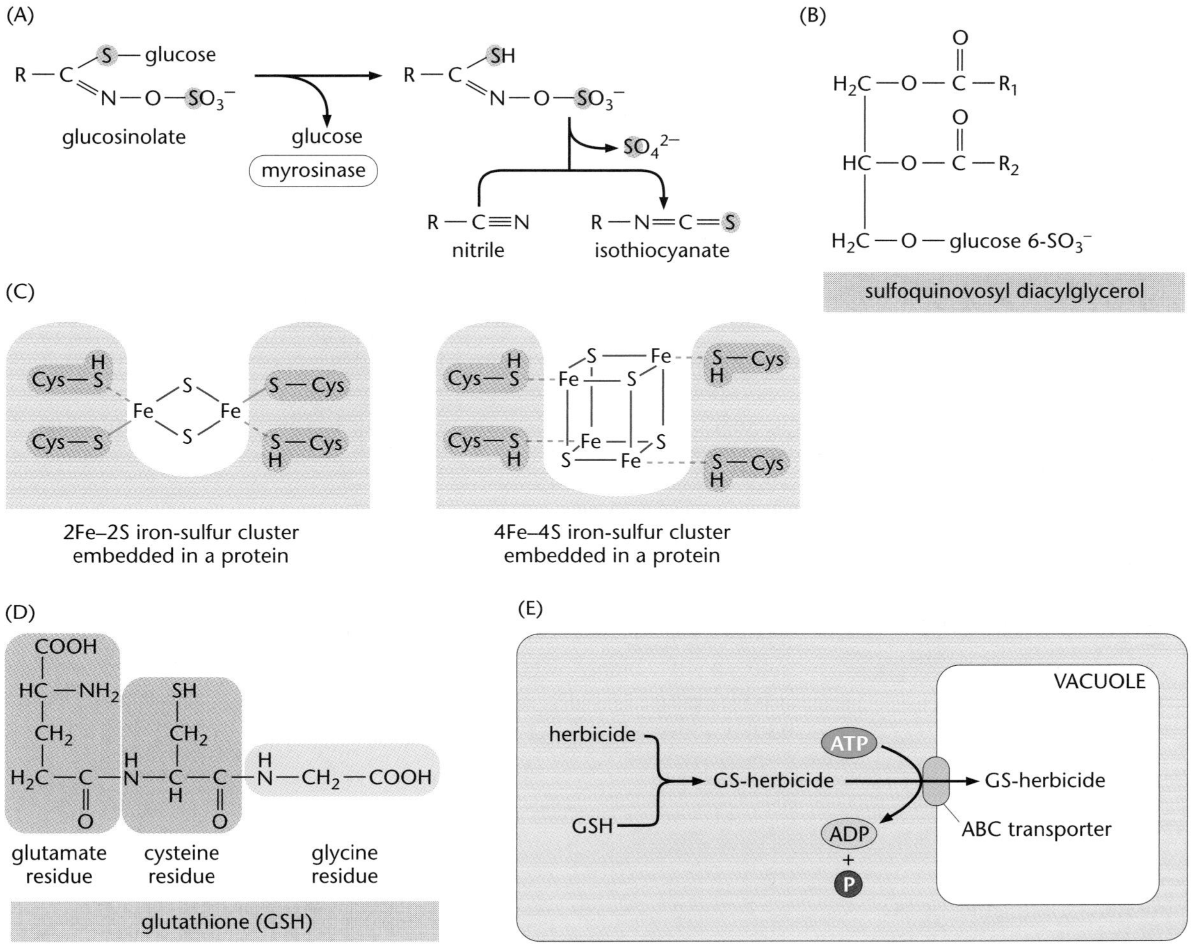

Figure 4–131.

Roles of sulfur-containing compounds. (A) The glucosinolates are important sulfur-containing metabolites found in many Brassicaceae (Cruciferae), including cabbage and its relatives (*Brassica* spp.) and Arabidopsis. Glucosinolates are considered to be primarily defense compounds. When tissue is wounded, stored glucosinolates are brought into contact with the enzyme myrosinase, which catalyzes the formation of compounds that are toxic or unpalatable to animal pests. In the example shown here, myrosinase removes a glucose residue to form an unstable intermediate that decays to highly reactive compounds: nitriles and isothiocyanates (R = short hydrocarbon chain or ring compound). (B) Sulfoquinovosyl diacylglycerol is a glycerolipid component of the thylakoid membrane. (C) Iron-sulfur clusters, coordinated to cysteine residues, occur in specific electron-transfer proteins. Shown here are the 2Fe–2S cluster found in the Rieske protein (*left*), which functions in electron transport on the lumenal side of the cytochrome b_6f complex of the thylakoid membrane; and a 4Fe–4S cluster (*right*) found in nitrite reductase (see Figure 4–114). (D) The tripeptide glutathione has a host of functions in the plant cell. It plays a major role in maintaining the redox balance in both the cytosol and the plastid and in scavenging reactive oxygen species (see Chapter 7). It exists *in vivo* as a mixture of the reduced form (GSH, shown here) and the oxidized form (GSSG) in which two molecules of glutathione are linked by a disulfide bond between their cysteine residues. Glutathione is also involved in storage and long-distance transport of sulfur, in the formation of phytochelatins, which bind potentially toxic heavy metals, and (D) in the conjugation of a variety of compounds for transport to the vacuole, such as flower pigments (anthocyanins), the degradation products of chlorophyll formed during leaf senescence, auxins, and (as shown here) some xenobiotic (foreign, biologically active) compounds such as particular classes of herbicides. Conjugation and removal to the vacuole of auxins and herbicides renders them biologically inactive; thus this mechanism contributes to control of active auxin levels in the cell and to detoxification of herbicides. Glutathione conjugates are transported across the tonoplast by a specific class of transporters (the ABC, or ATP-binding cassette, transporters). The energy for transport is provided by hydrolysis of ATP by the transporter.

sulfate transporters. These proteins have different affinities for sulfate and differ in their expression patterns; their relative roles in sulfate uptake are not yet understood.

The primary route of sulfate assimilation (Figures 4–132 and 4–133) is analogous to that of nitrate: reduction to sulfide (S^{2-}) via sulfite (SO_3^{2-}), followed by transfer onto an amino acid "backbone" to form cysteine. Most of the sulfate taken up by roots is transferred to leaves via the xylem, although some may be assimilated in root cells. Sulfate first enters plastids, where it reacts with ATP to form adenosine 5′-phosphosulfate, a reaction catalyzed by **ATP sulfurylase** (Figure 4–133). The sulfate group is then reduced by transfer of two electrons from glutathione, releasing free sulfite and AMP. Sulfite is reduced to sulfide by transfer of six electrons from reduced ferredoxin by **sulfite reductase**. Finally, sulfide is added to *O*-acetylserine to form cysteine, with the release of acetate. Cysteine is the starting point for most of the biosynthetic pathways involving sulfur-containing compounds. Sulfur assimilated in leaves may travel to other organs in the phloem, in the form of glutathione.

Some biosynthetic pathways proceed not from sulfide but from the first intermediate of the assimilation pathway, adenosine 5′-phosphosulfate (Figure 4–133). The enzyme **adenosine 5′-phosphosulfate kinase** adds a phosphate group to generate adenosine 3′-phosphate 5′-phosphosulfate. This compound is the sulfuryl donor in the synthesis of **glucosinolates**, sulfated lipids (Figure 4–131B), and sulfur-containing extracellular polysaccharides.

Regulation of sulfate uptake and metabolism reflects the plant's requirement for this nutrient both for amino acid and protein synthesis and for the synthesis of glutathione and other compounds essential for protecting the cell against oxidative stress. Expression of genes encoding some sulfate transporters is strongly up-regulated by low availability of sulfate in the soil, and is repressed by accumulation of glutathione or cysteine in the plant. Metabolites of sulfate reduction and assimilation—sulfide, *O*-acetylserine, cysteine, and glutathione—regulate the activities of proteins involved in these processes and in sulfate uptake. For example, *O*-acetylserine represses its own

Figure 4–132.
Overview of sulfur assimilation. Sulfate taken up from the soil is transported to the leaves for assimilation in chloroplasts, and may also be assimilated in nonphotosynthetic plastids in the root. Inside the plastid, sulfate is reduced to sulfite and then sulfide before its assimilation to form the amino acid cysteine. This is the starting point for the synthesis of most sulfur-containing compounds in the cell. Transport of assimilated sulfur from the leaves—the main site of assimilation—to other parts of the plant is usually in the form of glutathione.

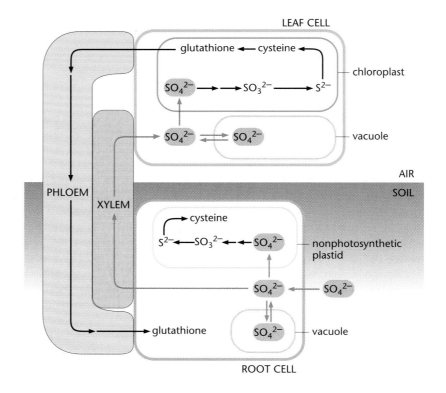

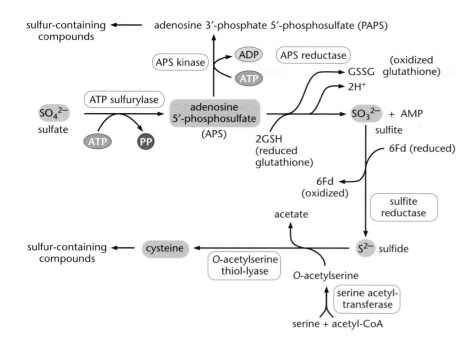

Figure 4–133.
Reactions of the sulfur-assimilation pathway. Sulfate is combined with ATP to give adenosine 5'-phosphosulfate (APS), catalyzed by ATP sulfurylase. APS is reduced in a reaction catalyzed by APS reductase, for which the reductant is reduced glutathione (GSH). This yields protons, oxidized glutathione (GSSG), AMP, and sulfite. Sulfite is reduced to sulfide in a reaction catalyzed by sulfite reductase, an enzyme with a 4Fe–4S cluster analogous to that of nitrite reductase. Six molecules of reduced ferredoxin (Fd) supply the reductant for conversion of one molecule of sulfite to sulfide. Sulfide is brought into organic combination by reaction with O-acetylserine to form cysteine. Cysteine is both a protein amino acid and the precursor or source of sulfur for many other sulfur-containing compounds, including methionine, S-adenosylmethionine, and glutathione. Some sulfur-containing compounds have a different origin (*upper part of diagram*): they are formed from a phosphorylated form of APS, adenosine 3'-phosphate 5'-phosphosulfate (PAPS).

synthesis by acting on serine acetyltransferase and O-acetylserine thiol-lyase, and activates sulfate reduction. Sulfide acts in a reciprocal manner, activating O-acetylserine synthesis and repressing sulfate reduction (Figure 4–134).

The requirement for reduced sulfur for amino acid synthesis is also reflected in the coordination between sulfate and nitrate assimilation. Expression of proteins involved in sulfate transport and reduction is controlled by the availability of nitrogen as well as sulfur. Nitrate itself can induce expression of these genes, and nitrogen levels also affect the levels of O-acetylserine and hence the extent to which this metabolite modulates sulfate reduction and assimilation. Expression of genes encoding enzymes of both nitrate and sulfate reduction is increased with increasing levels of sucrose, linking both of these processes to the overall availability of biosynthetic precursors.

Iron uptake requires specialized mechanisms to increase iron solubility in the soil

Iron, like phosphorus, exists in the soil primarily in forms not directly available to plants. In many soils, much of the iron is in the form of insoluble ferric (Fe^{3+}) hydroxides and complexes of ferric iron with organic compounds. Soluble inorganic iron, mostly in the ferrous (Fe^{2+}) form, is present in tiny concentrations, of the order of 0.01 to 1 nM. The plasma membrane of root cells has transporters for the Fe^{2+} ion, but concentrations of this ion in the soil are usually far lower than is necessary to maintain an adequate supply of iron to the plant. Sufficient iron is made available for uptake by two sorts of mechanisms (Figure 4–135). In grasses, the roots secrete specific compounds with a high affinity for ferric iron, known as **phytosiderophores**. The **chelation** of iron by these compounds provides a soluble form of the element that can be taken up by the roots. Eudicots and non-grass monocots obtain iron from soluble ferric chelates naturally present in the soil.

Relatively few phytosiderophores have been characterized. In oats and barley, the main compounds are the nonprotein amino acids avenic acid and mugenic acid (Figure 4–135A), respectively. They are synthesized from methionine, and the activities of the enzymes involved in their synthesis increase in response to reduced availability

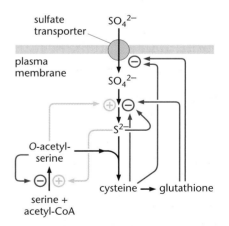

Figure 4–134.
Regulation of sulfate uptake, reduction, and assimilation. Products of sulfate reduction and assimilation—sulfide, cysteine, and glutathione—act as feedback inhibitors of sulfate uptake and reduction. Sulfide availability promotes synthesis of O-acetylserine (required for assimilation), and this compound in turn promotes sulfide synthesis and inhibits its own synthesis.

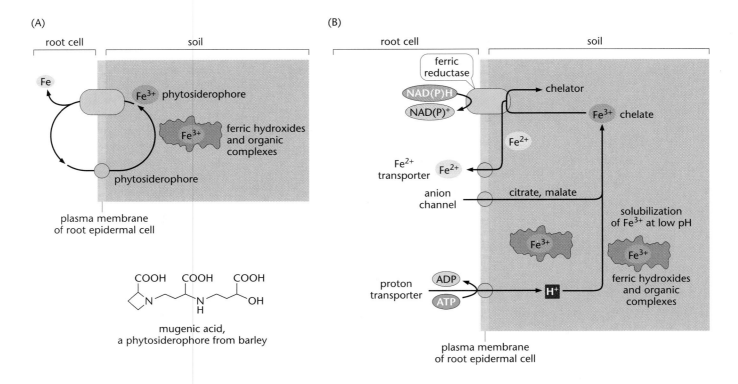

(A) root cell   soil

Fe

Fe³⁺ phytosiderophore

Fe³⁺ ferric hydroxides and organic complexes

phytosiderophore

plasma membrane of root epidermal cell

COOH COOH COOH

N N OH
H

mugenic acid,
a phytosiderophore from barley

(B) root cell   soil

ferric reductase

NAD(P)H
NAD(P)⁺

chelator

Fe³⁺ chelate

Fe²⁺

Fe²⁺ transporter Fe²⁺

anion channel citrate, malate

solubilization of Fe³⁺ at low pH

Fe³⁺

Fe³⁺

proton transporter

ADP
ATP

H⁺

ferric hydroxides and organic complexes

plasma membrane of root epidermal cell

Figure 4–135.
Mechanisms of iron uptake into root cells. (A) In grasses, efficient chelators of ferric (Fe³⁺) iron (phytosiderophores) are synthesized and secreted into the soil. These form complexes (chelates) with Fe³⁺, and the complexes are taken up into root cells by specific transporters. (B) In other groups of plants, Fe³⁺ in the soil is made available to the root in two ways. First, secretion of protons results in acidification of the soil immediately around the root (the rhizosphere) and hence solubilization of some Fe³⁺. Second, secreted organic acids such as citrate and malate act as chelators for solubilized Fe³⁺. A ferric reductase that spans the plasma membrane of the root cell uses NAD(P)H to reduce the chelated Fe³⁺·The resulting Fe²⁺ is taken up into the cell by a specific Fe²⁺ transporter.

of iron in the soil. The phytosiderophores are very efficient chelators of ferric iron. An iron transporter on the plasma membrane of grass root cells recognizes the ferric iron–phytosiderophore complex and transports it intact across the membrane.

In eudicots and in monocots other than grasses, two plasma membrane proteins are involved in iron uptake (Figure 4–135B). A **ferric reductase** reduces Fe³⁺ ions in soluble chelates. Ferric reductase is a member of the **flavocytochrome** family of proteins that transport electrons across membranes: it transfers electrons from NAD(P)H inside the cell to reduce Fe³⁺ ion outside the cell. The reduction releases Fe²⁺ ion from the chelate, and the Fe²⁺ can be taken up by the **ferrous iron transporter**. The abundance and activities of ferric reductase and the ferrous iron transporter are tightly controlled in response to both the local availability of iron in the soil and the iron status of the shoot. Studies in Arabidopsis have shown that expression of both the *FRO2* gene that encodes ferric reductase and the *IRT1* gene that encodes the major ferrous ion transporter is induced when iron is supplied to iron-starved plants. This iron-deficiency response is regulated by signals from the shoot: expression of both proteins is repressed when the shoot has sufficient iron. Expression of *FRO2* and *IRT1* is also influenced by levels of nitrogenous compounds and sugars in the plant, again illustrating the integration of nutrient assimilation with nutrient status and requirements for plant growth.

Most of the iron in the plant is used in the synthesis of heme (see Section 4.7) and in iron-sulfur clusters (Figure 4–131C). Free, unchelated iron is potentially damaging to the cell because its interaction with oxygen can generate **superoxides**. Levels of free iron are minimized by two mechanisms: first, by chelation with amino or organic acids; and second, by complexation with a specific protein. One of the main chelators of free iron in plants is the nonprotein amino acid nicotinamine (which is also an intermediate in the synthetic pathway of the grass phytosiderophores, mugenic and avenic acid). The main iron-protein complexes involve the protein **phytoferritin** (Figure 4–136). The phytoferritin complex consists of 24 protein subunits that form a hollow shell, which contains about 6000 iron atoms in the form of ferric oxide–phosphate complexes.

(A)

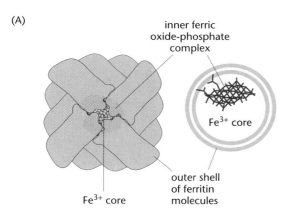

inner ferric
oxide-phosphate
complex

Fe^{3+} core

outer shell
of ferritin
molecules

Fe^{3+} core

(B)

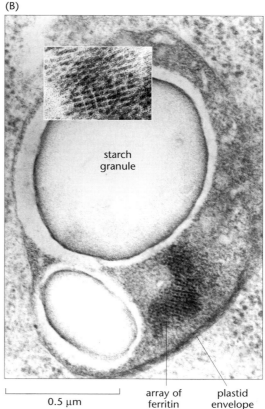

starch
granule

0.5 μm

array of
ferritin
complexes

plastid
envelope

Figure 4–136.
The phytoferritin complex. (A) The complex is a sphere, with an outer shell composed of 24 ferritin molecules and an inner core of a ferric oxide–phosphate complex (with Fe^{3+} ions). The complex is shown schematically from above (*left*) and in cross section (*right*). (B) Electron micrograph of an array of phytoferritin complexes in a plastid in a cell of sycamore (*Acer pseudoplatanus*) grown in tissue culture. The inset shows the complexes at a higher magnification.
(B, courtesy of Keith Roberts.)

Phytoferritin is encoded by small multigene families, with different members responsive to different developmental and environmental signals. In Arabidopsis, expression of three of the four genes increases in response to increased levels of iron. Expression of one of these is also up-regulated in response to treatment with hydrogen peroxide; this treatment imposes oxidative stress, and under these conditions the presence of free, unchelated iron is particularly damaging to cells. A fourth phytoferritin gene encodes a phytoferritin that forms an iron store in seeds. The mechanism and control of mobilization of iron from phytoferritins is not well understood.

4.10 MOVEMENT OF WATER AND MINERALS

There is a constant stream of water in plants, carrying not only water but also dissolved substances, including minerals, throughout the plant body. We describe here the mechanism of this water flow and its close connection with the movement of nutrients in the phloem.

Water moves from the soil to the leaves, where it is lost in transpiration

Water enters the roots, moves through the xylem to the leaves, and is lost to the atmosphere by evaporation (Figure 4–137A). Water in the xylem vessels forms a continuous column that links water in the soil, at the absorbing surfaces of the root, to the evaporating surface of the leaf and thus the atmosphere. This flow of water through the plant is known as the **transpiration stream**. For a C3 plant in a temperate climate, water evaporation from the leaves is of the order of 700 to 1300 moles per mole of carbon dioxide assimilated in photosynthesis.

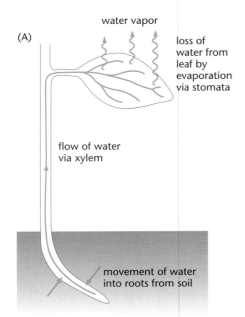

(A)

water vapor

loss of water from leaf by evaporation via stomata

flow of water via xylem

movement of water into roots from soil

(B)

water drawn from xylem into mesophyll cells down gradient created by evaporation

loading of sucrose into phloem causes movement of water from xylem to phloem

evaporation of water from mesophyll cell walls

movement of water vapor out of leaf down gradient of water vapor concentration

stomatal pore

atmosphere (low water vapor concentration)

upper epidermis of leaf

xylem elements

phloem complex

sucrose

substomatal cavity (high water vapor concentration)

lower epidermis

Figure 4–137.
Water movement from the soil to the atmosphere. (A) The transpiration stream. Loss of water from the leaf by evaporation through the stomata—the process of transpiration—draws water from the soil into the roots and up to the leaf via the xylem. (B) Water movement in the leaf. Two main forces draw water out of the xylem elements. First, loss of water vapor from the leaf down the water vapor gradient between the intercellular spaces and the atmosphere causes evaporation of water from the mesophyll cell walls. This lowers the water potential in the mesophyll cells, drawing water out of the xylem and into these cells. Although the major route of water loss is through open stomata, the waxy cuticle that covers the outer surface of the leaf is somewhat permeable to water in most species. Thus, even when the stomata are shut, some loss of water vapor from epidermal cells can still occur directly across the cuticle. Second, active loading of sucrose into the phloem creates a very high osmotic potential in the sieve elements that draws water out of the xylem and into the phloem. The resulting high pressure at the top of the phloem drives movement of sucrose to sink organs via pressure flow (see Section 4.4).

The flow of water is in part an inevitable consequence of the photosynthetic process. With the opening of stomata over the large, light-capturing surfaces of leaves to allow entry of carbon dioxide, much water is lost to the atmosphere, down the gradient of water vapor concentration between the internal spaces of the leaf (where water vapor concentration is at equilibrium with water inside cells) and the atmosphere (where water vapor concentration is usually much lower) (Figure 4–137B). But the transpiration stream not only replaces water lost through the stomata, it is also essential for many physiological processes: maintaining cell **turgor** (see Chapter 3, especially Section 3.5); maintaining cytoplasmic solute concentrations compatible with metabolic activity; transporting nutrients absorbed by the roots, and metabolites and phytohormones synthesized in the roots, to the shoot; cooling leaves by evaporation; and supplying water to the top of the phloem to enable movement of sucrose to nonphotosynthetic organs (see Section 4.4).

The mechanisms that control stomatal opening reflect the complex and potentially conflicting demands for carbon dioxide, for a flow of water through the plant to maintain physiological processes, and for conservation of water, which is frequently in limiting supply in the soil (see Section 4.3 and Chapter 7). Stomatal aperture is regulated in response to a host of environmental factors, including light intensity and quality, CO_2 concentration in the leaf, and water vapor concentration, and is also regulated by the circadian clock (see Chapters 3, 6, and 7).

Water moves from roots to leaves by a hydraulic mechanism

The main driving force for movement of water through the plant is the evaporative loss of water from the leaves. Evaporation lowers the water potential in the leaf cells and causes water to move out of the xylem vessels of the leaf; this reduces pressure in the xylem and thus "pulls" water up through the entire length of the xylem from the root. The distance can be as much as 100 m in the tallest trees (Figure 4–138). This remarkable phenomenon of long-distance upward movement of water in a continuous vertical column, caused by a negative pressure at the top, is attributable to the cohesive and adhesive properties of water. Given the strong mutual attraction between water molecules (cohesion) and their attraction to solid surfaces (adhesion), a narrow column of water will not rupture even under considerable negative pressure (i.e., water has high tensile strength). This hydraulic explanation for water movement in the plant is called the **cohesion-tension theory**. Although alternative explanations have been put forward, the cohesion-tension theory of water movement is currently the most widely accepted.

Water is taken up from the soil mainly in the apical regions of the root, where the surface area is greatly increased by the presence of root hairs (see, for example, Figure 5–29). The route by which water from the soil reaches the xylem vessels of the root is not fully understood. It is likely that three routes operate (Figure 4–139A) and that their relative importance varies diurnally and with developmental stage and environmental conditions. First, water may move from the soil to the xylem entirely in the apoplast—that is, within cell walls. Resistance to water movement in this apoplastic pathway is high, because the walls of the **endodermis** contain a band of suberin, a complex hydrophobic material consisting of lignin-like polymers and complex, long-chain fatty acids (see Section 3.6), linked to the cell wall. The radial band of suberin in the endodermis is called the **Casparian strip**. Movement of water in the apoplastic pathway is driven by the hydrostatic pressure gradient from the soil to the xylem, generated by the upward "pull" on water in the xylem vessels, as discussed above.

In a second route, water may be taken up into the symplast in root-hair, epidermal, and cortical cells, and then pass from cell to cell via plasmodesmata into the **stele**, where it moves out of cells into xylem vessels. The third route involves both the symplast and the apoplast: water passes from cell to cell via pores in the plasma membrane. Water can diffuse through lipid bilayers, but its diffusion through the plasma membrane is greatly enhanced by pores formed by a family of proteins called **aquaporins** (see Section 3.5). The capacity for water movement across the plasma membrane can be varied by changes both in the number of aquaporins present and in their channel activity. Expression of different members of the aquaporin gene family is altered in response to various environmental stimuli, and the channel activity of the aquaporin proteins is regulated by phosphorylation/dephosphorylation in response to water stress. Water movement in the two pathways that involve the symplast is influenced by osmotic as well as hydrostatic pressure gradients. Osmotic pressure gradients can exist across the root because mineral nutrients that enter the symplast of root cells from the soil are actively secreted (via plasma membrane transporters) from cells of the stele into xylem vessels. This can result in higher concentrations of solutes, and thus a higher osmotic pressure, in the xylem than in the soil and root cells.

The existence of different pathways for water movement across the root allows **root conductivity** (the permeability of the root to water, and hence the ease with which water can flow from the soil to the xylem) to change in response to the shoot's demands for water (Figure 4–139B). When the transpiration rate is high, water in the xylem is under tension and there is a considerable negative hydrostatic pressure (a strong "pull") in the root xylem; water is drawn in from the soil mainly via the apoplastic pathway, and root conductivity is relatively high. When transpiration is not occurring—for example, when the stomata are shut at night or in response to a water deficit—water in the xylem is under minimal tension (a weak "pull") and the hydrostatic pressure gradient across the root is minimal. The relatively high resistance of the apoplastic pathway means that water flow by this route is very limited. Most of the water flow in these low-transpiration conditions occurs via the cell-to-cell (symplastic) pathways, and root conductivity is low. The decline in root conductivity in response to a low transpiration rate is important in the conservation of water under drought conditions. This is described more fully in Section 7.3.

When transpiration is minimal and water is freely available in the soil, a positive pressure may build up in root xylem vessels. This occurs because water in the xylem is under minimal tension, as described above, so mineral nutrients secreted into xylem vessels are not carried away rapidly by the transpiration stream. Locally high solute concentrations can develop, drawing water into the xylem from surrounding cells. This phenomenon, known as **root pressure**, occurs in well-watered plants at night and

Figure 4–138.
A *Sequoiadendron giganteum* growing in California. Many of the oldest specimens of this tree (thought to be 2500–3000 years old) are more than 80 m tall. The world's tallest trees are specimens of the closely related *Sequoia sempervirens*, found on the Californian coast. Fifteen individuals of this species are more than 110 m high.
(Courtesy of Mike Murphy.)

Figure 4–139.
Movement of water from soil to root xylem. (A) Water can move by three routes: exclusively via the apoplast (within cell walls), exclusively via the symplast (through plasmodesmata), or via symplast and apoplast (crossing plasma membranes and cell walls). Movement by way of the apoplastic route is driven by negative hydrostatic pressure in the xylem. This route has a high resistance to water movement due to the relatively water-impermeable layer (the Casparian strip) in the walls of endodermal cells. Movement by way of the symplastic routes is driven by osmotic potential differences between the xylem vessels, the surrounding cells, and the soil. These differences are brought about by active secretion of mineral nutrients from the symplast of stele cells into the xylem vessels. (B) Different conditions favor the apoplastic and symplastic routes. The apoplastic route is favored when the transpiration rate is high and there is a strong negative hydrostatic pressure (a strong "pull") in the xylem vessels. Under these conditions, the apoplastic pathway allows a high rate of water movement from the soil to the xylem. The routes via the symplast are favored when the transpiration rate is low, with little hydrostatic pressure gradient from the soil to the xylem. Under these conditions, secretion of mineral nutrients from the symplast into the xylem results in a greater osmotic potential in the xylem than in surrounding cells and the soil, drawing water from the symplast into the xylem.

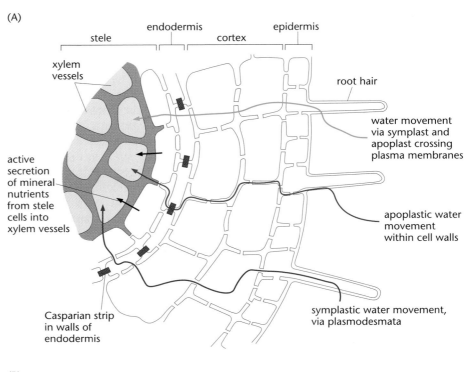

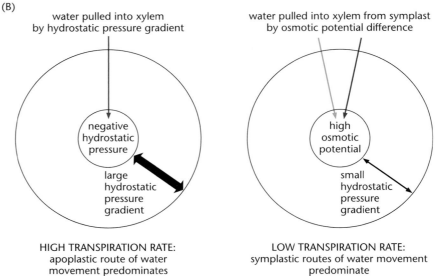

Figure 4–140.
Guttation at the margins of a leaf of lady's mantle (*Alchemilla*).

under conditions of very high air humidity (that is, when there is no gradient of water vapor concentration between the leaf air spaces and the atmosphere). The positive pressure in the root xylem vessels results in mass flow in the xylem (analogous to mass flow in the phloem; see Section 4.4), which may lead to exudation of xylem fluid from specialized structures on the margins of leaves known as **hydathodes**. The dewdrops seen on the tips of leaves as night ends are formed in this way: a process called **guttation** (Figure 4–140).

The movement of mineral nutrients in the plant involves both xylem and phloem

Most nutrient uptake takes place in the actively growing and expanding apex of the root and the adjacent root-hair zone. The forward growth of the apical region allows cells to exploit new sources of soil nutrients, as nutrients become depleted in the **rhizosphere** (the zone of influence of a root in the soil) adjacent to older parts of the root. Root hairs provide a very large surface area for nutrient uptake.

Nutrients are taken up into epidermal and cortical cells via specific transporters in the plasma membrane (see Sections 3.5 and 4.9). The Casparian strip in the walls of the endodermis is highly impermeable to most nutrients, so movement into cells of the stele adjacent to the xylem is symplastic. For most nutrients, export from the stele symplast to the xylem is brought about by transporters found specifically in the plasma membrane of cells adjacent to the xylem. For example, the potassium channels in the plasma membranes of these stele cells are different from those found in epidermal cells: different members of the potassium transporter gene family are expressed in the two cell types. The dominant potassium channel of stele cells favors potassium efflux (i.e., from the stele cell to the xylem), whereas that of epidermal and cortical cells favors influx (i.e., from the apoplast into the symplast) (Figure 4–141).

Nutrients entering the root xylem are carried to leaves in the transpiration stream. In the leaves they are taken up via transporters into the symplast of cells surrounding the xylem vessels. Some of the nutrients pass symplastically down concentration gradients to other leaf cells, to meet demands for growth and metabolic processes. The remainder are transferred to phloem elements, which lie in close proximity to xylem vessels in the minor vein endings of leaves. Nutrients entering the phloem are carried out of the leaf together with sucrose and travel to sink organs (Figure 4–142). The phloem, rather than the xylem, is the major nutrient supply route for many sink organs. Sinks such as vegetative apical meristems and developing fruits are likely to have a high demand for mineral nutrients. Because they have few stomata, the rate of evaporation from these organs is very low compared with that of leaves, hence inflow of water and nutrients via the xylem is limited; the phloem provides most or all of the mineral nutrients required.

SUMMARY

Plants obtain the major elements that make up the plant body—carbon, oxygen, hydrogen, and nitrogen—mainly as carbon dioxide, water, and nitrate. The organic compounds made from these inorganic materials are essential not only for the plants themselves but for almost all other forms of life that use plants directly or indirectly for food. Plant metabolism generates a vast diversity of organic compounds.

Metabolic pathways are regulated at two general levels. One is compartmentation—into cytosol, plastids, mitochondria, vacuoles, endomembrane system, and other organelles, and between soluble and membrane phases in these compartments—which

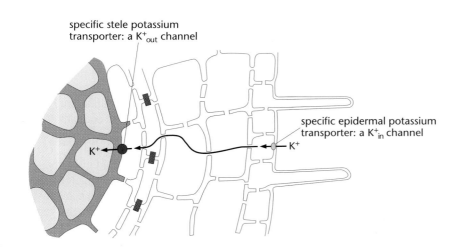

specific stele potassium transporter: a K^+_{out} channel

specific epidermal potassium transporter: a K^+_{in} channel

K^+

K^+

**Figure 4–141.
Two transport steps in the movement of potassium from the soil to the xylem.**
Uptake into the epidermal cells occurs via a transporter that forms an "inward" potassium channel (K^+_{in}), favoring movement of potassium into the cell. Movement from the point of uptake to the stele cells adjacent to the xylem is symplastic. Secretion into the xylem occurs via a transporter that forms an "outward" potassium channel (K^+_{out}), favoring movement of potassium out of the cell into the xylem vessel.

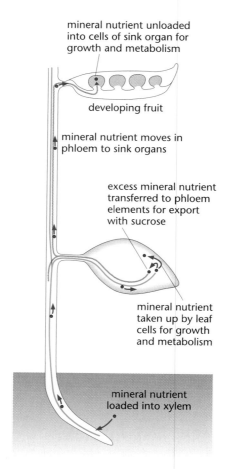

mineral nutrient unloaded into cells of sink organ for growth and metabolism

developing fruit

mineral nutrient moves in phloem to sink organs

excess mineral nutrient transferred to phloem elements for export with sucrose

mineral nutrient taken up by leaf cells for growth and metabolism

mineral nutrient loaded into xylem

Figure 4–142.
Transport of mineral nutrients to sink organs. Many sink organs have few or no stomata, so the capacity of the transpiration stream to deliver mineral nutrients for growth is severely limited. Instead, mineral nutrients reach these organs via the phloem. Nutrients arriving in the leaf from the root, via the xylem, are transferred to adjacent phloem elements; they are then exported to sink organs, together with sucrose synthesized in the leaf cells.

increases the potential for metabolic diversity. The other is the coordinated control of enzyme activities. "Coarse control" regulates the number of molecules of an enzyme in the cell; "fine control" determines the activity of these molecules. Fine control modulates enzyme activity by covalent modification (e.g., phosphorylation/dephosphorylation) of the enzyme protein or by noncovalent interactions between the enzyme and specific metabolites. Changes in enzyme activity determine the flux through a pathway and coordinate the activities of different pathways.

In photosynthesis, carbon dioxide is assimilated via the Calvin cycle in the chloroplast stroma, using energy (ATP) and reducing power (NADPH) supplied by light-driven reactions in the thylakoids. The light-harvesting processes involve absorption of light by chlorophylls followed by loss of excited electrons to electron carriers. Transfer of electrons along electron transport chains reduces $NADP^+$ and creates an electrochemical gradient across the thylakoid membrane that drives the phosphorylation of ADP by ATP synthases. The initial step in CO_2 assimilation, catalyzed by Rubisco, is carboxylation of an acceptor molecule (ribulose 1,5-bisphosphate). The cycle is "autocatalytic," with removal of an intermediate for net synthesis of product (sucrose) and regeneration of the CO_2 acceptor. Carbon assimilation and energy supply are coordinated by complex regulation of Calvin cycle enzymes. Sucrose synthesis is tightly controlled by the rate of photosynthesis and the demand for carbon by nonphotosynthetic parts of the plant.

Rubisco can also catalyze an oxygenation reaction that results in release of carbon dioxide (photorespiration) and has profound consequences for the plant's carbon and nitrogen economy. C4 plants have anatomical and biochemical adaptations that eliminate photorespiration.

Sucrose is transported from the leaves (source organs) to nonphotosynthetic parts of the plant (sink organs) in the phloem. Phloem loading and unloading may be apoplastic or symplastic. In the nonphotosynthetic cells, imported sucrose is used to provide biosynthetic precursors, ATP, and reducing power. Sucrose is metabolized to hexose phosphates, which are further broken down by glycolysis, the oxidative pentose phosphate pathway, and the Krebs cycle, ultimately providing NAD(P)H, ATP, and precursors. Most of the ATP is synthesized in the mitochondria, where the Krebs cycle generates reducing power that drives respiratory electron transport chains in the inner membrane. The resulting electrochemical gradient across the membrane drives the ATP synthesis. Nonphotosynthetic cells also convert imported sucrose to storage compounds, including starch and lipids.

The metabolic processes in plastids vary with cell type and developmental stage. Synthesis of fatty acids, starch, and chlorophyll occurs exclusively in plastids. Synthesis of membrane lipids occurs both in plastids (in a "prokaryotic" pathway) and elsewhere in the cell (in a "eukaryotic" pathway). Terpenoid synthesis occurs by different pathways in the plastid and the cytosol, forming different products; for example, tetrapyrroles, the precursors of chlorophyll, are synthesized in plastids.

Nitrogen is the most abundant element in plants after carbon, hydrogen, and oxygen. For most plants, the main source of nitrogen is soil nitrate, taken up by the roots via nitrate transporters, reduced to ammonium (via nitrate reductase in the cytosol and nitrite reductase in plastids), and assimilated into glutamate and glutamine. These amino acids are the sources of nitrogen for almost all other nitrogen-containing compounds. Nitrogen is stored as amino acids and specific storage proteins.

In addition to carbon, hydrogen, oxygen, and nitrogen, plants require 13 other elements for growth: macronutrients (e.g., phosphorus and sulfur) and micronutrients (e.g., iron), obtained mainly by uptake from the soil. The capacity for uptake and metabolism of a nutrient is controlled by its local availability in the soil and by the general nutrient status of the plant.

The transpiration stream in the xylem carries water and dissolved substances, including minerals, throughout the plant body by a hydraulic mechanism driven by evaporation of water from cell surfaces in the leaf and water loss via the stomata. This water flow is closely connected with movement of sucrose and nutrients in the phloem.

FURTHER READING

General

Heldt HW (2005) Plant Biochemistry, 3rd ed. Amsterdam: Elsevier Academic Press.

Plaxton WC & McManus MT (eds) (2006) Control of Primary Metabolism in Plants. Annual Plant Reviews, vol. 22. Oxford: Blackwell Publishing.

4.1 Control of Metabolic Pathways

ap Rees T & Hill SA (1994) Metabolic control analysis of plant metabolism. *Plant Cell Environ.* 17, 587–599.

Lunn JE (2007) Compartmentation in plant metabolism. *J. Exp. Bot.* 58, 35–47.

4.2 Carbon Assimilation: Photosynthesis

Long SP, Ainsworth EA, Rogers A & Ort DR (2004) Rising atmospheric carbon dioxide: plants FACE the future. *Annu. Rev. Plant Biol.* 55, 591–628.

Nelson N & Yocum CF (2006) Structure and function of Photosystems I and II. *Annu. Rev. Plant Biol.* 57, 521–565.

Raines CA (2004) The Calvin cycle revisited. *Photosynth. Res.* 75, 1–10.

4.3 Photorespiration

Reumann S & Weber APM (2006) Plant peroxisomes respire in the light: some gaps of the photorespiratory C_2 cycle have become filled—others remain. *Biochim. Biophys. Acta Mol. Cell Res.* 1763, 1496–1510.

Sage RF (2004) The evolution of C4 photosynthesis. *New Phytol.* 161, 341–370.

4.4 Sucrose Transport

Holbrook NM & Zwieniecki MA (eds) (2005) Vascular Transport in Plants. Amsterdam: Elsevier Academic Press.

Kehr J (2006) Phloem sap proteins: their identities and potential roles in the interaction between plants and phloem feeding insects. *J. Exp. Bot.* 57, 767–774.

4.5 Nonphotosynthetic Generation of Energy and Precursors

Fernie AR, Carrari F & Sweetlove SJ (2004) Respiratory metabolism: glycolysis, the TCA cycle and mitochondrial electron transport. *Curr. Opin. Plant Biol.* 7, 254–261.

Kruger NJ & von Schaewen A (2003) The oxidative pentose phosphate pathway: structure and function. *Curr. Opin. Plant Biol.* 6, 236–246.

4.6 Carbon Storage

Goepfert S & Poirier Y (2007) β-Oxidation in fatty acid degradation and beyond. *Curr. Opin. Plant Biol.* 10, 245–251.

Napier JA (2007) The production of unusual fatty acids in transgenic plants. *Annu. Rev. Plant Biol.* 58, 295–319.

Smith AM, Zeeman SC & Smith SM (2005) Starch degradation. *Annu. Rev. Plant Biol.* 56, 73–98.

4.7 Plastid Metabolism

D'Auria JC & Gershenzon J (2005) The secondary metabolism of *Arabidopsis thaliana*: growing like a weed. *Curr. Opin. Plant Biol.* 8, 308–316.

Dormann P & Benning C (2002) Galactolipids rule in seed plants. *Trends Plant Sci.* 7, 112–118.

Tanaka T & Tanaka A (2007) Tetrapyrrole biosynthesis in higher plants. *Annu. Rev. Plant Biol.* 58, 321–346.

Weber APM, Schneidereit J & Voll LM (2004) Using mutants to probe the *in vivo* function of plastid envelope membrane metabolite transporters. *J. Exp. Bot.* 55, 1231–1244.

4.8 Nitrogen Assimilation

Halkier BA & Gershenzon J (2006) Biology and biochemistry of glucosinolates. *Annu. Rev. Plant Biol.* 57, 303–333.

Lillo C, Meyer C, Lea US et al. (2004) Mechanism and importance of post-translational regulation of nitrate reductase. *J. Exp. Bot.* 55, 1275–1282.

Stitt M, Muller C, Matt P et al. (2002) Steps towards an integrated view of nitrogen metabolism. *J. Exp. Bot.* 53, 959–970.

4.9 Phosphorus, Sulfur, and Iron Assimilation

Briat JF, Curie C & Gaymard F (2007) Iron utilization and metabolism in plants. *Curr. Opin. Plant Biol.* 10, 276–282.

Hesse H, Nikiforova V, Gakière B & Hoefgen R (2004) Molecular analysis and control of cysteine biosynthesis: integration of nitrogen and sulphur metabolism. *J. Exp. Bot.* 55, 1283–1292.

Karandashov V & Bucher M (2005) Symbiotic phosphate transport in arbuscular mycorrhizas. *Trends Plant Sci.* 10, 22–29.

4.10 Movement of Water and Minerals

Holbrook NM & Zwieniecki MA (eds) (2005) Vascular Transport in Plants. Amsterdam: Elsevier Academic Press.

DEVELOPMENT

5

In Chapter 3 we considered the structure, development, and life cycle of plant cells: from formation of new cells by cell division, through expansion and growth, to adoption of specialized functions in the plant body. Here we turn to the development of the plant body as a whole: from the zygote to the mature, reproductive plant.

When multicellular organisms develop, single cells do not simply multiply—they adopt different roles (or fates) within the growing organism. This is a division of labor: the variety of cell types found in any multicellular organism reflects the specialized roles taken on by individual cells. Cell growth and division are coordinated to build organs (and in turn the whole organism) with a specific shape and function, and specialized cells and tissues appear in their characteristic locations and arrangements. Cells communicate and interact with each other to coordinate cell fate; in addition, the organism as a whole detects and responds to its surroundings. Reflecting the independent origins of multicellularity in plants and in animals, these groups use different molecules and structures to coordinate the development of specialized cells. There are, however, some striking similarities between animal and plant development, and this will be a recurring theme throughout this chapter. Nevertheless, many features of plant development are unique to plants, and we begin the chapter by highlighting the key concepts that under-pin the development of plant form. We start our discussion with an overview of the unique features of plant development. We then follow the progress of plant development from the embryo to the mature, flowering plant and reproduction—completing the cycle from seed to next-generation seed.

5.1 OVERVIEW OF PLANT DEVELOPMENT

One characteristic of plant development is that new organs are formed at the growing apices throughout the life of the plant: overall plant form is constantly changing. Not all the parts that make up a plant are visible in a mature embryo; roots, branches, leaves, and flowers are progressively added throughout the plant's life and in response to its surroundings. At the end of plant **embryogenesis** there are two distinct groups of cells (the meristems) from which the entire plant body is subsequently derived. The **shoot meristem** gives rise to the shoot and the **root meristem** goes on to form the root (Figure 5–1). In a mature mammalian embryo, in contrast, all body parts are already present in a recognizable form.

The reiterative production of new organs by the meristems results in the growth of shoots and roots that are made up of repeated structural units, or modules (Figure 5–2). For example, a shoot module comprises a **leaf**, a **lateral meristem**, and an **internode**. The way these modules develop is determined by a genetic program that is character-istic of each plant, but the environment has a strong influence on plant growth and on the time when meristems shift to producing a new type of module (for example, when instead of new leaves the plant starts producing flower buds, as described in detail later in the chapter).

Although plants cannot move to a different location in search of better conditions, the combination of continuous development and sensitivity to external factors allows them to adjust to changes in their environment, such as changing light and nutrient availabil-ity (Figure 5–3). For example, plants growing in soil with low levels of nitrogen have very few lateral roots, but if nitrogen content increases, many lateral roots are initiated near the source. Likewise, when gaps appear in forest canopies, lateral branches are ini-tiated in the shoot and grow to fill the available space. In both of these examples, growth remains modular; that is, new modules are made rather than preexisting mod-ules stretching to fill the gap.

Figure 5–1.
Embryo and mature plant of *Capsella bursa-pastoris*. (A) Longitudinal section through a mature embryo, with the shoot apical meristem (SAM) and root apical meristem (RAM) indicated. (B) The shoot of an adult plant (all derived from the shoot apical meristem). (A, courtesy of Judy Jernstedt.)

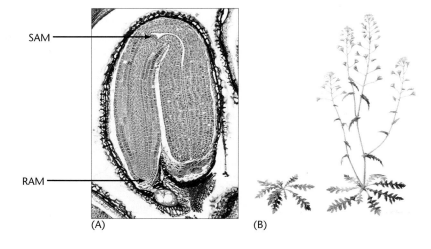

SAM

RAM

(A) (B)

Growth by continuous addition of repetitive units has consequences for plant reproduction. Isolated parts of plants are often able to survive and continue growing to reconstitute a new and clonally identical plant. For example, tubers, the shoot storage organs of potato, remain viable in the ground after other parts of the plant have died. In the next growing season, the tuber develops into a plant that is genetically identical to the plant on which it formed. The regenerative abilities of plants are revealed in an extreme form by the ability of differentiated cells to dedifferentiate and reconstitute entire plants, as seen in the regeneration of whole plants in **tissue culture** from a small number of leaf cells.

The ability to produce new plants from detached parts is possible in part because cell fate in plants can be more easily redirected than in animals. This flexibility includes the cells that eventually produce gametes. Early in the development of higher animals, a small number of cells destined to be the **germ line** are separated from the other (**somatic**) cells that will comprise the animal body. This does not happen in plants. Just as all the body parts are not preformed in a mature plant embryo, no defined group of cells is designated to become the germ line.

In animals, germ-line cells eventually undergo **meiosis** to form **gametes**. Another major difference in the way plants develop relates to the fate of the **haploid** cells produced by meiosis. In plants, meiosis takes place in **sporophytes**, and the haploid products of meiosis—the **spores**—develop into distinct gamete-producing organisms called **gametophytes**. The plant life cycle, therefore, alternates between a **diploid** sporophyte generation and a haploid gametophyte generation (Figure 5–4; see also Figure 5–86). In **angiosperms**, **gymnosperms**, and ferns, the sporophytes are the bodies of the plants we see around us, whereas the gametophytes are much smaller and, in the case of seed plants, grow within the reproductive structure of the sporophyte. In mosses, the gametophyte is the more visible generation and the sporophyte is dependent on the gametophyte.

The temporary existence of the plant body as a haploid organism has important genetic consequences. The haploid cells of plants have to perform many more functions than do animal gametes, such as functions necessary for cell division and specific types of growth. Some of the **genes** required for these functions are also used in the sporophyte. Because these genes are active as single copies in the gametophyte, **recessive mutations** are exposed in the haploid generation. This can act as a filter to prevent the accumulation of recessive, deleterious mutations, but it also reduces the opportunity

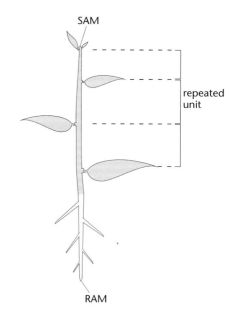

Figure 5–2.
Diagram showing repeated units that originate from the shoot apical meristem (SAM) and root apical meristem (RAM).

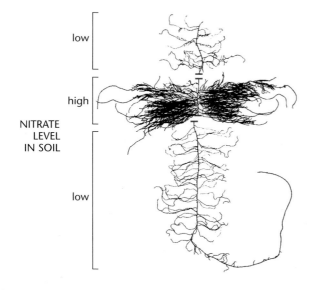

Figure 5–3.
Response of root development to changes in nutrient supply. Barley roots growing through a soil region with a high concentration of nitrate develop many more lateral roots. (Courtesy of Malcolm Drew.)

Figure 5–4.
Alternation of generations in the fern
Ceratopteris richardii. The diploid
sporophyte produces spores by meiosis.
A spore develops into a haploid (n) organism,
the gametophyte, which produces gametes by
mitosis. Fusion of gametes (fertilization)
produces the diploid zygote, which develops
into a new diploid (2n) sporophyte.

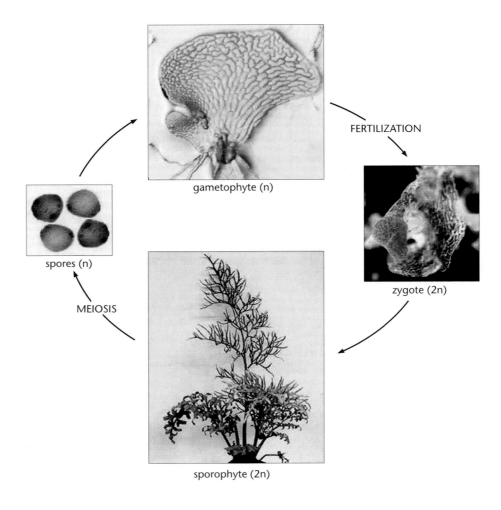

gametophyte (n)

FERTILIZATION

spores (n)

zygote (2n)

MEIOSIS

sporophyte (2n)

to accumulate genetic variability at these loci. It is possible that the reduction of haploid individuals during the evolution of higher plants minimized the genetically vulnerable generation.

Multicellularity evolved independently in plants and animals

As mentioned above, multicellular organisms are not simply collections of identical cells; they possess many cell types. These different cell types have acquired different functions; for example, **xylem** cells transport water and **guard cells** regulate CO_2 uptake. Once cells are no longer autonomous, communication networks are essential to regulate their development. The last common ancestor of plants and animals was unicellular; but even though multicellularity arose independently in plants and animals, similar mechanisms for intercellular communication have evolved in both kingdoms. For example, plant cells are connected by **plasmodesmata**, cytoplasmic channels allowing the movement of small molecules from cell to cell; and animal cells have **gap junctions**, which are similar in function to plasmodesmata though different in structure.

We cannot reconstruct the processes that drove the evolution of multicellularity, but there are a few examples from which we can infer how more complex life forms originated from unicellular ancestors. One such example is *Volvox*, a green alga whose progression can be followed from a single-celled organism to a colony of cells and ultimately to a multicellular organization in which the cells are so dependent on one another that they cannot live in isolation.

Volvox is a simple system in which to study the genetic basis of multicellularity

Volvox carteri develops from a single free-floating cell into a multicellular hollow sphere consisting of cells embedded in a gelatinous mass (Figure 5–5). The free-living *Volvox* cell divides repeatedly to form a colony of identical cells, and this becomes a multicellular organism as cells differentiate into two distinct cell types. Midway through *Volvox* development, asymmetrical cell divisions produce small cells that become sterile (somatic) cells and large cells that will form the reproductive cells, the **gonidia**. The gonidia accumulate at one end of the sphere, where they reproduce asexually. The flagellated somatic cells form a single layer and are connected by cytoplasmic bridges. The flagella can beat in synchrony, propelling the sphere of cells toward a source of light. Once differentiated, somatic cells can no longer divide. In order for *Volvox* to reproduce, therefore, both types of cells are necessary: the flagellar somatic cells for movement and the gonidia for reproduction. These two types of cells are dependent on each other and cannot live in isolation.

Cell size is the main determinant of cell fate in *Volvox*: if the number of large cells in a developing *Volvox* sphere is altered by genetic mutation, the number of gonidia formed also changes. Usually, a mature *Volvox* sphere is made up of about 3000 small somatic cells and 16 large gonidia. In *gonidialess* mutants, no asymmetrical cell divisions take place, so the spheres do not form gonidia and cannot reproduce. The wild-type GONIDIALESS protein is associated with the mitotic spindle in asymmetrically dividing cells and may be required for the asymmetrical positioning of the spindle. The transcriptional repressor SOMATIC REGENERATOR acts in the small somatic cells to repress the transcription of genes required for reproductive development, and a further protein, LATE GONIDIA, has been shown to repress somatic development in the large reproductive cells.

Analysis of *Volvox* development from a single cell to a multicellular organism has provided some clues about the genetic basis of multicellularity in this alga—the mechanisms by which cells divide asymmetrically to allow cell specialization and by which cell identity is maintained are beginning to unfold. How does this information help us investigate how *Volvox* might have evolved from a unicellular ancestor?

Before *Volvox* cells specialize, they are similar in appearance to the unicellular green alga *Chlamydomonas* (see Figure 5–6A). Early in the *Chlamydomonas* life cycle, cells have flagella and are motile. A transcriptional repressor similar to SOMATIC REGENERATOR is believed to repress reproductive development at this early stage. Later, the cells lose their flagella and group together. Cell division then produces new, free-floating *Chlamydomonas* cells and, at this point, a gene such as *LATE GONIDIA* is believed to stop the cells from developing as somatic cells.

Although similar regulatory genes are involved in the development of *Volvox* and *Chlamydomonas*, the major difference in gene regulation between the multicellular and unicellular systems is that in *Volvox* these genes are expressed simultaneously in different specialized cells, whereas in *Chlamydomonas* they are expressed one after the other in the same cell. Thus, one possibility is that the progression of a multicellular *Volvox* from a unicellular *Chlamydomonas*-like ancestor may have arisen when the program of regulatory gene expression altered from sequential changes in a single cell to parallel changes in different cell types (Figure 5–6). As noted above, in *Volvox* the ability to form different cell types depends on the ability to undergo asymmetrical cell division. Thus the key factors in the evolution of multicellularity in *Volvox carteri* are asymmetrical cell division and a change in pattern of regulatory gene expression.

The example of *Volvox* and *Chlamydomonas* illustrates how multicellular plants may have evolved from unicellular ancestors. *Volvox*, however, is not an ancestor of land plants; multicellularity not only arose independently in plants and animals, but most likely evolved multiple times within the plant kingdom. Nevertheless, the principles

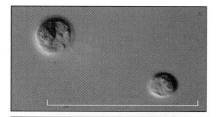
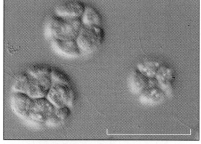
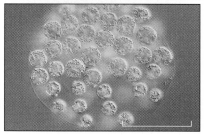
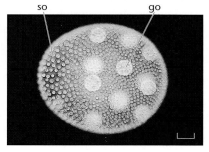

Figure 5–5.
Volvox carteri. The light micrographs show development from a single cell into a multicellular sphere (*top to bottom*). Labels on the mature *Volvox* indicate the small somatic cells (so) and large gonidia (go). Scale bar: 10 μm. (Coutesy of David Kirk.)

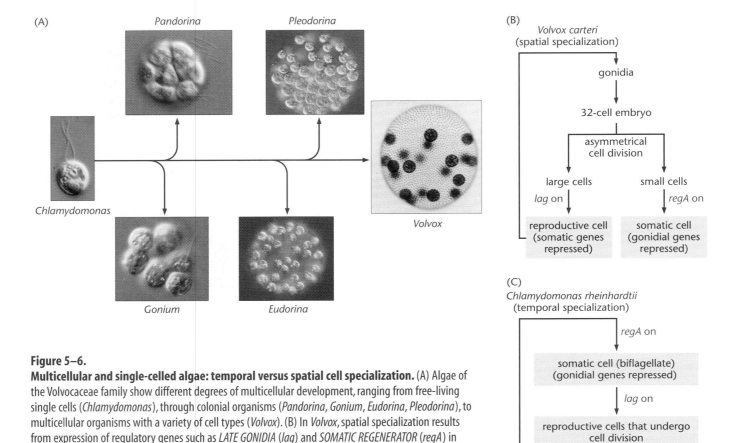

Figure 5–6.
Multicellular and single-celled algae: temporal versus spatial cell specialization. (A) Algae of the Volvocaceae family show different degrees of multicellular development, ranging from free-living single cells (*Chlamydomonas*), through colonial organisms (*Pandorina, Gonium, Eudorina, Pleodorina*), to multicellular organisms with a variety of cell types (*Volvox*). (B) In *Volvox*, spatial specialization results from expression of regulatory genes such as *LATE GONIDIA* (*lag*) and *SOMATIC REGENERATOR* (*regA*) in different cells of a single organism. (C) In *Chlamydomonas*, similar regulatory genes function sequentially to direct cell specialization at different stages of the life cycle. (A, courtesy of David Kirk.)

illustrated by *Volvox* and *Chlamydomonas* are probably important in the evolution of multicellularity in all plants. The ability of cells to remain attached after division was an obvious prerequisite. The ability of the unicellular ancestor to form specialized cells sequentially over time was also important. Asymmetrical cell division then probably had a role in allowing the alternative, specialized fates to be acquired by different cells in the same organism.

5.2 EMBRYO AND SEED DEVELOPMENT

In higher plants, the opposite ends of the mature embryo contain two populations of cells—the shoot and root meristems—that will provide new cells for shoot and root tissues throughout the subsequent growth of the plant. The position of a given cell in the embryo determines its fate. For example, in Arabidopsis, cells at the apical (top) end develop as **cotyledons** and a shoot **apical meristem**, whereas those at the basal (bottom) end develop as roots. Likewise, in the brown seaweed *Fucus*, the **zygote** divides asymmetrically to form a large apical cell that further divides to give rise to the leafy part of the alga, and a smaller basal cell that forms a **rhizoid**. The rhizoid subsequently divides and develops into a **holdfast** that binds the plant to its rocky substrate.

In both of these organisms, the embryos are polar—that is, the two ends of the embryo are different—and the dimension that runs between the top and the bottom poles of the embryo is called the **apical-basal axis**. Apical-basal polarity is exhibited from an early stage in higher- and lower-plant development (Figure 5–7). *Fucus* is a popular model system for studying the earliest stages of embryo formation; we begin here with a description of the formation of the apical-basal axis.

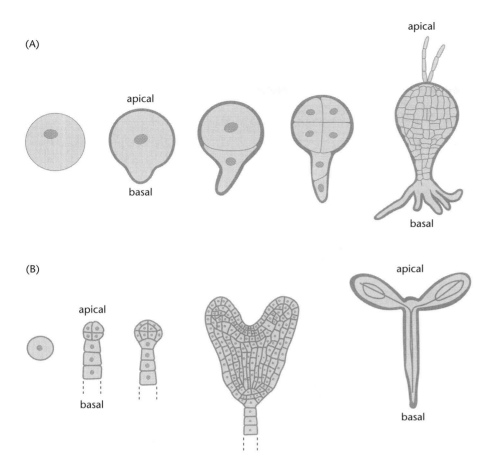

External cues establish the apical-basal axis in the *Fucus* embryo

An unfertilized *Fucus* egg lacks a **cell wall**, but soon after **fertilization** a uniform wall forms around the zygote, which then undergoes a predictable pattern of cell divisions to form the adult alga. The first division is asymmetrical, forming a larger apical cell and a smaller basal cell. The apical cell undergoes a series of longitudinal and transverse divisions to form a **thallus** and **stipes** (analogous to leaves and stems, respectively, in higher plants), and the smaller basal cell elongates to form a rhizoid, which divides to form the multicellular holdfast (analogous to roots).

The zygote becomes polarized even before the first cell division takes place, 6 to 12 hours after fertilization (Figure 5–7A). Just as a multicellular embryo is described as polar because top and bottom are different, the single-celled zygote is described as polar because some of its cellular components accumulate at one end and not at the other. Thus, the apical-basal axis is established very early in embryogenesis and is maintained throughout development.

Environmental signals affect polarization of the zygote during apical-basal axis formation. When the zygote is exposed to unidirectional light, an increase in cellular calcium ion (Ca^{2+}) concentration occurs on the nonilluminated side and marks the region where the rhizoid pole eventually develops (Figure 5–8). This is partly the result of localized activation of Ca^{2+} ion channels in the plasma membrane, allowing an influx of Ca^{2+} ions into the cell. Experimental disruption of the calcium gradient, for example, by adding or removing Ca^{2+} ions from the cytoplasm, results in a loss of polarity of the zygote. As embryos develop from polarized zygotes they develop rhizoids on the dark side and leaflike thalli on the illuminated side. If a polarized zygote is exposed to a second light pulse from a different direction, the apical-basal axis can also change

Figure 5–8.
The role of light and calcium ions in establishing apical-basal polarity in *Fucus*. (A) In the first 4 hours after fertilization, Ca^{2+} accumulates on the side of the cell facing away from light. The area with high Ca^{2+} becomes the basal pole and eventually develops rhizoids. (B) If the direction of light is changed during this early period, the Ca^{2+} ion distribution and thus the basal pole shift accordingly. (C) By 12 hours after fertilization, however, the Ca^{2+} ion distribution no longer responds to a change in the direction of light and the basal pole is fixed.

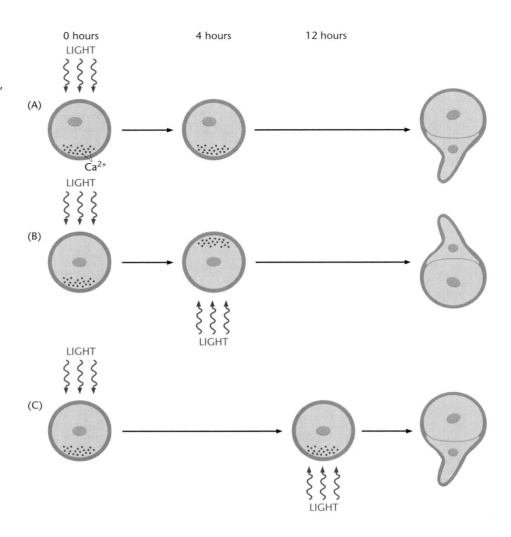

direction. The resulting rhizoid and thallus grow in line with the new axis, indicating that the polarity of the zygote is labile.

Around 12 hours after fertilization, the apical-basal axis becomes fixed and cannot be reversed by illumination from a second direction (Figure 5–8C). The local increase of Ca^{2+} concentration at the basal, rhizoid pole is maintained during axis fixation. In addition, a network of **actin** filaments accumulates at the basal pole. Experimental disruption of this network results in a loss of embryo polarity. It is likely that the microfilaments are required for the delivery of vesicles derived from the **Golgi apparatus** to the cell surface at the rhizoid pole. Molecules such as sulfated **polysaccharides** are exported in these vesicles (see Chapter 3) and make up part of the cell wall only at the basal pole. This localized secretion is required to maintain polarity. The cell wall is a source of signals that are necessary for establishing and maintaining the apical-basal axis: **protoplasts** (cells with their cell wall removed) derived from polarized embryos lose polarity.

The cell wall directs the fate of cells in the *Fucus* embryo

Cell walls contain at least some of the signals responsible for cell differentiation. As described above for the *Fucus* embryo, these signals are not uniformly distributed throughout the cell wall. This unequal distribution of signals is key to the development of neighboring cells. Factors in the apical wall determine apical cell fate, and different factors in basal walls determine basal cell fate.

A laser beam can be used to kill specific cells in the embryo (a process termed **ablation**). Laser ablation destroys the cell contents but leaves fragments of the cell wall in place, thus creating a space that is subsequently filled as neighboring cells divide. A study of this process reveals the effect of the cell wall on the fate of the incoming cells (Figure 5–9). This type of experiment shows that contact with a basal wall causes an incoming apical cell to shift fate and develop as a basal cell. If the apical cell divides without touching the basal wall, it remains an apical cell. Conversely, if an apical cell is ablated and its wall comes in contact with an invading basal cell, the cell assumes an apical fate. This suggests that factors in the apical and basal cell walls confer identity on cells in the corresponding regions of the embryo.

We now turn our attention to embryogenesis in seed plants, with Arabidopsis as the model plant.

Embryo development in higher plants occurs within the seed

Unlike the *Fucus* zygote, the zygotes of angiosperms and gymnosperms develop within a **seed**. For simplicity, we can describe development of the seed as including three main processes. One is embryogenesis, the process that forms an incipient plant (the embryo) from the unicellular zygote, through cell division, cell specialization, and organized growth. The second, occurring in parallel with embryogenesis, is specialization of tissues as nutritional reserves and protective layers around the embryo. The third important process in seed development prepares the embryo to survive a potentially long period of metabolic inactivity in a desiccated form. Here we focus on the first of these processes: embryogenesis.

As noted earlier, at the end of embryogenesis, two populations of cells are established at either end of the embryo—the shoot and root meristems—from which most of the shoot and root tissues arise after **germination**. Different plants reach this point of development in different ways; the stages of embryogenesis vary immensely among species, and the overall appearance of the final embryo is also species-specific. We describe here embryogenesis in Arabidopsis: the morphological changes and what is known about the factors determining cell specialization.

The first division of an Arabidopsis zygote is asymmetrical: a polarity along the apical-basal axis is clearly established as the zygote divides to form a small apical cell and a larger basal cell. The apical cell will form the majority of the embryo proper, and the basal cell will form a filamentous structure called the **suspensor**. The apical cell

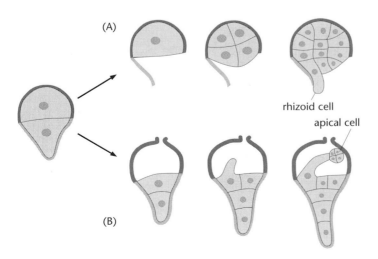

(A)

rhizoid cell

apical cell

(B)

Figure 5–9.
The cell wall carries information about apical-basal polarity in *Fucus*. (A) When the basal cell is destroyed by a laser beam (ablated) at the two-cell stage, if a descendant of the apical cell eventually touches the remaining basal cell wall, this descendant cell develops as a rhizoid cell. (B) Conversely, when the apical cell is destroyed, descendants of the basal cell acquire apical cell features if they touch the remnants of the apical cell wall. (From F. Berger et al., *Science* 263:1421–1423, 1994. With permission of AAAS.)

undergoes a combination of transverse and longitudinal divisions resulting in formation of the **globular embryo** (by about 60 hours after fertilization; Figure 5–10). The globular embryo has a distinct outer layer of cells, or **protoderm**. Further rounds of cell division result in the formation of a distinct **ground tissue** and a central **stele**, in which the vascular system of the embryo develops. The appearance of small mounds of cells (called **primordia**) that will become cotyledons marks the end of the globular stage and the emergence of the characteristic **heart stage**; it is at this point that the developing root and shoot meristems become visible. By 96 hours after fertilization, the **torpedo stage** is established, and the hypocotyl (see below) is clearly visible. After approximately 10 days the embryo is fully developed. The seed then undergoes a desiccation stage and dormancy is established, persisting until germination triggers growth.

The fate of embryonic cells is defined by their position

Cells in a growing Arabidopsis embryo continuously take from their surroundings the cues that direct their fate; this means that cell fate remains undefined until cell division has ceased and the cell's location in the embryo is fixed. The order in which cells divide and orient is mostly predictable, but not precisely the same in each developing embryo. These subtle differences from one embryo to another indicate that the pattern of cell division (lineage) is not important for cell fate in Arabidopsis.

A technique called **clonal analysis** (Box 5–1) has shown that the final location of a cell in a plant can determine cell specialization. In essence, a cell is marked with a genetic "tag" early in embryogenesis, and this tag marks the descendants of the cell through subsequent cell divisions and development. This technique has shown that the fate of a cell in the embryo is not dictated by the earlier position and role of its progenitor cell. For example, cells between the apical and basal meristems of a mature Arabidopsis embryo develop to form the **hypocotyl**, which is a boundary area between the root and

Figure 5–10.
Stages of Arabidopsis embryogenesis.
(A) Zygote. (B) One-cell stage. The one cell is the small apical cell (a) that results from the first zygotic division and develops into the embryo proper; the larger, basal cell (b) gives rise to the suspensor. (C) Two-cell stage, after the apical cell divides vertically. (D) Eight-cell stage. (E) Early globular (16-cell) stage. (F) Late globular stage. (G) Heart stage. (H) Mature embryo. (Courtesy of Gerd Jürgens. With permission from the Company of Biologists.)

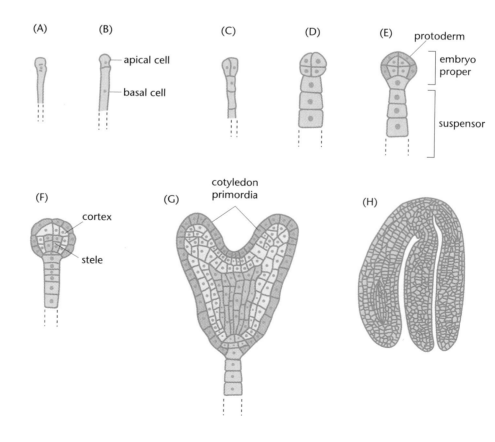

Box 5–1 Clonal Analysis

Clonal analysis is a technique used to trace how the descendants of a particular cell or group of cells contribute to the subsequent development of an organism. For example, as explained in Section 5.4, the meristem of angiosperms is made up of concentric L1, L2, and L3 layers, and the cells in one layer sometimes carry an albino mutation. It is possible to see how the descendants of a specific meristem layer contribute to making a leaf, because the descendant cells give rise to albino tissues (Figure B5–1).

Clonal analysis can also be performed using unstable mutations created by transposon excision (see Chapter 2). In this case, a mutant allele caused by insertion of a transposon disrupts the function of a gene that has a readily visible effect on tissue phenotype (for example, the gene is required to produce a pigment). When the transposon excises, not only does the cell where excision occurred become distinct from its neighbors, but the tissues formed by the progeny of that cell are also marked (see Figure 2–9).

Transposition usually occurs at random and so does not allow control over which cells are marked or at what developmental stage. One way to control the time of transposition is by introducing into the plant a **transgene** that expresses a **transposase** under the regulation of an inducible promoter. Another method is based on inducible expression of a sequence-specific **recombinase** such as the Cre recombinase, which originated from a bacterial phage (Figure B5–2). The recombinase catalyzes the excision of a transgene flanked by specific, short DNA sequences—in the case of Cre, the *loxP* sequences. As with transposon excision, excision by the recombinase activates a marker gene that was previously interrupted by the transgene.

Clonal analysis has provided important information on how cell division is regulated to build the different parts of a plant and on the question of whether differentiation depends on a cell's lineage or its position in the tissue—the answer being cell position.

Figure B5–1.
Variegated *Ficus* leaves. These leaves were produced by a single plant in which the L1 layer of the meristem consisted of cells unable to produce chloroplasts (albino mutant cells). The green color of the leaves is given by mesophyll cells, which originate mostly from the L2 and L3 layers. The edges of the leaves are albino because in these regions the L1 layer contributes cells to the mesophyll. The degree of this contribution varies from leaf to leaf; some leaves have a large albino sector (*arrow*), which arises from an early "invasion" of the L2 layer in the leaf primordium by descendants of the albino L1 layer.

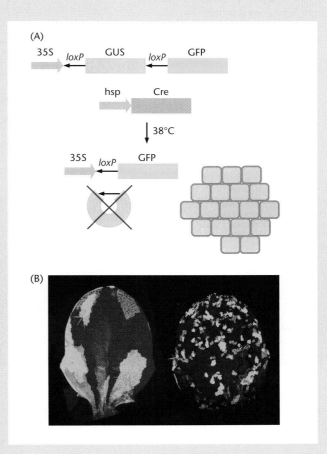

Figure B5–2.
Clonal analysis using the Cre-*loxP* system. (A) The plant contains two transgenes. One uses a promoter active in most plant tissues (the 35S promoter) to direct expression of the GUS reporter gene (β-glucuronidase gene; see Section 4.6), which is flanked by *loxP* sequences and followed by an inactive GFP (green fluorescent protein) reporter gene. In the second transgene, a heat-shock–inducible promoter (hsp) directs expression of Cre recombinase. Exposing the plants to high temperature (38°C) induces *Cre* expression, and the recombinase catalyzes removal of the GUS transgene; GFP is now under the control of the 35S promoter. If enough Cre recombinase is induced, GUS excision occurs in all cells; if the heat treatment is brief, GUS is excised in a random sample of cells. These cells and their descendants now express GFP instead of GUS. (B) Images of fluorescent GFP sectors induced in Arabidopsis leaves using the system described in (A), at an early stage (*left*) or later stage (*right*) of leaf development. Note that early induction results in larger GFP sectors, because the descendants of marked cells have more time to multiply. (Courtesy of Samantha Fox.)

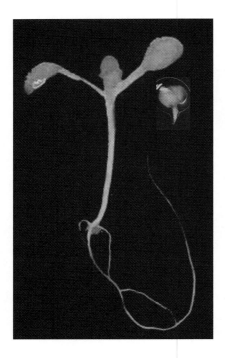

Figure 5–11.
Wild-type Arabidopsis and *monopteros*
mutant. A *monopteros* mutant seedling (*inset*)
is compared with a wild-type seedling. The
mutant has defective auxin signal transduction;
the root and hypocotyl are missing.
(Courtesy of Gerd Jürgens.)

the shoot in a young seedling. Clonal analysis has demonstrated that the cells that make up the hypocotyl can derive from progenitors in either the apical or the basal part of the embryo—again reinforcing the idea that assignment of cell fate is flexible in plants.

Progressive polarization of auxin transporters mediates formation of the basal pole in embryos

Auxin is a mobile signaling molecule involved in long-distance signaling. It is important in conveying positional information, both in the embryo and in the tissues of a growing plant. Auxin is synthesized near the shoot tip and transported to the root. Its long-distance effects have been well documented for decades—for example, as a signal produced by the shoot apex to inhibit growth of axillary meristems. In recent years, however, we have learned that auxin transport also provides short-distance signals involved in shaping tissues and organs. One of the best-studied examples is the role of auxin in shaping the early embryo.

Polarized auxin transport is required from the early stages of embryogenesis. When a developing Arabidopsis embryo is treated with inhibitors of auxin transport, it has defects in its basal pole. The resulting embryo either lacks roots or forms small peglike structures in place of the root and hypocotyl. Mutant plants such as *monopteros*, with defects in auxin signal transduction, have similar phenotypes; they form peglike structures instead of a root (Figure 5–11). The wild-type *MONOPTEROS* gene encodes a **transcription factor** involved in the auxin-response pathway.

Auxin transport is mediated by two types of carrier in the plasma membrane: efflux carriers at the sites of auxin export from the cell and influx carriers at the sites of auxin uptake. The PIN1 protein is believed to facilitate auxin efflux from cells in embryos and shoots (Figure 5–12). At the globular stage of embryogenesis, PIN1 protein is located on all plasma membranes of the embryo except those next to the outer cell walls of the protoderm (surface) layer. This suggests that, at this stage of development, auxin may be transported from cell to cell equally in all directions. As embryogenesis continues, however, the location of PIN1 is progressively limited to the basal membranes of most cells, suggesting that auxin transport has become polarized and now moves auxin from the apical to the basal region of the embryo. When this transport is inhibited, development of the basal pole stops.

(A)   (B)   (C)   (D)

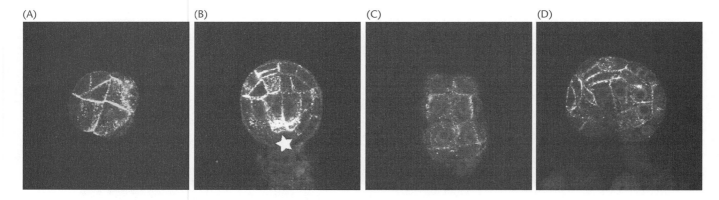

Figure 5–12.
Immunolocalization of the PIN1 protein in embryos of wild-type Arabidopsis and *gnom* mutant. (A) In the wild-type early globular embryo, PIN1 is found in the plasma membranes but is not oriented toward the basal part of the embryo. (B) By the late globular stage, PIN1 has accumulated on the side of the cells facing the basal region (*star*). (C, D) In the *gnom* mutant, PIN1 does not become oriented toward the basal region during the transition from early (C) to late (D) globular embryo; note also the abnormal shape of the mutant embryo. (From T. Steinmann et al., *Science* 286(5438):316–318, 1999. With permission of AAAS.)

Targeted vesicle fusion is required for the polar localization of PIN1 protein in the embryo and the establishment of polar auxin transport. The *GNOM* gene of Arabidopsis encodes a protein similar to the SEC 60 family of yeast proteins. Members of this protein family are required for the targeting of Golgi-derived vesicles to the plasma membrane. Embryos homozygous for the *gnom* mutation develop peglike structures at their basal pole. This phenotype resembles that of seedlings homozygous for the *monopteros* mutation and that of embryos treated with auxin-transport inhibitors, and thus the GNOM protein seems to be required for the regulation of auxin transport during embryogenesis. GNOM may act by transporting PIN1-containing vesicles to the plasma membrane. The requirement for GNOM in PIN1 localization was shown by determining the location of PIN1 protein in *gnom* mutant embryos. While PIN1 distribution is progressively polarized during embryogenesis in wild-type embryos, it remains unpolarized in embryos homozygous for the *gnom* mutation. Thus, GNOM is required for the polar localization of the PIN1 protein, which in turn is required to catalyze the polar auxin transport necessary for formation of the apical-basal axis in the embryo.

Radial cell pattern in the embryonic root and hypocotyl is defined by the SCARECROW and SHORT ROOT transcription factors

In addition to the apical-basal organization of the developing embryo described thus far, the embryonic cells are also arranged in a characteristic radial pattern. Mutants with defects in the development of the root and hypocotyl have provided information on how this radial pattern arises.

The radial organization of tissues in the hypocotyl is almost identical to that found in roots. The hypocotyl is composed of an outer epidermal layer (protoderm) and the three-layered ground tissue (with two cortical cell layers and an inner endodermal layer) surrounding two columns of cells from which the vasculature develops (Figure 5–13). The protoderm of the root comprises a lateral **root cap** layer that surrounds the **epidermis**. The ground tissue comprises a single layer of **cortex** and an **endodermis**, also surrounding the **diarch** stele. During development of the roots and hypocotyls, this radial organization is first visible with the formation of the protoderm layer in the globular embryo. As development proceeds, the ground tissue is formed, comprising the endodermis and cortex.

Since the radial patterns of the root and hypocotyl are similar, it is not surprising that genes regulating the patterning of cell layers in the root also function in the hypocotyl. SHORT ROOT and SCARECROW, putative transcription factors in the same protein family, are required for formation of the ground tissue, both in the embryo and during root development after embryogenesis. Plants homozygous for loss of function due to either *scarecrow* (*scr*) or *short root* mutations develop a ground tissue with a single cell layer instead of the normal two layers in the root, and a ground tissue of two cell layers rather than three in the hypocotyl (Figure 5–14). The cells that normally divide

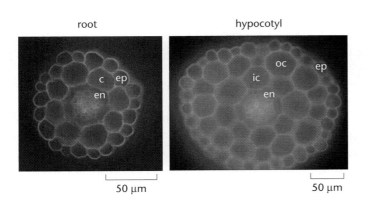

root   hypocotyl

50 μm   50 μm

Figure 5–13.
Transverse sections through Arabidopsis root and hypocotyl. The root (*left*) and hypocotyl (*right*) have a similar organization of concentric tissues. (en = endodermis; c = cortex; ic = inner cortex; oc = outer cortex; ep =epidermis) (With permission from the Company of Biologists and the author, courtesy of John Schiefelbein.)

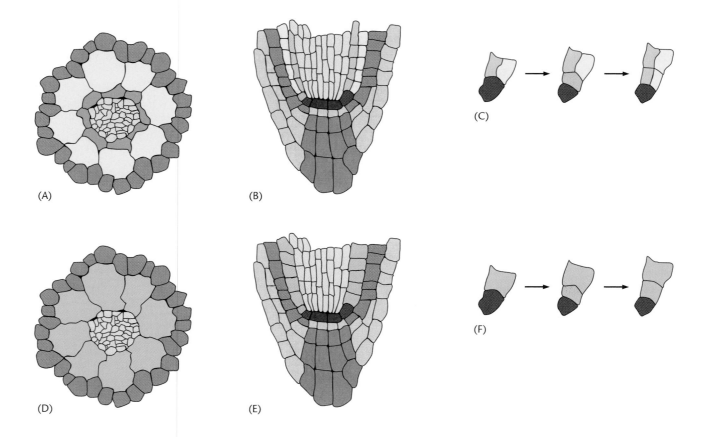

Figure 5–14.
Sections through primary root tip of wild-type Arabidopsis and *scarecrow (scr)* mutant. (A) Transverse and (B) longitudinal sections through a wild-type root, showing different tissues and cell types: epidermis (*purple*), cortex (*yellow*), endodermis (*blue*), stele (*pink*), lateral root cap (*light gray*), columella (*dark gray*), quiescent center (*red*), ground tissue initial (*brown*), and cortex/endodermis initial (*orange*). (C) Detail of how the cortex/endodermis initial (*orange*) originates from a transverse division of the ground tissue initial (*brown*), and then divides longitudinally to give rise to a cortical cell (*yellow*) and endodermal cell (*blue*). (D) Transverse and (E) longitudinal sections through a *scr* mutant root; note that the cortex and endodermis are replaced by a single cell layer of mixed identity (*green*). (F) Detail of how the *scr* mutant forms a single layer of ground tissue, due to failure of longitudinal division of the cortex/endodermis initial. (A–F, from K. Nakajima et al., *Nature* 413:307–311, 2001. With permission from Macmillan Publishers Ltd.)

longitudinally to form the cortex and endodermis in the wild-type plant fail to do so in the mutants. In roots, *SCARECROW* is expressed in the cells that divide to form the two cell layers, suggesting that the *SCARECROW* gene product is required for execution of the asymmetrical cell divisions that give rise to the cortex and endodermis. This gene is also expressed in the endodermis derived from the formative division, and thus it may play a role in maintaining endodermal identity.

SHORT ROOT is expressed in the stele tissues; that is, it is not expressed in the cells that undergo asymmetrical division in the formation of ground tissue throughout embryogenesis. This means that the function of SHORT ROOT is visible only in cells adjacent to those expressing *SHORT ROOT*. In cases like this, the function of the gene is said to be "non–cell autonomous," implying that a signal moves from the cells that express the gene to cells where its effect is seen. In the case of *SHORT ROOT*, the signal is the SHORT ROOT protein itself, which moves from the cells where it is synthesized to neighboring ground tissue precursors, where it controls **gene expression** (Figure 5–15).

Having described the establishment of apical-basal and radial asymmetry in the embryo, we now turn to a brief description of how meristems are established. The way meristems function throughout a plant's life cycle is discussed in more detail later in the chapter.

Clues from apical-basal and radial patterning of the embryo are combined to position the root meristem

The fully formed root meristem contains the progenitor cells for each of the root tissues. These **initials** divide frequently and give rise to regular files (orderly columns) of cells that differentiate into specialized tissues as new divisions displace cells from the

nuclei in endodermis

stele

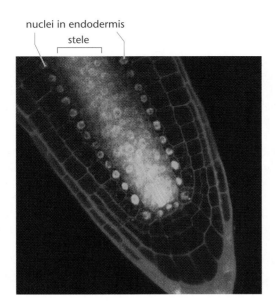

Figure 5–15.
Root tip of Arabidopsis expressing a fusion between SHORT ROOT and green fluorescent protein. The gene expressing the fusion protein (made visible by its fluorescent GFP tag) is transcribed only in the stele. The fusion protein is transported to the endodermis, where it accumulates in the nuclei. (From K. Nakajima et al., *Nature* 413:307–311, 2001. With permission from Macmillan Publishers Ltd, courtesy of Keiji Nakajima.)

meristem. The files of new cells are oriented both toward the more mature regions of the root (i.e., closer to the shoot) and toward the very tip of the root. The cells that accumulate at the very tip differentiate as the root cap, which protects the root meristem as it grows through the soil.

The initial cells of the root meristem surround a small group of slowly dividing cells called the **quiescent center (QC)**. As described more fully later in this chapter, the QC is believed to produce a signal that prevents the premature differentiation of initial cells. The QC and the surrounding cells originate from different parts of the early embryo. Most of the initials descend from the basal cells of the globular embryo. The QC and root-cap initials, however, descend from the topmost suspensor cell, which is incorporated into the developing embryo at the heart stage.

Because the initial cells cannot be maintained without the QC, establishment of the QC is central to the development of the root meristem. Although the QC has a different clonal origin from most of the initials, the mechanism that determines which cells become the QC is based not on which cells they descend from but on the position of the cells in the developing embryo. The position of the QC is determined by overlapping signals used for the apical-basal and radial patterning of the embryo. As mentioned above, auxin is transported to the basal region of the embryo. The higher auxin concentration at the base activates the regulatory *PLETHORA* (*PLT*) genes, which are necessary, but not sufficient, for QC development. The QC is formed by a subset of basal cells that, in addition to *PLT*, also express the *SCARECROW* gene, which, as explained above, participates in establishing the radial pattern of the embryo. Thus the inputs of apical-basal and radial patterning are combined to establish the position of the QC and consequently the root meristem (Figure 5–16).

The shoot meristem is established gradually and independent of the root meristem

The shoot apical meristem develops from a group of cells located in the area between the two forming cotyledons. The late globular embryo has an epidermal layer and a **hypodermal layer** overlying a group of cells that will form the vasculature of the hypocotyl. The hypodermal layer later divides to form a double layer. The embryonic

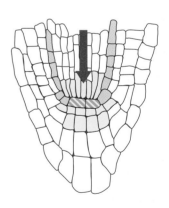

Figure 5-16.
Apical-basal and radial patterning are combined to set the position of the root meristem. The *SHORT ROOT* and *SCARECROW* genes (cells marked in *blue*) control the radial pattern of the root. Apical-basal patterning is controlled by auxin, which is transported from the shoot toward the root tip (*red arrow*); high levels of auxin accumulate at the root tip and activate the *PLETHORA* genes (cells marked in *yellow*). The overlapping activity of *PLETHORA*, *SHORT ROOT*, and *SCARECROW* specify the quiescent center (cells outlined in *red*), around which the root meristem is organized.

shoot meristem develops from epidermal, upper hypodermal, and lower hypodermal cells at the torpedo stage of embryogenesis (Figure 5–17). Frequent divisions make the cells in the developing meristem morphologically distinct from those in the surrounding area. Lower hypodermal cells divide in directions both parallel to the cell layer (periclinal divisions) and perpendicular to the cell layer (anticlinal divisions), whereas upper hypodermal and epidermal cells divide only anticlinally. This pattern of cell division persists in the seedling meristem, resulting in the characteristic concentric cell layers of the shoot meristem.

Screens for Arabidopsis mutants with defective meristems have identified genes involved in the development of the shoot meristem. One of these genes is *SHOOT MERISTEMLESS*. Plants homozygous for *shoot meristemless* (*stm*) mutations lack the shoot meristem when they germinate and show no signs of early meristem development in embryos. The root meristem forms normally in the *stm* mutant, showing that development of root meristem and of shoot meristem are controlled separately.

The wild-type *SHOOT MERISTEMLESS* (*STM*) gene encodes a DNA binding protein; this finding, in combination with the mutant phenotype, suggests that SHOOT MERISTEMLESS is required for the transcriptional activation of genes involved in formation of the shoot meristem. Meristem regulatory genes become active before the morphological changes that accompany meristem development. For example, *SHOOT MERISTEMLESS* is first expressed in a small group of cells in the globular embryo in the region where the future meristem will form (Figure 5–18). As embryogenesis progresses, additional regulatory genes come into play that have specialized roles in the meristem, such as the *CLAVATA* (*CLV*) genes, which control meristem size, and *CUP-SHAPED COTYLEDONS* (*CUC*) genes, which mark the boundaries of new organs that emerge from the meristem. Genes of this type are discussed in Section 5.4; we mention them here to illustrate the progressive refinement of gene expression during establishment of the meristems.

The endosperm and embryo develop in parallel

Embryo development does not take place in isolation. Angiosperms undergo **double fertilization**, giving rise to a diploid embryo and a **triploid** endosperm (see Figure 5–97). The embryo and endosperm then develop in parallel. Endosperm and embryo structures develop very differently, and endosperm development itself can vary widely, especially between **monocots** and **eudicots**. This difference depends, in part, on the relative importance of the endosperm in seed development. For example, in maize and coconut the endosperm is the principal source of nutrients during seed germination, whereas in some species, including Arabidopsis, it is a transient structure and other tissues serve as primary nutrient stores.

Figure 5–17.
Establishment of the shoot apical meristem during embryogenesis. Sections through Arabidopsis embryos at (A) the early torpedo stage and (B) the late torpedo stage. Cells of the embryonic shoot meristem are indicated in *red*. (A and B, with permission from the Company of Biologists.)

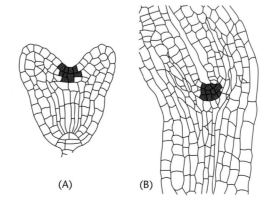

(A)   (B)

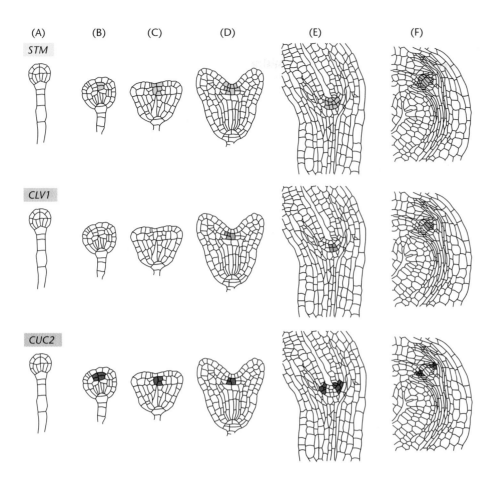

Figure 5–18.
Expression of shoot meristem regulatory genes in Arabidopsis. Expression of the regulatory genes starts early in embryogenesis and is gradually refined as the meristem develops. The diagrams show the expression pattern of *SHOOT MERISTEMLESS* (*STM*), *CLAVATA1* (*CLV1*), and *CUP-SHAPED COTYLEDONS 2* (*CUC2*) at different stages of embryogenesis. (A) Early globular stage; (B) late globular stage; (C) heart stage; (D) early torpedo stage; (E) late torpedo stage; and (F) maturing embryo, showing detail of the region containing the shoot meristem. (A–F, with permission from the Company of Biologists.)

The endosperm originates from a nuclear fusion event that occurs parallel to the fusion that forms the zygote. This process is described in more detail later (Section 5.6). Briefly, the **pollen grain** carries two sperm nuclei; when the **pollen tube** reaches the **ovule**, one of these nuclei fuses with the egg cell to form the zygote. The second sperm nucleus fuses with two other haploid cells of the female gametophyte to form the primary endosperm nucleus. This triploid nucleus then undergoes rounds of nuclear division without cytokinesis: the replicated chromosomes separate (**karyokinesis**) but no cell plate forms. The nuclei continue to divide until the developing embryo is surrounded by a **syncytium**, a continuous mass of cytoplasm shared by multiple nuclei (Figure 5–19).

Early in endosperm development, the nuclear divisions are synchronized: all cells divide at the same time. As development proceeds, nuclear divisions in the different regions of the endosperm become asynchronous. The free nuclei become surrounded by a plasma membrane and cell wall, forming discrete cells each containing a single nucleus when the embryo is at the globular stage. This process of **cellularization** begins at the chalazal pole of the endosperm, progressing toward the antipodal end. In Arabidopsis, the endosperm then degenerates and the developing embryo absorbs its contents. In species such as maize, barley, and wheat, the endosperm goes on to accumulate nutritional reserves and eventually makes up the bulk of the seed mass.

Division of the cells that give rise to endosperm is repressed until fertilization

In Arabidopsis, the two haploid nuclei of the female gametophyte fuse early to form the diploid **central cell nucleus**, which eventually fuses with a sperm nucleus delivered

Figure 5–19.
Endosperm development in Arabidopsis.
(A) Immediately after fertilization, the primary endosperm nucleus (*arrow*) and the zygote nucleus (z) are visible. (B) Syncytial endosperm, with multiple nuclei (*arrows*), surrounding the embryo (e) at the four-cell stage. (C) Cellularized endosperm surrounding a heart-stage embryo (e). (Courtesy of Frederic Berger.)

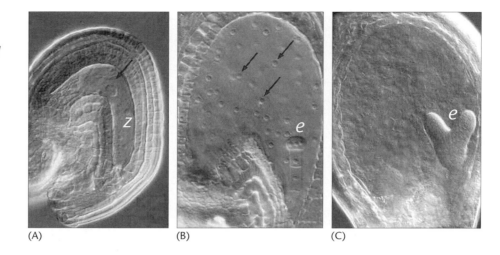

(A)   (B)   (C)

by the pollen tube to form the endosperm. Genetic analysis has indicated that cell division in the central nucleus is repressed until fertilization, when repression is released allowing the primary endosperm nucleus to divide.

Three proteins have been identified that are likely to form a complex that binds to **chromatin** to repress expression of the genes required for endosperm cell division and development. The three proteins were identified by analysis of fertilization in the endosperm mutants *medea* (*mea*), *fertilization independent seed2* (*fis2*), and *fertilization independent endosperm1* (*fie1*). In these mutants the central cell nucleus divides in the absence of fertilization to form an endosperm composed of diploid nuclei, instead of the wild-type triploid endosperm. When *mea* mutants are fertilized with wild-type nuclei, they have endosperm defects at later stages of development (Figure 5–20), suggesting that the MEDEA (MEA) protein is required throughout development of the endosperm.

The MEA and FIE1 proteins are conserved in animals. In *Drosophila*, a pair of proteins related to MEA and FIE1 interact within a protein complex (the Polycomb complex) that represses gene activity through chromatin remodeling (see Section 2.3). It is likely that a similar mechanism operates in Arabidopsis embryos to control cell division until fertilization has taken place.

Figure 5–20.
Endosperm development in the absence of fertilization in the *medea* (*mea*) mutant. (A) Developing seed of wild-type Arabidopsis, with an embryo at the four-cell stage (e) surrounded by the syncytial endosperm with multiple nuclei (*arrows*). (B) Unfertilized *mea* seed, with syncytial endosperm (*arrows*) but no embryo. (A and B, courtesy of Bob Fischer.)

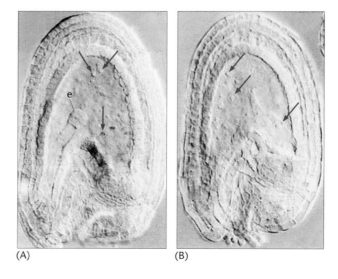

(A)   (B)

After embryo and endosperm development, seeds usually become dormant

When seed development is almost complete, the embryo usually becomes dormant. This typically involves the accumulation of storage products and desiccation. The ability of the embryo to survive in the seed for long periods without water, together with the availability of a stock of nutrients to sustain the early stages of germination, were probably crucial to the colonization of terrestrial environments by seed plants.

Maturation of the embryo is controlled, at least in part, by a signal, in the form of **abscisic acid (ABA)**, from the rest of the seed. In many species, when immature embryos are removed from seeds and grown *in vitro* they germinate and become seedlings. Thus dormancy is a condition that can be imposed on an embryo rather than a necessary stage in its development. This is also illustrated by the fact that the embryos of different species become dormant at different stages. At one extreme, orchid embryos can become dormant as early as the globular stage. In Arabidopsis, dormancy sets in when the shoot and root meristems are established but no leaf primordia have yet emerged. The embryos of grasses become dormant at a later stage, with several leaf primordia already emerging from the shoot apical meristem.

The onset of dormancy during seed development and the release from dormancy during germination are under the control of two **growth regulators** (or **phytohormones**), gibberellin and ABA, that act antagonistically. ABA promotes seed maturation and prevents seeds from germinating too early. Its importance has been demonstrated in studies of ABA-deficient mutants of Arabidopsis, tomato, and maize that lack particular enzymes involved in ABA biosynthesis. The ABA-deficient mutants of Arabidopsis and tomato do not complete the processes associated with maturation and dormancy, and in maize the mutant kernels germinate precociously on the ear (Figure 5–21). The maize *Vp1* gene encodes a transcription factor (Vp1) involved in the transduction of the ABA signal. In mutants lacking Vp1, **vivipary** (precocious germination) also takes place, mimicking the effects of ABA deficiency. Vp1 (and ABI3, the orthologous Arabidopsis gene product) regulates the expression of ABA-inducible genes during seed maturation.

Whereas ABA is required for repressing germination, **gibberellins** are required to initiate germination. The seed maturation program is usually completed when a seed develops desiccation tolerance, followed by desiccation itself. On **imbibition**, as the seed takes up water and rehydrates, it can enter the germination program. The first clear manifestation of germination is when growth of the **radicle** (embryonic root) enables it to penetrate the **testa** (seed coat) and the layers of the seed envelope (the **hyaline** and **aleurone** layers). *GA1* encodes an enzyme involved in early gibberellin biosynthesis in Arabidopsis. Mutant *ga1-3* plants with very low gibberellin levels are unable to germinate unless gibberellin is applied. When these gibberellin-deficient mutant embryos are removed from the testa and aleurone layers of the seed envelope,

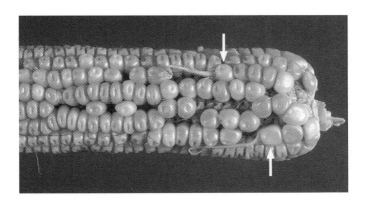

Figure 5–21.
**Viviparous seeds of the maize *Vp1*
mutant.** Seeds have germinated (*arrows*) while still attached to the cob.

Figure 5–22.
Gibberellin-deficient Arabidopsis mutant. *ga1-3* seeds do not germinate (i.e., do not overcome inhibition by the seed coat envelope layers) unless gibberellin (GA) is administered or the seed is removed from the seed coat or envelope. (WT, wild type)

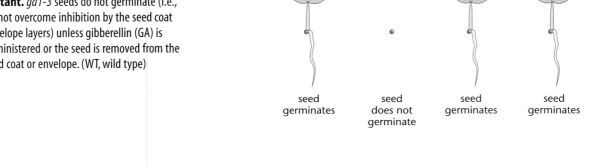

however, germination proceeds (Figure 5–22). Gibberellin is therefore necessary for the germinating seedling to overcome the restraint imposed by the seed coat and envelope layers.

The balance between gibberellin and ABA levels is an important factor in regulating seed germination, as shown experimentally. Double mutants have been made in Arabidopsis and maize that have low ABA and low gibberellin levels. In both instances the plants are still able to germinate, suggesting that ABA is responsible for the dormancy observed in gibberellin-deficient mutants such as *ga1-3*. This demonstrates that ABA and gibberellin act antagonistically in wild-type plants, and thus the balance between these two growth regulators seems to be important in the regulation of germination in many plant species (Figure 5–23).

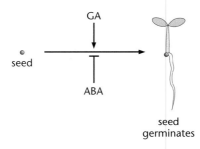

Figure 5–23.
Gibberellin (GA) promotes and abscisic acid (ABA) inhibits seed germination.

Another aspect of seed germination that is regulated by gibberellin is the mobilization of endosperm storage reserves that occurs during germination of cereal grains. In cereal grains, the aleurone cells form a specialized cell layer surrounding the endosperm. During germination, embryo-derived gibberellin stimulates the production of various hydrolyzing enzymes, including the enzyme **α-amylase**, in the aleurone layer. These enzymes are secreted into the endosperm, where they hydrolyze the food reserves (see Section 4.6) and thus make them available to the growing seedling.

It is likely that gibberellin interacts with a specific outward-facing receptor on the plasma membrane of aleurone cells. The signal-transduction chain that follows the cell's perception of gibberellin (described in Section 5.4) leads to transcriptional regulation of the genes encoding α-amylase. The **promoters** of the α-amylase genes contain **GA-boxes**, short regions of DNA sequence that confer gibberellin-induced transcription. A candidate transcription factor (GAMyb, of the **MYB family**) that binds to the GA-box has been identified in barley: GAMyb regulates the transcription of α-amylase genes and is itself gibberellin-inducible. However, GAMyb is not exclusively involved in the control of α-amylase genes; it also plays a role in regulating endosperm development.

5.3 ROOT DEVELOPMENT

The remainder of this chapter looks at how higher plants develop after embryogenesis—that is, how the shoot and root meristems give rise to the aboveground and belowground parts of the plant. Even though shoot and root development share features such as their origin from apical groups of actively dividing, undifferentiated cells, we describe their development separately for several reasons: their different evolutionary origins, their different structures, and control of their development by different genes. We begin, in this section, with the evolution and development of roots.

Plant roots evolved independently at least twice

Roots are generally responsible for water and nutrient uptake and, in the majority of land plants, are the site of symbiotic interactions between the plant and soil fungi and bacteria (see Section 8.5). Roots also serve to fix the plant into the ground, providing stability. They are found in all existing groups of vascular land plants, and they evolved independently at least twice (Figure 5–24). Rootlike organs growing from swollen stems are found in 350-million-year-old fossil quillworts (*Isoetes*); members of the quillworts are thought to be derived from rootless plants such as *Asteroxylon*. The rootlike structures of these ancient plants have a single vascular trace (containing xylem and phloem) surrounded by ground tissue, and have an epidermis with **root hairs**. It is likely that the roots of quillworts are modified **microphylls**, small leaves that develop in this group of plants. Quillworts, however, are not thought to be the ancestors of ferns and seed plants. Roots evolved at least once again in the progenitors of these plants, probably also around 350 million years ago. The roots of ferns and seed plants have a more complex vascular system than the roots of quillworts and may be derived from modified shoot structures.

Two types of roots are found in ferns. In the first type, all root cells are derived from a single initial cell; the pattern of cell divisions is regular and can be followed precisely. This type of development is found in the floating water fern *Azolla*. In the second type of fern root, the root cells cannot be traced from a single initial. The cellular pattern in the roots of conifers and flowering plants suggests that the roots of higher plants are derived from more than one cell and are more similar to the second root type in ferns. Roots have been lost during evolution in several lineages. The most dramatic example of this is *Psilotum*. Although it lacks roots, DNA sequence comparisons show that *Psilotum* is a true fern. A number of parasitic plants have also lost the ability to form roots, illustrating how organs may be lost as a result of morphological specialization during the evolution of parasitism.

Roots have several zones containing cells at successive stages of differentiation

The primary root, or radicle, is formed in the embryo and remains dormant until it emerges from the seed at germination. In some species the radicle forms the main root, from which all other roots arise as lateral branches. In others the **primary root** does not contribute much to the root system, which is instead derived mainly from **adventitious roots** (Figure 5–25). Adventitious roots can be loosely defined as roots arising from shoots; an example is the prop or brace roots of maize, which develop from the nodes of lower leaves on the maize stem and support the growing plant. Mutant maize plants lacking prop (brace) roots are unstable and topple over.

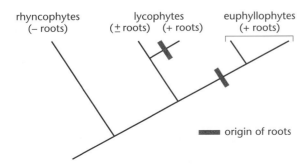

Figure 5–24.
Evolution of roots in vascular plants.
Roots evolved independently at least twice. The fossil record shows that roots originated separately in lycophytes (e.g., quillworts) and in the ancestors of ferns and seed plants (euphyllophytes).

Figure 5–25.
Adventitious roots. (A) Adventitious roots arising from nodes of a maize stem.
(B) Mutant maize plants that cannot form adventitious roots tend to topple over.
(A and B, courtesy of Frank Hochholdinger.)

(A)

(B)

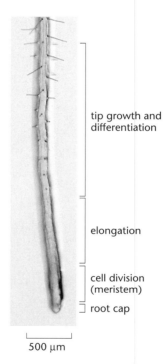

tip growth and differentiation

elongation

cell division (meristem)

root cap

500 µm

Figure 5–26.
The four growth zones of a root. The root tip is protected by a root cap and is organized into several zones with cells at different stages of development. (Courtesy of Seiji Takeda.)

The radicle is organized in four zones, from bottom to top (Figure 5–26). At the bottom is the root cap, which ensheaths the meristem and, as the name suggests, covers the very tip of the root, protecting it from mechanical injury. Above this is the meristem, or **zone of cell division**. Next, once cells stop dividing, they undergo rapid expansion in the **elongation zone**. In the uppermost zone—the **differentiation zone**—cell expansion has ceased and cells have acquired their final shape and form (with the exception of root-hair cells, which start growth at this stage). Secondary tissues such as **wood** develop some distance behind the root tip from a lateral meristem called the **cambium**. Cells derived from the inside of the cambium differentiate as xylem and cells formed on the outside of the cambium develop as **phloem**. In this way, a hard core of xylem can build up to form woody roots, as seen in trees.

The Arabidopsis root has simple cellular organization

The primary root of Arabidopsis is a convenient model system for root development because its cellular organization is simple and the patterns of cell division are predictable. The root is less than 100 µm in diameter and is composed of cells arranged in files in radially arranged cell layers. Details of the cellular organization are shown in Figure 5–14. The outermost layer at the very tip is the lateral root cap, which surrounds a single-celled epidermis. The root cap disintegrates as cells enter the elongation zone. The four remaining cell layers are the epidermis, cortex, endodermis, and stele (vascular) tissues.

In transverse view, usually 16 to 21 cells make up the circumference of the Arabidopsis root epidermis, and inside this are two 8-celled rings—the cortex and endodermis. The Arabidopsis root is particularly small and invariant; other plant species have more cell layers, and in some cases there is a lot more variation within the same species. The invariant cell numbers in the cortex and endodermis of the Arabidopsis root indicate that the patterns of cell division are under tight control during root development. The files converge to concentric rings of initials at the root tip, from which new cells in each file are derived. Clonal analysis has shown that one daughter cell of an initial undergoes a predictable set of divisions (see Figure 5–14C), while the other daughter cell becomes a new initial. Thus initial cells reproduce themselves while constantly contributing to new, differentiated tissues; this regenerative activity is a characteristic of **stem cells** (Box 5–2).

The initial cells at the root tip surround a small group of cells that divide at a much slower rate. These cells form the quiescent center (see Section 5.2). The cells in the QC

Box 5–2 Stem Cells in Plants and Animals

Stem cells have attracted widespread attention because of their potential use in medicine and the associated debate on the ethics of embryonic stem cell research. Stem cells have a much wider biological significance, however, because of their central role in the development of multicellular organisms, including plants.

A common definition of stem cells is that they provide the precursors for a variety of differentiated cell types while at the same time maintaining their own undifferentiated lineage. By this definition, stem cells have a prominent role in plant development: specific groups of cells in the meristems are the source of new cells that renew organs and sustain growth throughout the plant life cycle. In both the root and shoot meristems, an intercellular signal maintains the stem cells in an undifferentiated, proliferating state. In the root, the signal originates from the quiescent center (see Section 5.2). In the shoot tip, the stem cells reside in the central zone of the meristem, and the maintenance signal is produced by a small group of underlying cells, which in Arabidopsis express the *WUSCHEL* gene (see Section 5.4).

Maintenance of undifferentiated cells in a defined environment created by intercellular signals is a general feature of stem cells, and the microenvironment is called the "stem cell niche." In animals, stem cells function in an analogous way. For example, in the bone marrow, a signal produced by a specialized type of osteoblast maintains small numbers of long-term hematopoietic stem cells. In the *Drosophila* ovary, germ-line stem cells are maintained only within the range of an intercellular signal produced by specialized cells, called the "cap cells."

The similarities in the way stem cells function in plants and animals probably evolved independently under similar constraints of multicellular development. These constraints include the need to maintain a reserve of undifferentiated cells to replace cells that specialize at the expense of their ability to self-renew, and the need to impose external control on the location and number of these proliferating cells.

are thought to send signals to the initial cells to prevent them from differentiating; this is based on the finding that ablation of QC cells results in the premature differentiation of an adjacent **columella** initial. The nature of this signal, however, is as yet unknown.

Cell fate depends on the cell's position in the root

As a plant develops, cells continuously respond to positional cues and their identity reflects their position in the plant. Experiments using laser ablation of cells have shown that Arabidopsis root cells differentiate according to their position.

In one study, researchers ablated cortical initial cells to create a space and induce division of a neighboring endodermal cell. One of the products of this division filled the space left by the ablation. The newly divided cell differentiated as a cortical cell, as its position would direct, not as an endodermal cell as might be expected if lineage were directing cell fate in this region of the root (Figure 5–27).

Figure 5–27.
Dependence of cell fate on cell position in the root: cortex and endodermis.
(A) The cell layers of the cortex (*yellow*) and endodermis (*blue*) originate from asymmetrical division of a single cell layer (*brown*) in the root meristem. (B) When the cortical initial cell is ablated (*black*), the adjacent endodermal initial (*blue*) divides longitudinally. (C) When one descendant of the endodermal cell occupies the space left by the ablated cell, it becomes a cortical cell (*yellow*). For simplicity, only the two daughter cells of the endodermal initial shown in (B) are in color here. Note that these cells have been pushed away from the meristem, which constantly produces new endodermal and cortical initials.

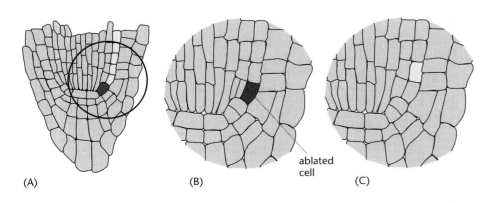

ablated
cell

(A)   (B)   (C)

(A)

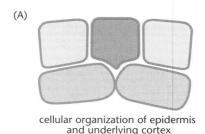

cellular organization of epidermis
and underlying cortex

(B)

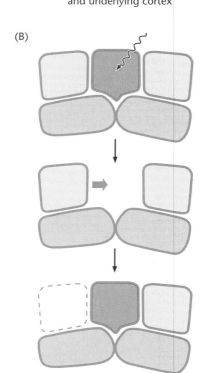

results of ablation

Figure 5–28.
Dependence of cell fate on cell position in the root: epidermis. (A) Schematic cross section through part of the epidermis and cortex. Non-hair cells (*beige*) flank a hair cell (*orange*), which overlies the junction between two cortical cells (*gray*). (B) After laser ablation of a hair cell, if a non-hair cell is displaced into the space overlying the junction between cortical cells, this displaced cell becomes a hair cell.

The root epidermis is made up of two kinds of cells: **hair cells** and non-hair cells. In Arabidopsis, the position of these cells relative to the underlying cortical cells determines their fate: cells overlying two cortical cells form hairs, and cells overlying a single cortical cell form non-hair epidermal cells (Figure 5–28). Epidermal cells of each type were ablated and the fates of neighboring cells and their derivatives were observed. If a neighboring cell or its derivative filled the space created by ablation, the new cell adopted the identity of the ablated cell; for example, if a non-hair cell filled the space left by an ablated hair cell, it differentiated as a hair cell. Thus, again, cells assumed the fate characteristic of their new position. The determinants of cell fate are most likely located in the cell walls between the epidermis and cortex.

Genetic analysis confirms the position-dependent specification of cell type

The simple organization of cells in the Arabidopsis root has made it easy to identify mutations in genes regulating the development of epidermal cells. In wild-type plants, the precursor cells that become epidermal hair cells are termed **trichoblasts**, and those that become non-hair cells are called **atrichoblasts**. One class of mutation has a "hairy" phenotype, meaning that cells in the atrichoblast position develop root hairs (Figure 5–29). Another class of mutation is characterized by a hairless phenotype or a decreased density of root hairs.

Three mutations have been characterized that produce a hairy phenotype. Plants that are homozygous for recessive *glabra2* mutations develop root hairs in every cell file. *GLABRA2* (*GL2*) encodes a DNA binding protein known to regulate transcription and is expressed in atrichoblasts but not in trichoblasts. Thus, GLABRA2 is probably required for the transcription of genes involved in non-hair cell development. The other two mutations that cause a hairy phenotype affect genes that control *GLABRA2*: *WEREWOLF* (*WER*) and *TTG1* encode proteins that interact to regulate *GLABRA2* expression, perhaps by binding to the *GLABRA2* promoter.

In contrast to these "hairy" mutants, plants homozygous for the *caprice* mutation develop few root hairs. Although *CAPRICE* (*CPC*) promotes the development of root hairs, it is expressed in neighboring non-hair cells. *CAPRICE* is thus another example of a gene with non–cell autonomous function. Like SHORT ROOT (see Section 5.2), the CAPRICE protein moves to a neighboring cell to regulate gene expression. The CAPRICE protein is a DNA binding protein that does not contain a transcription activation domain, and therefore is likely to act as a transcriptional repressor. Specifically, CAPRICE is thought to repress *GLABRA2* and prevent the cell from developing as a non-hair cell (see the model in Figure 5–30).

CAPRICE is activated by *WEREWOLF*. Thus, at the same time that *WEREWOLF* directs a cell to develop as a non-hair cell by activating *GLABRA2*, it sends a signal to neighboring cells (via CAPRICE) instructing them not to follow the same fate. As mentioned above, hair cells develop over the cell walls between the epidermis and cortex,

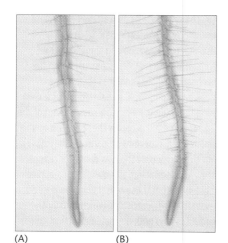

(A)   (B)

Figure 5–29.
Arabidopsis root hairs. (A) Wild-type and (B) *werewolf* mutant. Note the greater number of root hairs in the mutant. (A and B, courtesy of John Schiefelbein.)

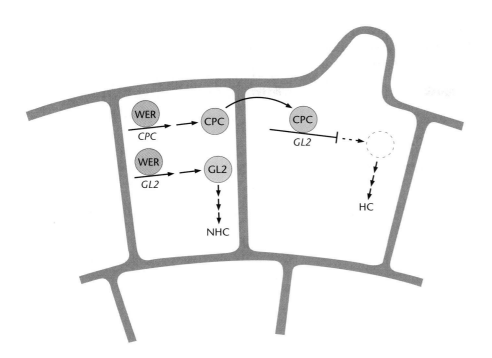

Figure 5–30.
Model for interaction between
WEREWOLF, CAPRICE, and GLABRA2
during development of the root
epidermis. Non-hair cells (NHC) express
WEREWOLF protein (WER; *orange*), which
activates the *CPC* and *GL2* genes. The
GLABRA2 protein (GL2; *blue*) promotes
differentiation as a non-hair cell. The CAPRICE
protein (CPC; *green*) is transported to an
adjacent cell, where it represses *GL2*. In the
absence of GL2 protein, the cell differentiates
as a hair cell (HC). (Solid straight arrow: gene
activation; blunted line: gene repression;
dotted arrows/lines: absence of activation;
curved arrow: movement of CPC protein)

so an additional signal is likely to originate from the cells below the epidermis and determine which epidermal cells express the WEREWOLF protein and inhibit hairless fate in their neighbors.

Lateral root development requires auxin

Arabidopsis has only a single root meristem during embryogenesis, so the majority of the mature root system results from the proliferation of lateral roots. The root system may therefore be thought of as repeating units of lateral roots, each with its own meristem and each capable of producing further lateral roots.

Lateral roots of Arabidopsis are derived from cells in the **pericycle** located near the xylem; these cells proliferate to form a new root meristem (Figure 5–31). The first divisions result in formation of a double layer of densely cytoplasmic cells, which continue to divide to form a lateral root primordium that contains all tissue layers (root cap, epidermis, cortex, endodermis, pericycle, and stele). The lateral root meristem expresses many of the same genes (for example, *SCARECROW* and *SHORT ROOT*) that are expressed in the embryo during formation of the primary root meristem, suggesting that similar regulatory mechanisms are involved in the formation of embryonic and lateral roots. During formation of the primordium, the expression of genes such as *SCARECROW* and *SHORT ROOT* is initiated at different times, indicating that formation of the lateral root meristem is a stepwise process.

Auxin controls lateral root development. There are two main streams of auxin flow in the root: one from top to bottom through the vascular system, and the other from bottom to top through the epidermis and ground tissue. Application of inhibitors of auxin transport to developing root systems has shown that it is the downward stream of auxin that is required for the formation of new lateral root primordia. Mutant plants that accumulate high levels of auxin also reveal the role of auxin in root development. For example, the auxin-accumulating Arabidopsis mutant *rooty* develops many more lateral roots than the wild type, and treatment of wild-type plants with auxin results in a proliferation of lateral roots. Few lateral roots develop in mutants such as *auxin resistant3* (*axr3*), which is defective in auxin-mediated signaling, and *auxin resistant1* (*aux1*), which is defective in auxin transport.

Figure 5–31.
Origin of lateral roots in Arabidopsis.
(A) The first sign that a lateral (secondary) root will develop is the longitudinal division of a pericycle cell (*arrow*). (B) The descendants of the pericycle cell divide to form a root primordium. (C) The root emerges through the epidermis of the parent root. (A–C, with permission from the Company of Biologists.)

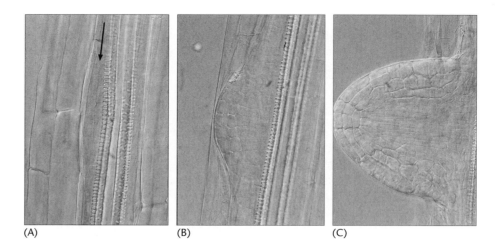

(A) (B) (C)

5.4 SHOOT DEVELOPMENT

The shoot originates from the shoot apical meristem, which arises during embryogenesis and generates almost all of the aboveground parts of the plant. Cells in the meristem divide more frequently than cells in any other part of the shoot, but the rapid addition of new cells in the meristem is balanced by a comparable number of cells that leave to form the stem or the lateral organs such as leaves. Thus the size of the meristem remains relatively constant, even when it is rapidly generating new plant tissue.

As well as providing new undifferentiated cells from which to build the plant body, the meristem establishes some basic features of the plant's geometry, such as leaf arrangement and branching pattern. Leaves begin to form on the flanks of the meristem as small mounds of cells, the leaf primordia (Figure 5–32). The arrangement of leaves relative to each other is determined by where and when the new primordia arise, and these parameters are modified by mechanisms that operate across different cell types and regions of the meristem. New branches can originate from lateral meristems left behind by the splitting of the apical meristem, or from initiation of new lateral meristems. To understand how the meristem maintains itself while regularly initiating

Figure 5–32.
Leaf primordia. Leaves originate as primordia on the flanks of the shoot apical meristem. Axillary meristems subsequently form at the junction between the leaf and the stem. The scanning electron micrograph shows a meristem and leaf primordia of tomato.

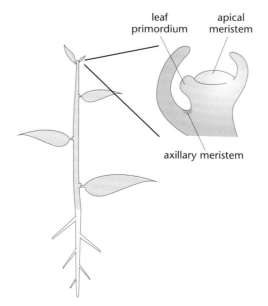

leaf primordium
apical meristem
axillary meristem

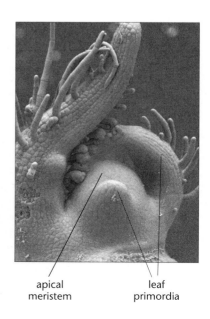

apical meristem
leaf primordia

new organs and branches, we must examine its structure in more detail. Most studies on meristem structure and function have been performed on higher plants. We focus on the shoot apical meristem of these plants first, and then highlight some of the variety of meristem structure across the plant kingdom. We then describe the details of shoot development: leaves, stems, and branches.

Cells in the shoot apical meristem are organized in radial zones and in concentric layers

The typical shoot apical meristem of a seed plant is radially symmetrical, measures between 100 and 250 μm across, and contains a few hundred cells. The meristem can be distinguished from the neighboring tissues by its characteristic cells. Reflecting their high rate of division, the cells are small with thin walls, and their cytoplasm is dense because they do not contain large vacuoles. Apart from relatively small differences in size, all meristem cells appear very similar.

Closer examination, however, reveals that the meristem is not a uniform mass of cells. Different regions can be distinguished by the orientation of cell divisions (Figure 5–33). In the outer (**tunica**) layers of the meristem, the divisions are parallel to the meristem surface (i.e., periclinal). New cell walls form in the plane perpendicular to the cell surface (anticlinally). These cells are aligned to form distinct layers. The number of layers varies between one and eight, depending on the plant species. The Arabidopsis meristem has two tunica layers, L1 and L2. The cells closer to the center of the meristem, which are enclosed by the tunica layers, divide in all directions (this group of cells is called L3 in Arabidopsis). For most cells, this separation is sustained throughout development of the mature organs and stem: cells from the outer layer of the meristem usually give rise to the epidermis, and those from the inner layers contribute to the inner tissues.

Different patterns of cell division, then, form the characteristic layers of cells in the meristem. We do not know whether the segregation of cells into layers is necessary for plant development, but the knowledge that this separation occurs has been exploited experimentally to show that meristem cells communicate with each other to coordinate their growth and differentiation. For example, by grafting diploid and **tetraploid** *Datura stramonium* (a plant in the nightshade family), researchers produced **chimeric** plants in which the meristem consisted of a mixture of diploid and tetraploid cells, which are readily distinguished by their different size. In the mixed meristems, the small diploid and the large tetraploid cells remained in separate layers, and the overall meristem structure stayed the same as in wild-type plants (Figure 5–34). The rates of cell growth and division in each layer had to adjust to accommodate the differences in cell size between layers and to maintain the shape of the meristem.

In addition to differences in the orientation of cell division, different parts of the meristem have different rates of cell division. Cells in the center and at the summit of the meristem (called the **central zone**; Figure 5–35) divide more slowly. The surrounding **peripheral**

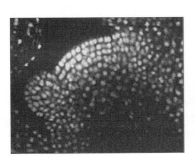
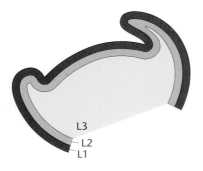

Figure 5–33.
Orientation of cell divisions in the Arabidopsis shoot apical meristem. In the micrograph of a longitudinal section through the meristem, the cell nuclei are stained with a fluorescent dye to show the alignment of cells in concentric layers. The schematic representation shows the cell layers: L1 and L2 are the two tunica layers; L3 is the group of cells enclosed by the tunica.

Figure 5–34.
Mixed meristems of chimeric *Datura* plants. The schematic longitudinal sections show cell layers of different ploidy. Note that tetraploid cells (*blue nuclei*) are larger than diploid cells (*unmarked nuclei*). The meristem layers are colored as in Figure 5–33: L1, *red*; L2, *orange*; L3, *yellow*. (A) All cells are diploid. (B) L1 is tetraploid. (C) L2 is tetraploid. (D) L3 is tetraploid.

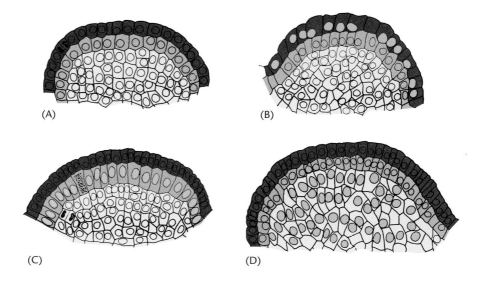

(A)　　　(B)

(C)　　　(D)

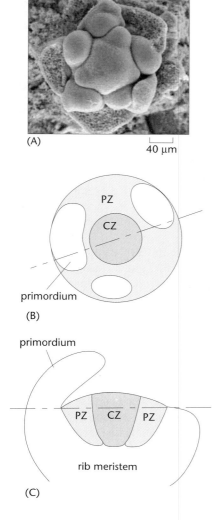

(A)　　40 µm

(B)

PZ

CZ

primordium

(C)

primordium

PZ　CZ　PZ

rib meristem

zone has higher division rates and is the area from which the primordia of new organs arise. In gymnosperms, the cells of the central zone are clearly larger, but in most angiosperms there is no obvious histological distinction between the central and peripheral zones. The different rates of cell division, however, are revealed by measuring the distribution of **mitosis** events across the meristem, either by looking at serial sections or by using labeled DNA precursors to show the synthesis of new DNA in individual cells.

New leaves originate from groups of cells from the peripheral zone of the meristem, which is then replenished both by divisions in the peripheral zone itself and by recruitment of cells from the central zone. The central zone also provides cells to the region at the base of the meristem (called the **rib meristem**; Figure 5–35), which gives rise to the tissues of the stem. This balance of cells between meristem zones and differentiating tissues has to be carefully controlled to maintain the structure and size of the meristem, and specific genes are dedicated to this control. By looking at which cells express these and other meristem-specific genes, studies have found that different areas of the meristem are even more specialized than suggested by the differences in orientation and rates of cell division. Localized gene activity underlies the mechanism that ensures the repetitive production of new organ primordia in a predictable pattern.

As described in Section 5.3 for the determination of cell fate in root tissues, regional specialization is imposed on cells by their position in the plant rather than their lineage. As cells are displaced into new regions, they must assume new functions (Figure 5–36). This has been demonstrated using chimeric plants in which the cells in different meristem layers are genetically distinct, similar to the *Datura* example described above. Although the marked cells usually remain in the same layer as their progenitors, occasionally a cell is displaced to an adjacent layer. The descendants of this "invading" cell are easily traced because of their genetic marker (for example, if they carry an albino mutation, with a defect in chlorophyll maturation, they give rise to a pale patch

Figure 5–35.
Central and peripheral zones of Arabidopsis shoot apical meristem. (A) Scanning electron micrograph (top view of meristem). (B) Schematic representation showing the central zone (CZ) and peripheral zone (PZ). (C) Longitudinal representation, with the section plane indicated by the diagonal line in (B).

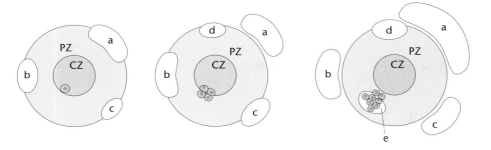

Figure 5–36.
Dependence of shoot cell specialization on position rather than lineage. During meristem growth, the descendants of a cell in the central zone (CZ) of the shoot apical meristem can be displaced to the peripheral meristem (PZ) and may eventually be recruited into leaf primordia. Labels a through e indicate the order of development of leaf primordia.

in green tissues; see Box 5–1). The invading cell assumes the fate corresponding to its new layer; the descendants of a cell displaced from L1 to L2, for example, differentiate as **mesophyll** cells instead of epidermal cells.

The type of shoot meristem described above is typical of angiosperms; that of lower vascular plants (mostly ferns and mosses) is different in important ways (Figure 5–37). For example, *Lycopodium* or ferns such as *Osmunda cinnamomea* typically have a surface layer of enlarged cells that divide asymmetrically, continuously producing smaller daughter cells that form the inner meristem layers. In mosses such as *Physcomitrella patens*, a single apical cell divides along alternating planes and provides new cells for the inner layers. The surface meristem layer in lower vascular plants may be functionally similar to the central zone in seed plants: both provide new cells to replenish the regions of the meristem in which new organs originate and differentiation begins. It is not known whether the morphological differences between the meristems of seed plants and lower vascular plants reflect a fundamental difference in meristem organization, or whether these differences mask an underlying similarity in the genes that control meristem functions. This type of question will become easier to answer as more meristem-related genes are discovered in higher plants and are used to identify their lower-plant counterparts.

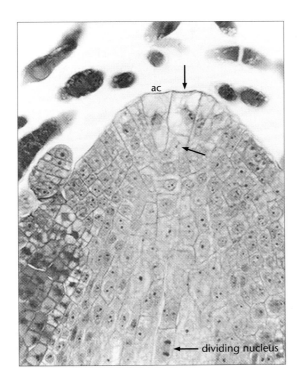

Figure 5–37.
Section through the shoot apical meristem of a fern (*Nephrolepis*). This longitudinal section shows a large apical cell (ac); a daughter cell produced by vertical division of the apical cell (*vertical arrow*); a cell produced by horizontal division of the daughter cell (*diagonal arrow*); and a dividing cell (dividing nucleus) below the apical meristem (*lower, horizontal arrow*). (Courtesy of James Mauseth.)

The number of new meristem cells is constantly balanced by the number that form new organs

The population of cells in the meristem is changeable, but the meristem structure itself remains constant (see Figure 5–36). Although the apical meristem remains at the tip of the shoot throughout development, some cells that made up the original meristem are left behind to differentiate and form new organs, while other cells respond to positional signals to maintain their meristematic roles. So, although the meristem is a permanent feature of a plant, and is still active even in a 3000-year-old bristlecone pine, the cells that make up the structure are constantly changing.

There is a set of genes that instruct cells to stay undifferentiated and to continue dividing while they are in the meristem (Table 5–1). The roles of these genes have been revealed by two main types of mutations: those that cause loss or decrease in size of the meristem, and those that cause a gradual increase in the meristem. Both types have been studied in Arabidopsis (Figure 5–38).

Some mutations that cause loss of the meristem correspond to genes that start functioning during embryogenesis to establish the shoot meristem and are required to maintain the meristem throughout the plant's life. These include *SHOOT MERISTEMLESS*, which is not only required to establish the shoot meristem in the embryo (as described above), but later is required to maintain cell division and delay differentiation of meristem cells.

Another gene with a central role in meristem maintenance is *WUSCHEL* (*WUS*). In *wuschel* mutants, the shoot apical meristem is quickly consumed by just a few leaf primordia (see Figure 5–38). New meristems are then reestablished on the **axils** of leaves, only to again stop functioning prematurely. This meristematic initiation and termination continues, generating a plant with leaves in a disorganized arrangement. Closer examination of *wuschel* meristems reveals that they cannot maintain the central zone, and this zone eventually shuts off the supply of cells that replenish the peripheral zone, from where the primordia emerge. Thus *WUSCHEL* is necessary to maintain the population of stem cells in the shoot meristem (see Box 5–2).

WUSCHEL encodes a transcription factor that is expressed just below the central zone in wild-type plants. Because the region of the meristem that is affected in the mutant is not the same region that expresses the gene, the cells that express *WUSCHEL* must be signaling to the cells above them to function as the central zone. This again highlights how cells in different regions of the meristem communicate and coordinate their behavior to maintain the overall structure.

Table 5–1 Genes that regulate shoot apical meristem development in Arabidopsis

Gene	Product	Initial stage of expression	Location of expression	Effect of mutation on shoot apical meristem
CLAVATA1 (CLV1)	Receptor kinase	Heart	Central zone	Larger
CLAVATA2 (CLV2)	Receptor-like protein	Unknown	Widespread	Larger
CLAVATA3 (CLV3)	Putative ligand	Heart	Central zone	Larger
SHOOT MERISTEMLESS (STM)	Transcription factor	Late globular	Throughout meristem	Absent
WUSCHEL (WUS)	Transcription factor	16-cell	Just below central zone	Repeatedly terminated and reinitiated

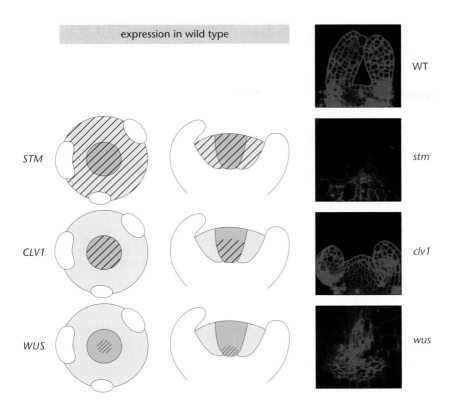

expression in wild type

STM

CLV1

WUS

WT

stm

clv1

wus

Figure 5–38.
Expression patterns of Arabidopsis meristem regulatory genes and corresponding defects in mutants.
The diagrams represent the areas of the meristem—top view (*left column*) and lateral view (*middle column*)—where the genes *STM*, *CLV1*, and *WUS* are normally expressed. The central zone, peripheral zone, and organ primordia are represented as in Figure 5–35B. The micrographs (made by confocal laser scanning microscopy) (*right column*) are longitudinal sections through the shoot apex of a wild-type seedling (WT) and *stm*, *clv1*, and *wus* mutant seedlings. Note that *stm* has no meristem and no organ primordia, *clv1* has an enlarged meristem, and *wus* has formed leaf primordia but the meristem is no longer present.

In *clavata* mutants, enlarged meristems produce more organs (such as leaves and floral organs) than in the wild type (see Figure 5–38). There are three different *CLAVATA* genes, and mutations in any of them produce similar effects. Two of them (*CLAVATA1* and *CLAVATA2*) encode components of a receptor, and the third (*CLAVATA3*) encodes an extracellular **ligand** that probably binds to the receptor. The *CLAVATA* genes are activated by *WUSCHEL* and act as a "brake" to limit that gene's expression. If *WUSCHEL* activity is excessive, the size of the central zone increases, but then *CLAVATA* activity also increases and represses *WUSCHEL*, bringing the meristem size back in balance (Figure 5–39). The *CLAVATA* pathway illustrates how cells within the meristem communicate to assess and control meristematic growth.

The genetics of meristem development has been best studied in Arabidopsis, but mutants with comparable meristem defects exist even in distantly related plants. For example, *knotted1* (*kn1*), the maize ortholog of *SHOOT MERISTEMLESS*, is necessary for meristem development (although the severity of the loss-of-meristem phenotype varies among maize plants with different genetic backgrounds). The *kn1* mutation was originally identified in mutants with altered gene-regulatory sequences that caused changes in the site of expression. The expression of genes in cells where they are not normally active is called **ectopic** gene expression. Ectopic *kn1* expression led to

Figure 5–39.
The size of the shoot meristem is maintained by a regulatory loop involving the *WUSCHEL* and *CLAVATA3* genes. *WUSCHEL* and *CLAVATA3* are expressed in different areas within the central zone of the meristem (indicated by the dashed white lines). Cells expressing *WUS* produce an unidentified signal that diffuses to the upper layers of the meristem to activate *CLV3* (arrow with positive sign). *CLV3*, in turn, produces a peptide signal that diffuses back to the inner cells of the meristem and represses *WUS* (arrow with negative sign). Excessive *WUS* activity would soon lead to its own repression, whereas a decrease in *WUS* levels would diminish *CLV3* expression and allow *WUS* expression to recover. Thus the interaction between *WUS* and *CLV3* stabilizes the levels of *WUS*, which in turn promotes meristem activity.

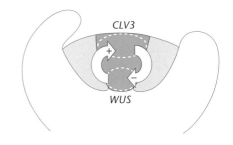

CLV3

WUS

deformed leaves (the "knots" that gave the gene its name), caused by increased cell division, suggesting that *kn1* can activate at least some characteristics of meristems when expressed ectopically (Figure 5–40).

Genes such as *SHOOT MERISTEMLESS* and *WUSCHEL* maintain division and prevent differentiation of meristem cells, but exactly how they do this is not well understood. One of the ways in which these genes control meristem cells, however, is through the production of phytohormones. The phytohormone cytokinin is important to maintaining cell division in the shoot meristem. This is shown, for example, by the rice mutant *lonely guy*, in which the shoot meristem is lost due to mutation of a gene required for cytokinin production. In Arabidopsis, *SHOOT MERISTEMLESS* activates genes involved in cytokinin synthesis, revealing a direct link between meristem regulatory genes and cytokinin function.

Organ primordia emerge from the flanks of the meristem in a repetitive pattern

One function of the shoot apical meristem is to initiate the formation of appendages such as leaves or flowers in specific positions and at regular intervals. These appendages are first visible on the edge of the meristem as primordia (Figure 5–32). The position and the timing of successive primordia determine the arrangement of leaves around the stem (**phyllotaxy**) that is characteristic of that species (Figure 5–41). In addition to being a source of evolutionary diversity, the pattern of emerging primordia can change in the life cycle of a single plant—for example, when the meristem starts to produce flowers instead of leaves. To understand how such basic features of plant architecture are set, we need to look at what causes primordia to appear at specific positions on the flanks of the meristem.

A common type of arrangement of primordia is spiral phyllotaxy (Figure 5–42; see also Figure 5–41B). In this arrangement, the average angle between a primordium, the center of the meristem, and the next youngest primordium is close to 137°. This angle has a special property: the ratio between the whole circumference and the larger sector is the same as the ratio between the larger sector and the smaller. Mathematical models show that spiral phyllotaxy provides the optimal even distribution of leaves around the apex when leaves are added one at a time and the total number of leaves to be added is not determined in advance (that is, the total number of leaves added is in response to the plant's environment rather than genetically predetermined). This optimal distribution may minimize competition for light between leaves.

Figure 5–40.
"Knots" on leaves of maize *knotted* mutant. (A) Wild-type maize leaf (*left*) compared with a *knotted* leaf (*right*). (B) Top view of a *knotted* leaf. The "knots" are caused by excessive cell division near the leaf veins. (A and B, courtesy of Sarah Hake.)

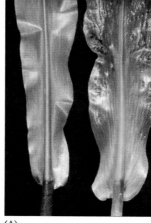

(A) (B)

(A)   (B)

Figure 5–41.
Two types of phyllotaxy. (A) Opposite phyllotaxy in *Crassula arborescens*. (B) Spiral phyllotaxy in *Aeonium lindleyi*.

Older primordia inhibit the emergence and growth of younger primordia in their vicinity: the closer to the preexisting primordium, the stronger is the inhibitory effect. This inhibition also depends on the age of the preexisting primordium: the older the primordium, the smaller its influence on new primordia. A primordium emerges in the region of the meristem where the combined negative effect of preexisting primordia is minimal (Figure 5–43). The point of minimal inhibition shifts to a new position every time another primordium emerges. Much of our knowledge about the predictable pattern of primordium emergence has come from early experiments in which primordia were separated from the meristem by incision, thus blocking communication between primordia. New primordia grew abnormally close to the older primordia from which they had been separated.

We still do not know exactly how leaf primordia inhibit the emergence of new primordia in their vicinity, but this effect is most likely mediated by changes in auxin transport. If auxin transport is inhibited either chemically or by mutation, primordia do not form; and if a small amount of auxin is applied to meristems in which auxin transport has been inhibited, a primordium is again initiated. In addition, in meristems, proteins that pump auxin out of cells (see Section 5.2) are oriented such that the auxin is pumped toward leaf primordia. Combining this evidence, it seems likely that primordia form in regions of the meristem with high auxin levels, and that existing primordia remove auxin from the surrounding meristem area. The emergence of each new primordium changes the distribution of auxin within the meristem and shifts the point where auxin is allowed to accumulate sufficiently to trigger the emergence of the next primordium.

Once the position of a new leaf has been set, a defined group of cells forms a primordium. The number of cells in this group varies with the species: approximately 100 to 200 in maize and tobacco, fewer in a plant with a smaller meristem such as Arabidopsis. This number is not set as a proportion of the total number of cells in the meristem, as has been demonstrated in mutants with altered meristem size. For example, *clavata* mutants have abnormally large meristems and many more primordia than the wild type, but the individual primordia are of normal size. Conversely, in meristems that become gradually smaller (for example, in *wuschel* mutants), primordia still have their characteristic number of cells until the meristem is consumed.

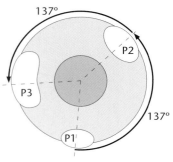

Figure 5–42.
Development of spiral phyllotaxy. Each new leaf primordium arises at an angle of approximately 137° from the previous primordium. The "P" labels refer to stages in primordium development: P1, P2, P3, and so forth, as development progresses.

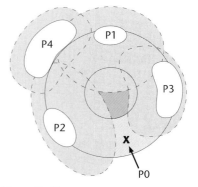

Figure 5–43.
Leaf primordia inhibit emergence of new primordia in their vicinity. The range of inhibition decreases with increasing primordium age. The combined inhibitory effect of all primordia determines where the next primordium can emerge (P0). The range of inhibition (represented by the gray area within the dashed line around each primordium) decreases with increasing primordium age.

Changes in gene expression precede primordium emergence

After a group of cells has been recruited to form a primordium, changes in the cells of this particular region of the meristem cause them to divide and grow in a different pattern. In cell division patterns, the first sign of emergence of a leaf primordium is an increase in the rate of periclinal cell division (that is, with new cell walls parallel to the meristem surface) several cell layers below the epidermis. There are also changes in gene expression at the site of primordium emergence, before any outgrowth is visible. For example, genes required for establishing the meristem (such as *SHOOT MERISTEM-LESS* in Arabidopsis and *knotted1* in maize) are turned off, and genes important for leaf development are turned on (for example, *PHANTASTICA* in *Antirrhinum* and *YABBY* genes in Arabidopsis, as described later in the chapter). All of these genes encode DNA binding proteins and are likely to control additional genes that ultimately modify growth patterns. The identity of these downstream genes, however, remains to be discovered.

The changes in cell division patterns seen early in primordium development suggest that the regulatory genes mentioned above somehow control the cell division machinery. However, another process that may be activated to trigger primordium growth is localized cell wall relaxation, which allows cell expansion driven by **turgor** pressure (see Section 3.5). This has been suggested by experiments using **expansin**, a protein that relaxes cell walls. In tomato meristems, leaf primordia emerged out of place in regions of the meristem where expansin was applied. If cell wall relaxation does allow primordia to bulge out of the meristem, then the changes in the orientation and rate of cell division could be a consequence of the increase in cell growth—that is, if cell division occurs when cells reach a specific size, then a localized increase in growth would eventually lead to a localized increase in the rate of cell division. The extent to which cell growth and cell division are used as control points in plant development is an area of active research.

Development of compound leaves is associated with expression of meristem genes during early leaf development

The repression of meristem regulatory genes in the leaf primordia is widespread but not universal, and plants with compound leaves are one exception. The blade of compound leaves is subdivided into smaller units called **leaflets**. Frequently, each leaflet resembles in shape the larger, compound leaf to which it belongs. Typical examples are the leaves of tomato, pea, and mimosa (Figure 5–44).

As in Arabidopsis and maize, the tomato ortholog of *KNOTTED1* and *SHOOT MERISTEMLESS* is expressed throughout the meristem. Unlike the maize and Arabidopsis genes, however, the tomato *KNOTTED1* ortholog is still active in the cells of young leaf primordia. The levels of *KNOTTED1* expression in the leaf primordium determine the degree of subdivision of the leaf. Mutations that reduce the levels of *KNOTTED1* expression during tomato leaf development simplify the leaf, while **transgenic** tomato plants with increased *KNOTTED1* expression develop leaves that are even more subdivided than those of the wild type.

The correlation between expression of meristem genes in leaf primordia and development of compound leaves is widespread across angiosperms and gymnosperms. In most cases, the genes expressed in primordia are related to *KNOTTED1* and *SHOOT*

Figure 5–44.
Simple and compound leaves. (A) Banana, a simple leaf. (B) *Mimosa pudica*, a compound leaf. (A and B, courtesy of Tobias Kieser.)

(A)   (B)

MERI-STEMLESS. The ability of *KNOTTED1*-like genes to promote the development of compound leaves can be explained if these genes cause the leaf primordia to retain some meristematic activity: the leaflets of compound leaves can be seen as derived from smaller primordia that develop on the edges of the primary leaf primordium.

Leaves are shaped by organized cell division followed by a period of cell expansion and differentiation

The changes in growth pattern that cause a leaf primordium to emerge from the meristem are just the first steps in a long developmental path. We now turn to the topic of how the small group of cells that forms the primordium develops into a large and complex leaf.

Typical leaves are flattened lateral organs whose primary function is **photosynthesis**. Unlike the meristem, the leaf has a limited potential for growth—in other words, it is a **determinate** organ. The variety of forms and functions of leaves, however, is staggering: the spines of cacti, the tendrils of vines, the **cataphylls** that form an onion, and the pitcher-shaped insect trap of *Nepenthes* are all modified leaves (Figure 5–45). Leaf type can also change during the life cycle of a single plant, and even flowers are made up of organs that have much in common with leaves. These organs are genetically determined, but little is known about which genes and processes have become modified during evolution to generate such variety.

Of the multitude of leaf types, we focus here on two examples: the eudicot leaves of Arabidopsis and monocot leaves of maize. A representative eudicot leaf arises from the meristem as a peglike primordium and forms a stalk-like **petiole** and a **blade** with reticulate **venation** (Figure 5–46). A typical monocot leaf is connected to the stem by a **sheath** and arises as a collar-shaped primordium that encircles the meristem. The venation of monocot leaves is usually parallel.

Different regions of the leaf primordium acquire different fates early in development

Leaves start their development as a group of cells recruited from the peripheral zone of the meristem. The formation of the genetically programmed size and shape of a leaf is brought about by growth in three dimensions: the **lateral axis** (width), the

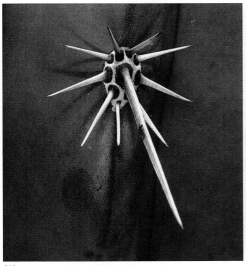

(A)

(B)

(C)

Figure 5–45.
Modified leaves. (A) Cactus spines. (B) Insect trap of *Nepenthes*. (C) Large floating leaves of *Victoria amazonica*. (A–C, courtesy of Tobias Kieser.)

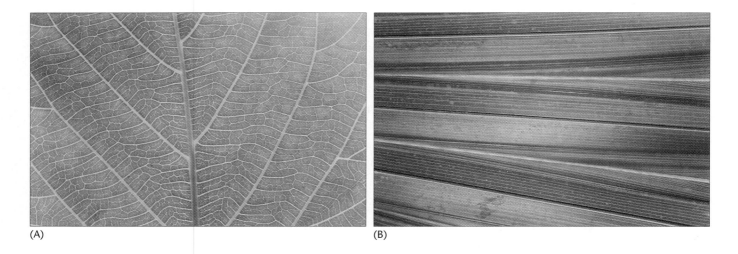

(A) (B)

Figure 5–46.
Leaf venation. (A) Reticulate venation of a typical eudicot leaf, *Acalypha hispida*.
(B) Parallel venation of a typical monocot leaf, *Sabal minor*.

proximo-distal axis (length), and the **abaxial-adaxial axis** (thickness) (Figure 5–47). Growth along each axis results from the combined effect of cell division and cell expansion, as explained below. At the same time, as the leaf grows, specialized cell types and tissues begin to form at specific positions in the leaf. These regional differences are established very early in development, before any signs of cell specialization, by regulatory genes that mark different regions along the growth axes.

We understand best how the abaxial-adaxial axis of the leaf is established. It is usually obvious which side of a mature leaf faces the light; this is called the "adaxial" side (**ad**jacent to the shoot **ax**is). The side that faced away from the meristem becomes the shaded side of the mature leaf and is called the "abaxial" side (away from the shoot axis). In many plants (**C3 plants**; see Chapter 4), the photosynthesizing mesophyll cells are densely packed as **palisade mesophyll** on the adaxial side; on the abaxial side, cells are loosely packed as **spongy mesophyll** and facilitate diffusion of gases through the tissues. Another difference is that stomata are more abundant on the abaxial epidermis.

Figure 5–47.
Differences between regions of the developing leaf along the abaxial-adaxial, lateral, and proximo-distal axes.
(A) Top view of the shoot apex; the lateral and abaxial-adaxial axes are indicated on a leaf primordium. (B) Lateral view, with proximo-distal and abaxial-adaxial axes indicated. (C) Mature leaf, showing lateral and proximo-distal axes. (D) Cross section of mature leaf, showing the abaxial-adaxial axis (ab/ad). (D, courtesy of David T. Webb.)

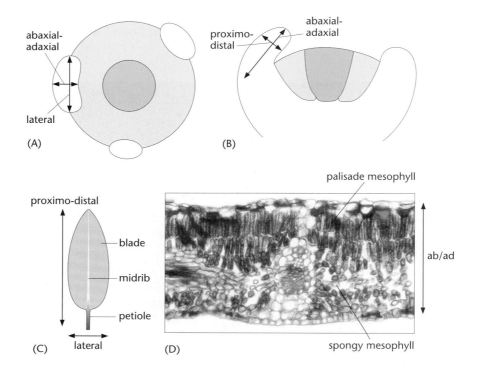

Although more clearly visible in a mature leaf, the differences between the adaxial and abaxial sides are established early in development.

It is likely that signals from the meristem are required for specification of adaxial identity in leaf primordia. If a very young leaf primordium is separated from the meristem by a vertical incision, the primordium develops into a radially symmetrical organ in which all cells express abaxial identity; that is, there is no difference between the adaxial and abaxial sides of the leaf. An alternative explanation for these results is that the cells forming the adaxial and abaxial parts of the primordium had already become distinct before the primordium was separated from the meristem, and that the incision specifically removed or interfered with the cells that would form the adaxial side (Figure 5–48). If only cells that are committed to one particular identity survived the incision, then no abaxial-adaxial asymmetry would develop and radial symmetry would result.

Specific genes regulate the differences between the two faces of the leaf

Another way to find out how leaves develop an abaxial-adaxial axis is to look for mutations that cause leaves to develop as radial organs (as in the surgical experiments described above) and identify the mutated genes. Mutants with radially symmetrical leaves have been found in various species, including *Antirrhinum* (snapdragon), Arabidopsis, and tobacco. In all cases, the mutations affect regulatory genes, mostly transcription factors. For example, the *PHANTASTICA* gene of *Antirrhinum* is needed to form the adaxial side of leaves and it functions at the beginning of primordium development, before the abaxial-adaxial axis is morphologically obvious. Other examples are the Arabidopsis *PHABULOSA* (*PHB*) and *PHAVOLUTA* (*PHV*) genes, which are normally expressed only in the adaxial side of developing leaves. Dominant mutations that cause expression of *PHB* or *PHV* throughout the leaf primordium cause the development of radially symmetrical leaves with only adaxial cell types. Mutants that have lost *PHB* and *PHV* expression would be expected to have the opposite effect— that is, to form leaves with only abaxial cell types. This is not seen, however, because of an earlier function of these genes in meristem development. Like the *stm* mutant mentioned earlier (Section 5.2), mutants that have lost *PHB* or *PHV* function cannot form a meristem and therefore do not produce leaf primordia. The relationship between meristem development and adaxial leaf development is discussed again later, in relation to the development of lateral meristems.

Figure 5–48.
Two interpretations of experiments in which an incision in the meristem caused development of radially symmetrical leaf primordia. (A) The asymmetry of primordia develops in response to a signal that emanates from the center of the meristem (*left column*). The incision prevents the signal from reaching the primordium (*right column*), which becomes radially symmetrical (ab = abaxial; ad = adaxial). (B) Primordia are asymmetrical because they are formed with cells recruited from different regions of the meristem (*left column*). Damage caused by the incision eliminates the primordium cells that normally form the adaxial side of the leaf (*right column*), causing the surviving part of the primordium to develop without abaxial-adaxial differences (i.e., as radially symmetrical).

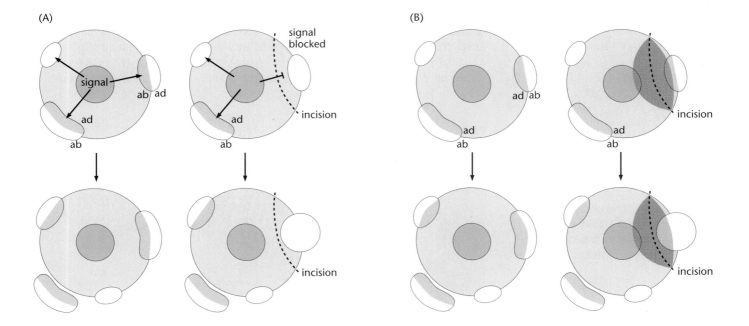

Radialized leaves with only abaxial cell types have been observed, however, and have revealed additional regulatory genes with functions and expression patterns complementary to those of *PHB* and *PHV*. One example is the *YABBY* gene family. In Arabidopsis, *YABBY3* is expressed only in the abaxial side of leaf primordia (Figure 5–49); if plants are engineered to express *YABBY3* in every cell of the developing leaf, the plant forms radially symmetrical organs with only abaxial cell types. In plants that carry mutations in *YABBY3* or in another *YABBY* gene called *FILAMENTOUS FLOWER*, floral organs develop with the opposite defect: they have mostly adaxial identity. This is significant because, as we discuss in more detail in Section 5.5, floral organs are essentially modified leaves. In leaves themselves, *yabby3* and *filamentous flower* mutations have much weaker effects, presumably because their missing functions are covered by the expression of additional *YABBY* genes. Together these results show that *YABBY* genes confer abaxial identity.

Lateral growth requires the boundary between the dorsal and ventral sides of the leaf

A striking observation from the experiments and mutants described above is that when an organ develops with only abaxial or only adaxial identity, it does not grow as a flat structure or blade, as is characteristic of leaves, but as a cylindrical organ. In other words, lateral growth does not occur in these leaves. This suggests that the boundary between abaxial and adaxial cells is necessary for lateral growth. This model is supported by work on the *phantastica* mutant of *Antirrhinum*. The wild-type *PHANTASTICA* gene is required for adaxial development, so the mutation transforms adaxial to abaxial identity. However, the mutant leaves are affected to variable degrees: in severe cases, the leaves become radial organs with only abaxial identity; in less affected leaves, only patches of the adaxial region acquire abaxial identity. At the boundaries between the adaxial cells and the mutant patches of abaxial cells, new leaf blades form, suggesting that contact between tissues with abaxial and adaxial identity is essential for lateral growth (Figure 5-50).

The examples described above imply that the leaf blade begins to develop only after the leaf primordium has emerged from the meristem, when the abaxial-adaxial boundary is established. In some cases, however, the final width of the leaf blade also depends on earlier events, when meristem cells are recruited to form the leaf primordium. In grasses such as maize, the leaf primordium emerging from the meristem already has a bladelike appearance, and the final width of the leaf is affected by the number of cells recruited to the edges of the leaf primordium. In the *narrow sheath* mutant of maize, the primordia do not extend around the meristem to the same extent as in the wild type, and the edges of mature leaves are deleted. As discussed above, one of the first signs of primordium development is the down-regulation of genes associated with meristem identity, such as *knotted1* (*kn1*) in maize. In the *narrow sheath* mutant, the area where *knotted1* is turned off is also narrower, suggesting that the failure to recruit a sufficient number of cells into the primordium may be caused by a reduced ability to down-regulate *kn1*.

Figure 5–49.
YABBY3 gene of Arabidopsis. *YABBY3* is expressed on the abaxial side of leaf primordia, as shown by *in situ* hybridization to *YABBY3* mRNA (cells stained *brown*). (A) Transverse and (B) longitudinal sections through the meristem and leaf primordia. (A and B, with permission from the Company of Biologists, courtesy of John Bowman.)

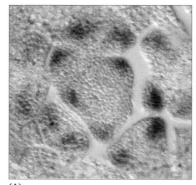
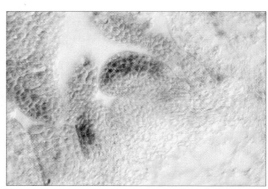

(A)　　　　　　　　　(B)

Figure 5–50.
Growth of the leaf blade along the axis defined by the boundary between the adaxial and abaxial sides of the leaf primordium. (A) In normal leaf development, the boundary between adaxial (ad) and abaxial (ab) sides is continuous, defining the two growing edges of the leaf blade. (B) Mutants that lose adaxial identity, such as the *phantastica* mutant, have no abaxial-adaxial boundary and no leaf edges, and the blade does not grow. (C) In some leaves of the *phantastica* mutant, only patches with abaxial identity appear, surrounded by regions that retain adaxial identity. Each patch defines a new abaxial-adaxial boundary and its associates leaf edges, causing extra leaf blades to grow. (D) Scanning electron micrograph (*top*) shows ridges formed on the leaf surface at the boundaries between patches with adaxial and abaxial identity, as in (C). A cross section through the ridges (*bottom*), which develop as leaf edges. (D, with permission from the Company of Biologists, courtesy of Andrew Hudson.)

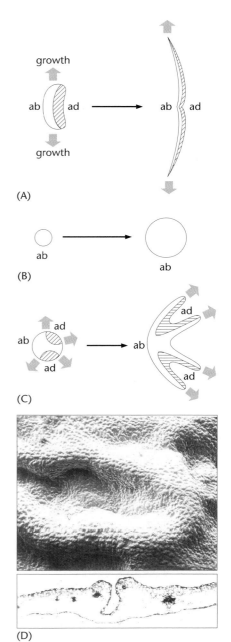

(A)

(B)

(C)

(D)

The leaf reaches its final shape and size by regulated cell division and cell expansion

Early in leaf development, cell divisions are distributed throughout the growing organ. At some point, cell division ceases and cell expansion becomes the major factor in leaf growth. The cessation of cell division occurs initially at the tip (distal end) and progresses toward the base (proximal end) of the leaf. This can be shown by following the proliferation of cells genetically marked at an early stage of leaf development. Albino cells can be created at random across the primordium, using mutation-causing irradiation. As this defect is heritable, the progeny of the irradiated cells form a pale patch as the leaves grow, and the size of the patch is proportional to the amount of cell division that has taken place since irradiation. These patches are generally larger at the bottom of the leaf than at the top, suggesting that cell division ceases at the tip of the leaf before it does at the base (Figure 5–51). Furthermore, if irradiation is carried out later in leaf development, pale patches form only at the base, indicating that cell division has already stopped at the tip.

After cell division stops, cell expansion continues for some time and then ceases too, again first at the tip of the leaf and then progressively toward the base. The continued cell expansion after division stops can be responsible for much of the final size of the leaf.

The time when cells stop dividing and grow only by expansion can vary across different layers of the leaf. This variation creates the differences in cell organization typical of the tightly packed palisade mesophyll and the air-filled spongy mesophyll. Growth stops earlier in the layer that becomes the spongy mesophyll, and air spaces are created when cells are pulled apart by continued growth of the leaf blade.

At maturity, the leaf achieves a final size and shape that is genetically determined. Little is known about how cell division and expansion are controlled to shape the leaf. There is genetic evidence, however, that cell expansion is regulated independently along the width and along the length of the leaf (Figure 5–52). In Arabidopsis, mutations in *ROTUNDIFOLIA3* cause the leaves to become shorter without changing their width,

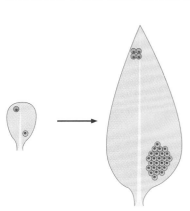

Figure 5–51.
Cell division in the developing leaf.
Cell division stops first near the leaf tip. If cells are marked (such as by conversion to albino cells) early in leaf development (*left*), by the time the leaf matures (*right*), the marked cells at the base of the leaf have proliferated more than the marked cells near the tip.

Figure 5–52.
Leaves of *angustifolia* and *rotundifolia3* Arabidopsis mutants. Wild-type (WT) leaves are compared with *angustifolia* (*an-1*) leaves (narrower than wild type, but same length) and *rotundifolia3* (*rot3-1*) leaves (shorter than wild type, but same width). (With permission from the Company of Biologists, courtesy of Tomohiko Tsuge.)

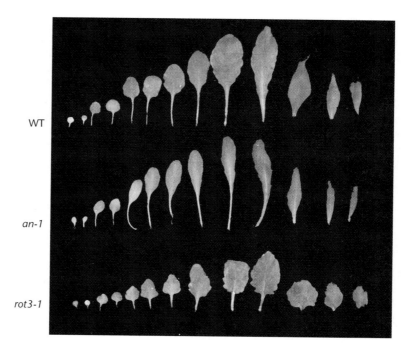

because the expansion of each cell is reduced specifically along the proximo-distal axis of the leaf. On the other hand, the *angustifolia* mutant of Arabidopsis forms narrow leaves with normal length, in this case due to a reduction of cell expansion along the lateral dimension of the leaf.

Leaf growth is accompanied by development of an increasingly elaborate vascular system, which is controlled by auxin transport

Leaves have an enormous variety of vein patterns. Not surprisingly, leaf venation is a classic criterion used in plant taxonomy. At maturity, a leaf vein contains xylem vessels (typically nearest the adaxial face of the leaf), phloem cells (normally near the abaxial side), and reinforcing lignified cells (Figure 5–53). The earliest morphological sign of vascular development is the appearance of elongated cells that are less vacuolated than surrounding cells in the primordium and are aligned in strands (**provascular strands**; Figure 5–54), which eventually become the paths of the mature veins. These early morphological changes are accompanied by changes in gene expression. For example, provascular strands express specific genes encoding **homeodomain** transcription factors (see Table 2–1): *ATHB8* in Arabidopsis and *Oshox1* in rice.

The first provascular strand to appear in a developing leaf becomes the **midvein**. In eudicots, this strand originates deep below the leaf primordium, as a branch of the provascular strands that are forming in the stem. This connection between the vascular system of the leaf and stem is called the **leaf trace**. Development of the leaf trace is a very early event in leaf formation and can precede the emergence of the corresponding primordium in the meristem. In monocots, the leaf trace also begins to form early, but not as a branch of the provascular strands in the stem. Instead, it appears between the stem and the leaf primordium and subsequently extends in both directions.

In both monocots and eudicots, the primary provascular strand gradually extends into the developing primordium and toward its tip. Secondary strands later extend from the primary vein toward the leaf margins (in eudicots) or appear parallel to the primary vein (in monocots). Further veins form (progressively narrower) between the older veins as the leaf grows. Leaf development is completed first at the tip and then proceeds toward

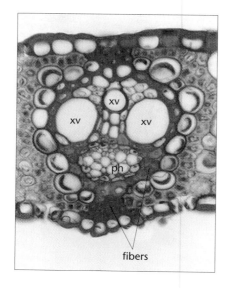

Figure 5–53.
Section through a mature vein in a sugarcane leaf. The section shows xylem vessels (xv), phloem (ph), and the fibers that provide mechanical support to the vein. (Courtesy of James Mauseth.)

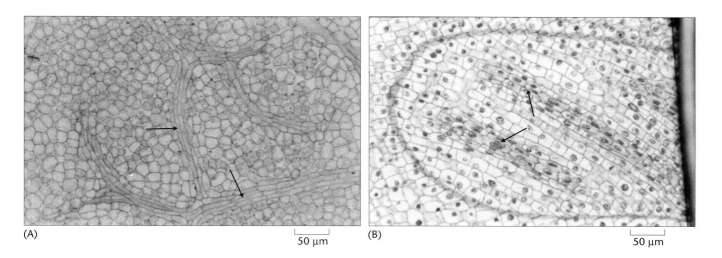

(A)

50 μm

(B)

50 μm

the base. This progression is seen in several processes—for example, as described above for the cessation of cell division and cell expansion. Vascular development is no exception: the minor veins appear in a tip-to-base progression.

The ordered formation of veins may be guided by a process called **auxin canalization**. According to this model, auxin is synthesized at the tip and edges of leaves and transported to the base, and it induces the cells to become more efficient auxin transporters. Because of this positive feedback loop, cells that initially have a slightly higher auxin flux become better and better transporters, draining auxin from surrounding cells and magnifying the difference in auxin transport (Figure 5–55). This can be visualized by thinking of what happens when a flooded field is drained: the receding water carves streams, these are reinforced by the increased flow, and eventually a network of merging streams is formed. According to the canalization model, the streams of auxin through the developing leaf set the path for the differentiation of veins.

The auxin canalization model was put forward after experiments showed that regeneration of the vasculature across a graft in pea stems was directed by a signal transported through the tissues in a polar fashion, from the shoot apex to the root. Furthermore, the newly formed veins tended to grow in the direction of older veins, suggesting that the older veins transported the inducing signal more efficiently, removing it from the surrounding tissues. Crucially, researchers found that externally applied auxin could replace the signal from the shoot apex.

The canalization model can explain the broad features of vein formation, but as yet cannot be used to explain the differences in vein patterns of individual species, and there

Figure 5–54.
Provascular strands. These sections of developing leaves show provascular strands (*arrows*). (A) Arabidopsis (a eudicot). (B) Maize (a monocot).

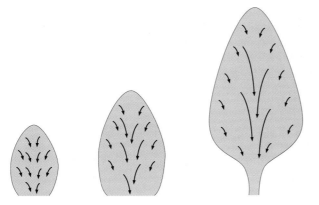

Figure 5–55.
Auxin channeling hypothesis. Early in leaf development (*left*), all cells have a similar capacity to transport auxin (*arrows*) from the tip and edges to the bottom of the primordium. A positive feedback loop causes the auxin flux to be gradually concentrated in fewer, stronger "streams" (*middle and right; larger arrows*). These paths of auxin transport eventually develop as leaf veins.

(A) (B)

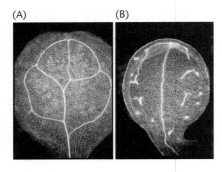

Figure 5–56.
Interrupted veins. (A) Continuous vein patterns in a wild-type Arabidopsis cotyledon. (B) Interrupted veins in the *scarface* mutant cotyledon. These images were produced by differential interference contrast microscopy. (A and B, with permission from the Company of Biologists.)

are some features of vascular development that are not easily explained. For example, some Arabidopsis mutants (*scarface, cotyledon vein patterning,* and *vascular network*) develop interrupted veins, in which the separate stretches of vein are aligned in the path that would be occupied by a normal vein (Figure 5–56). The auxin canalization model seems incompatible with the development of isolated pieces of vein: these would be comparable to disconnected stretches of a stream. To explain the interrupted veins in these mutants, it could be assumed that the path set for the developing veins is initially continuous but cells sporadically fail to differentiate as veins along that path. Alternatively, the mutants could be taken as evidence that the path of veins is established independent of continuity. A firm understanding of vein patterning will require identification of the molecules and genes involved in the process.

Cell communication and oriented cell divisions control the placement of specialized cell types in the leaf

In a mature leaf, about 10 different cell types are instantly recognizable by their distinctive morphology. These include vascular cells, photosynthetic mesophyll cells, and specialized epidermal cells such as trichomes and guard cells (Figure 5–57). The apparently low number of morphologically distinct cell types, however, masks a much higher degree of biochemical specialization. In most cases, we do not know how leaf cells differentiate and acquire their specialized roles.

Development of a functional leaf requires not only distinct cell types but also specific patterns of cell distribution. The best studied case of specification of cell fate and distribution in the epidermis is that of **trichomes** (hairlike structures protruding from the epidermis). They are a good model for several reasons: their distinctive shape means that mutants are easily identifiable, and mutants with defects at various points along the developmental pathway have been identified; their distribution across the epidermis is a relatively simple example of two-dimensional pattern formation; and because trichomes are not essential for survival under laboratory conditions, most mutants survive to be analyzed.

Trichomes can be multicellular, as in tobacco, or can consist of a single large cell, as in Arabidopsis (Figure 5–58). We describe the cellular processes involved in the formation of trichomes in more detail in Section 3.1. In essence, in Arabidopsis, the key stages in trichome formation are nuclear enlargement as DNA keeps replicating after cell division has stopped (a process called **endoreduplication**), increased cell size to form a spike-shaped outgrowth, and the formation of branches.

The mechanism that places trichomes at regular intervals on the leaf is similar to the mechanism for placing hair cells on the root epidermis (see Section 5.3), and several of

Figure 5–57.
Cross section of a typical eudicot leaf, showing the variety of cell types present in different tissues.
(Courtesy of David T. Webb.)

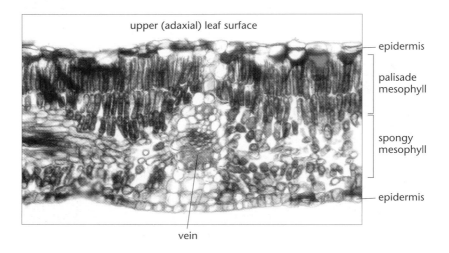

upper (adaxial) leaf surface

epidermis

palisade mesophyll

spongy mesophyll

epidermis

vein

the genes involved are either the same or closely related. In both cases, the mechanism leads to activation of the regulatory gene *GLABRA2* in only some cells of the epidermis. In the leaf, cells expressing *GLABRA2* develop as trichomes, whereas in the root, *GLABRA2* directs development of non-hair cells. The genes that regulate *GLABRA2* in the leaf and root are also similar: *WEREWOLF* and *GLABROUS1* encode closely related MYB transcription factors that activate *GLABRA2* in the root and leaf, respectively, and both function in a complex containing the same TTG protein and the same bHLH proteins GL3 and EGL3. Another similarity is that these complexes activate expression not only of *GLABRA2* but also of a transcriptional repressor that is transported across cells to prevent neighboring cells from acquiring the same fate. In the root the transported repressor is CAPRICE (see Figure 5–30), while in leaves it is the homologous protein TRYPTICHON. As discussed in Section 5.3, in the *caprice* mutant, non-hair cells form inappropriately where root hairs should be, resulting in roots with fewer hairs. Similarly, in the *tryptichon* mutant, failure to inhibit trichome fate in neighboring cells causes clusters of trichomes to appear where single trichomes normally form (Figure 5–59). The similarity in the mechanism that spaces *GLABRA2* expression in the leaf and root illustrates how a single mechanism for generating a pattern of regularly spaced cell types can be applied in different developmental contexts, even though the actual cell types formed (non-hair cells in the root, trichomes in the leaf) are very different.

In leaves, however, a distinctive mechanism is used to distribute stomata on the epidermis. **Stomata** are composed of pairs of specialized cells (the guard cells) that control the aperture of a pore through which gases are exchanged between the leaf and the environment. Each pair of guard cells is normally surrounded by non-stomatal cells (Figure 5–60). Studies in Arabidopsis have shown that oriented cell divisions are responsible for this patterning. The stoma and its surrounding cells originate through a series of asymmetrical cell divisions (Figure 5–61). After each of these divisions, the

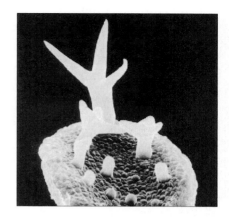

Figure 5–58.
Young Arabidopsis leaf with trichomes at different stages of development.

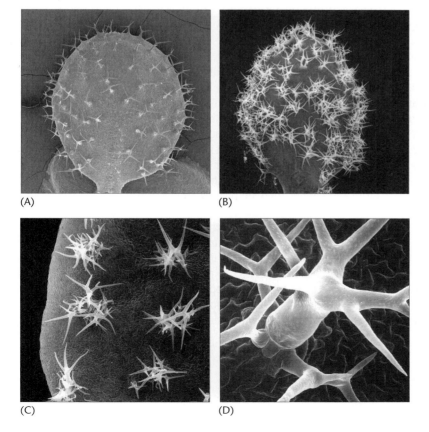

(A) (B) (C) (D)

Figure 5–59.
Trichomes on wild-type and mutant Arabidopsis leaves. These scanning electron micrographs show (A) the wild-type leaf, with evenly spaced, single trichomes; (B) a *tryptichon* mutant leaf, with clusters of trichomes; (C, D) close-ups of trichome clusters of a *tryptichon* leaf.

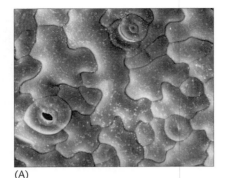

(A)

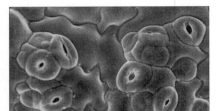

(B)

Figure 5–60.
Stomata. Scanning electron micrographs of leaf epidermis showing (A) evenly spaced stomata in wild-type Arabidopsis, and (B) clusters of stomata in the *too many mouths* mutant.

larger daughter cell becomes a **subsidiary cell** (non-stomatal cell). The smaller cell may repeat the asymmetrical division, or it may divide symmetrically to generate the pair of guard cells that form the stoma. Whenever an asymmetrically dividing cell is adjacent to a developing guard cell, the cell division is oriented to place the smaller daughter cell away from the flanking guard cell. The result is that at least one subsidiary cell is placed between each pair of stomata. Mutants such as *too many mouths* and *stomatal density and distribution1* fail to orient the cell divisions in response to preexisting guard cells, and as a consequence form clusters of guard cells. *TOO MANY MOUTHS* encodes a membrane-localized receptor similar to *CLAVATA1* (whose role in controlling meristem size was described earlier), suggesting that *TOO MANY MOUTHS* participates in cell-to-cell signaling to orient cell division and establish stomatal spacing.

Leaf senescence is an active process that retrieves nutrients from leaves at the end of their useful lifespan

Building a leaf requires input of energy and nutrients, until the leaf reaches sufficient photosynthetic capacity to become a net exporter of nutrients and energy to the rest of the plant. With time, however, the photosynthetic capacity of a leaf diminishes due to environmental damage, shading by further plant growth, or the end of the leaf's genetically programmed lifespan. When this stage is reached, leaf **senescence** is activated.

The old leaf is not simply shed or left to die. Leaf senescence is a carefully coordinated process in which nutrients are retrieved from the leaf for export to the rest of the plant (Figure 5–62). Some of the genes that become active in senescing leaves encode, for example, cysteine proteases, alkaline endopeptidase, ubiquitin, and ribonucleases—enzymes and other proteins involved in the breakdown of components of the old leaf for the recycling of carbon and nitrogen. At the same time, several other genes that are active during senescence (such as genes encoding metallothioneins, pathogenesis-related proteins, and peroxidases) participate in defending the cells against stresses such as wounding or pathogen attack. Activation of these genes may help prolong the life of the senescing leaf during nutrient retrieval or prevent the weakened, senescing leaf from becoming an entry point for infection.

The maintenance of cells and tissues during retrieval of nutrients may account for important differences between leaf senescence and other forms of genetically programmed death. In animal development, **apoptosis** is a prominent form of **programmed cell death**, during which an evolutionarily conserved set of **proteases** (called "caspases") and **nucleases** are activated to degrade the cellular contents. Similar processes can occur in plants, such as in response to pathogen attack. Senescing plant cells, however, do not show some of the typical features of apoptosis, such as early DNA breakdown or activation of caspases. The reason for this difference may be that during leaf senescence the salvage of nutrients depends on the ability to maintain the integrity of the dying cells and tissues for longer periods than are possible in apoptotic cells.

Leaf senescence can be accelerated by sugar depletion and darkness (both related to reduced photosynthetic productivity). **Ethylene** also controls the timing of senescence. This can be seen, for example, in the Arabidopsis *ethylene response 1* (*etr1*) mutants, which have a reduced ability to perceive ethylene, or in transgenic tomato that has reduced levels of 1-aminocyclopropane-1-carboxylic acid oxidase (the enzyme that catalyzes the last step in ethylene production; see Section 6.3). In these plants, leaf senescence is delayed, although not abolished.

Figure 5–61.
Cell divisions giving rise to stomata. Oriented cell divisions ensure that stomata are separated by at least one subsidiary cell. An asymmetrical division forms a small stomatal precursor (*light green*) and a larger, non-stomatal cell (*gray*). Divisions are oriented to avoid stomatal precursor cells developing next to each other. This results in evenly spaced pairs of guard cells (*dark green*).

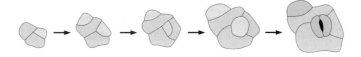

Figure 5–62.
Progressive loss of chlorophyll during leaf senescence. Note that the last areas to lose the green color are close to the veins, reflecting the fact that cells close to the veins need to remain active during nutrient export.

Increased levels of the growth regulator **cytokinin** also delay senescence. This has been shown in transgenic tobacco plants that express the *IPT* gene (which encodes isopentyl transferase, an enzyme from *Agrobacterium tumefaciens* that synthesizes cytokinin) under the control of the promoter from a senescence-activated gene; in this way, extra cytokinin is produced only when the leaf is about to begin the senescence program. The increased cytokinin levels were shown to block senescence. This ability to block senescence in transgenic plants can be put to practical use. Senescence has evolved in conditions of free competition between plants in the wild, but in agriculture it may unnecessarily reduce crop yield. In the cytokinin-overproducing tobacco plants, the yield of biomass and seeds was increased by 50%. It remains to be seen whether this effect can be exploited to increase agricultural yields.

Leaf senescence is often (but not always) accompanied by **abscission** (shedding). Like senescence, abscission is an active process. The cells in the abscission zone do not simply die, thus causing the leaf to break free. Instead, special cells develop in this region. During abscission, the **middle lamella** of their cell walls dissolves, causing the tissue between the cells to break (Figure 5–63). The cells left on the surface attached to the plant then finish the process as their walls become impregnated with **suberin** (the polyphenol-based polymer that makes tree bark and certain endodermal cells impermeable; see Section 3.6). Cells in the abscission zone also produce enzymes (such as polygalacturonidase and **cellulases**) that digest cell wall components. Another gene

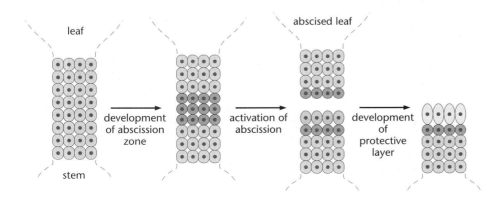

Figure 5–63.
Separation of the abscised leaf.
Separation occurs in the specialized tissues of the abscission zone. The cell layer that remains exposed on the shoot forms a protective, impermeable layer.

expressed specifically in the abscission zone encodes a transmembrane receptor kinase similar to CLAVATA1, suggesting that communication between cells is also essential for abscission.

Branches originate from lateral meristems whose growth is influenced by the apical meristem

In addition to the arrangement of leaves, another major parameter of shoot architecture is the pattern of branching. In lower plants (ferns, mosses), branches can originate by bifurcation of the shoot apex. In higher plants, however, branches usually originate from lateral meristems that form on the axil between the leaf and the stem. In most cases, the lateral meristems do not immediately form a branch, but instead remain temporarily inactive. This inactivity is imposed by a signal from the apical meristem (an effect known as **apical dominance**). Thus the branching pattern is determined not only by the positions where lateral meristems form, but also by control over when and at what rate they are allowed to grow.

Lateral meristems can develop in at least two different ways. In tomato, for example, they originate from meristematic cells that separate from the apical meristem together with the leaf primordium. In this species, the development of the leaf primordium and lateral meristem can be separated genetically. The *lateral suppressor* tomato mutant lacks lateral buds during vegetative growth, because the leaf primordia leave the apex without an associated group of meristematic cells at their base (Figure 5–64). Lateral meristems form differently in Arabidopsis. The cells at the base of the developing leaf begin to differentiate, then revert to meristematic after the primordium has separated from the apical meristem. The common origin of the leaf primordium and the lateral meristem in Arabidopsis has been demonstrated by clonal analysis (see Box 5–1). The cells that formed the lateral meristem were found to belong to the same group of cells that forms the central region of the primordium.

Regardless of whether lateral meristems develop from remnants of the apical meristem or are reinitiated in the axils of leaves, they are placed specifically near the adaxial side of the leaf. The association with the adaxial side is maintained in mutants with altered leaf polarity, as described above. In the Arabidopsis mutant *phabulosa-1d*, in which the

Figure 5–64.
***lateral suppressor* mutant of tomato.**
Close-up of the leaf axils of (A) the *lateral suppressor* mutant and (B) the wild type. (A and B, courtesy of Dörte Müller and Klaus Theres.)

(A)

(B)

entire leaf exhibits adaxial identity, the lateral meristem completely surrounds the base of the leaf. Conversely, expression of *YABBY* genes throughout the primordium not only produces primordia with only abaxial cell types but also eliminates the axillary meristems. The association of lateral meristems with the adaxial side of leaves is probably related to the fact that some of the genes that control adaxial identity in leaves also control meristem development. For example, as we have mentioned, *PHABULOSA* not only specifies adaxial identity but is also required (together with a small group of closely related genes) to form the shoot meristem in the first place.

Although most leaves have an associated lateral meristem, not all leaves produce new branches from their axils. This is because shortly after the lateral meristem is established, its growth is usually arrested. This arrest is triggered by a signal—believed to be auxin—emanating from the apical meristem. If the shoot apex is removed, the lateral meristems are activated in a matter of minutes or hours. This allows plants to reinitiate growth after the apex is destroyed—for example, by grazing animals or by pruning. The timing of the release of lateral meristem growth has a major effect on the final plant shape, as illustrated by a gene that has been central to the domestication of maize. The wild ancestor of maize is teosinte, a highly branched plant (see Section 9.1); in modern maize, the lateral buds are suppressed and all resources are concentrated in the main stem (Figure 5–65). A change in expression of the *teosinte branched 1* (*TB1*) gene is largely responsible for this difference. After the lateral meristems are established, *TB1* acts to inhibit their growth. The protein encoded by *TB1* belongs to a plant-specific family of transcription factors that includes other genes that modulate organ growth, such as *CYCLOIDEA* in snapdragon (described in Section 5.5).

Internodes grow by cell division and cell elongation, controlled by gibberellins

So far we have mainly discussed growth that results from the activity of meristems, but most plant growth is completed outside the meristems. An example is the cell division and cell growth that control final leaf shape and size, as described above. Another model of growth is internode elongation. The expansion of internodes in submeristematic regions is responsible for most of the length of a plant stem. The majority of cells in the internode are derived directly from the division of cells already located in the internode, rather than being produced in the meristem.

The growth regulator gibberellin has a major role in controlling growth in the submeristematic regions of the shoot. Mutations causing reduced gibberellin levels result in dwarfed plants, because both cell division and cell expansion in the internodes are reduced. Such dwarf mutants have been found in a wide variety of plant species, including Arabidopsis, maize, rice, and peas (see Section 9.2). Although growth is dramatically

(A)　　　　(B)　　　　(C)

Figure 5–65.
Teosinte and maize. (A) Teosinte, the ancestor of maize, is highly branched. (B) Maize has a single main stem. (C) The maize mutant *teosinte branched 1* has a highly branched stem, similar to that of teosinte.

stunted in these mutants, their stems have the same number of internodes as wild-type plants (Figure 5–66). Thus, the mutants are deficient in internode elongation but not in internode initiation. It is likely that gibberellin regulates the growth of other plant organs such as leaves in a similar way; the leaves of gibberellin-deficient mutants are smaller than those of wild-type plants.

One example of a dwarf mutation affects the pea gene *Le*, which encodes an enzyme (a hydroxylase) that converts gibberellin precursors to biologically active gibberellins. The mutant gene (*le*) encodes a nonfunctional enzyme, resulting in reduced levels of active gibberellin. Applying gibberellin to younger internodes of *le* mutants has no effect on internode growth or on the maintenance and structure of the meristem. When applied to older internodes, however, gibberellin increases both cell size and cell number and restores normal plant growth. As internodes develop beyond the age when they are responsive to gibberellin, it has no effect on internode size in dwarf mutants.

Internodal gibberellin levels are themselves influenced by the shoot apical meristem, via polar auxin transport. When the apical bud is removed from pea plants, internode elongation is greatly reduced, *Le* expression is down-regulated, and lower levels of active gibberellins are found in elongating internodes. These effects are reversed by the application of auxin, indicating that auxin is transported from the apical bud into the elongating internode. The auxin up-regulates *Le* expression, increasing the levels of active gibberellins and in turn increasing internode elongation (Figure 5–67).

Our understanding of how gibberellin controls internode growth comes from the analysis of dwarf mutants that look like gibberellin-deficient mutants (such as *le* mutants) but are *not* gibberellin-deficient. For example, the Arabidopsis *gai* mutant is dwarfed and dark green, and closely resembles gibberellin-deficient Arabidopsis mutants. But unlike the gibberellin-deficient mutants, *gai* fails to respond to treatment with gibberellin. This suggests that *GAI* does not encode an enzyme involved in gibberellin biosynthesis, but instead is involved at some point in the perception or signal transduction of this growth regulator (Figure 5–68). In fact, *GAI* encodes a nuclear transcription factor (GAI) that is likely to be part of the gibberellin signal-transduction chain.

GAI belongs to the DELLA protein family, named after a conserved amino acid sequence (**DELLA domain**) found in these proteins ("DELLA" is based on the single-letter code used to denote amino acid residues in proteins). DELLA proteins are thought to repress plant growth, and gibberellin promotes growth by overcoming the repressive effects of these proteins. It does this by targeting DELLA proteins for destruction by the proteasome. Thus gibberellin overcomes the growth-repressive effects of DELLA proteins by causing them to be removed from the nucleus. Mutations such as *gai* result in DELLA proteins that are resistant to gibberellin-induced degradation, which explains why growth of these mutants cannot be restored by treatment with gibberellin.

Mutations similar to *gai* have been very important in plant breeding. In particular, *gai*-like mutations affecting DELLA proteins in cereals have resulted in shorter plants with higher seed yields. In the late twentieth century this allowed a large increase in world food production, known as the "Green Revolution" (see Section 9.2).

The **proteasome** is a multisubunit enzyme complex that destroys proteins that are specifically targeted for destruction within it. Targeted proteins are labeled by attachment of a chain composed of **ubiquitin** monomers (such proteins are said to be **polyubiquitinated**). Eukaryotic cells contain a range of enzymes and enzyme complexes that polyubiquitinate specific proteins in response to signals. In the case of DELLA proteins, gibberellin acts through a gibberellin receptor to stimulate enzymes that specifically polyubiquitinate the DELLAs, thus targeting them for destruction. Proteasome-dependent protein destruction is fundamental to the signaling of several

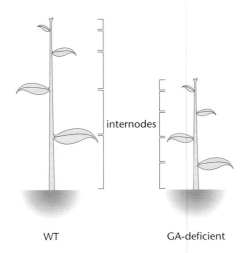

Figure 5–66.
Effect of gibberellin on internodes.
Gibberellin (GA)–deficient mutants are smaller than the wild type (WT) because the internodes are shorter, but the number of internodes is the same in the mutant and wild type.

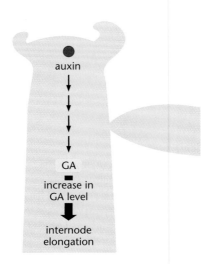

Figure 5–67.
Effect of auxin on internodes. Auxin produced in the shoot apex controls internode elongation by increasing gibberellin (GA) activity.

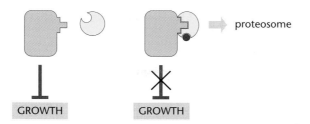

phytohormones, including ethylene (see Chapter 6) and auxin. The auxin receptor is now known to be a component of a multisubunit polyubiquitination complex that targets specific auxin signaling proteins for proteasome-dependent destruction in response to auxin.

A layer of meristem cells generates vascular tissues and causes secondary thickening of the stem

Thus far we have described how plants grow at the root and shoot apices, and how newly formed cells expand and differentiate into specialized tissues. This describes most of the growth seen in herbaceous plants such as Arabidopsis. A very large part of the world's plant biomass, however, is wood. The thick, woody stems of perennial plants are mostly formed by another type of meristem, the cambium (Figure 5–69), which lies between the phloem and xylem layers of the stem. (A cambium is also active in woody roots; see Section 5.3.)

Like the apical meristems, the cambium is made up of small, actively dividing cells and maintains itself while producing new tissues in a reiterative way. The continuous and repetitive activity of the cambium is visible in the form of rings in a cross section of a tree trunk. The **annual rings** are formed by the annual addition of a new layer of vascular tissue, followed by a period of reduced activity in winter or in the dry season.

The new cells produced by the cambium differentiate as vascular tissues: phloem toward the surface of the stem, xylem toward the inside (Figure 5–69). Both types of transport tissue have multiple cell types. For example, xylem contains **tracheary elements**, **parenchyma cells**, and **ray cells** (see Section 3.6). The tracheary elements are long, water-conducting tubes that originate from cells that undergo programmed death after their walls are thickened and reinforced with **lignin**. Parenchyma cells also have lignified walls, but these cells remain alive and store starch or fat. Ray cells facilitate movement of water and sugar across the radial dimension of the stem. Differentiation of these cell types can be observed in a cross section of the stem, with the cambium layer next to a layer of cell expansion, then a maturation zone where secondary cell walls and lignin are deposited, then a region of programmed cell death to form tracheary elements.

In spite of the abundance and enormous economic importance of wood, much less is known about the genetic basis of cambium development, and of the differentiation and programmed cell death in vascular tissues, than of other types of plant growth. In line with its role in vascular differentiation, auxin has a major role in the cambium. Auxin transport inhibitors block cambium growth, and direct measurements in different cell layers in pine show a steep auxin gradient around the cambium, with the highest level in the cambium and progressively lower levels across the cell-expansion zones of the phloem and xylem. This radial auxin gradient may provide positional information during vascular differentiation. Consistent with this idea, studies of poplar show that different *IAA* genes (which encode transcriptional regulators of the response to auxin) are expressed at different distances from the cambium in the differentiating vasculature.

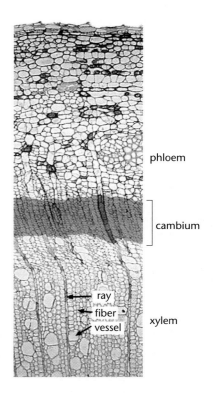

Figure 5–69.
Transverse section through poplar stem, showing the cambium. The cambium produces new phloem cells (toward the surface of the stem) and new xylem (toward the center of the stem).

5.5 FROM VEGETATIVE TO REPRODUCTIVE DEVELOPMENT

The transition from vegetative to reproductive growth is a major change in plant development. In many plants, the meristem structure changes when the plant shifts to reproductive development: the meristem enlarges and its cells show signs of increased metabolic activity (such as enlarged **nucleoli**). These changes have been observed in detail in Arabidopsis: the relatively flat vegetative meristem becomes dome-shaped and the meristem layers become more distinct. The traffic of small molecules through plasmodesmata from mature tissues to the meristem is temporarily restricted during the switch to flowering. The functional significance of these changes is not well understood, but they are the earliest detectable changes in response to the signals that initiate reproductive development.

Reproductive structures in angiosperms are produced by floral and inflorescence meristems

Two types of apical meristem can be distinguished in the reproductive phase of angiosperms. The **floral meristem** produces floral organs and gives rise to a single flower. Floral meristems usually terminate after producing a fixed number of organ primordia—that is, they are determinate (Figure 5–70). The **inflorescence meristem** gives rise to a cluster of flowers, or **inflorescence**. The small mounds of cells produced on the flanks of the inflorescence meristem develop as new meristems, often floral meristems. In many plants, the inflorescence meristem is **indeterminate**—its activity does not end after producing a fixed number of floral buds; others have determinate inflorescences.

The difference between inflorescence and floral meristems is illustrated dramatically in plants that are unable to shift from one type of meristem to the other. In Arabidopsis, this is seen in plants that carry mutations in two closely related genes, *APETALA1* (*AP1*) and *CAULIFLOWER*. In the double mutant, the mounds of cells on the flanks of the inflorescence meristem become new inflorescence meristems, instead of floral meristems. These new inflorescence meristems behave in the same way, until eventually the apex is occupied by a mass of inflorescence meristems, but no flowers. The same genetic defect is the basis for the cultivated cauliflower, in which the edible part is an accumulation of hundreds of thousands of inflorescence meristems (see Section 9.1 and Figure 9–10).

In plants that form more than a single apical flower, the earliest step in reproductive development is conversion of the vegetative meristem to an inflorescence meristem. This transition occurs in response to environmental signals such as a change in day length. These signals, and the genes that mediate their effects, are discussed in Chapter 6. Below, we focus on how the inflorescence meristem produces floral meristems and how these give rise to flowers.

Figure 5–70.
Floral meristems. During the reproductive phase, the apical and axillary shoot meristems become inflorescence meristems, which produce floral meristems. (A) Lateral view of Arabidopsis during the reproductive phase, showing the rosette (R), inflorescence (I), and lateral inflorescence (LI). (B) Top view of the Arabidopsis inflorescence apex; new floral buds (marked by *solid red circles*) arise in a spiral pattern around the apical meristem. (C) Higher magnification of the inflorescence meristem (*white triangle*) and young floral buds, with floral meristems (*red circles*) visible. (D) Schematic lateral view of the Arabidopsis inflorescence, corresponding to (A); *white triangles* and *red circles* represent inflorescence and floral meristems, respectively. (C, courtesy of Leonardo Alves Jr.)

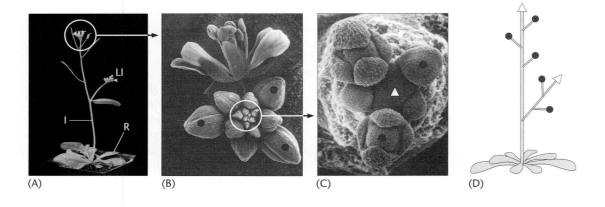

(A) (B) (C) (D)

Development of floral meristems is initiated by a conserved regulatory gene

The primordia produced on the flanks of an inflorescence meristem are converted to floral meristems by the action of a highly conserved gene—*LEAFY* in Arabidopsis, and its **homologs** across the plant kingdom. *LEAFY* encodes a transcriptional regulator whose role in floral meristem formation has been demonstrated by comparing wild-type, mutant, and genetically transformed Arabidopsis plants.

In the wild type, expression of *LEAFY* is the earliest sign that a group of cells derived from the apical meristem will form a flower. In *leafy* mutants, the cells that would normally form a floral meristem instead become a new shoot apex. Conversely, in transgenic plants that express *LEAFY* constitutively, the shoot meristem is converted to a single flower—although only after the plant has shifted from vegetative to reproductive development. Thus *LEAFY* expression is necessary and sufficient to commit meristem cells to flower formation during the reproductive phase of development. *LEAFY* performs this function by coordinating the transcription of additional regulatory genes, including *APETALA1* and *CAULIFLOWER*, and genes that direct the development of each type of floral organ, as described below.

The identification, in a wide variety of species, of genes with sequence and function similar to those of *LEAFY* indicates that the mechanism for the switch to floral development is conserved throughout the flowering plants (Figure 5–71). Indeed, the reproductive function of *LEAFY*-like genes probably extends beyond flowering plants: the expression of *LEAFY* homologs is also increased in gymnosperms in developing reproductive structures.

The expression pattern of *LEAFY*-like genes determines inflorescence architecture

Inflorescences can be determinate or indeterminate. Determinate inflorescences produce a limited number of flower buds, after which the inflorescence meristem is converted to a floral meristem. Typical examples of determinate inflorescences are a single flower at the end of a stalk and a **cyme**, with a terminal flower at the end of each branch. In indeterminate inflorescences, the apical meristems are not converted to terminal flowers and remain active throughout the life of the plant, as long as the environment is favorable. A typical inflorescence of this kind has an indeterminate number

Figure 5–71.
LEAFY-like gene of snapdragon. The *FLORICAULA* gene is the snapdragon homolog of the Arabidopsis *LEAFY* gene. Like *leafy* (*lfy*) mutants in Arabidopsis, the *floricaula* mutant (*left*) develops new shoot apices where the wild-type (*right*) normally develops a flower.

of flowers that grow along the stalk. This structure is found in Arabidopsis and *Antirrhinum* (Figure 5–72). Indeterminate inflorescences are believed to have evolved independently several times from the ancestral, determinate type.

In determinate inflorescences, the conversion of apical meristems to terminal flowers is probably due to the activity of *LEAFY*-like genes. Expression of these genes is repressed in the growing tips of indeterminate inflorescences. In Arabidopsis, *LEAFY* is repressed by *TERMINAL FLOWER 1* (*TFL1*). In *tfl1* mutants, *LEAFY* is no longer repressed in the inflorescence tip, so the inflorescence terminates prematurely forming a single flower (Figure 5–73). This is similar to the effect seen in plants with constitutive *LEAFY* expression. In transgenic plants that constitutively express *TFL1*, *LEAFY* expression is repressed everywhere and the plant resembles a *leafy* mutant. The role of *TFL1* in regulating *LEAFY* expression is also demonstrated in *tfl1:leafy* double mutants, which have inflorescence branches rather than a single flower: *TFL1* has no effect if there is no *LEAFY* function to be repressed. *TFL1* encodes a protein whose homologs in animals interact with kinases. This has led to the idea that *TFL1* is part of a **signaling cascade** that inactivates *LEAFY*.

Homologs of *TFL1* function similarly in other plants. For example, the *Antirrhinum* ortholog of *TFL* (*CENTRORADIALIS*) maintains the inflorescence meristem by repressing the *LEAFY* ortholog (*FLORICAULA*). In tomato, mutations in the *TFL* ortholog (*SELF-PRUNING*) cause the inflorescence to terminate prematurely after producing a smaller number of flowers than in wild-type plants (Figure 5–74). Plants containing this mutation are commercially important because their compressed growth facilitates mechanical harvesting. In the case of tomato, however, the *LEAFY* ortholog is expressed together with *SELF-PRUNING* in the inflorescence meristem, showing that *SELF-PRUNING* does not repress expression of the tomato *LEAFY* ortholog, although it might still antagonize its function.

Figure 5–72.
Indeterminate and determinate inflorescences. The photographs show the indeterminate inflorescences of (A) Arabidopsis and (B) *Antirrhinum*, and (C) the determinate inflorescences of several rice varieties. The diagrams indicate the positions of the inflorescence meristems (*white triangles*) and floral meristems (*red circles*). (Bi, courtesy of Enrico Cohen; Ci, courtesy of Junko Kyozuka.)

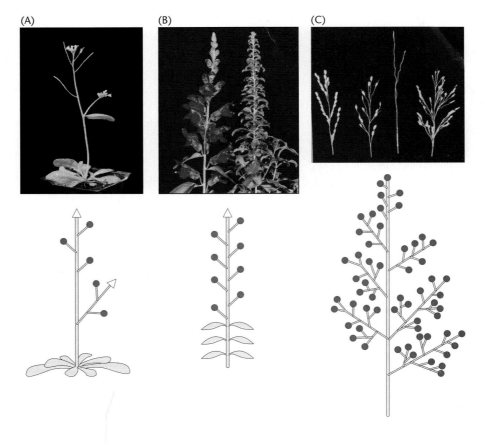

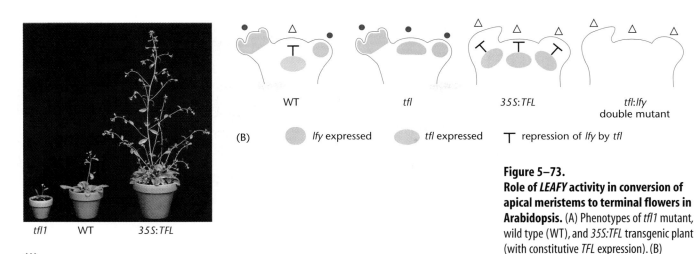

(B)

☐ *lfy* expressed   ☐ *tfl* expressed   T repression of *lfy* by *tfl*

(A)

Figure 5–73.
Role of *LEAFY* activity in conversion of apical meristems to terminal flowers in Arabidopsis. (A) Phenotypes of *tfl1* mutant, wild type (WT), and *35S:TFL* transgenic plant (with constitutive *TFL* expression). (B) Schematic diagrams of the inflorescence apex show the pattern of *TFL/tfl* and *LFY/lfy* expression in the wild type, the *tfl* mutant, the *35S:TFL* transgenic plant, and the *tfl:lfy* double mutant. *TFL* represses *LFY*, which directs development of floral meristems (*red circles*). Meristems that do not express *LFY* behave as inflorescence meristems (*white triangles*). (With permission from the Company of Biologists, courtesy of Desmond Bradley.)

Flowers vary greatly in appearance, but their basic structure is directed by highly conserved genes

Flowers have a wide range of shapes and forms, but all are made up from essentially the same components. In a typical flower, the female reproductive organs (**carpels**, which collectively make up the **gynoecium**) occupy the center and are surrounded by the male organs (**stamens**). Outer layers of sterile organs (the **perianth**) protect the reproductive organs and, in eudicots, often have special features to attract pollinators (Figure 5–75). In eudicot flowers, the perianth usually consists of **petals** and **sepals**. The flowers of grasses also have perianth organs, but the organs corresponding to petals are much reduced in size.

In spite of the enormous variety in the appearance of flowers, development of the floral organs is directed by a highly conserved set of genes. These genes were discovered through the study of **homeotic mutants**, in which floral organs of one type are replaced by another (for this reason, the corresponding genes are also called **floral organ identity genes**). For example, in a type of mutant called "double flower," the reproductive organs are replaced by additional perianth organs—a widely occurring mutation that forms attractive flowers in many species popular with plant breeders (Figure 5–76). The occurrence of homeotic mutations in widely different plant species was an early indication that the genetic basis of floral organ identity is evolutionarily conserved. Parallel studies of homeotic mutants in Arabidopsis and *Antirrhinum* revealed that similar mutants in the two species (Figure 5–77) carried mutations in homologous genes. These studies led to a unified model of how the identity of floral organs is genetically controlled, as described below.

Figure 5–74.
***Self-pruning* tomato mutant.**
(A) Wild-type tomato. The diagram represents the indeterminate growth, which alternates between producing shoot and determinate inflorescence branches (floral meristems represented by *red circles*) but maintains an indeterminate apical meristem (*white triangle*). (B) In the *self-pruning* mutant the inflorescence meristem is prematurely converted into a terminal floral meristem. (Ai and Bi, with permission from the Company of Biologists, courtesy of Eliezer Lifschitz.)

(A)

(B)

Figure 5–75.
Attraction of pollinators. In the flowering plants, a variety of features have evolved to attract pollinators. (A) In the tulip (*Tulipa*), showy petals are the main pollinator-attracting feature. (B) In *Passiflora*, the attractive, colored organs are modified stamens. (C) The flowers of the orchid *Ophrys insectifera* resemble female bees and are pollinated by male bees that try to mate with the flower. (A–C, courtesy of Tobias Kieser.)

(C)

In the ABC model of floral organ identity, each type of organ is determined by a specific combination of homeotic genes

Floral organs in eudicots are generally organized in four concentric whorls (ringlike arrangements). From the outermost to the innermost, the whorls are made up of sepals, petals, stamens, and carpels. Each organ type originates from a specific region in the young floral bud (Figure 5–78).

The fate of the cells that form each type of organ is controlled by floral organ identity genes, all of which encode transcription factors, in most cases with a DNA binding domain called the **MADS domain** (see Chapter 2). The genes that determine the four basic floral organ identities are grouped in three classes, A, B, and C. Note that each class does not always correspond to a single gene. For example, two Arabidopsis class B genes encode proteins that only function together in a multiprotein complex. Figure 5–79 shows the areas where each gene class is expressed in the wild-type flower. The domains of activity partially overlap, so each floral whorl has a unique combination of active identity genes. Class A genes alone are needed to form sepals in whorl 1; classes A and B together are necessary to form petals in whorl 2; classes B and C are

Figure 5–76.
Double flowers. (A) Wild-type flowers and (B) mutant flowers of *Nerium oleander.* The mutant flowers have additional petals. (A and B, courtesy of Tobias Kieser.)

(A)   (B)

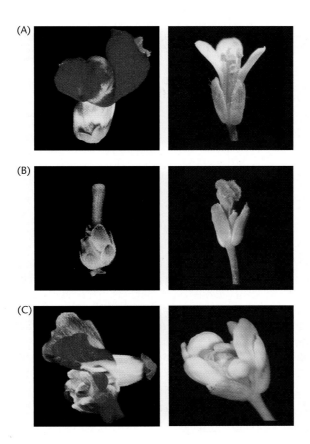

Figure 5–77.
Similar homeotic mutants of
***Antirrhinum* and Arabidopsis.** (A) The
wild-type flowers of *Antirrhinum* (*left*) and
Arabidopsis (*right*). (B) Mutants with petals
converted to sepals and stamens converted to
carpels: *deficiens* in *Antirrhinum* (*left*) and
apetala3 in *Arabidopsis* (*right*). (C) Mutants
with reproductive organs converted to sepals
and petals: *plena* in *Antirrhinum* (*left*) and
agamous in *Arabidopsis* (*right*).
(Ai–Ci, from E. Coen, *EMBO J.*
15(24):6777–6788, 1996. With permission
from Macmillan Publishers Ltd, courtesy of
Enrico Cohen.)

required for stamens to occupy whorl 3; and carpel formation needs class C genes only. Classes A and C genes are mutually exclusive; in the absence of one, the other extends to the whole flower. Together, these rules form the core of the **ABC model**.

The ABC model predicts the phenotypes when the expression of floral organ identity genes is manipulated. By using combinations of mutants and artificially expressed organ identity genes, flowers can be made to form any type of organ, in any position (Figure 5–80). If mutations in classes A, B, and C are combined in the same plant, the floral organs are similar to leaves. This indicates that floral organs and leaves are variants of the same basic type of organ. The similarity between leaves and floral organs is also revealed by the finding that mutations affecting leaf development can have a similar effect on floral organs. For example, the *phantastica* mutation (described in Section 5.4) that affects dorso-ventral identity in *Antirrhinum* leaves also causes floral organs to develop as radially symmetrical organs with only ventral cell types.

Figure 5–78.
Origin of floral organs from specific regions of the floral meristem. (A) Top view of the inflorescence apex of Arabidopsis, with the inflorescence meristem (*white triangle*) surrounded by young floral buds (*red circles*). (B) Section through a floral meristem, along the plane indicated in (A); the meristem has already generated sepal primordia (se) on its flanks. (C) Diagram representing the areas of the young floral bud that will give rise to sepals (*blue*), petals (*green*), stamens (*orange*), and carpels (*red*). (D) The mature flower. (A, courtesy of Leonardo Alves Jr.; B, courtesy of Elizabeth Lord.)

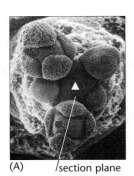

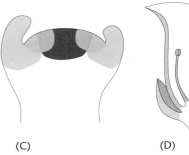
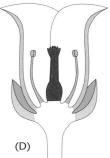

(A) /section plane (B) (C) (D)

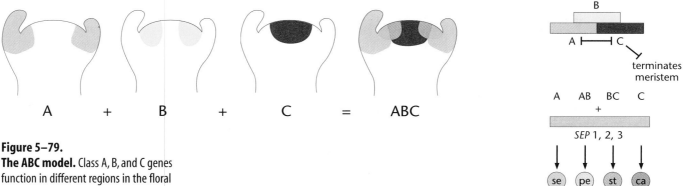

A + B + C = ABC

terminates meristem

A   AB   BC   C
+
SEP 1, 2, 3

se   pe   st   ca

Figure 5–79.
The ABC model. Class A, B, and C genes function in different regions in the floral meristem. Class A and C functions are mutually exclusive (A represses C; C represses A), as indicated by ⊢——⊣ (*upper right*). The region where class B genes function overlaps with the regions where A and C function. The different combinations of A, B, and C activity determine the identity of the organs that arise from each region of the meristem (se = sepals; pe = petals; st = stamens; ca = carpels). The ABC genes can direct floral organ formation *only* when combined with the *SEPALLATA* (*SEP*) genes, which are expressed throughout the flower (but not outside flowers).

Although manipulation of the ABC genes can be used to change floral organ identity within a flower, they alone do not convert leaves to floral organs when expressed outside flowers. Additional genes expressed throughout the flowers are necessary for the ABC genes to function. In Arabidopsis, these genes are *SEPALLATA1* through *SEPALLATA3* (see Figure 5–79), which, like the ABC genes, encode MADS domain proteins. Combined expression of SEPALLATA and ABC proteins converts leaves to floral organs, the identity depending on the particular combination of ABC proteins used.

Figure 5–80.
Changing the identity of floral organs in Arabidopsis by manipulating the expression of ABC genes. (A) Wild-type flower with the normal pattern of ABC gene expression. (B) Double mutant, with mutant (nonfunctional) B and C genes (B⁻, C⁻); all organs express only A genes and develop as sepals. (C) The same double mutant, but with B genes artificially expressed throughout (expanded); organ primordia develop as petals (A and B gene activities combined). (D) Mutant flower with C genes artificially expressed throughout (repressing A genes) and mutant B genes; all organs express C genes only and form carpels. (E) Mutant flower with mutant A genes (A⁻) and B genes expressed throughout; all organs develop as stamens (BC combination). (F) Triple mutant (mutant A, B, and C genes); floral organs are replaced by organs similar to leaves.

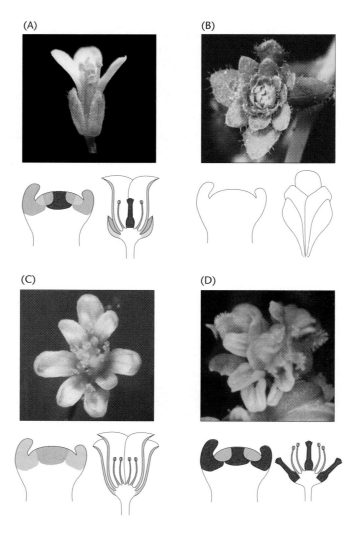

(A)   (B)

(C)   (D)

The SEPALLATA proteins physically associate with different ABC proteins. Each multi-protein complex composed of SEPALLATA and a different combination of ABC proteins is believed to control transcription of the genes required for development of a specific type of floral organ (Figure 5–81).

In addition to their functions in controlling floral organ identity, some of the ABC genes have other roles in floral development. As mentioned earlier, the floral meristem is normally determinate, producing a fixed number of organ primordia. In mutants affecting class C genes, however, the floral meristem becomes indeterminate and continues producing organs beyond the usual set of four whorls (as seen in the double flowers in Figure 5–77C). The class C genes limit the growth of the floral meristem in part by shutting down expression of *WUSCHEL*, which is necessary to maintain the undifferentiated cells at the center of the meristem (see Section 5.4).

Floral organ identity genes are conserved throughout the angiosperms

Following the isolation of floral organ identity genes from Arabidopsis and *Antirrhinum*, homologous genes were isolated from many other species. The expression patterns and functions of class B and C homologs are similar in most species examined (Table 5–2), although the function of class A homologs is more variable. In spite of some variation in detail, the essence of the ABC model is valid across the angiosperms: floral organ identity is specified by a set of similar genes that function according to a conserved combinatorial code.

Even in flowers, such as those of grasses, that look very different from "typical" flowers, the underlying expression patterns and functions of floral homeotic genes are similar to those described above. In grasses, the position of petals is occupied by smaller organs called **lodicules**, while the organs known as **palea** and **lemma** are loosely similar to sepals. As in eudicot flowers, mutations in the maize *APETALA3* homolog *SILKY* change stamens to carpels and lodicules to palea- or lemma-like organs (Figure 5–82).

The conservation of homeotic genes is also illustrated by the fact that even when expressed in distant species, the genes perform the same function. For example, expression of the class B *Antirrhinum* gene *DEFICIENS* in an Arabidopsis class B mutant

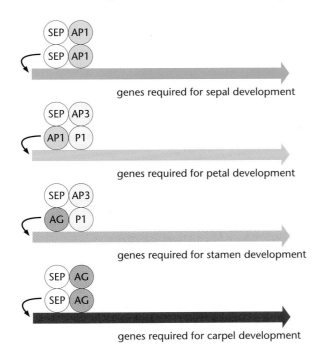

genes required for sepal development

genes required for petal development

genes required for stamen development

genes required for carpel development

Figure 5–81.
Combined action of ABC and SEPALLATA proteins. Each multiprotein complex is believed to control transcription of the set of genes required to form each type of floral organ. The proteins encoded by the ABC genes are AP3 (APETALA3), PI (PISTILLATA), AP1 (APETALA1), and AG (AGAMOUS). The SEP protein can be SEP1 (SEPALLATA1), SEP2 (SEPALLATA2), or SEP3 (SEPALLATA3).

Table 5–2 Homologs of class B and class C floral identity genes across model plant species

Species	Class B genes		Class C genes
Arabidopsis thaliana	*APETALA3*	*PISTILLATA*	*AGAMOUS*
Antirrhinum majus	*DEFICIENS*	*GLOBOSA*	*PLENA, FARINELLI*
Petunia hybrida	*PMADS1*	*FBP1*	*FBP6, PMADS3*
Oryza sativa (rice)	*OsMADS16*	*OsMADS2*	*OsMADS3*
Zea mays (maize)	*SILKY*	?	*ZAG1, ZMM2*

Note: The sequences, expression patterns, and mutant phenotypes of the genes in each column are similar. In some species, class C function is controlled by more than one gene.

(A)

(B)

Figure 5–82.
Homeotic gene mutation in maize.
(A) Male floret of wild-type maize, showing stamens (st), palea (pa), and lodicules (lo). (B) Male floret of the *silky1* mutant. The stamens (tst) and lodicules (tlo) are transformed into organs resembling carpels and palea/lemma.
(A, courtesy of C. Whipple and R. Schmidt.)

(*apetala3*) restores petal and stamen development: *DEFICIENS* and *APETALA3* are functionally equivalent.

Despite the conserved genetic control of organ identity, the final appearance of flowers is, of course, immensely varied. Much of this variation is either superimposed on the specification of floral organ identities (e.g., floral symmetry) or appears at steps after organ identity is specified (e.g., petal color). We now turn to these more varied features of flowers.

Asymmetrical growth of floral organs gives rise to bilaterally symmetrical flowers

Flower shape is not determined solely by organ identity; organ shape may vary in the same whorl. If shape is identical within each organ type, the resultant flowers are radially symmetrical (e.g., magnolia flowers). In other flowers (e.g., snapdragon and orchids), dorsal and ventral petals and sepals have distinct shapes, resulting in flowers with bilateral symmetry (Figure 5–83). Bilateral symmetry has evolved independently numerous times in the angiosperms, probably from radial symmetry.

Genes that establish bilateral symmetry have been identified in *Antirrhinum*. These flowers have clear differences between dorsal petals (those that emerge closest to the inflorescence meristem) and ventral petals (Figure 5–84). *CYCLOIDEA* and *DICHOTOMA* are genes encoding putative transcription factors, both of which are required for asymmetrical growth along the dorso-ventral axis in *Antirrhinum* flowers: dorsal organs look like the ventral organs in plants with mutations in either gene. In *cycloidea:dichotoma* double mutants, the organs are identical and the flowers are radially symmetrical: the plants have only ventral-type organs.

CYCLOIDEA and *DICHOTOMA* are expressed very early in the dorsal part of the developing flower and inhibit the growth of organ primordia that emerge in the dorsal region, so that dorsal petals are smaller than ventral. In later stages, the effects of *CYCLOIDEA* and *DICHOTOMA* on organ growth depend on the organ identity, revealing an interaction with the processes controlled by the floral homeotic genes.

Additional regulatory genes control later stages of floral organ development

The specification of organ identity is not restricted to the earliest stages of floral organ development; the floral organ identity genes continue to function until late stages. This

(A)   (B)

Figure 5–83.
Bilaterally symmetrical flowers.
(A) *Paphiopedilum* orchid. (B) *Viola tricolor*
(wild pansy). These are compared with
(C) the radially symmetrical flower of
Clematis × *jackmanii*.
(A–C, courtesy of Tobias Kieser.)

(C)

can be shown using a temperature-sensitive mutant allele of the class B gene *APETALA3* of Arabidopsis. By shifting mutant plants at different times from the temperature at which *APETALA3* is active to a temperature at which its function is lost, researchers showed that loss of activity even at late stages prevents the development of functional stamens. Thus floral organ identity genes are active in coordinating the

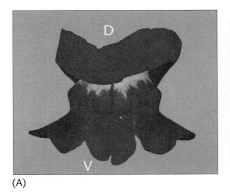

(A)   (B)

Figure 5–84.
Floral asymmetry in *Antirrhinum*.
(A) Front view of wild-type *Antirrhinum*
flower, with ventral (V) and dorsal (D) sides
indicated. (B) Radially symmetrical flower of a
cycloidea:dichotoma double mutant, in which
all organs are similar to the ventral organs of
the wild type. (A and B, from E. Coen, *EMBO J.*
15(24):6777–6788, 1996. With permission
from Macmillan Publishers Ltd, courtesy of
Enrico Coen.)

development of floral organs at all times in flower development. Very little is known about the processes and gene activities controlled by the homeotic genes.

Genes that control the development of specific parts of floral organs, however, have been identified. For example, in Arabidopsis, two such genes—*SHATTERPROOF1* and *SHATTERPROOF2*—are necessary for the development of specific tissues of the carpel, which eventually give rise to the structure that allows the fruits to burst open and release the seeds. *SHATTERPROOF* genes are closely related to *AGAMOUS*, a master regulator of carpel development. The role of *SHATTERPROOF* genes in a specific aspect of carpel development illustrates how regulatory genes can duplicate and acquire more specialized functions during evolution.

Another example of a gene that functions downstream of floral organ identity genes is the *Antirrhinum* gene *Mixta*, which is necessary for the development of a specific epidermal cell shape in petals (Figure 5–85). On the dorsal side of wild-type petals, the epidermal cells are conical; this shape is important in giving the petal its brightness, by refracting light toward the cell vacuoles where the pigments are. In *mixta* mutants, the cells are flat and the petal color appears dull. *Mixta* encodes a transcription factor of the MYB family, the same as *GLABRA1* and *WEREWOLF*, which as mentioned earlier, control the differentiation of other types of epidermal cells.

5.6 FROM SPOROPHYTE TO GAMETOPHYTE

In the life cycle of flowering plants, the developing stamens and carpels are the sites of meiosis. As mentioned in our overview of plant development (Section 5.1), a characteristic of plants is that the haploid products of meiosis are not gametes but spores (for this reason, the diploid plant is called the "sporophyte"). Unlike gametes, spores divide and develop into a haploid multicellular individual, male or female. It is within this new individual, the "gametophyte," that some cells differentiate as gametes (Figure 5–86). In this section, we focus on the development of angiosperm gametophytes: the pollen grain (male gametophyte) and the embryo sac (female gametophyte).

The male gametophyte is the pollen grain, with a vegetative cell, gametes, and a tough cell wall

In angiosperms, the pollen grain is a much-reduced male individual, made up of only two or three cells: one vegetative and one or two corresponding to the germ line (Figure 5–87). Some seed plants, however, have more complex pollen (with up to 40 cells in certain conifers), a reminder of the relation of pollen grains to the more complex gametophytes of lower plants.

Figure 5–85.
Scanning electron micrograph of the epidermis of *Antirrhinum* petal, showing flat-looking *mixta* mutant cells surrounding a patch of conical wild-type cells. The patch of wild-type cells developed because the *mixta* mutation is caused by a transposon, which can excise during development and give rise to a patch of cells in which the wild-type gene is restored (see Chapter 2 for transposons).

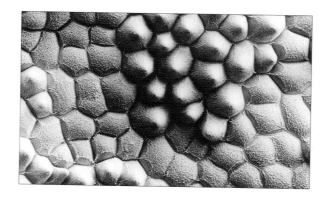

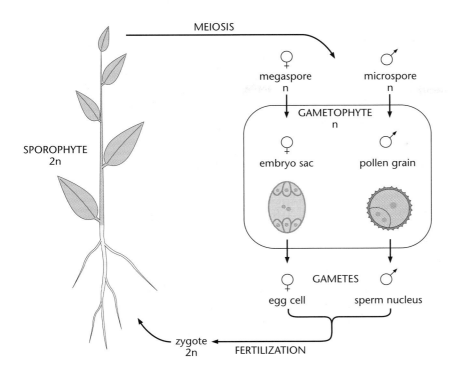

Figure 5–86.
Alternating generations in angiosperms.
In flowering plants, the large, conspicuous organism is the diploid (2n) sporophyte; meiosis takes place in the flowers. The resulting haploid cells (megaspores and microspores) give rise to the female and male gametophytes (embryo sac and pollen grain, respectively). These develop as small haploid organisms embedded in the sporophyte and produce the haploid gametes by mitosis. Fertilization produces a diploid zygote, which gives rise to a new sporophyte.

Pollen development begins in young **anthers** with the division of a subepidermal cell called the **archesporial cell** (Figure 5–88). Subepidermal cells are derived from the L2 layer of the meristem (see Section 5.4), so the origin of the male germ line from the L2 layer can be traced to this initial L2-derived cell. One of the cells derived from division of the archesporial cell eventually gives rise to the **tapetum**, a layer of cells that secrete essential substances to support pollen formation (see below); the other cell gives rise to the **microsporocyte** (also called the **pollen mother cell**). The microsporocyte is responsible for the transition from the diploid to the haploid phase of the life cycle. This transition can also be pinpointed genetically. In Arabidopsis, mutations in the *SPORO-CYTELESS* (*SPL*) gene (also called *NOZZLE*) specifically block the progression of the microsporocyte to meiosis; the same defect is seen in the female sporocyte (see below).

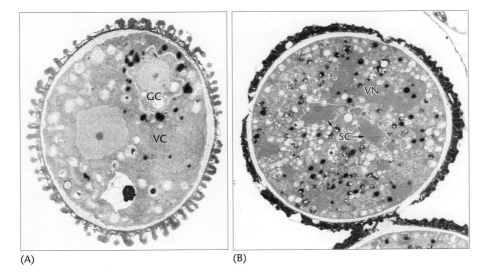

Figure 5–87.
Internal structure of Arabidopsis pollen grain. Electron micrographs of (A) an immature pollen grain, showing the vegetative cell (VC) and generative cell (GC), and (B) mature pollen, showing the nuclei of sperm cells (SC) and the vegetative nucleus (VN).

(A)   (B)

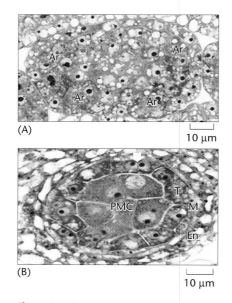

Figure 5–88.
Archesporial cells. (A) Section through a developing anther of Arabidopsis, showing the archesporial (Ar) cells. (B) At a later stage of anther development, the descendants of archesporial cells differentiate as the pollen mother cells (PMC), tapetum (T), and tissues surrounding the tapetum, known as the middle layer (M) and endothecium (En).

Many of the genes that are essential for gametophyte development (both male and female) function in meiosis. For example, the *SYNAPSIS/DETERMINATE INFERTILE 1* gene encodes a protein similar to **cohesins**, which are required in all eukaryotes for pairing of homologous chromosomes during meiosis. Another example is *AtSPO11*, which encodes a **topoisomerase**, an enzyme essential to initiating meiotic recombination. As in yeast and mammals, recombination in Arabidopsis is an essential part of the pairing and correct segregation of homologous chromosomes in meiosis. (See Section 3.2 for a discussion of meiosis and recombination.)

When meiosis is successfully completed, four haploid cells are formed. In the cycle of alternating diploid and haploid generations, these cells occupy a position equivalent to that of the haploid spores essential to the dispersal of mosses and ferns. In seed plants, however, the haploid cells form the gametophyte within the diploid plant. In the case of the male gametophyte, this requires only two mitotic divisions. The first division is unequal and generates a larger **vegetative cell** and a smaller **generative cell** (Figure 5–87). The nucleus of the generative cell has condensed chromatin and is less transcriptionally active than the nucleus of the vegetative cell. This is reflected, for example, in the finding that most of the genes expressed in pollen are active specifically in the vegetative cell. The vegetative cell is responsible for deposition of the inner wall of the pollen grain (the **intine**) and supports the growth of the pollen tube on germination of the pollen grain (see below).

The asymmetry of the division that separates the vegetative and generative cells is necessary to establish the fate of the generative cells. If the asymmetrical division is inhibited (for example, by treating developing *Tradescantia* pollen with **colchicine**, a drug that disrupts **microtubules**), only a vegetative cell develops. In the *GEMINI POLLEN* (*GEM*) mutant of Arabidopsis, the microspore divides in variable planes, and when the division is symmetrical, both cells develop as vegetative cells. Thus vegetative or generative fate may be controlled by the distribution of a factor in the cytoplasm of the microspore. Asymmetrical cell division coupled to specific cell fates is seen in both plant and animal development.

After separation from the vegetative cell, the generative cell divides again to form the two **sperm** nuclei, which participate in the double fertilization that is typical of seed plants (see the discussion of endosperm development in Section 5.2). The vegetative and sperm cells of the mature pollen grain are encased together in a coat made up of two cell walls. The internal one, the intine, is composed mainly of **callose** and **cellulose**. When the pollen grain germinates, the intine is continuous with the wall of the elongating pollen tube. The pollen grain is also surrounded by a tough external wall, called the **exine**, which is rich in phenolic compounds. This wall is interrupted by openings (apertures) through which the pollen tube germinates. The external surface of the exine frequently has ridges and spikes arranged in intricate patterns, which are genetically determined and can be used to identify the species that produced the pollen grain.

Pollen development is aided by the surrounding sporophyte tissues

The completion of pollen development does not depend only on gene activity within the developing pollen grain, but is intimately dependent on the surrounding sporophyte tissue. A specialized layer of sporophyte cells, the tapetum, is essential for the completion of pollen development. The tapetum develops from the sister cell of the archesporial cell (as described above) and surrounds the chamber where pollen grains develop. The tapetal cells produce proteins and other substances (such as **flavonols**) that form the outer coat (exine) of the pollen grain. The commitment of tapetal cells to supporting the development of the pollen grain culminates in suicide: at the end of pollen development, the tapetal cells undergo programmed cell death and release proteins, which are incorporated into the developing exine.

The essential role of tapetal cells in pollen development is reflected in the discovery that many of the mutations that cause male sterility affect primarily the tapetum. Male sterility is economically desirable in many crops, because preventing **self-fertilization** facilitates the production of more vigorous seed from crosses between different varieties (see Section 9.2 for more details on **hybrid vigor**). The tapetum-affected sterile male mutants with the highest economic importance are unusual in that they carry mutations not in the nuclear DNA but in the DNA of **mitochondria** (for this reason, they are called "cytoplasmic," rather than nuclear, mutations). Consequently such mutations are not inherited as Mendelian characters; instead, the phenotype of the progeny is determined by the female parent, which contributes the mitochondria to the progeny cells. Even though the defects caused by these mutations are present in the mitochondria of all cells, only the tapetal cells seem to be affected. This suggests that tapetal development is particularly sensitive to reduced mitochondrial function.

Cytoplasmic male sterility (CMS) mutations are used to facilitate the production of hybrid seeds, in combination with dominant nuclear mutations that restore fertility even in the presence of the defective mitochondria. One example is the CMS-T mutation in maize (Figure 5–89). The normally male-sterile CMS-T plants become fertile in the presence of the nuclear mutations *Restorer of fertility 1* and *Restorer of fertility 2* (*Rf1* and *Rf2*). Pollination of CMS-T plants with pollen from plants homozygous for *Rf1* and *Rf2* produces hybrid progeny that are fertile, self-pollinate, and set seed—in spite of inheriting mutated mitochondria from the female CMS-T parent. The mutated mitochondria of CMS-T plants produce an abnormal peptide of unknown activity (URF13). The *Rf2* gene encodes an aldehyde dehydrogenase, suggesting that *Rf2* prevents the accumulation of a toxic aldehyde produced by the defective mitochondria of CMS-T plants.

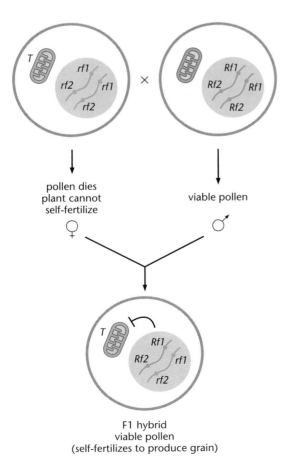

pollen dies
plant cannot
self-fertilize
♀

viable pollen
♂

F1 hybrid
viable pollen
(self-fertilizes to produce grain)

Figure 5–89.
Cytoplasmic male sterility (CMS) in maize. Plants that carry mitochondria with a mutant gene known as *T* cannot produce pollen and therefore can function only as female plants. The effect of *T*, however, can be suppressed by the combined action of the dominant *Rf1* and *Rf2* genes. After crossing the plant with *T*-mitochondria to a male plant homozygous for *Rf1* and *Rf2*, all the F1 progeny can produce pollen and self-fertilize to produce seeds, in spite of inheriting *T*-mitochondria from the female parent. CMS is used commercially to facilitate the production of highly vigorous F1 hybrid seeds (see Chapter 9).

Figure 5–90.
Ovule development in angiosperms.
(A, B) The archesporial cell in the ovule primordium undergoes the first round of meiotic division, while it becomes surrounded by two layers of protective integuments. (C) After the second meiotic division, three of the four haploid cells degenerate, leaving one megaspore. (D) In the mature ovule, the megaspore has undergone three rounds of mitosis to generate the haploid embryo sac, which contains the egg cell, central cell, synergid cell, and antipodal cells. The sporophytic, diploid integuments have now enveloped the embryo sac, leaving only a small opening (the micropyle). The funiculus attaches the ovule to the rest of the sporophyte. (E) Micrograph of an ovule primordium of Arabidopsis at the stage represented in (C); the megaspore is indicated by the *red arrow*. (F) Section through a mature Arabidopsis ovule, corresponding to (D). The nuclei of the central cell, egg cell, and two synergids are visible as white dots within the embryo sac. (E and F, courtesy of Gary Drews.)

The female gametophyte develops in the ovule, which contains gametes for the two fertilization events that form the zygote and the endosperm

The female gametophyte of seed plants is contained in the ovule, which is a composite structure made of both sporophyte (diploid) and gametophyte (haploid) tissues (Figure 5–90). Like its male counterpart, the female gametophyte is a much-reduced haploid individual. In most angiosperms, it consists of eight cells that form the **embryo sac**, which is surrounded and protected by two layers of (sporophytic) **integuments**. The integuments do not completely enclose the gametophyte, but instead leave an opening (the **micropyle**) through which the pollen tube will enter to deliver its sperm cells.

The angiosperm ovule begins its development as an outgrowth from the internal wall of the carpel. What determines the location of these outgrowths or directs their development as ovules is not certain. In *Petunia*, however, there is evidence that MADS domain proteins (in this case, FBP7 and FBP11) are again involved. In plants with reduced activity of these genes, the outgrowths do not develop as ovules but rather as carpel-like structures (Figure 5–91). Expression of FBP7 and FBP11 is also sufficient to trigger development of ovule tissue in other organs (such as sepals and petals).

At the tip of the outgrowth that will form the ovule, the female gametophyte originates from a subepidermal cell called the **archespore** (similar to what happens in the male gametophyte). In each outgrowth, only one subepidermal cell is normally selected to become the archespore. In the maize mutant *multiple archesporial cells 1* (*mac1*), each developing ovule contains multiple archespores, which may develop into ovules with several female gametophytes. The mechanism by which *MAC1* causes only one archesporial cell to develop may involve lateral inhibition—a process mentioned earlier in the

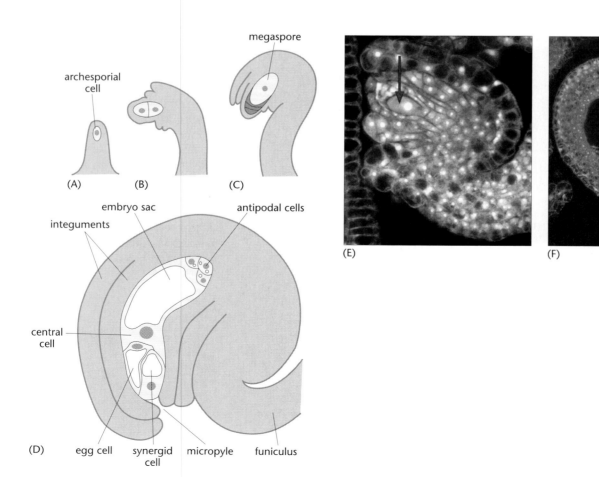

discussion of *TRYPTICHON* in trichome spacing (Section 5.4) and *CAPRICE* in root-hair development (Section 5.3).

The archesporial cell goes through meiosis, producing four haploid cells. Three of these undergo programmed cell death. The remaining cell multiplies to form the gametophyte (the embryo sac; Figure 5–90D). Three rounds of mitotic division generate the eight nuclei that form the typical angiosperm female gametophyte. The resulting nuclei initially form a syncytium (i.e., the nuclei share the same cytoplasm), but eventually the embryo sac becomes cellularized. Three of the nuclei remain in the center of the embryo sac. One corresponds to the **egg cell** (which during fertilization fuses with one of the sperm cells of the pollen tube to form the diploid embryo) and the two others form the **central cell** nuclei (which fuse with the second sperm nucleus to form the triploid endosperm). The remaining nuclei that form the female gametophyte migrate to either pole of the embryo sac before cellularization. At the end where the pollen tube will enter, two of the nuclei form the **synergid cells**, which have a role in attracting the pollen tube toward the ovule and degenerate after the pollen tube has entered. At the opposite end of the embryo sac, the remaining three nuclei form the **antipodal cells**, which after fertilization may undergo programmed cell death (in Arabidopsis) or may multiply (in maize).

Development of the female gametophyte is coordinated with development of the sporophyte tissues of the ovule

While the archesporial cell is forming at the tip of the ovule primordium, two collar-shaped outgrowths initiate at the base of the primordium. These grow to form the two sporophyte-derived integuments that envelop the developing gametophyte. Below the point where the integuments begin to form, the remaining cells of the ovule primordium form the stalk that connects the mature ovule to the carpel placenta.

As in pollen development, not only is the female gametophyte enclosed within the sporophyte, but it also requires genes that function in the surrounding sporophyte tissues to complete its development. In Arabidopsis, mutations that disrupt development of the integuments also affect development of the embryo sac. *INNER NO OUTER* (*INO*) and *AINTEGUMENTA* (*ANT*) are two examples of such genes: *ino* mutants fail to form the outer ovule integument, while in *ant* mutants, both integuments are defective. In both mutants, the embryo sac fails to develop. Because *INO* and *ANT* are expressed in the developing integuments but not in the cells that form the embryo sac, failure to complete development of the gametophyte in the mutants must be an indirect consequence of the defect in the integuments.

A pollen grain germinates on the carpel and forms a tube that transports the sperm nuclei toward the ovule

After completing their development, the semi-dried pollen grains are released from the anthers (a process called "anther dehiscence"). Depending on the reproductive strategy of each plant, the pollen is released directly on the gynoecium of the same flower or is carried by wind, insects, or other pollinators to a different flower. The gynoecium has a specialized surface, the **stigma**, to which the incoming pollen grain adheres. In some cases, pollen adhesion is encouraged by sticky substances secreted by the stigma. Other plants have "dry" stigmas, without sticky secretions; this category includes Arabidopsis. Even without sticky secretions, the stigmatic cells of Arabidopsis bind the incoming pollen rapidly and tightly (if extended to a surface of 0.1–0.5 m^2, the adhesion would be strong enough to lift a 100-kg weight). This adhesion, mediated by the exine, is selective: pollen from other species binds only weakly to Arabidopsis stigmas.

After binding to the stigma, the pollen grain hydrates. This process can also be species-selective and requires lipids present in the pollen coat. In Arabidopsis mutants that lack

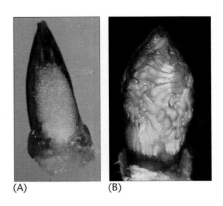

Figure 5–91.
Development of ovules. (A) Wild-type petunia; the carpel wall has been removed to reveal the ovules inside. (B) In petunia plants in which the genes producing FBP7 and FBP11 are silenced, the ovules are converted to structures resembling miniature carpels, compressed together in a spaghetti-like tangle.

(A)　　　　(B)

these lipids (*cer* mutants), the pollen fails to hydrate and is sterile. On rehydration, the pollen tube emerges from one of the apertures in the pollen grain. The pollen tube carries both the vegetative and sperm nuclei near its tip and grows rapidly toward the ovule, sometimes over a long distance (up to 10 cm in some lily species).

Pollen tube growth is functionally equivalent to cell migration. The pollen cell stays at the growing tip, while callose plugs deposited at regular intervals seal the empty tube left behind. The tube wall has an unusual composition, consisting mainly of callose and **pectin**. Growth of the pollen tube occurs by addition of new material at the tip, much like the growth of root hairs, fungal hyphae, and the axons of neurons. Accordingly, the tip has a high density of vesicles involved in **exocytosis** of new cell wall material (Figure 5–92).

Growth of the pollen tube is oriented by long-range signals in the carpel tissues and short-range signals produced by the ovule

The point of the pollen tube wall to which these vesicles are transported probably determines the direction of tube growth. Presumably this involves extracellular signal molecules, receptors on the pollen grain, and a signal-transduction pathway that controls vesicle targeting. Only a few components of this hypothetical chain have been identified, however, including a small **Rho-type GTPase** (in yeast and animals, these GTPases transduce signals from the cell surface to the actin **cytoskeleton**, which is involved in vesicle transport).

One signal that guides pollen tube growth is probably the extracellular matrix (ECM) inside the carpel. This type of guidance mechanism is analogous to that used in axon guidance during animal development: a tip-growing cell is guided to its target by a track marked by specific ECM molecules. In lily, a pectic polysaccharide and a small ECM protein have been implicated in pollen tube guidance.

When the pollen tube approaches the ovule, other, short-range guidance signals take over (Figure 5–93). These signals originate from the female gametophyte. Mutations that affect development of the female gametophyte disrupt the final stages of pollen tube guidance: tubes elongate down the carpel but fail to turn toward the ovule unless it contains a normal embryo sac. The specific source of this short-range signal may be the synergid cells. The best evidence for this comes from experiments in which pollen tubes are attracted *in vitro* to dissected embryo sacs from *Torenia fournieri*. Laser ablation of the synergids eliminated attraction of the pollen tube, while embryo sacs with the central cell or the egg cell ablated still attracted pollen tubes.

As soon as the pollen tube grows through the micropyle and reaches the female gametophyte, the ovule loses its ability to attract pollen tubes. This helps prevent **polyspermy**, fertilization by sperm cells from multiple pollen tubes.

Figure 5–92.
Schematic representation of the subcellular structure of the tip of a growing pollen tube. Note the high concentration of secretory vesicles at the tip and the callose plug that separates the tip from the older part of the pollen tube. (ER = endoplasmic reticulum)

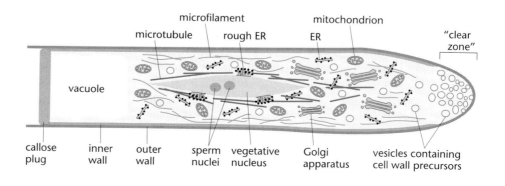

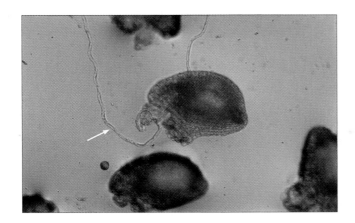

Figure 5–93.
Attraction of the pollen tube to the ovule. A pollen tube (*arrow*) grown *in vitro* is attracted to the micropyle of an isolated ovule of *Torenia fournieri*.
(Courtesy of Tetsuya Higashiyama.)

Plants have mechanisms that allow the growth only of pollen tubes carrying specific genes

Pollen grains landing on the stigma do not always grow through the carpel. Fertilization by pollen from other species, which would result in inviable progeny, can be prevented because foreign pollen fails to germinate or does not recognize the signals that attract the pollen tube to the ovule. Pollen that is genetically too close to the receiving plant can also be disadvantageous: self-fertilization can lead to "inbreeding depression" (the antithesis of **hybrid vigor**; see Section 9.2). Many plant species can avoid self-fertilization because, for example, the morphology of the flower or the timing of anther dehiscence prevents pollen from reaching receptive stigmas in the same flower. In other cases, even if the flower self-pollinates, the self-pollen is specifically recognized by the carpel and prevented from growing. This is called **self-incompatibility**.

In the best-studied examples of self-incompatibility, a protein expressed in the carpel recognizes a protein in the pollen grain and inhibits pollen growth. The genes encoding the pollen and carpel proteins are tightly linked and are inherited together (i.e., at a single locus, called the "S-locus"). The genes in the S-locus are highly polymorphic: there are many variants of the carpel and pollen proteins, which can recognize each other if they originate from the same S-locus allele, but cannot recognize proteins produced by other alleles. Since the pollen and carpel of the same individual express matching proteins, self-pollination triggers the self-incompatibility reaction. Pollen from a different individual, with a different pair of pollen- and carpel-expressed genes, is not recognized and grows unhindered (Figure 5–94).

To prevent self-pollination by means of these self-incompatibility systems, the matching carpel and pollen components must be inherited as a unit. This happens because recombination during meiosis is suppressed in the region between the matching self-incompatibility genes. How recombination is inhibited within the S-locus is not yet clear. Another intriguing but unanswered question is how so much allelic diversity has evolved at the S-locus, while still maintaining the match in each pair of carpel and pollen components.

Self-incompatibility can be gametophytic or sporophytic, depending on the origin of the pollen protein recognized

There are two types of self-incompatibility, based on the type of inheritance. The difference originates from the fact, noted earlier, that some proteins in the pollen grain are provided by the parent sporophyte (secreted by the tapetum) and others by the haploid gametophyte. This includes the pollen proteins that are recognized by the carpel. Thus self-incompatibility can be sporophytic or gametophytic, depending on whether the

Figure 5–94.
The molecular basis of self-incompatibility. Pollen and carpels of the same plant (*left*) contain the same alleles of the *SI* genes, *SM1* and *SF1* (*SM* = male-expressed *SI* gene; *SF* = female-expressed *SI* gene). The SI protein in pollen (*red blob*) is recognized by the SI protein in the carpel (*orange blob*), leading to inhibition of pollen tube growth. Pollen from a different plant (*right*), carrying a different *SI* allele (*SM2*, *SF2*), carries a protein (*green blob*) that is not recognized by the carpel protein, and pollen tube growth is not inhibited.

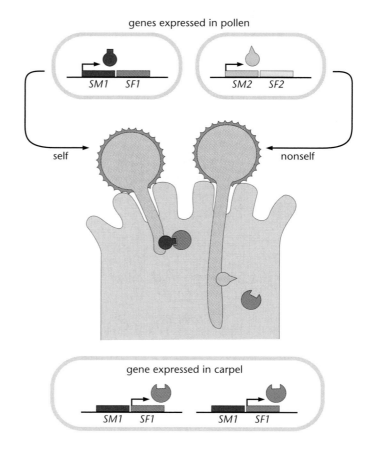

carpel recognizes pollen components deposited by the sporophyte or by the gameto-phyte. Genetically, this makes an important difference: in sporophytic incompatibility, rejection of the pollen grain is determined by both alleles in the diploid sporophyte; in gametophytic incompatibility, only the single allele present in the haploid pollen is relevant (Figure 5–95).

Gametophytic self-incompatibility is common in the Solanaceae (although in many cultivated plants of this family, such as tomato, potato, and petunia, the self-incompatibility genes have been eliminated during domestication). Sporophytic self-incompatibility has been studied mostly in Brassicaceae (which includes rapeseed and cabbage). In addition to the genetic differences mentioned above, the two types of self-incompatibility use different biochemical mechanisms to prevent growth of the self-pollen.

In the gametophytic self-incompatibility of Solanaceae, the carpel component is a **ribonuclease (RNase)**. The proteins encoded by the different S-RNase alleles have variable amino acid sequences, thought to interact with the matching pollen-expressed components, but the RNase catalytic site is invariable. Directed mutagenesis showed that the catalytic site is essential for the self-incompatible response and suggested that S-RNases cause the arrest of pollen tube growth by degrading RNA in self-pollen. The S-RNases enter pollen tubes grown *in vitro* or *in vivo*, regardless of their genotype, so the RNase activity most likely is normally inhibited inside the pollen, but the S-RNase is activated when it recognizes the matching pollen-expressed self-incompatibility component (Figure 5–96).

In the sporophytic self-incompatibility of *Brassica*, arrest of pollen growth happens at an earlier stage. The carpel protein is a receptor with kinase activity (SRK, **S-r**eceptor **k**inase), expressed in the epidermal cells of the stigma. The pollen component is a small cysteine-rich protein (SCR, **S-c**ysteine-**r**ich), present on the pollen coat. When the

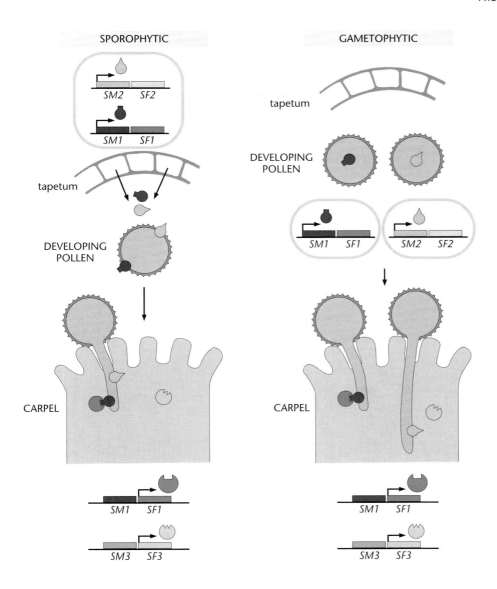

SPOROPHYTIC

GAMETOPHYTIC

tapetum

DEVELOPING
POLLEN

SM2 SF2

SM1 SF1

tapetum

DEVELOPING
POLLEN

SM1 SF1

SM2 SF2

DEVELOPING
POLLEN

CARPEL

SM1 SF1

SM3 SF3

CARPEL

SM1 SF1

SM3 SF3

Figure 5–95.
Sporophytic and gametophytic self-incompatibility. In sporophytic self-incompatibility (*left*), the pollen SI proteins (*red and green blobs*) are produced by the diploid sporophyte (in the tapetal cells) and deposited in the developing pollen grains. Each pollen grain therefore is marked by the two *SI* alleles present in the diploid, sporophytic tapetal cells. (*SM* and *SF* are explained in Figure 5–94.) If any one of the alleles is matched by the SI proteins in the receiving carpel (*orange and blue*), pollen tube growth is inhibited. In gametophytic self-incompatibility (*right*), the pollen SI proteins are produced by the haploid pollen cells. Each pollen grain is marked by only one of the two *SI* alleles present in the male sporophyte. Only a pollen grain containing a protein that matches the carpel SI proteins is inhibited.

pollen touches the stigma cell and the SRK recognizes the matching SCR protein on the pollen coat, a signal-transduction cascade is triggered in the stigma cell, which prevents it from sustaining germination of the pollen grain. The exact mechanism that arrests pollen germination, however, is still unknown.

Angiosperms have double fertilization

When the pollen tube reaches the ovule, it enters through the micropyle and bursts to release the sperm cells into the embryo sac. One of the sperm nuclei fuses with the egg cell, generating a diploid zygote, and the other fuses with the two polar nuclei to generate the triploid endosperm (Figure 5–97). This double fertilization is a defining feature of angiosperms. In nonflowering seed plants (e.g., conifers), the embryo-nourishing tissue in the seed is haploid, derived from the female gametophyte.

Two hypotheses have been proposed for the origin of double fertilization in angiosperms. One is that double fertilization originally produced two embryos, one of which was recruited to a nutritional role during evolution. If this is correct, both fertilization events originally formed diploid cells and the endosperm was later modified to triploid; consistent with this hypothesis, the endosperm of some basal angiosperms is diploid. The alternative hypothesis is that the endosperm evolved from tissues similar to the nour-

Figure 5–96.
RNase-based self-incompatibility of
Solanaceae. The SI protein produced by the carpel is an RNase (*orange blob*) that remains inactive until it binds to a matching pollen SI protein. The RNase is imported into the growing pollen tube. If the matching SI protein (*red blob*) is present in the pollen tube (*right*), the activated RNase destroys the cellular RNA that is essential to sustain pollen tube growth. If the SI protein (*green blob*) does not match (*left*), the RNA remains functional. (*SM* and *SF* are explained in Figure 5–94.)

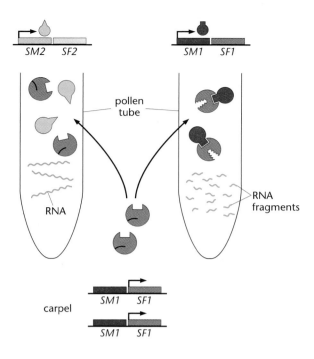

ishing tissues of nonflowering seed plants. In this case, the second fertilization event originated independent of the zygote-forming fertilization. Neither hypothesis has a decisive advantage at the moment.

Genes from the male and female gametes are not expressed equally after fertilization

Shortly after fertilization, even though each gene is represented by a copy from each parent, the female- and male-contributed gene sets do not function equally. The differential expression of female- and male-derived genes in the zygote, called **genomic imprinting**, is also seen in animals. In plants, this phenomenon has been shown to occur in the endosperm. For example, in maize and other grasses, studies have shown that the endosperm develops normally only when it contains two sets of maternal chromosomes for each set of paternal chromosomes. Hexaploid endosperm with two paternal and four maternal genomes (2:4 ratio) developed like the normal, triploid endosperm (ratio 1:2), whereas hexaploid endosperm with four paternal and two maternal genomes (ratio 4:2) was not viable. This illustrates the effect of imprinting: although both types of hexaploid endosperm had the same number of chromosomes, normal development depended on whether chromosomes were contributed by the male or female parent.

In Arabidopsis, imprinting occurs on the *MEDEA* (*MEA*) gene (mentioned in the earlier discussion of endosperm development; see Section 5.2). *MEA* is initially silenced by DNA **methylation** in both male and female gametophytes, but is specifically demethylated in the central cell, which is the female cell that fuses with one of the male gametes to give rise to the endosperm (see Section 2.3 on DNA silencing). The demethylation of *MEA* in the central cell, but not in the male gamete, results in expression of only the maternally inherited *MEA* during endosperm development. *MEA* expression in the central cell is important to prevent this cell from initiating endosperm development in the absence of fertilization.

Seed development without fertilization, as seen in *mea* mutants and in mutants that cannot demethylate *MEA* in the central cell, normally results in sterility. In some cases,

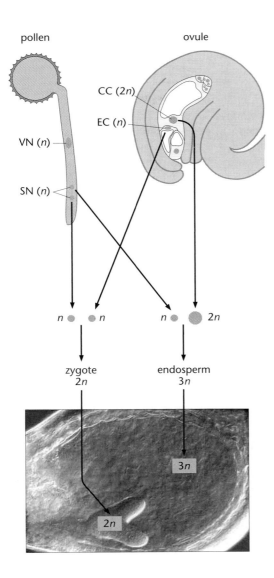

Figure 5–97.
Double fertilization in angiosperms.
One of the two sperm nuclei (SN) present in the pollen tube fuses with the egg cell (EC), giving rise to the diploid (2n) zygote. The other sperm nucleus fuses with the ovule's central cell (CC), which is a diploid cell produced by the fusion of two nuclei of the female gametophyte. This second fertilization produces a triploid (3n) nucleus, which gives rise to the endosperm. (VN = vegetative nucleus) (Courtesy of Frederic Berger.)

however, fertilization-independent seed development is part of a strategy for vegetative reproduction, as described below.

Some plants can form seeds without fertilization

Apomixis is asexual reproduction through seeds. Instead of originating from fertilization, the embryo in these seeds is a clone of the parent plant. This form of asexual reproduction is usually facultative—that is, the same plant can also produce sexual seeds. Apomixis occurs in an estimated 0.1% of angiosperms; familiar examples include the dandelion, citrus, mango, and blackberry. The main reason why apomixis attracts attention is economic. In major crops such as maize, the most productive genotypes are **F1 hybrids** (see Section 9.2). Their high productivity justifies the expense and extra work of producing hybrid seeds at every generation by crossing two different parental lines. Apomixis in these crops would be very desirable, because it would allow easy maintenance and propagation of genotypes that are very productive but highly heterozygous (and therefore mixed up by recombination during sexual reproduction).

Just as double fertilization is needed to initiate the embryo and endosperm in sexual seeds, so apomixis requires initiation of both embryo and endosperm development. Given that apomixis is found in many different plant families, it has most likely originated independently many times during evolution. Accordingly, the embryo and endosperm of apomictic seeds can originate in a variety of ways.

To bypass fertilization, the embryo must originate from a diploid cell. In some plants, this is achieved by forming a diploid egg cell, present in a diploid female gametophyte ("gametophytic" apomixis). To form the diploid embryo sac, the meiotic division that generates the **megasporocyte** is sometimes replaced by a mitotic division. In other cases, meiosis does occur, but the megasporocyte degenerates and is replaced by a neighboring sporophytic cell, which then develops into the diploid embryo sac. In yet other cases, a haploid embryo sac forms but is subsequently "invaded" by an embryo derived from the surrounding diploid sporophytic tissue. This is called **adventitious embryogeny** (the type of apomixis seen in citrus).

In most cases of apomixis, while the zygote-forming fertilization is bypassed, the endosperm-forming fertilization still occurs (that is, setting of seed still requires pollination). In other cases, the second fertilization event is bypassed, resulting in endosperm with no male contribution at all, and seed set without pollination.

The genetic basis of apomixis is not yet understood. Given that it occurs by modification of several crucial steps of sexual reproductive development (the embryo sac forms without meiosis; embryogenesis proceeds without fertilization; and endosperm-forming fertilization may be maintained or bypassed), we might expect the genetic basis of apomixis to be complex. Surprisingly, crosses between apomictic and nonapomictic plants showed that apomixis can segregate as only one or two gene loci, suggesting that a few genetic changes are sufficient to establish apomixis.

SUMMARY

Plants develop and grow throughout their life, forming new organs through the activity of meristems. Reiterative production of new organs by the meristems results in shoots and roots made up of repeated modules. The way these modules develop is determined by a genetic program characteristic of each plant, under strong environmental influences (e.g., on timing of the shift from vegetative to reproductive growth). Plants and animals became multicellular independently, so plants evolved their own mechanisms to coordinate cell behavior during development and to assign specialized cell functions according to their position in the organism. The fate of cells is determined by their position in the plant; positional information is conveyed by intercellular signals such as auxin.

An apical-basal and a radial axis are established early in plant development. The differences between apical and basal poles culminate in establishment of the shoot and root apical meristems; genes that establish the meristems during embryogenesis continue to function to maintain meristem function later in life. Embryogenesis is coordinated with development of the endosperm. Seed development includes a genetic program to make the embryo dormant and survive desiccation.

Roots have evolved more than once. The angiosperm root contains a group of self-renewing cells at the growing tip, followed by a succession of cells at different stages of growth and differentiation. New root meristems originate from pericycle cells to produce lateral roots. Intercellular communication establishes the arrangement of different cell types in the root.

Self-renewing stem cells are maintained in the center of the shoot apical meristem, while new leaf primordia emerge on the periphery of the meristem in a repetitive pattern that is organized in part by directed auxin transport. Primordium emergence is accompanied by down-regulation of genes involved in meristem development and activation of genes that direct the development of different regions of the leaf, such as the adaxial and abaxial regions.

Leaves grow initially by cell division, followed by a period of cell expansion. Intercellular communication and asymmetrical cell division are used to organize the pattern of specialized cell types in the leaf. The position of leaf veins is also established by a mechanism based on active auxin transport. Leaf senescence is an active process that retrieves nutrients from leaves at the end of their useful life span.

Branches originate from lateral meristems, under the influence of the apical meristem. Studies of dwarf mutants show that internodes grow by cell division and cell elongation, controlled by gibberellin. The cambium, a meristem between the phloem and xylem, generates secondary thickening of the stem and produces the woody stems of perennial plants.

In the transition from vegetative to reproductive growth, the angiosperm shoot apical meristem is converted to an inflorescence meristem, which produces floral meristems on its flanks. The transition from vegetative to reproductive development is controlled by a conserved set of regulatory genes. The development of floral organs is directed by floral organ identity genes, which function according to a conserved combinatorial code.

In plants, meiosis produces the cells that develop into the haploid gametophytes, which subsequently produce gametes by mitosis. The male gametophyte, formed in the anther, is the pollen grain. The female gametophyte, formed in the ovule, is the embryo sac. A pollen grain germinates on the stigma and forms a tube that transports the two sperm cells toward the ovule, with tube growth oriented by long- and short-range signals. In the embryo sac, one sperm nucleus fuses with the egg cell to form a diploid zygote and the other fuses with two polar nuclei to form the triploid endosperm, in the process of double fertilization. In self-incompatible plants, the carpel recognizes and disables pollen originating from the same individual, to avoid self-fertilization. In apomixis, a plant reproduces asexually by forming seeds with an embryo that is a clone of the parent plant.

FURTHER READING

General

Leyser O & Day S (2003) Mechanisms in Plant Development. Oxford: Blackwell Publishing.

Steeves TA & Sussex IM (1989) Patterns in Plant Development. Cambridge: Cambridge University Press.

5.1 Overview of Plant Development

David LK (2005) A twelve-step program for evolving multicellularity and a division of labor. *BioEssays* 27, 299–310.

Kirk DL (1997) The genetic program for germ-soma differentiation in *Volvox*. *Annu. Rev. Genet.* 31, 359–380.

Meyerowitz EM (2002) Plants compared to animals: the broadest comparative study of development. *Science* 295, 1482–1485.

Walbot V & Evans MMS (2003) Unique features of the plant life cycle and their consequences. *Nat. Rev. Genet.* 4, 369–379.

5.2 Embryo and Seed Development

Berger F, Taylor A & Brownlee C (1994) Cell fate determination by the cell wall in early *Fucus* development. *Science* 263, 1421–1423.

Friml J, Vieten A, Sauer M et al. (2003) Efflux-dependent auxin gradients establish the apical-basal axis of Arabidopsis. *Nature* 426, 147–153.

Gehring M, Choi Y & Fischer RL (2004) Imprinting and seed development. *Plant Cell* 16, S203–S213.

Hadfi K, Speth V & Neuhaus G (1998) Auxin-induced developmental patterns in *Brassica juncea* embryos. *Development* 125, 879–887.

Kaplan DR & Cooke TJ (1997) Fundamental concepts in the embryogenesis of dicotyledons: a morphological interpretation of embryo mutants. *Plant Cell* 9, 1903–1919.

Long JA & Barton MK (1998) The development of apical embryonic pattern in Arabidopsis. *Development* 125, 3027–3035.

Lovegrove A & Hooley R (2000) Gibberellin and abscisic acid signalling in aleurone. *Trends Plant Sci.* 5, 102–110.

McCarty DR (1995) Genetic control and integration of maturation and germination pathways in seed development. *Annu. Rev. Plant Physiol. Plant Mol. Biol.* 46, 71–93.

Steinmann T, Geldner N, Grebe M et al. (1999) Coordinated polar localization of auxin efflux carrier PIN1 by GNOM ARF GEF. *Science* 286, 316–318.

5.3 Root Development

Berger F, Haseloff J, Schiefelbein J & Dolan L (1998) Positional information in root epidermis is defined during embryogenesis and acts in domains with strict boundaries. *Curr. Biol.* 8, 421–430.

Dolan L, Janmaat K, Willemsen V et al. (1993) Cellular organisation of the *Arabidopsis thaliana* root. *Development* 119, 71–84.

Jiang K & Feldman LJ (2005) Regulation of root apical meristem development. *Annu. Rev. Cell Dev. Biol.* 21, 485–509.

Nakajima K, Sena G, Nawy T & Benfey P (2001) Intercellular movement of the putative transcription factor SHR in root patterning. *Nature* 413, 307–311.

Schiefelbein J (2003) Cell-fate specification in the epidermis: a common patterning mechanism in the root and shoot. *Curr. Opin. Plant Biol.* 6, 74–78.

van den Berg C, Willemsen V, Hendriks G et al. (1997) Short-range control of cell differentiation in the Arabidopsis root meristem. *Nature* 390, 287–289.

Vogler H & Kuhlemeier C (2003) Simple hormones but complex signalling. *Curr. Opin. Plant Biol.* 6, 51–56.

5.4 Shoot Development

Carles CC & Fletcher JC (2003) Shoot apical meristem maintenance: the art of a dynamic balance. *Trends Plant Sci.* 8, 394–401.

Doebley J, Stec A & Hubbard L (1997) The evolution of apical dominance in maize. *Nature* 386, 485–488.

Eshed Y, Baum SF, Perea JV & Bowman JL (2001) Establishment of polarity in lateral organs of plants. *Curr. Biol.* 11, 1251–1260.

Gan S & Amasino RM (1995) Inhibition of leaf senescence by autoregulated production of cytokinin. *Science* 270, 1986–1988.

McConnell JR & Barton MK (1998) Leaf polarity and meristem formation in Arabidopsis. *Development* 125, 2935–2942.

Nadeau JA & Sack FD (2002) Control of stomatal distribution on the Arabidopsis leaf surface. *Science* 296, 1697–1700.

Nelson T & Dengler N (1997) Leaf vascular pattern formation. *Plant Cell* 9, 1121–1135.

Peng J, Richards DE, Hartley NM et al. (1999) "Green revolution" genes encode mutant gibberellin response modulators. *Nature* 400, 256–261.

Quirino BF, Noh YS, Himelblau E & Amasino RM (2000) Molecular aspects of leaf senescence. *Trends Plant Sci.* 5, 278–282.

Reinhardt D, Pesce E, Stieger P et al. (2003) Regulation of phyllotaxis by polar auxin transport. *Nature* 426, 255–260.

Sablowski R (2007) The dynamic plant stem cell niches. *Curr. Opin. Plant Biol.* 10, 639–644.

Satina S, Blakeslee AF & Avery AG (1940) Demonstration of the three germ layers in the shoot apex of Datura by means of induced polyploidy in periclinal chimeras. *Am. J. Bot.* 27, 895–905.

Schmitz G & Theres K (1999) Genetic control of branching in Arabidopsis and tomato. *Curr. Opin. Plant Biol.* 2, 51–55.

Sinha N (1999) Leaf development in angiosperms. *Annu. Rev. Plant Physiol. Plant Mol. Biol.* 50, 419–446.

Szymkowiak EJ & Sussex IM (1996) What chimeras tell us about plant development. *Annu. Rev. Plant Physiol. Plant Mol. Biol.* 47, 351–376.

Tooke F & Battey N (2003) Models of shoot apical meristem function. *New Phytol.* 159, 37–52.

Tsuge T, Tsukaya H & Uchimiya H (1996) Two independent and polarized processes of cell elongation regulate leaf blade expansion in *Arabidopsis thaliana* (L.) Heynh. *Development* 122, 1589–1600.

Uggla C, Moritz T, Sandberg G & Sundberg B (1996) Auxin as a positional signal in pattern formation in plants. *Proc. Natl. Acad. Sci. USA* 93, 9282–9286.

Vollbrecht E, Veit B, Sinha N & Hake S (1991) The developmental gene Knotted-1 is a member of a maize homeobox gene family. *Nature* 350, 241–243.

Waites R & Hudson A (1995) Phantastica—a gene required for dorsoventrality of leaves in *Antirrhinum majus*. *Development* 121, 2143–2154.

5.5 From Vegetative to Reproductive Development

Jack T (2001) Relearning our ABCs: new twists on an old model. *Trends Plant Sci.* 6, 310–316.

Kempin SA, Savidge B & Yanofsky MF (1995) Molecular basis of the cauliflower phenotype in Arabidopsis. *Science* 267, 522–525.

Krizek BA & Fletcher JC (2005) Molecular mechanisms of flower development: an armchair guide. *Nat. Rev. Genet.* 6, 688–698.

Noda K, Glover BJ, Linstead P & Martin C (1994) Flower colour intensity depends on specialized cell shape controlled by a Myb-related transcription factor. *Nature* 369, 661–664.

Smyth DR (2005) Morphogenesis of flowers—our evolving view. *Plant Cell* 17, 330–341.

5.6 From Sporophyte to Gametophyte

Colombo L, Franken J, Koetje E et al. (1995) The petunia MADS box gene FBP11 determines ovule identity. *Plant Cell* 7, 1859–1868.

Higashiyama T, Yabe S, Sasaki N et al. (2001) Pollen tube attraction by the synergid cell. *Science* 293, 1480–1483.

Holdaway-Clarke TL & Hepler PK (2003) Control of pollen tube growth: role of ion gradients and fluxes. *New Phytol.* 159, 539–563.

Koltunow AM & Grossniklaus U (2003) Apomixis: a developmental perspective. *Annu. Rev. Plant Biol.* 54, 547–574.

Levings CS III (1993) Thoughts on cytoplasmic male sterility in cms-T maize. *Plant Cell* 5, 1285–1290.

Ma H (2005) Molecular genetic analyses of microsporogenesis and microgametogenesis in flowering plants. *Annu. Rev. Plant Biol.* 56, 393–434.

McCormick S (2004) Control of male gametophyte development. *Plant Cell* 16, S142–S153.

Nasrallah JB (2002) Recognition and rejection of self in plant reproduction. *Science* 296, 305–308.

Sheridan WF, Avalkina NA, Shamrov II et al. (1996) The mac1 gene: controlling the commitment to the meiotic pathway in maize. *Genetics* 142, 1009–1020.

Yadegari R & Drews GN (2004) Female gametophyte development. *Plant Cell* 16, S133–S141.

Zinkl GM, Zwiebel BI, Grier DG & Preuss D (1999) Pollen-stigma adhesion in Arabidopsis: a species-specific interaction mediated by lipophilic molecules in the pollen exine. *Development* 126, 5431–5440.

ENVIRONMENTAL SIGNALS

When you have read Chapter 6, you should be able to:

- Summarize the effects of environmental factors on seed germination and seedling development, noting the role of growth regulators.

- Distinguish between skotomorphogenesis and photomorphogenesis, and summarize their roles in plant development.

- Describe the structures, roles, and mechanisms of the five types of phytochrome in Arabidopsis.

- Define "phototropism," and outline the roles of photoreceptors in this plant response.

- List the roles of ethylene as a growth regulator, and outline its signaling pathway.

- Describe the roles of brassinosteroid in plant development, and outline its signaling pathway.

- Define "photoperiod," distinguish between short-day and long-day plants, and describe how plants detect and respond to environmental cues that induce flowering.

- Explain what circadian rhythms are and how they control the expression of some plant genes.

- Define "vernalization," and outline its role in plant development.

- Summarize how plant growth is affected by gravity and the mechanisms of gravitropism in hypocotyls, stems, and roots.

Plants constantly monitor their surroundings, and respond actively and continuously to signals that indicate environmental change, adapting their growth patterns, their form, and their progress through the life cycle to their immediate environment. As we described in Chapter 5, plants develop all of their organs after they germinate, through the activity of **meristems**, which remain active throughout the life of the plant. For example, in Arabidopsis and other eudicots, the **shoot apical meristem** gives rise to all the aboveground organs by repeatedly forming modules consisting of a segment of stem and one or more leaves or flowers. Because of this lifelong pattern of growth and development, plants can alter their form in response to changes in the environment.

The stimuli to which plants' developmental programs respond include temperature, day length, light, gravity, and the availability of water and nutrients, all of which we consider in this chapter. Although we consider each type of stimulus separately, plants do not respond to them individually; they are constantly integrating their responses to all of these stimuli. The effect of these signals can be to alter the growth of the stem or the

CHAPTER SUMMARY

6.1 SEED GERMINATION

6.2 LIGHT AND PHOTORECEPTORS

6.3 SEEDLING DEVELOPMENT

6.4 FLOWERING

6.5 ROOT AND SHOOT GROWTH

type of lateral organs formed, so that the form of the adult plant is governed by an interaction between intrinsic developmental programs and environmental signals. Local conditions therefore have a dramatic effect on the morphology of the adult plant. For example, plants grown in strong shade grow toward sunlight by altering their axis of growth. Similarly, plants switch from vegetative to reproductive growth in response to appropriate environmental conditions, and the time during shoot growth when this transition occurs greatly affects the morphology of the adult plant. Environmental responses such as these are an essential part of the sedentary lifestyle of plants. They are also crucial in agriculture: crops can be bred to be closely adapted to the environment in which they are cultivated, thereby ensuring maximum yield.

Chapter 7 deals with what happens when environmental factors are extreme and can potentially limit growth and development. Here we describe how environmental signals affect plant growth as an essential aspect of normal development.

In Chapter 5 we described how development is coordinated across the whole plant to produce a characteristic form. Environmental responses are also coordinated: signaling between organs means that stimuli perceived in one part of the plant can produce changes in another part. For example, signals produced in the leaf trigger flower development at the shoot apical meristem. Our discussion, then, focuses on how the plant perceives its surroundings and how this information is signaled to generate a response. The response might be a change in the pattern of development such as the induction of flowering or a modification of the speed or direction of growth. In many cases these responses are mediated by small signaling molecules called **growth regulators** (or **phytohormones**). There are eight generally recognized classes of growth regulators: auxins, cytokinins, brassinoßsteroids, gibberellins, ethylene, abscisic acid, jasmonic acid, and salicylic acid. All of them can probably move between cells to some extent, and some, such as auxin and jasmonic acid, can move long distances between plant organs. In this chapter we describe some effects of several of these growth regulators; others are discussed in Chapters 7 and 8.

Controlled experiments with defined environmental parameters are required to study the effect on development of particular environmental stimuli. In such experiments, most environmental factors are kept constant and the effect of one environmental variable on development is measured. Such approaches allow researchers to study single responses and the genetic pathways that control them, and to identify the proteins or growth regulators required for particular responses. However, this approach can be an oversimplification. In reality these pathways do not operate in isolation, and more than one pathway can influence the same response. Similarly, a single growth regulator may control many different responses. The real picture is likely to be one of complex cross-talk and networks of information transfer. A well-studied example of interacting pathways is provided by **ethylene**, which we will describe in some detail. For example, genetic analyses have revealed much about the role of ethylene in seedling development in the dark, but this is only one example of the role of ethylene signaling. There is extensive cross-talk between ethylene signal transduction and other hormonal pathways, and ethylene signaling even plays an important role in disease resistance.

The chapter begins where growth of the new plant begins: seed **germination**. We then consider the mechanisms by which plants detect and respond to light, and how these responses—and responses to temperature, gravity, and nutrient supply—affect seedling development, flowering, and shoot and root growth.

6.1 SEED GERMINATION

A **seed** can lie dormant or can germinate (see Chapter 5 on the formation and dormancy of seeds). Seed maturation often involves desiccation, and while rehydration is necessary for germination, the rehydrated seed may still remain dormant. **Dormancy** is probably an adaptive trait that ensures germination occurs at times and in environments likely to be favorable to the subsequent growth of the plant. Accordingly, seed dormancy is widespread in plants growing in the wild but is not found in many cultivated species (see below).

Seed dormancy in many, but not all, species is broken by light. For example, the dormancy of some varieties of lettuce seed is broken only after exposure to light. Other seeds require specific day lengths or intermittent exposure to light to trigger germination. Generally, in plants for which light is the stimulus for germination, dormancy is imposed by the **seed coat**; when the coat is removed, **embryos** germinate even in the absence of light. Seeds that depend on light to break dormancy tend to be relatively small. Seeds that do not require light tend to be larger and thus to have more energy reserves that allow the seedling to grow temporarily in the dark (e.g., when underground), in the absence of **photosynthesis**. Some light-dependent seeds are sensitive to shading by neighboring adult plants. Sunlight that has passed through other vegetation (e.g., a forest canopy) is relatively rich in far-red light (just a small portion of the light spectrum, described in more detail later). Plants recognize far-red–rich light as an indicator of shade and hence, in many cases, of an unsuitable environment for germination. Accordingly, far-red–rich light is known to inhibit seed germination in many light-dependent species; this may form part of the "shade avoidance response" (see Section 6.2).

Another factor that can release seeds from dormancy is low temperature. Many seeds germinate only after hydration followed by a period of cold (0–10°C), a trait that prevents germination until winter has passed, thus maximizing chances of survival. A common horticultural practice is to chill seeds before planting, thus breaking dormancy and minimizing the time the seeds spend in the soil before germinating.

Although dormancy is advantageous to many plants in the wild, it is a trait that has been selected against in crop domestication, beginning around 10,000 years ago. For farmers, it is easier if all the seeds of a crop germinate at the same time, thus maximizing the chances of synchronous maturity and shortening harvest time (see Section 9.1 on this and other facets of crop domestication).

Sometimes the environmental controls on seed germination are highly specific to the environmental niche that the organism occupies. For example, the seeds of some species are triggered to germinate by the heat and smoke produced by fires. This adaptation permits rapid seed germination and exploitation of the new environment that arises from the ashes of the old (hence this adaptation is sometimes called "fire following"). Some seeds respond specifically to the heat of the fire, others to the smoke. For example, recent work has shown that smoke, but not heat, induces the germination of yellow whispering bells (*Emmenanthe penduliflora*), an **annual plant** of the California chaparral (Figure 6–1). Smoke has many components, but a single one of these, nitrogen dioxide (NO_2), is sufficient to trigger germination.

Seed germination in parasitic plants is often dependent on a host-plant signal. **Obligate parasites**, such as *Striga* (witchweed), must attach themselves to a host within a few days of germination, or they will not survive. These plants detect the presence of a potential host by recognizing specific molecules released from the host-plant roots as a cue to seed germination. For example, *Striga* seeds germinate in response to four modified **hydroquinones** released from the roots of its maize or sorghum host (Figure 6–2). These compounds are relatively unstable, easily oxidized to biologically inactive forms, and found only in soil close to their source. Thus, *Striga* seeds germinate only when appropriate host roots are within reach of the germinating seedling (see Section 8.2).

For some seeds, as noted above, light is a major environmental regulator of germination. The importance of light reflects the dependence of **autotrophic** organisms such

Figure 6–1.
Flowers of *Emmenanthe penduliflora*.
This annual plant germinates and grows quickly after forest fires.
(Courtesy of Barbara J. Collins.)

Figure 6–2.
***Striga* growing on maize, one of its host plants.** The diagram shows the hydroquinone signal from the host root.
(Me = methyl group)
(Courtesy of Zeyaur R. Khan.)

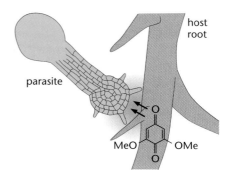

as plants on light to provide the energy for carbon fixation. Once a seed has germinated and the shoot emerges from the soil, light also plays a key role in the regulation of many further aspects of plant development, as described throughout this chapter. In the next section we begin a discussion of how the orderly development of plants is governed by light, and what we know about how plants detect light and translate this information into appropriate growth responses.

6.2 LIGHT AND PHOTORECEPTORS

Light quality and quantity regulate many aspects of plant growth and development, thus ensuring optimal exposure of photosynthetic plant parts to light and allowing maximal rates of **carbon assimilation**. Plants are therefore sensitive to, and regulate growth in response to, changes in the quantity (irradiance), quality (wavelength), and direction of light. Likewise, the timing of developmental processes such as flowering and onset of dormancy depends on a system of measuring and responding to changes in day length.

Seedlings of higher plants that develop in darkness have a completely different appearance from those that develop in the light. However, it is not just the presence or absence of light that influences development, but also the amount and wavelengths of light to which plants are exposed. For example, the structure of leaves varies depending on whether they develop in direct sunlight or are shaded by other leaves. Light is perceived through its absorption by **photoreceptors**, which convert the physical energy of light to chemical energy. Plants have a variety of photoreceptor systems that perceive the quality, quantity, duration, and direction of light. In this section we briefly explore how the various photoreceptors have been identified and describe what is known about the effects of light, from the early events in the transduction of light signals to the final plant responses.

Plant development proceeds along distinct pathways in light and dark

Seedlings of higher plants can adopt one of two developmental pathways, and the presence or absence of light determines which pathway is followed (Figure 6–3). When grown in the dark, seedlings exhibit **skotomorphogenesis** (*skotos* is Greek for "darkness") and become etiolated. **Etiolation** has several striking characteristics: embryonic stems (**hypocotyls** in **eudicots**, **epicotyls** in **monocots**) become very long and spindly, and the seedling is pale yellow or white because **chloroplast** formation and **chlorophyll** biosynthesis require exposure to light. In some eudicots, the **apical hook** (a curving of the hypocotyl that is thought to protect the apical meristem from damage during growth through the soil) is maintained unchanged throughout etiolation, the **cotyledons** remain unexpanded, and activity of the shoot apical meristem (from which most of the aboveground plant normally derives) is suppressed. In contrast, seedlings grown in the light exhibit **photomorphogenesis**: they have relatively short embryonic stems, lose the apical hook, develop expanded green cotyledons, and rapidly initiate the elaboration of **leaves** and **internodes** at the shoot apical meristem.

When the seedling shoot emerges from the soil surface and is exposed to light, its development switches from skotomorphogenesis to photomorphogenesis. This process is known as **de-etiolation**. In response to light, the seedling begins to turn green, the cells of the hypocotyl cease to expand, the apical meristem begins to initiate leaves, and a normal, light-grown seedling begins to grow. The differences between skotomorphogenesis and photomorphogenesis result from changes in the expression of specific genes: **gene expression** is activated or repressed by the presence or absence of light. Although the full suite of **genes** responsible for this difference is currently unknown, we do know that **transcription** of a large number of genes is induced by exposure to light. For example, de-etiolation in far-red light causes the increased expression of approximately 10% of Arabidopsis genes. Individual genes that are known to be light-regulated include those involved in the development of photosynthetic capacity, such as the nuclear-encoded genes for the small subunit of **Rubisco** and the chlorophyll *a/b* binding protein of **light-harvesting complexes** (see Section 4.2). The obvious difference between the two developmental programs (skotomorphogenesis and photomorphogenesis) provides us with a useful way to investigate light-signal transduction. Mutants have been identified that have impaired de-etiolation and defects in light perception or in light-signal transduction. In particular, the study of mutants with "dark-grown" phenotypes when grown in the light, or "light-grown" phenotypes when grown in the dark, has improved our understanding of the photomorphogenesis and skotomorphogenesis developmental pathways.

The environment of a growing seedling dramatically affects how a plant develops: etiolated plants are very sensitive to light, and skotomorphogenesis shifts to photomorphogenesis even after minimal light exposure. In many cases, only a few minutes of light each day is sufficient to induce photomorphogenic development. Skotomorphogenesis seems to be an adaptive, light-seeking response. For example, the elongated hypocotyl typical of skotomorphogenesis in eudicots pushes the shoot apex up into the light from the darkness of the soil.

Distinct photoreceptors detect light of different wavelengths

Plants possess a variety of photoreceptors, which are light-sensitive molecules that undergo a structural, energetic, or conformational change following the absorption of light energy (Figure 6–4). The function of photoreceptors is to detect the presence of light and then, via a linked signal-transduction cascade, to initiate the appropriate developmental response. Several different families of photoreceptors exist in plants, which by virtue of their different individual molecular structures are sensitive to light of different wavelengths. For example, the growth and development of plants is particularly affected by light in the far-red, red, blue, UV-A, and UV-B regions of the spectrum. Red light exerts the strongest influence on photomorphogenesis, and a family of photoreceptors known as the **phytochromes**, which respond especially to red and far-red light,

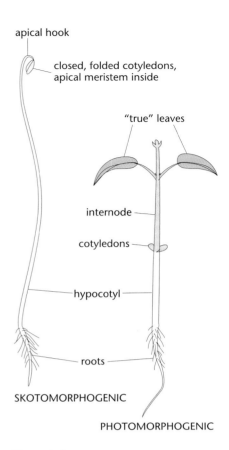

Figure 6–3.
Comparison of the morphologies of "dark-grown" (skotomorphogenic) and "light-grown" (photomorphogenic) plants. Note the elongated hypocotyl and inhibited meristematic program (development of internode and "true" leaves) of the skotomorphogenic plant.

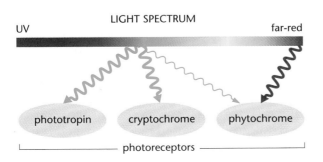

Figure 6–4.
Light and photoreceptors. The diagram shows the wavelengths of light that activate three families of plant photoreceptors.

are widespread in the plant kingdom. The **cryptochromes** detect and mediate responses to light in the blue and ultraviolet regions of the spectrum. Another family of blue-light–activated photoreceptors is the **phototropins**, which are responsible for blue-light–activated phototropism (described in more detail below) and mediate the photoreception of blue light in **guard cells**, thus controlling the blue-light–induced opening of **stomata**. Yet another blue-light receptor is FKF1, a protein that mediates the photoperiodic control of flowering. The functions of these photoreceptors are considered in more detail below.

Phytochromes are converted from an inactive to an active form by exposure to red light

The phytochromes of higher plants are homodimeric proteins, consisting of two identical polypeptide chains (**apoproteins**), each with an attached **chromophore**. In essence, the chromophore captures a light signal and undergoes isomerization, resulting in a conformational change in the protein. Subsequently, the protein initiates changes in gene expression that cause appropriate growth responses in the plant. Arabidopsis has five distinct phytochrome apoproteins (PHYA, PHYB, PHYC, PHYD, and PHYE), encoded by genes *PHYA* through *PHYE*. When we use the term "phytochrome," we are referring to a dimerized pair of apoproteins and their attached chromophores—that is, the form in which the phytochrome acts to perceive light and cause the appropriate growth response. When we refer to "PHY" we mean the apoprotein gene product of a *PHY* gene.

The phytochrome chromophore is phytochromobilin, a **tetrapyrrole** (see Section 4.7), which is attached to the apoprotein monomer at a specific cysteine residue. Much of our understanding of phytochromes has come from the study of plants with impaired phytochrome function. For example, phytochrome mutants can be isolated simply by screening for plants with altered hypocotyl elongation: light-grown mutant hypocotyls grow longer than those of wild-type plants, and the mutant plants therefore resemble etiolated (dark-grown) plants, even in the presence of light. Mutants isolated using this screen are collectively called "*hy* mutants" (Table 6–1). The Arabidopsis wild-type genes *HY1* and *HY2* encode enzymes involved in the synthesis of phytochromobilin (Figure 6–5). The *hy1* and *hy2* mutants have abnormally long hypocotyls because, lacking chromophore and thus complete phytochromes, their light detection is impaired.

Phytochromes provide a developmental switch that is controlled by their sensitivities to different wavelengths of light: red light (650–680 nm) induces many types of growth responses, and far-red light (710–740 nm) inhibits them. A classic example is

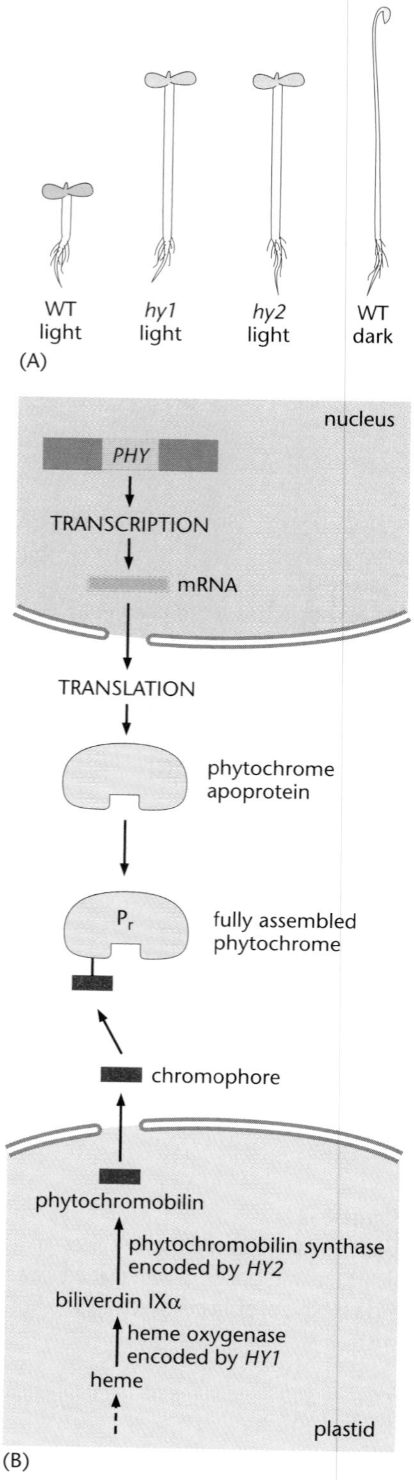

Figure 6–5.
Assembly of active phytochrome from apoprotein and phytochromobilin. The *HY1* and *HY2* genes encode enzymes that function in chromophore biosynthesis. (A) These genes were first identified as loss-of-function mutations, *hy1* and *hy2*, that result in elongated hypocotyls in the light. Note that the light-grown *hy1* and *hy2* mutants have longer hypocotyls than light-grown wild-type (WT) plants, approaching the length of those of dark-grown wild-type plants. The elongated hypocotyls of *hy1* and *hy2* mutants in the light are due to impaired phytochrome function. (B) Key steps in the formation of active phytochrome. The phytochrome apoprotein is translated from mRNA in the cytosol, following transcription of the *PHY* gene and nuclear export of the mRNA. Meanwhile, the chromophore, phytochromobilin, is synthesized in plastids from heme via steps catalyzed by enzymes encoded by *HY1* and *HY2*. Covalent linkage of apoprotein to chromophore in the cytosol completes formation of the phytochrome monomer (P_r; see the text and Figure 6–6). The pathway from heme to phytochromobilin is reduced in *hy1* or *hy2* loss-of-function mutants, causing a reduction in chromophore levels, a consequent reduction in functional phytochrome levels, and the characteristic elongated hypocotyl of light-grown *hy1* or *hy2* mutants.

Table 6–1 Long hypocotyl (*hy*) mutants

Mutant gene name (Wild-type gene name)	Alternative mutant gene name (Alternative wild-type gene name)	Wild-type gene product	Function
hy1 (*HY1*)		HEME OXYGENASE	Chromophore synthesis
hy2 (*HY2*)		PHYTOCHROMOBILIN SYNTHASE	Chromophore synthesis
hy3 (*HY3*)	*phyB* (*PHYB*)	PHYTOCHROME B	Encodes phytochrome B apoprotein
hy4 (*HY4*)	*cry1* (*CRY1*)	CRYPTOCHROME	Encodes cryptochrome 1 apoprotein
hy5 (*HY5*)			Basic leucine zipper (bZIP) transcription factor

Note: As the mutated genes are isolated and characterized, some *hy* mutants have acquired new names, corresponding to the function of the wild-type gene. Both names are given here.

the control of lettuce seed germination: red light stimulates germination and far-red light inhibits it. The final stimulus in a succession of exposures to red and far-red light determines whether or not germination occurs (for example, red, far-red, red causes germination; red, far-red, red, far-red does not). The explanation for these observations is that phytochrome exists in two interconvertible forms (Figure 6–6): P_r, which

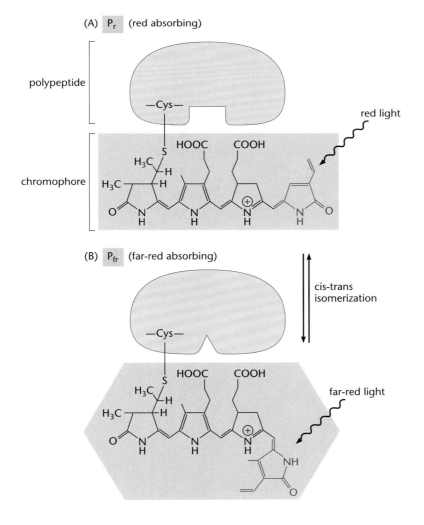

Figure 6–6.
Interconversion of the P_r and P_{fr} forms of phytochrome. (A) The P_r (red-absorbing) form of phytochrome consists of phytochrome apoprotein covalently bound to the tetrapyrrole chromophore. (B) Absorption of red light by the chromophore results in structural (cis-trans) isomerization of the chromophore and altered conformation of the apoprotein, producing the P_{fr} form. Absorption of far-red light by P_{fr} restores the phytochrome to the P_r form.

absorbs red light, and P_{fr}, which absorbs far-red light. Phytochromes are synthesized in the P_r form, which is converted to P_{fr} by the absorption of red light; P_{fr} is converted back to P_r by the absorption of far-red light. As many light responses are triggered by red light, P_{fr} is generally assumed to be the active form of phytochrome.

Exactly how phytochromes work remains unclear. For example, phytochromes may act as light-regulated **protein kinases**, as transcriptional regulators, or both. The protein kinase activity of phytochromes is differentially and reversibly responsive to red and far-red light. However, the importance of this kinase activity is unclear, since the N-terminal half of the phytochrome protein (lacking the kinase domain) is able to complement phytochrome-deficiency **mutations**. Ultimately, when activated by light, phyto-chromes A and B are known to move into the nucleus. The other members of the phytochrome family are likely to do the same, although this has not yet been demonstrated. One pathway of phytochrome action is as follows (outlined in Figure 6–7). Absorption of red light by P_r converts it to P_{fr} and is thought to activate phytochrome **serine/threonine protein kinase** activity and phytochrome **autophosphorylation**. P_{fr} then enters the nucleus and interacts with PIF3, a **basic helix-loop-helix (bHLH)** transcriptional regulator. *In vitro* experiments suggest that the PIF3-phytochrome interaction is dependent on the phytochrome being in the P_{fr}, rather than the P_r, form. In summary, phytochrome is activated in the cytoplasm by absorption of light, enters the nucleus, and interacts with PIF3 to alter the regulation of gene transcription. PIF3 action is covered in more depth in a subsequent section of this chapter. Phytochrome probably also acts as a signal in the cytoplasm, independent of its effects on nuclear gene transcription. Such cytoplasmic signaling might occur via substrate proteins phosphorylated by the phytochrome, such as PKS1 (PHYTOCHROME KINASE SUBSTRATE 1).

Phytochromes are not restricted to higher plants; phytochrome-like proteins have been found in lower plants, algae, cyanobacteria, and bacteria. Indeed, it is likely that all photosynthetic organisms contain phytochromes. Phytochrome-like proteins have also been found in nonphotosynthetic bacteria (e.g., *Deinococcus radiodurans* and *Pseudomonas aeruginosa*). Comparisons of amino acid and DNA sequences of phytochromes and their genes across this range of organisms reveal similarities in two key domains in all phytochromes so far analyzed (Figure 6–8): first, the phytochrome sequences of higher plants carry C-terminal serine/threonine kinase domains, and bacterial phytochromes carry C-terminal **histidine kinase** domains. Second, near the N-termini of the proteins, higher-plant phytochromes carry a chromophore-binding domain, and bacterial phytochromes contain a different but analogous attachment site for chromophores.

Figure 6–7.
A phytochrome signaling pathway that involves PIF3. Red light activates phytochrome. The activated P_{fr} autophosphorylates; phosphorylated P_{fr} enters the nucleus and, in combination with transcription factors such as PIF3, regulates the transcription of specific genes. In addition, interaction of P_{fr} with cytoplasmic substrates such as PKS1 may trigger rapid changes in cellular physiology in response to light.

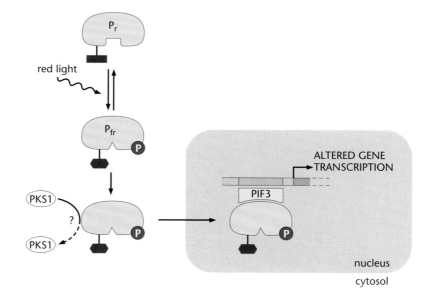

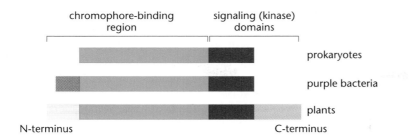

Figure 6–8.
Related domains in phytochromes of a range of organisms. The C-terminal signaling domain is a serine/threonine kinase domain in plant phytochromes and a histidine kinase domain in bacterial phytochromes. The chromophore-attachment domains are analogous in plants and bacteria.

Distinct forms of phytochrome have different functions

As described above, Arabidopsis has five different phytochromes: A, B, C, D, and E (Figure 6–9). Phytochrome A differs from the others in that it accumulates to high levels in dark-grown, etiolated plants. Phytochrome A levels decline in light-grown Arabidopsis and other plants, by three mechanisms: *PHYA* transcription is repressed (observed in monocots only); *PHYA* **messenger RNA (mRNA)** is degraded; and phytochrome A in the P_{fr} form is ubiquitinated (by an ubiquitin ligase enzyme) and degraded by the **proteasome** (see Section 5.4). Phytochromes B, C, D, and E are present in low but stable levels in light- and dark-grown plants; their gene transcription and protein stability are not affected by light conditions. The high levels of phytochrome A found in dark-grown monocots made these plants an ideal source of phytochrome for much of the defining work on phytochrome structure and function. In all light-grown plants, the five phytochromes are present in low and approximately equal abundance. The properties of phytochrome A are consistent with its important role in the transition from skotomorphogenesis to photomorphogenesis: as soon as the dark-grown plant encounters light, a rapidly decaying burst of phytochrome A–associated signaling triggers photomorphogenic development.

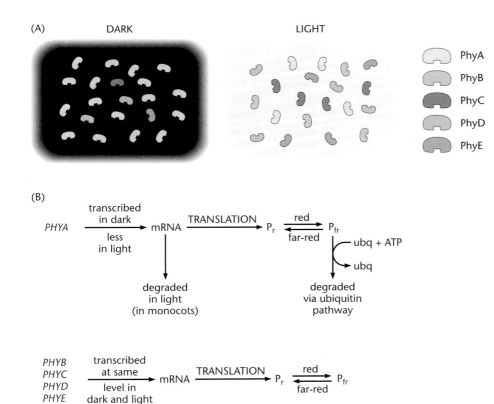

Figure 6–9.
Regulation of expression of phytochrome genes. (A) Phytochrome A accumulates to relatively high levels in dark-grown plants. In light-grown plants, the levels of phytochromes A, B, C, D, and E are more alike. (B) Summary of differences in the light regulation of levels of phytochrome-encoding mRNAs and phytochrome proteins, comparing phytochrome A with phytochromes B, C, D, and E. (ubq = ubiquitin)

Our understanding of the relative complexity of the phytochrome proteins, and of the sometimes overlapping and sometimes discrete functions of different phytochromes, has advanced considerably through the combination of two approaches. First, the family of genes encoding phytochrome apoproteins has been isolated from Arabidopsis. Comparison of the structures and expression patterns of different members of a family of phytochromes in this single species helps us understand how the different family members might have evolved and taken on specialist roles. Second, detailed physiological studies of Arabidopsis mutants with defective light responses, coupled with identification of the mutated genes, have allowed the attribution of particular functions to individual phytochromes in wild-type plants.

The predicted amino acid sequences of PHYA, PHYB, and PHYC are equally divergent from each other, with about 50% homology. The three genes encoding these proteins are probably derived from a common ancestor by **gene duplication** (see Section 2.4), and this separation is believed to have taken place at about the same time that flowering plants emerged (in the Cretaceous Period, around 130 million years ago; see Chapter 1). PHYD and PHYE are more closely related to each other and to PHYB, so are likely to have derived from PHYB more recently (also by gene duplication).

As noted above, plants with aberrant hypocotyl growth (e.g., the Arabidopsis *hy* mutants) provide a powerful genetic screen for plants that are less responsive to light than the wild type. Specifically, *hy1* and *hy2* mutants have elongated hypocotyls when grown in white light because they are deficient in the phytochromobilin chromophore; reduced levels of active phytochrome are produced since reduced levels of the chromophore are available. Other *hy* mutants have mutations in the apoprotein-encoding genes. So, if a plant has a mutation in *PHYA*, it should still be able to synthesize PHYB through PHYE; if a plant is mutated in *PHYB*, it should still be able to synthesize PHYA and PHYC through E. Analysis of these mutants under different light conditions, therefore, helps to distinguish the roles of individual phytochromes in detecting and responding to different types of light (Figure 6–10).

Arabidopsis *phyA* mutants were isolated under continuous far-red light on the basis of their insensitivity to a regimen of far-red light. *phyB* mutants are similarly insensitive to red light. In both of these examples, mutant hypocotyls are longer than those of wild-type plants grown under the same conditions. Phytochrome A is responsible for the detection of and response to far-red light (far-red detection becomes particularly important when seeds germinate in the shade of other plants, because chlorophyll in the shading leaves filters out much of the red light), and phytochrome B is responsible for the detection and response to red light. Phytochrome B tends to take over from phytochrome A following de-etiolation, because phytochrome A is light-labile whereas phytochrome B is light-stable.

Figure 6–10.
Light responses in mutants lacking phytochrome A or B. (A) The *phyA* mutant (which lacks phytochrome A) has a short hypocotyl (like that of the wild type, WT) in red or white light, but the *phyB* mutant (which lacks phytochrome B) has a long hypocotyl in these conditions. (B) In far-red light, however, the *phyA* mutant has a long hypocotyl, and the WT and *phyB* mutant do not (see text for further explanation).

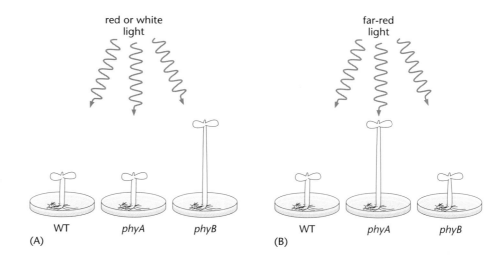

Although the roles of phytochromes A and B often counterbalance one another, they can sometimes overlap (Figure 6–11). For example, phytochrome B is the predominant phytochrome involved in the induction of *CAB* mRNA in response to a red-light pulse; however, phytochrome A also contributes to this response (this *CAB* gene response is discussed below). The involvement of phytochrome A is apparent only in the absence of phytochrome B (i.e., in a phytochrome B–deficient mutant), however, in which red-light responses are still observed, albeit at a lower level. These responses are further reduced in double mutants lacking phytochromes A and B (Figure 6–11). These two phytochromes also act cooperatively in other light responses. A wide range of genes show altered expression patterns in response to pulses of red light; phytochromes A and B mediate these responses, which are stronger when both phytochromes are present than when either is present alone. In other light responses, phytochromes A and B seem to act in opposite directions. For example, phytochrome A promotes germination in far-red light, whereas phytochrome B inhibits it; phytochrome A contributes to induction of flowering by long days, whereas phytochrome B inhibits flowering in a manner that seems to be insensitive to day length. We described above how phytochrome modulates gene transcription via interaction with PIF3. However, PIF3-dependent signaling is but one phytochrome signaling pathway. It may be that the different functions of phytochromes A and B reflect differences in the signal-transduction systems into which they feed; under some conditions these signaling pathways provoke the same growth response, and under others they produce opposite responses. It is the balance between the levels of the two active phytochromes (in their P_{fr} forms) that defines the plant response.

Phytochrome plays a role in shade avoidance

Many plants ("sun plants") have the capacity to detect the shade cast by other plants and to alter their growth accordingly by investing more resources into the elongation of stems and less into the expansion of leaves (Figure 6–12). This "shade avoidance response" enables plants to grow out of the shade and into direct sunlight, thus increasing their capture of energy for photosynthesis. Because chlorophyll absorbs red light but is relatively transparent to far-red light, light filtered through vegetation has relatively high far-red content, as does light reflected from neighboring plants. Several different phytochromes act during the shade avoidance response by detecting far-red–rich light, moving to the nucleus, and directly affecting gene expression. Phytochrome B, together with its close relatives phytochromes D and E, plays the major role in shade avoidance by established plants.

Studies on Arabidopsis with mutated *PHYB* genes have allowed researchers to infer the role of phytochrome B in wild-type plants. When *phyB* mutants (deficient in functional phytochrome B) are grown in white light, they develop longer **petioles** (the stem that joins a leaf to the shoot) and flower earlier than wild-type plants. These phenotypes mimic the actions of wild-type plants exhibiting the shade avoidance response, indicating that phytochrome B normally acts in white light to repress shade responsiveness. Another response, known as the "end-of-day response," is found in unshaded plants as the day draws to an end and incident light becomes higher in far-red content. When either shaded or end-of-day conditions are mimicked in the lab, *phyB* mutants exhibit greatly reduced responsiveness compared with wild-type controls. This suggests that phytochrome B is the major phytochrome involved in mediating shade avoidance and end-of-day responses (at high ratios of far-red to red light). It also illustrates that the roles

Figure 6–11.
Overlapping roles of phytochromes A and B. In dark-grown wild-type (WT) seedlings, *CAB* mRNA (encoding the chlorophyll *a/b* binding protein) is induced to detectable levels following a pulse of red light (*CAB* mRNA levels, 4 hours after the pulse, are represented by the *red bars*). Light induction of *CAB* mRNA expression is reduced in *phyA* or *phyB* mutants (which lack phytochrome A or phytochrome B, respectively) and is further reduced in *phyA:phyB* double mutants (which lack both phytochromes).

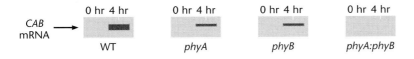

Figure 6–12.
The shade avoidance response. (A) Plants grown in full sun have short internodes and petioles. (B) Plants shaded by other plants exhibit the classic shade avoidance response: extended internodes and petioles.

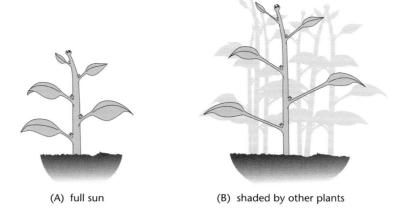

(A) full sun   (B) shaded by other plants

of individual phytochromes are separate in different parts of the plant and at different developmental stages. Recall that earlier we described how phytochrome A is responsible for inhibition of hypocotyl elongation by far-red–rich light; here, we see that phytochrome B mediates an effect of far-red–rich light on the growth of leaves and shoots.

The end-of-day and shade responses have been characterized at the genetic level mainly in Arabidopsis. Although less well characterized in other plant species, it is likely that phytochrome B (perhaps with other closely related phytochromes analogous to D and E in Arabidopsis) mediates the shade avoidance response in many other plants. As PHYB, PHYD, and PHYE are more closely related to one another than they are to PHYC or PHYA, this subgroup of phytochromes perhaps evolved as an adaptation to growth in shaded habitats.

Cryptochromes are blue-light receptors with specific and overlapping functions

Blue-light responses are easier to define than the phytochrome-mediated responses, because they are simply those responses that take place in the presence of blue light— they have no equivalent to red–far-red reversibility. Often plants respond to blue light in combination with other wavelengths. For example, inhibition of hypocotyl elongation, the transition from vegetative to reproductive growth (flowering), and **circadian clock** entrainment (discussed in Section 6.4) are all responses that involve blue, red, and far-red light. Sometimes, however, plant responses are dictated by blue light alone; examples are **phototropism** (growth toward a light source) and stomatal opening. Two classes of photoreceptors, the cryptochromes and phototropins, confer responses to blue light.

Cryptochromes were the first blue-light photoreceptors to be characterized. They were identified in Arabidopsis and subsequently discovered in a diversity of organisms ranging from cyanobacteria, ferns, and algae, to *Drosophila*, mouse, and human. Cryptochromes have important functions in the circadian clocks of animals, and in plants they are involved in the entrainment of the circadian clock (see Section 6.4) as well as in the activation of light-induced gene expression and in the repression of hypocotyl elongation by light. All cryptochromes so far identified have homology to **photolyases** (Figure 6–13), enzymes originally characterized in bacteria that act as photoreceptors and mediate light-dependent DNA repair (see Section 7.1). Based on sequence similarities, cryptochromes can be classified into three groups: plant cryptochromes, animal cryptochromes, and cryptochrome-DASH proteins (CRY-DASH, so-named because they are found in **D**rosophila, **A**rabidopsis, **S**ynechocystis, and **H**omo sapiens). The presence of CRY-DASH in cyanobacteria indicates that these photoreceptors evolved before the

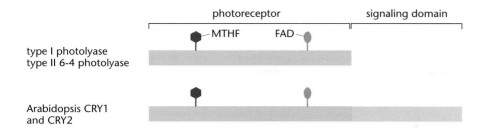

Figure 6–13.

Comparison of photolyases and cryptochromes of Arabidopsis. Photolyases are enzymes that repair pyrimidine dimers in DNA. This repair activity is activated by exposure to blue or UV-A light. Pterin methyltetrahydrofolate (MTHF) and flavin adenine dinucleotide (FAD), attached to the photolyase protein, are involved in the absorption of blue light. The N-terminal regions of CRYPTOCHROME 1 (CRY1) and CRYPTOCHROME 2 (CRY2) (*green*) are related to photolyases and bind the same chromophores. However, the cryptochrome photoreceptors also contain a C-terminal region (*blue*), proposed to be involved in light signal transduction.

divergence of prokaryotes and eukaryotes. In contrast, animal cryptochromes are most similar to a family of photolyases found in plants and animals, and plant cryptochromes (CRY1 and CRY2; see below) are distantly related to this family of photolyases (Figure 6–14), suggesting that plant and animal cryptochromes evolved independently after divergence of the two kingdoms.

Cryptochromes are similar to phytochromes in that they are made up of apoproteins and chromophores that absorb light. However, the chromophore of cryptochromes is attached noncovalently to the apoprotein, whereas that of phytochromes is attached covalently. Although similar in sequence to photolyases, cryptochromes do not have photolyase activity. They do, however, bind the same chromophores as photolyases and probably absorb blue light by a related mechanism.

The Arabidopsis cryptochrome family has three members, CRY1, CRY2, and CRY3 (CRY-DASH). The gene encoding CRY1 was isolated by screening mutant Arabidopsis populations for plants with reduced inhibition of hypocotyl elongation (resulting from the *hy* class of mutations described above) during treatment with blue light; *cry1* mutants have elongated hypocotyls when germinated in blue light. *CRY2* was identified in the Arabidopsis genome sequence because it is closely related in sequence to *CRY1* (at least in the region encoding the N-terminal chromophore-binding domain). The gene encoding *CRY3*, identified later, is less closely related to *CRY1* and *CRY2* than these genes are to each other and its functions have not yet been defined. The functions of CRY1 and CRY2 in mediating blue-light responses overlap, but are not identical. For example, CRY2 is rapidly degraded in etiolated seedlings exposed to blue light, whereas CRY1 is light-stable, suggesting that CRY1 is more likely to be involved in high-irradiance responses (Figure 6–15). The individual roles of the two proteins are described more fully below.

The chromophores of cryptochromes and photolyases are **flavin adenine dinucleotide (FAD)** and pterin methyltetrahydrofolate (MTHF). The mechanism by which blue light triggers photolyase activity is understood in some detail (Figure 6–16), and cryptochromes probably act in a similar way. MTHF absorbs blue light and transfers excitation energy to the FAD. Electron transfer from FAD to either a signaling-partner protein or to an amino acid residue in the cryptochrome protein itself is then presumed to activate signal transduction. Both CRY1 and CRY2 are phosphorylated when plants are exposed to blue light. Furthermore, CRY1 produced artificially in insect cells

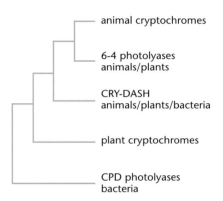

Figure 6–14.

Phylogenetic relationship between photolyases and cryptochromes. This superfamily of proteins is divided into four groups. Animal cryptochromes and the 6-4 photolyases are in one group; the other groups consist of plant cryptochromes, CRY-DASH proteins, and CPD photolyases.

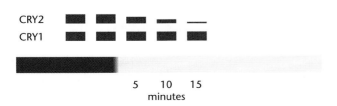

Figure 6–15.

Instability of CRY2 protein in etiolated seedlings exposed to light. CRY2 protein accumulates in etiolated seedlings, but within 1 hour of transfer from darkness to white or blue light the protein is not detected (protein level indicated by *purple bars*). In contrast, the CRY1 protein is stable under the same conditions. The bar at the bottom represents the transition from dark to light, and time after lights on is indicated.

Figure 6–16.
Repair of thymidine dimers in DNA by photolyases. The pterin methyltetrahydrofolate (MTHF) of the photolyase absorbs blue light, and the resulting excitation energy (EE) is transferred to the flavin adenine dinucleotide (in its reduced form, FADH$_2$). The excited flavin transfers an electron to the pyrimidine dimer, allowing repair of the DNA.

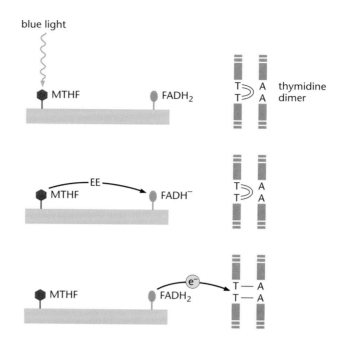

is phosphorylated *in vitro* when exposed to blue light, suggesting that the phosphorylation is mediated by CRY1 itself and is triggered by blue light. These results imply that phosphorylation may also be important in the function or regulation of cryptochromes in plants. Signaling by plant cryptochromes is likely to occur in the nucleus, because both CRY proteins are found in the nucleus in Arabidopsis. A major role for plant cryptochromes is the regulation of light-induced gene expression by inhibiting the activity of the ubiquitin ligase CONSTITUTIVE PHOTOMORPHOGENESIS 1 (COP1), as discussed in more detail in Section 6.3.

Arabidopsis mutants with impaired functioning of single cryptochrome photoreceptors have been used to study the individual contributions of the CRY proteins. The CRY1 photoreceptor promotes **anthocyanin** accumulation and suppression of hypocotyl elongation, and synchronizes the circadian clock to the day–night cycle (see Section 6.4). CRY2 has a smaller role in suppressing hypocotyl elongation in blue light and is important only in very low-intensity blue light where maximal photoreceptor sensitivity is required. However, CRY2 has an important role in promoting the transition from vegetative growth to flowering, and mutations that impair CRY2 function delay the transition to flowering.

In addition to their effects under blue light, CRY1 and CRY2 trigger responses to shorter-wavelength UV-A light. **Transgenic plants** containing more CRY1 than wild-type plants are more sensitive both to blue and to UV-A light than the wild type, suggesting that CRY1 acts as a photoreceptor in both regions of the spectrum. Mutations in *CRY1* only weakly reduce the suppression of hypocotyl elongation that normally occurs with UV-A treatment, but if both CRY1 and CRY2 are inactivated, a strong effect on hypocotyl elongation can be seen. This shows that the roles of CRY1 and CRY2 in mediating responses to UV-A light overlap.

Phototropins are blue-light receptors involved in phototropism, stomatal opening, and chloroplast migration

Phototropism is the directional growth of plants toward a light source. This pattern of growth increases the opportunity for light capture and hence photosynthesis in the leaves. In most plant species, phototropism is triggered by blue light. In the laboratory, phototropism is most often studied in etiolated seedlings, which bend toward a unidirectional light source. This approach also provides a convenient method to identify Arabidopsis mutants with impaired phototropism. Plants impaired in the phototropic response are called *nonphototropic hypocotyl* (*nph*) mutants. Roots show the opposite response, growing away from a unilateral light source. Arabidopsis plants mutated in the *NPH1* gene are impaired in phototropism when exposed to low-intensity blue light, but show a normal response to high-intensity blue light. The hypocotyls of *nph1* plants do not bend toward low-intensity blue light, and their roots do not grow away from it. In contrast, *cry1* and *cry2* mutants are unaffected in their phototropic responses (Figure 6–17).

NPH1 encodes the blue-light photoreceptor phototropin 1 (phot1), and therefore the gene was renamed *PHOT1*. This protein contains a serine/threonine kinase domain at the C-terminus and two repeated domains, each approximately 100 amino acids long, at the N-terminus. These domains are found in proteins from bacteria to mammals, and were named "LOV domains" because proteins containing them are regulated by **l**ight, **o**xygen, and **v**oltage. As described for cryptochromes, phot1 is a flavoprotein, but in this case **flavin mononucleotide (FMN)** is the chromophore; one molecule of FMN is bound to each LOV domain.

A second gene in Arabidopsis encodes another protein with two LOV domains and a kinase domain, phototropin 2 (phot2). Genetic experiments indicate that phot2 and phot1 have related functions, because *phot1:phot2* double mutants lack phototropic responses in both low- and high-intensity blue light (Figure 6–17), whereas *phot1* mutants retain the response to high-intensity blue light.

The *phot1:phot2* double mutant is also impaired in other blue-light responses. Chloroplasts normally migrate in response to changes in light intensity. At low light intensities, chloroplasts are distributed within the cell to maximize exposure to light, but at high light intensities they are positioned to minimize exposure and thereby reduce **photodamage**. The *phot1:phot2* double mutant does not show chloroplast movement in response to low or high light intensities, demonstrating the importance of phototropin in these responses. Exposure to light also normally induces stomatal opening, allowing uptake of carbon dioxide for photosynthesis. The *phot1:phot2* double mutant shows no stomatal opening in blue light, demonstrating that phototropins are also essential for this response (Figure 6–17).

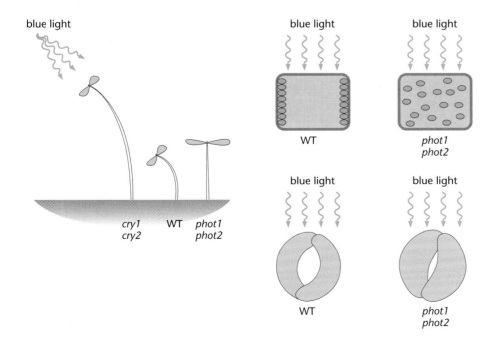

Figure 6–17.
The roles of phototropins in blue-light responses. The wild-type (WT) seedling shows a phototropic response, growing toward a unidirectional light source. The *cry1:cry2* double mutant shows a wild-type phototropic response, but its hypocotyl is longer because inhibition of hypocotyl elongation by blue light requires CRY function. The *phot1:phot2* double mutant shows no phototropic response, and hypocotyl elongation is inhibited as in wild-type plants. In addition, *phot1:phot2* shows neither chloroplast movement away from strong light (*upper right*) nor opening of stomatal guard cells in response to light (*lower right*).

The first step in phototropin signaling probably involves autophosphorylation. Phosphorylation of membrane proteins has long been known to play an early role in phototropism, as shown in the growing regions of etiolated seedlings exposed to light. Furthermore, this phosphorlyation is greatly reduced in a *phot1* mutant. PHOT1 protein made artificially in insect cells undergoes autophosphorylation in response to light, suggesting that this is an early step in the signaling mechanism of PHOT1.

The nature of the signal-transduction chain activated by phototropin is unknown. One way to investigate the mechanism is to screen for more *nph* mutants in Arabidopsis and to study those that do not carry mutations in the gene encoding phototropin. These mutations are likely to define proteins that are required for phototropin signal transduction. The *NPH3* gene is affected by one of these mutations and encodes a membrane-associated protein that interacts directly with phototropin in *in vitro* assays. Thus the NPH3 protein and phototropin may exist *in vivo* as a complex associated with the plasma membrane. Mutations in a gene closely related in sequence to *NPH3* prevent root phototropism, which suggests that different members of this class of protein may be involved in distinct aspects of phototropin signaling. The bending of the hypocotyl that occurs during phototropism is caused by higher accumulation of **auxin** on one side of the hypocotyl, as in gravitropism (see Section 6.5). Auxin concentration is higher on the side of the hypocotyl furthest from the light and is responsible for the increased growth of the hypocotyl on that side. Mutations in the *PIN3* gene, which encodes a component of the auxin efflux carrier related to PIN1 and PIN2 (see Section 6.5), reduce phototropism of Arabidopsis seedlings, indicating that altered distribution of the PIN3 protein may be responsible for the differential accumulation of auxin during phototropism. How phototropin alters the distribution of PIN3 is not known.

Some photoreceptors respond to red and blue light

So far we have focused on phytochromes, which control plant responses to red and far-red light, and cryptochromes and phototropins, which control responses to blue and UV-A light. Some photoreceptors, however, seem to regulate responses in both the red/far-red and blue regions of the spectrum. For example, in addition to their red and far-red responses, phytochromes can also control responses to blue light under specialized conditions. This has been demonstrated in mutants incapable of making fully functional phytochrome A. Suppression of hypocotyl elongation under low-irradiance blue light does not occur in *phyA* mutants, and hypocotyls grow significantly longer than in wild-type seedlings.

Another example of a photoreceptor that is thought to control responses to blue and red/far-red light is PHY3, a protein identified in the fern *Adiantum*. This is a hybrid photoreceptor: its N-terminal amino acids show homology to the phytochrome chromophore-binding domain (Figure 6–18). Recombinant protein, expressed in *Escherichia coli* and reconstituted with a **phycocyanobilin** chromophore, shows the red–far-red reversibility characteristic of phytochrome. The remainder of the protein is very similar to phototropin and contains the chromophore-binding domain and the kinase domain. This single protein, therefore, has features of both red/far-red–light and blue-light photoreceptors and may control responses to light in both regions of the spectrum. This is in agreement with the observation that red and blue light act cooperatively in the phototropism of *Adiantum*, in contrast to Arabidopsis, in which phototropism is regulated only by blue light and mediated via phototropin.

Biochemical and genetic studies provide information on the components of the phytochrome signal-transduction pathway

Photoreceptors are responsible for the initial perception of light, but additional molecules are needed for signal transduction. We have described in brief what is known about signal transduction from phytochromes A and B, cryptochromes, and phototropins. Here

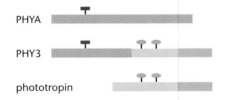

Figure 6–18.
A photoreceptor of *Adiantum* with features of both phytochromes and phototropins. The PHY3 photoreceptor of the fern *Adiantum* contains regions of homology with phytochrome in the N-terminal region (marked in *pink*) and with phototropin at the C-terminus (marked in *blue* and *purple*). The regions of homology contain the chromophore binding site of phytochrome (elevated *red rectangle*) and the LOV domains to which the chromophores of phototropin attach (elevated *blue oval* shapes). The kinase domain of phototropin is also conserved (*purple rectangle*). Phytochrome A (PHYA) and phototropin of Arabidopsis, along with their attached chromophores, are shown for comparison.

we describe in more detail the key biochemical and genetic studies that have provided information on these signal-transduction pathways, concentrating on phytochromes, which are the best-characterized system. We also discuss what the studies indicate about signal transduction downstream of the photoreceptors.

One way of identifying signal-transduction components downstream of the photoreceptors is to screen for mutants that exhibit phenotypes similar to those of mutants lacking individual photoreceptors. For example, *phyA* mutants have an elongated hypocotyl in far-red light but not in red or white light. Plants with functional phytochrome A but impaired phytochrome A–associated signal-transduction machinery show phenotypes identical to those of *phyA* mutants. Analysis of such mutants helped to identify the components of signal transduction. Several different genes specifically affect phytochrome A signaling, and findings on the different phenotypes caused by mutations in these genes paint a picture of a complex chain, or even network, of events. Similarly, mutants with a fully functional *PHYB* gene but phenotypes similar to those of *phyB* mutants have helped to identify the signaling components between PHYB perception of red light and altered gene expression.

By comparing the results from these two approaches, researchers have identified some genes that are specific to phytochrome A or B and other genes that are common to both pathways (summarized in Table 6–2). It is possible that the genes isolated by just one of the approaches represent early steps in phytochrome signal transduction, and that genes isolated by both approaches represent common steps; that is, the two pathways may converge.

Table 6–2 Genes of the phytochrome signal-transduction pathways

Gene name	Gene product and its function	Phytochrome signal-transduction pathway(s) affected
FHY1	Nuclear protein that amplifies PHYA signal	A
FHY3	WD 40 protein	A
SPA1	WD 40 protein	A
FAR1	Nuclear protein	A
PAT1	Cytoplasmic member of the GRAS family of signal-transduction proteins	A
FIN219*	Cytoplasmic member of the GH3 family of proteins	A
RED1		B
PEF2		B
PEF3		A & B
PEF1		A & B
PSI2		A & B
PIF3		A & B
PKS1		A & B
NDPK2	Kinase that interacts with phytochromes	A & B

*May define a link between PHYA and auxins in the regulation of growth.

As Table 6–2 shows, many of the proteins identified as involved in phytochrome A or B signal transduction are of as yet unknown function. The functions of other proteins have been more fully characterized, and several of these proteins seem to be common to both phytochrome A and phytochrome B pathways. Three such proteins known to interact with phytochromes A and B are PIF3, a nuclear-localized bHLH **transcription factor**; PKS1, a novel cytoplasmic protein that is phosphorylated by phytochrome; and the enzyme nucleoside diphosphate kinase 2, whose activity is regulated by phytochrome. These proteins are not structurally or functionally related, and they interact with different phytochrome domains; thus they are not regulated by phytochrome via the same mechanism. The observation that substrates for the light-regulated kinase activity of phytochrome are located in both nuclear and cytoplasmic compartments, and the observation that phytochromes themselves can move from the cytoplasm to the nucleus after red-light activation, together suggest that phytochromes can phosphorylate a number of substrates in a variety of subcellular locations. This begins to explain how phytochromes can regulate a large number of developmental processes throughout plant development.

As described previously, the best-understood branch of phytochrome signaling involves the nuclear transcription factor PIF3 (Figure 6–19). This protein was originally identified by virtue of its ability to associate with phytochrome in yeast. The interaction of phytochrome with PIF3 is light-dependent: red light induces rapid binding of phytochrome (P_{fr}) to PIF3, and far-red light causes the complex to dissociate. PIF3 binds to a sequence known as the **G-box motif** (CACGTG), which is found in the **promoter** of several light-regulated genes. Some of these genes are involved in chloroplast development and photosynthesis, and induction of their expression by light is reduced in *pif3* mutants compared with wild-type plants. Therefore, phytochrome probably regulates expression of these genes by interacting with PIF3 and enabling PIF3 to activate gene expression. In fact, PIF3 is just one of a family of PIF3-related transcription factors, some of which act as activators and some as inhibitors of phytochrome signaling.

Among the best-studied light-regulated genes are the *CAB* (*LHCB*) genes, which encode the chlorophyll *a/b* binding proteins of the light-harvesting complex required for photosynthesis (see Section 4.2). Arabidopsis contains three of these genes, and their expression has been studied extensively. Transcription of these genes is rapidly induced by activated phytochrome (Figure 6–19), and *CAB* mRNAs are detected in high abundance approximately three hours after exposure to red light. By taking progressively shorter fragments of the upstream region of these genes and fusing them to a **marker gene** such as the β-glucuronidase or luciferase gene (for more on the latter marker see Section 6.4), a 78-bp fragment was shown to be sufficient to invoke expression of the marker gene in response to light. This suggests that proteins involved in regulating or activating *CAB* gene expression can recognize and bind to DNA sequences within this fragment.

Experiments that combined protein extracts from plants with radiolabeled DNA fragments derived from the promoter identified a protein complex capable of binding these sequences. (Binding of proteins to DNA can be evaluated based on the altered mobility of the protein-DNA complex compared with unbound DNA in **polyacrylamide gels**.) To test whether binding of this protein complex is required for light induction of the promoter, a series of mutations were introduced into the promoter that prevent binding of the complex, and the mutated promoter was inserted upstream of a marker gene (β-glucuronidase). The mutations indeed abolished the promoter's responsiveness to red light. The CIRCADIAN CLOCK ASSOCIATED 1 (CCA1) protein was subsequently shown to bind to the same DNA sequences as this protein complex, and therefore is likely to be part of the complex. Transcription of the *CCA1* gene itself responds to red light, and although the mechanism has not been defined it is probably related to that described above for activation of light-induced genes by PIF3 (Figure 6–19). The identification of a CCA1 binding site required for phytochrome induction of *CAB*

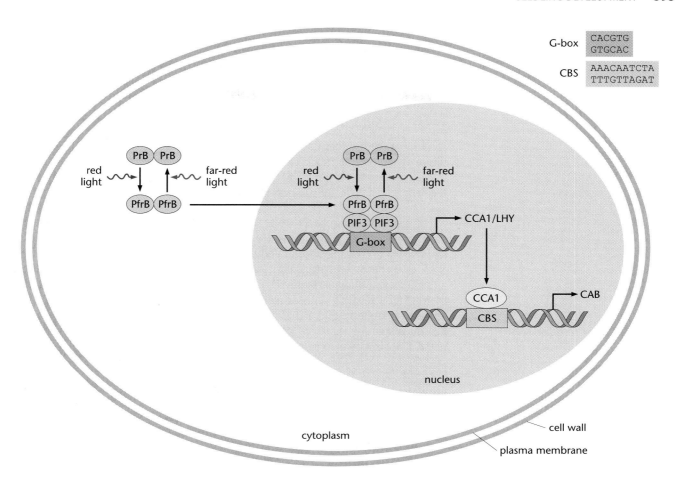

G-box `CACGTG`
`GTGCAC`

CBS `AAACAATCTA`
`TTTGTTAGAT`

expression may define the end point of a short signal-transduction chain extending from the phytochrome. Although a single CCA1 binding site seems to be essential for induction of *CAB* expression by red light, this is an unusual situation. A single transcription factor often binds to several positions within a promoter; such redundancy means that each of these sites must be altered before a reduction in gene expression can be detected.

6.3 SEEDLING DEVELOPMENT

Many seedlings begin their growth underground, in darkness. Growth in the dark is probably controlled by a balance of positive and negative regulation: that is, some of the genes involved are activated in the absence of light, others (those involved in photomorphogenic growth) are inhibited. Likewise, during photomorphogenesis, genes necessary for etiolated growth are repressed and those required for photomorphogenic responses are activated. Neither photomorphogenesis nor skotomorphogenesis is the default developmental pathway: they are alternative fates that a plant can follow, depending on the environmental cues that it perceives and processes.

Seedlings growing underground encounter physical resistance from the soil. On encountering obstacles, they produce the gaseous growth regulator ethylene. The ethylene signaling pathway is one of the best-understood signaling pathways in plants, and in this section we describe its role in growth and, wherever possible, how its role has been determined experimentally. Keep in mind, however, that ethylene also plays important roles in other developmental processes: the regulation of fruit ripening, seed germination, abscission, senescence, and responses to pathogen attack. Ethylene (like other

Figure 6–19.
Activation of *CAB* gene transcription by phytochrome B. In response to red light, the P$_{fr}$ form of phytochrome B (PfrB) is formed in the cytoplasm and moves to the nucleus. In the nucleus, PfrB interacts with the transcription factor PIF3 and the complex binds to the G-box in the promoter of the *CCA1* gene, activating its transcription. The CCA1 transcription factor (or LHY; see Section 6.4) then binds to the CCA1 binding site in the *CAB* promoter (CBS) and activates its transcription.

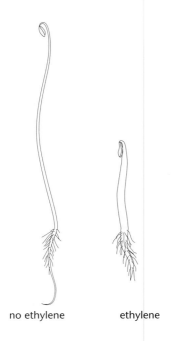

no ethylene   ethylene

$$H_3C - S - CH_2 - CH_2 - CH - COO^-$$
$$|$$
$$NH_3^+$$

methionine

ATP

$PP_i + P_i$

$$H_3C - \overset{\oplus}{\underset{|}{S}} - CH_2 - CH_2 - CH - COO^-$$

adenine ribose NH_3^+

S-adenosylmethionine (SAM)

ACC synthase

$$\underset{H_2C}{\overset{H_2C}{\diagdown}} \overset{NH_3^+}{\underset{COO^-}{\diagup}}$$

1-aminocyclopropane-
1-carboxylic acid (ACC)

ACC oxidase ⟶ CO_2 + NH_4^+

$CH_2 = CH_2$

ethylene

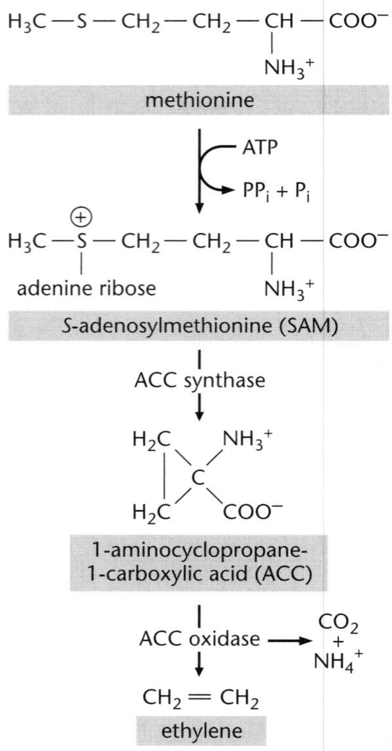

Figure 6–21.
Steps in the biosynthetic pathway from methionine to ethylene.

Figure 6–20.
Ethylene regulation of dark-grown seedlings. The untreated seedling has a long, slender hypocotyl, elongated root, and closed apical hook. The ethylene-treated seedling has an exaggerated apical hook and stunted root and hypocotyl.

growth regulators) does not act alone. For example, ethylene and gibberellin interact in the regulation of internode elongation in deepwater rice; ethylene and **abscisic acid** interact in the regulation of seed germination; and ethylene and auxin interact in the regulation of a characteristic growth response called the **triple response**.

The triple response probably protects delicate structures, such as the shoot apical meristem, from damage as they force through the soil. The three elements of the response are: hypocotyls become shorter and thicker; roots become shorter and thicker; and the apical hook becomes exaggerated—all relative to what is observed in plants growing in a medium that offers little physical obstruction. The triple response can be mimicked by treating a seedling with exogenous ethylene (Figure 6–20), and a combination of ethylene treatments and analyses of mutant plants with defective ethylene responses has elucidated the signal-transduction pathway that connects the plant's perception of physical obstructions to its altered growth responses.

Ethylene is synthesized from methionine in a pathway controlled by a family of genes

Ethylene is produced from methionine via the synthesis of 1-aminocyclopropane-1-carboxylic acid (ACC), catalyzed by the enzyme ACC synthase. ACC is then oxidized by the enzyme ACC oxidase (Figure 6–21). The induction of ethylene at different times during development (including growth in darkness) and in response to different environmental variables (for example, stresses such as those occurring during pathogen attack, abscission, and senescence) involves the controlled expression of individual members of a family of ACC synthase genes. In tomato, for example, which has six **isozymes** of ACC synthase, only two of the ACC synthase–encoding genes show increased expression during fruit ripening.

Not all ACC found in plant tissue is converted to ethylene; it can also be converted to nonvolatile conjugated forms. Stores of conjugated ACC are thought to play an important role in the control of ethylene production (that is, ethylene production is not simply controlled by the activity of the biosynthetic enzymes). ACC itself may also be used as a signal under **anaerobic** (or **anoxic**) conditions (see Section 7.6).

Genetic analysis has identified components of the ethylene signal-transduction pathway

The triple response is used as an assay to study the genetic basis of the ethylene responses of plants. Most of this work has been carried out in Arabidopsis. Mutagenized populations of seedlings are screened to identify mutants with a perturbed ethylene response. In essence, detection of two classes of mutant might be expected from such a screen (Figure 6–22). The first class consists of mutants that fail to exhibit the triple response when exposed to ethylene. These ethylene-insensitive mutants have long rather than short hypocotyls and roots when grown in an ethylene-containing atmosphere. The mutants are deficient in the perception or transduction of the ethylene signal. The *etr1* mutants are an example of this first class. The second class consists of constitutive triple-response mutants, which exhibit a triple response in the absence of ethylene exposure; that is, they have a constitutively activated ethylene-signaling pathway. This can result either because the mutant seedlings produce more ethylene than wild-type plants, or because the signaling pathway is activated in the absence of ethylene, as in the *ctr1* mutants.

Ethylene receptors are characterized by an N-terminal ethylene binding domain and a histidine protein kinase–like domain. They are similar to a class of receptors found in bacteria called "two-component histidine kinase receptors." Two-component regulation usually involves (1) a sensor molecule carrying a histidine kinase domain that autophosphorylates in response to an environmental stimulus, and (2) a response regulator molecule carrying a receiver domain with an aspartate residue that is phosphorylated by transfer of the phosphate from the histidine of the sensor molecule (Figure 6–23). The Arabidopsis and tomato ethylene receptor families each have five members, and **homologs** have been reported in a wide variety of plants. The genome of the cyanobacterium *Synechocystis* also contains a homologous gene whose product is capable of binding ethylene, suggesting that ethylene receptors may originally have evolved in bacteria.

The five ethylene receptors of Arabidopsis are ETR1 and ETR2 (ETHYLENE RECEPTOR 1 and 2), ERS1 and ERS2 (ETHYLENE RESPONSE SENSOR 1 and 2), and EIN4 (ETHYLENE INSENSITIVE 4) (Figure 6–24). Loss of any single receptor has little obvious effect on plant phenotype, suggesting that the functions of individual receptors in Arabidopsis overlap, and one can take the place of another if the full complement is not present. Of course, this level of redundancy might just apply to ethylene receptors involved in the triple response, and individual receptors might have slightly different roles during a plant's lifetime. For example, the five ethylene receptor genes in tomato have distinct expression patterns throughout development, indicating that each receptor may have a different tissue- and stage-specific role in ethylene signaling. Expression of some receptor genes is inducible—by flooding or ripening, for example.

ETR1 was the first gene to be identified in screens for ethylene-insensitive mutants; *etr1* mutations are genetically dominant, and in addition to abolishing the seedling triple response they abolish the response of adult plants to endogenously produced ethylene. These mutations are dominant because they render the mutant receptor insensitive to ethylene; thus, in the presence of ethylene, the mutant receptor continues to signal (unlike the wild-type receptor; see below). ETR1 and other ethylene receptors exist as dimers that span the **lipid bilayer** of membranes (Figure 6–25). The N-terminal, membrane-spanning domain of the ETR1 protein contains the ethylene binding site; the histidine kinase domain is cytoplasmic; and the receiver domain is also cytoplasmic and found toward the C-terminus. The ethylene binding site is a hydrophobic pocket embedded in the plasma membrane, and binding requires the presence of copper ions, probably supplied by RAN1, the product of the *RESPONSIVE TO ANTAGONIST 1 (RAN1)* gene; this protein is a homolog of copper-transporting **ATPases** in yeast and humans.

In addition to evidence provided by mutant phenotypes, experimental data also support the idea that ETR1 is an ethylene receptor. Leaves of *etr1* mutants have a significantly reduced capacity for binding ethylene—only about 20% of that of wild-type controls.

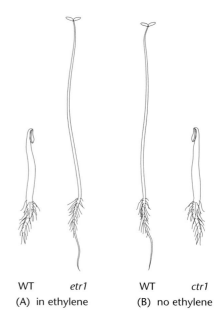

WT   *etr1*   WT   *ctr1*
(A) in ethylene   (B) no ethylene

Figure 6–22.
Comparison of wild-type plants and ethylene triple response mutants.
(A) The wild-type (WT) plant exhibits a classic triple response (short root, short hypocotyl, exaggerated apical hook) when grown in ethylene, whereas the *etr1* mutant shows growth like that of a wild-type plant in the absence of ethylene. (B) In the absence of ethylene, the *ctr1* mutant exhibits a constitutive triple response.

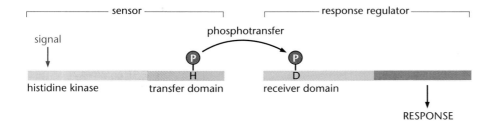

sensor   response regulator
signal   phosphotransfer
P   P
H   D
histidine kinase   transfer domain   receiver domain
RESPONSE

Figure 6–23.
Bacterial two-component regulation.
A signal causes autophosphorylation of the sensor; transfer of a phosphate group (phosphotransfer) from a histidine residue (H) in the sensor to an aspartate residue (D) in the response regulator causes phosphorylation of the response regulator (see text for details).

Figure 6–24.
The Arabidopsis family of ethylene receptors. ETR1, ETR2, and EIN4 contain features of both sensor and response regulator molecules, whereas ERS1 and ERS2 lack a receiver domain. ERS1 and ERS2 might recruit the ETR1, ETR2, or EIN4 receiver domains, or might use other response regulators.

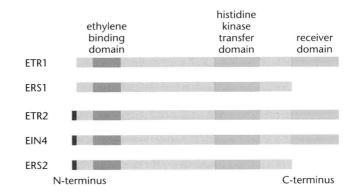

Also, when *ETR1* is expressed in yeast it confers high-affinity ethylene binding, but when mutated *ETR1* genes are expressed in yeast (i.e., those that confer ethylene insensitivity in plants) no ethylene binding takes place.

The ethylene response is negatively regulated by binding of ethylene to its receptors

As we have noted, the roles of the various members of the ethylene receptor family overlap (at least during the triple response). That is, most Arabidopsis plants with single or double loss-of-function mutations in genes encoding ETR1 and related receptors still display a triple response similar to that of wild-type plants. Arabidopsis plants that are triple or quadruple ethylene-receptor mutants, however, have a constitutive triple-response phenotype (Figure 6–26). In these mutants, ethylene is not perceived but the downstream signaling pathway is constitutively active. This means that in wild-type

Figure 6–25.
The domains of ETR1. ETR1 spans the lipid bilayer of the plasma membrane. Ethylene (C_2H_4) binding, which occurs in a hydrophobic pocket in the plasma membrane, requires the binding of copper (Cu^{2+} ions).

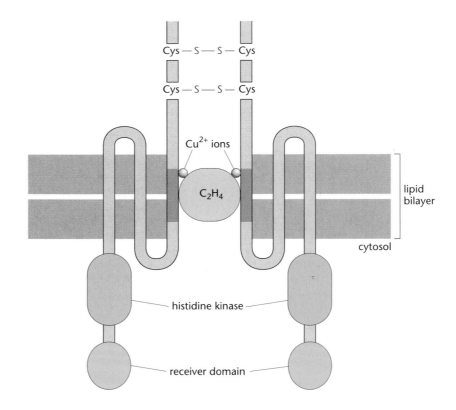

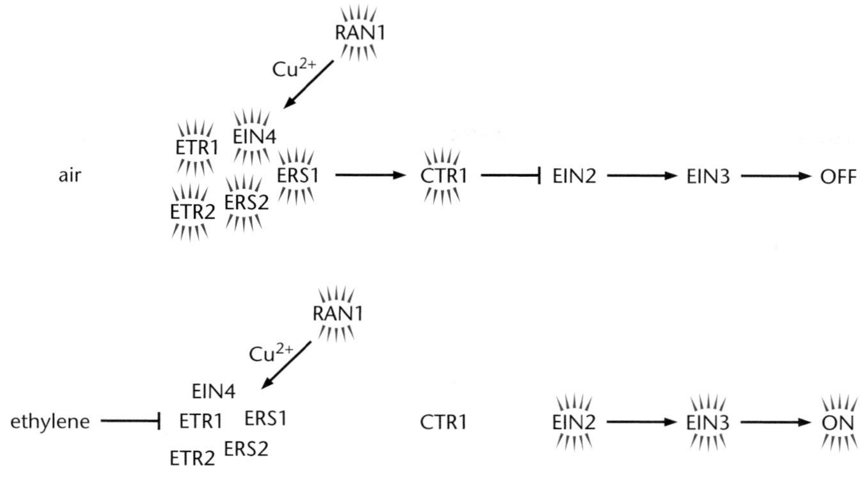

Figure 6–26.
Negative regulation of ethylene signaling. ETR1 and related ethylene receptors are active in the absence of ethylene and act as negative regulators of the ethylene response. Components CTR1, EIN2, and EIN3 are downstream of the ethylene receptors in ethylene signaling (as discussed in the text). Components are shown as active (*red spikes*) or inactive (*no spikes*) in the presence of ethylene, resulting in an activated (ON) or repressed (OFF) ethylene response.

plants, ethylene receptors normally inhibit ethylene responses until ethylene binds to them, at which point inhibition stops and the signaling cascade proceeds unimpeded. The negative regulation of ethylene signaling is an important concept to grasp: before these experiments were carried out, the intuitive assumption would have been that the pathway was positively controlled.

Active ETR1 (i.e., ETR1 in the absence of ethylene) activates CTR1 (CONSTITUTIVE TRIPLE RESPONSE 1), a kinase related to the mammalian Raf family of protein kinases that interacts with the cytoplasmic portion of ETR1 *in vitro*. Raf kinases are serine/threonine protein kinases. They are activated by Ras, a small GTP-binding protein, and then regulate the **mitogen-activated protein kinase (MAPK) cascade**. MAPKs are serine/threonine protein kinases that regulate mammalian cell responses (including gene expression, mitosis, differentiation) to extracellular stimuli (mitogens). The stimuli activate the MAPK via a signaling cascade involving MAPK, MAP kinase kinase (MAP2K), and MAP kinase kinase kinase (MAP3K). Thus, it is possible (although not proven) that CTR1 acts as a negative regulator of ethylene signaling through the activation of a MAPK cascade.

Inactivation of CTR1 allows activation of downstream components of the ethylene signaling chain

Beyond CTR1 inactivation, we have only an incomplete understanding of the signal-transduction pathway that ultimately results in ethylene responses. Analyses of double mutants have revealed an order of gene action in the ethylene-signaling pathway. The order suggests that at least two gene products, EIN2 and EIN3 (described below; see also Figure 6–26), act downstream of the *CTR1* gene product.

Arabidopsis EIN2 is an integral membrane protein (i.e., a protein component of the plasma membrane) with a membrane-spanning N-terminal portion homologous to a mammalian family of transporter proteins known as "NRAMP" metal-ion transporters. EIN2, however, lacks detectable ion-transport activity, so its biochemical role in ethylene signaling is not clear.

Other components downstream of CTR1 are EIN3 and ERF1, transcription factors that bind to the promoter of ethylene-inducible genes, activating their transcription (Figure 6–27). EIN3 is a key regulator of ethylene response: in the absence of ethylene, degradation of EIN3 (via the ubiquitin-proteasome system) is promoted by interaction with a specific E3 ubiquitin ligase. In the presence of ethylene, EIN3 dimers bind to a

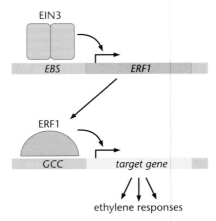

Figure 6–27.
The EIN3 to ERF1 transcription factor cascade. Ethylene, by inhibiting CTR1, causes increased levels of the EIN3 dimer. EIN3 binds to the *EBS* site in the promoter of the *ERF1* gene and activates its transcription. The resulting increase in the level of ERF1 transcription factor activates target genes that carry GCC-box promoter elements, resulting in ethylene responses.

defined target sequence in the promoter of the ethylene-inducible gene *ERF1*. The ERF1 protein belongs to a family of DNA binding proteins known as the "AP2-like" proteins, which show sequence similarities with the **AP2 (APETALA2)** transcription factor family (other members of this family are involved in transcriptional responses to drought, salt, and cold stresses; see Chapter 7). ERF1 binds to a GCC-box, a motif found in the promoters of several ethylene-responsive genes, including those involved in disease resistance. The EIN3 to ERF1 transcription factor cascade thus links ethylene production induced during pathogen attack to the induction of defense-related genes.

Ethylene interacts with other signaling pathways

Ethylene does not act alone in the triple response. There is considerable evidence that auxin and abscisic acid are also involved in this and other ethylene responses. For example, auxin-resistant *axr1* Arabidopsis mutants have reduced root and apical-hook sensitivity to ethylene, and *aux1* and *eir1* mutants, affected in the polar transport of auxin, show ethylene resistance specifically in root growth. In addition, exogenous ethylene inhibits the polar transport of auxin. Many ethylene-signaling mutants also show altered seedling responses to abscisic acid, and DELLA function (traditionally associated with gibberellin action; see Section 5.4) also regulates the triple response.

Ethylene interacts with other growth regulators in other plant responses. We have described briefly here (and do so in more detail in Section 8.4) its role in plants' defense against disease. In these responses, ethylene sometimes acts in concert with defense signals to amplify responses, and sometimes acts antagonistically to signals (e.g., jasmonic acid) to fine-tune the response to specific predators.

The light responses of seedlings are repressed in the dark

Seedlings retain the capacity to enter photomorphogenic or skotomorphogenic development, and it is the presence or absence of light, respectively, that determines which developmental pathway a seedling will follow. As noted above, many mutants with impaired light responses show aspects of skotomorphogenic development even when exposed to light. These plants have nonfunctional photoreceptors or altered proteins that block the usual light-signal-transduction pathways. During skotomorphogenesis, photomorphogenesis is actively repressed, and there are many regulatory proteins that act to prevent photomorphogenesis in the dark. The presence of these proteins was initially deduced from the isolation of mutants that exhibit characteristic features of photomorphogenic development when grown in the dark (Figure 6–28). Such features include short hypocotyls, expanded cotyledons, partial leaf development, partial chloroplast differentiation, and increased expression of light-regulated genes such as the *CAB* (*LHCB*) gene family, which encodes light-harvesting chlorophyll *a/b* binding proteins, and the gene encoding chalcone synthase, an enzyme involved in synthesis of anthocyanin pigments. Many mutants with this sort of phenotype have been described and are classified as *de-etiolated* (*det*), *constitutive photomorphogenic* (*cop*), or *fusca* (*fus*) mutants.

The *det*, *cop*, and *fus* mutations are recessive, so gene activity is lost in plants homozygous for the mutant genes. This suggests that in wild-type plants grown in the dark, the *DET*, *COP*, and *FUS* genes act to repress photomorphogenesis. If *cop1* or *det1* mutations are combined with mutations that impair light perception (such as *phyB*, *phyA*, or *cry1*), the double mutants also show the photomorphogenic phenotype associated with the *cop/det/fus* single mutations. Since light perception is impaired in these plants, the *cop/det/fus* mutations allow photomorphogenic growth in the absence of photoreceptor signaling. This strongly suggests that wild-type *COP/DET/FUS* genes negatively regulate photomorphogenesis and act within the signaling pathways controlled by phytochromes and cryptochromes (Figure 6–29).

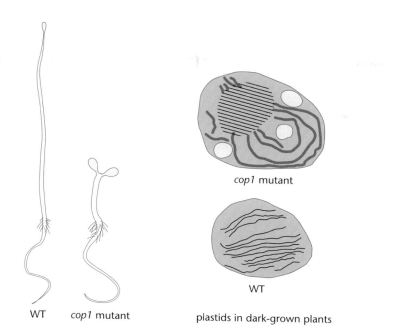

WT   *cop1* mutant   plastids in dark-grown plants

Figure 6–28.
Comparison of dark-grown wild-type and *cop1* mutant seedlings. The dark-grown wild-type (WT) seedling shows the elongated hypocotyl and closed cotyledons characteristic of skotomorphogenic development. In contrast, the dark-grown *cop1* mutant has a shorter hypocotyl and open cotyledons, as typically seen in wild-type plants exposed to light and characteristic of photomorphogenic development. The plastids of dark-grown wild-type plants contain prolamellar bodies (aggregates of disorganized membranes) and lack thylakoid membranes (*lower right*). However, in dark-grown *cop1* mutants the plastids show signs of chloroplast development; prolamellar bodies are absent, and parallel thylakoid membranes are formed (*upper right*).

In addition to acting as negative regulators of photomorphogenesis in the dark, the *COP/DET/FUS* genes have other important effects on plant development. For example, all mutations in these genes cause severe defects in plant growth in the light. Strong *det1* or *cop1* mutations that abolish protein function are lethal, whereas weaker *det1* or *cop1* mutations reduce the stature and fertility of light-grown plants. Moreover, some *det1* and *cop1* mutations cause defective differentiation of particular cell types in light-grown plants—for example, chloroplast differentiation can occur in root cells, and expression of the enzyme chalcone synthase (restricted to the epidermis in wild-type plants) can occur in leaf mesophyll cells. Thus the wild-type *COP1* and *DET1* genes not only repress photomorphogenesis in the dark but are also important for the proper development of light-grown plants and for the correct spatial pattern of light-regulated gene expression.

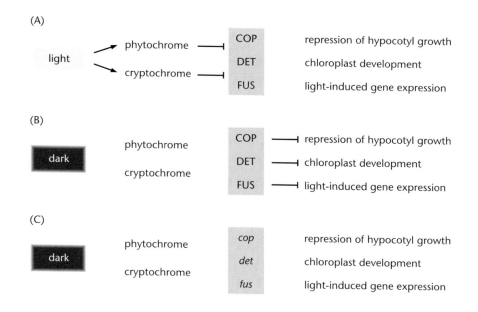

Figure 6–29.
The role of COP/DET/FUS proteins in regulating photomorphogenesis.
(A) Light-grown wild-type seedlings. Photomorphogenic development is represented by chloroplast development, light-induced gene expression, and repression of hypocotyl growth. Exposure of the seedling to light activates the phytochrome and cryptochrome photoreceptors, and these repress the activity of the COP/DET/FUS proteins. (B) Dark-grown wild-type seedlings. Skotomorphogenic development requires repression of photomorphogenesis by the COP/DET/FUS proteins. Photoreceptors are not activated by exposure to light and thus do not repress the activity of COP/DET/FUS proteins, and therefore these proteins repress photomorphogenesis. (C) Dark-grown *cop*, *det*, or *fus* mutant seedlings. The repression of photomorphogenesis shown in (B) does not occur, because one component of the COP/DET/FUS system is absent. Photomorphogenic growth occurs even in the absence of light.

COP1 and the COP9 signalosome function by destabilizing proteins required for photomorphogenesis

The COP1 protein represses photomorphogenesis in the dark by binding to several proteins required for photomorphogenesis and targeting them for degradation. COP1 carries four WD 40 repeat domains (a structural domain made up of β protein folds and responsible for protein-protein interactions in many eukaryotic proteins); these domains are essential for repression of photomorphogenesis and directly bind to several proteins, including HY5, a **basic leucine zipper (bZIP)** transcription factor. HY5 forms a heterodimer with a related protein, HYH, and binds to the promoters of genes activated by light. (*hy5* mutants are impaired in photomorphogenesis, showing an elongated hypocotyl in white light; see Table 6–1 for a reminder of *hy* gene nomenclature.)

In wild-type Arabidopsis seedlings, HY5 is 15-fold less abundant in dark-grown than in light-grown seedlings, and this difference is caused by regulation of the protein's stability rather than regulation of *HY5* transcription. Furthermore, in *cop1* mutants HY5 accumulates to the same extent in light-grown and dark-grown seedlings, demonstrating that COP1 is required for reduced HY5 levels in darkness. Inhibitors of proteasome activity prevent HY5 degradation in the dark, and therefore COP1 may destabilize HY5 by catalyzing the attachment of several copies of ubiquitin to HY5 and thus targeting it for destruction by the proteasome (Figure 6–30).

Two mechanisms ensure that COP1 promotes the degradation of HY5 in the dark but not in the light. The first involves subcellular compartmentalization of the COP1 protein. In dark-grown plants COP1 is present in the nucleus, whereas in light-grown plants it is found in the cytoplasm (Figure 6–31). HY5 is located in the nucleus. Thus the two proteins are present in the same cellular compartment only under the environmental conditions (darkness) in which they are thought to interact. The presence of COP1 in

Figure 6–30.
COP1-catalyzed ubiquitination of HY5 transcription factor in the dark but not the light. The RING finger, coiled-coil, and WD 40 repeat domains of COP1 are shown. (A) In the light, the HY5 protein (a bZIP transcription factor) accumulates and activates the expression of target genes. (B) In the dark, HY5 binds to the WD 40 domain of COP1. COP1 then catalyzes the attachment of a chain of ubiquitin (ubq) molecules to HY5. Ubiquitinated HY5 is degraded by the proteasome.

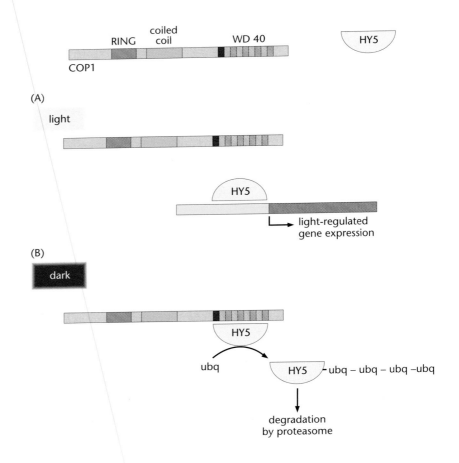

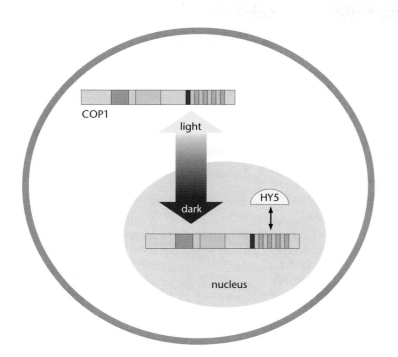

Figure 6–31.
Change in cellular location of COP1 on exposure of plants to light. In the dark, COP1 accumulates in the nucleus, where it targets HY5 for destruction. In light-grown plants, COP1 is found in the cytosol.

the cytoplasm of light-grown plants suggests that its location within the cell is regulated by light, and that this in turn regulates the stability of HY5 in response to light.

The second mechanism that reduces COP1 protein activity in the light is its inhibition by cryptochromes, the blue-light photoreceptors described in Section 6.2. Transgenic plants expressing high levels of the C-terminal part of CRY1 or CRY2 have a phenotype similar to that of *cop1* mutants. In particular, when grown in the dark they show characteristics of plants grown in the light, such as a short hypocotyl. The C-terminal part of CRY1 also physically interacts with COP1 protein in yeast cells, whereas in plant cells, COP1 and CRY1 physically interact in both the light and dark. These observations suggest that CRY1 binds to COP1, and when the plant is exposed to blue light, CRY1 undergoes a change in conformation that leads to the inhibition of COP1. The inhibition of COP1 then allows HY5 and other proteins involved in photomorphogenesis to accumulate in the light and photomorphogenesis to proceed.

The *COP1* gene is one of 11 *COP*, *DET*, and *FUS* genes that have been shown (by mutation studies) to be required for the repression of photomorphogenesis in the dark. Many of the other genes encode proteins that form a large protein complex called the COP9 **signalosome**. One of these genes is *COP9*; in *cop9* mutants, the COP9 signalosome does not form. There are at least eight other proteins in the COP9 signalosome, and as the complex does not accumulate in plants carrying mutations in any one of the other *COP/DET/FUS* genes, these probably also encode components of the complex. The COP9 signalosome is thought to activate COP1 activity, given that HY5 also accumulates in the *cop9* mutant when grown in the dark. The function of the COP9 signalosome is probably conserved in a wide range of organisms, because a protein complex containing very similar subunits is found in mammals and in the yeast *Schizosaccharomyces pombe*.

Brassinosteroids are required for repression of photomorphogenesis in the dark and other important functions in plant development

Brassinosteroids are another type of growth regulator required to prevent photomorphogenesis in etiolated seedlings. Their importance was demonstrated by the study of *de-etiolated* mutants. As described above, COP1 and the COP9 signalosome destabilize proteins required for photomorphogenesis. However, other *COP/DET/FUS* gene products probably act by independent mechanisms, as illustrated by differences in the effects of mutations in the three genes. For example, although *det1* and *cop1* mutants grown in the dark contain partially differentiated chloroplasts, the plastids of *det2* mutants remain as **etioplasts** when grown in the dark. Also, unlike *det1* and *cop1* mutations, the *det2* mutation does not alter the spatial pattern of light-regulated gene expression when plants are grown in the light. These observations suggest that COP1 and DET2 act independently to maintain etiolation and repress photomorphogenesis by different mechanisms. Characterization of the function of DET2 demonstrated the importance of brassinosteroids in repressing photomorphogenesis.

DET2 encodes an enzyme involved in the biosynthesis of **brassinolide**, the most active of the brassinosteroids (Figure 6–32), which promotes plant growth. Brassinolide levels in *det2* mutants are less than 10% of that found in wild-type plants, and the phenotype (photomorphogenic growth in the dark) can be corrected by exogenously applied brassinolide. Several mutants showing phenotypes similar to that of *det2* have been isolated and can also be corrected by exogenous brassinolide. The corresponding genes are all involved in brassinosteroid synthesis, and all such mutations have dramatic effects on light-grown plants in addition to causing the de-etiolation of dark-grown plants. Mutant light-grown plants are extremely dwarfed, with short internodes, abnormally dark-green curled leaves, and reduced fertility and **apical dominance**. These findings indicate that brassinosteroids inhibit photomorphogenesis in the dark and facilitate several aspects of plant development in the light.

In Arabidopsis, brassinolide is synthesized from campesterol, a sterol component of the plasma membrane, via a forked pathway (Figure 6–33). Three steps convert campesterol to campestanol, and the 5α-reductase encoded by *DET2* catalyzes the second of these steps. The pathway branches after campestanol, with both branches leading to formation of brassinolide. In one branch, C-6 oxidation occurs early in the pathway, and in the other it occurs late. Both branches have seven intermediates between campestanol and brassinolide.

Although most brassinolide mutants described thus far are in Arabidopsis, dwarf mutants with altered brassinolide synthesis also have been described in tomato and pea. Analysis of these mutants reveals the same forked biosynthetic pathway, but the late C-6 oxidation pathway seems to predominate in tomato.

All of the mutants with impaired brassinosteroid biosynthesis can be restored to wild-type phenotype by the application of brassinolide. Analysis of a second class of mutants with similar phenotypes, but which cannot be corrected by exogenous brassinolide, enabled the identification of proteins required for brassinosteroid signal transduction. Root growth of wild-type plants is reduced in the presence of exogenous brassinolide, and *brassinosteroid insensitive* (*bri*) mutants were first isolated by identifying plants insensitive to this effect of brassinolide. In a second approach, researchers isolated

Figure 6–32.
Brassinolide and its effect on Arabidopsis seedlings. (A) Structure of brassinolide. (B) A comparison of wild-type and *de-etiolated2* mutant seedlings that were grown in the dark for 10 days. The wild-type (*left*) is etiolated; the *de-etiolated2* mutant seedling (*middle*) shows the short hypocotyl and open cotyledons characteristic of photomorphogenic development. Addition of brassinolide to the *de-etiolated2* mutant corrects the mutant phenotype (*right*). (C) Seedlings of the same genotypes, shown in the same order as in (B), but grown in the light for 12 days. Addition of brassinolide corrects the dwarf phenotype observed in the leaves and petioles of the *de-etiolated2* mutant. (B and C, from J. Li et al., *Science* 272:398–401, 1996. With permission from AAAS, courtesy of Jianming Li.)

(A)

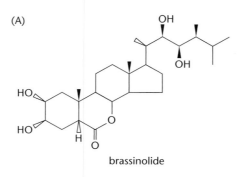

brassinolide

(B)   (C)

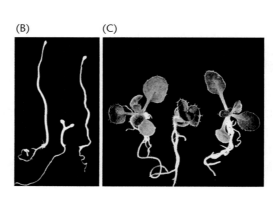

Figure 6–33.
Major steps in the brassinolide biosynthetic pathway. Campesterol, a phytosterol found in membranes, is formed by a sterol-specific pathway. Steps subsequent to campesterol represent the brassinolide-specific pathway. The reaction catalyzed by the DE-ETIOLATED2 (DET2) enzyme is one of the early steps specific to brassinolide synthesis. The major division into the early and late C-6 oxidation pathways is shown.

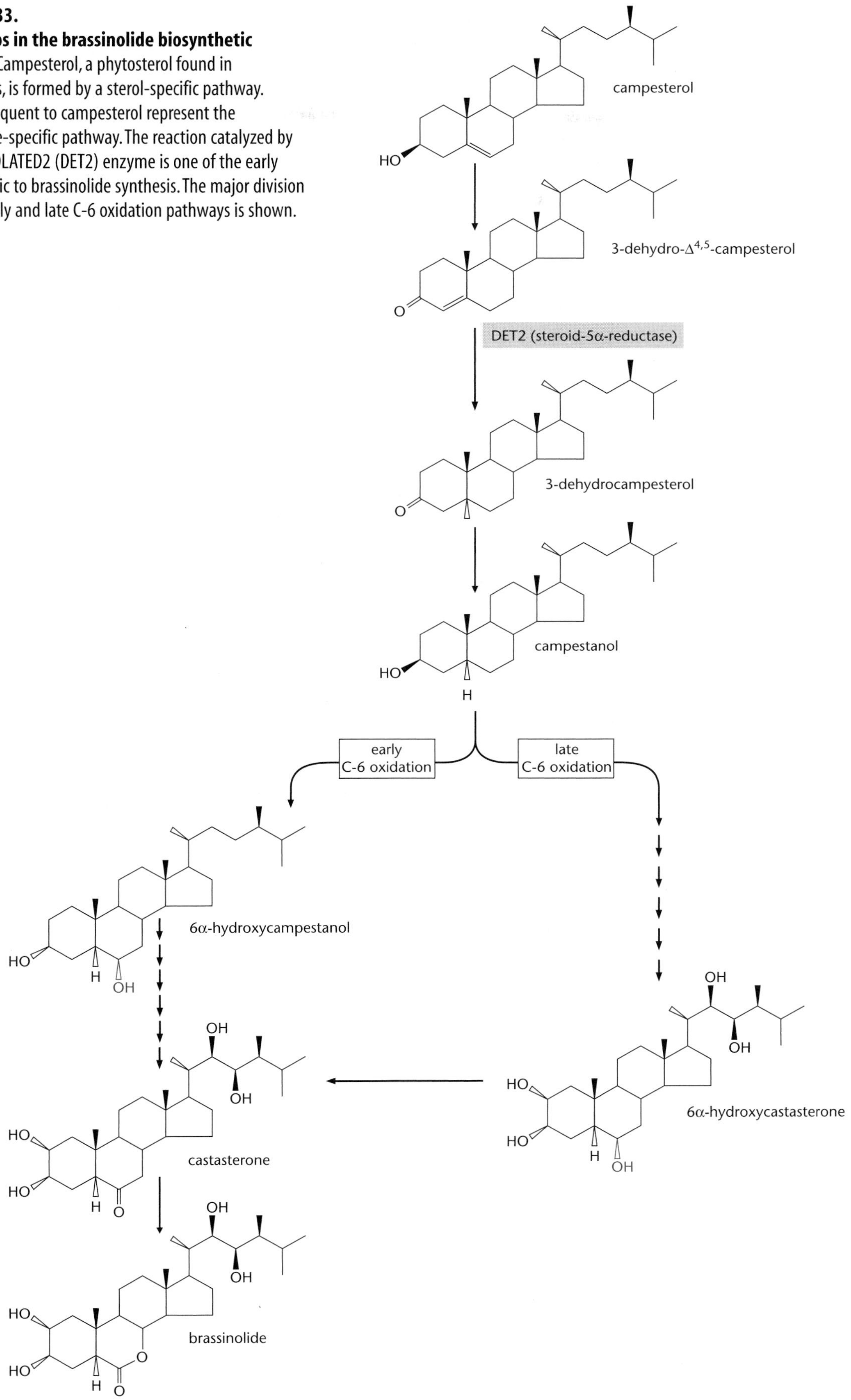

mutants with a phenotype similar to that of plants with impaired brassinosteroid biosynthesis, and then tested them to identify those whose phenotype was not corrected by applied brassinolide. Many independent mutant alleles of the *BRASSINOSTEROID INSENSITIVE 1* (*BRI1*) gene were identified using each of these approaches. In *bri1* mutants, brassinolide accumulates to higher levels than in wild-type plants, and the expression of CPD, an enzyme involved in brassinolide biosynthesis, is increased. These observations suggest that **feedback inhibition** by brassinolide represses its own synthesis and that this requires *BRI1*. Similar mechanisms of hormone **homeostasis** have been described for other phytohormones, such as gibberellins.

BRI1 encodes a **leucine-rich repeat (LRR)** receptor kinase that spans the plasma membrane (Figure 6–34). The extracellular domain of BRI1 contains 25 LRRs, similar to those found in proteins that confer disease resistance (see Section 8.4) and in CLAVATA1 (see Section 5.4); an unusual domain of 70 amino acids, called the "island domain," is inserted between LRRs 21 and 22. The intracellular (cytoplasmic) domain of BRI1 is a serine/threonine protein kinase. Mutations that impair BRI1 activity are mainly clustered in the island domain and in the intracellular kinase domain, emphasizing the importance of these domains for BRI1 activity. The plasma membranes of wild-type plants bind brassinolide, and this binding is increased in plants containing more copies of BRI1, which suggests that BRI1 may directly bind brassinolide. Furthermore, a part of the BRI1 protein comprising the island domain and a neighboring LRR that was made in *E. coli* cells binds brassinolide. In addition, BRI1 self-phosphorylates in the presence of brassinolide. These observations indicate that BRI1 is likely to be a membrane-bound receptor for brassinolide that autophosphorylates when brassinolide binds to the extracellular island domain. A BRI1-related protein, BAK1, consisting of five extracellular LRRs and an intracellular kinase domain, may participate with BRI1 in a membrane-bound receptor complex. The BRI1 and BAK1 proteins interact, and the two proteins can phosphorylate each other. However, BAK1 mutations only weakly impair brassinolide signaling and do not affect the binding of brassinolide to BRI1. Therefore the autophosphorylation of BRI1 together with transphosphorylation of BAK1 probably initiates an intracellular signal-transduction chain that is activated by brassinolide.

Activation of BRI1 by brassinolide leads to the transcriptional up-regulation of more than 400 genes and the down-regulation of approximately 300 genes. The signal-transduction pathway required for these transcriptional changes (Figure 6–35) was defined by isolating mutations that impair or activate the pathway. BRASSINOSTEROID INSENSITIVE 2 (BIN2) is a serine/threonine protein kinase that is a negative regulator

Figure 6–34.
Structure of the BRASSINOSTEROID INSENSITIVE 1 (BRI1) protein. BRI1 is located in the plasma membrane. The N-terminal, extracellular domain consists of 25 leucine-rich repeats (LRRs; shown as coils) and an "island domain" (70 amino acids) between LRRs 21 and 22. The C-terminal, intracellular domain consists of a serine/threonine protein kinase. BRI1 autophosphorylates in the presence of brassinolide and this initiates a signal-transduction chain.

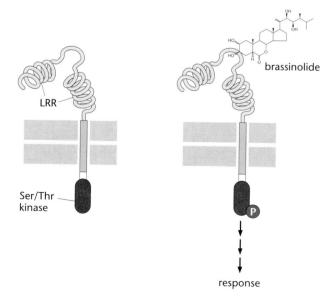

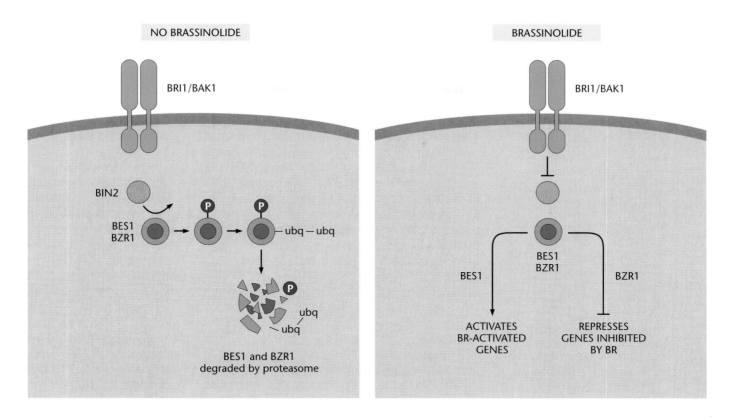

of brassinolide signaling. In the absence of brassinolide, BIN2 phosphorylates two transcription factors, BES1 and BZR1, and the phosphorylated forms of these transcription factors are ubiquitinated and degraded by the proteasome. However, in the presence of brassinolide, BIN2 activity is inhibited by BRI1, allowing BES1 and BZR1 to accumulate. BES1 then activates the expression of brassinolide-responsive genes, whereas BZR1 represses the expression of genes—such as the gene encoding the enzyme CPD—that are repressed by brassinolide. Thus brassinolide alters gene expression patterns by binding to BRI1 outside the cell and initiating signaling steps that involve protein phosphorylations that result in stabilization of key transcription factors.

6.4 FLOWERING

The transition from vegetative to reproductive development—that is, to flowering—is the first step in angiosperm sexual reproduction; the end result is the formation of seeds. The timing of this transition is strictly regulated to ensure that as many seeds are formed as possible and that seed development can be successfully completed in suitable environmental conditions. Developing seeds act as strong **sink organs**, requiring transfer of sugars from the leaves through the **phloem** (see Section 4.4). To ensure successful seed development, plants need to maintain a period of vegetative development, during which leaves are formed and photosynthetic products (**photosynthate**) are accumulated, before floral development is initiated. If the transition to flowering occurs too early, when there are too few leaves to produce photosynthate, seed production is compromised.

Similarly, flowering must occur under optimal environmental conditions to ensure seed development. Flowering is often regulated so as to ensure that seed development occurs when plants are exposed to optimal conditions of temperature, water availability, and light. Regulation of flowering can also ensure that plants in a population flower synchronously. A striking example is the mass flowering of trees in the tropical forests

Figure 6–35.
The signal-transduction pathway induced by brassinolide. In the absence of brassinolide (*left*), the BIN2 protein kinase is active and phosphorylates the transcription factors BES1 and BZR1. These phosphorylated transcription factors are substrates for attachment of ubiquitin and are degraded by the proteasome. In the presence of brassinolide (*right*), the BRI1/BAK1 receptor complex is activated and inhibits the BIN2 protein kinase. Inhibition of BIN2 stabilizes the BES1/BZR1 transcription factors. BES1 activates brassinolide-responsive (BR-activated) genes, whereas BZR1 represses genes whose expression is inhibited in the presence of brassinolide.

of Southeast Asia. Many trees of the Dipterocarpaceae family flower at irregular intervals, with an average of once every four years. The environmental cues and molecular mechanisms that trigger these reproductive episodes are not known. This strategy enables cross-fertilization between individuals and increases the opportunity for seed survival, because animals that scavenge seeds are not able to consume the huge number of seeds formed.

Many plant species go through a juvenile phase in vegetative development during which they will not flower even if exposed to environmental conditions that would induce flowering later in vegetative development. This juvenile phase probably ensures that plants produce sufficient numbers of vegetative leaves and therefore have enough photosynthetic capacity to support flower and seed development. After the juvenile phase, plants enter an adult phase of vegetative development in which they are able, or competent, to flower if exposed to the appropriate environmental conditions. The juvenile and adult phases are often associated with distinct vegetative characteristics. For example, during its juvenile phase, English ivy (*Hedera helix*) produces leaves that are unlobed and relatively small and produced in alternate **phyllotaxy**, whereas in the adult phase the leaves are lobed, larger, and produced in spiral phyllotaxy. Flowers are formed only during the adult phase.

Seasonal cues control flowering in many plants, thereby ensuring that flowering occurs at the time of year that coincides with the most advantageous environmental conditions. For example, flowering may take place when day length becomes longer or after exposure to an extended period of low temperature characteristic of winter. These responses ensure that flowering occurs in spring or early summer, and therefore seed development can be completed before the onset of the next winter. The environmental cues that trigger flowering vary widely among species, and even varieties of the same species can show different flowering responses to environmental stimuli. Varieties with different flowering responses often originate from locations at different latitudes or altitudes, suggesting that altering the flowering response to match environmental conditions is an important adaptation to life at different locations.

The mechanisms of flower development are described in Section 5.5. Here we are concerned with earlier stages in the control of flowering: the mechanisms by which plants detect and respond to environmental cues that induce flowering, such as day length and temperature. We describe genetic pathways in Arabidopsis that induce flowering in response to different environmental stimuli, and how these pathways activate genes known to confer floral identity on the developing primordium.

Reproductive development in many plants is controlled by photoperiod

In many plant species, developmental transitions are controlled by day length. The initiation of flowering is the most widespread of these responses, but other examples are the formation of tubers in potato and the onset of bud dormancy in trees. Plants measure and respond to the length of the day by a process called **photoperiodism**, which was first described in detail by W. W. Garner and H. A. Allard in the United States in 1920. They described a variety of tobacco, Maryland Mammoth, that flowered during the short summer days in southern U.S. states but not in the longer days farther north. However, Maryland Mammoth plants would flower in the north if they were moved into a darkened building during the afternoon, thereby reducing the duration of daylight to which they were exposed. Later, Garner and Allard characterized the day-length responses of many plant species and showed that they could be classified into three major photoperiodic response types: long-day, short-day, and day-neutral plants (Figure 6–36).

A **short-day plant** flowers when the daylight is shorter than a critical length. For example, *Xanthium strumarium* (the common cocklebur) remains vegetative when grown in

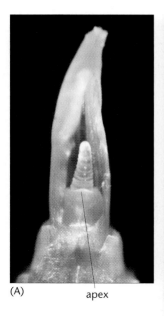

(A) apex

(B)

(C)

(D)

Figure 6–36.
Flowering in the photoperiodic plants
Lolium temulentum* and *Xanthium
strumarium. (A) The vegetative apex of
Lolium, a long-day plant. (B) Within 15 days of
exposure to long days the apex forms many
florets and anther initials. (C) Seeds begin to
mature within 40 days of exposure to long
days. (D) *Xanthium*, a short-day plant,
producing flower buds.
(A–C, courtesy of Rod King; D, courtesy of Jim
Lewis and Lucy Rubino.)

day lengths of 16 hours but flowers when the day length is shortened to 15 hours, indicating that the critical day length for these plants falls between 15 and 16 hours. **Long-day plants** show the reverse response, flowering only when daylight exceeds a critical length. Varieties of the classical long-day plant *Lolium temulentum* (a species of ryegrass) have a critical day length of 14 hours and flower when the length of daylight exceeds this. Critical day length can vary greatly among species or among varieties of the same species. Some plants have an obligate requirement for photoperiod and remain vegetative indefinitely in noninductive conditions, whereas others show a facultative photoperiodic response, in which case they will flower under all conditions but much later under noninductive conditions. **Day-neutral plants** show no flowering response to day length.

Other photoperiodic response types have been described in addition to long-day, short-day, and day-neutral plants. Some plants require exposure to both long days and short days, and the order of exposure is important. For example, long-day–short-day plants flower only if exposed to long days followed by short days, whereas short-day–long-day plants flower only if exposed to short days followed by long days. Another photoperiodic response type is represented by **intermediate-day plants**, which flower only if the day length is neither too long nor too short.

An important issue in understanding the underlying mechanism of photoperiodism is to determine whether the duration of light or of darkness is being measured. Short-day plants measure the duration of darkness, but for long-day plants the situation is less clear. One approach to determine whether night or day is the determining factor is to grow plants in light–dark cycles that are longer or shorter than 24 hours, thus enabling day and night length to be varied independently. This method was used to demonstrate that the short-day *X. strumarium* measures the length of the night (Figure 6–37). As mentioned above, *X. strumarium* has a critical day length between 15 and 16 hours,

Figure 6–37.
Flowering of *Xanthium strumarium* in response to the length of night. Flowering of *X. strumarium* occurs when day length is short and night length is long (*top*). Flowering is repressed (vegetative growth only) when day length is long and night length is short; when the plant is exposed to light for 5 minutes during the night under short-day/long-night conditions; or when the plant is exposed to a shorter daily cycle of 16 hours with a short day and a short night (*2nd to 4th rows*). Plants do flower if exposed to a longer daily cycle of 32 hours with a long day and a long night (*bottom*).

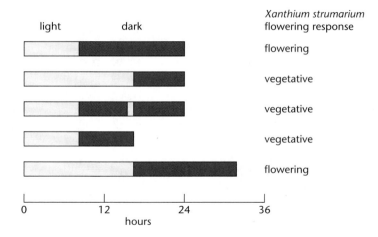

flowering when day length is shortened to 15 hours, but from another perspective *X. strumarium* could be described as flowering when the night is lengthened to 9 hours. That this species does indeed measure night length was shown experimentally. For example, disruption of the dark period with 5 minutes of light (a treatment termed "night breaks") is enough to prevent flowering, and combining long days with long nights triggered flowering. Short-day plants might therefore be better thought of as long-night plants.

Although short-day plants are extremely sensitive to night breaks, long-day plants are generally much less sensitive to them. Nevertheless, some long-day plants will flower in short-day conditions with night breaks of 30 to 60 minutes, suggesting that exposure to long nights inhibits flowering. In other long-day plants, however, the inhibitory effects of long nights are less clear and there seems to be a requirement for exposure to long days. In sum, whether day or night length is measured is less clear for long-day plants than for short-day plants.

Phytochromes and cryptochromes act as light receptors in the photoperiodic control of flowering

The control of flowering by day length requires photoreceptors to distinguish the light and dark phases of the daily cycle. Phytochromes and cryptochromes are the photoreceptors required for photoperiodism, but their relative importance varies among species. The demonstration that flowering of short-day plants was inhibited by short night breaks provided an experimental approach to identify which wavelengths of light and which photoreceptors are involved in photoperiodic control of flowering. Short night breaks of red light (wavelengths 600–660 nm) were shown to inhibit flowering of the short-day plants *X. strumarium* and *Glycine max* (soybean). If the red-light night breaks were rapidly followed by exposure to far-red light (700–760 nm), the plants flowered, indicating that the repression of flowering by red light was reversible by far-red light. These findings suggested the involvement of phytochrome in detecting the night breaks that prevent flowering of short-day plants.

An alternative approach to identifying the photoreceptors involved in photoperiodism is to assess the control of flowering in response to photoperiod in mutants that lack particular photoreceptors. This approach was employed in pea, which is a long-day plant. A mutant in which the gene encoding phytochrome A is inactivated was found to flower at the same time under long and short days, and is therefore insensitive to photoperiod. Phytochrome A therefore plays a crucial role in photoperiodism in this long-day plant and is required to distinguish between long and short days.

Mutations that impair phytochrome A function have a less pronounced effect on photoperiodism in another long-day plant, Arabidopsis: phytochrome A mutants flower only slightly later than wild-type plants under long days. Some artificial long-day conditions that trigger flowering in wild-type plants fail to provoke flowering in *phyA* mutants, however. For example, artificial daily cycles with 8 hours of white light followed by 8 hours of low-intensity far-red–rich light promoted flowering in wild-type Arabidopsis plants but not in *phyA* mutants. These observations indicate a role for phytochrome A in photoperiodism of Arabidopsis, but its effect is obscured in white light. This is because other photoreceptors also control the response to photoperiod in Arabidopsis. These include the blue-light photoreceptor cryptochrome 2 (CRY2): Arabidopsis *cry2* mutants show a reduced photoperiodic response, flowering later than wild-type plants under long days. Blue light promotes flowering in members of the Brassicaceae (Cruciferae), including Arabidopsis, more strongly than in other long-day plants. In Brassicaceae, therefore, both cryptochrome 2 and phytochrome A have major roles in responding to photoperiod, whereas in other long-day plants, such as pea, phytochrome A has the major role.

Circadian rhythms control the expression of many plant genes and affect the photoperiodic control of flowering

Many physiological and developmental processes show **circadian rhythms**. The first of these rhythms to be recognized was the leaf movements of mimosa plants, opening and closing at a particular time each day. The time taken for a rhythm to make one complete cycle and return to the starting point is known as the "period" of the rhythm, and these rhythms were called "circadian" because the period is approximately one day (*circa*, "approximately," and *dia*, "day"). Although circadian rhythms were first discovered in plants, they were later found also in some prokaryotes, fungi, insects, and mammals, including humans. Plant processes regulated by circadian rhythms include,

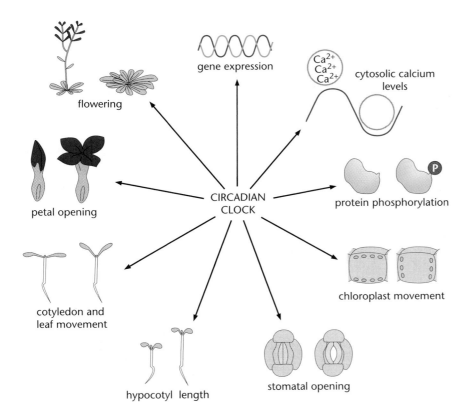

Figure 6–38.
Control of plant processes by the circadian clock. These circadian processes include control of flowering by day length and of daily rhythms in hypocotyl elongation, petal opening, cotyledon or leaf movements, stomatal opening, and intracellular chloroplast movements. Biochemical processes controlled by the circadian clock include daily rhythms in gene expression, in phosphorylation of certain proteins, and of calcium levels in the cytosol.

besides leaf movements, hypocotyl extension, stomatal opening, and gene expression (Figure 6–38).

A crucial feature of circadian rhythms is that the rhythmic activity continues when plants are moved from daily light–dark cycles to continuous darkness or continuous light (Figure 6–39). The rhythms that continue under constant conditions, called "free-running rhythms," demonstrate that circadian rhythms are not simply a direct response to daily changes in environmental conditions. Plants must contain an internal system capable of measuring 24-hour time intervals and generating these rhythms. This system is often called the "circadian clock" or **circadian oscillator**.

Analysis of global gene-expression rhythms in Arabidopsis has demonstrated the extent of circadian regulation in plants. Monitoring of the abundance of more than 8000 mRNAs over a 24-hour cycle revealed that approximately 6% of them showed circadian rhythms. An important feature of plant circadian rhythms, as illustrated by comparing the peaks and troughs of gene expression, is that a circadian rhythm can show a peak in activity at any time of day. Some mRNAs are most abundant at dawn; others peak in the middle of the day or in the evening.

Circadian rhythms of gene expression typically peak just before the time of day when the protein encoded by the gene is most needed. Circadian rhythms can therefore anticipate environmental conditions in which the gene activity is required, and this probably explains how these rhythms confer a selective advantage. For example, more than 20 genes encoding enzymes involved in phenylpropanoid biosynthesis peak in activity around 4 hours before dawn. **Phenylpropanoids** are a large class of **secondary metabolites**, some of which protect plants from the effects of UV light (see Section 7.1). Circadian rhythms in the expression of these genes enable all of the necessary enzymes to be present simultaneously, and thus the synthesis of phenylpropanoids proceeds in anticipation of the onset of dawn.

Just as plants time circadian rhythms to coordinate diurnal leaf movements, so they use these rhythms to measure day length during the photoperiodic control of flowering (Figure 6–40). Exposure of short-day plants to light during the night inhibits flowering, but the inhibitory effect is much more effective if plants are exposed to light at certain times during the night rather than at other times. For example, *Chenopodium rubrum* (red goosefoot) plants are induced to flower by 72 hours of continuous darkness. The time that these plants are most sensitive to night breaks was investigated by exposing them to 4 minutes of red light at different times during the 72 hours of darkness. The degree of sensitivity to these night breaks varied in a circadian rhythm. When a short light exposure was provided 5 to 10 hours or 35 to 40 hours after the start of the dark period, flowering was almost completely prevented, but if the light was given between 20 and 25 hours after the start of the dark period, no effect on flowering response was observed. These experiments led to the conclusion that in photoperiodism, a circadian rhythm regulates flowering and this rhythm is sensitive to light at certain times in the cycle. In short-day plants, exposure to light at these times represses flowering. The time of flowering is therefore determined by whether exposure to light coincides with

Figure 6–39.
Characteristics of circadian rhythms.
Two circadian rhythms that peak in activity at different times (or phases) of the cycle are shown in *red* and *blue*. During daily cycles of day and night (entraining conditions), one of these rhythms (*red*) peaks in activity (represented by gene expression) at dawn, and the other (*blue*) peaks in the evening. Under entraining conditions, each cycle lasts exactly 24 hours, referred to as the "period" of the rhythm. Under continuous light or continuous darkness (free-running conditions), the rhythms continue in the same phases and with a period close to 24 hours.

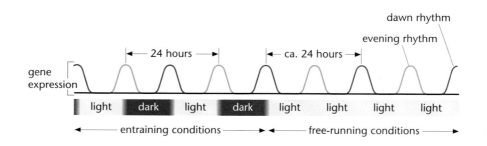

(A)

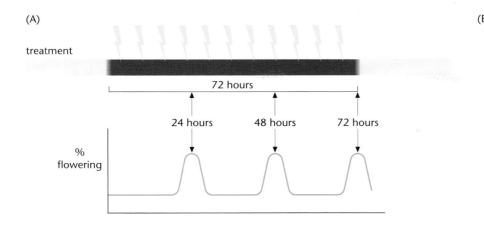

(B)

a particular phase of the rhythm, and this is therefore described as the **external coincidence model of photoperiodic flowering**. The light-sensitive rhythms are often called **photoperiodic response rhythms**. Long-day plants also contain a photoperiodic response rhythm, but in these plants coincidence of light with the rhythm at appropriate times promotes rather than suppresses flowering.

Circadian rhythms in plants result from input of environmental signals, a central oscillator, and output of rhythmic responses

The systems that generate circadian rhythms in plants and other organisms are often conceptually divided into three parts (Figure 6–41). The first comprises input pathways that transmit signals triggered by environmental cues to the circadian clock. These input pathways ensure that the clock is synchronized with the daily day–night cycle. The second, central part generates the rhythm and is the circadian clock or oscillator. The third part consists of a wide range of rhythmic outputs, such as the expression of genes involved in photosynthesis or phenylpropanoid biosynthesis, that are controlled by the circadian clock but are not required to generate the rhythms.

Circadian rhythms are synchronized with, or "entrained to," the daily day–night cycle by photoreceptors and input pathways. In this way, all rhythms are synchronized to dawn or dusk, ensuring that the peak of the rhythm occurs at the appropriate time in the day–night cycle. Under natural cycles of night and day, rhythms are entrained to a period of 24 hours. However, under free-running conditions of continuous light or continuous darkness, the period of a circadian rhythm is not 24 hours but can vary between 19 and 29 hours. Thus, in day–night cycles, the circadian clock is entrained to 24 hours by the light-to-dark or dark-to-light transitions at dusk and dawn, respectively, in every cycle, and consequently circadian rhythms are responsive to alterations in day length as the seasons change. The importance of input pathways, then, is that they entrain the circadian oscillator to the daily cycle of day and night, ensuring that output rhythms occur at characteristic times within the daily cycle and remain coupled to the daily cycle during the changing seasons as the relative duration of day and night changes.

Light and temperature are the major environmental signals that entrain circadian rhythms to the daily cycle. The photoreceptors required for entrainment have been examined by studying the effects of mutations that impair photoreceptor function on the circadian rhythm of expression of the fusion gene *CAB:LUC*, formed by fusion of the promoter of the gene encoding chlorophyll *a/b* binding protein and the marker gene luciferase (Figure 6–42). Luciferase is an enzyme found in the firefly that causes the emission of light by catalyzing the ATP-dependent decarboxylation of its substrate luciferin. Plants carrying *CAB:LUC* emit light when treated with luciferin and ATP, and

Figure 6–40.
Circadian-rhythm control of the response of flowering to photoperiod.
(A) Demonstration of a photoperiod response rhythm in *Chenopodium rubrum*. Flowering of this species occurs under short days and is inhibited under long days. In the laboratory, the plant flowers if grown in continuous light and exposed to a single dark period of 72 hours. The effect of light on flowering was tested by exposing plants to 4-minute flashes of red light (*yellow pulse symbols*) at different times in the 72-hour dark period. When light was given about 24, 48, and 72 hours after the start of the dark period, flowering occurred; when given at other times flowering was repressed. This experiment demonstrated that flowering is not triggered only by the duration of darkness, but is controlled by an interaction between a circadian rhythm and exposure to light. (B) Flowering *C. rubrum*, a model species for studying the physiology of photoperiodism. (Courtesy of Tobias Kieser.)

Figure 6–41.
Components of a circadian system. The circadian system can be divided into three parts. Input pathways synchronize, or entrain, the oscillator to the daily day–night cycle. Light is the major input signal in plants and entrains the oscillator through the phytochrome (PHY) and cryptochrome (CRY) photoreceptors. Temperature will also entrain the circadian oscillator (*not shown*). The central oscillator is the core of the circadian system and is the molecular machinery that generates the 24-hour time-keeping mechanism. Output pathways, controlled by the central oscillator, are individual clock-controlled biological processes, such as hypocotyl elongation, flower opening, and stomatal opening.

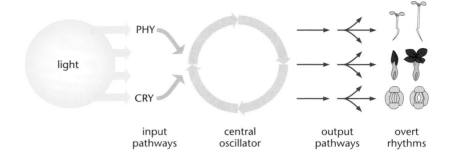

input pathways | central oscillator | output pathways | overt rhythms

this can be detected using a camera. Thus, circadian rhythms in transgenic plants expressing *CAB:LUC* can be monitored in intact seedlings by following the emission of light from seedlings during a 24-hour cycle, or for much longer times under free-running conditions.

When wild-type plants were kept in continuous darkness, the period of the circadian rhythm in *CAB:LUC* expression was longer than when the same plants were held in continuous light. Impairment of photoreceptors in light-grown plants mimics this effect of darkness by lengthening the period of the circadian clock. This approach was used to demonstrate that under red light, both PHYA and PHYB entrain the circadian clock, although PHYB does this only under high-irradiance red light and PHYA only under low-irradiance red light (Figure 6–43). Cryptochromes and PHYA entrain the circadian clock under blue light, with CRY1 acting at all irradiance levels of blue light, PHYA acting only under low irradiance, and CRY2 having only a minor effect. Thus the period length of the circadian clock is regulated by a variety of photoreceptors that act together to monitor different qualities and intensities of light.

The circadian system in different tissues of the plant seems to be independent and can be entrained independently. For example, when different leaves of a plant are exposed to distinct light–dark cycles, circadian rhythms entrain to different phases in the treated tissues. This contrasts to the situation in mammals, where a central pacemaker controls the circadian system and light entrainment occurs through the eyes and a region of the brain known as the "suprachiasmatic nucleus."

The central oscillator, the central part of circadian-rhythm generation, probably acts on a principle similar to the oscillators of mammals and *Drosophila*, which have been described in more molecular detail. In all three systems, proteins acting within the oscillator feed back to repress their own activity or the transcription of their own mRNA (Figure 6–44). One cycle of synthesis and repression of these proteins, followed by synthesis again, lasts 24 hours and generates the period of the circadian rhythm. Such a cycle is described as an "autoregulatory" **negative feedback** loop. A major feature of

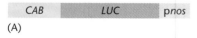

(A)

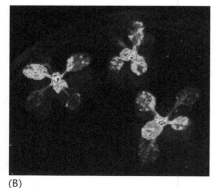

(B)

Figure 6–42.
Visualizing circadian rhythms in transgenic plants. (A) The promoter of the *CHLOROPHYLL A/B BINDING PROTEIN* (*CAB*) gene is fused to the open reading frame encoding the firefly luciferase enzyme (*LUC*). At the end of the open reading frame, a polyadenylated (p*nos*) sequence—isolated from the *NOPALINE SYNTHASE* (*NOS*) gene of the T-DNA of the *Agrobacterium tumefaciens* Ti plasmid—is inserted to ensure that mRNA is properly processed in the transgenic plants. Transcription of the *CAB* gene is controlled by the circadian clock, and the *CAB* promoter confers circadian clock regulation on luciferase (*LUC*) expression. The presence of luciferase causes plants to emit light, which can be detected with a camera and used to follow the activity of the circadian clock. (B) Arabidopsis seedlings emitting light due to the expression of luciferase.

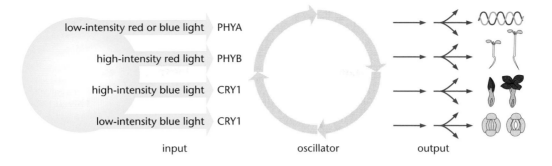

Figure 6–43.
Photoreceptors that entrain the circadian clock under different light conditions. Entrainment of the circadian clock to light signals involves input pathways activated by several phytochrome (PHYA, PHYB) and cryptochrome (CRY1) photoreceptors. These photoreceptors are activated by different wavelengths and intensities of light, ensuring that the circadian clock is entrained under most ambient light conditions.

such a feedback loop is that it incorporates a 24-hour delay into the cycle, and this seems to be caused largely by control of import of the proteins into the nucleus and the rate at which they are degraded there.

Only some of the components of the circadian oscillator in plants have been definitively identified. One of these is TIMING OF CAB 1 (TOC1). Mutations that impair TOC1 activity shorten the period of the circadian rhythm in *CAB:LUC* expression when the oscillator is entrained by light or temperature cycles. The abundance of *TOC1* mRNA shows a circadian rhythm, and this mRNA encodes a nuclear protein with sequence motifs similar to those found in two-component signal-transduction systems (as described above) common in bacteria.

Two other probable oscillator proteins are the nuclear **MYB**-related transcription factors LHY and CCA1 (Figure 6–44). These are closely related proteins whose abundance shows a circadian rhythm. When the genes are expressed constantly using a heterologous promoter, all circadian rhythms tested are abolished, suggesting that rhythmic expression of these genes is essential for function of the circadian clock. In addition, mutations that impair the function of CCA1 or LHY cause circadian rhythms to cycle with a short period, and inactivating both genes *CCA1* and *LHY* causes circadian rhythms to stop prematurely in free-running conditions of constant light.

A model has been proposed in which LHY and CCA1 interact with TOC1 to create a negative autoregulatory feedback loop (Figure 6–44). Expression of *LHY* and *CCA1* mRNA begins around dawn. These transcription factors then repress *TOC1* expression and eventually, as their proteins accumulate, repress their own expression. As the levels

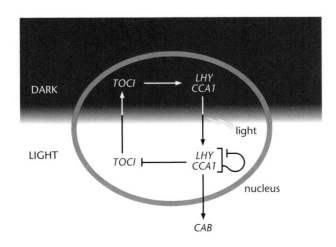

Figure 6–44.
An autoregulatory feedback loop proposed as the circadian oscillator in Arabidopsis. The genes encoding the transcription factors LHY and CCA1 are expressed in a circadian rhythm so that their mRNAs accumulate in the morning. In addition to regulation by the circadian clock, transcription of the *LHY* and *CCA1* genes is activated by light at dawn. The LHY and CCA1 proteins repress their own transcription and that of the *TOC1* gene, which encodes a pseudo-response regulator. As LHY and CCA1 protein levels fall during the day, *TOC1* mRNA levels rise, peaking at the end of the day. The TOC1 protein activates *LHY* and *CCA1* expression, starting a new cycle. The duration of one cycle is about 24 hours. LHY and CCA1 are also thought to activate the expression of other circadian-clock–controlled genes, such as *CAB*.

of LHY and CCA1 proteins fall, the expression of *TOC1* mRNA rises, with a peak of expression at the end of the day. TOC1 protein then activates the expression of the *LHY* and *CCA1* genes, thereby starting another cycle.

In addition to being self-regulatory, the circadian oscillator controls the expression of genes that are not part of the oscillator. Many of these genes also act in the third part of the circadian-rhythm system: the output pathways involved in biochemical or developmental processes. Examples of such output pathways include those controlling the expression of genes required for phenylpropanoid synthesis or for photosynthesis. Increasing or decreasing the expression of genes in a single output pathway does not have a general effect on circadian rhythms of other pathways, but alters the activity of only the single pathway affected. The oscillator also feeds back to regulate the expression of input pathways. For example, *PHYB* is a circadian clock–controlled gene, and its product also acts as a photoreceptor in input pathways to the oscillator. Similarly, the *EARLY FLOWERING 3* gene is controlled by the circadian clock and therefore is formally part of an output pathway, but its protein product modulates the effects of the input pathway activated by phytochrome B (Figure 6–45). Therefore, a strict separation between input and output pathways is not always clearly defined.

Substances produced in leaves can promote or inhibit flowering

Exposure of plants to appropriate day lengths triggers floral development at the apex. This poses the question of whether day length is detected directly at the apex or in other tissues, with signaling from these tissues to the apex triggering flowering. The answer in many species is that day length is detected in the leaf, and a signal translocated from the leaf triggers flowering at the apex. This can be shown by grafting experiments in which a leaf exposed to one photoperiod is attached to a shoot exposed to a different photoperiod (Figure 6–46). An example of the use of this approach is in the short-day plant *Perilla*. Grafting of a single leaf from a short-day–grown plant onto the shoot of a long-day–grown plant is sufficient to induce flowering. The signal translocated, often referred to as the **floral stimulus** or **florigen**, is transported in the phloem. The time required for the leaf to export the floral stimulus can also be measured in these experiments. When the leaf of a short-day–grown *Perilla* plant was attached to a long-day–grown plant for 24 hours, this was sufficient to induce flowering in 50% of recipient plants. Thus in these plants, sufficient floral stimulus to trigger flowering was translocated from the leaf over 24 hours.

A transmissible signal has also been suggested to induce flowering in maize. Flowering of maize induces the development of the "tassel," the male reproductive structure formed at the apex of the plant, and the "ear," the female reproductive structure formed in a leaf axil (Figure 6–47). Flowering is greatly delayed in *indeterminate (id)* mutants. The *id* mutant does eventually form the ear and tassel, but much later than wild-type plants, and the tassel is defective. The *ID* gene was isolated by **transposon tagging** (see

Figure 6–45.
Feedback of some circadian-clock–controlled genes to influence input pathways. The *EARLY FLOWERING 3* (*ELF3*) gene is controlled by the circadian clock, and its protein product is expressed in a circadian rhythm, indicating that it is part of a circadian output pathway. However, ELF3 also feeds back to modulate the input signaling pathway activated by phytochrome B (PHY).

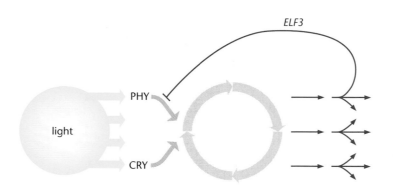

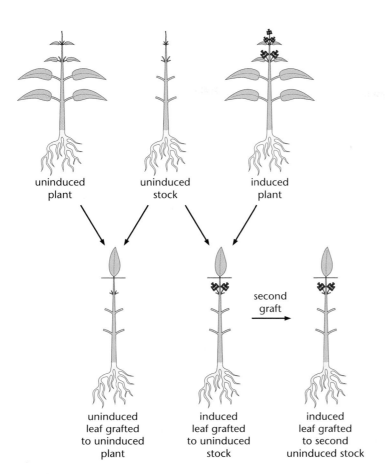

uninduced   uninduced   induced
plant   stock   plant

second
graft

uninduced   induced   induced
leaf grafted   leaf grafted   leaf grafted
to uninduced   to uninduced   to second
plant   stock   uninduced stock

Figure 6–46.
A graft-transmissible floral stimulus produced in *Perilla* leaves in response to photoperiods that induce flowering.
A *Perilla* plant grown in long days is not induced to flower (*upper left*). Removal of the leaves from this plant generates the uninduced stock (*upper middle*), a recipient for leaf grafts. A *Perilla* plant grown in short days produces flowers in the axils of leaves (*upper right*). Grafting a leaf from an uninduced plant onto an uninduced stock does not cause flowering (*lower left*). However, grafting of a leaf from an induced plant onto an uninduced stock does cause flowering (*lower middle*). Therefore, a substance formed in the induced leaf causes flowering of the uninduced stock. Grafting the same leaf onto a different stock is sufficient to induce flowering a second time (*lower right*). This process can be repeated up to seven times.

Section 2.5) and was found to encode a protein containing **zinc finger motifs** related to transcription factors, suggesting that *ID* encodes a regulatory protein. Although the *id* mutation has dramatic effects on flowering time and the morphology of the tassel, the *ID* gene is not expressed in the shoot apical meristem but is expressed in young leaves. In addition, small sectors of *ID* activity on the leaves of an *id* mutant are sufficient to accelerate flowering. *ID* expression in the leaves has a non–cell-autonomous effect on the flowering behavior of the apex, implying that the gene product is involved in the synthesis or transport of a transmissible signal.

The floral stimulus that is formed in the leaf and induces flowering at the apex was extremely difficult to identify by biochemical means. The difficulties led to the proposal that the floral stimulus might be a complex mixture of substances. Genetic approaches may provide information on the identity of these substances (as described below). In pea, a long-day plant, many mutations or naturally occurring alleles that either accelerate or delay flowering have been identified. These different plant varieties have also been used in grafting experiments to determine whether the mutations affect synthesis of, transport of, or response of the apex to transmitted substances formed in the leaves. Grafting experiments in pea provide evidence both for a floral stimulus and for transmitted substances that delay flowering: a floral repressor (Figure 6–48).

In pea, the *STERILE NODES* (*SN*) and *DIE NEUTRALIS* (*DNE*) genes are required for the synthesis of substances that inhibit flowering. For example, *dne* mutants flower much earlier than wild-type plants under short-day conditions. In grafting experiments, flowering of long-day–grown wild-type plants is delayed by grafting onto stocks (roots and base of the shoot) of short-day–grown wild-type plants, suggesting that an inhibitor of flowering is formed in the short-day–grown stocks. This effect is abolished

Figure 6–47.
Induction of flowering in maize by activity of the *INDETERMINATE* gene of leaves. In a wild-type maize plant (*upper left*), *INDETERMINATE* (*ID*) is expressed in young leaves (*red*) and induces flower development at the shoot apical meristem, which forms the male flower (the tassel), and in axillary meristems, which form the female flower (the ear). In *id* mutants (*upper middle*), flowering is delayed so that more leaves are formed before flowering, no ear develops, and the tassel is malformed. Some (unstable) *id* mutations are caused by insertion of a transposon (*upper right*), and excision of the transposon in parts of the leaf is sufficient to restore ear and tassel development and to induce earlier flowering than in stable *id* mutants. The photographs show, left, a wild-type maize plant (*left*) compared with an *id* mutant (*right*); middle, a wild-type tassel (*left*) compared with a malformed *id* mutant tassel (*right*); and right, the modified tassel of an *id* mutant in which a shoot has emerged from a spikelet.

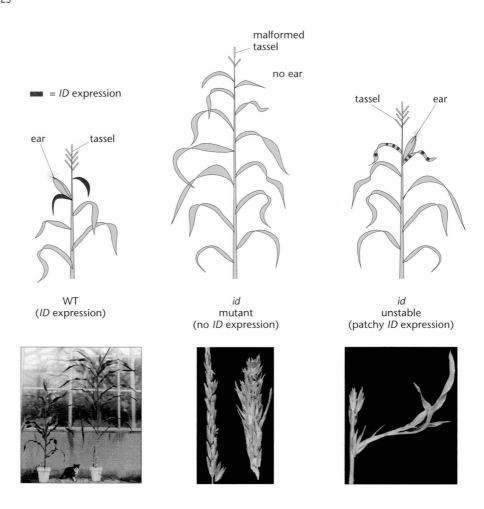

in the *dne* mutant, consistent with the idea that the *DNE* gene is required for the synthesis of an inhibitor.

Similar experiments suggest that the *GIGANTEA* gene of pea is required for the synthesis of a floral stimulus. In contrast, the *LATE FLOWERING* gene delays flowering, but this effect is not graft-transmissible from the stock to the shoot, which indicates that it probably acts in the shoot at or after the perception of the floral stimulus. Isolation of these pea genes should provide information on the identity of the translocatable substances that affect flowering and on the proteins required in the apex to respond to them.

Similar groups of genes are involved in photoperiodic control of flowering in Arabidopsis and in rice

Genes that regulate flowering time have been studied most extensively in Arabidopsis. Flowering of this model species is regulated by environmental signals such as photoperiod and **vernalization**. Arabidopsis flowers much earlier when plants are exposed to long days (16 hours of light) rather than short days (8 hours of light). Vernalization treatments—extended exposures to low temperatures that mimic winter conditions—soon after germination also accelerate flowering. Genes that control the response to these environmental signals were initially identified by studying mutations that alter flowering time or by analyzing genetic differences between naturally occurring varieties that flower at different times. Such approaches identified classes of genes required to respond to distinct environmental signals. For example, one class of genes specifically promotes flowering when plants are exposed to long photoperiods, while another class

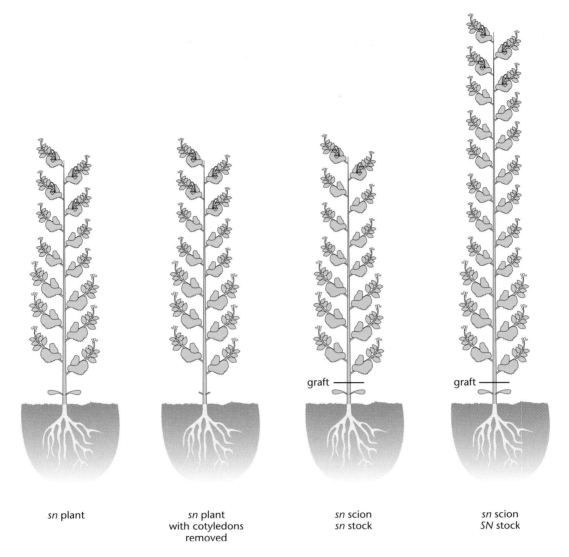

sn plant

sn plant
with cotyledons
removed

sn scion
sn stock

sn scion
SN stock

acts independent of photoperiod but is related to the vernalization response. We describe here how these groups of genes define regulatory pathways that control flowering time (Table 6–3), and how these pathways interact. We also discuss an orthologous system in rice, a short-day plant.

Mutations that impair the photoperiodic response in Arabidopsis were identified because they delay flowering under long days but not under short days (Figure 6–49). The genes identified by these mutation studies are required in wild-type plants to promote flowering specifically in response to long days. These genes include *GIGANTEA* (*GI*), *CONSTANS* (*CO*), and *FLOWERING LOCUS T* (*FT*).

CO is a circadian clock–controlled gene that encodes a nuclear protein containing zinc finger domains involved in protein-protein interactions and a second plant-specific domain similar in sequence to the DNA binding domain of the yeast protein HAP2. The abundance of *CO* mRNA is reduced by mutations in the *GI* gene, suggesting that *gi* mutations delay flowering by decreasing the expression of *CO*, but the mechanism by which the GIGANTEA protein (GI) regulates *CO* transcription is unknown; the biochemical function of GI has not yet been described. In turn, the CONSTANS protein (CO) activates expression of the *FT* gene, which encodes a small protein (FT) related to the Raf kinase inhibitor proteins of animals, and to the TERMINAL FLOWER protein of Arabidopsis. These observations define a hierarchy of gene function within the photoperiodic response and define a circadian clock output pathway that controls flowering.

Figure 6–48.
Control of a graft-transmissible inhibitor of flowering by the *SN* gene of pea.
On the left is a flowering plant that carries an *sn* mutation. Removal of cotyledons from an *sn* plant does not affect flowering (*2nd from left*), nor does grafting the shoot (or scion) of an *sn* plant onto the root and lower shoot, including the cotyledons, of an *sn* stock (*3rd from left*). Grafting the scion of an *sn* plant onto the stock of a plant carrying an active *SN* gene, however, delays flowering, so the scion produces more vegetative nodes before flowering (*right*). This shows that the stock of the *SN* plant produces a graft-transmissible substance that delays flowering.

The detailed pattern of expression of *CO* mRNA suggests that the protein may define a photoperiodic response rhythm similar to those described earlier (Figure 6–50). The level of *CO* mRNA shows a circadian clock–controlled peak late in the day, and this coincides with exposure of plants to light in long photoperiods. In contrast, in short

Table 6–3 Genes encoding proteins in the regulatory pathways that control flowering time

Gene name	Abbreviation	Flowering pathway	Putative function of encoded protein
CONSTANS	CO	Photoperiodic	Transcriptional activator
GIGANTEA	GI	Photoperiodic	Unknown
FLOWERING LOCUS T	FT	Photoperiodic	Transcriptional regulator related to Raf kinase inhibitor
SUPPRESSOR OF OVEREXPRESSION OF CONSTANS 1	SOC1	Photoperiodic/ vernalization	MADS box transcription factor
FLOWERING LOCUS C	FLC	Vernalization	MADS box transcription factor
FRIGIDA	FRI	Vernalization	Unknown
VERNALIZATION INSENSITIVE 3	VIN3	Vernalization	Related to histone deacetylation
VERNALIZATION 1	VRN1	Vernalization	Unknown
VERNALIZATION 2	VRN2	Vernalization	Related to histone methylation
FLD	FLD	Autonomous	Related to histone deacetylation
FVE	FVE	Autonomous	Related to histone deacetylation
FY	FY	Autonomous	Polyadenylation of mRNA
FCA	FCA	Autonomous	RNA binding

Figure 6–49.
Effect of environmental conditions on the flowering response of wild-type and mutant Arabidopsis. When grown under short days characteristic of winter, the Columbia variety, a summer annual widely used in the laboratory (*upper row*), flowers late after forming many vegetative leaves. Under long days characteristic of summer, Columbia plants flower early, and flowering is not accelerated by cold treatment (vernalization) that simulates winter conditions. Mutations that change the flowering behavior of Columbia have been isolated. Mutants in which the photoperiodic pathway is impaired (*middle row*) flower at the same time as wild-type plants under short days and later than wild-type plants under long days, and the late-flowering phenotype under long days is not corrected by vernalization treatments. Mutants with an impaired autonomous pathway (*lower row*) flower later than wild-type plants under long and short days, but flower early if exposed to vernalization treatments.

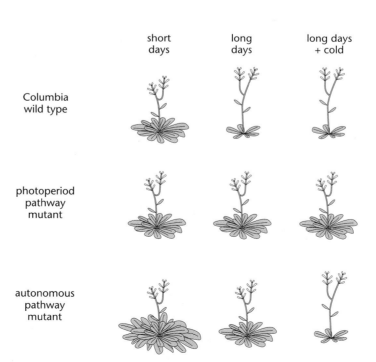

photoperiods, the peak in *CO* mRNA occurs only in darkness. Exposure of plants to light stabilizes the CO protein, and therefore the protein activates *FT* expression and promotes flowering only under long days, when there is a coincidence between *CO* mRNA expression and light. Exposure to blue and far-red light stabilizes the CO protein, and these are the wavelengths previously shown to promote flowering in Arabidopsis, suggesting that at least part of the effects of light quality on flowering can be explained by CO protein stabilization. Thus the regulation of CO protein fits an external coincidence model, in which light (an external signal) must be present at the same time as—must coincide with—the peak of the circadian rhythm in *CO* expression.

The identification of a molecular pathway required for the flowering response to photoperiod suggested that the systemic signal transmitted from the leaf to the apex might be encoded by genes within this pathway. *CO* expression occurs in the **companion cells** of the phloem (see Section 4.4), and the CO protein activates transcription of *FT* in these cells. Furthermore, expression of *CO* or *FT* under the control of promoters specific to phloem companion cells complemented the *constans* mutation, and expression of *FT* complemented the *ft* mutation. Therefore, the phloem-transmissible signal must occur downstream of *FT* transcription. An attractive possibility is that the signal is a product of the *FT* gene, either the *FT* mRNA or FT protein (Figure 6–51). In support of this possibility, FT protein has been detected in the phloem, and FT protein fused to green fluorescent protein (GFP) expressed in phloem companion cells moved through the phloem to the shoot apex. In addition, although *FT* promoter activity is detected only in the leaf, FT protein functions at the shoot meristem by interacting with FD protein, a bZIP transcription factor, to activate transcription of the floral meristem identity gene *APETALA1* (*AP1*) and probably of genes that promote flowering, such as *SUPPRESSOR OF OVEREXPRESSION OF CONSTANS 1* (*SOC1*). These observations support a model in which FT protein is transmitted from the leaf to the apex during floral induction.

In rice, a short-day plant, orthologs of the *CO* and *FT* genes promote flowering in response to short days, suggesting that the same genes act in rice and Arabidopsis to confer short-day and long-day responses, respectively. The rice orthologs of *CO* and *FT*

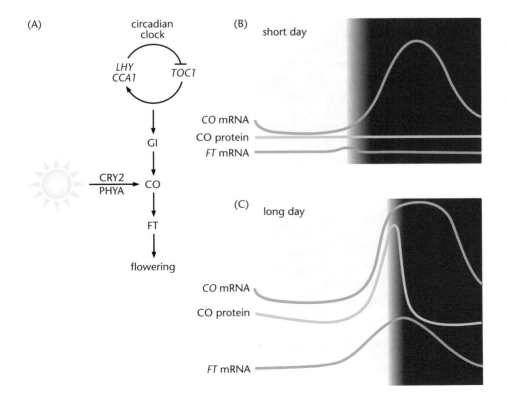

Figure 6–50.
Regulation of the Arabidopsis photoperiodic flowering pathway.
(A) The hierarchy of gene action in the pathway. The circadian clock (see Figure 6–44) controls the transcription of *GI*, *CO*, and *FT*. The GIGANTEA protein (GI) increases transcription of *CO*, and the CONSTANS protein (CO) activates *FT* transcription. CO protein is stabilized by activity of the photoreceptors CRY2 and PHYA. (B) and (C) Regulation of the pathway by day length. Under short days, *CO* mRNA rises during the night. In the dark, CO protein is ubiquitinated and degraded and therefore *FT* transcription is not activated. In contrast, under long days, *CO* mRNA expression coincides with exposure of plants to light. CO protein is then stabilized through the activity of the CRY2 and PHYA photoreceptors. CO protein accumulates and activates *FT* transcription. FT protein then promotes flowering.

Figure 6–51.
Induction of flowering by movement of FT protein. CO protein activates *FT* transcription in the phloem system of the leaf. FT protein moves to the meristem through the phloem. At the meristem, FT binds to the bZIP transcription factor FD. The FD/FT complex binds to the promoter of the floral meristem identity gene *AP1* and activates transcription. The FD/FT complex probably also activates flowering-time genes in the meristem.

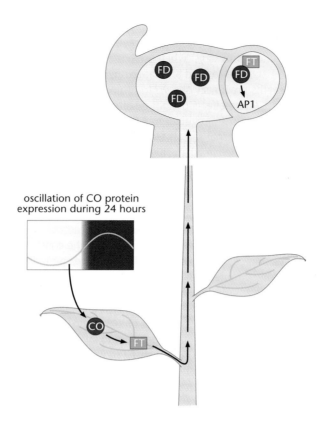

oscillation of CO protein expression during 24 hours

are *HEADING DATE 1* (*Hd1*) and *HEADING DATE 3a* (*Hd3a*), respectively. These findings suggest that the same proteins control photoperiodic flowering in rice and Arabidopsis, but that they confer different responses in these two species.

The mechanisms regulating *FT* and *Hd3a* transcription in Arabidopsis and rice, respectively, differ. *FT* transcription is activated under long days and does not occur under short days. The reverse happens in rice: *Hd3a* mRNA is expressed under short days but not long days (Figure 6–52). Differences in the biochemical function of CONSTANS and HD1 seem to be responsible for the differential regulation of *FT* and *Hd3a*. In rice, HD1 represses *Hd3a* transcription under long days and thereby delays flowering. Therefore, whereas the CO protein acts as a transcriptional activator of *FT* under long days in Arabidopsis, HD1 is a repressor of *Hd3a* under long days in rice. Furthermore, whereas CO does not regulate flowering under short days in Arabidopsis, HD1 promotes flowering under short days by activating *Hd3a* transcription in rice. Thus in rice, HD1 has a dual function: it acts as a transcriptional activator of *Hd3a* under short days and as a transcriptional repressor of *Hd3a* under long days. These functions of HD1 confer a short-day flowering response by ensuring that *Hd3a* is expressed only under short days.

The mechanism underlying the dual function of HD1 in rice is not known. However, an external coincidence model similar to that described for Arabidopsis has been proposed. According to this model, when *Hd1* expression coincides with exposure to light, then HD1 is modified by phytochrome signaling so that it acts as a transcriptional repressor. However, when the gene is expressed in darkness under short days, HD1 acts as a transcriptional activator.

Vernalization is detected in the apex and controls flowering time in many plants

Many plants require vernalization before flowering. This promotion of flowering by cold generally occurs with exposure to temperatures between 1 and 7°C. Plants that

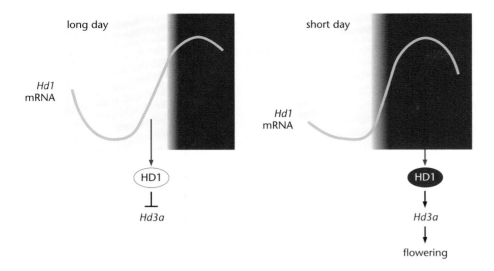

Figure 6–52.
Regulation of photoperiodic flowering in rice. *Hd1* transcription is controlled by the circadian clock. Under long days, *Hd1* mRNA rises at the end of the day and peaks during the night. HD1 protein that is formed while plants are exposed to light under long days acts as an inhibitor of *Hd3a* transcription, and thus flowering is repressed. In contrast, HD1 protein formed in the dark under short days activates *Hd3a* transcription and flowering is promoted.

require vernalization are exposed to these conditions during the winter and then flower the following spring. This ensures that the plants do not flower prematurely in late summer or autumn, but only after exposure to winter temperatures.

The widespread cultivation of winter wheat is an example of the agricultural importance of the vernalization response. Winter wheat varieties are sown in autumn and require vernalization for flowering. The longer period of vegetative growth ensures that these varieties produce higher yields than spring wheat, which is sown in spring and does not require vernalization. Some plants show a facultative requirement for vernalization; for example, varieties of Arabidopsis that require vernalization do eventually flower if they are not vernalized, but they flower much more rapidly if given a vernalization treatment soon after germination. Other plants, such as *Hyocyamus niger* (black henbane), have an absolute requirement for vernalization and will not flower without it.

Vernalization typically requires exposure to low temperatures for 1 to 3 months. Within this time period vernalization is a quantitative response: shorter exposures promote flowering, but to a lesser extent than longer exposures. For example, a vernalization-requiring (winter-annual) variety of Arabidopsis will flower late, after forming approximately 40 vegetative leaves, if it is not vernalized, whereas the corresponding non-vernalization-requiring (summer-annual) variety will flower early, after forming approximately 12 leaves (Figure 6–53). Vernalization for 5 weeks causes the winter-annual variety to flower earlier, after forming about 12 leaves. Since vernalization is dependent on the duration of cold treatment, vernalization for 2 weeks causes plants to flower at a time intermediate between that of nonvernalized and fully vernalized plants. In this respect, vernalization differs from induction of flowering by photoperiod, which usually requires exposure to inductive day lengths for only one or a few days to trigger early flowering; vernalization typically requires exposure to cold for several weeks.

The effect of vernalization can be stably inherited through many cell divisions (**mitosis**). This aspect of vernalization is apparent from experiments performed with *H. niger*, which requires both vernalization and exposure to long days before flowering. By vernalizing *H. niger* plants and then growing them in short days for different lengths of time before exposure to long days, researchers found that plants retained the effect of vernalization for about 300 days, by which time all the leaves and leaf primordia present during the vernalization treatment had senesced and fallen from the plant. Thus all the primordia and leaves had been formed after the treatment, and vernalization had been stably inherited through mitosis for many cell divisions.

Figure 6–53.

Vernalization response. Vernalization induces early flowering of winter-annual varieties of Arabidopsis and of summer-annual varieties carrying a mutation that impairs the autonomous pathway. Summer-annual varieties flower rapidly if exposed to long days characteristic of summer; exposure of these plants to 8 to 12 weeks of low-temperature conditions characteristic of winter (vernalization treatment) has little effect on the time of flowering (*upper diagram*). In contrast, winter-annual varieties do not flower (they grow vegetatively) for many months if grown in long-day conditions; after exposure to vernalization treatment, however, they flower rapidly (*lower diagram*). This shows that winter-annual varieties require vernalization for early flowering. The photographs show different genotypes exposed to vernalization for 0, 1, 3, or 6 weeks. The summer-annual variety Landsberg *erecta* (L-*er*) flowers at the same time under all conditions. Flowering of the autonomous pathway mutant *fca-1* is accelerated by 3 or 6 weeks of vernalization. Flowering of winter-annual plants, which carry an active *FRIGIDA* allele (*FRI*), is accelerated by 3 or 6 weeks of vernalization. An *fca-1* mutant carrying a *vrn1* mutation does not respond to vernalization, demonstrating that *VRN1* expression is required for the vernalization response. (With permission from the Annual Review of Genetics, courtesy of Caroline Dean and Josh Mylne.)

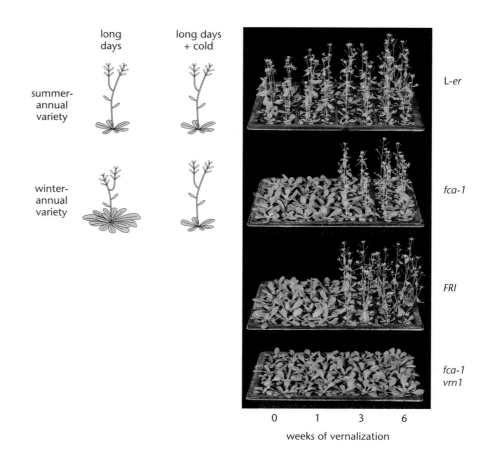

This conclusion is further supported by experiments in which *Lunaria* (honesty, or money plants) or *Thlaspi arvense* (field pennycress) were vernalized, then leaves were removed and used to regenerate plants *in vitro*. The regenerated plants behaved as if they had been vernalized, indicating that the effect was stably maintained through the regeneration process. The effects of vernalization are not transmitted through **meiosis**, however, and the progeny of vernalized plants must be exposed to cold to promote flowering.

Vernalization is detected at the shoot apex in intact plants. This was demonstrated, for example, in celery plants by locally cooling or heating the apex. When the apex was cooled and thus exposed to a vernalization treatment, but the rest of the plant was exposed to ambient greenhouse temperatures, the plant behaved as if it had been vernalized. In the reciprocal experiment, in which the apex was kept warm but the other tissues were exposed to low temperatures, flowering occurred as if the plant was not vernalized.

Grafting experiments with *H. niger* also provide evidence for apical perception of vernalization. When the apices of vernalized plants were grafted onto stocks of nonvernalized plants, the grafted tissues flowered early as if they had been vernalized. When apices of nonvernalized plants were grafted onto vernalized stocks, no early flowering occurred.

Genetic variation in the control of flowering may be important in the adaptation of plants to different environments

The flowering response to environmental cues can differ widely among varieties of the same species. The response often depends on the geographic location in which that variety grows, suggesting that variation in the genetic control of flowering time is an important adaptation to local environmental conditions. (Varieties collected at particular locations are often called **ecotypes** or **accessions**.) The control of flowering by day length and temperature synchronizes the life cycle of plants with the annual cycle of the changing seasons. For example, plants with a requirement for vernalization often show a winter-annual growth habit; they germinate in late summer and flower the following spring after extended exposure to low winter temperatures. Similarly, some short-day plants flower during the spring in response to short days, whereas others flower in late summer or autumn when hours of daylight are again below the critical day length. For many plant species, genetic variants with different flowering responses have been isolated in different locations.

Photoperiod is a consistent indicator of seasonal changes. However, day length also varies with latitude and shows the most dramatic seasonal differences at high latitudes (i.e., nearer the poles). Plant species with a wide distribution over a range of latitudes often show different photoperiodic responses in different parts of their range. This was demonstrated by the analysis of more than 40 strains of *X. strumarium* isolated in different locations of North America, extending from Mexico, at a latitude of 20.4° north, to Quebec, Canada, at 45.5° north. This species is a short-day plant that flowers in late summer when day length falls below the critical value. Comparison of the flowering responses of these strains demonstrated a trend in critical day length between the southern and northern varieties (Figure 6–54).

Plants collected in the southern part of the range require shorter day lengths to initiate flowering than those collected in the northern part. The flowering of the northern varieties while day length is longer means that they flower soon after the summer solstice, which increases the probability of completing seed development before the onset of winter temperatures. The southern varieties flower in response to the shorter day lengths later in the summer, and because the onset of low winter temperatures occurs later in the year at these latitudes, plants setting seeds are not damaged by exposure to cold. Similar variation has been described for other plant species and suggests that photoperiodic responses play an important role in adapting plants to their geographic location.

The vernalization response is also likely to be important in the adaptation of plants to their environment. The requirement for vernalization often varies within a species, so that some varieties require vernalization before flowering while others do not. For

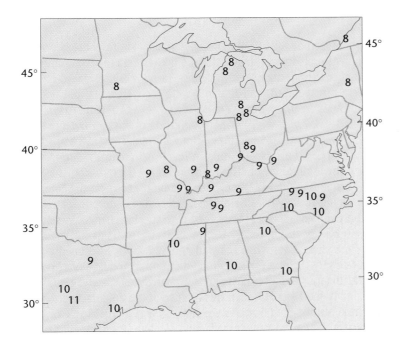

Figure 6–54.
Geographic distribution of varieties of *Xanthium strumarium* with different photoperiodic responses. *Xanthium strumarium* is a short-day plant that grows in many parts of the United States (side scales show latitude in degrees north; numbers on the map indicate night length in hours). Varieties found in the northernmost part of the range are induced to flower when night length is relatively short (8 hours), whereas those in the south flower when night length is longer (10–11 hours). The northern varieties therefore flower in high summer, ensuring that seed development is complete before the start of winter.

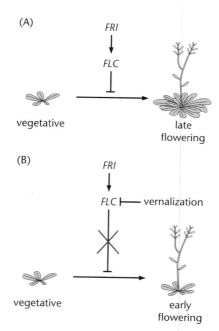

Figure 6–55.
Molecular control of vernalization response in Arabidopsis.
(A) In winter-annual varieties, the *FRIGIDA* (*FRI*) and *FLOWERING LOCUS C* (*FLC*) genes confer a vernalization response. *FLC* encodes a MADS box transcription factor that represses flowering, so these varieties flower late, or remain vegetative indefinitely, under long-day, summer conditions. (B) On exposure to vernalization, *FLC* expression is repressed; this repression removes the floral repressor and flowering can occur.

example, some wild isolates of Arabidopsis collected from natural environments, such as the Stockholm variety, are very late flowering when they are not vernalized but will flower rapidly after vernalization. This is in contrast to early flowering varieties of Arabidopsis, such as Landsberg *erecta*, which do not show a vernalization requirement and flower rapidly irrespective of whether they are vernalized.

When varieties that require vernalization are crossed with those that do not, subsequent generations inherit the requirement for vernalization. Winter-annual, vernalization-requiring varieties contain active copies of two genes, *FRIGIDA* (*FRI*) and *FLOWERING LOCUS C* (*FLC*), whereas in summer-annual varieties that do not require vernalization one or both of these genes are inactive. The *FLC* gene encodes a **MADS box** transcription factor that represses flowering. The function of the FRI protein is not known, but it acts to increase the abundance of the *FLC* mRNA. Thus, plants carrying both *FRI* and *FLC* show high levels of expression of the floral repressor *FLC* and flower very late (Figure 6–55). The principal way in which vernalization promotes flowering in Arabidopsis is to decrease the transcription of *FLC*; this is based on the finding that vernalization of plants carrying active *FRI* and *FLC* genes causes a reduction in *FLC* mRNA, and this correlates with earlier flowering.

Summer-annual varieties that flower early and do not require vernalization often contain naturally occurring mutations in the *FRI* gene. Several different types of *fri* mutation have been found, suggesting that *FRI* function was lost independently in different early-flowering varieties. Many varieties carrying active *FRI* alleles originated from northern Europe, where the summers are shorter and where *FRI* function probably provides a selective advantage.

Vernalization response in Arabidopsis involves modification of histones at the *FLC* gene, which is also regulated by the autonomous flowering pathway

We can now summarize the characteristics of vernalization: relatively long exposures (1–3 months) to low temperatures are required to induce a vernalization response; the effect of vernalization is stable through mitosis but reset by meiosis; and the vernalization response in Arabidopsis requires transcriptional repression of the *FLC* gene, which encodes a repressor of flowering. These features can be explained by the finding that vernalization in Arabidopsis causes alterations to the **histones** that are part of the **chromatin** at the *FLC* gene (Figure 6–56). (See Sections 2.1 and 2.3 for descriptions of histones and their role in gene regulation.)

The importance of histone modification in the vernalization response became clear after the identification of Arabidopsis mutants that did not respond to vernalization. Plants that flower rapidly after vernalization, such as those carrying active alleles of *FLC* and *FRI* (*FLC:FRI*), were mutagenized, and mutants were recovered that did not flower soon after the vernalization treatment: the *vernalization insensitive* (*vin*) or *vernalization* (*vrn*) mutants (Figure 6–53; Figure 6–57). Characterization of the *vin/vrn* mutants and isolation of the *VIN/VRN* genes provided insight into the molecular mechanisms controlling vernalization. In *FLC:FRI* plants, *FLC* mRNA levels fall gradually during vernalization, and this allows flowering to occur. The reduction in *FLC* expression is associated with changes to the histones at the *FLC* gene. Specific amino acids near the N-terminus of histones can be modified *in vivo*—methylated, acetylated, phosphorylated, or ubiquitinated—and these modifications are associated with specific changes in gene expression. For example, in one of the histones, H3, methylation of lysine-27 is associated with repression of gene expression, while acetylation of lysine-9 is associated with gene transcription (Figure 6–56). During vernalization, the histone modifications in the chromatin at the *FLC* gene undergo a series of characteristic changes. The acetyl groups are removed from lysine-9 and lysine-14 of H3, then two methyl groups are attached to lysine-27 and to lysine-9. These histone modifications are likely to play a crucial role in vernalization, because *vin/vrn* mutations prevent vernalization and alter these modifications.

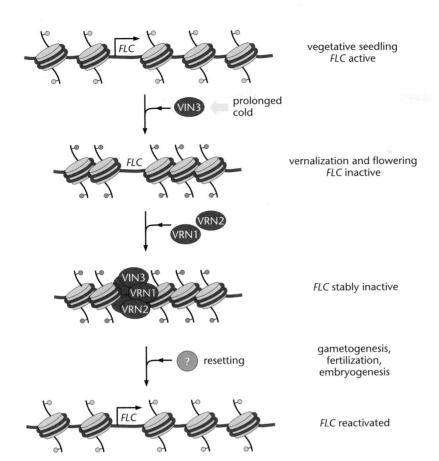

vegetative seedling
FLC active

prolonged
cold

vernalization and flowering
FLC inactive

FLC stably inactive

gametogenesis,
fertilization,
embryogenesis

resetting

FLC reactivated

Figure 6–56.
Repression of *FLC* expression by vernalization via histone modification.
Before vernalization (*top*), *FLC* is expressed and represses flowering. *FLC* chromatin has an open structure; histone H3 modifications associated with active transcription, such as acetylation of lysine-9 and lysine-14, are present at *FLC*. During vernalization, *VIN3* is expressed and the VIN3 protein contributes to deacetylation of histone H3 at *FLC*; the *FLC* chromatin becomes closed and *FLC* transcription is repressed. The repressed state is maintained by the VRN1 and VRN2 proteins and is associated with methylation of lysine-27 and lysine-9 of histone H3. VRN1 and VRN2 prevent transcription of *FLC* when plants are returned to normal growth temperatures after vernalization. During gametogenesis, *FLC* expression is reset (by an unknown mechanism) so that *FLC* chromatin returns to the open state and *FLC* is expressed in the seedlings of the next generation. In each diagram the *blue disks* represent histones, the red line represents DNA, and the *colored circles* represent different modifications present at the N-terminus of histone H3.

The *vin3* mutation prevents the reduction of *FLC* mRNA levels during vernalization in *FLC:FRI* plants, and alters the pattern of histone modifications observed in *FLC* chromatin. In *vin3* mutants, deacetylation of lysine-9 and lysine-14 and dimethylation of lysine-9 and lysine-27 of H3 at *FLC* do not occur during vernalization. This suggests that the role of VIN3 is to initiate histone modifications at *FLC* during vernalization. Consistent with this suggestion, the VIN3 protein contains a plant **homeodomain** that in other proteins is known to be involved in initiating changes in chromatin structure. Furthermore, the abundance of the *VIN3* mRNA rises during vernalization, between 1 and 2 weeks after the start of cold treatment, indicating that VIN3 may have an early role in mediating the response to vernalization.

Mutations in the *VERNALIZATION 1* (*VRN1*) and *VERNALIZATION 2* (*VRN2*) genes also affect *FLC* chromatin, but with a different effect from *vin3* mutations (Figure 6–57). In *vrn1* and *vrn2* mutants, *FLC* expression is reduced by vernalization treatment, as observed in wild-type plants; but after vernalized plants are returned from the cold to

Figure 6–57.
Levels of *FLC* and *VIN3* mRNA during and after vernalization in plants of different genotypes. The abundance of *FLC* or *VIN3* mRNA during and after vernalization is represented by the thickness of the bars. In *FLC:FRI* plants, *FLC* mRNA levels decline during vernalization and are maintained at low levels after vernalization. *VIN3* mRNA levels rise during vernalization. In *vin3* mutants, *FLC* mRNA levels do not decline during vernalization. In *vrn2* or *vrn1* mutants, *FLC* mRNA levels decline during vernalization but are not maintained at a low level, and increase again after vernalization.

plants tested, mRNA measured		weeks of vernalization							weeks after vernalization				
		0	1	2	3	4	5	6	1	2	3	4	5
FLC:FRI	*FLC* mRNA	▬	▬	▬	—	—							
FLC:FRI	*VIN3* mRNA		—	▬	▬	▬	▬	▬	—	—			
FLC:FRI vin3	*FLC* mRNA	▬	▬	▬	▬	▬	▬	▬	▬	▬	▬	▬	▬
FLC:FRI vrn2	*FLC* mRNA	▬	▬	—	—	—				▬	▬	▬	▬
FLC:FRI vrn1	*FLC* mRNA	▬	▬	—	—	—				▬	▬	▬	▬

normal growth temperatures, *FLC* expression rises again. *VRN1* and *VRN2* are therefore required to maintain repression of *FLC* expression, but not for the initial reduction in expression. Consistent with this conclusion, deacetylation of lysine residues of H3 at the *FLC* locus occurs during vernalization in *vrn1* and *vrn2* mutants. However, dimethylation of lysine-9 and lysine-27 does not occur.

VRN2 encodes a protein that shares homology with the *Drosophila* protein Suppressor of Zeste 12, or Su(z)12, that acts as part of the Polycomb complex to stably repress gene expression during embryogenesis. The Polycomb system also acts by methylating lysine residues of H3. *VRN1* encodes a plant-specific DNA binding protein that has an effect similar to that of VRN2 on *FLC* expression and histone modification. Therefore, the role of VRN1 and VRN2 is to stably repress *FLC* expression after vernalization and ensure that its expression does not increase again when plants return to normal growth temperatures.

Identification of histone modifications as the basis of the vernalization response helps explain the classical observation that the vernalized state is stable during mitosis, because histone modifications are stably inherited through cell division. Furthermore, histone modifications can be reset at meiosis, which explains why the vernalized state is not inherited by the progeny of a vernalized plant.

Besides its role in the vernalization response, *FLC* expression is also important in the **autonomous flowering pathway** of Arabidopsis. Mutations that impair the activity of this pathway increase *FLC* expression and thereby cause late flowering. Autonomous pathway mutants flower later than wild-type plants under long and short days. However, the late-flowering phenotype can be corrected if the plants are exposed to vernalization. These mutants are not impaired in photoperiodic response, because they flower later under short days than under long days, as does the wild-type plant.

The common feature of all the proteins in the autonomous pathway of Arabidopsis is that they are required to regulate *FLC* expression and to keep *FLC* expression low in early-flowering varieties. Two genes in the autonomous pathway, *FLD* and *FVE*, encode proteins that are involved in histone deacetylation at *FLC*. In *fld* and *fve* mutants, the histones at *FLC* are hyperacetylated, *FLC* mRNA is expressed at high levels, and flowering is delayed. In contrast, the FCA and FY proteins, which are also in the autonomous pathway, are involved in **RNA processing**. FCA contains RNA binding motifs and binds to the FY protein, which is related to a yeast protein required for RNA polyadenylation (addition of a **poly-A tail**). Therefore, the FCA–FY complex might be involved in processing *FLC* mRNA and maintaining it at low levels. In summary, genes in the autonomous pathway are involved in diverse biochemical processes that maintain *FLC* mRNA at low levels. Mutations in these genes cause late flowering because *FLC* mRNA levels rise and repress flowering.

Photoperiodic and vernalization pathways of Arabidopsis converge to regulate the transcription of a small set of floral integrator genes

In addition to the pathways described above, flowering in Arabidopsis is also regulated by **gibberellins**: they accelerate flowering, and mutations that strongly reduce gibberellin biosynthesis delay flowering. These mutations have their strongest effect under short days, and initiation of flowering via the gibberellin pathway may be responsible for the initiation of eventual flowering of wild-type plants under short days.

The photoperiodic, autonomous, and gibberellin pathways independently promote flowering. However, ultimately they increase the expression of the same set of genes, including *LEAFY* (*LFY*) and *APETALA1* (*AP1*), that confer floral identity on primordia (Figure 6–58; see also Section 5.5). These distinct flowering pathways must eventually converge to regulate the expression of the same genes. The gibberellin and photoperiodic

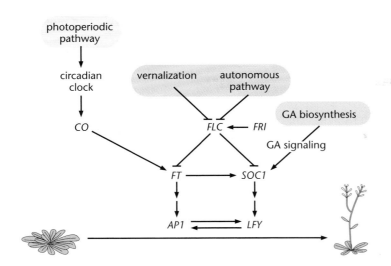

Figure 6–58.
Integration of Arabidopsis flowering pathways to regulate flowering time and the expression of floral meristem identity genes. The photoperiodic, vernalization, autonomous, and gibberellin (GA) pathways promote flowering of Arabidopsis. All of these pathways converge on expression of the flowering-time genes *FT* and *SOC1*. These genes are required, in turn, for expression of the *APETALA1* (*AP1*) and *LEAFY* (*LFY*) genes, which act in the floral primordium to confer floral meristem identity.

pathways promote expression of *LFY* through different sequence elements in the promoter, suggesting that convergence of these flowering-time pathways occurs on the *LFY* promoter. Similarly, expression of *SOC1* and of *FT* is increased by the photoperiodic and gibberellin pathways and repressed by FLC. These observations indicate that the photoperiodic, gibberellin, and vernalization pathways all converge on the regulation of transcription of the *FT*, *SOC1*, and *LFY* genes. These genes are therefore sometimes described as **floral integrator genes**, because their regulation integrates the information from different flowering pathways.

6.5 ROOT AND SHOOT GROWTH

So far, the chapter has focused on various aspects of the ways in which growth and development are regulated by the presence or absence of light in the environment. However, many other environmental variables also influence growth, and in this section we examine a few well-studied examples.

Plant growth is affected by gravitational stimuli

Unlike other environmental stimuli, gravity is constant in both intensity and direction. Gravity is thus a vectorial stimulus, and plants orient their growth with respect to the constant orientation of the gravitational vector—a response known as **gravitropism**. In general, shoots grow upward, away from the soil and toward the sun (negative gravitropism), and roots grow downward, deeper into the soil (positive gravitropism), providing increased anchorage and greater exposure to soil-borne nutrients and water.

Growth directions are more varied than simply positive and negative with respect to gravity, however. For example, lateral roots grow at an angle to the gravitational vector and do not show the complete positive gravitropism exhibited by primary roots. Many recent advances in the understanding of how plants respond to gravitational signals have come from the analysis of mutants that are affected in the simple positive or negative gravitropisms of roots and shoots. These advances have confirmed the importance of **statoliths** (a specialized type of **amyloplast**, a starch-containing plastid; see Section 3.3) in graviperception, of **columella** (cells of the **root cap**) in root graviperception, and of cells of the **endodermis** in shoot graviperception (see Chapter 5 for descriptions of root and shoot structure). The findings have also revealed the crucial role of the growth regulator auxin in mediating gravitropic responses in roots.

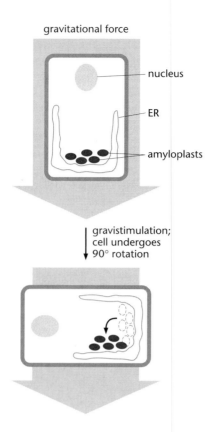

Figure 6–59.
Role of statoliths (amyloplasts) in gravitropism. When the orientation of statolith-containing cells with respect to gravity is altered, the statoliths re-sediment to lie on the (new) bottom surface of the cell. Statolith re-sedimentation triggers rechanneling of auxin flux and a resultant growth curvature (see the text discussion). (ER = endoplasmic reticulum)

Statoliths are key to graviperception in stems, hypocotyls, and roots

Both shoots and roots exhibit responses to gravity. All parts of the elongating stem, in isolation, have the ability to respond to the gravitational stimulus, so the gravitational response of plant stems is not governed by a single site of graviperception. Rather, graviperception occurs along the length of the elongating stem.

As noted above, the starch-containing statoliths are central to the process of graviperception. **Starch** is denser than cytoplasm, and as a result the statoliths tend to sediment with respect to the gravitational vector, congregating on the lower side of the cell (Figure 6–59). When the orientation of the cell changes, the statoliths re-sediment and congregate on the new lower side. The cell detects changes in the sedimentation of the statoliths, and the plant uses this information to effect changes in the relative growth rates of cells in the growing regions of organs, thus restoring the gravisensing cells to the correct orientation with respect to the gravitational vector.

The exact mechanism by which the cell detects changes in statolith sedimentation is poorly understood, but there is experimental evidence supporting this role of statoliths in graviperception. For example, Arabidopsis *phosphoglucomutase* (*pgm*) mutants lack the ability to synthesize starch (phosphoglucomutase is an enzyme of the starch biosynthetic pathway). These mutants have reduced graviresponses in stems, hypocotyls, and roots. However, these starch-deficient mutants are not completely agravitropic, suggesting that although statoliths are important for graviperception, plants probably have additional mechanisms for detecting and responding to the gravitational vector.

Columella cells of the root cap are the site of graviperception in the growing root

Although roots, hypocotyls, and stems all exhibit graviresponses, which are thought to be elicited primarily by changes in the sedimentation of statoliths (amyloplasts), there are clear differences in the mechanism of the graviresponse in these organs.

A variety of experiments, including surgical- and laser-ablation studies, have indicated that the primary site of root graviperception is the root cap (the developmental origin and function of the root cap are described in Chapter 5). It contains a group of central columella cells that are continuously replaced through division of columella initials in the root apical meristem. Columella cells are highly specialized: they are relatively depleted in the microfilaments and microtubules that compose the **cytoskeleton** (see Chapter 3), contain little, if any, vacuole, and contain amyloplasts. The reduced cytoskeleton of the columella cells enables their amyloplasts (statoliths) to rapidly re-sediment as the cell changes orientation with respect to the gravitational vector.

The crucial role of the columella cells in the perception of gravitational stimuli is clear, but other factors are also likely to be involved in graviperception. For example, laser ablation of entire Arabidopsis root caps does not completely abolish the gravitropic responses of roots. The presence of residual responses, although substantially reduced compared with control plants, implies that roots must have additional ways of responding to gravitational stimuli.

Following the initial perception of the gravitational stimulus by the columella cells, a signal passes from the site of perception to the "growth region" of the root (the **elongation zone**; see Section 5.3). This signal results in differential growth of cells on the opposing sides of the growth region, and consequent reorientation of the columella cells (and the rest of the root) with respect to the gravitational vector. The nature of this signal is not completely understood, but there is considerable evidence for the involvement of auxin.

The endodermal cell layer is the site of graviperception in growing stems and hypocotyls

In Arabidopsis, the site of graviperception in hypocotyls and stems is the endodermis, a single cell layer that contains statoliths and forms a cylinder surrounding the **stele**. Loss-of-function alleles at the *SCARECROW* (*SCR*) and *SHORT ROOT* (*SHR*) loci cause absence of the endodermis from root, hypocotyl, and stem. Mutants lacking the endodermis lack shoot and hypocotyl gravitropic responses, but they have normal root-cap columella cells and normal root gravitropic responses. As in roots, a signal must be produced in the gravisensing endodermal cells and transferred to the cells (the overlying epidermal and cortical cells) that effect the changes in stem growth. As with root gravitropism, there is evidence that auxin is involved in the process, as described below.

Mutations in auxin signaling or transport cause defects in root gravitropism

There is clear evidence, from both physiological and genetic studies, that auxin is involved in relaying the message from the site of graviperception to the site of the growth response, and auxin probably mediates the physiological signal generated in response to gravistimulation in roots. Some indirect evidence also suggests the involvement of Ca^{2+} ion gradients and electrical currents in this process.

Auxin is transported to the root tip via cells of the stele, the central core of the root. Once it reaches the root tip, the auxin is thought to be redistributed to tissues surrounding the stele and transported basipetally (i.e., back up the root) into the root elongation zone. Here, the auxin regulates cell elongation. According to one classical theory (the Cholodny-Went hypothesis), gravistimulation can be explained in terms of the establishment of an auxin gradient across the root tip (Figure 6–60), which results in different concentrations of auxin, and thus different degrees of inhibition of elongation, in the elongation zones on opposite sides of the root. The differential growth that results reflects the auxin concentration gradient across the root and causes changes in the direction of root growth.

The role of auxin in signaling graviresponses is supported by evidence from physiological and genetic studies. Inhibitors of polar auxin transport block gravitropic responses at concentrations that are too low to affect root growth itself. Furthermore, gravistimulation results in the redistribution of radiolabeled auxin across the root tip, with accumulation of the auxin at the real (physical) bottom of the root.

The genetic evidence for the involvement of auxin in root gravitropism is particularly strong. Increasing concentrations of exogenous auxin are proportionately inhibitory to

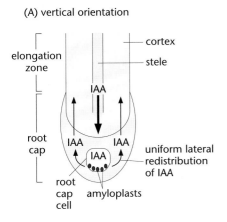

(A) vertical orientation

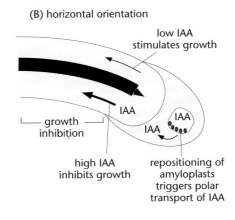

(B) horizontal orientation

Figure 6–60.
The Cholodny-Went hypothesis. (A) When a root is in a vertical orientation, equal auxin (IAA) flux on both sides of the root maintains the direction of growth. (B) When the root is moved into a horizontal position, the statoliths (amyloplasts) re-sediment, resulting in a high auxin level on the (new) bottom side of the root. This locally high auxin level causes growth inhibition in the bottom part of the root, and a resultant downward curvature.

the growth of roots. This provides the basis for selecting mutant plants with roots relatively unaffected by (resistant to) the exogenous auxin. Such plants have longer roots than wild-type plants when subjected to auxin concentrations that are usually inhibitory to root growth. A number of genes whose products are involved in auxin signaling have been identified in this way, and the products of many of these genes affect plant responses to gravity.

For example, the Arabidopsis *AUX1* gene encodes a transmembrane protein that is related to amino acid **transporters** and seems to play a role in the influx of auxin into cells (see Section 5.2 on auxin influx and efflux carriers). Mutations in *AUX1* cause root growth that is resistant to the auxins IAA and 2,4-D, but not to another auxin, 1-NAA. Both IAA and 2,4-D require an "influx carrier" to enter plant cells, whereas 1-NAA is freely permeable. *AUX1* function is clearly necessary for gravitropism, as mutants lacking this function are agravitropic. Normal gravitropism can be restored to these mutants by exogenous 1-NAA, suggesting that although *AUX1* function is necessary for gravitropism, it is not involved in determining the auxin gradients that are thought to generate gravitropic growth.

In addition to transmembrane proteins that regulate auxin influx, plant cells have auxin efflux carriers that are involved in transport of auxin out of cells. The Arabidopsis *PIN2* gene is required for root gravitropism. *PIN2* encodes a transmembrane protein that is structurally related to bacterial transporters and to the Arabidopsis *PIN1* gene product (PIN1; an auxin efflux facilitator). It seems likely that PIN2 is also an auxin efflux carrier, or (like PIN1) a closely associated protein. PIN2 is probably involved in establishment of the auxin concentration gradients that are associated with root gravitropism.

Thus there is strong evidence that auxin is involved in root gravitropism, and the requirement for the various influx and efflux carriers is compatible with the role of the auxin concentration gradient in the Cholodny-Went hypothesis. It is also possible, however, that although auxin is necessary for gravitropism, it does not itself provide the signals enabling root cells to undergo the differential elongation that drives the change in direction characteristic of the root gravitropic response.

The extent of lateral root elongation varies in response to soil nutrient levels

One of the most dramatic examples of the plasticity of plant architecture in response to environmental signals is the shape and form of root systems. Internal genetic controls set the basic parameters of root branching (see Section 5.3), but the final form of a root system is hugely affected by the nature of the soil in which it grows. For example, the root systems of many plant species are highly responsive to variation in the concentration of soil nutrients (such as nitrate, ammonium, and phosphate; see Chapter 4). When a root system encounters localized nutrient-rich regions of soil, it responds with a proliferation of lateral roots into these regions, thus presumably increasing nutrient assimilation (Figure 6–61).

The signaling pathway by which roots detect and respond to soil nitrate levels is still being characterized, although one component of a likely signaling system has been identified in Arabidopsis (Figure 6–62). The *ANR1* gene was identified by virtue of its nitrate-inducible expression. *ANR1* mRNA is found specifically in roots; it is undetectable in nitrate-starved roots but is rapidly induced in nitrate-starved roots that are exposed to nitrate. *ANR1* expression is nitrate-specific and is unaffected by levels of other nutrients, such as potassium or phosphate. *ANR1* encodes a member (ANR1) of the MADS box family of nuclear transcription factors, but the downstream genes on which it acts are not yet characterized. The root systems of plants with down-regulated *ANR1* transcription either completely fail to respond or have a marked reduction in response to localized nitrate-rich regions. Thus ANR1 is involved in regulating the developmental plasticity of roots that is mediated by nitrate availability.

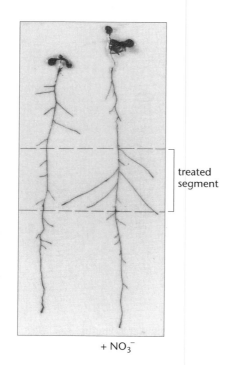

treated
segment

+ NO₃⁻

Figure 6–61.
Effect of nitrate on root growth. Lateral root growth is promoted in regions that receive nitrate (NO₃⁻; treated segment) (*right*) but not in other regions of the root, and not in a control plant (*left*). (From H. Zhang and B.G. Forde, *Science* 279:407-409, 1998. With permission from AAAS, courtesy of Brian G. Forde.)

How might ANR1 perform this function? The growth of the primary root of Arabidopsis is unaffected by nitrate concentration. When plants are grown in soil that is uniformly nitrate-rich, root branching is inhibited. Only when roots are exposed to nitrate-rich pockets of soil does lateral root proliferation occur, specifically within the region that is nitrate-rich. Nitrate affects elongation of lateral roots, not their initiation. All these observations can be incorporated into a model of nitrate-directed root growth: the tips of lateral roots perceive nitrate via a nitrate receptor; the information passes through a signal-transduction chain, which includes ANR1; and activation of this signaling chain promotes the growth of lateral roots. However, if the overall nitrogen status of the plant is not limiting its growth, a signal from the shoot inhibits the growth of lateral roots and overrides the nitrate signal.

Nitrate is one of the most mobile nutrients in the soil. (In contrast to phosphate, for example, which most plants assimilate via symbiotic relationships with fungi; see Section 8.5). Given the mobility of nitrate, the selective advantage of the capacity for localized root proliferation in nitrate-rich pockets of soil is unclear. In natural, unfertilized soils, however, decaying organic matter is a major source of nitrogen. In aerobic conditions this organic matter releases both the highly mobile nitrate ion and the relatively immobile ammonium ion. Under these circumstances, the faster diffusion of nitrate through the soil solution means that it is the first nutrient to be perceived by nearby roots. Thus a plant that can respond to nitrate by localized root proliferation will also be able to exploit the additional, less mobile nutrients that may be located in the nitrate source.

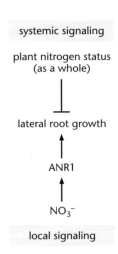

Figure 6–62.
Model explaining the regulation of lateral root production by NO_3^-.
Systemic signals of the plant's general nitrogen status (of unknown nature) inhibit lateral root growth. Localized concentration-sensitive detection of NO_3^- at a lateral root tip promotes lateral root growth via an ANR1-dependent signaling pathway.

SUMMARY

The form of the adult plant is governed by an interaction between its intrinsic developmental programs and environmental signals, including temperature, day length, light, gravity, and the availability of water and nutrients. Environmental responses are coordinated, so that stimuli perceived in one part of the plant body can produce changes in another part. Many of these responses are mediated by growth regulators (phytohormones).

In many plants, seed dormancy is broken by light and/or by low temperature. The seedling shoot shows two distinct types of growth in darkness and light: when the shoot emerges from the soil, development switches from skotomorphogenic (etiolated) to photomorphogenic. Both types of growth are probably controlled by a gene-orchestrated balance of positive and negative regulation.

Photoreceptor systems perceive the quality, quantity, duration, and direction of light and, via linked signal-transduction pathways, initiate developmental responses. The phytochromes respond to red and far-red light (e.g., in the shade avoidance response); the cryptochromes and phototropins respond to blue light (e.g., in phototropism and stomatal opening). Phytochromes are converted from an inactive to an active form by exposure to red light. The interconversion acts as a developmental switch: red light induces growth responses, and far-red light inhibits them. Phytochromes may act as light-regulated protein kinases and/or as transcriptional regulators.

Ethylene plays important roles in seed germination and seedling growth, and in many other plant responses. It acts in concert with other phytohormones, such as with abscisic acid and auxin in the triple response of seedlings. Induction of ethylene at different times during development and in response to different environmental signals involves the controlled expression of a family of genes.

The timing of the transition from vegetative to reproductive development (flowering) is strictly regulated in response to seasonal cues. Through their photoreceptors and signaling systems, plants measure and respond to day length. The major groups of plants are classified as long-day, short-day, or day-neutral. Short-day plants measure the duration of darkness rather than light; the situation is less clear for long-day plants. Circadian rhythms control the expression of many plant genes and affect the photoperiodic control of flowering and other responses. Circadian rhythms result from input of environmental signals that ensure the clock is synchronized with the night–day cycle, generation of a circadian oscillator, and output of rhythmic responses (e.g., expression of genes involved in photosynthesis).

In many species, day length is detected in the leaf and triggers flowering at the apex; the signal is transported in the phloem. Vernalization, detected in the shoot apex, also controls flowering time in many plants. In Arabidopsis, photoperiodic and vernalization pathways converge to regulate the transcription of floral integrator genes. Genetic variation in the control of flowering may be important in the adaptation of plants to different environments.

Many other environmental variables also influence growth. Graviperception by the root (by statoliths and the columella) and shoot (by the endodermis) results in auxin-mediated gravitropic responses. Lateral root elongation varies in response to levels of nutrients, such as nitrate, in the soil.

FURTHER READING

General

Abeles FB, Morgan PW & Saltveit ME (1992) Ethylene in Plant Biology. San Diego, CA: Academic Press.

Hall AJW & McWatters HG (2005) Endogenous Plant Rhythms. Oxford: Blackwell Publishing.

Schäfer E & Nagy F (2006) Photomorphogenesis in Plants and Bacteria. Dordrecht, The Netherlands: Springer.

Thomas B & Vince-Prue D (1997) Photoperiodism in Plants. San Diego, CA: Academic Press.

Weigel D & Jurgens G (2002) Stem cells that make stems. *Nature* 415, 751–754.

6.1 Seed Germination

Bewley JD (1997) Seed germination and dormancy. *Plant Cell* 9, 1055–1066.

Joel DM, Steffens JC & Matthews DE (1995) Germination of weedy root parasites. In Seed Development and Germination (J Kigel, G Galili eds), pp 567–599. New York: G. Marcel Dekker.

Keeley JE & Fotheringham CJ (1997) Trace gas emissions and smoke-induced seed germination. *Science* 276, 1248–1250.

Penfield S, Gilday AD, Halliday KJ et al. (2006) DELLA-mediated cotyledon expansion breaks coat-imposed seed dormancy. *Curr. Biol.* 16, 2366–2370.

Peng J & Harberd NP (2002) The role of GA-mediated signalling in the control of seed germination. *Curr. Opin. Plant Biol.* 5, 376–381.

6.2 Light and Photoreceptors

Ahmad M & Cashmore AR (1993) Hy4 gene of *A. thaliana* encodes a protein with characteristics of a blue-light photoreceptor. *Nature* 366, 162–166.

Briggs WR & Christie JM (2002) Phototropins 1 and 2: versatile plant blue-light receptors. *Trends Plant Sci.* 7, 204–210.

Briggs WR & Olney MA (2001) Photoreceptors in plant photomorphogenesis to date: five phytochromes, two cryptochromes, one phototropin, and one superchrome. *Plant Physiol.* 125, 85–88.

Kohchi T, Mukougawa K, Frankenberg N et al. (2001) The Arabidopsis *HY2* gene encodes phytochromobilin synthase, a ferredoxin-dependent biliverdin reductase. *Plant Cell* 13, 425–436.

Lin C, Yang H, Guo H et al. (1998) Enhancement of blue-light sensitivity of Arabidopsis seedlings by a blue light receptor cryptochrome 2. *Proc. Natl. Acad. Sci. USA* 95, 2686–2690.

Martinez-Garcia JF, Huq E & Quail PH (2000) Direct targeting of light signals to a promoter element-bound transcription factor. *Science* 288, 859–863.

Muramoto T, Kohchi T, Yokota A et al. (1999) The Arabidopsis photomorphogenic mutant *hy1* is deficient in phytochrome chromophore biosynthesis as a result of a mutation in a plastid heme oxygenase. *Plant Cell* 11, 335–348.

Quail PH (2002) Phytochrome photosensory signalling networks. *Nat. Rev. Mol. Cell Biol.* 3, 85–93.

Reed JW, Nagpal P, Poole DS et al. (1993) Mutations in the gene for the red/far-red light receptor phytochrome B alter cell elongation and physiological responses throughout Arabidopsis development. *Plant Cell* 5, 147–157.

Shalitin D, Yang HY, Mockler TC et al. (2002) Regulation of Arabidopsis cryptochrome 2 by blue-light-dependent phosphorylation. *Nature* 417, 763–767.

Whitelam GC, Johnson E, Peng J et al. (1993) Phytochrome A null mutants of Arabidopsis display a wild-type phenotype in white light. *Plant Cell* 5, 757–768.

6.3 Seedling Development

Chang C, Kwok SF, Bleecker AB et al. (1993) *Arabidopsis* ethylene-response gene *ETR1*: similarity of product to two-component regulators. *Science* 262, 539–544.

Chory J, Peto C, Feinbaum R et al. (1989) *Arabidopsis thaliana* mutant that develops as a light-grown plant in the absence of light. *Cell* 58, 991–999.

Deng XW, Matsui M, Wei N et al. (1992) COP1, an Arabidopsis regulatory gene, encodes a protein with both a zinc-binding motif and a G-beta homologous domain. *Cell* 71, 791–801.

Guo H & Ecker JR (2004) The ethylene signaling pathway: new insights. *Curr. Opin. Plant Biol.* 7, 40–49.

Kieber JJ, Rothenberg M, Roman G et al. (1993) *CTR1*, a negative regulator of the ethylene response pathway in *Arabidopsis*, encodes a member of the raf family of protein kinases. *Cell* 72, 427–441.

Osterlund MT, Hardtke CS, Wei N & Deng XW (2000) Targeted destabilization of HY5 during light-regulated development of Arabidopsis. *Nature* 405, 462–466.

Potuschak T, Lechner E, Parmentier Y et al. (2003) EIN3-dependent regulation of plant ethylene hormone signaling by two *Arabidopsis* F-box proteins: EBF1 and EBF2. *Cell* 115, 679–689.

Serino G & Deng XW (2003) The COP9 signalosome: regulating plant development through the control of proteolysis. *Annu. Rev. Plant Biol.* 54, 165–182.

Vert G, Nemhauser JL, Geldner N et al. (2005) Molecular mechanisms of steroid hormone signaling in plants. *Annu. Rev. Cell Dev. Biol.* 21, 177–201.

Wang HY, Ma LG, Li JM et al. (2001) Direct interaction of Arabidopsis cryptochromes with COP1 in light control development. *Science* 294, 154–158.

6.4 Flowering

Bäurle I & Dean C (2006) The timing of developmental transitions in plants. *Cell* 125, 655–664.

Colasanti J, Yuan Z & Sundaresan V (1998) The indeterminate gene encodes a zinc finger protein and regulates a leaf-generated signal required for the transition to flowering in maize. *Cell* 93, 593–603.

Corbesier L, Vincent C, Jang S et al. (2007) FT protein movement contributes to long-distance signaling in floral induction of Arabidopsis. *Science* 316, 1030–1033.

Harmer SL, Hogenesch JB, Straume M et al. (2000) Orchestrated transcription of key pathways in Arabidopsis by the circadian clock. *Science* 290, 2110–2113.

Harmer SL, Panda S & Kay SA (2001) Molecular bases of circadian rhythms. *Annu. Rev. Cell Dev. Biol.* 17, 215–253.

Hayama R & Coupland G (2004) The molecular basis of diversity in the photoperiodic flowering responses of Arabidopsis and rice. *Plant Physiol.* 135, 677–684.

Henderson IR & Dean C (2004) Control of Arabidopsis flowering: the chill before the bloom. *Development* 131, 3829–3838.

Ishikawa R, Tamaki S, Yokoi S et al. (2005) Suppression of the floral activator Hd3a is the principal cause of the night break effect in rice. *Plant Cell* 17, 3326–3336.

Johanson U, West J, Lister C et al. (2000) Molecular analysis of FRIGIDA, a major determinant of natural variation in Arabidopsis flowering time. *Science* 290, 344–347.

Michaels SD & Amasino RM (2000) Memories of winter: vernalization and the competence to flower. *Plant Cell Environ.* 23, 1145–1153.

Reid JB, Murfet IC, Singer SR & Weller JL (1996) Physiological-genetics of flowering in Pisum. *Semin. Cell Dev. Biol.* 7, 455–463.

Zeevaart JAD (1976) Physiology of flower formation. *Annu. Rev. Plant Physiol.* 27, 321–348.

6.5 Root and Shoot Growth

Morita MT & Tasaka M (2004) Gravity sensing and signaling. *Curr. Opin. Plant Biol.* 7, 712–718.

Muller A, Guan C, Galweiler L et al. (1998) *AtPIN2* defines a locus of *Arabidopsis* for root gravitropism control. *EMBO J.* 17, 6903–6911.

Zhang H & Forde BG (1998) An Arabidopsis MADS box gene that controls nutrient-induced changes in root architecture. *Science* 279, 407–409.

ENVIRONMENTAL STRESS

Plant species have diverged to inhabit an enormous range of environments. They may survive extremes of temperature (as low as −40°C or as high as +50°C), water deficits so severe that no water is available from the soil, prolonged periods of water-logging and oxygen deprivation, or high levels of external salts. Under these different environmental conditions, plants that are adapted to life under extremes of environment may continue to assimilate carbon through **photosynthesis** and maintain basic cellular processes. Some have developed specialized metabolism or morphology that facilitate the supply of light, water, and oxygen for these vital processes. Others have specialized life cycles that allow opportunistic use of more favorable conditions and survival during less favorable conditions by different types of **dormancy**.

Successful growth and colonization in environmental extremes is the prerogative of a limited number of plant species (Figure 7–1). Yet, fixed in their environment, all plants must be able to tolerate transitory environmental change. Tolerance of transitory stress depends on a range of physiological responses that **acclimate** a plant's growth, development, and metabolism to less favorable conditions. In fact, most plants are subject to

(A)

(B)

Figure 7–1.
Plants inhabit extreme environments.
(A) Bristlecone pine (*Pinus aristata*) growing
in hot, dry conditions in Bryce Canyon National
Park, Utah. (B) Sea lily (*Pancratium maritinum*)
growing in the hot, dry, salty conditions of
Lara Beach, Cyprus. (A and B, courtesy of
Tobias Kieser.)

some form of environmental stress for some periods of their lives. In this chapter we
explore how the basic processes of plant metabolism and development, outlined in the
preceding chapters, are modified to acclimate plants to transient extremes in light availability and irradiance, water deficit or excess, high external salt concentrations, and
extremes of temperature. As we use the term here, a "stressful environment" is any
environment that is less than optimal for plant growth.

Appreciation of how plants acclimate to changing environmental conditions provides a
good foundation for understanding how some species are **adapted** to life under
extreme environmental conditions. Often adaptations involve constitutive operation of
acclimating responses to stress. However, most plants that compete effectively in continuously extreme environments also show additional metabolic or developmental
modifications that assist their survival. We consider how plants acclimate to transient
environmental conditions and compare these processes to those that adapt particular
plant species to long-term survival in environmental extremes.

In nature, environments are rarely the source of a single type of stress for plants. For
example, soils with high salt levels present the dual challenges of increased ion toxicity and water deficit. Low temperatures are often accompanied by stress from too much
light because, at these temperatures, light energy exceeds the needs of photosynthesis.
Low temperatures that induce freezing of water in the soil involve water deficit in the
plant's tissues as well as the problems of low temperature. Consequently, survival of a
single species in a particular environment usually involves combinations of different
adaptations. In this chapter we consider the primary environmental challenges of light
availability, temperature, water availability, salinity, and oxygen availability as individual stresses, but we emphasize the common components of the responses of plants to
these different stresses.

Responses to environmental conditions require, first, that the organism perceives the
stress. This signal then has to be transmitted to invoke the appropriate responses, both
metabolic and developmental. This sequence of events—stress perception, signal transduction, and response induction—underlies acclimation responses to all types of environmental challenge.

7.1 LIGHT AS STRESS

Plants require light for photosynthesis. In habitats with limited light, plants must have
forms of development, metabolism, and life cycle that optimize the capture and use of
light energy. Too much light, however, can damage plants. We focus first on the stress

effects of too much light and the protective mechanisms that plants have developed, then discuss some adaptations to low light. We then consider the mechanisms of damage by ultraviolet (UV) light and how plants counteract them.

Photosystem II is highly sensitive to too much light

Chloroplasts are especially sensitive to light stress, in particular **photosystem II (PSII)**, which catalyzes the water-splitting reaction of photosynthetic energy conversion. The light-harvesting pigment complexes of the PSII **reaction center** antennae absorb light, causing excitation of **chlorophyll**. Under normal photosynthetic conditions, this energy is transferred to a chlorophyll molecule at the reaction center and drives the photochemical reactions of photosynthesis. If incident light is too high, the excitation of chlorophyll exceeds the energy dissipated through photosynthesis. This problem is most marked under conditions of high light and low temperature, when **photoexcitation** proceeds at a high rate but the chemical reactions of photosynthesis are slow. Much of our discussion here builds on foundations laid in Chapter 4 (see Section 4.2), which details the mechanisms of photosynthesis.

High light induces nonphotochemical quenching, a short-term protective mechanism against photooxidation

Light is sometimes stressful even in moderate environments, such as with the transition from cloud cover to full sun, and rapidly induced responses are required to protect the plant from damage. Too much light gives rise to excess electrons in the chloroplasts, and the energy from these excited electrons must be dissipated to prevent **photooxidative damage**. The most rapid response to high light intensity (occurring in seconds to minutes) is a decrease in the pH of the **thylakoid** lumen caused by the excess electrons, resulting in an increase in the pH gradient across the thylakoid membrane.

If the flow of electrons into the photochemical reactions of photosynthesis exceeds the flow out, the excess of electrons in the **electron transport chains** is dissipated by transfer of electrons from **photosystem I (PSI)** to oxygen rather than to **NADP** (Figure 7–2). The oxygen is reduced to superoxide anion radicals (see Table 7–1).

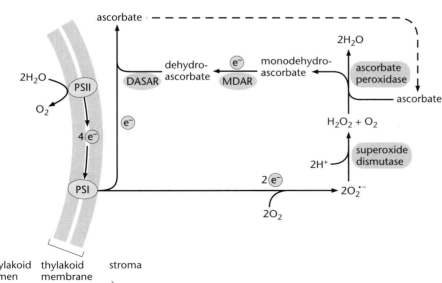

Figure 7–2.
The water–water cycle. If electron flow (e^-) into the photochemical reactions, from photosystem II (PSII) to photosystem I (PSI), exceeds outflow through ferredoxin (see Figure 4–12), oxygen is reduced to form superoxide ($O_2^{\cdot-}$) in the chloroplast stroma. The action of superoxide dismutase, ascorbate peroxidase, monodehydroascorbate reductase (MDAR), and dehydroascorbate reductase (DASAR) removes the superoxide and further electrons. The net output is a pH gradient across the thylakoid membrane that induces additional pathways for energy dissipation.

thylakoid lumen   thylakoid membrane   stroma

pH gradient

Superoxide radicals are eliminated by the enzyme superoxide dismutase (SOD) to yield hydrogen peroxide, which is, in turn, reduced to water by reaction with ascorbate, catalyzed by ascorbate peroxidase. The oxidized ascorbate is reduced (recycled) to ascorbate by two electrons from the electron transport chain. Because this process is the reverse of the oxidation of water occurring in PSII (see Figure 4–12), this mechanism for dissipating excess electrons is referred to as the **water–water cycle**.

The water–water cycle allows the flow of electrons to continue under light stress, and the electron flow, in turn, sets up the pH gradient across the thylakoid membrane (see Section 4.2). The pH gradient slows the overall rate of electron transfer through the electron transport chain and elicits an extra means of energy dissipation within minutes of light stress, through the activation of xanthophyll synthesis via the **xanthophyll cycle** (Figure 7–3).

Xanthophylls are a class of **carotenoids**. The xanthophyll cycle interconverts three xanthophylls: zeaxanthin, antheraxanthin, and violaxanthin. In low or limiting light, zeaxanthin epoxidase converts zeaxanthin to violaxanthin via the intermediate antheraxanthin. In excess light, violaxanthin de-epoxidase converts violaxanthin back to antheraxanthin and thence to zeaxanthin. Two of these three xanthophylls act to remove excess excitation energy from chlorophyll. Zeaxanthin and antheraxanthin effectively transfer energy from the excited chlorophyll and restore the chlorophyll to its ground state: since the lowest excited energy states for both these xanthophylls are at or below that of excited chlorophyll, energy can flow down this gradient (Figure 7–4) and is eventually dissipated as heat as the pigments return to their ground states. This process is called **nonphotochemical quenching**.

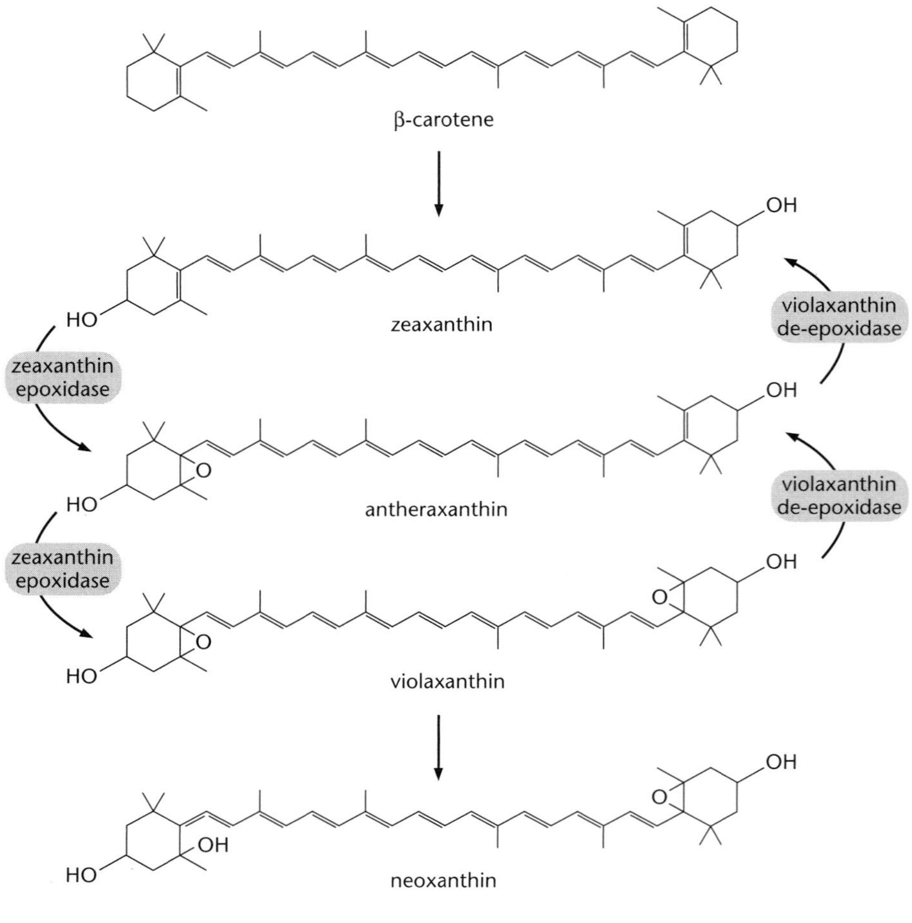

Figure 7–3.
The xanthophyll cycle. This cycle is part of the β-carotene branch of carotenoid biosynthesis in plants.

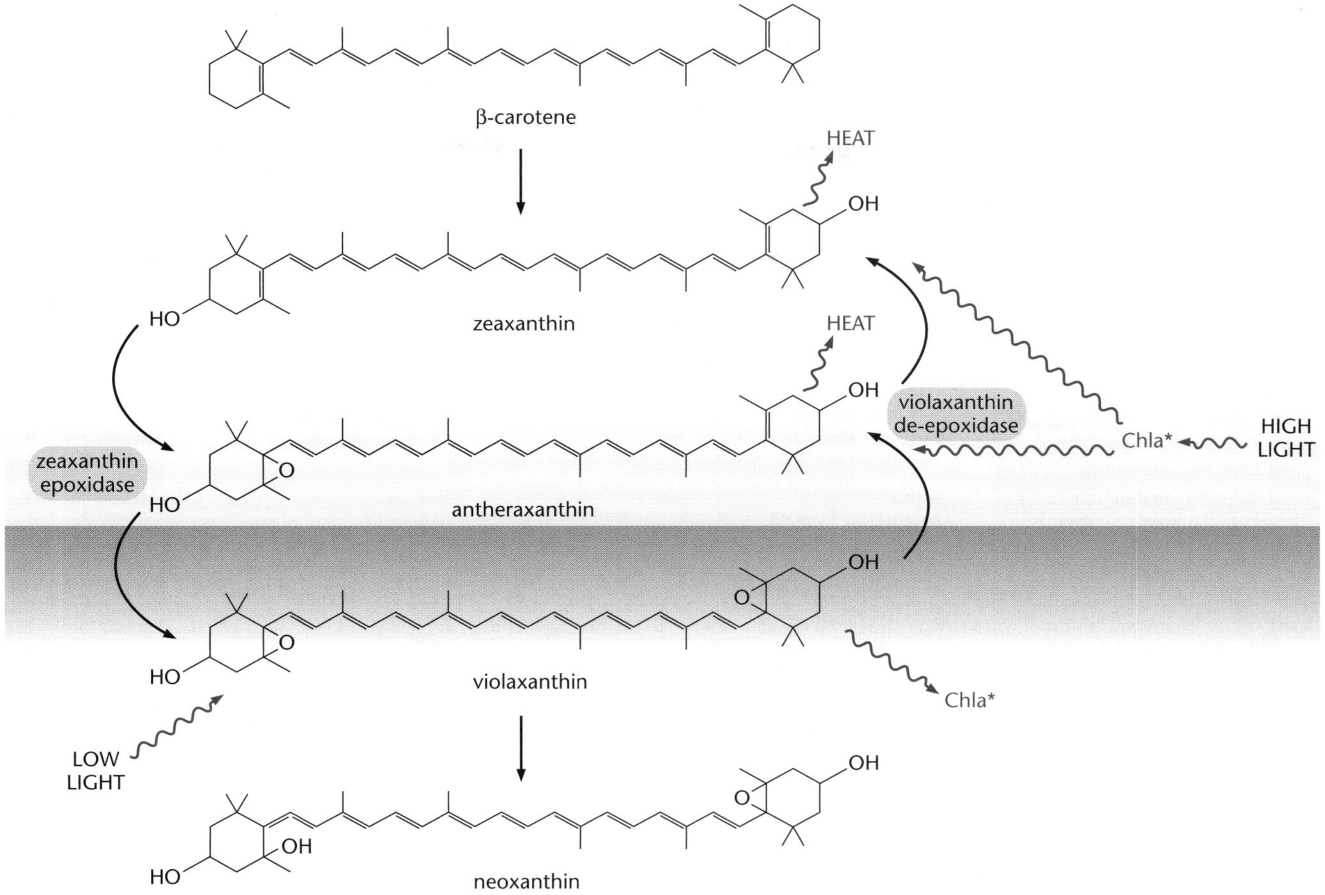

Figure 7–4.
Changing outputs of the xanthophyll cycle under high and low light conditions. In low light, violaxanthin absorbs light energy and transfers it to chlorophyll (Chla), acting as an accessory light-harvesting pigment. In high light, violaxanthin de-epoxidase activity is induced and this enzyme converts violaxanthin to antheraxanthin and zeaxanthin, which take excess excitation energy from chlorophyll and dissipate it as heat. (Chla* indicates activated chlorophll.)

Since the lowest excited energy state of violaxanthin is higher than that of chlorophyll, this xanthophyll cannot dissipate excess energy from chlorophyll. Instead it may function as an accessory light-harvesting pigment. In high light conditions, the enzyme violaxanthin de-epoxidase is induced by the large pH gradient generated across the thylakoid membrane and converts violaxanthin to the energy-dissipating zeaxanthin (Figure 7–5), thus providing more photoprotection to PSII. Violaxanthin de-epoxidase is located in the thylakoid lumen, whereas the xanthophyll pigments are bound to the **light-harvesting complex (LHC)** proteins in the **antennae** of the photosystems. As the thylakoid lumen becomes acidified, violaxanthin de-epoxidase associates with the thylakoid membrane, where its xanthophyll substrates are located. The affinity of the enzyme for ascorbate (see Section 7.7), which acts as a cofactor, is also greatly increased by low pH.

The importance of the xanthophyll cycle to light stress was shown by the finding that plants with **mutations** in violaxanthin de-epoxidase have an increased sensitivity to light stress. Loss of another carotenoid, lutein, also increases sensitivity to high light, and double mutants with impaired violaxanthin de-epoxidase and impaired lycopene-ε-cyclase (the enzyme that synthesizes lutein) undergo photooxidative bleaching and premature senescence under high light intensity (Figure 7–6).

The relative importance of different quenching mechanisms for photoprotection varies among species. For example, in Arabidopsis, zeaxanthin contributes about 85% of the plant's nonphotochemical quenching capacity, whereas in *Chlamydomonas* its contribution is only 25%.

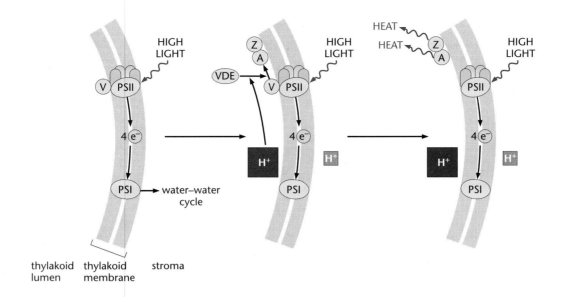

Figure 7–5.
The role of violaxanthin de-epoxidase in nonphotochemical quenching. Under high light conditions, the energy of the excess excited electrons is dissipated by the water–water cycle (*left*), setting up a pH gradient across the thylakoid membrane (see Figure 7–2). Violaxanthin (V) is associated with the thylakoid membrane on the lumenal side. The pH gradient causes movement of violaxanthin de-epoxidase (VDE) to the membrane (*middle*), where it encounters its substrate (violaxanthin) and converts it to antheraxanthin (A) and zeaxanthin (Z). Antheraxanthin and zeaxanthin absorb excess excitation energy from chlorophyll and dissipate it as heat (*right*), protecting the photosynthetic machinery through nonphotochemical quenching.

phytoene

↓

ζ-carotene

↓

lycopene

lycopene-ε-cyclase

β-carotene

α-carotene

zeaxanthin

lutein

antheraxanthin

violaxanthin de-epoxidase

violaxanthin

zeaxanthin epoxidase

loroxanthin

neoxanthin

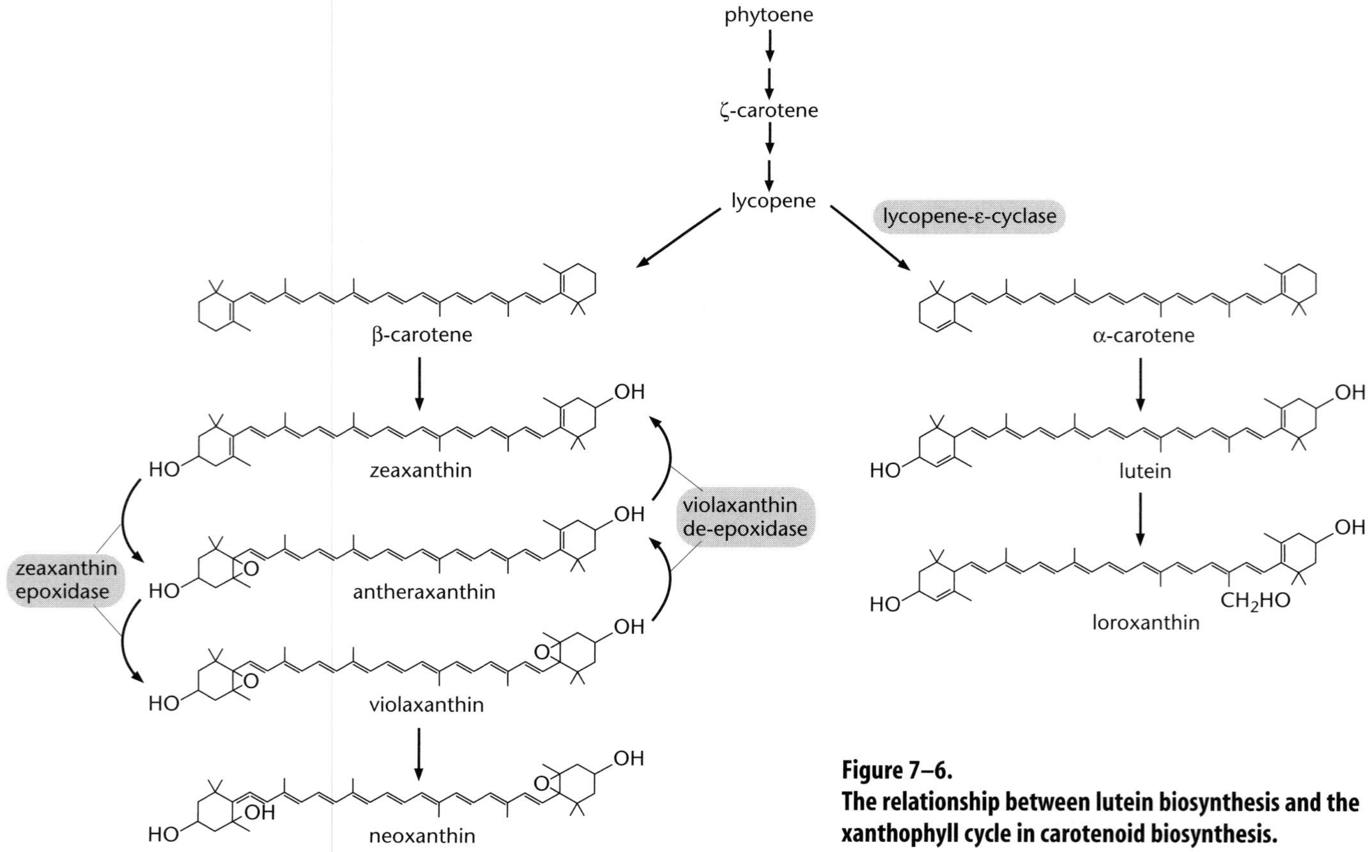

Figure 7–6.
The relationship between lutein biosynthesis and the xanthophyll cycle in carotenoid biosynthesis.

Vitamin E–type antioxidants also protect PSII under light stress

Tocopherols are vitamin E–type compounds that are powerful **antioxidants**, chemical scavengers that can quench singlet oxygen—a highly reactive oxygen species (Table 7–1). They are synthesized in plastids (Figure 7–7) via the **isoprenoid** and **shikimate pathways** (see Sections 4.7 and 4.8). Their levels increase in plants under high-light stress.

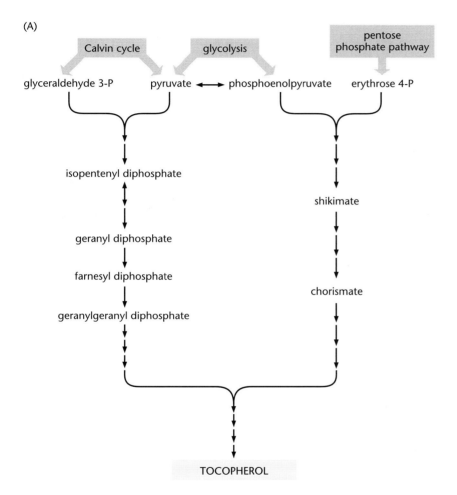

Figure 7–7.
Tocopherol synthesis. (A) Convergence of the biochemical pathways for the synthesis of tocopherol (vitamin E) in plants: the isoprenoid (*left*) and shikimate (*right*) pathways. (B) Structural similarity between tocopherol and plastoquinone.

Table 7–1 Major types of reactive oxygen species

Reactive species	Symbol	Half-life at 37°C (sec)
dioxygen biradical	•O-O•	>100
singlet oxygen	O-O:	1×10^{-6}
superoxide radical	•O-O:	1×10^{-6}
hydrogen peroxide	H:O-:H	
hydroxyl radical	H:O•	1×10^{-9}

The tocopherols are lipid-soluble, **amphipathic** molecules, with a hydrophobic tail and polar (hydrophilic) head group, that are located in plastids. In chloroplasts, their hydrophobic tails are inserted in the thylakoid membrane. This location enables tocopherols to protect the membranes from photooxidative damage such as lipid peroxidation, thus maintaining membrane stability during light stress and, in turn, reinforcing the pH gradient across the thylakoid membrane that drives the other mechanisms of nonphotochemical quenching such as the water–water cycle and violaxanthin de-epoxidase activity.

Tocopherols are produced constitutively at similar levels to **plastoquinones** (components of the photosynthetic electron transport chain), which are similar in structure (see Figure 7–7B). These structural similarities suggest that tocopherols may also function in electron transport during photosynthesis. *In vitro*, the product of tocopherol quenching of singlet oxygen, tocopherol quinone, can efficiently oxidize the cytochrome components of the **cytochrome b_6f** complex that links PSII and PSI (see Figure 4–12) and can allow cyclic electron flow and energy dissipation.

Photodamage to photosystem II is quickly repaired in light stress–tolerant plants

In all plants, a key polypeptide of the PSII reaction center, **protein D1** (see Figure 4–13), is damaged by exposure to strong light, disrupting photosynthesis and limiting growth. Plants that have the ability to replace damaged D1 relatively quickly can often survive high light. Protein D1 normally has a high turnover rate (50–80 times higher than that of other membrane-bound photosynthetic proteins), and in many plants (pea is a well-studied example), turnover is accelerated under strong light. The D1 protein, encoded by *psbA* in the chloroplast genome, contains binding sites for the chlorophyll molecules of PSII (including **P680**, the primary electron donor) and for electron acceptors such as plastoquinone. When D1 is damaged by singlet oxygen that forms as a result of excess light, it becomes a target for proteolytic degradation. In plants tolerant of high irradiance, new D1 is rapidly synthesized and assembled with the other PSII polypeptides, thus restoring photosynthetic function.

Proteolysis of damaged D1 follows two distinguishable steps: (1) endoproteolytic cleavage by a serine/threonine protease, which requires **GTP**, and (2) further degradation by a different, **ATP**-requiring protease. Replacement of the D1 protein that is destroyed under conditions of excess excitation energy requires increased synthesis of D1 protein and its transfer to the non-appressed (nonstacked) thylakoid membranes. The synthesis of D1 is thought to respond to the oxidation state of plastoquinone after light stress, which signals to induce transcription of *psbA* and other genes encoding plastid-encoded reaction center proteins. Translation of *psbA* mRNA into D1 protein is also controlled by the redox status of the plastid through the plastidial **ferredoxin–thioredoxin system**. The other components of PSII reassemble around D1, and the entire complex migrates back to the appressed thylakoid membranes of the **grana** where PSII is concentrated.

The replacement of damaged D1 protein and restoration of PSII are important in reducing photoinhibition in plants that grow well in full sun ("sun plants"), and there is evidence for a pool of precursor D1 protein available for rapid repair in sun-adapted plants. However, some plants that normally grow in shady environments ("shade plants"), such as *Tradescantia albiflora*, can grow in higher light (upward of 300 μmol m^{-2} sec^{-1}) as well as in shade (50 μmol m^{-2} sec^{-1}), despite considerable levels of photoinhibition under high irradiance. *Tradescantia* has a slow PSII repair cycle. Similarly, the shade plant *Oxalis argona* shows no D1 degradation after photoinhibition. Its photoinhibited PSII reaction centers remain physically intact, retaining the damaged D1 in the appressed thylakoids and providing increased protection against the damaging by-products of excess light by dissipation of excitation energy.

Some plants, such as winter evergreens, have mechanisms for longer-term protection against light stress

Nonphotochemical quenching is a rapid protective response to high-light stress, but many plants also use a variety of more slowly induced mechanisms that can provide longer-term protection. For example, a group of proteins termed the "early light-induced proteins" (ELIPs) are induced by high light intensity. They are nuclear-encoded and belong to the same family of proteins as the light-harvesting complex proteins that bind to the pigments (such as chlorophyll) that capture light for photosynthesis (Figure 7–8). ELIPs are thought to dissipate energy under high light, by some as yet unknown mechanism. ELIPS and LHC proteins may have diverged during evolution from a common ancestral protein, a high-light–induced protein, whose original function might have been energy dissipation. Early plants probably encountered high light more often than low light as an environmental stress. As plants diversified and their numbers increased, shading became more common, and the function of some diverged ELIPs may then have evolved toward light harvesting.

Other longer-term adaptations to light stress are exemplified in many evergreens, which keep their leaves and overwinter in environments with high light intensity and so are particularly vulnerable to light stress at the lower seasonal temperatures. During the winter months, synthesis both of LHC proteins and of chlorophyll decreases, thus reducing the capacity of these plants to capture light energy, so avoiding overload. This is seen in conifers in alpine environments, which often become **chlorotic** (having yellowed leaves) during winter months. In contrast, energy-dissipating pigments such as zeaxanthin build up in these evergreens. These changes take place without being triggered by an increased pH gradient across the thylakoid membranes, and probably involve the formation of stable energy-dissipating chlorophyll–xanthophyll–protein complexes.

Excess **reactive oxygen species (ROS)** generated by light stress (see Table 7–1) can cause debilitating photooxidative damage to plants. Plant defenses against such damage include inducible ROS-scavenging systems, such as superoxide dismutase, ascorbate peroxidase, glutathione-*S*-transferases, and catalases (see Section 7.7). The chloroplast

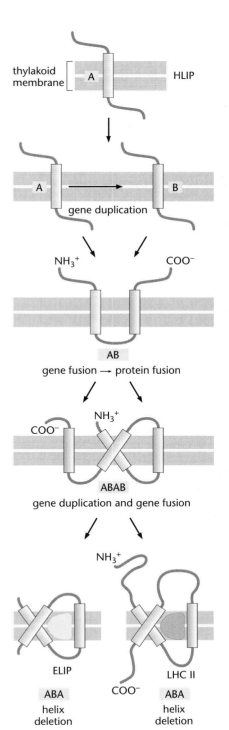

Figure 7–8.
Scheme for the evolution of proteins associated with the photosystems. The ancestral protein is suggested to have been induced by high light intensity (HLIP, high light–induced protein). The gene sequences encoding the transmembrane domain, A, may have been duplicated to form B. Further gene fusions gave a fusion protein ABAB linked by hydrophilic regions of peptide. Thereafter, deletion of the sequence encoding one helix gave rise to proteins with three helical domains, in the order ABA, which evolved functions either in energy dissipation under high light (ELIPs, early light-induced proteins) or as accessory proteins facilitating light capture in low light (LHCs, light-harvesting complex proteins).

is the primary site of ROS production during light stress, so scavenging systems located in the plastids are particularly significant in photoprotection.

Under light stress, some of the light energy (5–50%) is still passed on to the photochemical processes of the plastid. **Photorespiration**, a cycle of reactions through which the plant loses carbon dioxide (see Section 4.3), also acts as a sink for some of this energy. The protective role of photorespiration against photooxidation can be demonstrated experimentally. If the capacity of the photorespiratory pathway is increased by overexpression of glutamine synthase in plastids, photoprotection is increased. If the capacity is decreased by inhibition of glutamine synthase, the plant becomes more sensitive to light stress. Photorespiration may have this effect by decreasing the levels of ROS entering the plastidial scavenging pathways (see Section 7.7), thus offering a degree of long-term protection against light stress, especially under conditions of low carbon dioxide availability (Figure 7–9).

Several changes in leaf morphology also occur in response to changing irradiance levels. For example, in the shade plant *Tradescantia*, strong light promotes the growth of thicker leaves with 30% less chlorophyll than in plants grown in low light. This thickening results from elongation of **palisade cells** and formation of extra layers of palisade **mesophyll**. The plants have 3 to 5 layers of mesophyll under higher light intensity, but only 2 to 4 layers under low light. The **epidermal cells** of leaves grown under strong light are thicker than those grown under shade. *Tradescantia* also shows an altered composition and alignment of chloroplasts in conditions of light stress (Figure 7–10). The thylakoid membrane area is greatly decreased, resulting in reduced grana formation. The chloroplasts move from near the surface of palisade cells to align along the vertical walls, decreasing light absorption capacity by 10%. Chloroplast realignment is signaled by the **phototropin** blue-light receptors (see Section 6.2). Together, these adaptations provide a significant photoprotective effect when plants are exposed to high light.

Figure 7–9.
Fates of absorbed photons in photosynthesis and photorespiration.
A network of reactions allows photorespiration to protect against photooxidation by removal of reactive oxygen species (ROS). Under high light, the increased flow of electrons can be used to fix photorespiration-produced carbon dioxide in the carbon-assimilating reactions of photosynthesis. This reduces excess electrons that might otherwise produce ROS and photooxidative damage. (Z = zeaxanthin; V = violaxanthin.)

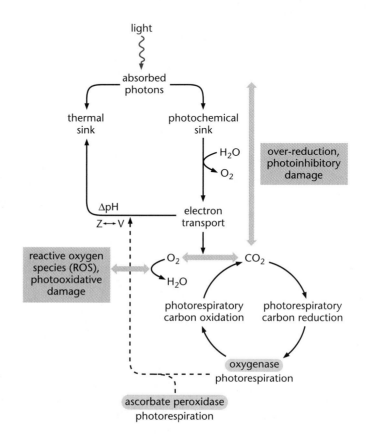

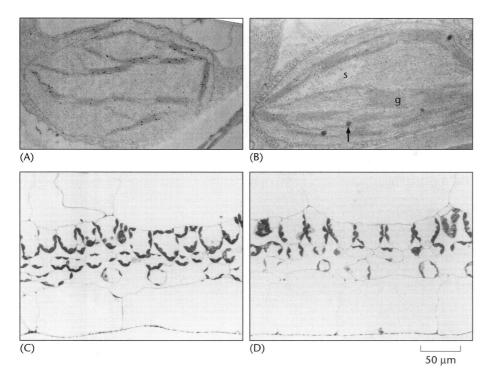

(A)　　　　　　　　　(B)

(C)　　　　　　　　　(D)

50 µm

Figure 7–10.
Effects of light on chloroplast development. Electron micrographs (*top*) show chloroplasts of *Tradescantia* grown (A) under high light (100% irradiance) and (B) in shade (12.5% irradiance). More grana develop in the shade-grown chloroplasts, thus improving light-capture efficiency (g = thylakoid grana; s = starch granule; arrow indicates osmophilic bodies). Light micrographs (*bottom*) show transverse sections of leaves of *Tradescantia albiflora* grown under (C) low light and (D) high light. Under strong light the chloroplasts align along the vertical edges of the palisade cells to reduce light absorption and photooxidative damage. (A and B, courtesy of Satoshi Yano.)

(A)

(B)

In many plants, such as the shade-loving basal angiosperm *Amborella trichopoda*, full sunlight induces changes in the orientation of the leaves. They become folded along their midvein, reducing light interception (Figure 7–11).

Low light leads to changes in leaf architecture, chloroplast structure and orientation, and life cycle

Growth patterns in many plants change under shady conditions. Commonly, stem growth is faster and flowering is accelerated (Figure 7–12). Both of these responses increase the chances of survival: the first by outgrowing the competing plants of the canopy, the second by promoting earlier production of **seeds**, which can survive until conditions improve. Shade is characterized by high ratios of far-red light to red light, caused by the absorption of red light by chlorophyll in the leaf canopy. The intensity of the far-red signal depends on the proximity of neighboring plants and on population density. A plant detects the ratio of far-red to red light by photoreceptors known as **phytochromes** (see Section 6.2). In Arabidopsis, a rosette plant, the most ecologically significant result of the **shade avoidance response** is early **bolting** (rapid stem elongation) and accelerated flowering. In white clover (*Trifolium repens*), which competes principally with grasses, the primary response to shade is extension of the petiole, which projects leaves above the canopy where they can harvest more light (see Figure 7–12B). The sensitivity of different plants to far-red light varies, often depending on their natural habitat. In forest tree successions, species typical of early colonizers are often more responsive to far-red light than are late successional species, which tolerate more shade before shade avoidance responses are triggered.

Changes from high to low irradiance invoke changes in the positioning of chloroplasts and in the light-harvesting machinery of photosynthesis. Under low light, chloroplasts

Figure 7–11.
Changes in leaf morphology of *Amborella trichopoda*. These plants are from the tropical montane rain forest of New Caledonia. (A) Flattened leaves of shade shoot. (B) Folded leaves of sun shoot.

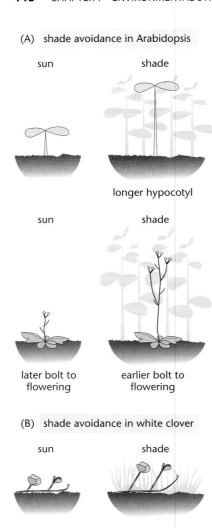

(A) shade avoidance in Arabidopsis

sun shade

longer hypocotyl

sun shade

later bolt to flowering

earlier bolt to flowering

(B) shade avoidance in white clover

sun shade

Figure 7–12.
Different components of the shade avoidance response. (A) Hypocotyl extension and accelerated flowering in Arabidopsis. (B) Petiole extension in white clover (*Trifolium repens*).

of both sun and shade plants orient perpendicular to the direction of the light, increasing the plant's light absorption by 10%. In Arabidopsis, chloroplast reorientation is signaled by phototropin photoreceptors (phototropins are also the photoreceptors responsible for blue-light–activated phototropism; see Section 6.2). In sunflower and other sun plants, the volume of the thylakoids increases when the plants become shaded. The non-appressed thylakoids are the first to increase in volume, then the appressed thylakoids, together resulting in a net widening of the thylakoid stacks (**grana**). In low light, extra recruitment of LHC proteins to the photosystems also enhances light capture, while proteins involved in carbon fixation are decreased. LHC gene expression is induced by low light, and the signal for this induction may be the redox state of plastoquinone, although the molecular mechanism for this signal-transduction pathway is not understood.

Some shade plants have characteristic leaf morphology. For example, in the shaded conditions under which *A. trichopoda* usually grows, its leaves are flat and arranged on pendant shoots in a non-overlapping pattern that increases the area of light interception (see Figure 7–11). *A. trichopoda*, as is typical of shade plants, has no distinct palisade layer in its leaf mesophyll, in contrast to the extended palisade cells of light-adapted plants described above. The spongy cells of *A. trichopoda* leaves are irregular in shape, promoting internal light scattering and thus facilitating the harvesting of diffuse and far-red–enriched light.

Light influences other aspects of shade plant development such as the switch from juvenile to adult phases of vegetative growth, a switch that is particularly marked in woody climbing plants. Such plants start growth in the understory but can reach the top of the canopy, and thus high irradiance, in their adult phase. Changes in morphology from the juvenile to the adult phase include changes in leaf anatomy, morphology, and size, **internode** length, **phyllotaxy** (arrangement of leaves around the stem), wax content, and **trichome** (leaf and stem hair) formation.

In the woody climber *Artabotrys hexapetalus*, the first-formed, lateral buds develop as thorns, whereas later buds produce leafy branches or leafy flowering branches. Low irradiance promotes the formation of thorns, whereas exposure of the same buds to high irradiance produces leafy branches (Figure 7–13). The production of thorns under low light intensity facilitates climbing to reach the upper surface of the forest canopy, where the production of leafy branches (with hooks) helps increase light capture. Thorns also defend the base of the plant against herbivorous predators.

Some plant species of tropical rain forests (e.g., *Ficus barbata*, *Cissus discolor*, *Begonia* spp., and *Anthurium* spp.) have specialized, cone-shaped epidermal cells on the **adaxial** (upper) leaf surface. The bulging outer walls of these cells reflect incident light onto the photosynthetic cells of the mesophyll below (Figure 7–14). Some species (e.g., *Fagus sylvatica*) have conical epidermal cells in both sun and shade leaves. The degree of reflection varies with the height and steepness of the sides of the cone and the refractive properties of the cell wall, and is more pronounced in the epidermal cells of shade leaves.

Some plants grow predominantly in conditions of low irradiance but take advantage of limited periods of high irradiance. Such species are common in the ground flora of deciduous woods. Examples are *Anemone nemorosa* and the bluebell, *Endymion non-scriptus*. These plants grow very rapidly in the spring, increasing their vegetative growth and flowering before the leaves develop on the trees above. During this period of rapid growth, they synthesize carbohydrates and store them in subterranean organs such as bulbs. During the summer months, in full shade, the plants are in a resting state. Some obligate shade plants of deciduous forests, such as *Oxalis acetosella* and *Viola sylvestris*, may have a second period of vigorous assimilation in the autumn (when irradiance increases due to leaf fall from the canopy above) and grow actively throughout the winter.

Figure 7–13.
Architecture of mature *Artabotrys hexapetalus*. Three main morphological forms are related to position within the canopy. In the lowest, shadiest region, axillary buds grow as thorns; in the middle regions, axillary branches grow as vegetative branches; and in the upper, lightest regions, axillary branches grow as reproductive branches with hooks.

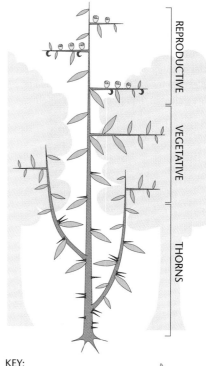

KEY:
= flower ꙮ = hook = thorn = leaf

Ultraviolet irradiation damages DNA and proteins

Besides the stress of increased light intensity, plants are subjected to stress from the ultraviolet wavelengths in incident irradiation. There are three ultraviolet components in solar radiation: UV-A (wavelengths 315–400 nm), UV-B (280–315 nm), and UV-C (200–280 nm) (Figure 7–15). UV-C is the most potentially damaging component but does not reach the earth's surface, due to filtering by **stratospheric** ozone. Thus the most damage to living organisms is caused by UV-B, which makes up 1 to 5% of incoming solar radiation. The proportion of UV-B in incident radiation varies with atmospheric path length (dependent on latitude), altitude, cloud cover, reflectance from the ground, thickness of the stratospheric ozone layer, and other factors. UV-B reaches its highest levels in regions at low latitudes and high altitudes. The increasing depletion of the ozone layers over the polar ice caps, however, is allowing more UV-B radiation to reach these areas. UV-A generally makes up about 6% of incoming solar radiation; its effects are considerably less damaging than those of UV-B.

Ultraviolet irradiation damages DNA and proteins, and plants use both restorative and repair mechanisms to counter this stress. DNA damage is caused when UV light induces the formation of **pyrimidine dimers** between adjacent pyrimidine residues, which block DNA synthesis and **transcription** (Figure 7–16). All organisms seem to have mechanisms to repair such damage: most excise the modified bases and then repair the resulting gap; bacteria and plants have an additional mechanism, "photoreactivation," which can reverse some mutations by restoring the original nucleotides. In organisms capable of photoreactivation, this tends to be the major mechanism of rectifying UV damage to DNA.

Photoreactivation is catalyzed by **photolyases**, enzymes that absorb energy from visible light and use it to break the carbon–carbon bonds between pyrimidine residues in

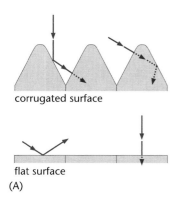

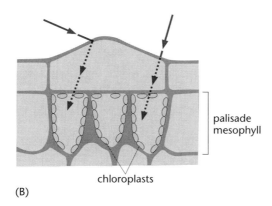

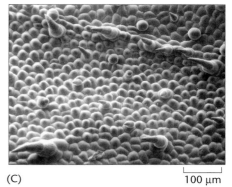

corrugated surface
flat surface
(A)

palisade mesophyll
chloroplasts
(B)

(C)   100 μm

Figure 7–14.
Light focusing by conical cells. (A) Comparison of reflection and penetration of light on corrugated and flat surfaces. More light, especially that arriving at low angles of incidence, is reflected into tissues with a corrugated surface. (B) Enhanced light capture by corrugated leaf surfaces results in greater light reception by the chloroplasts of the mesophyll cells. (C) Scanning electron micrograph of conical cells on a leaf of *Coleus hybridus*.

Figure 7–15.
Components of ultraviolet radiation in sunlight. Of the three types, UV-B causes the most damage in biological systems. In the classification of the three types of UV light based on their biological effects, the ranges are shifted to slightly longer wavelengths.

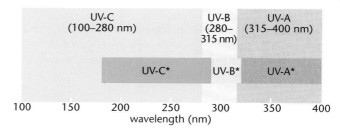

pyrimidine dimers (Figure 7–17); the enzymes recognize and bind specifically to these dimers. Photolyases have two associated cofactors: one transiently donates an electron that breaks the pyrimidine dimer double bond, and the other acts as an antennal protein, harvesting light in the 360 to 420 nm range and thus exciting the electron donor. Photolyases are an ancient family of protective repair enzymes found in all kingdoms and are activated by light in the UV-A to blue range. Genes encoding photolyases are induced by light in this range but generally not by red light or UV-B. Plants have three subgroups of protein families in the photolyase superfamily. Two subgroups repair different kinds of pyrimidine dimers, and Arabidopsis mutants deficient in enzymes of these subgroups are more sensitive to UV light, which illustrates the protective role of some photolyases during UV stress. Photolyases of the third subgroup are not involved in DNA repair but act as blue-light photoreceptors (this is the **cryptochrome** family; see Section 6.2), illustrating the evolutionary relationships between blue-light receptors and the DNA-repair photolyases.

Some plants use nucleotide excision and repair to rectify DNA damage in the absence of light. The damaged DNA strand is nicked on the 5′ and 3′ sides of the site of damage, the damaged oligonucleotide is removed, and the undamaged DNA strand is used as a template to repair the damaged strand (Figure 7–18).

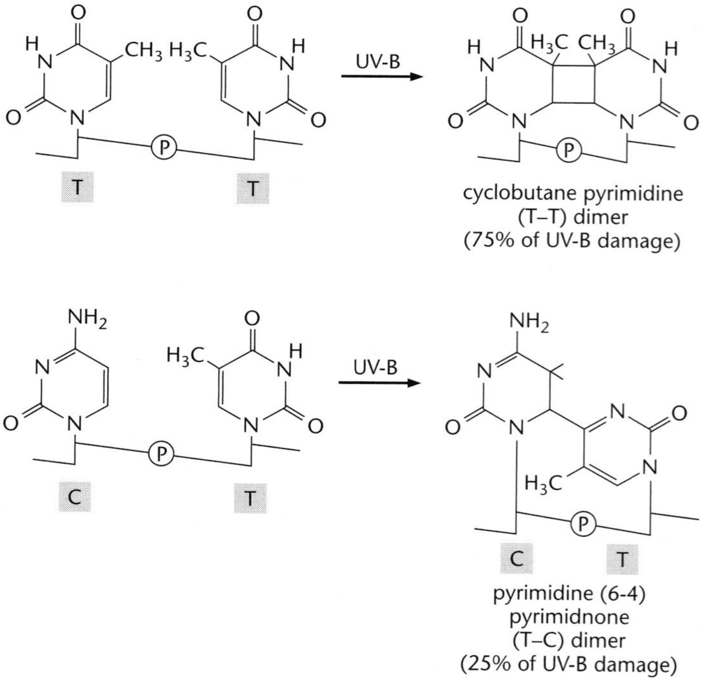

Figure 7–16.
Formation of two types of pyrimidine dimer in DNA, induced by UV irradiation. (T = thymine; C = cytosine.)

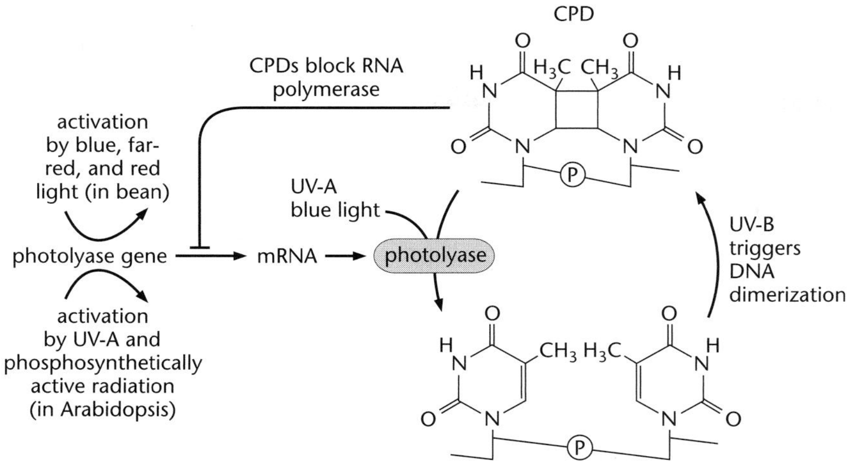

Figure 7–17.
Regulation of photoreactivation of DNA damaged by formation of a cyclobutane pyrimidine dimer (CPD). Photoreactivation, catalyzed by photolyases, restores the pyrimidine residues. Transcription of genes encoding photolyases is induced by light of all wavelengths except UV-B. Photolyase activity requires blue or UV-A irradiation.

Ultraviolet irradiation also damages plant proteins, particularly those involved in photosynthesis. **Rubisco**, violaxanthin de-epoxidase, and the PSI and PSII reaction centers are all particularly susceptible to such damage. UV-B light induces more rapid repair of the damaged protein D1 of PSII by the mechanism described earlier. This may serve as a protective mechanism under high irradiance, by removing and replacing photodamaged D1.

It is not only damaged protein that adversely affects photosynthetic capacity: morphological changes in response to UV irradiation, as described below, also reduce productivity (in addition to reducing UV light capture).

Resistance to UV light involves the production of specialized plant metabolites, as well as morphological changes

Ultraviolet irradiation, even at low levels, stimulates the synthesis of **phenylpropanoids** (see Section 4.8), such as flavonoids and sinapoyl esters. These products of **secondary metabolism** serve as particularly effective sunscreens (absorbing UV radiation between 280 and 340 nm and reducing the penetrating UV more than 90%) but do not reduce the absorption of photosynthetically active radiation. In many plants, **flavonoids** accumulate in response to exposure to UV-B. In Arabidopsis, **sinapoyl esters** and flavonoids accumulate in the epidermal layers exposed to UV irradiation and allow light of the wavelengths used in photosynthesis to penetrate the epidermis.

There is believed to be a specific UV-B receptor with an absorption maximum at 290 nm. UV-B induces the synthesis of certain phenylpropanoids, and the genes encoding the biosynthetic enzymes are induced coordinately (Figure 7–19). Transcriptional repressors of these genes are down-regulated in response to UV-B, suggesting that de-repression of **gene expression** is an important component of the UV-B signal. **Transcriptional activators** may also be important in coordinating the sequence of gene inductions (see Section 2.3 and Box 2–1 for information on **transcription factors**). Flavonoid biosynthesis is induced by the cooperative interaction of **MYB**-like and **basic helix-loop-helix (bHLH)** transcription factors. Expression of these transcriptional activators is induced by light and probably also by UV-B. In many plant species there is a very strong correlation between flavonoid accumulation and UV-B irradiance: flavonoid levels are lower in plants growing in canopies and shaded areas and higher in plants of the same species (different **ecotypes**) growing at higher altitudes.

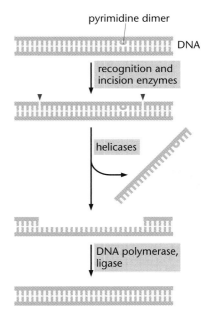

Figure 7–18.
Nucleotide excision and repair. Damaged DNA is excised as an oligonucleotide of 23 to 32 nucleotides; this requires the action of excision enzymes and enzymes known as helicases. The excised region is repaired by the action of the enzymes DNA polymerase and ligase, using the 3′ end of the broken strand as a primer and the undamaged strand as a template.

Figure 7–19.
Transcriptional responses to UV-B light that induce changes in phenylpropanoid metabolism in Arabidopsis. The response occurs through both transcriptional activators and repressors. The result is an accumulation of phenylpropanoids (sinapoyl esters, hydroxycinnamic acid esters, anthocyanins) that act as sunscreens by absorbing UV-B.

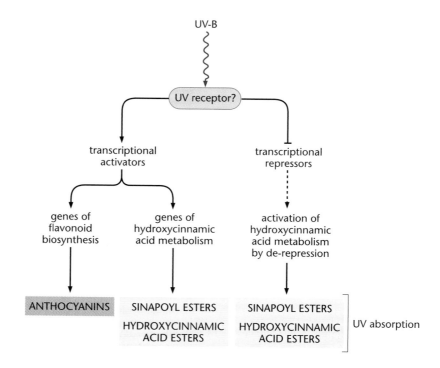

Other secondary compounds, such as polyamines and alkaloids, provide alternative UV screens in some species. These protective secondary metabolites also have antioxidant properties, which offer further protection against stress (see Section 7.7).

Ultraviolet light also induces morphological responses. Exposure to UV light promotes leaf thickening, which reduces the penetration of UV to the photosynthetic cells. This response is similar to the response to high irradiation levels. Another UV-induced change is the redistribution of chloroplasts away from the adaxial surface of the cell, thus reducing damage. These morphological changes may be brought about by UV-dependent modulation of the activity of the phytohormone **auxin** (see Chapters 5 and 6). UV light induces peroxidase activity, which may accelerate auxin degradation and thus modulate its effect on leaf development. In some species, such as *Dudleya*, specialized waxes on the leaves reflect UV-B while having little effect on photosynthetically active radiation. Arabidopsis mutants with impaired wax production show no change in UV tolerance, however, so the contribution of these **cuticular waxes** (see Section 3.6) to UV protection may vary from species to species.

Reductions in leaf size and leaf curling also reduce photosynthetic capacity (in addition to reducing incident UV irradiation). There has been concern about the potential reduction in plant productivity that might result from the increased UV irradiation accompanying depletion of the stratospheric ozone layer in recent years. So far, however, ozone depletion is estimated to have had only a marginal effect on the photosynthetic yields of crop plants.

7.2 HIGH TEMPERATURE

Most growing plants cannot survive air temperatures much over 45°C, and temperatures greater than 30°C are generally stressful. Some Mediterranean species can survive temperatures as high as 48 to 55°C, tropical trees can survive at 45 to 55°C, and subtropical woody plants (such as palms) at 50 to 60°C (Figure 7–20). Even plants that survive in hot climates, however, are temperature-stressed when ambient temperatures exceed 30°C. Leaf temperatures can rise 5 to 10°C higher than air temperature,

(A)

(B)

(C)

particularly if concurrent water shortage has triggered stomatal closure. Desiccated cells and tissues, such as seeds and **pollen grains**, can withstand much greater heat than growing plants. For example, dry alfalfa seeds can remain viable after exposure to temperatures of up to 120°C, and red pine pollen can survive up to 70°C.

Heat stress often accompanies periods of low water availability, and plants adapted to hot environments can generally survive both heat and water stress. We discuss water stress in detail in Section 7.3; here we focus on heat stress. Net carbon gain by plants is particularly sensitive to high temperatures: as ambient temperature increases, the rate of photosynthetic carbon fixation increases less rapidly than the rate of photorespiration, so net carbon gain is reduced as temperature increases. The point at which carbon assimilation by photosynthesis equals carbon loss through photorespiration is called the **compensation point** (see Section 4.3). As net carbon gain decreases with increasing temperature toward or beyond the compensation point, growth slows or stops altogether.

Photosynthesis itself is also sensitive to high temperature. The photosynthetic protein complexes are distributed between the stacked (appressed) and nonstacked (non-appressed) regions of thylakoid membranes. The organized structure of these membranes is destroyed at high temperatures, preventing efficient electron transfer, disconnecting reaction centers from antennae pigments, and uncoupling **photophosphorylation** (the process by which ATP is synthesized; see Section 4.2).

High temperature induces the production of heat shock proteins

Prokaryotic and eukaryotic organisms have a generalized physiological response to heat stress: a decrease in the production of certain proteins accompanied by synthesis of a new set of proteins called **heat shock proteins (HSPs)**. Higher than normal, but sub-lethal, temperatures induce the production of HSPs, which protect the organism from severe damage, increase the temperatures at which it can survive, and allow cellular and metabolic activities to resume (acclimation). Tissue temperatures of 45 to 55°C are considered the normal limits for most plants. The temperature at which the **heat shock response** is invoked depends on the optimal growth temperature for the particular species; for example, the response occurs at 45°C for tropical cereals (e.g., sorghum and millet) and at 35°C for temperate species (e.g., rye grass).

The heat shock proteins from all kinds of organisms fall into five major categories according to molecular weight (in kilodaltons, kDa): HSP100, HSP90, HSP70, HSP60, and small HSPs. All HSPs are molecular **chaperones**, proteins that recognize and bind to proteins that are in an unstable or inactive state. Protein **denaturation**, the loss of the three-dimensional structure required for full activity, is the main consequence of high temperatures. HSPs stabilize and control cycles of protein refolding and prevent the production of inactive, misfolded proteins. Studies have shown that many plants can survive otherwise lethally high temperatures if they have previously been exposed

Figure 7–20.
Heat-tolerant plants. (A) Purple coneflower (*Rudbeckia purpurea*), a heat-tolerant perennial. (B) *Cynoglossum magellense*, a member of the borage family. This plant grows in hot, dry regions of Europe. Its silvery-blue foliage reflects light and reduces the heat load of the leaves. (C) Wild olive (*Olea europaea*), growing in Cyprus. Leaf hairs give the foliage a silvery appearance. The hairs reflect light and reduce heat load.
(A–C, courtesy of Tobias Kieser.)

to smaller temperature increases. This pretreatment induces the synthesis of HSPs, giving the plant acquired **thermo-tolerance**. Note that HSPs, despite their name, are also induced by other environmental stresses such as cold and drought, and acquisition of thermo-tolerance following heat stress often affords cross-protection against other environmental stresses, because the heat shock proteins act to stabilize protein structures.

Molecular chaperones ensure the correct folding of proteins under all conditions

Molecular chaperones play an essential role in cells, both stressed and nonstressed. They prevent incorrect folding of newly synthesized proteins, and classes of chaperones related to HSP90, HSP70, and HSP60 function at normal temperatures in all types of organism. Molecular chaperones generally contain domains that bind to adenine nucleotides (ATP and ADP), and the hydrolysis of ATP to ADP is essential for chaperone activity. We show the characteristics of HSP70 and HSP100 family members, which serve as good examples to illustrate the general features of molecular chaperones (Figure 7–21). HSP70s have a peptide binding domain adjacent to the ATPase domain; in addition, they interact with proteins called "co-chaperones," which promote the activity of HSPs in maintaining target proteins in their native conformation.

As newly synthesized polypeptides emerge from **ribosomes**, more than 50% become associated with chaperones. This delays the folding of the nascent proteins until they reach a critical size, at which point the correct, native conformation forms. Proteins targeted to different subcellular compartments must be moved across organellar membranes, and chaperones are also involved in **protein targeting**.

Families of heat shock proteins play different roles in the heat stress response in different species

Although a wide variety of organisms carry genes that encode members of all five groups of HSPs, the importance of particular HSP families in the heat shock response or in acquired thermo-tolerance differs between species. In Arabidopsis, for example, members of the HSP100 family are particularly important in the development of thermo-tolerance. This has been demonstrated by analysis of mutant plants with an altered HSP100-encoding gene. Wild-type Arabidopsis plants grow well at temperatures of 45°C if pretreated at 38°C. In contrast, mutant plants have impaired growth at 45°C, even after pretreatment at 38°C. If the wild-type HSP100-encoding gene is overexpressed in **transgenic** Arabidopsis, increased thermo-tolerance results. The other HSPs found in Arabidopsis are likely to play roles in maintaining growth at higher temperatures, rather than in the specific adaptations involved in acquired thermo-tolerance.

In other plants, small HSPs have important roles in thermo-tolerance. For example, wheat varieties differ in degrees of thermo-tolerance, and there is a strong correlation between the accumulation of small HSPs and thermo-tolerance. In transgenic carrot and tobacco cells, overexpression of a gene encoding a small HSP also confers improved thermo-tolerance.

Figure 7–21.
Characteristic protein domain structures of heat shock proteins. (A) HSP70s, which have an ATPase domain followed by a peptide binding domain. (B) HSP100s, which consist of nucleotide binding domains.

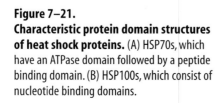

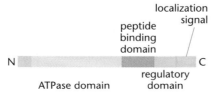

(A) characteristic primary structure of HSP70s

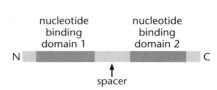

(B) characteristic primary structure of HSP100s

Synthesis of heat shock proteins is controlled at the transcriptional level

The production of HSPs in response to heat stress is controlled principally through the regulation of HSP gene transcription, and this regulation is mechanistically similar in all eukaryotic organisms. The HSP genes contain multiple copies of a regulatory sequence motif, the **heat shock element (HSE)**, within their **promoters**. At least three such elements are required for efficient binding of the **heat shock factor (HSF)**, a transcriptional activator that associates as a trimer on target gene promoters (Figure 7–22). In plants, a large family of genes encodes members of the HSF family (Arabidopsis has 21 HSFs), whereas other eukaryotes have a single gene (in yeast) or a small gene family (four genes in vertebrates) encoding HSFs. The plant HSF proteins share a conserved DNA binding domain and an adjacent hydrophobic region involved in **oligomerization** of the proteins through the association of coiled-coil regions of the monomers. Plant HSF proteins fall into three subgroups—HSFA, HSFB, and HSFC—defined by their protein structures. The HSFA proteins contain a transcriptional activation domain and function in inducing HSP gene expression in a manner analogous to HSF proteins in other eukaryotes. The roles of the HSFB and HSFC proteins are not well defined, but they may function as co-activators or co-repressors of HSFA proteins through complex formation.

Expression of HSFA under nonstressed conditions leads to induction of HSPs and improved thermo-tolerance. The regulation of HSF activity in response to heat shock is multifaceted (Figure 7–23). The C-terminal domain of the HSFA protein promotes transcriptional activation. Under normal, nonstressed conditions, the ability of this domain to activate transcription is repressed. Also, HSF trimer formation is negatively regulated

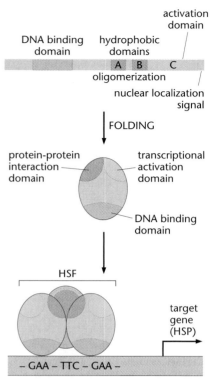

Figure 7–22.
Generalized structure of the heat shock factor (HSF) protein of plants. The HSF polypeptide forms a trimer by association through the hydrophobic domains (*pink*). The trimer binds to heat shock elements, via its DNA binding domains (*orange*), in the promoters of genes that encode heat shock proteins (HSPs). Transcriptional activation requires the activation domain (*blue*).

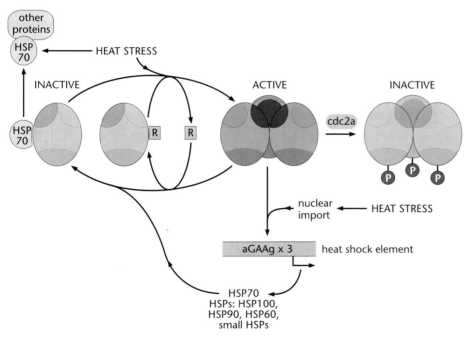

Figure 7–23.
Model of HSF regulation of HSP transcription in response to heat shock. Heat stress results in dissociation of a negative regulatory molecule (R) from the heat shock factor (HSF) polypeptide, allowing HSF to oligomerize (forming the trimer) and bind to the heat shock element and thus induce production of heat shock proteins (HSPs). HSF is negatively regulated by phosphorylation, which may be catalyzed by cyclin-dependent protein kinase 2 (cdc2a) during the cell cycle (see Section 3.1). Induction of HSP70 also negatively regulates HSF. During heat shock, HSP70 proteins associate with other proteins to stabilize their structure, so less HSP70 is available to associate with HSF. After heat shock, HSP70 accumulates and interacts with HSF to inhibit oligomerization, thus negatively regulating its own synthesis.

Figure 7–24.
Xanthostemon aurantiacum, **growing in the Rivière Bleue Reserve of New Caledonia.** Angled leaves help to reduce heat load. (Courtesy of Pete Lowry.)

under normal conditions, and this prevents high-affinity binding to the heat shock element. HSF is further negatively regulated by association with chaperones of the HSP70 class. Following heat shock, the demand for chaperones increases rapidly, and association of HSP70 with HSF is reduced as the HSP70 pool is depleted. This freeing of HSF allows it to multimerize and bind to the HSE, so activating HSP gene transcription. HSF itself, or HSF interaction with HSP70, may be the primary sensor of temperature stress. Nuclear import of HSF is also induced by heat stress, adding another layer to the temperature-induced control of HSP transcription.

Some plants have developmental adaptations to heat stress

Developmental adaptations are found in plants that live in hot climates. Many of these adaptations are also beneficial during water stress (see Section 7.3). However, one mechanism that cools leaves and protects against high-temperature injury is a high **transpiration** rate, which dissipates heat as water evaporates. This response is not suitable for drought conditions, however, and is found in tropical plants that have a high water supply but not in plants of dry environments that are experiencing heat stress. Overheating of leaves may also be reduced by orienting leaves at steep angles to the incident light; this can reduce the temperature of leaves by 3 to 5°C compared with leaves at right angles to sunlight. This trait is common in plants adapted to hot, dry Mediterranean summers (Figure 7–24).

The heat load of leaves can also be reduced by reflectance of incident light by a whitish leaf surface and by the copious production of trichomes, epidermal hairs that reflect incident light. The importance of trichomes in reducing the heat load of leaves in species such as olive (*Olea europaea*; see Figure 7–20C) can be demonstrated simply by shaving the leaves and observing the reduced heat tolerance. Some plants of hot, dry climates also have a thick corky bark (Figure 7–25), which reduces heat absorption and insulates the **phloem** and **cambium** from water loss.

7.3 WATER DEFICIT

Water deficit occurs as a result of drought, salinity, and low temperature

Water is a fundamental requirement for life. The most prevalent environmental stresses for plants are those that limit water supply, either through a lack of available water in the environment (in drought or freezing conditions) or through a lowering of the external water potential (in saline conditions), which restricts water entry into the plant.

Figure 7–25.
Thick corky bark of the coral tree (*Erythrina latissima*). A thick bark helps reduce water loss in hot, dry climates. (A) Close-up of a tree trunk where the cork has been partially stripped away. (B) View of a stand of cork oaks; the cork has been stripped from the trunk and lower regions of the branches. (Courtesy of Arthur Gibson.)

(A)   (B)

The stresses of drought, salinity, and low temperature are linked by their effects on water availability, and the responses of plants to these stresses show many similarities in signaling mechanisms and in the biochemical and metabolic responses invoked. Cellular water deficits stop growth and can result in protein denaturation, membrane damage, loss of **turgor**, and changes in solute concentration. Plant responses operate to ameliorate these biochemical changes and to limit water loss through transpiration to prevent further damage.

Plants use abscisic acid as a signal to induce responses to water deficit

The first step in a plant's response to stress is its recognition of the signal from the environment. Under conditions that lower external water potential, early signal recognition is thought to involve cellular water loss, although how this is perceived is not yet known. Likely mechanisms involve responses to loss of cell turgor, changes in membrane area, changes in membrane stretch, changes in cellular **water potential**, or alterations in cell wall–plasma membrane connections. (See Section 3.5 for more details of these cellular characteristics.)

The perception of water deficit is well characterized in yeast, and related signal-recognition and signal-transduction pathways may occur in plants. Yeast has two pathways that sense external osmotic conditions (Figure 7–26). The first is constitutive during normal osmotic conditions and is inhibited by high external osmolarity (low-water conditions); the second pathway is activated by increases in external osmolarity. Both pathways regulate the same phosphorylation cascade, the "HOG (high osmolarity glycerol) pathway," which controls the response of yeast to osmotic stress (yeast respond by synthesizing glycerol).

The constitutive pathway comprises a complex of a transmembrane **histidine kinase** (the Sln1p histidine kinase), a signal transducer, and a response regulator. The histidine

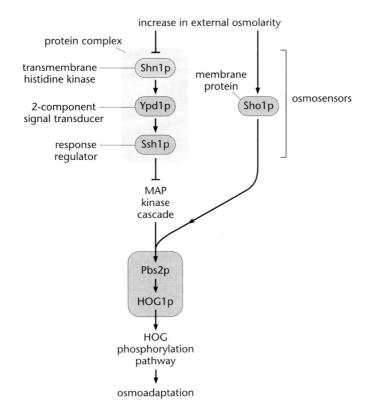

Figure 7–26.
The two osmosensing pathways in yeast.
The Sln1p pathway negatively regulates the HOG (high osmolarity glycerol) pathway and is repressed by increases in external salt concentration. Sln1p, Ypd1p, and Ssh1p act together in a protein complex that negatively regulates a MAP kinase cascade that activates the HOG pathway. The Sho1p pathway is activated by increasing external salt and activates the HOG pathway.

kinase transfers a phosphate via the signal transducer to the response regulator protein. Phosphorylation of the response regulator inhibits its ability to activate the HOG pathway. When external osmolarity rises, the Sln1p histidine kinase is inactivated and the dephosphorylated response regulator can activate the HOG pathway. In the alternative pathway, another membrane protein, Sho1p, interacts with and activates a positive regulator of the HOG pathway. Either mechanism of activating the HOG pathway induces the transcription of genes required for adaptation to high-osmolarity stress—for example, the genes involved in glycerol synthesis. Functional **homologs** of the yeast Sln1p histidine kinase have been identified in plants, suggesting that plants might employ a similar mechanism to sense water deficit.

Another mechanism that might be used by plants to detect water deficit involves tensional changes to the **cytoskeleton** caused by loss of cellular turgor. Some plant genes induced by water stress are also induced by mechanical stimulation (touch), suggesting induction by a common mechanism, although the function of these genes is not yet known. This overlap in responses suggests that changes in cytoskeletal tension could provide a mechanism through which plants detect both touch and water shortage.

The signal-transduction pathways activated in response to water deficit are relatively well-characterized in plants and provide a good example of the complexity of signal transduction and the physiological responses invoked by an environmental stress. We describe these processes in detail to provide a conceptual model for how responses to other stresses might be organized.

Perception of water deficit gives rise to increases in the internal concentrations of **abscisic acid (ABA)** in many plants. ABA, often termed the "stress hormone," is induced by many different types of environmental stress, including drought, salinity, freezing, chilling, wounding, and **hypoxia** (low oxygen). It also controls developmental processes that involve desiccation of plant tissues, such as seed maturation, during which as much as 90% of the original cellular water content may be lost.

Under conditions of water stress, ABA is synthesized *de novo* via the **terpenoid** pathway (Figure 7–27). This is thought to occur in roots, which are usually the first organs to perceive water deficits. Most of the genes encoding enzymes of ABA biosynthesis are induced by water deficit. In particular, the activity of the 9-*cis*-epoxycarotenoid dioxygenase (NCED) enzyme is thought to limit ABA availability, and expression of the *AtNCED3* gene is induced in Arabidopsis by water deficit. Turnover of ABA is also important in determining its activity, and the enzyme ABA 8-hydroxylase is a P450 enzyme, the activity of which is induced rapidly on rehydration, following dehydration stress, to reduce ABA levels.

ABA signals from the roots to the other tissues. Increases in ABA are translated into changes in gene expression via a signal-transduction pathway. Several receptors for ABA have been identified, including the RNA binding protein FCA (involved in controlling flowering time), the H subunit of Mg^{2+} chelatase (an enzyme involved in chlorophyll synthesis), and a **leucine-rich repeat (LRR)** receptor-like kinase (RPK1) of Arabidopsis. Of these, there is genetic evidence to support the operation of Mg-chelatase and RPK1 in the early stages of ABA perception that modulate seed **germination**, post-germination growth, and stomatal movement. The signal-transduction pathway seems to involve a **protein kinase/phosphatase cascade** in which calcium participates.

Several genes are known to encode components of this signal-transduction pathway in Arabidopsis: *ABI1* and *ABI2* encode serine/threonine phosphatases, suggesting that they regulate target proteins by dephosphorylating them. Both ABI1 and ABI2 phosphatases play a role in stomatal closure, and expression of the *ABI1* and *ABI2* genes changes in response to water deficits. A third gene, *ERA1*, encodes a farnesyl

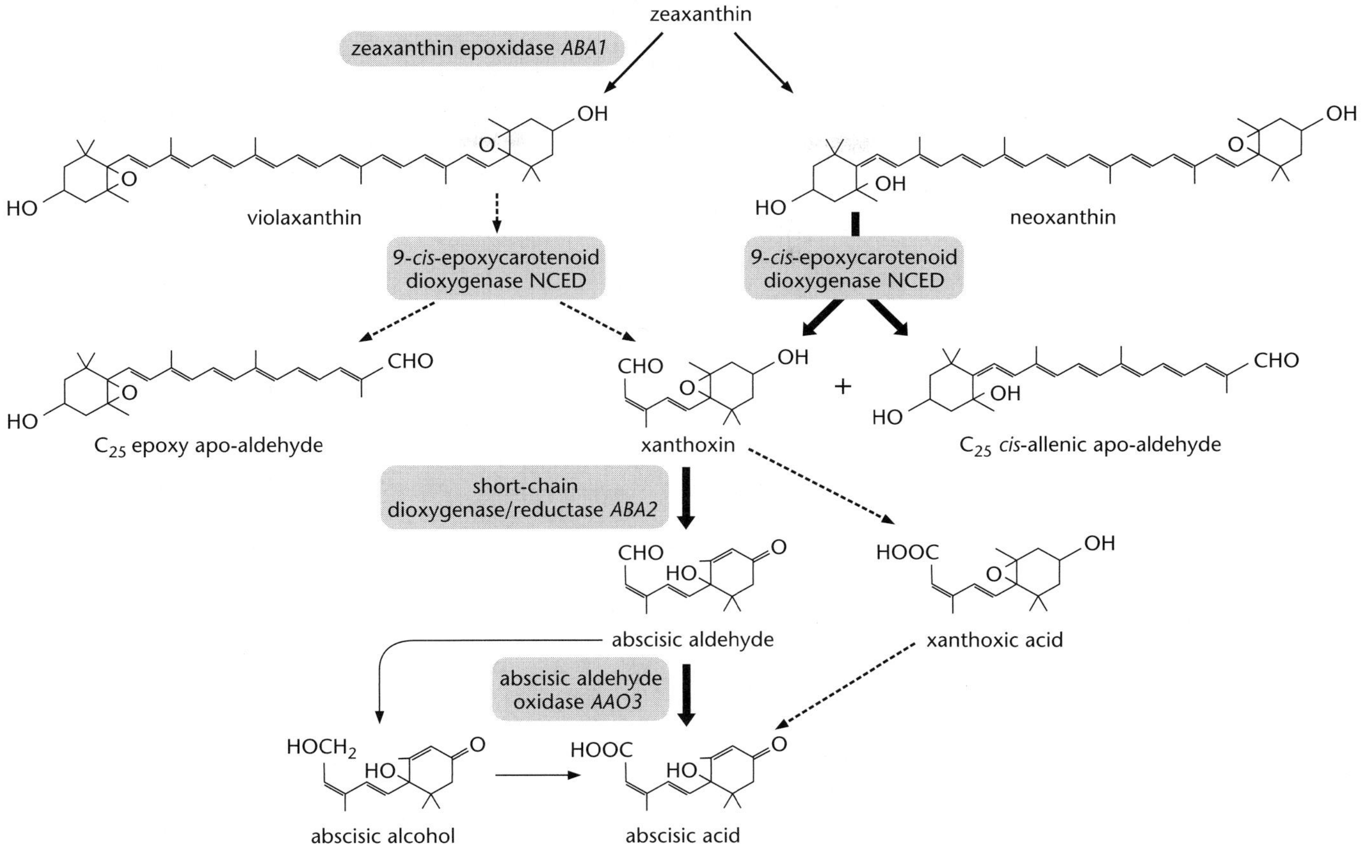

Figure 7–27.
Biosynthesis of abscisic acid from xanthophylls. The names of the genes that encode the enzymes in Arabidopsis are included where known. Shown here are the major pathway (*heavy arrows*), minor pathways (*light arrows*), and hypothetical pathways (*dashed arrows*).

transferase. The transfer of a farnesyl group (a linear group of three **isoprene units**) to a protein generally enables membrane localization of that protein. Plants with a mutation in the *ERA1* gene show enhanced responses to ABA. Thus ERA1 is likely to be involved in the farnesylation of a negative regulator of ABA sensitivity. Likely candidates for such a regulator include a receptor or a component of the signal-transduction pathway that requires membrane localization.

Abscisic acid signaling during water deficit gives rise to changes in expression of water-deficit response genes. Responses that depend on ABA signaling are of two types: those requiring and those not requiring new protein synthesis (Figure 7–28). Genes responding to ABA without induced protein synthesis include those with **ABA-response elements (ABREs)** in their promoters. Many genes with ABREs were first identified as induced during seed maturation (a natural water deficit condition), including *Em1* in wheat and *rab* genes in rice. Other genes that contain ABREs are induced during drought stress, such as *rd29A* and *rd29B* in Arabidopsis. Some genes encoding **late embryo abundant (LEA) proteins** carry ABREs in their promoters and are induced in mature seeds. Transcription factors of the **basic leucine zipper (bZIP)** transcription factor family (e.g., EMBP-1 in wheat) bind to the ABRE, and the promoters of many ABA-responsive genes contain additional elements, which are probably the binding sites for other transcription factors.

Abscisic acid also induces changes in gene expression through pathways that do require protein synthesis. The *rd22* gene of Arabidopsis, which encodes a protein structurally similar to seed storage proteins, is up-regulated by water deficit and ABA, but its promoter does not contain an ABRE. In Arabidopsis, transcription factors AtMYC2/rd22BP1 and AtMYB2, belonging to the bHLH and MYB families, respectively

Figure 7–28.
Signal-transduction pathways between perception of water-stress signals and changes in gene expression. Water stress may result from drought, high salinity, or cold temperatures. At least four signal-transduction pathways operate; some are ABA-dependent, some ABA-independent. The relationship of these signaling pathways to the cold-responsive pathway is also shown. bHLH, MYB, bZIP, NAC, and ZFHD are transcription factor families, members of which respond to the signals induced by water stress and modulate gene expression (see Section 2.3 and Box 2–1 on transcription factors). DREB2A–B and CBF1–3 belong to the AP2 family of transcription factors. HOS1 and ICE are transcription factors involved in cold responses (see Section 7.5).
(ABRE = ABA-response element; DRE/CRT = drought-response element.)

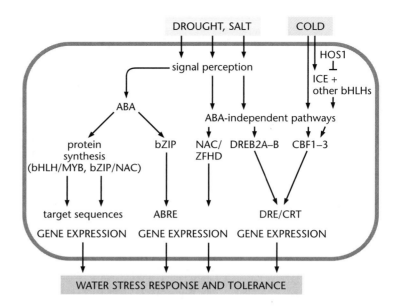

(see Section 2.3), have been shown to bind to the promoter of this gene. The expression of both transcription factors is induced by water deficit (dehydration stress or salt stress), and AtMYB2 synthesis is induced by ABA. It is likely that AtMYC2/rd22BP1 and AtMYB2 are synthesized in response to water stress via a signal from the ABA-dependent pathway and cooperatively induce the expression of other ABA-inducible genes, such as *rd22*. AtMYC2/rd22BP1 also negatively regulates some transcriptional responses to **jasmonic acid** involved in defense responses to biotic challenge (see Section 8.4). Cross-talk between plant responses to abiotic and biotic stress may operate through the activity of this common regulatory component.

A NAC-domain transcription factor, RD26, is induced by drought stress and by ABA in Arabidopsis. Reduced expression of *RD26* limits the sensitivity of Arabidopsis plants to ABA, suggesting that RD26 may regulate gene expression in response to ABA during water deficit.

Plants also use ABA-independent signaling pathways to respond to drought

Several genes induced by drought do not require ABA for expression; these genes have been identified because they are still induced in response to water stress in mutants with impaired ABA synthesis or perception. This suggests that, in addition to the ABA-mediated signaling pathways described above, ABA-independent signaling pathways also induce responses in gene expression when plants are subject to water stress (see Figure 7–28). Some of these genes can also be induced by salt stress or low temperature. Many of the genes that are activated independently of ABA encode proteins of unknown function. For example, the function of the protein product of the *KIN2* gene has not yet been established, but the protein has some structural similarities to **antifreeze proteins** of animals and may serve a role in protecting cytoplasmic proteins from damage during water deficit.

Expression of a conserved sequence known as the **drought-response element (DRE)**, present in the promoters of many genes, is up-regulated in response both to water stress and to low temperature. The DRE is bound by members of the plant-specific **AP2 (APETALA2)** transcription factor family. Three AP2 family members, CBF1, CBF2, and CBF3, have been shown to bind to the DRE and increase expression of cold-inducible genes (see Section 7.5). Other AP2 transcription factors, DREB2A and DREB2B, activate gene transcription via the DRE in response to water stress.

The ABA-dependent and ABA-independent pathways may converge in signaling for changes in gene expression in response to water deficit. Genetic evidence for this comes from the observation that some genes are regulated by an ABA-independent pathway in response to water stress or low temperature but are also activated by ABA signaling. One means of convergence could be the interaction between transcription factors that recognize ABREs and those that recognize DREs and thus modulate expression of genes. For example, *rd29A* (which encodes a protein of unknown function) is activated by ABA-dependent and ABA-independent pathways, and contains both DRE and ABRE sequences in its promoter. The product of the *rd29A* gene, therefore, is likely to act after the two signaling pathways have merged.

Some genes are activated in response to water stress but do not respond to cold treatment or ABA signaling. These genes may be controlled by a fourth signaling pathway, one that responds specifically to water stress. In Arabidopsis, transcription factors of the NAC-domain family (ANAC019, ANAC055, and ANAC072) and from the **zinc finger** homeodomain group (ZFHD1) may operate to regulate responses in gene expression specific to water stress. Several target genes in this category (e.g., *rd19*, *rd21*, and *erd1* of Arabidopsis) encode proteases, or regulatory subunits of proteases. Protease induction in response to drought stress may increase the turnover of damaged proteins and thus provide amino acids for the synthesis of new proteins.

Abscisic acid regulates stomatal opening to control water loss

During water stress, transpirational loss of water is reduced as stomata close in response to the signal from ABA. (See Section 3.5 for details of the mechanisms of the opening and closure of **stomata**.) Stomatal movement depends on changes in **guard cell** turgor, which is controlled by fluxes of K^+ ions into or out of the cells, with either malate or Cl^- as counter-ions. K^+ fluxes are controlled by ion channels in the guard cell plasma membrane and in the **tonoplasts**, principally a plasma membrane channel that moves K^+ into the **cytoplasm** (K^+ inward rectifying channel) and a channel that moves K^+ out (K^+ outward rectifying channel). ABA is detected externally, probably at the guard cell plasma membrane, and perhaps also internally. The receptor might be a LRR receptor-like kinase, RPK1, in Arabidopsis. Perception of ABA gives rise to an increase in cytosolic Ca^{2+}, through release from intracellular stores and movement of Ca^{2+} from the **vacuole** to the cytoplasm. Increases in cytosolic Ca^{2+} inhibit the K^+ inward rectifying channel and activate the K^+ outward rectifying channel (through depolarization of the plasma membrane and inhibition of the plasma membrane **H^+-ATPase**). This results in a net export of K^+ ions from the guard cells, a decrease in cell turgor, and stomatal closure (Figure 7–29).

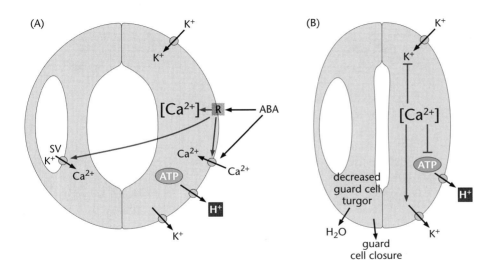

Figure 7–29.
Model for responses of guard cells to the ABA signal. (A) Perception of abscisic acid (ABA) by a guard cell receptor (R) induces calcium (Ca^{2+}) efflux from the vacuole through slow vacuolar channels (SV) and Ca^{2+} influx into the cell through calcium-permeable ion channels in the plasma membrane. (B) Increased cytosolic calcium level, [Ca^{2+}], inhibits potassium (K^+) inflow through the plasma membrane and promotes K^+ outflow from the cell. The increased cytosolic Ca^{2+} also inhibits the plasma membrane H^+-ATPase, so depolarizing the plasma membrane. These changes result in a decrease in guard cell turgor and stomatal closure.

Drought-induced proteins synthesize and transport osmolytes

One of the most widespread responses to water stress, whether caused by drought or high salinity, is the synthesis of compatible **osmolytes** (osmotically active metabolites) such as **polyols** (polyhydric alcohols, or sugars), the amino acid proline, and **betaines** (quaternary ammonium compounds, or **onium compounds**) (Table 7–2). These solutes are "compatible" in that they do not interfere with cellular structure and function. The accumulation of organic solutes in stressed cells lowers the cellular water potential, thus allowing increased water uptake.

Osmolytes probably have additional roles in making cells more tolerant to water stress. For example, when levels of cytosolic osmolytes are increased in transgenic plants, moderate improvements in tolerance to water stress result, even though the increases in osmolyte concentration are not enough to mediate the changes in **osmotic potential** that occur during acclimation to water deficit (osmotic adjustment). The increased osmolyte levels may have several **osmoprotective** effects: stabilizing protein and membrane structure, scavenging free radicals and thus protecting against oxidative

Table 7–2 Osmoprotective compounds in plants

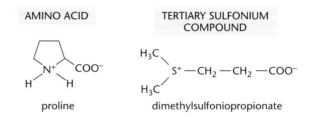

damage, and providing a cellular store of carbon, nitrogen, and reducing power to fuel metabolism once the stress has been relieved.

Figure 7–30 illustrates a possible mechanism for the role of sugars (polyols) as osmolytes in stabilizing proteins during water stress.

Many plants, including Arabidopsis, accumulate the osmolyte proline in response to water deficit. Proline is synthesized from glutamate (Figure 7–31) by the action of Δ^1-pyrroline-5-carboxylate synthase (an enzyme with both kinase and dehydrogenase activities), a spontaneous cyclization, and a reduction catalyzed by Δ^1-pyrroline-5-carboxylate reductase. Expression of the carboxylate synthase is strongly induced by water deficit, and this accelerates the rate of proline synthesis. At the same time, degradation of proline by proline dehydrogenase in the **mitochondria** slows down; this enzyme is dependent on respiratory electron transport and ATP generation, both of which are reduced during conditions of water stress. Thus proline levels are increased both by increasing synthesis and by reducing breakdown.

Proline levels are also increased by the increased efficiency of proline transport. One effect of the perception of water deficit by some plants is the up-regulation of genes that encode proline **transporters**. Arabidopsis proline transporters have been identified that mediate loading of proline into the phloem from source tissues and unloading into sink tissues (see Section 4.4). One of the proline-transporter–encoding genes, *ProT2*, shows differential expression in response to water deficit and salt stress. It is induced slowly over a long time period in response to drought, but very rapidly in response to salt stress. Its principal role, therefore, is likely to be in the redistribution of osmolytes during salt stress.

In addition to the osmolytes listed above, LEA proteins also accumulate in water-stressed cells. LEA proteins are hydrophilic globular proteins, originally characterized as accumulating in seeds during maturation and desiccation. Some LEA proteins accumulate in vegetative tissues after water stress, and their protective effect has been demonstrated in transgenic plants. Rice plants engineered with a gene that enables biosynthesis of an LEA protein (HVA1) from barley maintain higher growth rates in stressed conditions than do control plants.

There are several different groups of LEA proteins, most of which are localized in the cytosol. These highly hydrophilic proteins tend to be rich in alanine and glycine and lacking in cysteine and tryptophan residues. The high number of polar residues suggests that LEA proteins coat intracellular macromolecules with a protective layer of water. On further dehydration they may provide a layer of their own hydroxylated residues that interact with the surface groups of other proteins, acting as "replacement water." Other LEA proteins may form salt bridges between their own charged amino acids (which are arranged in regions of amphipathic **α helix**) and other charged proteins, so stabilizing and protecting these proteins under conditions of water deficit.

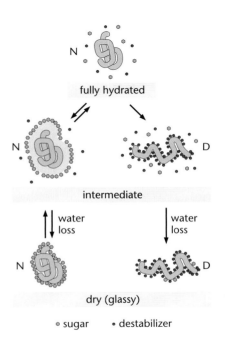

Figure 7–30.
Scheme for possible role of sugars in stabilizing proteins at different stages of water loss.
N is the native folded form of the protein; D is the denatured, unfolded form. Molecular crowding during water loss increases the interaction of solutes with the protein surface. In water-stress–tolerant cells with compatible solutes such as sugars, the solutes interact with the protein surface preferentially, excluding destabilizing solutes and maintaining protein folding. In the absence of compatible solutes, the destabilizing solutes interact with the protein surface, causing the protein to denature. At very low water contents, sugar molecules replace water around the protein via hydrogen bonding, so stabilizing the protein in the dry (glassy) state. Compatible solutes other than sugars fail to stabilize proteins in the dried state. The reversibility of the processes through dehydration and rehydration is indicated by arrows.

Figure 7–31.
Pathways for proline biosynthesis and degradation. Proline is synthesized from glutamate by three enzymes. It is broken down to Δ^1-pyrroline-5-carboxylate (P5C) by proline dehydrogenase. (ABA = abscisic acid.)

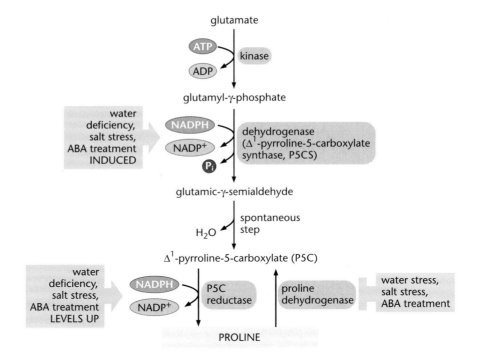

Ion channels and aquaporins are regulated in response to water stress

Drought and high salinity both induce expression of water-deficit response genes, such as genes that encode ion channels and water channels. Induction of plasma membrane potassium channels by salinity stress permits an increase in uptake of Na$^+$ ions and thus osmotic adjustment. In both drought-stressed and salt-stressed plants, expression of specific genes that encode **aquaporins** (Figure 7–32) is also induced. These water channels facilitate the movement of water into and out of cells (as well as between the vacuole and cytoplasm; see Sections 3.5 and 4.10). The water-transporting activity of aquaporins is modified by phosphorylation and by changes in oligomerization. Some aquaporins show reduced phosphorylation under water stress, a mechanism that may serve to limit water loss by reducing the activity of specific aquaporin channels.

Figure 7–32.
General structure of aquaporins. The first, cytosolic loop and the third, extra-cytosolic loop each contain a conserved Asn-Pro-Ala sequence that probably dips into the vacuolar membrane (tonoplast), collectively forming a seventh transmembrane structure. Water can pass through the middle. Conserved serine (Ser) residues are phosphorylated to regulate water transport activity.

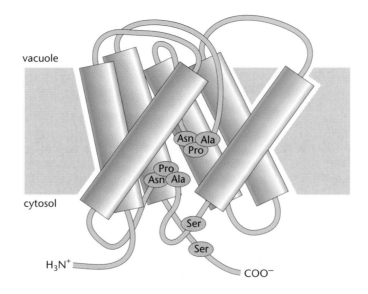

Many plant species adopt metabolic specialization under drought stress

Water stress, whether induced by drought or salinity, brings concomitant problems in maintaining carbon fixation. If stomata are closed to reduce transpiration rates, the supply of carbon dioxide becomes limiting for growth and productivity. A specialized metabolism, **crassulacean acid metabolism (CAM)**, is found in many plants tolerant of arid or saline conditions. **CAM plants** can fix carbon dioxide into inorganic acids at night. This allows water-stressed plants to open their stomata only at night, limiting transpirational water loss. CAM is a modification of the C3 photosynthetic pathway. It is similar in some respects to the C4 pathway that prevents photorespiration (see Section 4.3 for a discussion of **C3** and **C4 photosynthesis**), but there are important differences. In C4 plants, the reactions of photosynthesis and carbon fixation are spatially separated by a specialized anatomy; in CAM plants, the reactions are temporally separated (Figure 7–33).

In CAM plants, the stomata open at night and carbon dioxide is captured by **phosphoenolpyruvate (PEP) carboxylase** in the cytosol, reacting with PEP to form oxaloacetate. The oxaloacetate is reduced to malate, which is stored in the vacuole. With the dawn, the stomata close, stored malate is transported from the vacuole to the cytosol and decarboxylated, and Rubisco fixes the released carbon dioxide into 3-phosphoglycerate (3PGA), which can be further metabolized via the **Calvin cycle**. PEP carboxylase activity is suppressed during the day and activated at night to avoid a futile cycle of carboxylation and decarboxylation. Like C4 metabolism, CAM may have evolved when plants were faced with limited supplies of atmospheric carbon dioxide.

Some plants are **obligate** CAM utilizers, with net nocturnal carbon dioxide fixation and daily fluctuations in organic acids whatever the availability of water. Others are **facultative**, inducing CAM only when needed under conditions of water stress. In the

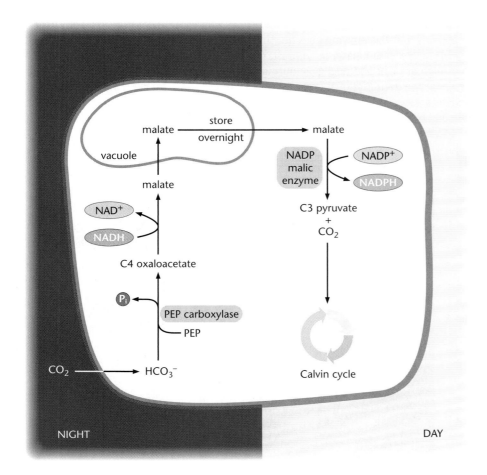

Figure 7–33.
Separation of photosynthesis and carbon fixation in CAM plants. In the dark, the stomata open and CO_2 enters the leaves to be fixed by PEP (phosphoenolpyruvate) carboxylase. The oxaloacetate so formed (four carbons) is converted to malate, which is transported into the vacuole. In the light, malate is mobilized from the vacuole and the CO_2 is released by malic enzyme, forming pyruvate (three carbons). The CO_2 is then reassimilated by the Calvin cycle.

ice plant, *Mesembryanthemum crystallinum* (Figure 7–34), for example, expression of a gene encoding a CAM-specific **isoform** of PEP carboxylase is highly induced by salt or drought stress in plants more than 5 weeks old. Water stress is probably perceived by the roots and signaled to the leaves, enabling the switch from C3 photosynthesis to CAM. Some plants are neither obligate nor facultative but show "weak CAM," with daily fluctuations in organic acids but no net carbon dioxide fixation in the dark, a process called **CAM cycling**. CAM cycling allows a plant to survive with no net loss of carbon to the environment. Plants exhibiting CAM cycling tend to grow in habitats with unpredictable daily water supplies. Generally, CAM plants are more successful in environments having long periods with no free water. Carbon fixation by CAM is limited by the amount of malate that can be stored in the vacuole. Under conditions where some free water is available, C4 photosynthesis provides a more productive metabolic strategy than CAM.

PEP carboxylase and the organic acid decarboxylase (which catalyze opposing metabolic reactions) are both located in the cytosol, yet function at different times of the day. The regulation of PEP carboxylase is an important factor in the control of CAM activity (Figure 7–35). PEP carboxylase is controlled at several levels. First, it is controlled transcriptionally; expression of CAM-specific isoforms increases rapidly during drought or water stress. Recognition motifs for MYB-like transcription factors have been identified within the promoters of the genes encoding the CAM enzymes and are important for induction in response to salt stress. Second, PEP carboxylase is **allosterically inhibited** by malate. Thus its activity is reduced in the light as malate is released from the vacuole. Third, the sensitivity of PEP carboxylase to malate is modulated by phosphorylation. The "night form" of PEP carboxylase is phosphorylated and is less sensitive to inhibition by malate, whereas the "day form" is dephosphorylated and much more (tenfold) sensitive to malate inhibition. This regulation amplifies the activation–inactivation cycle of PEP carboxylase.

A phosphatase that dephosphorylates PEP carboxylase has been identified in *Kalanchoe*, but this enzyme does not show diurnal fluctuations in activity. The activity of PEP carboxylase kinase, in contrast, does show significant diurnal periodicity: activity is high at night and negligible during the day. The kinase activity is controlled principally at the transcriptional level, with expression of its gene very high in the dark but undetectable in the middle of the day. This **circadian** regulation of expression of the PEP carboxylase kinase gene seems to be the cornerstone for efficient CAM operation. In *M. crystallinum*, PEP carboxylase kinase expression is also induced by salt stress.

Figure 7–34.
The ice plant (*Mesembryanthemum crystallinum*), a CAM plant.
(Courtesy of Tobias Kieser.)

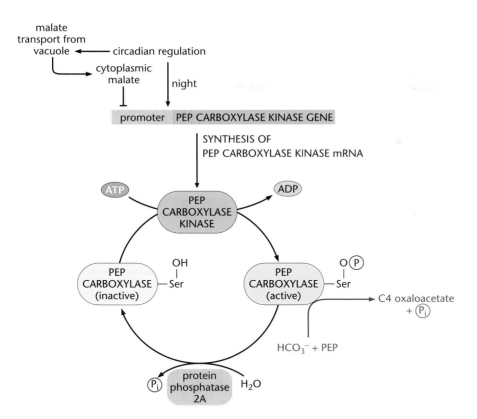

Figure 7–35.
Model for the control of PEP carboxylase activity in relation to day–night cycles in CAM plants. At night, malate is transported from the cytosol to the vacuole and expression of PEP carboxylase kinase is activated. PEP carboxylase kinase activates PEP carboxylase by phosphorylating it, so stimulating CO_2 fixation to form oxaloacetate, which is subsequently converted to malate. As malate accumulates, PEP carboxylase kinase expression is repressed, so switching off PEP carboxylase activity.

Regulation of expression of the gene encoding PEP carboxylase kinase may work through repression by malate. Thus, when malate is exported from the vacuole at the start of the day, it regulates PEP carboxylase kinase gene expression and reduces activity. Consequently, PEP carboxylase undergoes net dephosphorylation and becomes more sensitive to malate inhibition. PEP carboxylase activity is lowered until malate efflux into the cytoplasm falls off, when PEP carboxylase kinase activity increases with consequent increases in PEP carboxylase activity. In this way, malate transport across the tonoplast could be the primary target of the circadian oscillator regulating CAM.

For CAM to operate efficiently in water-stressed plants, and to allow carbon fixation under conditions of reduced transpiration, stomatal behavior must also be reversed relative to rhythms normally operative in C3 and C4 plants. Evidence from facultative CAM plants such as *M. crystallinum* suggests that the blue- and red-light photoreceptors controlling stomatal opening become inactivated with the onset of CAM. Control of stomatal opening may then be transferred to ABA or to changes in carbon dioxide concentration.

Plants that tolerate extreme desiccation have a modified sugar metabolism

Some plants can tolerate severe desiccation. Most plants in this group belong to the lower plant orders, including mosses, lichens, and ferns. The moss *Tortula ruralis* desiccates within 1 to 2 hours but can resume a fully hydrated state within 90 seconds. Some angiosperms can be dried to within 2 to 5% of their normal water content, and yet recover rapidly on rehydration. For example, the resurrection plant *Craterostigma plantagineum* loses water in a slow and controlled manner over several days and can complete water uptake and recover from a dried state in 12 to 15 hours (Figure 7–36).

It is not known precisely which special features of *C. plantagineum* adapt it to tolerate extreme desiccation. Dehydration induces massive changes in gene expression by ABA-dependent and ABA-independent signaling pathways, giving rise to *de novo* synthesis of many of the same proteins induced in response to water stress. LEA protein

untreated

dried

rehydrated

Figure 7–36.
A resurrection plant (*Craterostigma plantagineum*) in its natural (untreated), dried, and rehydrated forms. (B, courtesy of Dorothea Bartels.)

and aquaporin levels increase, as do the levels of many proteins involved in sugar metabolism. All of these proteins make contributions to the desiccation tolerance of *C. plantagineum*. In addition, in fully hydrated plants as much as 50% of the dry weight of leaves is made up of an unusual eight-carbon sugar, 2-octulose. Upon dehydration stress, 2-octulose is converted very rapidly to **sucrose**. The accumulation of sucrose in water-stressed plants generally serves an osmoprotective role, and the high levels of 2-octulose characteristic of hydrated *C. plantagineum* may allow for much more rapid conversion to sucrose during water stress than occurs in other plants.

Sugars protect cellular components from dehydration damage by forming a "glass state" (much like the sugar remaining in a cup after the tea has been drunk). In this state, rates of molecular diffusion and chemical reaction are greatly reduced. The "glass" prevents cellular collapse, and the cytoplasm resembles a solid, brittle material but retains the disorder and physical properties of a liquid. The composition and concentration of the sugars affect formation of the glass state, and sucrose is particularly suited to glass formation. Other sugars, including trehalose, are believed to stabilize and protect membrane structures.

Rehydration is often more damaging to plants than dehydration. Water replaces the sugar at the membrane surface, and cellular components are repartitioned from the membranes into the cytoplasm. This leads to ion leakage and membrane disruption. In species that recover from desiccation very rapidly, such as *T. ruralis*, these effects on membranes are transient, and a set of biochemical modifications buffers and protects membranes during rehydration. An array of novel genes is also expressed. In less rehydration-tolerant plants, it is often the **imbibition** (uptake of external water) accompanying rehydration that is lethal.

Physical mechanisms governing rehydration of resurrection plants are important determinants of the tolerance of these plants to desiccation stress. In *C. plantagineum*, all the induction of novel gene expression occurs during desiccation, including the synthesis of many protective proteins. The events during rehydration contribute only to metabolic recovery.

Many plant species adapted to arid conditions have specialized morphology

Plants that grow in arid environments are called **xerophytes**, from the Greek meaning "liking dry conditions." In many arid environments, the dominant vegetation is composed of drought-tolerant **perennials**. These are plants with specialized morphology of two major types: succulents and nonsucculent perennials.

Succulent plants (Figure 7–37) tolerate the water deficit caused by either drought or salinity stress by storing water. Succulence may occur in roots (e.g., *Fouquieria* spp.),

Figure 7–37.
Succulent plants. (A) The ponytail palm (*Nolina recurvata*), with succulent roots.
(B) The candelabra tree (*Euphorbia candelabrum*), with succulent stems.
(C) Living stones (*Lithops gesincoe*), with succulent leaves.
(A–C, courtesy of Tobias Kieser.)

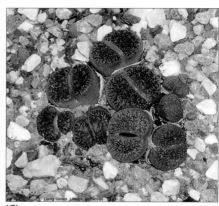

(A) (B) (C)

in stems (e.g., cacti and spurges), or in leaves (e.g., *Agave, Aloe, M. crystallinum*, house leeks, and stonecrops). Most succulents store water in parenchymatous cells with enlarged vacuoles. Transpiration rates are low, and thick **cuticles** further reduce water loss. Crassulacean acid metabolism is common in succulent plants.

Cacti have shallow root systems. They differ in this respect from many other xerophytes, which have extensive root systems to tap deep reserves of water (*Mesquite* roots may be 65 feet, or 20 m, long or more). The shallow roots of cacti allow them to tap the transitory surface moisture of the soil, a resource not used by other desert perennials. In some species the roots are **drought deciduous**: the rootlets drop off (abscise) during drought, then regrow rapidly after rain.

Nonsucculent perennials also show distinctive morphological adaptations to arid environments (Figure 7–38). These adaptations include thick cuticles that reduce water loss, stomata sunk in grooves or pits, copious production of trichomes (hairs), and reduction of leaf surface area.

Thickened cuticles not only reduce water loss by transpiration but also prevent leaf damage and breakage under wilting. Some plants have shiny cuticles, which reflect light and reduce the heat load of the leaves. Increased numbers of trichomes may play a role in water retention by trapping a layer of stationary air next to the leaf surface. This air-trapping function is probably more significant in species with hairs concentrated in recessed stomata (e.g., *Oleander*). Hairs lining the cavity above sunken stomata trap a column of damp air in the microclimate of the guard cells, limiting transpirational water loss when the stomata open for carbon dioxide uptake. Trichomes are also important because they reflect light, thus reducing leaf temperatures and so transpiration rates.

The leaves of many nonsucculent perennials of arid environments are highly reduced in surface area. This limits water loss through the leaf surface. Some xerophytes have no leaves, and carbon fixation is restricted to stems or petioles. Asparagus has reduced leaves (scales) from which photosynthetic leaflike branches arise. The water-use efficiency of a stem or petiole structure is greater than that of a flattened leaflike structure

Figure 7–38.
Morphological adaptations to drought.
(A) Section of agave (*Agave americana*) leaf, showing thickened cuticle and sunken stomata. (B) Micrographs showing increased numbers of trichomes on the underside of leaves of wild Brassica species (*B. incana*) in (iii) compared to *B. rapa* (i) and a cross (*B. rapa* x *B. incana*) shown in (ii). Varieties with increased drought tolerance have more trichomes on their older leaves, which may reduce water loss. (C) Section of oleander (*Nerium oleander*) leaf, showing thick cuticles and stomata confined to pits lined by trichomes (hairs). (D) Photograph showing small leaves on a spiny tree (*Alluaudia procera*) of the forests of Madagascar. (B, courtesy of Ruth MacCormack; D, courtesy of Guenther Eichhorn.)

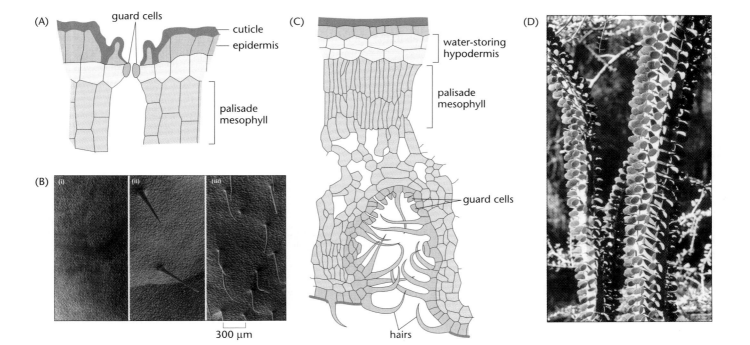

under conditions of water deficit. Tissues are often reinforced with **sclerenchymatous cells**, which make the tissues more resistant to shrinkage, an important feature when plants are under threat of wilting. Spines are also common on xerophytic plants, and in species such as gorse (*Ulex europaeus*) the density of spines may be related to water availability; more spines form under water deficit, and more foliage leaves form when water is available.

Many desert plants are drought deciduous, losing their leaves under dry conditions. For example, *Salvia mellifera* loses 90% of its leaves under water stress. Those that are retained are smaller than those lost, and the retained leaves are rotated so that their white undersides become exposed. Again, the white undersides reflect light, lowering plant temperature and conserving water. Some species lose all their leaves under dry conditions. For example, *Fouquieria splendens*, growing in the deserts of the southwestern United States and Mexico, has no leaves for most of the year. The tissue of the stem lying beneath the cork between leaf bases contains chloroplasts, which fix carbon dioxide. Net carbon fixation takes place only as the leaves rapidly regrow after rainfall. The major function of the stem chloroplasts is probably to refix carbon dioxide lost in **respiration**.

Some desert plants retain their leaves during drought stress, but the shape or orientation of the leaves is altered. Surface area can be reduced by leaf folding, as in many desert legumes and grasses. For example, in Kentucky bluegrass (*Poa pratensis*), longitudinal furrows on the leaves contain enlarged cells with high water content (Figure 7–39). As water is lost, leaves lose turgidity, the furrow collapses, and the leaf folds. Other grasses with shallow root systems completely dry out in a dry season, a process called "aestivation." The leaves die back but are not abscised, and the mulch formed by the dead leaves protects young buds at the soil surface.

Some species growing in very arid environments obtain water through their leaves as well as through their roots. Atmospheric bromeliads such as *Tillandsia landbeckii*, found in the Atacama Desert of Peru and Ecuador, grow in regions that have no rainfall but have very dense fogs, as damp air from the sea encounters the dry air of the high, arid desert. These plants have no roots. They absorb water from the fog through specialized scale-like trichomes on their leaves. *Tillandsia usneoides* (Spanish moss), an **epiphyte** (i.e., a plant that grows on another plant), also absorbs water from aerial sources through dense, specialized trichomes. The trichomes are nail-shaped, consisting of a multicellular stalk that carries the flattened dead cells of the shield (Figure 7–40). Two "foot cells" are settled deep in the epidermis of the base of the stalk, making direct contact with the mesophyll cells, and a "dome cell" is situated at the top of the stalk. Water is drawn from the atmosphere through the dead cells of the shield and absorbed into polysaccharide in a chamber (separate from the cytoplasm) above the dome cell. Water is conveyed osmotically through the dome and stalk cells and distributed to the mesophyll by the foot cells. The walls of the stalk cells are heavily suberinized to restrict water flow to the **symplast** (the intracellular volume bounded by the plasma membrane and interconnected among

Figure 7–39.
Folding of leaves of Kentucky bluegrass (*Poa pratensis*). The leaf folds when the cells in the grooves lose turgidity.

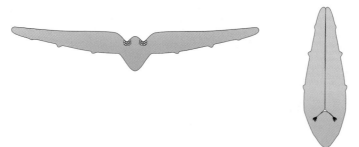

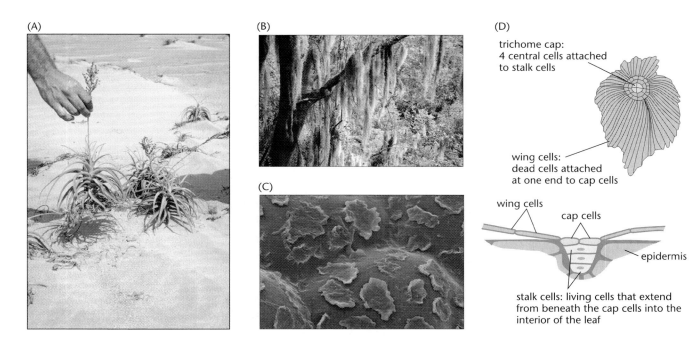

cells by **plasmodesmata**), and extra plasmodesmata facilitate the symplastic flow between cells of the trichome and the mesophyll.

Rapid life cycling during water availability is common in plants of arid regions

In arid regions with periods of higher rainfall, there is an abundant flora of small **annual plants** that complete their life cycles very rapidly. They often germinate only after several successive imbibitions during rain and when temperatures are most favorable for successful germination. The time from seed germination to seed set may be just a few weeks; many of these plants produce only a single pair of leaves before flowering. Typically these annuals have shallow root systems that tap surface moisture only. Some species survive periods of drought not as seed but as subterranean perennating (i.e., surviving from one season to another) organs (e.g., *Welwitschia*, found in the Namibian Desert; Figure 7–41). As with the seeded plants, these storage organs produce only two leaves during periods of water availability, before returning to the dormant state.

In the warm deserts of the southwestern United States and northwestern Mexico, many annuals complete their life cycles in the cool, wetter winters. They germinate between September and December, and their vegetative leaves tend to be organized in a **rosette**. The rosette creates a microenvironment that is both warmer and more humid than the aboveground air (which in winter months may be 0–10°C). As temperatures start to rise in spring, the stems bolt (elongate) and produce **cauline leaves** (leaves growing on an inflorescence). The rosette leaves die, and photosynthesis is supported by the green stems and cauline leaves.

The reproductive cycles of many plant species that survive arid conditions can also respond quickly to changes in climate. For example, species of *Lupinus* and *Eriogonum* enter into prolonged vegetative growth during favorable conditions. Even at the early stages of vegetative growth these plants produce a few flowers and fruits; thus the plant can reproduce even if the favorable conditions do not last. (Most annual plants undergo a developmental switch from vegetative to reproductive growth rather than being able to do both concurrently; see Section 5.5.) *Perityle emoryi* is a facultative annual that under favorable conditions produces alternately **determinate** inflorescences and lateral vegetative apices, a developmental strategy that increases reproduction under extreme but changeable conditions.

Figure 7–40.
Plants without roots. These plants can obtain water from mist. (A) *Tillandsia purpurea*, sitting on sand in Peru. (B) The epiphyte Spanish moss, *Tillandsia usneoides*. (C) Micrograph of water-absorbing trichomes of *Tillandsia usneoides*. (D) Diagram of a leaf scale used for water absorption in *Tillandsia usneoides*.
(A, courtesy of Michael O. Dillon; C, courtesy of Julian Collins and H. A. Sadaqat.)

Figure 7–41.
Welwitschia mirabilis, **growing in the Namibian desert.** (Courtesy of Dierk Wanke.)

7.4 SALT STRESS

Much of the world's land is saline and unsuitable for most plants. Saline soil has been estimated to make up about 6% of the earth's surface, though the figure depends very much on the definition of "salinity" being used. Salinity is often of profound local importance: for example, in Australia as much as 30% of the land is saline, and in Pakistan 26% of cultivated land is salt-affected. Salt is also the most significant toxicity problem for rice, the world's primary source of human dietary carbohydrate and protein. Modern agricultural practices, land clearance, inadequate or poor irrigation, and lack of drainage often increase soil salinity, so on a global scale the area of land subject to salinity problems is likely to increase to an estimated 50% of all arable land by 2050.

Salinity restricts the agricultural potential of land because almost all modern crops are derived from plants lacking the genetic basis for salt tolerance. Clearly, understanding the processes of plant responses to salt stress and the adaptive mechanisms of salt-tolerant species (**halophytes**) is an important objective in attempts to stem the loss of arable land and to increase food productivity through the use of marginal lands.

Salt stress disrupts homeostasis in water potential and ion distribution

In high concentrations, salt creates the problem of low water availability, and plant responses are similar to those for water deficit. These include the synthesis of compatible osmolytes (organic solutes; Table 7–2), LEA proteins, and aquaporins, as described in Section 7.3. High salt levels also impose ion stress on susceptible plants. Most plants (unlike most animals) do not require sodium, but potassium is an essential plant nutrient, and specific K^+ transporters usually maintain high cellular levels of this ion. Na^+ ions are similar to K^+ ions in radius and ion hydration energy; when external salt concentrations are high, the two ions compete for uptake via cellular transporters. High Na^+:K^+ ratios reduce plant growth and are eventually toxic. Many plants tolerant to high salt levels (in excess of 300 mM) can maintain a high cytoplasmic K^+:Na^+ ratio by sequestering Na^+ ions in their vacuoles (Figure 7–42).

Salt stress is signaled by ABA-dependent and ABA-independent pathways

Sometimes plants are subject to occasional or periodic salt stress, to which they respond with massive changes in gene expression and a series of physiological changes that keep salt out of the cells. Responses to salt stress, like those to drought, involve ABA-dependent and ABA-independent signaling pathways.

The perception of salt stress may involve the same mechanisms as the perception of water stress. The signal-transduction pathway controlling ion homeostasis has been

Figure 7–42.
Relative concentrations of organic solutes and inorganic ions in the vacuole and cytosol of a plant cell under salt stress.

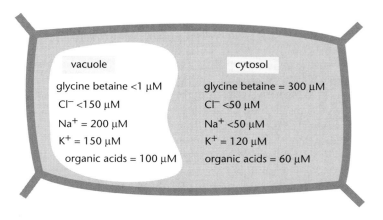

vacuole	cytosol
glycine betaine <1 µM	glycine betaine = 300 µM
Cl^- <150 µM	Cl^- <50 µM
Na^+ = 200 µM	Na^+ <50 µM
K^+ = 150 µM	K^+ = 120 µM
organic acids = 100 µM	organic acids = 60 µM

characterized in Arabidopsis through the study of mutant plants that are overly sensitive to salt (*sos* mutants) (Figure 7–43). An early response to salt stress is an increase in intracellular calcium levels, and *SOS3*, which encodes a calcium-binding protein related to the calcium sensors of animals, is the first gene in the signal-transduction pathway. The SOS3 protein binds calcium and undergoes a conformational change that facilitates its association with SOS2, a **serine/threonine protein kinase**. SOS2 contains a negative regulatory domain, but interaction with SOS3 relieves the auto-inhibition of SOS2 and induces its kinase activity. Although many kinases are induced in response to general stress or to osmotic or cold stress, SOS2 has a specific role in adaptation to high sodium (low potassium) stress.

The SOS3–SOS2 complex phosphorylates and activates SOS1, a plasma membrane Na^+/H^+ **antiporter** that exports Na^+ ions from the cell. *SOS1* is expressed principally in cells surrounding the xylem, suggesting that SOS1 may function to load Na^+ ions into the xylem for long-distance transport during salt stress. The SOS3–SOS2 complex can also activate a low-affinity Na^+ transporter on the plasma membrane to assist in Na^+ efflux, and the Na^+/H^+ antiporter on the tonoplast to increase vacuolar sequestration of Na^+ ions.

The Arabidopsis *sos* mutants demonstrate the importance of ion homeostasis during salt stress, in particular the role of SOS1. The short, calcium-dependent signal-transduction pathway is similar to signaling pathways governing ion homeostasis in animals and yeast, although some details are specific to plants. Similar signal-transduction pathways may link other forms of environmental stress (e.g., water deficit or low temperature) with their cellular response processes. Arabidopsis has numerous genes that encode proteins homologous to SOS2 and SOS3, and these are likely to signal other forms of stress.

Adaptations to salt stress principally involve internal sequestration of salts

Between 5000 and 6000 species of plants worldwide are halophytes, plants adapted to growth in saline conditions. Most other plants are **glycophytes**, which lack the genetic basis for salt tolerance. Salt tolerance has evolved independently many times: halophytes have developed within about half of the extant families of higher plants. Despite the diversity of origins, the salt-tolerance mechanisms that have evolved are remarkably similar.

Uptake of water against the low external water potential of saline soil is achieved through a controlled net uptake of Na^+ ions counterbalanced by other ions (such as Cl^-)

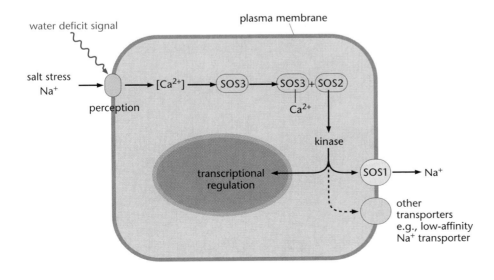

Figure 7–43.
The signal-transduction pathway for responses to salt stress in Arabidopsis.
Water deficit, as a result of high salinity, induces increased levels of intracellular calcium. Calcium is bound by the Ca^{2+}-binding protein SOS3, which undergoes a conformational change that facilitates its interaction with the kinase SOS2. The SOS3–SOS2 interaction relieves autoinhibition of SOS2 and induces kinase activity. The SOS3–SOS2 complex phosphorylates and activates SOS1, an Na^+/H^+ antiporter of the plasma membrane. Increased activity of SOS1 reduces the level of cytosolic Na^+ in saline conditions. The SOS3–SOS2 complex probably activates other transporters, including a low-affinity Na^+ transporter on the plasma membrane and an Na^+/H^+ antiporter on the tonoplast that acts to sequester Na^+ ions in the vacuole.

into cell vacuoles. This drives water uptake into the cells. The overall osmolarity of halophytic shoots is maintained at two to three times the osmolarity of the external (soil) solution. Despite these high levels of Na^+ (and usually Cl^-) flux through the cytoplasm, cytoplasmic concentrations of ions in halophytic plants are maintained at relatively normal, nontoxic levels. This is achieved by sequestering the salt in the vacuoles through the action of Na^+ and Cl^- importers on the tonoplast (Figure 7–44). To counterbalance the increased osmolarity of the vacuole, organic solutes accumulate in the cytoplasm (see Figure 7–42). These solutes are also believed to act as free-radical scavengers, osmoprotectants, and stabilizers of proteins and membranes.

There are several mechanisms by which the cells of halophytes take up Na^+ ions across the plasma membrane. Uptake of both Na^+ and Cl^- is most likely to occur via ion channels, but it may be supplemented by vesicular import following **pinocytosis** (uptake by invagination of the plasma membrane). Na^+ can also enter via a low-affinity K^+ transporter. From the cytoplasm, Na^+ is actively transported into the vacuole by a Na^+/H^+ antiporter in the tonoplast. The driving force is provided by tonoplastic H^+-ATPase and H^+-pyrophosphatases. Cl^- may follow the Na^+ passively into the vacuole by a specific **uniport** channel in the tonoplast. Some halophytes express the Na^+/H^+ antiport system constitutively, whereas in some salt-tolerant glycophytes it is induced in response to high external salinity; this induction is very rapid and may involve the activation of

Figure 7–44.
Ion movements and consequent water movements in plants. These diagrams compare responses to salt stress in a halophyte (*left*) and a salt-sensitive glycophyte (*right*).

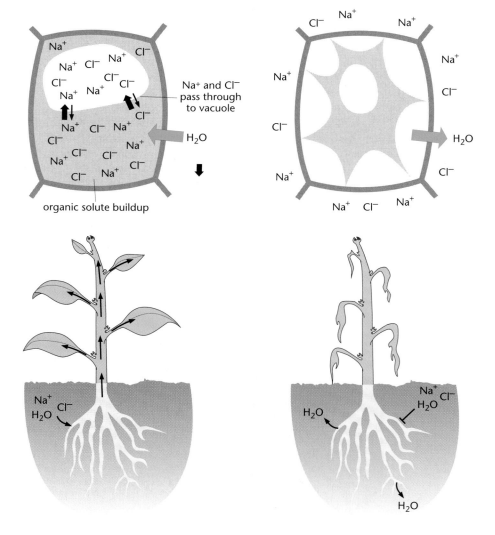

preexisting antiport proteins. Halophytes have specialized tonoplast membranes that withstand high vacuolar salt concentrations without ion leakage into the cytoplasm. Cation channels are usually closed to prevent leakage, and in some halophytes, such as *Suaeda maritima*, tonoplast membrane lipids contain high levels of **saturated fatty acids** that minimize permeability to NaCl.

The importance of the tonoplast antiport system to salt-tolerant plants has been demonstrated in Arabidopsis (a glycophyte) in plants that overexpress a Na^+/H^+ antiporter protein. High levels of this protein permit unimpaired growth on 200 mM NaCl, coupled to internal accumulation of Na^+ ions in the vacuoles. In contrast, in wild-type Arabidopsis the protein is not induced by saline conditions, less Na^+ is accumulated, and growth is considerably impaired at 200 mM NaCl. These results suggest that a primary difference between halophytes and salt-sensitive glycophytes may be the signaling pathways that control the activity of this antiport system.

Halophytes may also be able to expel Na^+ and Cl^- from the cytoplasm via a Na^+/H^+ antiport and H^+-ATPase system on the plasma membrane. This mechanism of salt tolerance is used in response to sudden salt shocks rather than for long-term adjustment to high salt (since it operates in opposition to the mechanisms promoting sequestration of salt in the vacuole). Because the cytoplasm of halophytic plants is not particularly high in anion or cation concentrations, their cytoplasmic enzymes and proteins are not especially tolerant of high salt levels.

The cytoplasmic accumulation of organic solutes in response to salt accumulation in vacuoles is another hallmark of halophytes and salt-tolerant glycophytes. The primary difference between accumulators and nonaccumulators is the level of such solutes that the plant can synthesize and whether synthesis is induced in response to water deficit. As noted earlier, the accumulation of nontoxic, "compatible" osmolytes also occurs in plants exposed to dry conditions, to provide osmotic balance and to protect cytoplasmic contents (see Section 7.3). Osmoprotective compounds fall into three major categories: quaternary ammonium (onium) compounds, amino acids, and polyols/sugars (Table 7–2).

Glycine betaine is the most commonly synthesized onium compound. It accumulates to high levels in a wide range of plant families, suggesting synthesis is a ubiquitous plant feature. Indeed, even species that do not accumulate glycine betaine synthesize trace amounts of the compound, suggesting that the biosynthetic pathway is intact in all plants. Glycine betaine is synthesized from serine via a choline intermediate (Figure 7–45). The enzymic steps involved in the conversion of serine to choline vary from one species to another, but the synthesis of glycine betaine from choline follows a common route: a two-step oxidation via betaine aldehyde as an intermediate. Glycine betaine has osmoprotective and **cryoprotective** effects because, unlike salt, it decreases osmotic potential without disturbing macromolecule–solvent interactions.

Other betaines accumulate in some plant species in response to salt stress; for example, *Limonium* species synthesize β-alanine betaine from β-alanine. *Wedelia biflora* synthesizes tertiary sulfonium compounds (e.g., β-dimethylsulfoniopropionate, DMSP) from methionine and accumulates them in the cytosol. Some halophytes are able to accumulate more than one osmoprotectant, depending on their nutritional status. For example, species of *Spartina* may accumulate DMSP in response to high salinity, but if nutrient levels, especially nitrogen levels, are high, the plants will preferentially accumulate glycine betaine.

Not all halophytic species accumulate onium compounds in response to salt stress. Some, such as the ice plant (*M. crystallinum*), accumulate polyols such as methylated inositols: pinitol and ononitol. Synthesis of pinitol and ononitol in *Mesembryanthemum* is induced by both salt stress and low temperature. Other plants, including Arabidopsis, accumulate proline (see Section 7.3).

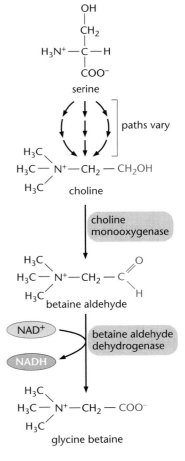

Figure 7–45.
Biosynthesis of glycine betaine from serine in plants. All plants share a common pathway from choline to glycine betaine.

Although synthesis and accumulation of organic solutes in response to salt stress is common to all halophytes and salt-tolerant glycophytes, genetically engineered increases in such solutes in salt-sensitive glycophytes lead to only marginal increases in salt tolerance. These observations suggest that osmolyte accumulation is not a major bottleneck limiting salt tolerance in glycophytes.

Other mechanisms that facilitate water uptake in halophytes include adjusted expression of the genes encoding water channels (aquaporins). Researchers have compared the expression of aquaporin genes in the responses of a glycophyte (Arabidopsis) and a halophyte (*Mesembryanthemum*) to salt stress. In Arabidopsis some genes encoding plasma membrane aquaporins are induced immediately upon salt stress. In *Mesembryanthemum*, aquaporin transcripts initially decline in response to salt stress, and increase only as leaves regain turgor and increase their cytosolic pinitol levels. These differences imply that water permeability is closely regulated in halophytic cells and that halophytes and glycophytes differ in their perception of stress signals or in the processing of such signals.

Physiological adaptations to salt stress include modulation of guard cell function

Halophytic plants accumulate salt, but many do not have obvious morphological adaptations to deal with this extra load. Salt accumulates in shoots, and the rate of transpirational water loss from leaves influences this accumulation. **Apoplastic** accumulation of salt (i.e., occurring through the region outside the plasma membrane, mainly through cell walls), which could result from high transpiration rates, could cause cell dehydration and collapse. Reduced rates of transpiration and water loss from shoots therefore facilitate the accumulation of salt in vacuoles of the shoot, where it causes less damage to the plant. Several halophytic species have modified stomatal responses to saline conditions, which, in turn, influence transpiration rates. In glycophytes, Na^+ ions stimulate stomatal opening, but in halophytes Na^+ ions promote stomatal closure.

Comparison of guard cells from closely related glycophytic and halophytic species has revealed comparable sets of plasma membrane ion channels, particularly K^+ inward and outward rectifying channels (see Section 3.5), that regulate stomatal aperture. Prolonged exposure of guard cells of halophytic plants to Na^+ ions results in inhibition of the K^+ inward rectifying channel, thus promoting stomatal closure. The difference in response of halophytes and glycophytes to Na^+ ions results from differences in the signaling between Na^+ ions and the K^+ inward rectifying channel. This signaling pathway probably involves increases in cytosolic Ca^{2+}, which inhibit the K^+ inward rectifying channel—much like the action of ABA in promoting stomatal closure in response to water deficit (see Section 7.3).

Morphological adaptations to salt stress include salt-secreting trichomes and bladders

Some plants have morphological features that facilitate their survival in high-salt environments, especially high-salt soils. Many halophytes are succulents and have adapted leaf morphology to facilitate water storage. This stored water can be used in times of water deficit to keep the cells, particularly photosynthetic cells, hydrated to maintain carbon fixation. In the leaves of halophytic succulents such as prickly saltwort (*Salsola kali*; Figure 7–46) and sea-blite (*Suaeda maritima*), most of the leaf tissue consists of large thin-walled cells in the center of the leaf that store water. The photosynthetic mesophyll cells are arranged around the periphery of this aqueous tissue.

In the silver goosefoot (*Obione portulacoides*), a central cellular water reserve in the leaf is supplemented with modified epidermal hairs that swell to form bladders (Figure 7–47).

Figure 7–46.
Prickly saltwort (*Salsola kali*), growing on sand dunes in the Gironde, France.
(Courtesy of Erick Dronnet.)

In times of drought or exceptionally high external salinity, the adjacent photosynthetic cells withdraw moisture from both the central aqueous tissue and the surface bladder cells. The bladder cells contract. When water availability increases or external salt concentrations fall, the bladder cells refill with water to regenerate the reservoir. Some plants, such as salt-tolerant asparagus, develop succulence only under saline conditions.

Halophytes may also be adapted to highly saline environments by secreting salts through specialized secretory hairs called **salt glands**. The mechanisms by which these glands secrete salts vary; salts may be secreted directly through pores to the outside or into the vacuole of the gland cells. Salt glands are common among mangrove plants, including black mangrove (*Avicennia germinans*), but they are not universally present in halophytic plants; for example, red mangroves, which are equally salt-tolerant, lack salt glands.

In *Atriplex* species, multicellular (1–3 cells) hairs develop into salt glands (Figure 7–48). The terminal cell becomes a bladder cell and develops a large vacuole into which salts are actively transported. The cell is supported by one to two support cells. These cells are waterproofed by thick, cutinized walls and do not contain vacuoles, and thus they maintain directional salt and water flow. Salts accumulate in the vacuole of the basal epidermal cell by active transport of Cl^- ions. From here the salts pass in vesicles from the basal cell, through the support cells, to fuse with the vacuole of the bladder cell.

Other forms of salt gland secrete salts directly to the outside. The salt glands of halophytic grasses are simple in structure: they consist of two cells, a basal cell and a cap cell (Figure 7–49). The cuticle overlying the cap cell becomes detached from the cell wall to form a cavity. The basal cell is enlarged and constricted into a neck that protrudes slightly from the surface of the leaf and supports the cap cell. The wall of the neck region is thickened and highly lignified, making it impermeable to water and salts. The plasma membrane of the basal cell is highly invaginated and folded on the surface adjoining the cap cell. The basal cell is connected by numerous plasmodesmata to the adjacent mesophyll and epidermal cells. Water and salt pass into the basal cell cytoplasm from the surrounding cells by active uptake from the apoplast through the plasma membrane. Further apoplastic flow through the gland is prevented by the thickened walls of the neck. Salt solutions move into the cytoplasm of the cap cell by diffusive flow, and then are actively pumped into the space (the collecting chamber) between the cell wall and the cuticle. From there, the solution moves out through cuticular pores, driven by the pressure of accumulating salt solution in the collecting chamber.

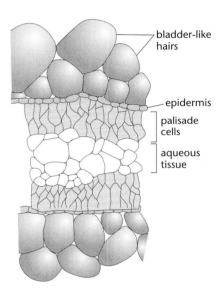

Figure 7–47.
Bladder cells in silver goosefoot (*Obione portulacoides*). The diagram shows a section of the leaf of this halophyte. The bladder cells provide a reservoir of water.

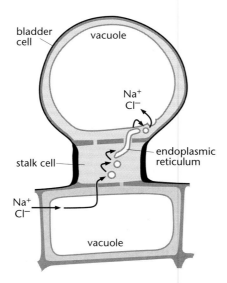

Figure 7–48.
Salt gland of the salt bush (*Atriplex lentiformis*). Cl^- ions and accompanying Na^+ ions pass by membrane transport, then by vesicular transport, into the vacuole of the bladder cell.

In halophytic eudicot species, salt glands tend to have a more complex, multicellular structure. In mangrove (*Avicennia* spp.; Figure 7–50), the salt gland is composed of 2 to 4 collecting or basal cells supporting a disk-like stalk cell, beneath 8 to 12 radially arranged secretory cells. A cuticle with numerous pores lies at the top of the gland over the secretory cells. The side walls of the stalk cell are heavily cutinized, whereas the transverse wall connecting with the basal cells contains many plasmodesmata. Salts accumulate in the basal cell, moving down a concentration gradient from the xylem, across the leaf. From the basal cell, salts move via the symplast of the stalk cell: apoplastic flow (and backflow) is prevented by the heavy cuticle of the stalk cell walls. Salts are actively secreted from the secretory cells into the space beneath the cuticle. This space contains material, probably **pectin**, that functions as a flow channel, absorbing the salts as they are eliminated from the cytoplasm of the secretory cells and directing them out through cuticular pores.

In some halophytes, such as sea lavender (*Statice*), salt may be exuded via **hydathodes** (see Section 4.10). Hydathodes are special organs that develop on the leaf margins, often at the end of principal veins. They form early during leaf development and exude excess water that might otherwise damage the young leaves in the highly humid environment of the developing bud. In some halophytic plants this water-exudation mechanism has been adapted to provide a means of removing excess salts.

Not all halophytes have salt glands, but in those that do there is a strong association between density of salt glands, Na^+ secretion rate, and salinity tolerance. In these

Figure 7–49.
Salt gland of Bermuda grass (*Cynodon dactylon*). This diagram shows the movement of salt from basal cell to cap cell. The basal cell collects salt by apoplastic flow and through numerous plasmodesmatal connections with the surrounding leaf mesophyll cells. Sodium and chloride ions move passively through the apoplast (*black arrows*) to the partitioning membrane of the basal cell. Here, they cross into the cytoplasm of the basal cell by energy-requiring transmembrane flux (*blue arrows*). The salts enter the cap cell by passive flow through the symplast (*red arrows*), then move by pressurized flow from the collecting chamber to the outside.

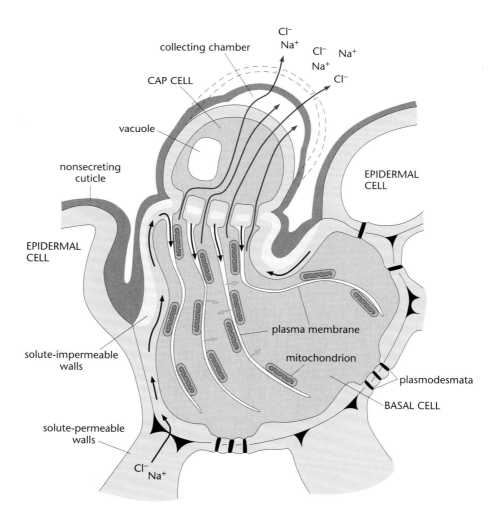

species, salt excretion may also perform subsidiary adaptive roles, such as light reflection by surface salts on the leaves and stems of desert halophytes, which reduces plant temperature.

Osmotic stress stimulates reproduction in some halophytes

Plants growing on saline soils often limit their periods of active growth to times when salinity is at a minimum. In the rainy season, rainfall dilutes the salt concentration in the soil and washes salts downward (provided there is adequate drainage), reducing the effective water deficit imposed by salt stress. When annual halophytes such as *Lasthenia glabrata* are exposed to periods of increased osmotic stress (dry periods), reproductive activity is stimulated. The osmotic stress induces hormonal changes that direct a shift in emphasis from vegetative to reproductive development a few weeks after the onset of osmotic stress.

For most plant species, germination is inhibited by high salt concentrations or low osmotic potential of the external medium. This may be a consequence of seed dormancy being linked to tissue desiccation. Seed germination of halophytic species is usually restricted to periods of minimum salinity, such as times of high rainfall. Another mechanism that overcomes the difficulty of germination in salty conditions is **vivipary**, in which seedlings germinate before the fruit has abscised; thus the new plants encounter their first saline soils as seedlings (Figure 7–51). This mechanism also overcomes problems of habitats with high water content and is common in mangrove shrubs, such as *Rhizophora mangle*, which grow in salt water (sea water is typically 350 mM NaCl).

7.5 COLD

Low temperature is similar to water deficit as an environmental stress

For most plants outside the tropics, low temperatures that present a real problem are those that induce freezing. Ice formation starts in the intercellular spaces, because the extracellular fluid has a lower solute content than cells. This causes a fall in the water potential outside the cell, promoting the movement of unfrozen water down the gradient and out of the cells. At −10°C, more than 90% of the osmotically active water moves out of a plant cell. Consequently, the major challenge for plants at temperatures low enough to induce freezing is water deficit. Many of the physiological adjustments

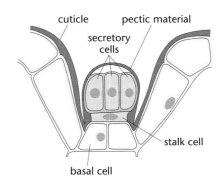

Figure 7–50.
Salt gland of the mangrove (*Avicennia marina*). Salt is collected by the basal (collecting) cells and passes by symplastic flow through the stalk cell to the secretory cells. Between the cuticle and the secretory cells is dense pectic material that acts as a flow channel for the secreted salt solution.

(A)

(B)

Figure 7–51.
Viviparous seedling of a mangrove shrub (*Rhizophora mangle*). (A) Radicles emerging from fruit still attached to the parent tree. (B) Maturing seedling. (A and B, courtesy of Gavin W. Maneveldt.)

that plants make to survive periods of low temperature are very similar to adaptations to drought or salt stress, discussed in Sections 7.3 and 7.4. Indeed, many of the genes induced by acclimation to low temperature are also induced by drought or salt stress.

Temperate plants acclimated by prior exposure to low temperatures are resistant to freezing damage

Many temperate plants can survive low temperatures (below freezing) by prior exposure to low, nonfreezing temperatures, a process known as **cold acclimation**. For example, the temperature at which 50% of plants die (LT_{50}) is improved from $-6°C$ for nonacclimated winter rye to $-21°C$ for acclimated rye, from $-6°C$ to $-10°C$ for spinach, and from $-3°C$ to $-10°C$ for Arabidopsis acclimated by exposure to $4°C$ for 2 days, before freezing.

Cold acclimation allows plants to survive longer spells of cold and colder conditions; it is a cumulative process that can be stopped, reversed, or restarted. For example, the threshold temperature for onset of cold acclimation in cereals is $10°C$, with an optimum of $3°C$. As temperatures fall below $10°C$ there is an increase in the rate of cold acclimation. Different parts of a plant may cold-acclimate independently. Once acclimation has been completed, cold hardiness remains as long as temperatures remain below freezing. Temperatures above $10°C$, however, lead to a rapid loss of cold hardiness. This is why summer frosts can be so damaging, even for cold-tolerant species.

The principal cause of damage in plants exposed to low temperatures without acclimation is membrane damage. Acclimation involves a series of physiological changes that have an additive effect to protect cellular membranes from freezing damage. Membrane damage arises from expansion-induced lysis (which occurs on thawing), phase transition (a sharp change in the physical properties of membranes, including fluidity and permeability), and loss of osmotic responsiveness. The plasma membrane typically contains high proportions of **phospholipids** (phosphatidylcholine and phosphatidylserine), **sterols** (free and glycosylated forms), and **cerebrosides** (lipids composed of ceramide and a single sugar residue). During cold acclimation there is an increase in the proportion of phospholipids and a concomitant decrease in cerebrosides in the plasma membrane. These compositional changes result in a decreased incidence of expansion-induced lysis and a decreased tendency for membrane lipids to fuse and rearrange to form pores. The changes also increase membrane hydration and decrease the curvature of membrane monolayers. Cumulatively, these changes reduce the likelihood of freezing-induced membrane damage and cellular impairment or death.

Some proteins induced during cold acclimation reduce the incidence of phase transition by stabilizing membranes likely to serve as nucleating sites for such changes (e.g., the chloroplast inner envelope). Levels of enzymes involved in the synthesis of sucrose and other simple sugars are raised; the resulting increased sugar levels stabilize membranes and protect against freezing-induced damage. In addition, some novel hydrophilic proteins and LEA proteins are induced during cold acclimation and protect membranes and cellular proteins against the effects of water deficit, as they do under drought and salt stress. In particular, these proteins reduce the likelihood of general protein denaturation at low temperatures, which occurs for reasons similar to those that destabilize proteins at high temperatures. The induction of chaperonins, a type of molecular chaperone (see Section 7.2), during acclimation also reduces cellular protein denaturation and consequent loss of function.

Osmolytes are synthesized during acclimation, a response analogous functionally to the synthesis of osmolytes under drought or salt stress (see Sections 7.3 and 7.4). Arabidopsis synthesizes and accumulates proline in response to low temperature. The *eskimo* mutant of Arabidopsis, which is constitutively freezing-tolerant (i.e., it does not require prior acclimation), accumulates enhanced levels of proline (see Section 7.3), suggesting that osmolyte induction is a mechanism contributing to cold acclimation. Other species accumulate different osmolytes, including glycine betaine (see Section 7.4) and soluble sugars (see Table 7–2), in response to low temperatures.

Exposure to low temperatures induces cold-regulated (*COR*) genes

Cold acclimation induces the expression of a battery of genes, collectively termed **cold-regulated genes (COR)**, which include many genes that are also induced by other forms of water deficit. It is likely that each *COR* gene contributes to freezing tolerance to a small degree. Some *COR* genes are also induced by stresses other than cold or water deficit (e.g., those involved in **anthocyanin** biosynthesis), and these genes probably play an indirect role in protecting against cold stress.

The function of some of the proteins encoded by *COR* genes has not been established biochemically, but many are predicted to be highly hydrophilic proteins. These probably have similar functions to LEA proteins; the effects of hydrophilic COR proteins on electrolyte leakage suggest that they stabilize the plasma membrane against freezing injury. Several are also induced in response to water deficit, where they may function in a similar fashion to reduce cellular damage. Overexpression of an LEA protein and a novel hydrophilic COR protein from spinach in transgenic tobacco slowed the rate of freezing-induced cellular damage. Proteins known as "antifreeze proteins" are also synthesized during cold acclimation. These proteins are secreted into the apoplast and inhibit the nucleation of ice crystals or their re-formation after a freeze–thaw cycle. Generally there is a good correlation between cold tolerance and accumulation of antifreeze proteins, although it is unlikely that antifreeze protein accumulation alone can determine the lower temperature limits for plant survival.

The *COR15a* gene of Arabidopsis plays a direct role in promoting freezing tolerance. Constitutive expression of *COR15a* in transgenic plants increases the freezing tolerance of nonacclimated plants by as much as 2°C in the range −4°C to −8°C. The COR15a protein is targeted to chloroplasts, where it is thought to decrease the rate at which the membranes of the chloroplast inner envelope undergo phase transition at low temperature.

Some *COR* genes encode molecular chaperones (chaperonins) such as HSP90 and HSP70-12, which act to prevent protein denaturation at low temperatures (see also Section 7.2). Other *COR* genes include those encoding proteins potentially involved in low-temperature signaling, such as MAP kinase (MAPK) and MAP kinase kinase kinase (MAP3K) (see Section 6.3), calmodulin-related proteins, and a series of transcription factors belonging to the AP2 family of DNA binding proteins. These transcription factors have been shown to play a central role in inducing expression of *COR* genes in response to cold; overexpression of these transcription factors induces expression of *COR* genes and enhances freezing tolerance in nonacclimated Arabidopsis plants.

Expression of the transcriptional activator CBF1 induces *COR* gene expression and cold tolerance

Three transcription factors have been characterized in Arabidopsis that induce *COR* gene expression. CBF1 is a DNA binding protein of the AP2 family of transcription factors specific to plants. It binds to a GCC motif (the CRT/DRE motif; Figure 7–52) in the promoters of many cold- and drought-induced genes, and functions as a transcriptional activator. Two other structurally related Arabidopsis proteins, CBF2 and CBF3, also bind to the same recognition motif, and the expression of all three genes is induced by low temperature.

Overexpression of *CBF1* in Arabidopsis results in constitutive expression of several *COR* genes, including *COR6.6*, *COR15a*, *COR47*, and *COR78*. Nonacclimated plants overexpressing *CBF1* show increased freezing tolerance (a 3.3°C increase in freezing

Figure 7–52.
Role of the CBF1 transcriptional activator in Arabidopsis. This scheme shows the domain structure of CBF1 and how it activates *COR* gene expression by binding to the CRT/DRE (GCC) motif in the promoters of *COR* genes.

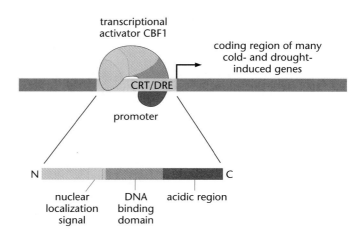

tolerance over a 7- to 10-day period). Similar results have been obtained by overexpression of the related genes *CBF2* and *CBF3*. Plants overexpressing *CBF1* are also more tolerant to drought, probably because many of the induced *COR* genes also serve protective functions in drought-induced water deficit. *CBF1*, *CBF2*, and *CBF3* are induced rapidly on exposure of plants to low temperature, and increases in their transcript levels can be detected within 15 minutes of cold treatment. The induction of *CBF* gene expression is thought to be transcriptional and mediated by other transcriptional activators, including ICE (**i**nducer of **C**BF **e**xpression). ICE is a bHLH transcription factor that activates *CBF3* expression, although there is evidence that other bHLH proteins induce the expression of the other *CBF* genes. Overexpression of *ICE* results in increased *CBF3* expression, but only under low-temperature conditions, suggesting that ICE is subject to post-transcriptional regulation responsive to low temperature in its regulation of cold responses. The induction of *CBF* gene expression in response to cold is negatively regulated by HOS1. *HOS1* encodes a RING finger protein probably associated with targeted protein turnover by the **proteasome**. The most likely target of HOS1 regulation is ICE itself, and because *HOS1* is constitutively expressed under normal conditions but is rapidly switched off in plants exposed to low temperature, it may regulate ICE activity at the post-transcriptional level in response to cold.

Although *CBF* overexpression gives rise to both cold and drought tolerance, *CBF* expression is not normally induced by drought. The drought response seems to be mediated by a related group of AP2 transcription factors, DREB2A and DREB2B (see Section 7.3). These are induced specifically in response to drought and also bind to the GCC (CRT/DRE) motif, thereby inducing *COR* gene expression in response to water deficit. Thus, independent signals operate through separate but related transcription factors to induce a common set of genes involved in protection of cells during water deficit, in response to low temperature or drought (Figure 7–53).

Figure 7–53.
Role of transcription factors in responses to cold and drought. This scheme shows the relationship between cold and drought signals and the transcription factors that induce activation of drought-responsive and cold-responsive genes. "p" indicates promoter regions of cold- and drought-responsive genes.

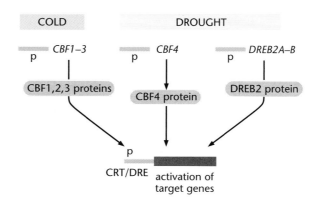

Low-temperature signaling involves increases in intracellular calcium

The signal-transduction pathway responding to low temperature is not well characterized, but there is a rapid increase in cytosolic free calcium in response to low temperature, due mainly to an influx from extracellular sources. This increase in cytosolic calcium is necessary for induction of *COR* genes and for freezing tolerance.

Increases in cytosolic free calcium may lead to changes in protein phosphorylation that induce the acclimation response. Some MAP kinases are specifically activated by low temperature, indicating that a MAPK cascade may form part of the signaling pathway for low-temperature acclimation.

ABA-dependent and ABA-independent pathways signal in response to cold

The CBF pathway operates independently of abscisic acid, but ABA levels do increase in response to low temperatures in a range of species, suggesting that, as with other water-deficit responses, an ABA-dependent pathway may also induce gene expression changes in response to low temperature. Application of ABA can enhance freezing tolerance, and, in Arabidopsis, mutations that affect ABA synthesis or ABA perception impair cold acclimation. However, the role of ABA in cold acclimation may be indirect. ABA levels increase only transiently in response to low temperature, whereas the acclimation response continues over a much longer period. Application of ABA to plants induces expression of water-deficit response genes that also protect plants against low temperature. Although some *COR* genes are dependent on ABA for their expression during low-temperature challenge, these tend to be genes that are induced only weakly in response to low temperature but much more strongly in response to water deficit. On balance, the role of ABA in cold response is less significant than its role in drought or salt stress responses.

Plant species of warm climates are chill-sensitive

Temperate species are challenged as temperatures drop below freezing point. However, many warm-climate species are damaged by low, nonfreezing temperatures (0–12°C). Included in this group are crops that originated in tropical climates such as maize, tomato, cucumber, and soya (soybean), many of which are now grown in temperate climates. The chill sensitivity of these crops can significantly compromise their productivity. Photosynthetic carbon fixation is limited at low temperatures in chill-sensitive plants, in part because low temperature reduces the reaction rates for carbon dioxide fixation and this limits the efficacy of sinks for the absorbed excitation energy from the light reactions. A rapidly reversible down-regulation of photosystem efficiency protects the photosynthetic machinery from photooxidative damage involving interconversion of xanthophyll pigments and formation of a trans-thylakoid electrochemical potential difference (see Section 7.1) (Figure 7–54). This can be seen in maize and tomato plants grown at low temperature in the light; additional absorbed light energy is dissipated as heat, and photosynthetic carbon fixation is limited. Carbohydrate metabolism itself is particularly sensitive to low temperature, and the primary limitation to photosynthetic carbon fixation between 0 and 10°C in crops such as tomato may result from a decrease in activity of two Calvin cycle enzymes, sedoheptulose 1,7-bisphosphatase and fructose 1,6-bisphosphatase, in the chloroplast stroma (see Section 4.2).

Vernalization and cold acclimation are closely linked processes in wheat and other cereal crops

Some plant species are adapted for periods of growth at low temperatures through a requirement for **vernalization**, exposure to a period of cold that is essential for later flowering and reproduction (see Section 6.4). Vernalization provides a mechanism to

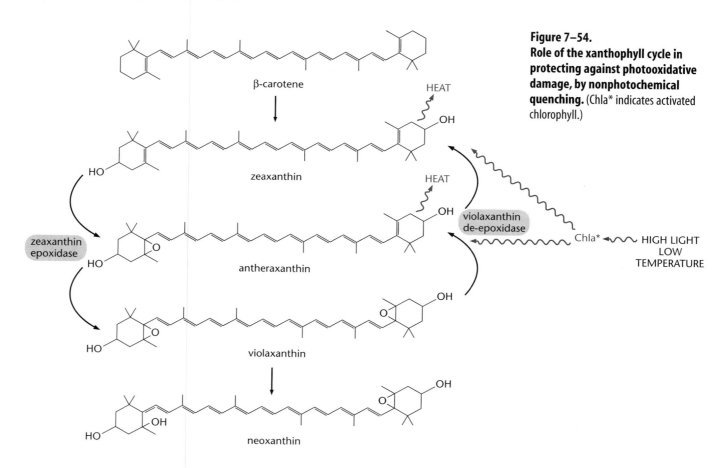

Figure 7–54.
Role of the xanthophyll cycle in protecting against photooxidative damage, by nonphotochemical quenching. (Chla* indicates activated chlorophyll.)

ensure that plants growing in temperate conditions flower at the appropriate time. This is important because flowering is particularly sensitive to low-temperature damage.

Vernalization is very important in wheat. Winter wheat will not flower unless exposed to a period of low temperature. Spring wheat has no vernalization requirement and will flower irrespective of its earlier growth conditions. Vernalization is governed by the gene *VRN1*, which is located on chromosome 5A in wheat. This part of the wheat genome also governs the plant's response to low temperature, which suggests that the processes of vernalization and cold acclimation are functionally related. Indeed, cold acclimation and vernalization are induced by temperatures in a similar range. Many cold acclimation traits map to this same chromosomal region, including antifreeze protein accumulation, sugar accumulation, increased ABA in response to low temperature, unsaturated phospholipid synthesis, prostrate growth habit, and flowering time.

Prolonged exposure of wheat to low temperature eventually leads to "vernalization saturation," and no further advancement in flowering results from longer exposure. Prolonged exposure of winter wheat to low temperature also results in a loss of cold tolerance. This also suggests a direct association between the ability of wheat to cold-acclimate and the completion of vernalization. It is possible that tolerance to low temperature is a function of the degree and duration of *COR* gene expression and that vernalization genes determine the duration of *COR* gene expression.

7.6 ANAEROBIC STRESS

Too much water creates an environmental stress for terrestrial plants in the form of oxygen deprivation. Wetlands account for 6% of the world's terrestrial habitats. Very wet soils are common in large parts of the world, and **anaerobic** stress is as much a problem for agriculture and forestry on flood plains and following irrigation as it is for natural plant populations. Rice, the most important crop of tropical and semi-tropical climates, is cultivated predominantly on flooded, anaerobic soils. In northern latitudes, water-logging is common during the winter, and low oxygen levels may be reinforced by ice crusts that prevent oxygen diffusion—for example, in tundra. Oxygen shortage can also be a problem in warmer climates where higher rates of respiration by soil microflora generate anaerobic conditions more rapidly. Permanent water-logging prevails in bogs and swamps such as the Okefenokee Swamp of Georgia, the Bijou of Louisiana, the cypress stands of Florida, and the Great Dismal Swamp of Virginia and North Carolina (Figure 7–55). These swamps are dominated by woody species such as the bald cypress (*Taxodium distichum*) and there are few **herbaceous** plants. Here we look at how plants cope with temporary or permanent shortages of soil oxygen.

Water-logging is a cause of hypoxic or anoxic stress for plants

When soils are flooded, air spaces and air pockets in the soil are lost. For plants, hypoxia (oxygen limitation) starts when gaseous oxygen levels in the soil fall below 50 mmol/m^3. Generally, after the onset of flooding (i.e., submergence), the levels of available oxygen decline by 60% in an hour and 95% in a day, although the precise rates are influenced by soil type and prevailing temperatures. Gaseous oxygen is displaced from the soil because of its low solubility and low rate of diffusion in water, and because soil microorganisms tend to consume all the readily available oxygen. Although all plants are obligate aerobes (i.e., they cannot live without oxygen), some species can survive or even grow with some or all of their organs in **anoxic** (oxygen-free) surroundings. Most plant species are tolerant to short-term hypoxia (low oxygen). Pretreatment, or acclimation, of plants to hypoxia creates greater tolerance to sustained hypoxia but not to anoxia (complete lack of oxygen).

Hypoxia is signaled by a Rop-mediated signaling pathway involving transient induction of ROS

Low oxygen availability is signaled by Rop (RHO of plant) **G proteins** that are active in their GTP-bound forms and inactive in the GDP-bound forms, in a pathway similar to the Rho-GTP-dependent (**Rho-GTPase**) pathway in mammals. Despite hypoxia involving reduced oxygen availability, the output of the Rop-signaling pathway is probably a transitory increase in reactive oxygen species, which may result from inhibition of mitochondrial respiration. A transitory increase in ROS is necessary for induction of alcohol dehydrogenase gene (*ADH*) expression. Reactive nitrogen species may also signal under hypoxic conditions.

Figure 7–55.
Swamps. (A) The Okefenokee Swamp in Georgia. (B) The Great Dismal Swamp in North Carolina. (A, courtesy of John A. Lawrence, Jr.; B, courtesy of Brian Thomas, TRC.)

(A)   (B)

Anoxia induces shifts in primary metabolism

In plants facing short-term anoxia, the major problem is blocking of the final oxidative step of respiration, the addition of electrons to oxygen to form water. This limits ATP production and the recycling of NAD/NADP intermediates that are essential for continued metabolism, and hence limits growth (see Section 4.5). Under these conditions, plants generate ATP and recycle **NAD** by switching from respiration to **fermentation**. There are three major fermentation products: lactate, ethanol, and alanine (Figure 7–56). All

Figure 7–56.
Metabolic pathways active under short-term anoxic or hypoxic conditions in plants. The pyruvate produced in glycolysis undergoes fermentation reactions and NAD is recycled. The three major fermentation products are ethanol, lactate (lactic acid), and alanine. Ethanol is the primary product during anoxic stress. Enzymes: SS, sucrose synthase; GPI, glucose phosphate isomerase; F1,6P ALD, fructose 1,6-bisphosphate aldolase; GAPDH, glyceraldehyde 3-phosphate dehydrogenase; LDH, lactate dehydrogenase; PDC, pyruvate decarboxylase; ADH, alcohol dehydrogenase; AlaAT, alanine aminotransferase.

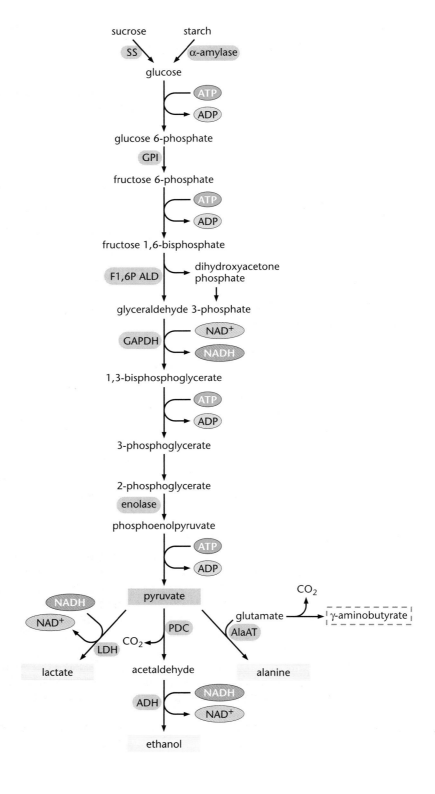

are formed from pyruvate, the end product of **glycolysis**, and formation of all three uses NADH (directly or indirectly), thus recycling NAD.

Lactate (lactic acid) is formed from pyruvate by lactate dehydrogenase. Lactate accumulation is potentially toxic, as it can acidify the cytosol. Consequently, **lactate fermentation** is usually an early response to anoxia and is soon replaced by ethanol production. Indeed, in some species the cytosolic acidification caused by lactate accumulation induces the first enzyme of ethanol biosynthesis, **pyruvate decarboxylase**, and reduces lactate dehydrogenase activity. Ethanol is the principal fermentation product in plants exposed to anoxic stress. It readily passes through membranes and can be lost from cells to the soil, reducing problems of cellular toxicity. Ethanol is synthesized in two steps: first, conversion of pyruvate to acetaldehyde by pyruvate decarboxylase, then reduction of acetaldehyde to ethanol by alcohol dehydrogenase, with the concomitant regeneration of NAD. The third fermentation product, alanine, is synthesized from pyruvate and glutamate, a reaction catalyzed by alanine aminotransferase. Alanine accumulates in plants experiencing low oxygen conditions, but not to the same extent as ethanol.

The principal metabolic adjustment to short-term flooding is the induction of the **ethanol fermentation** pathway. Although in some species lactate signals a switch to ethanol fermentation, in other species lactate fermentation continues alongside. An alternative switch may depend on the concentration of pyruvate in the tissues. Pyruvate decarboxylase (the first enzyme of ethanol synthesis) has a lower affinity for pyruvate than does pyruvate dehydrogenase (which under normal, aerobic conditions channels pyruvate into the **tricarboxylic acid cycle**; see Section 4.5). Therefore, under aerobic conditions, the affinity of pyruvate decarboxylase for pyruvate is too low for ethanol fermentation to be significant. When the tricarboxylic acid cycle is blocked under anaerobic conditions, the concentration of pyruvate increases and pyruvate decarboxylase activity becomes significant, switching the output of glycolysis to ethanol. In rice, a rapid switch to pyruvate decarboxylase and ethanol fermentation is favored by high basal pyruvate levels.

Of the two enzymic steps in ethanol fermentation, that catalyzed by pyruvate decarboxylase is thought to be rate-limiting. Various attempts have been made to improve the tolerance of crops to anoxia by increasing the activity of pyruvate decarboxylase, because many flood-tolerant species are known to have constitutively high levels of fermentation enzymes. Overexpression of pyruvate decarboxylase in tobacco did not enhance survival under oxygen stress, however, perhaps because tobacco roots already have considerable levels of the enzyme under anoxic conditions. In contrast, overexpression of pyruvate decarboxylase in rice roots did significantly increase ethanol production and plant survival after submergence. These data suggest that ethanol fermentation can play an important role in strategies to improve submergence tolerance, at least for some plant species.

In Arabidopsis, expression of several genes involved in fermentation pathways is induced by hypoxia, and the importance of fermentation to survival during anoxia has been demonstrated by mutational analysis. The gene encoding alcohol dehydrogenase (*ADH1*) is induced during anoxia, and mutation of this gene severely reduces the tolerance of plants to anaerobiosis. This underscores the importance of the ethanol fermentation route to survival during anoxia/hypoxia. Under hypoxic stress, the roots of Arabidopsis lines that have a mutated gene encoding alcohol dehydrogenase cannot acclimate to oxygen stress, but the shoots can. This suggests that acclimation in shoots can involve a mechanism not based on ethanol fermentation, but the significance of this non-ethanol-based acclimation pathway to shoots under normal conditions is not clear.

Arabidopsis has two genes that encode pyruvate decarboxylase: *PDC1*, most strongly expressed in roots and induced by hypoxia, and *PDC2*, expressed at a low level in both leaves and roots and not induced by hypoxia. Genes involved in the other fermentation pathways are also induced by low oxygen: lactate biosynthesis is up-regulated by increased expression of the genes encoding lactate dehydrogenase, and alanine biosynthesis is up-regulated by increased expression of genes encoding alanine aminotransferase.

The promoter of the Arabidopsis *ADH* gene includes a sequence motif called the **anaerobic response element (ARE)**, which controls the transcriptional response to low-oxygen stress. This element may be involved in general stress responses, as it may also regulate responses to cold and dehydration. In maize, the *ADH1* gene promoter contains a motif closely associated with the ARE, known as the **G-box**, which is also involved in the induction of gene expression in response to hypoxia. A protein that binds to the G-box is associated with hypoxic stress, and a G-box binding factor gene (*GBF1*) is induced just before *ADH1* induction. The activity of GBF1 is probably, in turn, regulated by phosphorylation. The signals inducing responses in gene expression to hypoxic or anoxic stress are not well defined, but are likely to involve calcium signaling. For example, calcium fluxes could induce changes in a signal-transduction cascade that targets phosphorylation of transcription factors that modulate *ADH1* gene expression.

Most varieties of rice are intolerant of complete submergence and die after a week of anoxic conditions. However, some varieties of Indica rice are more tolerant of complete submergence owing to a major quantitative trait locus, *Submergence 1* (*Sub1*). Three genes encoding AP2-type transcription factors (*Sub1A*, *Sub1B*, and *Sub1C*) lie at the *Sub1* locus. Expression of *Sub1A* is strongly induced by submergence, and polymorphisms in this gene correlate with submergence tolerance or intolerance in different varieties. Sub1A induces expression of *ADH* and confers increased submergence tolerance to less tolerant varieties. There is evidence that submergence-tolerant varieties were selected independently in flood-prone areas of India and Sri Lanka for enhanced activity of the *Sub1A* gene.

The genes that encode the enzymes involved in fermentation are not the only genes induced by hypoxic stress. About 20 different genes are induced by hypoxia in maize, and these genes encode about 70% of the new protein synthesized in hypoxia-stressed plants. These new proteins are predominantly enzymes of glycolysis and sugar phosphate metabolism. Glycolysis (which is one-eighteenth as efficient at producing ATP as oxidative respiration) is stimulated by increases in glucose indirectly derived from cleavage of sucrose by **sucrose synthase** (Figure 7–57), which is itself strongly induced by anoxia.

If sucrose is cleaved principally by sucrose synthase rather than **invertase** in anoxic tissues, the products are **UDP-glucose** and fructose rather than glucose and fructose. Glucose 1-phosphate is synthesized from UDP-glucose and **inorganic pyrophosphate** by UDP-glucose pyrophosphorylase. In addition, pyrophosphate can be used instead of ATP to convert fructose 6-phosphate to fructose 1,6-bisphosphate, by the action of pyrophosphate-dependent phosphofructokinase. Using these three enzymes—sucrose synthase, UDP-glucose pyrophosphorylase, and pyrophosphate-dependent phosphofructokinase—cells can achieve a net gain of 4 ATP per glucose over conventional glycolytic metabolism of sucrose using invertase, hexokinase, and phosphofructokinase. There is evidence for this shift toward sucrose synthase/pyrophosphate-based glycolysis under anoxic conditions in both maize and rice.

Aerenchyma facilitates long-distance oxygen transport in flood-tolerant plants

Plants tolerant to longer-term anoxia undergo distinctive developmental adjustments that do not occur in most nontolerant species. Principal among such changes is the formation of internal long-distance gas-transport pathways. These are created by cortical tissues that form gas-filled spaces known as **aerenchyma**. Many species adapted to

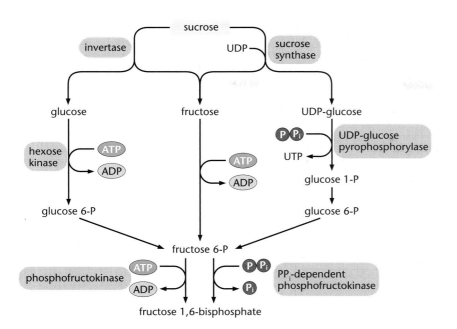

Figure 7–57.
Metabolism of sucrose under normal versus anoxic conditions. This scheme of sucrose metabolism and hexose activation (phosphorylation) emphasizes the importance of sucrose synthase, UDP-glucose pyrophosphorylase, and pyrophosphate (PP$_i$)–dependent phosphofructokinase in the metabolism of sucrose under anoxic conditions. This pathway has a higher net gain of ATP per glucose molecule than if sucrose is metabolized by invertase.

wetlands form aerenchyma constitutively in roots, leaves, and stems. In other species, such as maize, aerenchyma formation is induced in response to poor aeration. Aerenchyma consists of extracellular, gas-filled spaces within tissues. It is formed either by cell separation or by selective cell death. In addition, barriers to apoplastic water ingress are synthesized, such as impermeable **exodermis**, thickened cells surrounding the aerenchyma, and thickened **endodermis**.

Aerenchyma is formed by two different routes (Figure 7–58), "schizogenous" and "lysigenous." Schizogenous aerenchyma is formed by regulated patterns of cell expansion and cell separation that create spaces between the cells, as seen in the submerged

Figure 7–58.
Structure and development of aerenchyma. (A) Schizogenous aerenchyma in roots of *Acorus calamus*. An aerenchymatous space is indicated by the *white arrow*. (B) Development of lysigenous aerenchyma in maize roots: (i) 0.5-day-old root in 21% oxygen; (ii) 0.5-day-old root in 3% oxygen (*red arrow* shows developing aerenchyma); (iii) mature root in 21% oxygen; (iv) mature root with aerenchyma in 3% oxygen. (A, courtesy of Jean Armstrong.)

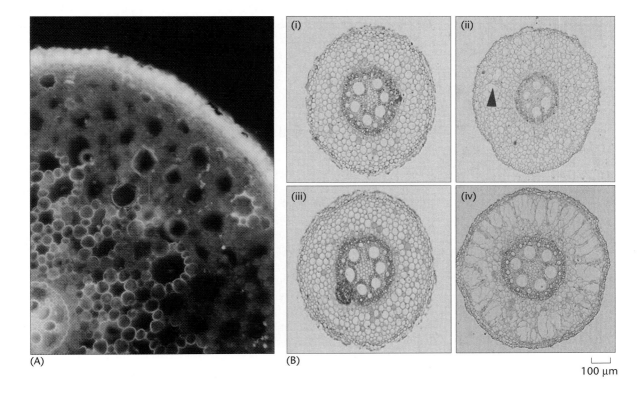

(A) (B)

100 μm

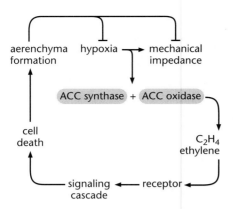

Figure 7–59.
Induction of cell death during formation of lysigenous aerenchyma. This cell death program of plants involves an ethylene signal-transduction pathway.

aquatic plant *Potamogeton pectinatus*. The mechanisms underlying this regulated cell growth and separation are not understood. Usually, schizogenous aerenchyma is formed constitutively in wetland species, but its formation can be promoted by hypoxia. Lysigenous aerenchyma arises from spatially controlled cell death that creates spaces in tissues. In maize and rice, for example, hypoxia induces death and lysis of cells in the midcortex of the root, and this expands radially to form spaces with bridges of cells between.

The cell death that occurs during lysigenous aerenchyma formation is programmed and is marked by early cellular changes, including invagination and shrinkage of the plasma membrane. Subsequently, chromatin condensation and DNA nicking are observed in dying cells. The organellar membranes remain intact. Finally the cell walls degrade, forming the air space, and wall degradation is associated with increases in carboxymethyl cellulase and xyloglucan endotransglycosylase activity. Although these cellular changes bear some similarity to **programmed cell death** (**apoptosis**) in animals, the ordering of events suggests that lysigenous aerenchyma is formed by a novel cell death program, partly resembling apoptosis and partly resembling cytoplasmic cell death in animals (Figure 7–59). The signals that mark some cells to die and others to survive are not known, but the result is a regular ordering of cells and spaces across the tissue.

The development of aerenchyma results in higher oxygen concentrations and deeper soil penetration of roots. Oxygen moves down its concentration gradient to the roots. Aerenchyma may initiate in the roots but usually extends into the shoots and leaves, thus maintaining a gas-space continuum. Generally, **secondary growth** inhibits aerenchymal function, so the efficacy of aerenchyma is limited to herbaceous species. However, some woody species adapt their secondary cortex to become aerenchymatous; for example, in the legume *Aeschynomene aspera*, secondary xylem is modified for this function. As an alternative to aerenchyma, many woody species of wetlands have extensive root systems to tap the more aerobic regions of the soil.

Production of lysigenous aerenchyma is induced in response to **ethylene** (see Figure 7–59). Ethylene concentrations rise under flood conditions, for at least three reasons: ethylene does not diffuse away from roots, oxygen deprivation stimulates expression of the genes required for ethylene synthesis, and abiotic production of ethylene increases in water-logged soils. Production of ethylene from *S*-**adenosylmethionine** is catalyzed by two enzymes: ACC (1-aminocyclopropane-1-carboxylic acid) synthase and ACC oxidase. Oxygen is required for ACC oxidase activity, so ethylene synthesis is inhibited by anoxia. Under hypoxic conditions, however, ACC oxidase is active and ethylene levels increase.

Water-logging is associated with other developmental adaptations that increase plant survival

The rhizomes of wetland species often remain dormant, enabling survival during prolonged periods of anoxia during the winter months. These rhizomes store large amounts of starch, **fructans** or free sugars, proteins, and amino acids. The large carbohydrate reserves and the amino acids provide translocatable sources of carbon and nitrogen that can be used when rapid growth resumes in the spring. Toxic gases that accumulate under water-logging are detoxified: ammonia by fixation as amino acids, which are stored, and hydrogen sulfide by assimilation of sulfide and its storage as **glutathione**. The accumulation of glutathione also facilitates the rapid onset of growth in spring by providing reserves of antioxidant. Flood-tolerant species may also accumulate superoxide dismutase, which prevents free-radical damage during regrowth (see Section 7.7). A comparison of two species of iris with different tolerance to flooding (Figure 7–60) showed that the tolerant species, *Iris pseudacorus*, synthesizes SOD anaerobically, but the sensitive species, *Iris germanica*, does not and consequently suffers significant free-radical damage to membranes on re-aeration.

(A)   (B)

Figure 7–60.
Iris species with different tolerance to flooding. (A) *Iris pseudacorus*, flood-tolerant. (B) *Iris germanica*, flood-intolerant. (A and B, courtesy of Tobias Kieser.)

Many species tolerant to hypoxia undergo rapid shoot expansion in response to slight oxygen stress, and this trait is correlated with ecologically identifiable species variations in tolerance to submergence. The primary stimulus for this type of expansion growth is ethylene, which increases as a result of both entrapment in flooded plants and increased synthesis in response to hypoxia. Ethylene promotes wall loosening by stimulating xyloglucan endotransglycosylase and cellulase activity. It may also induce wall expansion by promoting proton excretion from cells. In rice seedlings, rapid extension growth is restricted to the **coleoptile**; other organs show inhibited growth under anoxia. Ethylene may operate in conjunction with other phytohormones, especially gibberellic acid, in promoting shoot expansion.

An extreme example of anaerobic shoot extension is that of the fennel-leaved pondweed *Potamogeton pectinatus* (Figure 7–61). Elongation is powered by sugar release from starch-filled tubers. Growth may be as rapid as 120 mm over 6 days and occurs through a combination of cell expansion and cell division. Besides ethylene, auxin may be involved in promoting this growth response. Carbon dioxide from respiration acidifies the environment and further promotes extension growth.

Many plants synthesize **adventitious roots** in response to anoxia (Figure 7–62A, B). In some species, this involves outgrowth of preexisting **primordia**; in others, initiation of new primordia. These additional roots replace roots damaged by anoxic conditions, and they are more effective at transporting oxygen because they develop aerenchyma and grow in the surface areas of the soil where oxygen levels are generally higher. Adventitious roots may also provide mechanical support against the pressure from flood water. In maize and willow, flooding promotes outgrowth of preexisting primordia, and ethylene promotes the formation of these adventitious roots. In sunflowers, adventitious roots form by initiation of new primordia induced by auxin.

In species that grow in permanently flooded conditions, such as black mangroves (*Avicennia* spp.), extensions of the roots called "pneumatophores" grow vertically to protrude above the water level (Figure 7–62C). Pneumatophores allow oxygen to enter and diffuse to the roots down its concentration gradient via the large air spaces. Cypress trees (*T. distichum*), such as those growing in the Okefenokee Swamp or the Great Dismal Swamp, also develop outgrowths of their roots, known as "cypress knees." It is

(A)

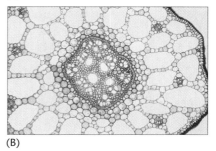

(B)

Figure 7–61.
The fennel-leaved pondweed (*Potamogeton pectinatus*). (A) Leaves and (B) cross section through root showing aerenchyma. (A, courtesy of Petr Krasa; B, courtesy of David T. Webb.)

Figure 7–62.
Morphological adaptations to flooding. (A) Adventitious roots of willow (*Salix europaea*). (B) Adventitious roots of red mangrove (*Rhizophora mangle*). (C) Pneumatophores of the black mangrove (*Avicennia germinans*). (B and C, courtesy of Wayne Armstrong.)

not known whether these structures are sites of enhanced oxygen diffusion. In some species adapted to flooding, such as willow and alder, oxygen transmission to the roots is enhanced by diffusion through cracks in the bark known as **lenticels**.

Limitations on shoot and leaf growth and on metabolism help promote survival under anoxic conditions. For example, flooding tends to induce stomatal closure in shoots, because water is less readily available via the anoxic-damaged roots. Stomatal closure is probably promoted by ABA under water-logged conditions, and ABA levels may increase in leaves as a result of reduced export of ABA by the phloem. Leaves of solanaceous plants (the nightshade family) often show **epinastic** leaf curvature (varying during the day) (Figure 7–63), slower leaf expansion, and accelerated leaf **senescence**. These changes reduce transpiration rates under water-logged conditions, thus increasing survival rates: since water conductivity in the roots is reduced as a result of tissue damage, high transpiration rates are incompatible with survival. Ethylene, which is synthesized in the shoots, induces these developmental changes. In anoxic roots, ACC synthesis increases through the induction of ACC synthase, but ethylene production is limited by the lack of oxygen for ACC oxidase. The extra ACC is transported from the roots to the shoot, where it acts as a signal of anoxic conditions (see Figure 7–63). Once in the aerobic tissues of the shoot, ACC is oxidized to ethylene by ACC oxidase and ethylene promotes changes in leaf movements, expansion, and senescence.

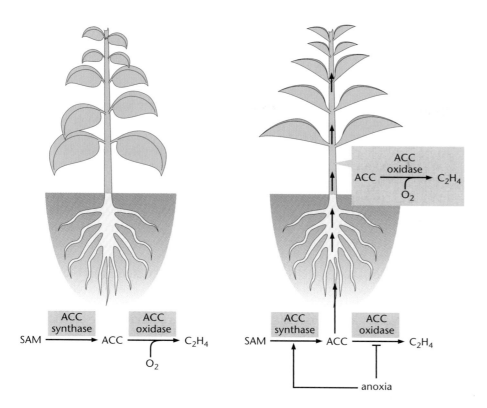

Figure 7–63.
Developmental responses of aerial tissues to flooding. These responses, including epinastic leaf movements, are controlled by ethylene (C_2H_4) produced from 1-aminocyclopropane-1-carboxylic acid (ACC). (SAM = S-adenosylmethionine.)

Plants synthesize oxygen-binding proteins under hypoxic conditions

All plants synthesize hemoglobin-like proteins that serve functions in regulating oxygen availability. The role of **leghemoglobin** in legume nodules in the binding of oxygen and its diffusion to nitrogen-fixing symbionts is well known (see Section 8.5). These leghemoglobins bind oxygen reversibly, but plants also synthesize nonsymbiotic hemoglobins that undergo stable oxygen binding and are unlikely to serve as oxygen carriers. Nonsymbiotic leghemoglobins can act as nitric oxide (NO) scavengers, by catalyzing the conversion of NO to nitrate in an NAD(P)H-dependent reaction. However, it is not known whether the role of nonsymbiotic leghemoglobins in hypoxia is to provide tolerance through removal of reactive nitrogen species (peroxynitrite, $ONOO^-$) or through suppression of NO signaling. It may be that the binding of oxygen by nonsymbiotic hemoglobins also maintains the energy status of cells in which energy demand exceeds supply by mitochondrial respiration.

Genes encoding nonsymbiotic hemoglobins are induced in response to hypoxic stress, and some of these genes are also induced by other stresses, such as osmotic stress. Some genes encoding nonsymbiotic hemoglobins are induced by other stresses such as cold but are not induced by anaerobiosis. This implies that nonsymbiotic hemoglobins may serve general roles in stress tolerance in addition to more specific roles in tolerance of hypoxia.

7.7 OXIDATIVE STRESS

Reactive oxygen species are produced during normal metabolism, but also accumulate under a range of environmental stress conditions

Reactive oxygen species are involved in all major areas of aerobic metabolism. Although they are produced during the normal operation of respiratory and photosynthetic electron transport, they can be toxic. This is because they initiate a cascade of reactions that can result in the production of highly destructive species such as hydroxyl radicals and lipid peroxides. Efficient antioxidant systems are therefore present that prevent initiation of these cascades.

Most types of stress result in increased production or accumulation of ROS. In particular, the combination of high light intensity and low temperature generates high levels of ROS as a result of the imbalance between increased light-driven excitation of the photosynthetic reaction centers and decreased energy dissipation through carbon fixation. Other stresses, including drought, salinity, and too much UV light, also result in disruption of electron transport chains (probably through membrane damage) and the generation of ROS. Exposure to strong oxidizing agents such as ozone generates ROS directly.

Reactive oxygen species are produced during the reduction of molecular oxygen. This occurs in three steps (Figure 7–64). The first step generates the short-lived and relatively nondiffusible hydroperoxyl radical ($HO_2^\bullet$) and the superoxide radical ($O_2^{\bullet-}$). Superoxide radicals are highly reactive and oxidize amino acids (histidine, methionine, and tryptophan) and lipids, thus causing protein and membrane damage. The second step reduces superoxide further to hydrogen peroxide (H_2O_2). This relatively long-lived species oxidizes sulfhydryl groups in particular. The third reduction gives rise to the most toxic species, the hydroxyl radical ($HO^\bullet$), which has a short half-life but extremely high oxidizing potential. Hydroxyl radicals have a very high affinity for all biological molecules; once formed, these radicals are too reactive to be controlled by any biological molecules and consequently cause massive cellular damage.

Hydroxyl radical formation is prevented by the action of superoxide dismutase (Figure 7–65), which acts by eliminating the superoxide precursors of hydroxyl radicals. Plants produce several different isoforms of SOD that use different divalent cations as cofactors. Different isoforms are active at different stages of plant development and under different environmental stresses. However, the activity of SOD only converts a more reactive species (superoxide) to a less reactive but longer-lived species (hydrogen peroxide). The hydrogen peroxide is removed by other enzymes: catalases and peroxidases. Catalases, which are localized in **glyoxysomes** and **peroxisomes**, scavenge most of the hydrogen peroxide, while ascorbate peroxidase removes hydrogen peroxide in other subcellular compartments, particularly chloroplasts.

Ascorbate metabolism is central to the elimination of reactive oxygen species

Hydrogen peroxide is eliminated from plastids by the activity of ascorbate peroxidase, which converts ascorbate to monodehydroascorbate and water. Monodehydroascorbate is further reduced to dehydroascorbate by monodehydroascorbate reductase, and then to ascorbate by dehydroascorbate reductase, using reduced glutathione (GSH) as a source of reducing power. This reaction also produces glutathione disulfide (GSSG), which can be converted back to GSH by glutathione reductase. These reactions complete the **ascorbate–glutathione cycle** (Figure 7–66), which eliminates excess hydrogen peroxide and regenerates the small antioxidant molecules glutathione and ascorbate.

Figure 7–64.
Production of reactive oxygen species (ROS). Oxygen is reduced in three stages, producing superoxide, hydrogen peroxide, and hydroxyl radical. The principal cellular targets of each ROS are shown.

Three steps of oxygen reduction	Product	Relative energy compared with molecular oxygen	Cellular targets
1 $O_2 + e^- \longrightarrow O_2^{\cdot-}$	superoxide	+7.6	specific enzymes, chlorophyll
2 $O_2^{\cdot-} + e^- + 2H^+ \longrightarrow H_2O_2$	hydrogen peroxide	−21.7	specific enzymes, chlorophyll, unsaturated fatty acids
3 $H_2O_2 + e^- + H^+ \longrightarrow HO^\bullet + H_2O$	hydroxyl radical plus water	−8.8	DNA, all proteins, lipids

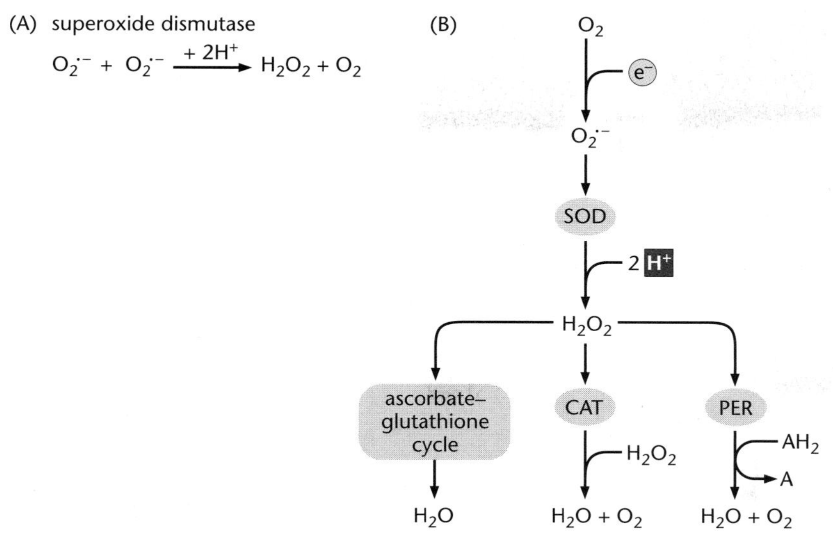

(A) superoxide dismutase

$$O_2^{\cdot -} + O_2^{\cdot -} \xrightarrow{+ 2H^+} H_2O_2 + O_2$$

(B)

$$O_2$$

$$\downarrow e^-$$

$$O_2^{\cdot -}$$

$$\downarrow SOD$$

$$\downarrow 2\ H^+$$

$$H_2O_2$$

ascorbate–glutathione cycle → H_2O

CAT → H_2O (H_2O_2) → $H_2O + O_2$

PER → AH_2 / A → $H_2O + O_2$

Figure 7–65.
Scavenging of superoxide by superoxide dismutase. (A) Superoxide dismutase (SOD) converts superoxide to hydrogen peroxide and molecular oxygen. (B) Scheme showing the action of SOD in scavenging reactive oxygen species and the removal of its reaction product, hydrogen peroxide, by three different routes: the ascorbate–glutathione cycle, catalase (CAT), and peroxidase (PER).

Ascorbate is the major antioxidant in plants and plays a central role in the removal of hydrogen peroxide. It can interact directly with and detoxify hydroxyl radicals, superoxide radicals, and singlet oxygen. It is present in large amounts (millimolar concentrations) in chloroplasts and other subcellular compartments. In addition to its function as a direct antioxidant in the ascorbate–glutathione cycle, ascorbate also reduces the oxidized form of α-tocopherol (an important antioxidant in nonaqueous structures such as membranes; see Section 7.1) and so maintains levels of vitamin E–type antioxidants. Finally, ascorbate is also present in significant amounts in the apoplast, where it plays an important role as antioxidant in response to direct oxidative stress such as exposure to ozone.

The importance of the ascorbate–glutathione cycle in limiting ROS accumulation has been demonstrated in transgenic plants with increased glutathione reductase activity, which increases tolerance to oxidative stress. Transgenic plants in which glutathione reductase activity has been reduced by **gene silencing** are less tolerant of oxidative stress.

At the heart of ROS scavenging is the activity of ascorbate peroxidase (APX). In Arabidopsis the expression of the gene encoding APX is induced by ZAT12, a zinc finger transcription factor. Overexpression of *ZAT12* enhances oxidative and light stress–responsive gene expression and results in enhanced tolerance of light, cold, and oxidative stresses. *ZAT12* expression is, in turn, positively regulated by heat shock factors (HSFs), which are directly sensitive to redox balance in their ability to activate *ZAT12* transcription.

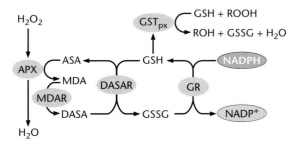

Figure 7–66.
The ascorbate–glutathione cycle. This cycle removes hydrogen peroxide from plastids by converting it to water. Intermediates: ASA, ascorbate; MDA, monodehydroascorbate; DASA, dehydroascorbate; GSH, reduced glutathione; GSSG, oxidized glutathione. Enzymes: APX, ascorbate peroxidase; MDAR, monodehydroascorbate reductase; DASAR, dehydroascorbate reductase; GR, glutathione reductase; GST$_{px}$, peroxisomal glutathione-S-transferase.

Hydrogen peroxide signals oxidative stress

Hydrogen peroxide levels signal excess accumulation of reactive oxygen species in cells. This has been demonstrated in engineered plants with reduced activity of class I catalase (which removes hydrogen peroxide from the palisade cells of the leaf mesophyll). These plants have elevated levels of hydrogen peroxide and show constitutive expression of stress-related genes such as those encoding heat shock proteins and pathogen-related proteins. If hydrogen peroxide levels increase following environmental stress, they signal the induction of ROS-scavenging enzymes (superoxide dismutases, ascorbate peroxidase, other peroxidases, and catalases) and other protective proteins. The hydrogen peroxide signal acts in a self-enhancing loop in collaboration with salicylic acid, and also interacts with the jasmonic acid pathway (see Section 8.4) and ethylene signal-transduction pathway (see Section 6.3). The pathway linking hydrogen peroxide to changes in gene expression has not yet been elucidated. The finding that it is possible to distinguish between molecular responses to hydrogen peroxide and to superoxide suggests that superoxide, and possibly other ROS, also signal oxidative stress.

Ascorbate metabolism is central to responses to oxidative stress

The relative importance of a plant's responses to increased ROS and its response to oxidative stress is a subject of some debate. The activity of the scavenging enzymes (including SOD, catalase, ascorbate peroxidase, monodehydroascorbate reductase, dehydroascorbate reductase, glutathione reductase, and glutathione peroxidase) increases following oxidative stress. When these enzymes are expressed at a high level in transgenic plants, however, improvements in tolerance to ROS are usually relatively minor. This suggests that the availability of key antioxidant molecules may be more important than the activity of the ROS-scavenging enzymes in determining a plant's tolerance to oxidative stress.

The two major plant antioxidants are reduced glutathione and ascorbate, with ascorbate being quantitatively the more important. The pool size of glutathione increases in response to oxidative stress, following induction of glutathione synthesis. In plants that are adapted to stressful conditions, glutathione pool sizes are constitutively high. Overexpression of glutathione reductase in engineered plants results in higher levels of ascorbate in leaves (via the ascorbate–glutathione cycle) and improved tolerance to oxidative stress.

The reduced form of glutathione (GSH) is synthesized from three amino acids, glutamate, cysteine, and glycine (Figure 7–67), through the action of two enzymes: γ-glutamylcysteine synthetase and glutathione synthetase. Oxidative stress causes an increase in the activity of this biosynthetic pathway by a combination of metabolic de-repression and transcriptional induction. The first step, catalyzed by γ-glutamylcysteine synthetase, is inhibited by GSH. The lowering of GSH levels immediately after exposure to oxidative stress relieves this inhibition and so increases synthesis. The expression of γ-glutamylcysteine synthetase and glutathione synthetase genes is increased in response to oxidative stress.

The pathway for biosynthesis of ascorbate is not yet fully resolved, but there is good evidence from experiments using radiolabeled substrates and from analysis of mutants and transgenic plants that ascorbate is synthesized from glucose via GDP-mannose (Figure 7–68). Mutant plants with reduced GDP-mannose pyrophosphorylase activity have lower levels of ascorbate (at about 30%) relative to wild-type plants. The mutant plants are significantly more sensitive than wild-type plants to oxidative stress induced by exposure to ozone. In wounded or stressed wild-type plants, the synthesis of ascorbate is increased by induction of the biosynthetic pathway. The activity of the final enzyme in the pathway, galactose 1,4-lactone dehydrogenase, increases following stress.

In summary, the most effective means of protection against oxidative stress seems to be the induced synthesis of the small-molecule antioxidants glutathione and ascorbate

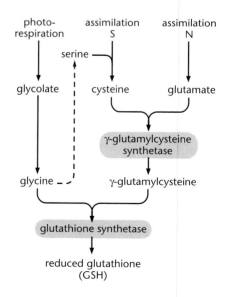

Figure 7–67.
Biosynthesis of reduced glutathione from its precursor amino acids.

following signaling from ROS. Induction of antioxidant synthesis may be a common component of responses to many different types of environmental stress, as well as to direct oxidative stress itself.

SUMMARY

All plants have ways of acclimating to stress through physiological responses that modify their growth, development, and metabolism. Stress perception, signal transduction, and response induction underlie acclimation to all types of environmental stress. In contrast, some plants have adapted to extreme environments.

Although plants depend on light to grow, light supply can represent a significant stress. High light produces excess energy from excited electrons, which must be dissipated to prevent photooxidative damage. Mechanisms of dissipation include the water–water cycle, xanthophyll cycle, and increased photorespiration. Plants also undergo morphological adaptations to high-light stress. At the other extreme, low light intensity, detected by phytochromes, results in morphological, growth, and life cycle changes. Ultraviolet irradiation, especially UV-B, damages DNA and proteins. Damaged DNA can be repaired by processes such as photoreactivation catalyzed by photolyases. Resistance to UV damage involves specialized plant metabolites acting as sunscreens (e.g., flavonoids and sinapoyl esters) and morphological adaptations.

High temperatures induce heat shock proteins, which are molecular chaperones that direct the correct refolding of denatured proteins. Developmental adaptations to growth at high temperatures include changes in leaf orientation and morphology.

The most prevalent environmental stresses for plants are those that limit water supply, and include drought, salinity, and low temperature. Plant responses to these stresses involve similar signaling mechanisms and metabolic responses. Abscisic acid (ABA) has a central signal-transduction role, although ABA-independent signaling pathways are also used. Responses to water stress include regulation of stomatal closure and the synthesis of osmolytes, which lower cellular water potential, stabilize protein and membrane structure, and protect against oxidative damage. Ion channels and aquaporins are also regulated during water stress. Some plants adopt crassulacean acid metabolism, which provides a mechanism for maintaining carbon dioxide fixation while keeping stomata closed during daylight. Plants adapted to arid environments have a range of morphological adaptations (e.g., water storage in succulent plants; distinctive morphology and life cycles in nonsucculents).

High salt levels impose both water stress (with plant responses including the synthesis of osmolytes and aquaporins) and ion stress. Adaptations in halophytes mainly involve ways of sequestering or secreting salts. Osmoprotective compounds include glycine betaine, amino acids, and polyols. Plants also respond by modulating guard cell function and changes in life cycle.

Outside the tropics, the main stress caused by low temperatures is freezing, which damages cellular membranes and causes stress through water deficit. Physiological adjustments are similar to those for drought or salt stress; many of the genes induced by cold acclimation are also induced by drought and salt stress. Cold acclimation, which includes the synthesis of osmolytes, increases resistance to freezing damage. Products of cold-regulated genes stabilize plasma membranes and prevent protein denaturation.

Too much water—flooding and water-logging—causes oxygen deprivation in plants. In short-term anoxia the final step of respiration is blocked and plant metabolism switches to fermentation. In flood-tolerant plants, aerenchyma allows long-distance oxygen transport. Other adaptations to hypoxia include ethylene-induced rapid shoot expansion and formation of adventitious roots and "pneumatophores."

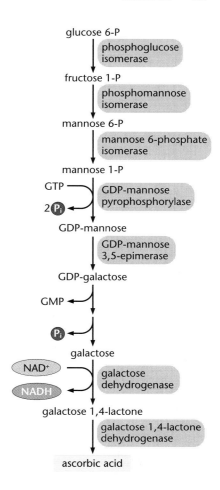

**Figure 7–68.
Biosynthesis of ascorbate from hexose phosphates in higher plants.**

High levels of reactive oxygen species (ROS) are common to growth in many stressful environments. High levels of ROS, especially the hydroxyl radical and hydrogen peroxide, are highly destructive to plant cells. ROS accumulate under a range of stresses: high light intensity, low temperature, drought, salinity, and too much UV light, or under oxidative stress such as exposure to ozone. Plant responses include induction of ROS-scavenging enzymes (superoxide dismutases, catalases, and peroxidases) and the ascorbate–glutathione cycle, which eliminates excess hydrogen peroxide.

FURTHER READING

General

Buchanan B, Gruissem W & Jones RL (eds) (2000) Biochemistry and Molecular Biology of Plants. Somerset, NJ: American Society of Plant Biologists (*especially chapters 17, 22, & 24*).

Encyclopedia of Life Sciences. Chichester, UK: John Wiley & Sons. www.els.net/

Fritsch FE & Salisbury EJ (1946) Plant Form and Function. London: G. Bell and Sons.

Hallahan DL, Gray JC & Callow JA (eds) (2000) Plant trichomes. In Advances in Botanical Research, vol. 31. New York: Academic Press.

7.1 Light as Stress

Gutschick VP (1999) Biotic and abiotic consequences of differences in leaf structure. *New Phytol.* 143, 3–18.

Jansen MAK, Gaba V & Greenberg BM (1998) Higher plants and UV-B radiation: balancing damage, repair and acclimation. *Trends Plant Sci.* 3, 131–135.

Niyogi KK, Grossman AR & Bjorkman O (1998) Arabidopsis mutants define a central role for the xanthophyll cycle in the regulation of photosynthetic energy conversion. *Plant Cell* 10, 1121–1134.

Sultan SE (2000) Phenotypic plasticity for plant development, function and life history. *Trends Plant Sci.* 5, 537–542.

7.2 High Temperature

Wang WX, Vinocur B, Shoseyov O & Altman A (2004) Role of plant heat-shock proteins and molecular chaperones in the abiotic stress response. *Trends Plant Sci.* 9, 244–252.

7.3 Water Deficit

Hoekstra FA, Golovina EA & Buitink J (2001) Mechanisms of plant desiccation tolerance. *Trends Plant Sci.* 6, 431–438.

Mulroy TW & Rundel PW (1977) Annual plants: adaptations to desert environments. *Bioscience* 27, 109–114.

Nimmo HG (2000) The regulation of phosphoenolpyruvate carboxylase in CAM plants. *Trends Plant Sci.* 5, 75–80.

7.4 Salt Stress

Glenn EP, Brown JJ & Blumwald E (1999) Salt tolerance and crop potential of halophytes. *Crit. Rev. Plant Sci.* 18, 227–255.

Zhu JK (2002) Salt and drought stress signal transduction in plants. *Annu. Rev. Plant Biol.* 53, 247–273.

7.5 Cold

Chinnusamy V, Zhu J & Zhu J-K (2006) Gene regulation during cold acclimation in plants. *Physiol. Plantarum* 126, 52–61.

van Buskirk HA & Thomashow MF (2006) Arabidopsis transcription factors regulating cold acclimation. *Physiol. Plantarum* 126, 72–80.

Yamaguchi-Shinozaki K & Shinozaki K (2006) Transcriptional regulatory networks in cellular responses and tolerance to dehydration and cold stresses. *Annu. Rev. Plant Biol.* 57, 781–803.

7.6 Anaerobic Stress

Dolferus R, Klok EJ, Delessert C et al. (2003) Enhancing the anaerobic response. *Ann. Bot.* 9, 111–117.

Drew MC, He C-J & Morgan PW (2000) Programmed cell death and aerenchyma formation in roots. *Trends Plant Sci.* 5, 123–127.

Xu K, Xu X, Fukao T et al. (2006) Sub1A is an ethylene-response-factor-like gene that confers submergence tolerance to rice. *Nature* 442, 705–708.

7.7 Oxidative Stress

Foyer CH & Noctor G (2005) Redox homeostasis and antioxidant signaling: a metabolic interface between stress perception and physiological responses. *Plant Cell* 17, 1866–1875.

Noctor G & Foyer CH (1998) Ascorbate and glutathione: keeping active oxygen under control. *Annu. Rev. Plant Physiol. Plant Mol. Biol.* 49, 249–279.

INTERACTIONS WITH OTHER ORGANISMS

8

When you have read Chapter 8, you should be able to:

- Explain "coevolution" by using examples of plant–pathogen and plant–pollinator interactions.

- Explain why monocultures of crops are susceptible to disease epidemics, with specific examples.

- Summarize the selective pressures on the evolution of plant pathogens and plants' defenses against them.

- Distinguish between biotrophic and necrotrophic pathogens, outline the routes by which pathogens enter plants, and describe the roles of effector molecules in the infection process.

- Summarize how *Agrobacterium* transfers its T-DNA into plant cells and how this system is used in biotechnology.

- Outline the gene-for-gene model of plant–pathogen interactions, including the role of *avr* genes.

- Summarize the types of fungal and oomycete pathogens.

- Summarize the roles of insects as pests and as vectors of viral pathogens.

- Give an overview of the plant viruses, describing the four main families, and explain the role of RNA silencing in plants' resistance to viral infection.

- Describe the basal and constitutive defense mechanisms of plants.

- Describe the two main ways in which pathogen molecules can be recognized by plants and result in defense activation.

- Define "symbiosis," and give details of the formation of nodules and mycorrhizae on plant roots, the types of plants, bacteria, and fungi involved, and how the symbionts benefit.

- Describe the main class of R proteins and their roles in plant defense, and summarize the two main types of systemic resistance in plants.

CHAPTER SUMMARY

8.1 MICROBIAL PATHOGENS

8.2 PESTS AND PARASITES

8.3 VIRUSES AND VIROIDS

8.4 DEFENSES

8.5 COOPERATION

In Chapters 6 and 7 we focus mainly on plants' interactions with physical, or abiotic, aspects of their environment: light, oxygen, water, minerals, and so forth. But we cannot fully understand plants' growth, development, and diversity without also understanding their interactions with their biotic environment: with other organisms. Plants are the source of organic carbon—that is, are the food—for almost all of the nonphotosynthetic organisms on earth. Many, though not all, of these interactions with other organisms are deleterious to the growth of the plant, because of wholesale removal of plant tissues, or plant disease. In this chapter, our focus is on interactions with nonhuman organisms: bacteria, oomycetes, fungi, insects and other herbivores, viruses, and even other plants—and we begin with the deleterious interactions. Many of these organisms can cause plant disease and damage. Generally, microbial disease-causing organisms are called **pathogens**, and herbivorous insects, mammals, and birds that eat vegetative tissues and seeds are referred to as **pests**.

Most plant pathogens and many pests have coevolved with their host plants since long before the domestication of plants as crops. However, the onset of agriculture around 10,000 years ago brought about dramatic and important changes in the relationship between domesticated species and their pathogens and pests. During domestication and agricultural mechanization, the size, density, and genetic uniformity of crop populations have all increased. Today, large areas of the planet are occupied by **monocultures** of millions of genetically identical crop plants. In this situation, natural selection strongly favors genetic variants of pathogens and pests that can overcome the defense mechanisms of a crop plant in order to exploit it as food. If such a genetic variant arises—through **mutation** or **genetic recombination**—it will reproduce rapidly and may cause massive crop losses.

The losses caused by epidemics of pathogen-induced diseases in crop plants can influence human history. The Irish potato famine of 1845–1848 was triggered by an epidemic of potato late blight caused by the **oomycete** *Phytophthora infestans* (Figure 8–1); this reduced the Irish population, through starvation and emigration, from 8 to 5 million people in three years. For millions of people living in developing countries today, hunger is exacerbated by crop losses through disease and insect damage. In developed countries, where food is plentiful, crop disease is still of considerable economic importance. In the United Kingdom, late blight control is estimated to cost about $400 per hectare each year, which translates to $55,000,000 for the country as a whole. Worldwide, between 40 and 50% of crops are lost each year because of pests, weeds, and pathogens. Plant pathogens and pests also affect the value of crops by causing blemishes on fruits and vegetables: these can severely reduce market value. Pathogens and pests can also attack plant material after harvest, during transport and storage, causing **post-harvest losses**.

Interactions between pathogens and their host plants can be represented as consisting of four phases, reflecting a sequence of events in the evolution of host–pathogen relationships; these are shown as a zigzag diagram in Figure 8–2. We can summarize the phases, and the **selective pressures** underlying their evolution, as follows:

1. Most microorganisms have surface molecules (often **cell wall** components) that are recognized by **receptors** on plant cells, triggering the activation of plant defenses referred to as **basal defense mechanisms**. These defenses can be regarded as a form of plant immunity. They are also referred to as **PAMP-triggered immunity** (PAMPs are **pathogen-associated molecular patterns**).

2. The existence of plant basal defense mechanisms places selective pressure on potential pathogens for genetic variants that produce proteins or other compounds—known as **effector molecules**—that interfere with and suppress these mechanisms, enabling the pathogen to attack the plant. This phenomenon is known as **effector-triggered susceptibility**.

3. The existence of effector molecules that allow successful colonization by the pathogen places selective pressure on the plant population for genetic variants that can recognize pathogen effector molecules and respond to their presence by triggering further defense mechanisms. These variant plants have receptors for the effector molecules, encoded by genes known as **resistance (R) genes**. This form of defense is known as **R gene–mediated defense**, or **effector-triggered immunity (ETI)**. It usually involves stronger activation of defense mechanisms than does PAMP-triggered immunity.

Figure 8–1.
An economically important plant pathogen. Leaf of potato infected with potato late blight, caused by the oomycete *Phytophthora infestans*. White downy sporangiophores (sporangium-bearing branches) surround necrotic areas. (Courtesy of Willmer Perez.)

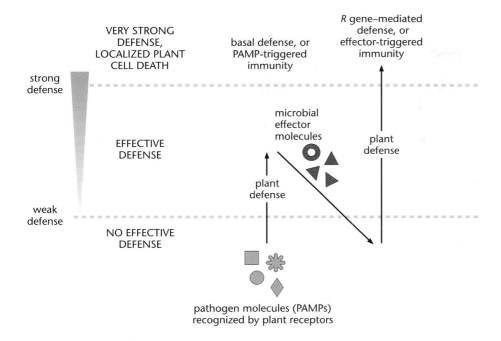

Figure 8–2.
The "zigzag" explanation of events in coevolution of host plants and their pathogens. The diagram shows the level of plant defense (*left*) in response to challenge by microbial pathogens (*right*). When a pathogen comes into contact with a potential host plant, surface molecules on the pathogen are recognized by receptors in the plant, triggering defense responses. This basal defense, also called PAMP-triggered immunity, favors the selection of genetic variants of the pathogen that produce effector molecules able to overcome the plant defense. In turn, the evolution of effector molecules favors the selection of genetic variants of the host plant in which the presence of effector molecules triggers further defense mechanisms. This is known as *R* gene–mediated defense, or effector-triggered immunity; it is usually a stronger form of defense than PAMP-triggered immunity. The evolution of *R* gene–mediated defense places selective pressure on the pathogen for variants that produce modified effector molecules, no longer recognized by the defense system. These concepts are further explored in Section 8.4.

4. The presence of R proteins in the plant places selective pressure on potential pathogens for genetic variants that no longer make the effectors that these receptors recognize. These variant pathogens will not trigger the *R* gene–mediated defense mechanism and so will be able to attack and exploit the plant.

Thus there is an intricate pattern of **coevolution** between plants and their pathogens. A successful pathogen that can overcome basal defense mechanisms in the plant provides a selective pressure on the plant species for variants that can recognize pathogen effector molecules and defend themselves against the pathogen, via R proteins. An R protein that recognizes a widespread pathogen effector molecule provides a selective pressure on the pathogen for variants that can avoid detection and thus avoid triggering defense mechanisms. This is a very important concept: we will return to it several times in this chapter.

Not all interactions with other organisms are harmful to plants. Some are mutually beneficial: these are referred to as **symbiotic interactions**. For example, some insects use nectar from flowers as food and also facilitate the dispersal of **pollen** to other flowers. Bacteria that can convert nitrogen gas to ammonia form symbiotic relationships with plants, in which the bacteria receive organic carbon for energy and supply the plant with ammonia for amino acid synthesis. Animals that eat fruits aid the dispersal of **seeds** to suitable sites for germination.

In this chapter we examine examples from the range of interactions between plants and other organisms. Our focus is on interactions that have been well characterized using biochemistry and genetics. We first discuss the mechanisms through which various types of pathogen attack and exploit plants as food, then describe the range of mechanisms by which plants detect and resist these attacks. Finally we examine some symbiotic relationships between plants and other organisms.

8.1 MICROBIAL PATHOGENS

Many pathogens are specialized to thrive on one particular plant species and cannot attack and cause disease in others. Others can attack many, often unrelated species of plants. In order to exploit a particular species as food, a pathogen must be able to overcome the species' defense mechanisms. Most plant species are resistant to most pathogens. As we describe above, the production by the pathogen of effector molecules that can suppress plant defenses without triggering recognition by R proteins is crucial to pathogen success. In this section we describe mechanisms by which microbial (and viral) pathogens attack and exploit their host plants; attack and exploitation by other types of organism is described in Section 8.2, and viruses are covered in more detail in Section 8.3. The mechanisms by which plants detect and resist attack are discussed in Section 8.4.

Most pathogens can be classified as biotrophs or necrotrophs

On entry into the host plant, pathogens either kill plant cells and feed on the resulting dead material (necrotrophy), or they grow in association with living tissue until they are ready to reproduce (biotrophy). **Necrotrophs** kill plant tissues by producing chemicals toxic to plant cells and hydrolytic enzymes that break down the polymers of the plant cell wall. Many necrotrophs can exploit a broad range of plant species. For example, the necrotrophic **fungus** *Botrytis cinerea* can infect at least 1000 plant species. Necrotrophic bacteria of the genus *Erwinia* cause rots in a range of fruit and vegetables (Figure 8–3A, B). In contrast, **biotrophs** are often highly specialized for growth on one particular host plant, and invaded cells must stay alive for the biotroph to complete its normal life cycle. For example, the powdery mildew fungus (*Blumeria graminis*) occurs in genetically distinct forms—called *formae specialis* ("forms of the species"; f. sp.)—that infect either barley or wheat, but not both. Infection by biotrophic pathogens often has large effects on the metabolism and development of the host plant, both as a result of loss of metabolites from the host to the pathogen and because of alterations in levels of **phytohormones** involved in growth and development. Effects can include delayed **senescence** of infected regions of leaves (see Section 5.4) and stunting and abnormal growth patterns (Figure 8–3C, D). Biotrophic pathogens include mildew and rust fungi, viruses, and nematodes.

Figure 8–3.
Examples of effects of necrotrophic and biotrophic pathogens on host plants.
(A) The nectrotrophic fungus *Botrytis cinerea* causing mold on grapes. The gray coloration of the grape surface is caused by a mat of sporulating (spore-forming) fungal hyphae. (B) The necrotrophic bacterium *Erwinia* (*Pectobacterium*) causing rot in potato, seen in potato tuber slices. (C) Delayed senescence in regions of a maize leaf infected with the biotrophic rust fungus *Puccinia sorghi*. The green regions ("green islands") are areas where the fungus is present; the dark regions in the centers of the islands are fungal spores on the leaf surface. (D) Distortion and variegation in tobacco leaves infected with *Tobacco mosaic virus*. (A, courtesy of M. Schuster; B, courtesy of Allan Collmer; C, courtesy of Tony Pryor; D, courtesy of Michael Shintaku and Rick Nelson.)

(A)
(B)
(C)
(D)

Some pathogens act as biotrophs in the early stages of plant infection, then become necrotrophic as the infection proceeds. An example is the agent of potato late blight, *P. infestans* (see Figure 8–1). The potato leaf remains alive during the initial phases of infection, but tissues are killed and colonized by this oomycete (see Figure 8–8) as the infection proceeds. Many bacteria in the genus *Pseudomonas* also have an initial phase of colonization of the intercellular spaces of the host-plant leaf in which the adjacent leaf cells remain alive. As in potato blight, leaf cells collapse and dead patches (lesions) develop as the infection progresses. This type of pathogen is known as a **hemibiotroph**.

Pathogens enter plants via several different routes

There are three main routes of entry for pathogens into plants: direct penetration through intact surfaces, entry through natural openings (such as **stomata**), or opportunistic entry through wound sites (Figure 8–4). Fungi can enter plants by all three routes; some species use just one route, and others can enter in more than one way. Bacteria rarely enter plants by direct penetration. If present on a film of water covering the plant surface, they can simply swim in through openings. Some bacteria depend on insects for entry into plants. These are mainly bacteria that infect vascular tissues (**phloem**, **xylem**, and associated cell types). When the mouthparts of an insect penetrate the phloem during feeding, they become contaminated with bacteria infecting this region. The bacteria are then transmitted to a new vascular region when the insect next feeds. **Viruses** and **viroids** (viroids are very short, circular, single-stranded RNAs; see Section 8.3) tend to enter through wounds made by the feeding insects or **nematodes** by which they are transmitted (their "vectors") or by mechanical damage to the plant.

Most microbial pathogens attack a specific part of the plant. Root-infecting fungi and bacteria persist in a dormant state in the soil until they detect compounds exuded by plant roots (such as sugars, amino acids, or other chemicals produced only by certain species

direct penetration

fungal spore on plant surface   appressorium

penetration peg piercing cuticle and cell wall

intercellular mycelium

penetration through natural openings

fungal hypha entering through stomatal opening

bacteria moving into leaf via stomatal opening

Figure 8–4.
Methods of invasion of plant organs by fungi and bacteria. Some fungi have mechanisms allowing direct penetration through the cuticle and cell wall of the plant. Other species of fungi, and pathogenic bacteria, enter through natural openings or through existing wounds and cracks in the plant surface.

penetration through wounds

fungal hypha entering leaf through a wound

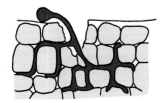

fungal hypha entering a root through natural cracks between main root and lateral root

of plants). After germination of the microbial spore or dormant form, entry of the microbe can occur either through wounds, or at the points where lateral roots emerge from the main root (Figure 8–4), or by direct penetration of the walls of root **epidermal cells**. Fungi and bacteria that infect the aerial parts of plants can enter through natural openings such as stomata, **hydathodes** (structures that exude water from leaf edges; see Section 4.10), and **nectaries**, or through wounds, or by direct penetration of epidermal cell walls. Unlike the root epidermis, the epidermis of the aerial parts of plants is covered by a waxy **cuticle** (see Section 3.6); pathogens entering by direct penetration must be able to attach to and penetrate this layer, as well as the cell wall itself.

Cladosporium fulvum is an example of a pathogen that uses natural openings to enter its host plant (Figure 8–5). This biotrophic fungus causes leaf mold disease of glasshouse tomatoes. Spores produced by asexual reproduction—which are called **conidia**—germinate on the leaf surface, and the **hyphae** that emerge grow across the leaf surface and enter through stomata. Hyphae then ramify through the intercellular spaces of the leaf. The fungus uses nutrients that leak from surrounding plant cells as its source of food. After 10 to 14 days of growth within the leaf, specialized hyphae called "conidiophores" emerge from the stomata to release conidia, which continue the life cycle (see Figure 8–22). The conidiophores prevent closure of the stomata, thus preventing control of water loss from the plant and giving rise to the disease symptoms.

Magnaporthe grisea (the cause of rice blast disease; Figure 8–6) is a hemibiotrophic fungus that enters its host by direct penetration. When a *Magnaporthe* conidium germinates on the leaf surface it produces a short hypha (a germ tube), which differentiates into a structure called an **appressorium**. The appressorium is a flattened bulblike structure, which attaches very tightly to the cuticle of the leaf by exuding adhesive proteins (hydrophobins) and mucilage (Figure 8–6B, C, D; see also Figure 8–4). The appressorium produces a penetration peg, which is much narrower than most hyphae. The growing point of the penetration peg advances like a needle into and through the cuticle and the cell wall. Several features of the penetration peg are important in enabling this penetration to occur.

First, the interior of the fungal cell maintains a very high **turgor** pressure (or hydrostatic pressure), making it a rigid structure. The turgor pressure is generated by breakdown of storage compounds—**lipids** and a starchlike glucose polymer called **glycogen**—contained in the conidium, to produce glycerol. Accumulation of glycerol draws water into the fungus, raising the outward pressure against the fungal cell wall (see Section 3.5 for more explanation of turgor pressure). Second, to prevent rupture of the fungal cell wall by the high internal pressure, the wall is strengthened by deposition of melanin, an extremely strong, cross-linked derivative of the amino acid tyrosine. Mutants of *M. grisea* that do not accumulate high concentrations of glycerol, or that cannot synthesize melanin, cannot sustain high turgor pressure in the infection peg, and they show reduced pathogenicity because penetration of the cell surface is less likely to be successful. Third, entry of the infection peg is probably also assisted by the production and secretion of fungal enzymes that degrade plant cell walls. The fungal cell wall has

Figure 8–5.
The *Cladosporium fulvum* life cycle on tomato. After germination of the spore on the leaf surface, hyphae run across the surface until they encounter a stomatal opening through which they can enter the intercellular spaces of the leaf. Hyphae then ramify through these intercellular spaces. Approximately 14 days after infection, sporulation structures, the conidiophores, emerge through the stomatal openings.

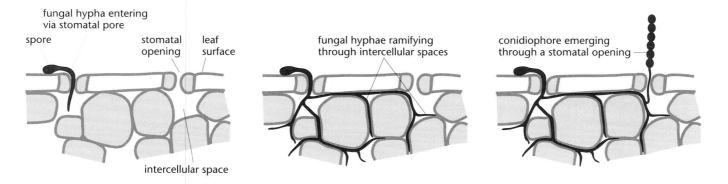

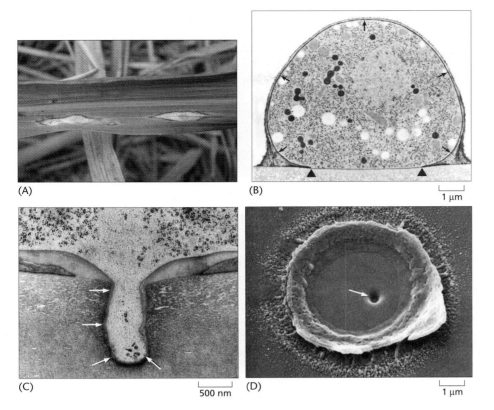

Figure 8–6.
Infection of rice leaves by the rice blast fungus *Magnaporthe grisea*. (A) Rice leaf infected with *M. grisea*, showing regions of cell death caused by the fungus. (B) Transmission electron micrograph showing an *M. grisea* appressorium cut perpendicular to the substrate. The melanin layer (*arrows*) surrounds the cell, except for an area in contact with the substrate, comprising the appressorium pore from which the penetration peg will emerge (*bottom, between arrowheads*). (C) Infection peg penetrating cellophane, an artificial substrate. The peg cell wall (*arrowed*) is continuous with the pore cell wall. (D) Impression of penetration peg left on an artificial substrate. The circular dent represents the site of the appressorium: the arrow shows where the penetration peg entered the substrate. (A, courtesy of Nick Talbot; B-D, from *Annu. Rev. Microbiol.* 50:491–512, 1996. With permission from Annual Reviews, courtesy of Richard J. Howard.)

a very different structure from the plant cell wall—it consists mostly of a polymer called **chitin**—so production by the fungus of enzymes (such as pectinases and cellulases) that degrade plant cell walls has no effect on its own cell wall.

After entry, the hyphae of *Magnaporthe* ramify throughout the leaf. Following a brief biotrophic phase, the fungus kills plant cells in the infected area, giving rise to lesions.

The rapid degradation of cell walls that can occur during invasion by a necrotrophic pathogen is dramatically illustrated during the infection of bean leaves by *Botrytis fabae*. Following enzyme-assisted penetration of the cuticle, *B. fabae* secretes polygalacturonases, enzymes that cause the surrounding cell wall to expand and allow rapid growth of hyphae. The fungus actually grows within the degraded cell wall and kills cells by the hydrolysis of integral pectic components.

Blumeria graminis (the cereal powdery mildew fungus) is an example of a biotroph that makes an appressorium and infection peg to penetrate the plant cell wall and then obtains nutrients by forming specialized feeding structures called **haustoria** (Figure 8–7). Haustoria are extensions of hyphae that project into the plant cell and provide a large surface area across which the pathogen can absorb nutrients. The haustoria of *B. graminis* have an exceptionally large area of contact with the plant cell cytoplasm because of their many fingerlike projections. Although the haustoria penetrate into the interior of the plant cell, the **plasma membranes** of both the host and the fungal cell remain intact. Between the plasma membranes of fungus and host lies an extracellular matrix of cell wall materials through which nutrients and signaling molecules are exchanged. Unlike most pathogens, powdery mildew fungi only infect epidermal cells. Haustoria are formed by many distantly related species of fungi and oomycete pathogens, and it is thought that they have evolved independently several times. Note that oomycetes are not fungi, and indeed are more closely related to the brown algae (Phaeophyceae) and to the malarial parasite *Plasmodium*, and even to plants, than they are to fungi (Figure 8–8).

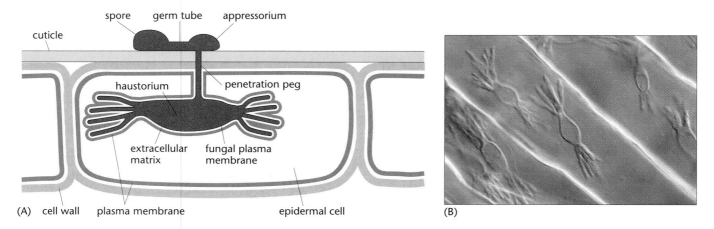

Figure 8–7.

Formation of haustoria inside plant cells by a biotrophic fungus.
(A) An epidermal cell of a cereal leaf infected by the powdery mildew fungus *Blumeria graminis*. After germination of the fungal spore, a germ tube and an appressorium form on the leaf surface. A penetration peg then penetrates the cuticle and cell wall. These processes are similar to those described for the necrotrophic fungus *Magnaporthe grisea* (see Figure 8–6). However, in this case, rather than resulting in the death of the plant cell, penetration by

B. graminis produces a feeding structure, the haustorium, that takes up nutrients from the plant cell while the cell remains alive. Note that the haustorium remains outside the plant plasma membrane. The fingerlike projections of the haustorium are surrounded by plant plasma membrane, giving a large area of contact across which nutrients can pass to the fungus. (B) Surface of a cereal leaf infected with *B. graminis*. The leaf has been treated to reveal fungal haustoria in the epidermal cells. (B, micrograph from MycoAlbum CD by George Barron.)

Pathogen infections lead to a broad range of disease symptoms

Pathogens bring about a variety of symptoms in infected plants. These symptoms are sometimes specific to a particular pathogen, allowing it to be identified by the symptoms alone, but different pathogens can cause similar effects (Figure 8–9). Some biotrophs can grow for lengthy periods in their host plants without causing any obvious disease symptoms. Diseases are often named after the symptoms caused, rather than

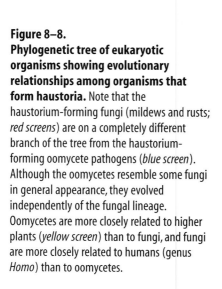

Figure 8–8.
Phylogenetic tree of eukaryotic organisms showing evolutionary relationships among organisms that form haustoria. Note that the haustorium-forming fungi (mildews and rusts; *red screens*) are on a completely different branch of the tree from the haustorium-forming oomycete pathogens (*blue screen*). Although the oomycetes resemble some fungi in general appearance, they evolved independently of the fungal lineage. Oomycetes are more closely related to higher plants (*yellow screen*) than to fungi, and fungi are more closely related to humans (genus *Homo*) than to oomycetes.

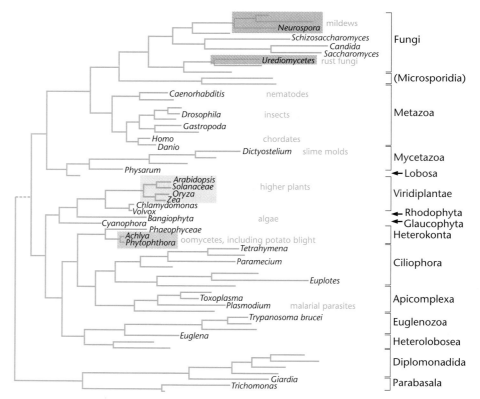

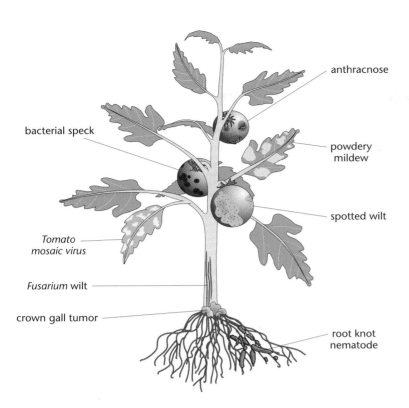

Figure 8–9.
Diseases caused by different types of pathogens on tomato. The diagram illustrates symptoms caused by bacteria (bacterial speck, spotted wilt, crown gall), fungi (*Fusarium* wilt, powdery mildew, anthracnose), viruses (*Tomato mosaic virus*), and nematodes (root knot nematode).

after the causative agent; for example, bacteria, fungi, and oomycetes can cause diseases called **blights**, which are characterized by rapid browning and death of the affected areas of the plant (see Figure 8–1). Bacteria and fungi can cause **vascular wilts**, which result from colonization and blockage of the xylem by pathogens, resulting in reduced water flow to the leaves. Bacteria, fungi, insects, and nematodes can cause reprogramming of host development, for example by altering the levels of phytohormones, resulting in **galls** (in which plant cell proliferation is locally deregulated), root knots, cysts, witches' brooms (with profuse upward branching of twigs), or leaf curls (Figure 8–10A, B). Club root of crucifers, caused by *Plasmodiophora brassicae*, leads to massive enlargement of the roots (Figure 8–10C).

Many pathogens produce effector molecules that influence their interactions with the host plant

Plant pathogens synthesize a wide spectrum of molecules that can enhance their ability to gain nutrients from their host plant and hence reproduce successfully. We use the

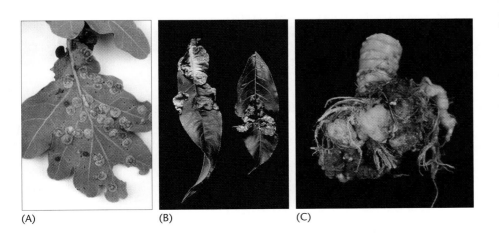

(A)   (B)   (C)

Figure 8–10.
Examples of disease symptoms involving disturbance of phytohormone levels in host plants. (A) Spangle galls on an oak leaf, caused by the gall midge *Cecidomyia poculum*. The adult midge lays eggs in leaf tissues. This results in hormonal disturbances in the leaf, giving rise to massive proliferation of cells at the leaf surface to form a gall, in which the insect larvae grow and feed. (B) Curling, thickening, and pigmentation of a peach leaf caused by infection by the fungus *Taphrina deformans*. (C) Massive enlargement and distortion of the roots of a crucifer (brassica)—a disease called club root—caused by infection by the unicellular organism *Plasmodiophora brassicae*. (B, courtesy of J. Pscheidt; C, courtesy of Marc Cubeta.)

general term "effector molecule" to include any molecule made by a pathogen that can enhance its ability to colonize, and grow and reproduce on or in, its host plant. Effector molecules are sometimes called "compatibility factors" because interactions between plants and pathogens that result in disease are described as "compatible"; interactions that do not lead to disease are termed "incompatible." Many effector molecules probably function to suppress basal defense mechanisms that are triggered by pathogen attack (in other words, to overcome PAMP-triggered immunity; see Figure 8–2). Isolating and identifying effector molecules can provide clues about how the pathogen achieves successful colonization and growth. Production of these molecules usually occurs only in the presence of a potential host plant. We consider here three important classes of effector molecules: enzymes, toxins, and growth regulators. Additional classes of effector molecule are considered in Sections 8.2 and 8.3.

Enzymes that degrade components of plant cell walls, such as cutinases, cellulases, xylanases, pectinases, and polygalacturonases, are secreted by both fungal and bacterial pathogens. Cutinases hydrolyze the cuticle that covers the aerial surfaces of plants, and the other enzymes hydrolyze major components of the cell wall: **cellulose**, **hemicellulose**, and **pectin** (see Chapter 3). The actions of these enzymes weaken plant cell walls and may separate the cells, opening up routes by which the pathogen can enter and colonize the plant. Most pathogens produce several different cell-wall–degrading enzymes, which probably act in concert to weaken the plant cell wall. Mutations that eliminate the production of any one of these enzymes usually have little effect on the overall success of the pathogen (i.e., on its pathogenicity), but their combined actions make a significant contribution to pathogenicity.

Some pathogenic bacteria, for example *Erwinia carotovora*, produce cell-wall–degrading enzymes only when bacterial populations on the host plant reach a certain density. This monitoring of bacterial population density, known as **quorum sensing**, uses mechanisms that were first discovered during studies on a symbiotic interaction between the luminescent marine bacterium *Vibrio fischeri* and marine animals such as certain squid and fish species. *V. fischeri* bacteria accumulate in specialized light organs in their host animals (Figure 8–11). When the bacterial population reaches a sufficient density, the cells become luminescent.

In *E. carotovora*, quorum sensing allows the production of cell-wall–degrading enzymes only when population density is high enough to maximize chances of a successful invasion. If the enzymes were produced at low pathogen densities, they might trigger the defense mechanisms of the host plant (see Section 8.4) without being present in sufficient quantity to overwhelm the plant's defenses. Quorum sensing in both *E. carotovora* and *V. fischeri* is mediated by *N*-acylhomoserine lactones, small molecules released from the bacterial cells into the surrounding medium (Figure 8–12). At high bacterial densities, the concentration of *N*-acylhomoserine lactone is sufficient to induce the expression of genes encoding luminescent proteins in *V. fischeri* and cell-wall–degrading enzymes in *E. carotovora*.

Some fungal pathogens secrete enzymes that detoxify host-plant molecules that inhibit fungal growth. These are discussed in Section 8.4, where we consider the defense mechanisms of the host plants.

Pathogens make a huge array of **toxins**, chemicals that are either generally toxic to plants or toxic only to a particular plant species. Regardless of whether they are generally or specifically toxic, many toxins act by inhibiting the action of specific proteins in host plants. We consider three examples of toxins produced by important fungal and bacterial pathogens. The examples illustrate the very different types of host proteins that may be targeted: in these cases, a nuclear protein involved in structuring chromatin, a proton pump on the plasma membrane, and an enzyme of nitrogen assimilation.

Figure 8–11.
Bobtail squid, *Euprymna berryi*. Its light-emitting organs are inhabited by the bacterium *Vibrio fischeri*. (Courtesy of Scubazoo/Science Photo Library.)

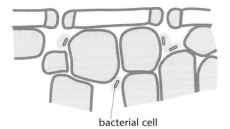

bacteria enter the leaf through stomatal openings; bacterial numbers low; *N*-acylhomoserine lactone concentration low

bacterial population density increases; *N*-acylhomoserine lactone concentration rises, triggering production of cell-wall–degrading enzymes and other effectors by the bacteria

high bacterial population density and high production of cell-wall–degrading enzymes and other effectors; localized death of plant cells

bacterial cell

cell-wall–degrading enzymes and other effectors

The fungus *Cochliobolus carbonum*, the cause of Northern leaf blight disease (Figure 8–13A), makes a toxin called "HC toxin." Like many other fungal toxins, HC toxin is a cyclic peptide (Figure 8–13B). It inhibits the activity of the enzyme histone deacetylase in the maize plant. The deacetylation of **histones**, the proteins associated with **DNA** in **chromatin** in the nucleus, affects gene expression (see Section 2.3). It is thought that alteration of gene expression in the maize plant when histone deacetylase is inhibited by HC toxin compromises the plant's defense against invasion by *C. carbonum*. The importance of HC toxin to the pathogenicity of *C. carbonum* is shown by the finding that mutants with impaired HC toxin biosynthesis are not pathogenic. Maize plants carry the *Hm1* gene, which encodes an enzyme that can metabolize and thus detoxify HC toxin (Figure 8–13C). This gives the plant resistance to leaf spot

Figure 8–12.

Quorum sensing during growth of a pathogenic bacterium in leaf intercellular spaces. Bacteria enter the leaf through stomatal apertures. Early in colonization, the bacterial population inside the leaf is sparse and the *N*-acylhomoserine lactone concentration is low. As the density of the bacterial population rises, the elevated *N*-acylhomoserine lactone concentration leads to expression of cell-wall–degrading enzymes. Production of high concentrations of enzymes and other effector molecules results in death of the plant cells.

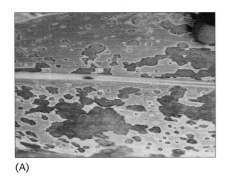

(A)

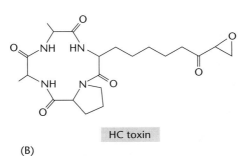

HC toxin

(B)

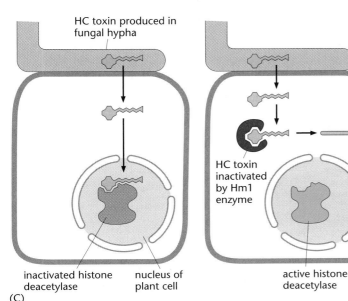

HC toxin produced in fungal hypha

HC toxin inactivated by Hm1 enzyme

inactivated histone deacetylase

nucleus of plant cell

active histone deacetylase

(C)

Figure 8–13.

Infection of maize plants by *Cochliobolus carbonum*. (A) Northern leaf blight disease of maize plants, caused by *C. carbonum*. (B) Structure of the HC toxin, a cyclic peptide produced by *C. carbonum*. (C) In a susceptible plant (*left*), HC toxin produced by fungal hyphae enters the cell nucleus and inhibits histone deacetylase, compromising the expression of genes required for resistance to the fungus. A resistant plant (*right*) produces the Hm1 protein, an NADPH-dependent oxidoreductase enzyme that modifies the structure of HC toxin and renders the toxin unable to inhibit histone deacetylase. (A, courtesy of Guri Johal.)

disease. In the United States, leaf spot disease is found in areas with moist climates that favor the spread of *C. carbonum*. Traditionally, the disease has caused only minor crop losses, but it became a major problem when a maize line developed in a dry area was used as a parent for a variety to be grown in a moist area. The new variety proved to be extremely susceptible to leaf spot disease. It was discovered that the parent line from the dry area carried a mutation in the *Hm1* gene, so could not metabolize HC toxin. This mutation had been transferred to the new variety.

An example of a nonspecific toxin is fusicoccin, which is produced by the fungus *Fusicoccum amygdali*. Fusicoccin constitutively activates the **H⁺-ATPase** (**proton pump**) of the plasma membrane. The action of the H⁺-ATPase is a major determinant of the electrical potential across the plasma membrane (see Chapter 4). The control of this electrical potential is of particular significance in stomatal guard cells, where changes in its magnitude bring about changes in cell turgor and hence changes in the aperture of stomata (see Sections 3.4, 3.5, and 7.3). In the presence of fusicoccin, the rate of proton pumping across the plasma membrane of guard cells cannot be modulated. The cells no longer respond to the environmental signals that normally bring about loss of turgor and hence stomatal closure. The irreversible opening of the stomata leads to abnormally high water loss, wilting, and eventual death of the plant. Although stomatal opening brings about obvious disease symptoms in the plant, whether it is directly advantageous to the fungus is unclear. It is possible that stimulation by fusicoccin of the H⁺-ATPase in other types of plant cells brings about release of nutrients that promote the growth and reproduction of the pathogen.

Some toxins produced by fungal pathogens (called **mycotoxins**) cause major health problems for humans and domesticated animals that consume infected plants. Outbreaks of a disease known as St Anthony's fire in Europe in the Middle Ages were caused by ergot (*Claviceps purpurea*), which infects grasses, including the cereal crop rye (Figure 8–14). Toxins from the fungus—including compounds related to the hallucinogen lysergic acid diethylamide (LSD)—contaminated rye flour and hence the bread made from it. *Aspergillus flavus*, a fungus that infects peanuts, produces aflatoxin, a potent carcinogen in humans. *Fusarium* and *Aspergillus* species that cause diseases of the ears of wheat and the cobs of maize, and also infect damp stored grain, produce several dangerous toxins. For example, *Fusarium* species make zearalenones (steroid hormone mimics that have been found to cause testicular feminization in pigs) and fumonisins (associated with high rates of esophageal cancer in humans). Unlike the toxins from *Claviceps* and *Aspergillus*, the zearalenones have an important role in the pathogenicity of the fungus.

The bacterium *Pseudomonas syringae* pv. *tabaci* produces the toxin tabtoxin ("pv." is an abbreviation for **pathovar**, or disease-causing strain). Tabtoxin is a dipeptide, a conjugate of the amino acid threonine and the nonprotein amino acid tabtoxinine (Figure 8–15A). Tabtoxin itself is not responsible for the toxic effects in the host plant. The plant contains a peptidase, an enzyme that cleaves the peptide bond of tabtoxin to

Figure 8–14.
Infection by *Claviceps purpurea*. (A) An ear of the grass *Agropyron repens* infected by *C. purpurea*; the large black grains contain spores of the fungus. (B) Structure of ergotamine, a toxin produced by *C. purpurea*, which is harmful to humans.

ergotamine

(A) (B)

Figure 8–15.
Tabtoxin and its effects in plants infected by *Pseudomonas syringae*. (A) Structure of tabtoxin, showing the tabtoxinine and threonine components. Cleavage of the peptide bond between the two amino acids by a plant peptidase releases the toxin tabtoxinine. (B) Soybean leaf infected with *P. syringae* pv. *tabaci*, showing the characteristic pale halos where chlorophyll is destroyed around points of infection. (B, courtesy of George N. Agrios. Fig.12.9, p.328 reprinted from Plant Pathology 5th ed, by George Agrios. © 2005 Elsevier Ltd. Reprinted with permission.)

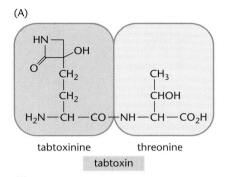

(A)

tabtoxinine   threonine

tabtoxin

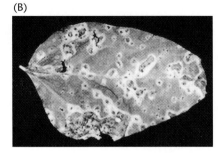

(B)

release threonine and tabtoxinine. Tabtoxinine is a potent inhibitor of **glutamine synthetase**, an enzyme essential for the assimilation of ammonia into organic molecules in plants (see Section 4.8). In leaves, glutamine synthetase is part of the mechanism that reassimilates the very large amounts of ammonia released in the **photorespiratory cycle** (see Section 4.3). Inhibition of this enzyme in leaves infected by *P. syringae* pv. *tabaci* leads to destruction of **chlorophyll** and yellowing of affected tissues to form a characteristic halo around lesions (Figure 8–15B). Other pathovars of *P. syringae* make the toxins syringomycin and syringopeptin. These have a very different action on the host plant: they form pores in the host plasma membranes, which may facilitate leakage of nutrients for the bacteria and may also weaken the host's defense responses.

A third class of effector molecules includes those that can control plant growth. These either act as **growth regulators** (phytohormones) themselves or inhibit the production or the action of existing phytohormones in the host plant. This type of effector molecule may result in abnormal growth in infected plants. For example, infection by the fungus *Gibberella fujikuroi* causes rice shoots to grow much faster than their uninfected neighbors (Figure 8–16). The Japanese name for this disease translates as "foolish seedling" disease. Analysis of infected rice plants revealed that the fungus was producing a molecule that accelerated plant growth: this was named **gibberellin**, after the fungus. Subsequent research showed that plants themselves produce gibberellins. This class of phytohormone is a very important part of the mechanisms that control growth in healthy plants (see Chapter 5). The best-studied instance of modification of plant growth by a pathogenic bacterium is that of *Agrobacterium tumefaciens*, which causes crown gall and hairy root disease in many eudicot plant species. We describe this interaction in detail below.

Agrobacterium transfers its DNA (T-DNA) into plant cells to modify plant growth and feed the bacterium, and this transfer system is used in biotechnology

Crown gall and hairy root diseases are caused by the closely related bacteria *Agrobacterium tumefaciens* and *Agrobacterium rhizogenes*, which engage in a highly unusual and complex biotrophic interaction with plant cells. After infection of wounded plant tissue by the bacterium, galls or tumors emerge. Galls caused by *A. tumefaciens* support growth of the invading bacteria: the cracked and open structure of the galls creates a haven for bacterial proliferation (Figure 8–17A). *A. rhizogenes* infects plant roots, causing tumors that give rise to a proliferation of roots (Figure 8–17B). The plant cells in the galls produce **opines**, conjugates of amino acids and sugars, that provide a food source for *Agrobacterium* but cannot be metabolized by the plant cells or by other bacteria.

The unusual feature of *Agrobacterium* infections is that the bacterium engineers the plant to make molecules that promote successful bacterial colonization and growth, rather than producing effector molecules itself. It does this by transferring some of its own DNA into the plant cells (Figure 8–18). Cells that receive bacterial DNA are said to be "transformed." The bacterial transferred DNA (**T-DNA**) encodes proteins that confer two main properties on the transformed cells. First, the cells are stimulated to divide more rapidly: genes on the T-DNA encode enzymes involved in the synthesis of the phytohormones **auxin** and **cytokinin**, which promote cell division. Second, the

Figure 8–16.
Symptoms of "foolish seedling" disease caused by infection of rice by *Gibberella fujikuroi*. The tall pale seedling (*arrow*) is infected by the pathogen.
(Courtesy of Yuji Kamiya.)

(A)

(B)

Figure 8–17.
Symptoms of infection of plants by
Agrobacterium spp. (A) Crown gall on a
Pelargonium stem, caused by *A. tumefaciens*
infection. (B) "Hairy roots" caused by
A. rhizogenes infection of the hypocotyl of a
poplar (*Populus tremuloides*) seedling.
(A, courtesy of Halvor Aarnes.)

Figure 8–18.
Main processes involved in
transformation of plant cells by
Agrobacterium tumefaciens. In the
bacterium, the Ti plasmid gives rise to a
T-DNA. The T-DNA is transferred to the plant
cell in a process mediated by expression of
other genes carried on the Ti plasmid. Inside
the plant cell, the T-DNA integrates into a
plant chromosome. Expression of genes on the
T-DNA leads to abnormal production of
phytohormones in the plant cell—and hence
cell proliferation to form a gall—and to
synthesis of opines, which can be taken up
and metabolized by the bacterial cells. Opines
are the main source of carbon and nitrogen for
the bacteria living in the gall.

transformed cells produce opines; again, T-DNA genes encode the enzymes needed to synthesize opines. Opines are not made by untransformed plants; the enzymes needed for opine synthesis are not encoded in plant genomes.

Within the bacterium, the T-DNA that will be transferred to the host and the genes required for its transfer are present on a **plasmid**, a circle of DNA separate from the main bacterial chromosome. In *A. tumefaciens* this plasmid is called the "tumor-inducing plasmid" or **Ti plasmid** (Figure 8–19); in *A. rhizogenes*, it is called the "root-inducing plasmid" or **Ri plasmid**. The Ti and Ri plasmids are similar in structure but encode different enzymes of phytohormone synthesis. The hormonal perturbations caused by expression of the Ti-plasmid genes in a plant cell lead to gall formation, whereas perturbations caused by expression of the Ri-plasmid genes lead to root proliferation. These functions are interchangeable: if the Ti and Ri plasmids are swapped between *A. tumefaciens* and *A. rhizogenes*, then disease symptoms are also swapped.

The proteins required to mediate T-DNA transfer from the bacterium to the plant cell nucleus are encoded by the *virulence* (*vir*) genes of the Ti and Ri plasmids. These genes form seven groups, or **operons**, *virA* to *virG*. The T-DNA is bordered by 24-bp DNA sequences, directly repeated on both the right and left borders of the T-DNA (Figure 8–19). These sequences are targets for the machinery that conjugates the DNA into the plant chromosome.

Agrobacterium usually enters the plant and causes disease only at wound sites. Wounded plants secrete an array of phenolic compounds, including acetosyringone (Figure 8–20), that induce the expression of genes in *Agrobacterium* needed for infection of plant cells. Bacteria detect acetosyringone through proteins encoded by *virA* and v*irG*. The VirA protein can bind acetosyringone. When binding occurs, VirA activates a **transcription factor** encoded by *virG*. VirG then activates expression of VirB, VirC, VirD, VirE, and VirF proteins, which are all components of the mechanisms required to

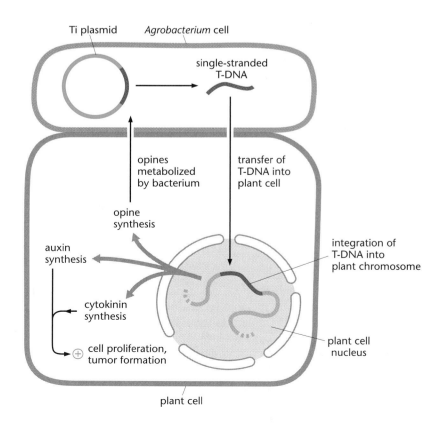

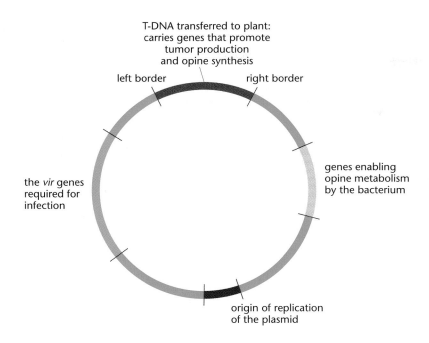

left border

right border

T-DNA transferred to plant:
carries genes that promote
tumor production
and opine synthesis

the *vir* genes
required for
infection

genes enabling
opine metabolism
by the bacterium

origin of replication
of the plasmid

Figure 8–19.
Structure of a Ti plasmid. The *Agrobacterium* Ti plasmid is a circular DNA strand carrying genes necessary for successful plant infection. The *vir* genes are required to promote transfer of the T-DNA into the plant cell and its integration into a plant chromosome. The T-DNA is demarcated by left and right border sequences. Integration of the T-DNA into a chromosome is initiated at the right border and terminated at the left border. Ti plasmids also carry genes encoding enzymes that metabolize the opines produced by infected plant cells, providing the bacterium with carbon and nitrogen.

conjugate the T-DNA from the Ti plasmid into the plant cell and integrate the DNA into the plant genome.

The T-DNA transfer process involves creation of **pilus** structures on the bacterial surface. The T-DNA leaves the bacterium through these pili and enters the plant cell. Once in the cell, the T-DNA is integrated into the host DNA at random locations in the host chromosomes. The mechanism by which *Agrobacterium* delivers T-DNA from the bacterium to the host-plant genome evolved from bacterial **conjugation**, a process by which bacterial plasmid DNA can be transferred from one bacterium to another. The process is initiated in the bacterium, when proteins encoded by *virD* genes cleave the 24-bp border sequences on one strand to release a single-stranded T-DNA molecule. This molecule binds to specific VirD and VirE proteins and is then delivered to the plant cell via the pore complex. Inside the plant cell, the T-DNA is targeted to the nucleus by the bound VirD and VirE proteins, which carry amino acid sequences recognized by the machinery that imports proteins into the nucleus. Integration of the T-DNA into the host chromosomes is promoted by very short regions of homology between T-DNA and plant DNA.

The natural transfer of DNA from *Agrobacterium* to plants has led to the organism being dubbed "nature's genetic engineer." This natural process has been exploited by biotechnologists to develop a system with which to introduce any new piece of DNA into plant cells without causing tumors, a process called plant **transformation** (see Chapter 9). The system separates the *vir* genes and the T-DNA components of the Ti plasmid onto two different plasmids that can be propagated in the same bacterial cell, creating a **binary vector system** (Figure 8–21). The genes encoding enzymes of opine and phytohormone biosynthesis are deleted.

Figure 8–20.
Role of acetosyringone in *Agrobacterium* infection. Acetosyringone is released from wounded plant cells and induces expression of *Agrobacterium vir* genes, thus initiating T-DNA transfer. Acetosyringone binds the VirA protein, enabling it to activate the VirG transcription factor. This in turn activates expression of the *vir* genes—*virB* through *virF*—that encode the proteins that permit entry of the T-DNA into the plant cell and its integration into a plant chromosome.

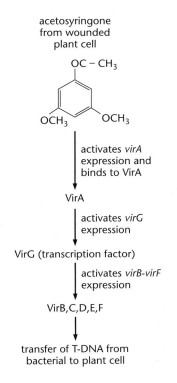

acetosyringone
from wounded
plant cell

activates *virA*
expression and
binds to VirA

VirA

activates *virG*
expression

VirG (transcription factor)

activates *virB-virF*
expression

VirB,C,D,E,F

transfer of T-DNA from
bacterial to plant cell

Figure 8–21.

Use of a binary vector system for plant transformation. Using standard techniques for DNA manipulation in the bacterium *Escherichia coli*, researchers engineer a T-DNA that contains the foreign gene of interest, a selectable marker gene, and the left and right borders necessary for integration into a plant chromosome. The selectable marker gene encodes a protein that enables detoxification of a compound normally poisonous to plant cells. (For example, a commonly used selectable marker gene encodes an enzyme that modifies and detoxifies the antibiotic kanamycin.) A plasmid containing this engineered T-DNA is transferred to a strain of *Agrobacterium* containing a second, "helper" plasmid with the *vir* genes necessary for T-DNA transfer to the plant cell and integration into a plant chromosome. The *Agrobacterium* strain containing the two plasmids is used to infect plant tissue. The infected tissue is grown on a selective medium—that is, a medium containing the toxic compound (e.g., kanamycin) that can be detoxified by the protein encoded by the selectable marker gene in the T-DNA. Plant cells that have not been infected are killed by the toxic compound in the medium. Cells successfully infected by *Agrobacterium*—and thus containing the foreign gene of interest—can detoxify the toxic compound and are able to grow. Transformed plants expressing the foreign gene can be regenerated from these cells.

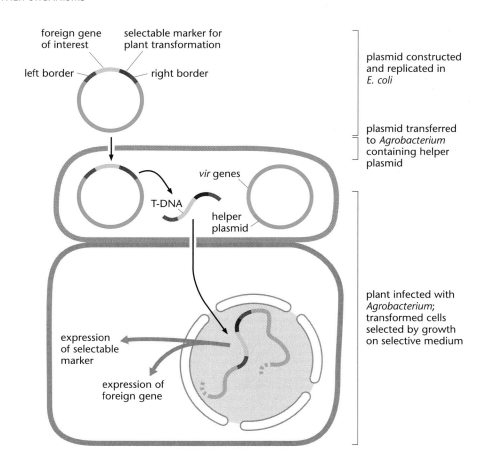

The T-DNA component is carried on a plasmid that can replicate in *Agrobacterium* and in *Escherichia coli*, the bacterium used in laboratories for DNA manipulation and replication. The T-DNA retains its left and right borders (LB and RB), and carries between them an introduced "selectable" **marker gene**, which encodes a protein that confers resistance to an antibiotic or another compound that is normally lethal to plant cells. This marker enables researchers to specifically select plant cells that have been transformed: the transformed cells will be able to grow on substrates containing the toxic compound but untransformed cells will be killed. Additional genes or other DNA sequences can be cloned into the T-DNA region between the left and right borders and next to the selectable marker. The second plasmid in the binary vector system is a Ti plasmid lacking the T-DNA region but still carrying the *vir* genes.

The binary vector system is assembled in a strain of *Agrobacterium* that carries a Ti plasmid derivative that provides *vir* genes but lacks its own T-DNA, and the *Agrobacterium* is allowed to infect plant cells. The *vir* genes carried on one of the two plasmids enable the T-DNA carrying the selectable marker and a foreign gene on the other plasmid to be transferred into a plant cell and integrated into its genome. Growth on a substrate containing the appropriate toxic compound, followed by regeneration of whole plants from the transformed cells, allows the impact of the expression of the foreign gene on plant growth to be studied. This technique has been widely exploited both in biotechnology (see Section 9.3) and in curiosity-driven research (see Section 2.5).

Agrobacterium strains carrying binary vectors can also be infiltrated into leaves, resulting in transient delivery and expression of genes in the T-DNA in the infiltrated region of the leaf. This can provide rapid tests for gene function without the need to regenerate **transgenic plants**. For example, as described below, transient expression was used to show that bacterial effector proteins are recognized inside the plant cell.

Some pathogen effector molecules are recognized by the plant and trigger defense mechanisms

We have described how pathogens make a diverse set of effector molecules that promote their pathogenicity. The effector molecules considered so far were all discovered through chemical and biochemical studies of diseased plants. Additional effector molecules were first identified through genetic studies, which revealed the existence of single plant genes—the *R* (resistance) genes—that render a plant resistant to a particular pathogen. Within a given species, plants carrying a particular *R* gene can recognize one specific effector molecule produced by a pathogen. This recognition triggers defense responses that prevent growth of the pathogen (effector-triggered immunity; see Figure 8–2). Plants that lack this *R* gene cannot recognize the effector molecule and so do not initiate defense responses, and thus they become diseased. Conversely, a strain of the pathogen that does not make the specific effector molecule recognized by the plant will be able to infect plants carrying that particular *R* gene, because the pathogen does not trigger the defense response. This situation is described by the **gene-for-gene model**.

The gene-for-gene model was originally put forward in the 1940s by Harold Flor. He studied the genetic basis of the ability of some races (genetic variants) of the rust fungus *Melampsora lini* to overcome the resistance conferred by an *R* gene in the flax plant (*Linum usitatissimum*), and thus to cause disease. Flor found that virulent races could overcome *R* gene–mediated resistance and cause infection, and that the virulence was genetically recessive. He inferred that the virulent races lacked a functional gene that was present in races of the rust fungus that did not cause disease in resistant flax plants. Races that do not cause disease in plants carrying a particular *R* gene are designated "avirulent races," and the functional gene that results in avirulence in the fungus—and that is absent from virulent races—is called an "avirulence gene," or *avr* gene (Figure 8–22).

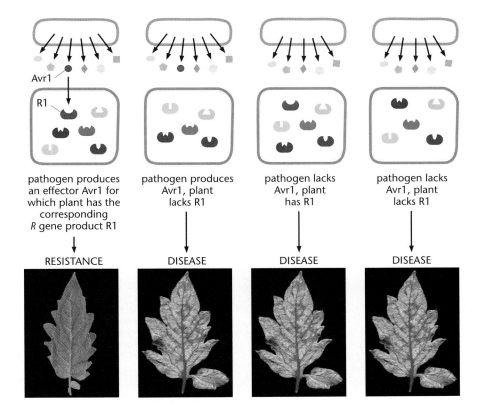

Figure 8–22.
Illustration of the gene-for-gene model.
The four strains of pathogen (*top*) each produce a distinct range of effector molecules. Two of the four carry the *avr1* gene and so produce effector molecule Avr1 (*red*). The four genotypes of the host plant each have a range of *R* genes, and two of the four have the *R1* gene, which encodes the R1 protein that enables the plant to recognize the pathogen Avr1. A plant is resistant to the pathogen only when a pathogen carrying the *avr1* gene infects a plant carrying the *R1* gene, thus triggering *R* gene–mediated defense mechanisms (*left*). In all other cases the plant is unable to recognize the pathogen and thus is susceptible to infection and disease. The photographs show resistant and susceptible interactions between the fungus *Cladosporium fulvum* and a tomato leaf.

Avr1

R1

pathogen produces an effector Avr1 for which plant has the corresponding *R* gene product R1

pathogen produces Avr1, plant lacks R1

pathogen lacks Avr1, plant has R1

pathogen lacks Avr1, plant lacks R1

RESISTANCE

DISEASE

DISEASE

DISEASE

Many bacterial *avr* genes have been identified, by a process in which plasmids carrying pieces of bacterial chromosome from an avirulent strain are transferred into a virulent strain of the same species. The resulting bacteria are screened to find which have become avirulent—that is, which are unable to infect a host plant carrying the appropriate *R* gene. These bacteria must be carrying the *avr* gene from the avirulent strain (Figure 8–23). On further analysis of the transferred DNA responsible for avirulence, the *avr* gene can be defined.

Sixty years after it was first put forward, the gene-for-gene model is still a useful basis for understanding many plant–pathogen interactions. The model provided the first clue to the existence of a set of effector molecules—produced in pathogens that carry *avr* genes—whose function is to promote the pathogen's successful colonization of the plant by overcoming basal resistance mechanisms, but which can potentially be recognized by the plant as a signal to activate defenses. We discuss these effector molecules below; in Section 8.4 we explain how *R* genes enable plants to recognize effector molecules and trigger defenses against the microbes that produce them.

The gene-for-gene model explains the interactions of many biotrophic and hemibiotrophic pathogens with plants. Examples include interactions of rust fungi with maize, wheat, and flax; *Tobacco mosaic virus* with tobacco and *Potato virus X* with potato; powdery mildew with barley and wheat; downy mildew with *Brassica* species; late blight with potato; *Pseudomonas* and *Xanthomonas* bacteria with cereals and legumes; and nematodes with tomato and potato. In each of these examples, for resistance to occur, an *R* gene must be present in the host plant and the corresponding *avr* gene must be present in the pathogen. Importantly, if a corresponding *R* gene is not present in the host, effector molecules encoded by the pathogen *avr* genes contribute to the pathogen's proliferation and reproductive success. Pathogen races that carry functional *avr* genes are often more pathogenic than races that do not.

The products of some bacterial *avr* genes act inside the plant cell

Early evidence that Avr proteins act inside cells of the host plant was obtained by using *Agrobacterium* to transiently express the gene encoding an Avr protein from

Figure 8–23.

Identification of bacterial *avr* genes. This method is used to identify a bacterial *avr* gene that has been defined on the basis of the inability of the strain carrying the gene to cause disease on a particular plant genotype (i.e., a plant carrying the corresponding *R* gene). DNA from the bacterium containing the *avr* gene of interest is digested to produce fragments. Individual fragments are transferred into bacteria lacking the *avr* gene, which are able to cause disease on the plant with the corresponding *R* gene. This produces a set of strains carrying different fragments of DNA from the original bacterium. Most of the fragments will not carry the *avr* gene of interest, and the strains containing these fragments will continue to cause disease on a plant carrying the corresponding *R* gene. However, a strain containing the fragment carrying the *avr* gene of interest will be recognized by the plant and so will not cause disease. The DNA fragment can be recovered from this strain and sequenced to reveal the nature of the *avr* gene.

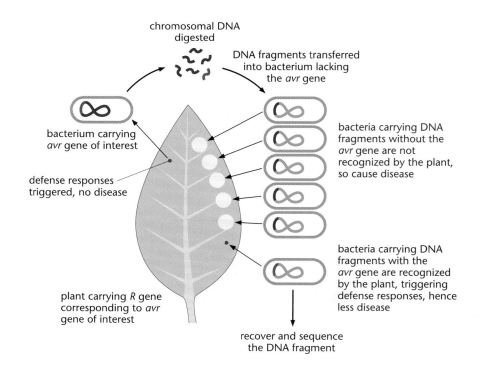

Xanthomonas, AvrBs3, in the leaves of pepper (*Capsicum*) plants. This technique ensures that the Avr protein is expressed inside the plant cell. The protein triggered a defense response in plants carrying the corresponding *R* gene; it did not produce a defense response in plants that lacked the *R* gene. This result suggests that during an interaction between an avirulent strain of *Xanthomonas* (i.e., a strain expressing *avrBs3*) and a plant carrying the corresponding *R* gene, detection of the AvrBs3 protein by the host takes place inside plant cells rather than in the intercellular space where the bacteria multiply.

Many bacterial *avr* genes have now been identified. Most encode hydrophilic proteins without obvious homology to other proteins, but some encode proteins with known functions. For example, some encode **proteases** and others encode enzymes that promote phosphorylation of host proteins, or remove phosphate groups from host proteins. Protein **phosphorylation/dephosphorylation** is an important means of regulating protein activity (see Section 4.1).

Plant-pathogenic bacteria usually deliver 10 to 30 different types of effector molecule into host cells. Although it is not known precisely how these effector molecules facilitate infection by the bacterium, they most likely interfere with signaling pathways that would normally allow the host cell to detect and respond to PAMPs. Some Avr proteins are located at the host-cell plasma membrane. Others are targeted to the nucleus, where they are thought to bind to DNA and thus alter **transcription** of host genes. The AvrBs3 protein of *Xanthomonas* acts in this way.

We have described how bacterial Avr proteins can act in plant cells both as effectors and, paradoxically, as triggers that alert host plants with the appropriate *R* genes to their presence. But how do they enter the plant cell? Plant-pathogenic bacteria have a specialized mechanism, the **type III secretion system**, for delivering effector molecules. This mechanism is also used by many bacterial pathogens of animals. Many of the bacterial effector molecules we described earlier—enzymes, toxins, and phytohormones—are secreted into the extracellular space of the host plant, but the type III secretion system delivers specific effector proteins into the cytoplasm of host cells.

The type III secretion system involves a protein complex that spans both inner and outer bacterial membranes and a pilus, a projection that penetrates the host cell (Figure 8–24). In plant-pathogenic bacteria, these structures are encoded by the *hypersensitive response and pathogenicity* (*hrp*) genes. Bacteria carrying mutations in *hrp* genes are unable to deliver effector molecules, including Avr proteins, into plant cells. This results in loss of both pathogenicity and the capacity to elicit a defense response in host plants carrying appropriate *R* genes.

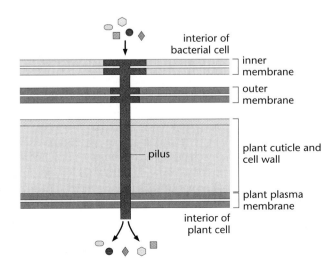

Figure 8–24.
Transfer of effector proteins from pathogenic bacteria into plant cells via a type III secretion system. The type III secretion system forms a tube that extends from the interior of the bacterial cell through both bacterial membranes, the plant cuticle, the plant cell wall, and the plant plasma membrane, into the interior of the plant cell. The tube consists of protein complexes that span the inner and outer bacterial membranes and a pilus that projects into the plant cell. Several types of bacterial effector molecules can enter the plant cell through this structure.

The structural proteins of type III secretion systems show homology to components of the flagellar apparatus used in bacterial motility, and the mechanism for their assembly presumably evolved from the flagellar assembly pathway. Both mechanisms export proteins from the bacterial cell in an ordered way to make a complex structure (the flagellum or the pilus) that extends from the cell membrane through the wall to the outside of the cell. Exactly how the pilus penetrates the plant cell wall and plasma membrane to deliver the bacterial effector molecules into the plant cytoplasm is not yet clear.

The functions of fungal and oomycete effector molecules are poorly understood

Identification of *Avr* genes and effectors from fungal and oomycete pathogens is more difficult than in bacteria, because most of these organisms are more difficult to transform with foreign DNA and many cannot be grown in the laboratory outside their host plants. Nevertheless, some fungal *Avr* genes have been identified. Avr proteins from *Cladosporium fulvum* are secreted into the intercellular spaces of infected leaves. They can be isolated and characterized by infiltrating the intercellular spaces with water and then recovering the fluid by low-speed centrifugation (Figure 8–25). Although the functions of many Avr proteins recovered in this way are not yet known, the functions of some have been elucidated. Avr4, for example, is a chitin-binding protein that is believed to mask the chitin of the fungal cell wall at the hyphal tip, preventing the plant's recognition of and response to the pathogen. Avr2 is a protease inhibitor that binds with high affinity and inhibits a plant protease, Rcr3, found in the intercellular spaces of tomato leaves. The *Cf-2* resistance gene of tomato confers resistance to *C. fulvum* carrying *Avr2*. Resistance also requires the presence of Rcr3, expression of which is normally induced in response to *C. fulvum* infection. Plants that carry the *Cf-2* gene but do not express Rcr3 are not resistant to *C. fulvum* strains carrying the *Avr2* gene. Thus it seems that the product of the *Cf-2* gene enables the host plant to recognize when Rcr3 has been inhibited by Avr2 and to activate defenses in response to this event (an effector-triggered immune response; see Figure 8–2).

In addition to these methods of directly identifying proteins from infected leaves, genetic and genomic approaches are beginning to provide information about fungal and oomycete effector proteins. As with bacteria, genetic studies are being used to identify *Avr* genes whose products can be recognized by the host plant. Once such genes are known, searches of the pathogen genomes can identify related genes that may encode additional effectors. We describe here several examples of effector proteins identified by these approaches.

The *Bgh* genes of the powdery mildew pathogen *B. graminis* f. sp. *hordei* encode Avr proteins that trigger resistance in barley plants carrying specific **alleles** at the *Mla* locus. The function of these Avr proteins is not known. It is also not clear how they move from the fungus into the host cytoplasm, because they do not carry **signal peptides** for export from the fungal cells. Effectors recognized by specific host *R* genes have also been identified for oomycete pathogens, including *Phytophthora infestans*, the causal agent of potato blight. All those identified so far are small proteins that carry a signal peptide for export from the pathogen and also a small conserved sequence of amino acids, or **motif**, containing arginine (Arg) and leucine (Leu) (Arg-X-Leu-Arg, where X is any amino acid), that is required for uptake from the space between haustorium and plant cell, into the plant cell. A related motif (Arg-X-Leu) is also found in secreted effectors of the malaria pathogen *Plasmodium falciparum*, and is necessary for uptake of the effectors into human erythrocytes. Remarkably, then, this mechanism is conserved among oomycetes and their apicomplexan relatives such as *P. falciparum* (Figure 8–8). Genomic DNA sequences from oomycete pathogens are revealing genes encoding hundreds of additional small proteins with secretion signal peptides and Arg-X-Leu-Arg motifs that probably act as effectors.

Most effectors of fungal and oomycete pathogens are likely to be proteins. However, the *Avr* gene product of *Magnaporthe grisea*, Ace1, is similar to proteins involved in

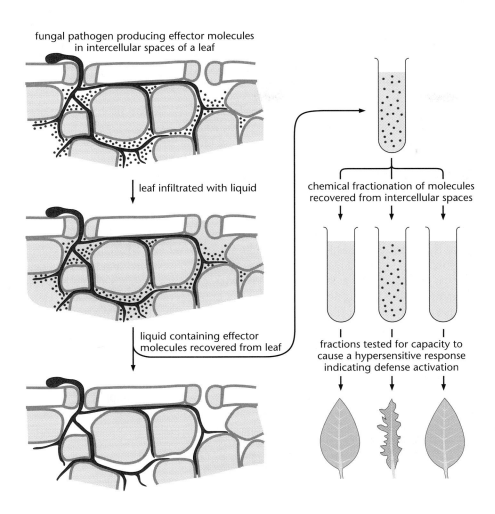

fungal pathogen producing effector molecules
in intercellular spaces of a leaf

leaf infiltrated with liquid

liquid containing effector
molecules recovered from leaf

chemical fractionation of molecules
recovered from intercellular spaces

fractions tested for capacity to
cause a hypersensitive response
indicating defense activation

Figure 8–25.
Identification of fungal Avr proteins.
Molecules secreted from fungal hyphae into the intercellular spaces of an infected leaf can be recovered by infiltrating the spaces with liquid then centrifuging the leaf gently to recover a solution of intercellular compounds. Individual effectors can then be identified by chemical fractionation of the solutes (e.g., by liquid chromatography) followed by testing of individual fractions for their ability to activate a defense response (a hypersensitive response; see Section 8.4) in a plant lacking the corresponding *R* gene, causing disease-like symptoms.

the synthesis of **polyketides**, a class of compounds that includes several important antibiotics. It seems likely that the Ace1 protein synthesizes a toxin that acts as an effector molecule.

Fungi have no equivalent of the bacterial type III secretion system, and effector proteins must be secreted into the intercellular spaces of the host plant and then taken up into host cells. In addition, some fungal pathogens have specific plasma membrane transporters that aid in the secretion of nonprotein toxins into the intercellular spaces of the host plant. For example, in *M. grisea* the expression of genes encoding ATP-dependent membrane transporters (**ABC transporters**; see Section 3.5) is induced by antifungal compounds produced by the host plant (see Section 8.4). The transporters probably play a role in exporting toxins that act as effector molecules: strains of *M. grisea* that cannot produce a vital component of the ABC transporter show greatly reduced pathogenicity.

8.2 PESTS AND PARASITES

Parasitic nematodes form intimate associations with host plants

Many different genera of nematodes feed on plants, causing disease. Feeding occurs through a hollow **stylet** that can penetrate plant cell walls. Some nematodes, the **ectoparasites**, feed at the surface of roots. They are mobile in the soil and can move from one plant to another. Other nematode species are **endoparasites** that invade root

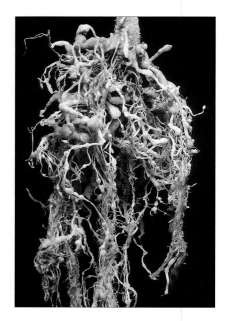

Figure 8–26.
Disease of tomato roots caused by the root knot nematode *Meloidogyne*. Some nematode species invade the root system and cause root cells to enlarge and divide. The resulting root distortions and swellings reduce the capacity of the root system to take up water and nutrients.

Figure 8–27.
Life cycle of root-infecting nematodes. (A) The free-living juvenile nematode is attracted to the plant by secretions from the root (*not shown*). It penetrates the root and starts to feed from cells in the vasculature. Feeding stimulates changes in the anatomy of the infected region to form "feeding structures." Feeding by cyst nematodes induces formation of a syncytium, which may involve hundreds of plant cells. Feeding by root knot nematodes induces endoreduplication of DNA and the consequent formation of giant cells. Female nematodes produce eggs internally; the female swells and protrudes from the root surface. The female then dies, forming a cyst that protects the eggs until they are shed. (B) Cross section of a mature female *Heterodera trifolii*, a cyst nematode, containing eggs and protruding from the surface of the root of a clover plant. Females are typically about 0.5 mm long. (C) Cross section of a *Meloidogyne incognita* (N), a root knot nematode, inside a tobacco root; it is feeding from giant cells formed from cells in the host vascular tissue. (FC = feeding cell)

tissues and feed inside the root. They depend on living host plants to complete their life cycle—that is, they are biotrophs. Two genera of endoparasitic nematodes are the source of much crop damage: the cyst nematodes (*Heterodera* spp. and *Globodera* spp.) and root knot nematodes (*Meloidogyne* spp.) (Figure 8–26).

Mature endoparasitic nematodes release eggs into the soil. These remain dormant until they detect specific molecules secreted from the roots of a potential host plant, which stimulates hatching. The juvenile nematodes are motile. They swim to and penetrate plant roots and migrate to the vascular tissue, where they initiate feeding and become sedentary (Figure 8–27).

Juvenile cyst nematodes, including those of *Heterodera*, produce a stylet that penetrates the plant cell wall but not the plasma membrane. Chemicals and enzymes released by the stylet have a profound effect on the anatomy of the infected part of the plant. **Symplastic** connections between the infected cell and adjacent cells become more extensive, eventually leading to cellular fusion. Up to several hundred plant cells may be recruited into a feeding structure with extensive symplastic continuity (a **syncytium**). The feeding nematode then grows until it is ready to produce and release its eggs (Figure 8–27B). Root knot nematodes use a different feeding strategy. The presence of juveniles induces DNA **replication** in host cells that is uncoupled from cell division (**endoreduplication**; see Section 3.1), which results in abnormal growth and the formation of giant cells (Figure 8–27C).

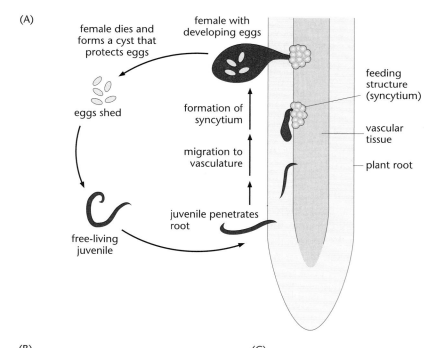

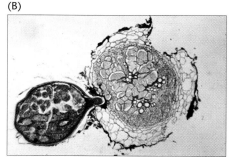

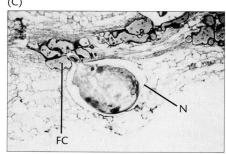

Both the syncytial feeding structures induced by cyst nematodes and the giant cells induced by root knot nematodes become closely associated with the host phloem. **Transfer cells**, with extensive wall invaginations that increase the surface area for nutrient transfer (see Section 4.4), develop between the feeding structures and phloem elements. This allows the feeding nematode to divert substantial quantities of sugars and amino acids from the host plant for its own growth and reproduction, resulting in substantial yield losses to the plant.

Nematodes produce a large number of effector molecules that alter the growth and metabolism of the host plant. As is the case for microbial pathogens, some of these molecules are detected by the host plant, producing a defense response. For example, a gene-for-gene relationship has been demonstrated between a cyst nematode and its potato host. If the potato has the *R* gene *H1*, it can detect and produce a defense response to nematodes with a particular *Avr* gene. Several plant *R* genes that confer resistance to particular nematode strains have been isolated. Interestingly, all of these encode proteins of the same class as the *R* gene products involved in resistance to bacterial pathogens. These *R* gene products are discussed in Section 8.4.

Insects cause extensive losses in crop plants, both directly and by facilitating infection by pathogens

Many insects depend on plants for food or for shelter in which to complete stages of their life cycle. They can cause immense destruction to crops. Chewing insects that damage crops by large-scale consumption of plant tissues include the locust, the Colorado potato beetle, the European corn borer, and the cotton boll weevil (Figure 8–28). Sap-sucking insects such as aphids, whiteflies, leafhoppers, and thrips insert specialized mouthparts (stylets) into the phloem of the host plant and take in the phloem contents as food (Figure 8–29). The damage to crops is thus less immediately obvious than that caused by chewing insects, although heavy infestations can lead to significant yield losses.

(A) (B) (C) (D)

Figure 8–28.
Chewing and sap-sucking insects.
(A) Soybean aphid on a soybean leaf.
(B) Colorado beetle on a potato stem.
(C) Cotton boll weevil. (D) Longitudinal section of an uninfested maize stem (*top*) and an infested maize stem carrying a pupa of the European corn borer (*bottom*).
(A, courtesy of Bob O'Neill.)

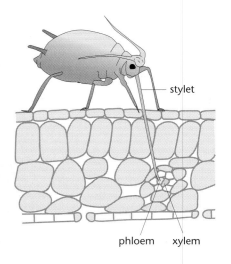

Figure 8–29.
Aphid feeding from leaf phloem. The stylet is inserted through the cell layers of the leaf and into the sieve element of the phloem, which contains very high concentrations of sucrose—food for the aphid.

Figure 8–30.
Parasitic plants. (A) European mistletoe (*Viscum album*). (B) Desert dodder (*Cuscuta denticulata*) swamping a bush of *Ambrosia dumosa* in the western United States. (C) Flower (the only aboveground structure) of *Rafflesia arnoldii*. The flowers may be up to 100 cm in diameter and weigh more than 10 kg. (C, courtesy of Troy Davies.)

Some insects, particularly sapsuckers, are also indirectly responsible for plant diseases because they transmit plant viruses from one host to another. Such insects act as **vectors** for the virus. Viruses transmitted by insects are either stylet-borne (adhering in a specific manner to an insect's mouthparts) or contained in the insect's saliva (**circulative viruses**). When circulative viruses are ingested by the insect, they pass through the gut wall to the salivary glands via the blood system (hemocoel), and can then be reintroduced to a plant during feeding. Some circulative viruses persist and replicate inside their specific insect vectors, and these may be regarded as insect viruses that have adapted to plants. These virus–insect interactions are highly specialized; indeed, many viruses are adapted to one particular insect species but can infect a broad range of plants. Vector insects deliver the virus to either epidermal or **mesophyll cells** or directly to the vascular system of the host plant. Viruses introduced into the phloem can rapidly spread through the plant.

The mechanical damage caused by chewing insects indirectly facilitates infection of plants by a range of pathogens. Some bacterial, fungal, and viral pathogens gain entry to plants primarily through wounds such as those caused by insect feeding. For example, mechanical damage caused to maize cobs by European corn borer infestation leads to higher levels of cob colonization by necrotrophs such as *Fusarium* species and higher levels of the associated mycotoxins.

Plants have a sophisticated array of defense mechanisms against attacking insects. Many of these mechanisms are related to those triggered by microbial pathogens. We describe some plant defense mechanisms against insects, and some specialist interactions in which insects overcome such defenses, in Section 8.4.

Some plants are plant pathogens

About 4000 plant species in several different families are biotrophic pathogens of other plants. The ability to parasitize other plants has evolved independently several times in angiosperm evolution. Parasitic plants differ greatly in the extent to which they depend on their host plants and in the route by which they tap resources. They also differ in the degree to which they have lost the capacity for **photosynthesis**. Many plant parasites, such as mistletoe (*Arceuthobium* and *Viscum*; Figure 8–30A), contain chlorophyll and can photosynthesize, but have no roots. They rely on the host plant primarily for minerals and water. Others, such as dodder (*Cuscuta* spp., also known as strangleweed; Figure 8–30B) and the bird's-nest orchid (*Neottia nidus-avis*), which parasitizes beech trees (*Fagus sylvatica*), cannot photosynthesize and have no true roots. They rely on the host plant for organic carbon and nitrogen as well as minerals and water. Some plant parasites (e.g., mistletoes and dodder) attach to stems of the host plant, whereas others attach to roots; examples of the latter are bird's-nest orchid and *Rafflesia arnoldii*, a parasitic plant found in the jungles of Southeast Asia that has one of the world's largest flowers (Figure 8–30C).

The most destructive parasitic plant of crops is witchweed (*Striga* spp.; Figure 8–31A), which takes water and minerals from the roots of a wide range of plant species. Its

(A)

(B)

(C)

effects in tropical Africa are devastating and very difficult to control. Witchweed seeds close to host roots germinate in response to chemicals present in root exudates, and the seedling roots grow toward the host in response to the chemical concentration gradient. Upon contact with the host, the *Striga* root swells into a haustorium, which rapidly penetrates the host root (Figure 8–31B). More than two-thirds of the 73 million hectares cultivated in sub-Saharan Africa are infested with *Striga*, and in extreme cases complete failure of maize, sorghum, millet, and legume crops can result. Each *Striga* plant can produce up to 500,000 seeds, allowing rapid spread in favorable conditions.

The haustoria of plant parasites are multifunctional organs that attach the parasite to the host, invade host tissues, and provide the vascular continuity between host and

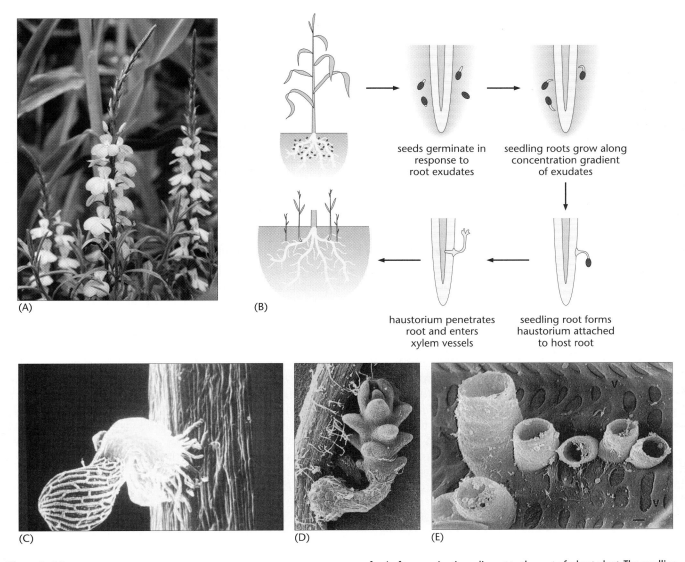

Figure 8–31.
***Striga*, a devastating parasitic weed.** (A) *Striga* plants flowering close to the base of a maize plant on which they are parasites. (B) Development of *Striga* plants. Seeds germinate close to the roots of a host plant. The *Striga* roots grow toward the root surface of the host, where they form haustoria. Haustorial cells grow into the root, penetrating the xylem vessels. Mature *Striga* plants take water and minerals from the xylem, drastically reducing the host plant's yield. (C–E) Scanning electron micrographs showing *Striga* infection and development. (C) A *Striga* seed (the structure with a netted surface) after germination adjacent to the root of a host plant. The seedling root has formed a haustorium on the host root surface; the haustorium is surrounded by root hairs that promote its attachment to the root surface. (D) A later stage in *Striga* development. The shoot is starting to develop, and attachment to the xylem of the host root is established. (E) The inner face of a xylem vessel wall in a host-plant root, showing the tube-like water-conducting vessels (oscula) of a *Striga* plant penetrating the wall. (D and E, from I. Dorr, *Annals Bot.* 79:463–472, 1997. With permission from Oxford University Press.)

parasite through which the parasite gains water and nutrients. Analysis of haustorium development is helped greatly by the ability to study it in the laboratory in the absence of a host plant. The roots of *Striga* seedlings respond rapidly to application of exudates from host-plant roots. Localized swelling of *Striga* root tips to form haustoria occurs within hours, caused initially by expansion and later by division of cortical cells. At the same time there is a proliferation of **root hairs** over the swollen zone (Figure 8–31C); these hairs are different from typical root hairs in that their surface is coated with hemicellulose-rich protrusions (papillae). The papillae promote attachment of the hairs, and hence the haustoria, to host-cell surfaces.

Although *Striga* haustoria can attach to the roots of non-host species, successful penetration occurs only on host species. Attachment of haustoria to the roots of non-host plants elicits defense responses similar to those to microbial pathogens (see Section 8.4). On host plants, haustorial cells at the parasite–host interface elongate and divide, pushing through the **epidermis** and **cortex** of the host root. When *Striga* seedlings are placed on sorghum roots, they can penetrate through the cortex within two to three days (Figure 8–31D).

Once a haustorium has penetrated through to the **stele**, the leading cells enter host xylem vessels through the **pits** in the xylem walls (see Section 3.6). Haustorial cells lose their cell walls at the tips, and these cells and adjacent cortical cells in the parasite root differentiate to form a continuous water-conducting system that links the host xylem with the vasculature of the parasite root (Figure 8–31E).

8.3 VIRUSES AND VIROIDS

Viruses and viroids are a diverse and sophisticated set of parasites

Viruses and viroids are major causes of plant disease. Viral particles typically comprise a **nucleic acid** surrounded by a protein coat, called a **capsid**; viroids are short, circular, single-stranded **RNA** molecules. Both are obligate biotrophs, in that they can replicate only within living cells. Plant viruses and viroids utilize cellular components from the host plant for their replication, but in many cases the mechanisms by which they disrupt cellular processes and cause disease symptoms are not clear. Viruses and viroids rarely kill their hosts, but they weaken the plant and reduce growth and seed yield. The disease symptoms caused by plant viruses are often included in viral names: for example, *Tobacco ringspot virus*, *Potato leaf roll virus*, *Beet curly top virus*, *Tobacco mosaic virus*, and *Tomato bushy stunt virus*. Viruses can be transmitted from plant to plant by a variety of other organisms, including insects, nematodes, and even fungi. They are also transmitted when plants are vegetatively propagated. This is a particular problem for crops that are propagated by cuttings or vegetative organs rather than by seed, such as banana or potato. Viruses are sometimes transmitted from one sexual generation of a plant to the next in the seed—for example, *Pea seed-borne mosaic virus*—but this is less common.

The genomes of plant viruses encode small numbers of proteins. These allow viral replication, facilitate spread of the virus from cell to cell, and suppress host defense responses. All viruses (except geminiviruses and nanoviruses) encode **replicases** (RNA or DNA polymerases), which replicate the viral genome inside the host cell. They also encode coat proteins, which associate to form the capsid, and this defines the viral shape. Most viruses also encode one or more proteins called **movement proteins**. In all cases movement proteins modify the structure and/or function of **plasmodesmata**, channels that interconnect plant cells. This modification is necessary because plasmodesmata do not normally allow the passage of molecules as large as viral particles or viral nucleic acids. The structural modification can occur in the form of new tubular (strawlike) structures that allow the particles to pass through, or the movement proteins bind viral nucleic acids to form complexes that can move through plasmodesmata that have been modified in less obvious ways. Many viruses also encode suppressors of the host's RNA-silencing mechanisms (see Section 8.4). The genomes of viruses that are transmitted by insects, fungi, or nematodes often encode "helper proteins" that promote interaction of the viral particle with the vector. Some viral genomes encode proteases that enable post-translational processing of the **translation** product of the viral nucleic acid (see below). All of these virus-encoded proteins can be thought of as effector molecules, analogous to those of microbial pathogens.

Associated with the compact organization of viruses, two different mechanisms enable multiple proteins to be produced from a single nucleic acid molecule contained in the capsid. Some viral RNAs direct the synthesis of a single large protein (a polyprotein) that is then processed by virally encoded proteases to form multiple proteins with distinct functions. Other viral RNAs can be replicated inside the host cell to produce one or more separate, smaller RNAs. Each of these **subgenomic RNAs** can direct the translation of a subset of viral proteins.

Viral classification is based on a range of factors that include viral particle size and shape, the nature of the viral genome, and the different strategies for viral replication and protein production. Plant viruses are most often spherical (isometric), or rodshaped with a straight or wavy appearance. The nucleic acid of plant viruses may be RNA or DNA, and it may be single-stranded (ss) or double-stranded (ds), depending on the type of virus. The majority (about 70%) of plant viruses have genomes of ssRNA with coding potential (+ sense); viral particles containing complementary (−) strand RNA also occur. In this section we describe the general features of three major groups of viruses: single-stranded (+) RNA viruses, exemplified by tobamoviruses and potyviruses; dsDNA viruses, exemplified by caulimoviruses; and ssDNA viruses, exemplified by geminiviruses. We discuss the salient features of their structures and replication mechanisms, and how the viruses might cause disease symptoms. Plant defenses against viruses are described in Section 8.4.

Different types of plant viruses have different structures and replication mechanisms

The type member of the **tobamovirus** group is *Tobacco mosaic virus* (TMV). This is a simple, single-stranded (+) RNA, straight rod-shaped virus (Figure 8–32A) that is transmitted mechanically rather than via a vector. The multiplication of the virus occurs in several steps (Figure 8–32B). After entry into the cell the virus must be uncoated before the RNA can be translated. For TMV this process is coupled with the first step of translation such that the ribosome binds to the 5′ end of the RNA and displaces coat protein subunits as the RNA is read. The first protein to be produced is the RNA polymerase, which uses the viral RNA molecule as a template on which to synthesize complementary (−) RNA. This proceeds from the 3′ end of the RNA. The (−) strands produced by the RNA polymerase serve two purposes. First, parts of the (−) strand are copied again by RNA polymerase, but starting at an internal site and so producing smaller subgenomic (+) RNA molecules, each of which encodes a subset of the viral proteins. These subgenomic RNAs are in turn translated to give specific viral proteins. Second, the whole (−) strand is used as a template by RNA polymerase to produce new, complete (+) strands. These associate with the coat proteins produced by translation of the subgenomic viral RNA to form new viral particles.

The TMV genome encodes only four proteins: the two largest (126 kDa and 183 kDa) are components of the RNA polymerase; the other two are a coat protein (18 kDa) and a movement protein (30 kDa). The movement protein changes the size limitations of plasmodesmata (see Section 3.4) to allow passage of larger particles, facilitating the movement of a TMV RNA–movement protein complex from cell to cell through the host plant. Because the virus encodes so few proteins, it is dependent on components of the host cell for many aspects of its replication.

Figure 8–32.
Structure and replication of *Tobacco mosaic virus*. (A) Electron micrograph of *Tobacco mosaic virus* particles. Each rod is 18 nm in diameter. (B) Cycle of synthesis of new viral particles inside a host-plant cell. (1) The viral RNA strand (+) is uncoated and is translated by plant cell ribosomes (*green blob*) to produce an RNA-dependent RNA polymerase (*blue oval*). (2) The RNA polymerase uses the viral RNA as a template to make the complementary (−) strands (*purple*). (3) Partial replication of (−) strands, from promoters within the strands, produces shorter, subgenomic (+) strands (*blue*). Full-length (+) strands are also produced. (4) The subgenomic strands are translated on plant ribosomes to produce both viral coat proteins and movement proteins. (5) Some full-length (+) strands are packaged with movement proteins for movement through plasmodesmata into other plant cells. (6) Other full-length (+) strands are packaged with viral coat proteins to produce new viral particles.

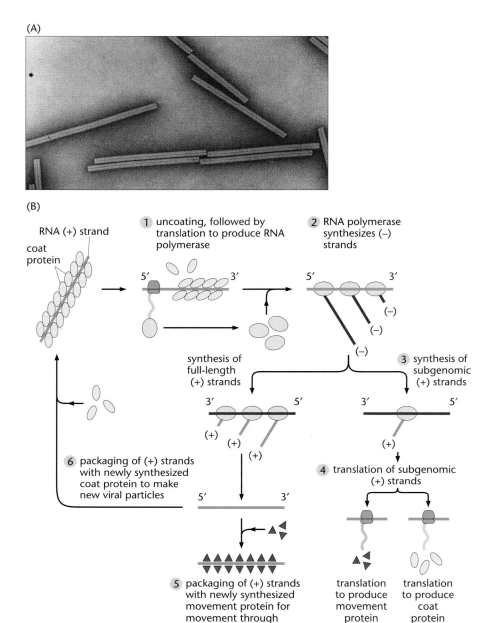

The type member of the **potyvirus** group is *Potato virus Y* (PVY). The particles of this virus are flexuous rods and, like TMV, have a single-stranded (+) RNA genome. Potyviruses are more complex than tobamoviruses. The extra complexity is of two sorts. Potyviruses encode more proteins than TMV, and their RNAs are translated to produce large polyproteins that are cleaved to form the active viral proteins. For example, the approximately 10-kb genome of PVY RNA is translated into a single polyprotein of about 360 kDa that is subsequently cleaved at specific points to produce, ultimately, about 10 proteins, each of which has separate and often multiple functions. In addition to the components of the viral RNA polymerase, coat proteins, and movement proteins, potyviruses encode proteases (the role of one protease in suppressing host-plant defense responses is considered in Section 8.4), an aphid transmission factor, and the protein VPg, which is attached to the end of the RNA molecule and is required for initiation of replication. Although the basic principles of RNA replication for potyviruses are similar to those for tobamoviruses, potyviruses do not produce subgenomic RNAs—hence the need for the different translation strategy.

The type member of the **caulimovirus** family is *Cauliflower mosaic virus* (CaMV). This virus is spherical and has a genome composed of circular dsDNA. The caulimovirus family falls within the larger badnavirus family, which collectively is unusual in that the dsDNA genome is replicated via a viral RNA intermediate in a process called **reverse transcription**. When CaMV particles enter a plant cell, the coat proteins are removed and the dsDNA enters the host-cell nucleus (Figure 8–33). Here it becomes twisted and coiled to form a circular mini-chromosome. This is transcribed by the host-cell RNA polymerase into two ssRNAs of different lengths, the 35S RNA and the 19S RNA, both of which are transported into the plant cytoplasm and translated. The 35S RNA encodes five proteins: coat protein, movement protein, a protein facilitating aphid transmission of the virus, a capsid-modifying protein that aids both intercellular movement of the virus and insect transmission, and the enzyme **reverse transcriptase**. The smaller 19S RNA is translated into a protein that activates translation of the 35S RNA. The cytoplasmic viral RNA also serves as a template for reverse transcription to form viral DNA. The CaMV reverse transcriptase can copy both RNA and DNA. Hence, it synthesizes the first complementary DNA strand from the RNA, and this strand becomes the template to create dsDNA. The dsDNA associates with coat proteins to form new viral particles. The DNA promoter for the 35S RNA of CaMV has been widely used in plant transformation experiments. It allows high-level expression of introduced genes in many different plant species and organs (see Section 9.3).

The **geminiviruses** are a large family of viruses that have small, circular ssDNA genomes of between 2.7 and 5.4 kb. Within this family are various subgroupings that differ in the number of unit-sized (~2.7 kb) DNAs that comprise the complete infectious virus. The name "gemini" (Latin for "twins") refers to the appearance of the viral particles, which consist of two fused, partially isometric subparticles (Figure 8–34A). Evidence suggests

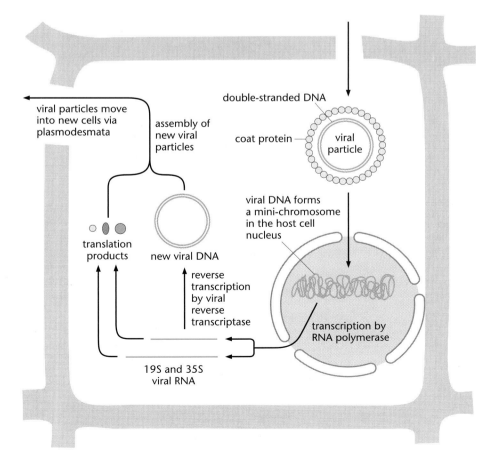

Figure 8–33.
Replication of *Cauliflower mosaic virus*. When the viral particle enters a plant cell (*top right*), coat proteins are removed and the double-stranded DNA enters the nucleus, where it forms a mini-chromosome. This is transcribed to produce a 19S and a 35S RNA. The RNAs are translated to produce new proteins (as described in the text), including a movement protein for cell-to-cell transmission, a reverse transcriptase, and coat protein required for assembly of new viral particles. The reverse transcriptase synthesizes the complementary DNA strand from the viral RNA, and then converts this to dsDNA, which is assembled into new viral particles.

that these paired particles contain only one of the unit-sized DNA molecules. Since the collective functions of the smaller and larger geminiviruses are similar, the smaller viruses have evolved a more compact organization with multifunctional proteins.

Geminiviruses encode factors required to initiate replication of their ssDNA genomes, but—unlike the viruses discussed above—they do not encode replicases (nucleic acid polymerases). They rely instead on host-plant nuclear DNA polymerases to replicate their genomes. However, differentiated, nondividing cells normally lack DNA polymerase and the associated factors required for replication, and thus the geminiviruses induce the expression of these host factors in order to replicate. This induction is achieved by inhibition of **retinoblastoma protein (RBR)**, a negative regulator of the cell cycle (see Section 3.1). Inhibition of RBR function allows expression of transcription factors that induce differentiated cells to reenter the cell cycle. This allows the viral genome to be replicated by the induced DNA replication machinery of the host plant, in a manner that shows parallels with animal tumor viruses.

When geminivirus DNA enters a cell, the ssDNA is converted to dsDNA in the nucleus (Figure 8–34B). The dsDNA is used as a template for transcription, and for DNA replication by a **rolling-circle replication** mechanism to produce more ssDNA genomes for encapsidation. Proteins are encoded on both strands of the DNA, hence transcription must be bidirectional to produce RNAs for all the encoded proteins.

Geminiviruses are found mostly in warmer climates, probably because of the geographic restrictions in the distribution of their leafhopper and whitefly insect vectors. Some examples are *Wheat dwarf virus* and *Maize streak virus* (Figure 8–34A), both transmitted by leafhoppers; *Tomato yellow leaf curl virus*, a whitefly-transmitted

Figure 8–34.
Replication of a geminivirus. (A) Electron micrograph of *Maize streak virus* particles. Note that each particle consists of two joined, incomplete icosahedra, hence the name "geminivirus." (B) The single-stranded viral DNA enters the plant cell nucleus, where a complementary DNA strand is synthesized. This forms a mini-chromosome, which is transcribed and translated to produce the proteins necessary for replication of the viral DNA and its assembly into new viral particles. Replication of viral DNA from the circular dsDNA occurs by "rolling-circle" replication. This process produces a succession of new ssDNA circles, which are packaged with coat proteins to form new viral particles. Although virally encoded proteins are required for rolling-circle DNA replication, the DNA polymerase is of plant origin.
(A, courtesy of Margaret Boulton.)

(A)

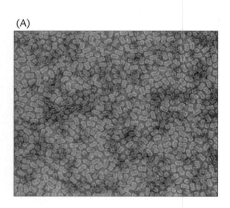

(B)

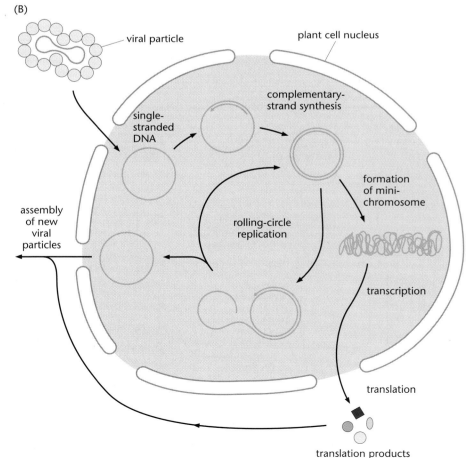

virus that causes a serious disease of tomato crops; and the whitefly-transmitted *African cassava mosaic virus*, which causes devastating losses in cassava crops.

What is the basis for viral symptoms in infected plants? Yield losses occur in part because the synthesis of viral particles reduces the availability of sugars and particularly amino acids to the plant. TMV particles can accumulate to levels of up to 1g/kg fresh weight of an infected tobacco leaf. However, these effects are not a sufficient explanation for most viral symptoms. Symptoms also result from more subtle perturbations of host biology. For example, many viruses encode proteins that suppress gene silencing in the host plant. **Gene silencing** is a complex phenomenon that is involved both in a plant's defense mechanisms (**RNA silencing**; see Section 8.4) and in the control of gene expression during normal plant growth and development. Viral suppressors of gene silencing may disable host defense mechanisms and interfere with the control of development in the host plant. Some of the symptoms that result from infection resemble the disturbed developmental phenotypes that result from mutations in the plant RNA-silencing machinery (see Section 8.4).

8.4 DEFENSES

Pathogens and pests have diverse mechanisms for attacking plants and exploiting them as food. It is remarkable that most plants are resistant to most plant pathogens, and the nature of plant defense against pathogens is the subject of this section.

A pathogen can fail to cause disease in a particular plant species for many different reasons. The structure of the cell wall of a non-host plant can present a physical barrier that prevents successful attack. The non-host plant may be insensitive to toxins produced by necrotrophic pathogens, or it may be able to metabolize and thus detoxify these toxins. Some plants contain chemicals that deter attack by pests and pathogens, or poison them once they start to attack. Some of these general or "non-host" mechanisms that prevent pathogens from attacking non-host species are **constitutive**: they are present continuously in the plant whether or not it is being attacked. Many other resistance mechanisms are **inducible**, expressed only when an attack occurs. Inducible responses include the basal defense and *R* gene–mediated defense mechanisms described in the chapter introduction and in Section 8.1. For example, a fungal pathogen may be prevented from attacking a non-host plant in part because the cell walls resist penetration (a constitutive defense mechanism), but also because new cell wall material is synthesized and deposited at the site where the pathogen attempts penetration (an inducible defense mechanism).

When a pathogen penetrates a potential host plant, it triggers basal defense mechanisms (see Figure 8–2). Basal defense mechanisms are elicited by pathogen molecules that are shared by many pathogenic species. Such molecules include chitin and other **glucans** that make up the cell walls of fungi; **flagellin**, the protein that makes up bacterial flagella; and a component of bacterial ribosomes, elongation factor–Tu (EF-Tu). These molecules are collectively referred to as PAMPs (pathogen-associated molecular patterns). The basal defenses they elicit include deposition of extra cell wall material by the plant, production of toxic molecules, and sometimes death of the plant cell under attack by the pathogen. As we described in the chapter introduction in discussing effector-triggered immunity, some pathogens produce specific effector molecules that can attenuate or overcome the basal defenses of the plant, thus promoting successful colonization by the pathogen. The effectors themselves may be triggers for a second type of plant defense mechanism, mediated by *R* genes (see Figure 8–2 and Section 8.1).

Here we describe the constitutive, basal, and *R* gene–mediated defense mechanisms by which plants resist attack by pathogens. We first discuss how the presence of pathogens is recognized by plants, and how this recognition process activates signaling pathways that lead to defense responses. We then consider four types of plant defense mechanism: detoxification of toxic compounds produced by pathogens, production of antimicrobial compounds, restriction of pathogen invasion by cell death (the **hypersensitive response**, or **HR**), and systemic resistance. Finally we give two further examples of defense mechanisms against particular categories of pathogens: the response to herbivores and the response to infection by viruses.

Basal defense mechanisms are triggered by pathogen-associated molecular patterns (PAMPs)

Basal plant defense responses triggered by conserved **elicitor molecules** made by microbes—that is, by PAMPs—are the "first line of defense" against pathogens (see Figure 8–2). We describe here how PAMPs are perceived by the plant and the downstream signaling events that lead to a defense response.

Discovery of the chain of events leading to the plant defense responses elicited by infecting microbes is difficult, because not all cells simultaneously experience the pathogen, and events may occur at different rates and levels in different cells. Researchers have been able to make progress in this area in several ways: by eliciting responses with chemically pure PAMPs in solution rather than with whole microbes; by looking for mutant plants in which PAMPs no longer elicit defense responses; by studying cell-suspension cultures rather than intact plants or plant organs; and, recently, by using genomic information to study the changes in gene transcription in response to elicitation.

The sequence of events can be summarized as shown in Figure 8–35. The microbial elicitor is recognized by a receptor protein, usually located in the plasma membrane of the plant cell. This recognition triggers an influx of calcium (Ca^{2+}) ions into the cell and

Figure 8–35.
Summary of events in recognition of a microbial elicitor (PAMP) by a plant cell.
The elicitor is recognized by the external domain (*green*) of a receptor protein kinase (*red* indicates cytoplasmic protein kinase domain). Recognition triggers activation of a protein kinase signalling cascade, AND an NADPH oxidase, and also activates a calcium channel allowing Ca^{2+} ions into the cell. The NADPH oxidase on the plasma membrane catalyzes production of superoxide (O_2^-) which is converted to hydrogen peroxide (H_2O_2). The protein kinase domain of the receptor protein, and elevated levels of Ca^{2+}, superoxide and hydrogen peroxide, activate a signalling pathway consisting of a series of protein kinases, resulting in activation of defense against microbes.

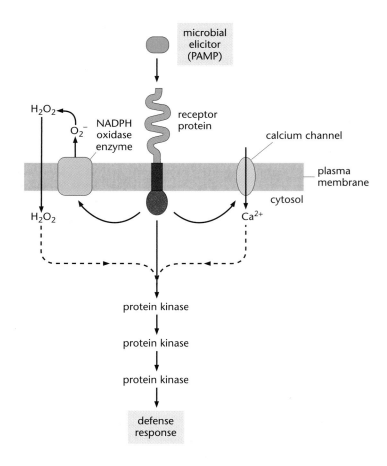

production of **reactive oxygen species (ROS)**, superoxide (O_2^-) and hydrogen peroxide (H_2O_2), which are antimicrobial and also act as signaling molecules. These early events activate a signaling pathway that consists of a series of **protein kinases**, and this in turn activates the expression of genes that bring about the defense response.

By searching for mutant plants that lack the ability to respond to PAMPs, researchers have identified plant receptors that recognize the elicitors and trigger the cascade of events leading to the defense response. The bacterial protein flagellin is a particularly well-studied PAMP. Its action is attributable to Flg22, a sequence of 22 amino acids; this part of the protein is the most highly conserved across bacterial species. Arabidopsis mutants that do not show defense responses when challenged with Flg22 are found to carry mutations in the *FLS2* gene, which encodes a transmembrane protein of the plasma membrane. The external part of the FLS2 protein (i.e., on the outside of the plasma membrane) consists of repeated short amino acid sequences with a high frequency of leucine residues (Figure 8–36). This type of domain, known as a **leucine-rich repeat (LRR)**, is common in proteins involved in defense against pathogens in both plants and animals (as discussed below). The function of the LRR region is to recognize the Flg22 sequence. On the inner side of the plasma membrane, the FLS2 protein has a protein kinase domain involved in a signaling pathway that activates downstream events when the outside LRR domain binds Flg22.

The use of cell-suspension cultures and purified PAMPs to produce synchronized defense responses has allowed detailed study of the initial events in the signaling pathway. Early studies of this kind were made with suspensions of cultured cells of parsley (*Petroselinum*), using PEP13, a PAMP that is part of a transglutaminase enzyme from the oomycete pathogen *Phytophthora megasperma*. More recent studies using other cell-suspension and elicitor systems have shown that many of the early events are conserved across plant species and types of PAMP. In cultured parsley cells, addition of PEP13 results within 1 to 2 minutes in the activation of ion channels, leading to an influx of Ca^{2+} ions into the cells. The importance of the influx of calcium is illustrated by the finding that addition of a calcium chelator—a compound that binds and removes Ca^{2+} ions from solution—abolishes defense responses in the cell culture. Within 5 minutes of addition of PEP13, the cells start to produce reactive oxygen species such as superoxide and hydrogen peroxide. These molecules are antimicrobial, and they also act as secondary signaling molecules to elicit subsequent defenses. The rapid production of ROS following elicitation requires oxygen (O_2) and is known as the "respiratory burst."

An important contribution to ROS synthesis is made by the enzyme **NADPH oxidase**, which transfers an electron from NADPH across the plasma membrane to convert extracellular O_2 to O_2^- (Figure 8–37; see also Figure 8–35). Superoxide is usually rapidly converted to H_2O_2 either spontaneously or through the action of **superoxide dismutase**. The Arabidopsis genome contains eight genes encoding NADPH oxidase (*Atrboh* genes: *A. thaliana* **r**espiratory **b**urst **o**xidase **h**omologs), two of which have been shown to be required for the production of ROS in response to Flg22 and during incompatible interactions (i.e., host–pathogen interactions that do not result in disease; see below) with *Hyaloperonospora parasitica* and *Pseudomonas syringae*.

At least two classes of protein kinases are rapidly activated on elicitation: the **mitogen-activated protein (MAP) kinases** and the **calcium-dependent protein kinases (CDPKs)**. MAP kinases participate in signaling pathways in all eukaryotes, whereas CDPKs are specific to plants. Both classes are involved in a cascade of events that leads to the extensive induction of gene expression in response to PAMPs (Figures 8–35 and 8–38). Expression of some genes is induced very quickly (within 15–30 minutes), and this induction is independent of ROS production. Rapidly induced genes typically encode regulatory and signaling proteins, such as further protein kinases and transcription factors. These proteins contribute additional complexity to the signaling

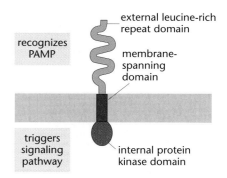

Figure 8–36.
Structure of FLS2 protein. The external, leucine-rich repeat (LRR) domain binds to the Flg22 sequence of amino acids in bacterial flagellin (*not shown*). Binding activates the internal protein kinase domain, triggering the signaling pathway that activates defense responses.

Figure 8–37.
Production of reactive oxygen species in response to PAMPs. (A) Conversion of oxygen to superoxide (O_2^-) and hydrogen peroxide (H_2O_2) by the plasma membrane enzyme NADPH oxidase and the intracellular enzyme superoxide dismutase. (B) Activity of the NADPH oxidase. Electrons from intracellular NADPH are donated to extracellular oxygen to form superoxide, via electron-carrying FAD and heme groups in the membrane-spanning enzyme.

(A)

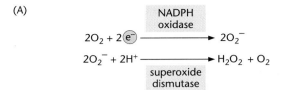

(B)

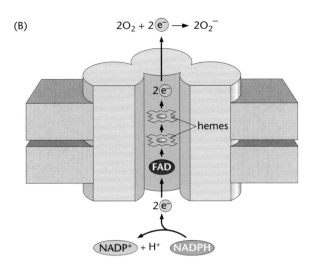

cascade and bring about a second phase of induction of gene expression that is dependent on ROS production. The genes expressed in this second phase encode enzymes that synthesize plant defense components (e.g., toxins that deter pathogen attack) or small signaling molecules (e.g., salicylic acid, ethylene, and jasmonic acid)

Figure 8–38.
Downstream events in the defense response triggered by a microbial elicitor. Signaling cascades of MAP kinases and calcium-dependent protein kinases activate expression of a first set of genes that encode regulatory and signaling proteins. These proteins in turn activate expression of a second set of genes. The second set of genes encode defense proteins that act to prevent microbial attack; components of the proteasome (ubiquitin ligases) that degrades components of the signaling pathway; and proteins that synthesize further signals that enhance and extend the defense response.

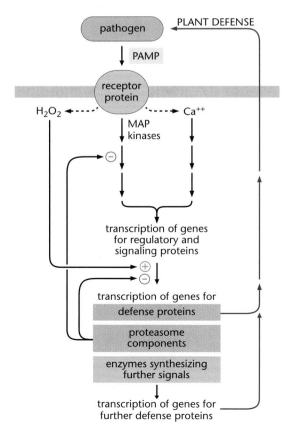

that further enhance plant defense mechanisms via other signaling pathways (both of these second-phase processes are considered below). Later-expressed genes also encode proteins responsible for targeting other proteins for degradation via the **proteasome** (see Section 5.4 for a description of the proteasome-ubiquitin system). Protein turnover is important to the proper functioning of signaling pathways. A protein that has been activated by phosphorylation by a protein kinase in a signaling cascade, or that is newly synthesized as a signaling component, must be inactivated once the response to the initial signal is fully operational. Inactivation is often achieved by the addition of ubiquitin molecules to the protein (ubiquitination), making it a specific target for degradation by the proteasome.

The events described above constitute the sensing and signaling pathways of the basal defense mechanism. They seem to be conserved across different species of plants and in response to different PAMPs. It is likely that plants have a multitude of different receptors for PAMPs produced by pathogens, which trigger a complex network of signaling events.

A similar set of defense responses is activated during *R* gene–mediated defense responses: effector molecules produced by pathogens are recognized by receptor proteins and trigger signaling networks that resemble those of the basal defense response. Although their sensing and signaling pathways are similar, the downstream defense mechanisms triggered in *R* gene–mediated defense are generally stronger than those of the basal defense response. *R* gene–mediated defense results in both higher levels of resistance to the pathogen and more extreme consequences for the plant, including selective cell death (see the zigzag scheme in Figure 8–2 and further discussion below).

An example of this difference in the strength of the two types of response comes from studies of plant defense against the obligately biotrophic fungus *Uromyces phaseoli* var. *vignae*, cowpea rust (Figure 8–39). Spores of the fungus were inoculated onto leaves of plants not normally infected by the rust (non-hosts) and onto resistant and susceptible varieties of cowpea. In the susceptible varieties, the rust penetrated through stomatal pores to produce intercellular hyphae and penetrated mesophyll cells to produce haustoria. On some non-hosts such as pea, fungal growth was restricted soon after stomatal penetration, indicating the presence of antifungal compounds in the plant before challenge (see below). In most other non-hosts, such as broad bean and French bean, the fungus failed to penetrate into mesophyll cells. The failure to form haustoria was associated with the rapid accumulation of a papilla of material inside the plant cell wall at sites of attempted penetration. (For a further example of this type of defense mechanism see Figure 8–40.) In varieties of cowpea with *R* gene–mediated resistance to the rust fungus, wall-based defenses did not develop and haustoria began to form, but the penetrated cells then rapidly collapsed and died—a hypersensitive response (discussed later in this chapter). In this interaction the biotrophic rust died within cells undergoing the hypersensitive response, either through simple starvation or through the accumulation of antifungal

Figure 8–39.
Interactions between cowpea rust fungus and host and non-host plants.
(A) Interaction with a susceptible host plant. The fungal hyphae successfully invade leaf cells and form haustoria, enabling further spread of the fungus and eventual sporulation. (B) On some non-host plant species, spread of the fungus is restricted by the constitutive production of antifungal compounds in the leaf. (C) On other non-host plant species, attempted penetration of leaf cells by fungal hyphae triggers local massive production of new cell wall material (papillae) on the inside of the existing wall, preventing further invasion. (D) On a host plant carrying the corresponding *R* gene, fungal invasion of a leaf cell triggers the death of that cell, preventing the fungus from gaining nutrients and restricting its spread within the leaf.

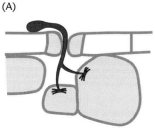

susceptible host plant:
infection

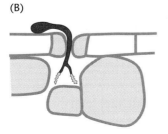

non-host plant:
toxin production restricts
haustorial growth

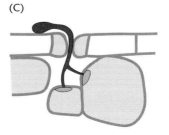

non-host plant:
formation of papillae
prevents cell penetration

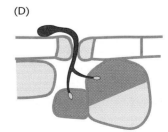

host plant with *R* gene–
mediated resistance:
penetrated cells die,
other cells remain healthy

Figure 8–40.
Cell-wall located resistance of Arabidopsis to barley powdery mildew, *Blumeria graminis* f. sp. *hordei*. Arabidopsis is not a host for this fungus. The germination of a spore on the leaf surface triggers basal defense mechanisms including the rapid deposition of new cell wall material at the attempted site of penetration. Genes involved in this response include *PEN1*, encoding a syntaxin (see Chapter 3) necessary for secretion of material into the cell wall; *PEN2*, encoding an enzyme that catalyzes the release of an antimicrobial indole compound from a glucosinolate inside the leaf cell (described later in this Chapter); and *PEN3*, encoding a plasma membrane transporter involved in the secretion of the indole compound to the outside of the cell. The photograph shows the production of reactive oxygen species (stained brown) at the site of attempted penetration of the germ tube, in response to recognition of fungal elicitors (see Figure 8–35).

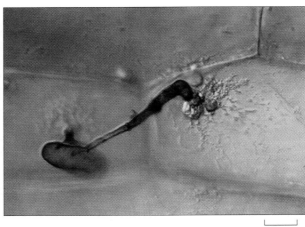

12 μm

compounds after infection. The important general lesson here is that basal defense responses often act to restrict the growth of the fungus by the presence of antimicrobial compounds or by highly localized alterations to the plant cell wall, or both. By contrast, *R* gene–mediated defense is usually linked to localized host-cell death and the consequent death of the fungus.

Below we describe in more detail the nature and functions of some key components of the sensing and signaling pathways that operate in PAMP-triggered and *R* gene–mediated defense responses. We then discuss downstream defense mechanisms, including synthesis of antimicrobial compounds and activation of the hypersensitive response.

R proteins and many other plant proteins involved in defense carry leucine-rich repeats

Many proteins involved in plant defense, including many R proteins, rely for their function on the recognition of other molecules, especially other proteins. These interactions are necessary, for example, in recognizing pathogen molecules that then trigger a plant defense response, and in direct binding to and inhibition of pathogen proteins that are harmful to the plant. Such recognition functions are most frequently carried out by a specific domain of the protein, the leucine-rich repeat. LRR-containing proteins include FLS2, the receptor that recognizes bacterial PAMP flagellin (described above); polygalacturonase-inhibiting proteins (PGIPs), which are induced during activation of plant defenses and inhibit microbial enzymes that attack plant cell wall components; and R proteins that recognize pathogen effector molecules. LRR proteins are also important in other aspects of plant growth and development. For example, the Arabidopsis proteins CLAVATA1, required for maintenance of the shoot apical meristem (see Section 5.4), and BRI1, required for brassinosteroid perception (see Section 6.3), carry LRR domains.

LRR domains typically contain many repeats of leucine-rich amino acid sequences that are 23 to 29 residues in length. Within part of these repeats, leucine residues occur at every second or third position: X-Leu-X-X-Leu-X-Leu-X, where X is another amino acid. This part of the repeats forms a structure within the protein called a **parallel β sheet**, in which the leucine residues are in the interior of the protein and the X residues are exposed to the exterior, where they form the surface that interacts with other proteins (Figure 8–41). The structure allows for very specific interactions and for the generation of differently interacting surfaces via single amino acid changes. This specificity is illustrated by studies of the PGIPs from bean. Two different PGIPs are secreted from the cells of the bean plant in response to fungal attack. Both consist of 10 LRRs, with short flanking regions. Only eight amino acid variations distinguish the two proteins: five of them are within the LRR structure and two are adjacent to it. In spite of the great similarity of the proteins, they recognize polygalacturonases differently. PGIP-1 interacts with and inhibits the polygalacturonase from the fungus *Aspergillus niger* but not that from *Fusarium moniliforme*, whereas PGIP-2 interacts with and inhibits both proteins. We discuss the importance of the variability and specificity of LRRs in the evolution of plant defense responses below.

R genes encode families of proteins involved in recognition and signal transduction

The role of an *R* gene product (an R protein) is to recognize an effector molecule from a pathogen and then activate a signaling pathway that results in activation of defense responses. *R* genes can activate resistance to a very wide range of pathogens, including bacteria, fungi, oomycetes, viruses, and nematodes. Despite this diversity of pathogens, *R* genes encode proteins with only a limited range of structures, which are often conserved across a very wide range of plant families. This conservation suggests that the processes underlying recognition by an R protein of unrelated pathogens are mechanistically highly conserved and probably activate a relatively small number of defense pathways. Figure 8–42 shows examples of the main classes of R proteins; we describe here and below some examples that illustrate important features of these proteins.

Many R proteins contain LRR domains (Figure 8–42). Extracellular LRR domains are responsible, for example, for pathogen recognition by the tomato Cf proteins. These proteins confer resistance to strains of the fungus *Cladosporium fulvum* that carry the corresponding *Avr* genes, which encode small, secreted cysteine-rich apoplastic peptides with multiple disulfide bridges. For example, the tomato gene *Cf-4* confers recognition of and resistance to *C. fulvum* carrying *Avr4*, and *Cf-9* confers recognition of *C. fulvum* carrying *Avr9*. *Cf-4* and *Cf-9* encode closely related (90% identical) proteins, and small differences in their LRRs specify whether Avr9 or Avr4 is recognized. Cf proteins have only small intracellular domains. It is not known how pathogen recognition mediated by the extracellular LRR domain triggers a signaling pathway inside the cell.

The biggest class of LRR-containing R proteins consists of intracellular proteins rather than those located in the plasma membrane (Figure 8–42), although the intracellular proteins may be associated with the inner face of the plasma membrane. This type of R protein is involved in resistance to pathogen effectors that act inside the cell. The proteins consist of an LRR domain that is involved in recognizing—directly or indirectly—pathogen effector molecules that have entered the cell and a domain that binds nucleotides (e.g., ATP) and is thought to activate a signaling pathway.

In addition to LRR domains, some R proteins carry other conserved motifs involved in protein-protein interactions. These include coiled-coil and TIR motifs. TIR motifs are also found in animal LRR-containing receptor proteins, the Toll-like receptors, which are involved in recognizing pathogen-associated molecules—including flagellin and components of bacterial cell walls—and triggering immune responses. The similarities between the receptors of the innate immune response in animals and the receptors of PAMP- and effector-triggered immunity in plants suggest that these proteins are evolutionarily very ancient, probably predating the split between animals and plants (see Chapter 1).

Some R proteins do not contain LRR domains (Figure 8–42). The *Pto* gene of tomato, for example, encodes an intracellular **serine/threonine protein kinase**. It confers resistance to strains of *P. syringae* that carry the *avrPto* gene. *RPW8* is an Arabidopsis *R* gene that confers resistance to powdery mildew. It encodes a plasma membrane protein with a coiled-coil domain but no LRR domain.

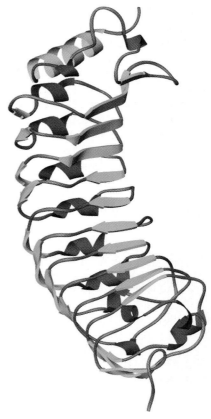

Figure 8–41.
Structure of an LRR domain in an R protein. Repeated leucine-rich repeats of about 23 to 29 amino acid residues fold within the protein to form parallel β sheets. The inner faces of the sheets (facing the interior of the protein molecule) are formed by leucine residues; the outer faces recognize and bind to other proteins. Single amino acid variations on the outer face can drastically affect the ability of the LRR domain to bind target proteins. The structure shown is a polygalacturonase-inhibiting protein (PGIP) of *Phaseolus vulgaris*.

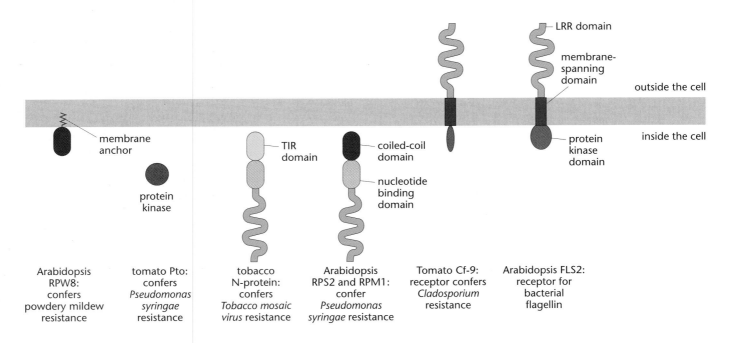

Figure 8–42.
Examples of the main types of R protein.
(*Left to right*) The Arabidopsis RPW8 protein belongs to the signal-anchor coiled-coil (SA:CC) class, consisting of a coiled-coil domain (*brown*) that mediates interactions with other proteins and an anchor that attaches the protein to a membrane. The tomato Pto protein is an intracellular serine/threonine protein kinase. The tobacco N-protein and Arabidopsis RPS2 and RPM1 proteins belong to the nucleotide binding (NB) LRR class; they are intracellular proteins with LRR domains (*green*) and nucleotide binding domains (*pink*), plus domains that mediate interactions with other proteins (TIR or coiled-coil domains). The tomato Cf-9 and Arabidopsis FLS2 proteins (see also Figure 8–36) span the plasma membrane and have external LRR domains. FLS2 has a protein kinase domain inside the cell; Cf-9 has a short domain inside the cell for which the specific function is not known.

Most R proteins do not directly recognize pathogen effector molecules

The simplest hypothesis to explain the mechanism by which R proteins trigger defense responses is that they bind directly to the effector molecules produced by pathogens and that this binding triggers activation of a signaling pathway. However, there is increasing evidence that this is not correct in most cases. Instead it seems that most R proteins recognize plant molecules that have been modified by the pathogen effectors. As explained above, plants respond to pathogen attack by activating basal defense mechanisms, and some pathogens can overcome these defenses by producing effector molecules that modify—and thus reduce the severity of—the plant response. It is believed that most R proteins recognize and respond to these modifications rather than to the effector molecules themselves. Thus each R protein can be regarded as a "guard" of a specific component of the basal defense pathways. If that component is modified by a pathogen effector molecule, the modification is recognized by the R protein and the *R* gene–mediated defense response is activated.

Evidence for this mode of action of R proteins comes from genetic analysis of *R* gene–mediated resistance in Arabidopsis. Infection of Arabidopsis plants by *P. syringae* triggers basal defense mechanisms via a pathway that involves RIN4 or RIN4-related proteins (Figure 8–43). Strains of *P. syringae* that can overcome this basal defense do so by producing effector molecules encoded by the *avrB*, *avrRpm1*, or *avrRpt2* genes, which inactivate RIN4 and thus suppress basal defenses and allow infection. The Avr proteins are secreted into the plant cell via the *P. syringae* type III secretion system (see Section 8.1). The AvrB and AvrRpm1 proteins cause phosphorylation of RIN4, and the AvrRpt2 protein is a protease that degrades RIN4. Arabidopsis plants that carry the *RPM1 R* gene are resistant to infection by *P. syringae* carrying the *avrRpm1* or *avrB* genes, because RPM1 recognizes phosphorylated RIN4. When RIN4 is phosphorylated, RPM1 activates a signaling pathway that leads to defense responses and hence suppression of *P. syringae* infection. Similarly, Arabidopsis plants that carry the *RPS2 R* gene are resistant to infection by *P. syringae* carrying the *avrRpt2* gene. RPS2 is normally found in a complex with RIN4. Degradation of RIN4 by AvrRpt2 protease releases RPS2, allowing it to activate a defense response. Thus, RIN4 is "guarded" by the products of two *R* genes that monitor its status and respond to specific types of effector-mediated damage by activating further defense mechanisms.

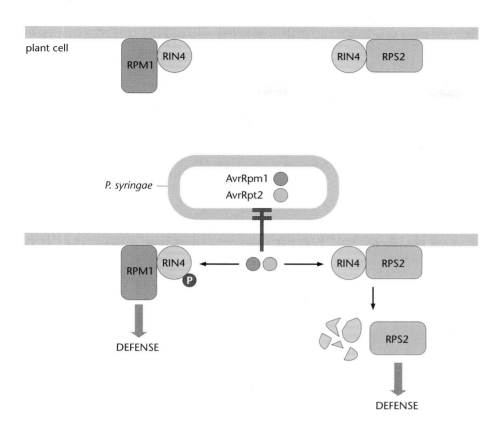

Figure 8–43.
Mode of action of Arabidopsis R proteins in defense against *P. syringae*. In the absence of *P. syringae*, the RIN4 protein is "guarded" by R proteins RPM1 and RPS2 (*top*). Bacterial effectors that modify RIN4 alter the interaction between RIN4 and the R proteins, causing the R proteins to trigger defense responses (*bottom*). If the bacterium secretes AvrRpm1 protein into the plant cell, this protein phosphorylates RIN4. The phosphorylation event is recognized by RPM1, which then activates a defense response. If the bacterium secretes AvrRpt2 into the plant cell, this protein degrades RIN4. This releases RPS2, which then activates a defense response.

A further example is provided by the mode of action of the *Cf-2 R* gene of tomato, described above. The presence of the *Cf-2* gene enables the plant to recognize when the Avr2 protease inhibitor of *C. fulvum* has bound to the plant intercellular protease Rcr3, thus triggering defense mechanisms.

Although many R proteins probably indirectly recognize the consequences of effector action, rather than the effectors themselves, some are known to interact directly with pathogen effector molecules. These include the protein kinase encoded by the tomato *Pto* gene, which directly binds to the product of the *avrPto* gene of *P. syringae*. This direct recognition triggers a further *R* gene–mediated response of the type described for the RIN4 system above. The interaction between the Pto and AvrPto proteins is recognized by the R protein Prf, which then activates a defense response (Figure 8–44).

Progress in understanding the signaling pathways activated by *R* genes is being made by searching for mutant plants in which defense responses are not triggered even when the appropriate *R* gene and effector molecule are present. These studies show that, in many cases, several different *R* genes feed into a common signaling pathway. For example, in Arabidopsis, mutations in the gene *NDR1* prevent activation of defense responses via both RPM1 and RPS2, the two R proteins that "guard" the RIN4 component of the basal defense mechanism against *P. syringae*. The NDR1 protein spans the plasma membrane and interacts with RIN4, but the mechanism by which it receives signals from the R proteins and transmits them into a signaling pathway is not yet known.

Similarly, mutations in the Arabidopsis *EDS1* (*ENHANCED DISEASE SENSITIVITY 1*) gene prevent activation of defense responses via a class of R proteins that includes RPP1, a protein that confers resistance to races of downy mildew. *eds1* mutants become susceptible to races of downy mildew that normally infect species of *Brassica* but not Arabidopsis. This suggests that RPP1 contributes resistance in Arabidopsis to downy mildew races that infect *Brassica*.

Figure 8–44.
Mode of action of tomato Pto protein.
The AvrPto effector protein of *P. syringae*, when secreted into a tomato cell, binds to the Pto protein of the tomato. The Pto-AvrPto complex is recognized and bound by the tomato Prf protein. This binding triggers defense responses in the tomato cell.

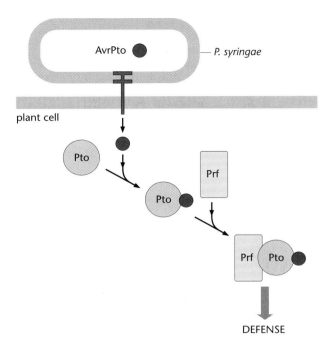

EDS1 is a member of a family of proteins that interact with each other and are involved in both basal and *R* gene–mediated defense against biotrophic pathogens. The importance of these proteins is demonstrated by the susceptibility to pathogens of mutants lacking the PAD4 and SAG101 members of the family. Arabidopsis is normally not susceptible to the barley powdery mildew *Blumeria graminis* f. sp. *hordei*, because very few fungal hyphae penetrate the cell wall (see Figure 8–40) and further development of those that do is prevented by defense responses inside the cell. In *pad4/sag101* mutants the incidence of penetration is not increased but hyphae that penetrate into cells are able to develop further. If the cell wall defenses of the *pad4/sag101* mutant are also reduced—for example, by introduction of the *pen2* mutation that prevents the secretion of an antifungal compound from the plant cells—the Arabidopsis plant becomes a full host to barley powdery mildew, allowing hyphal ramification and sporulation of the fungus.

R gene polymorphism restricts disease in natural populations

R genes are remarkable for the rapid rate at which they diversify under selective pressure from pathogens. As new *avr* or *Avr* genes appear in pathogen populations, so new *R* genes able to recognize the Avr proteins and activate defense responses are selected in host plant populations. It was once believed that a unique and dedicated mechanism must operate at *R* gene loci to generate novel sequences at a high rate, as a basis for rapid evolution under a selective pressure. However, no such mechanism has been found. It seems more likely that there is selection for a high level of **polymorphism** at *R* loci—in other words, that many variants of a particular *R* gene are maintained within a given plant population. Natural plant populations are rarely subject to epidemic disease caused by virulent strains of pathogens. This is probably because polymorphism increases the likelihood that individual plants in a population have different *R* genotypes. To understand how such high levels of polymorphism arise, we need to look at the structure of *R* loci.

R loci can be classified as simple or complex (Figure 8–45), depending on how they achieve polymorphism. "Simple *R* loci" consist of a single copy of the *R* gene, and polymorphism arises because many different alleles of the gene exist in the population. Although an individual plant in a population of a diploid species has a maximum of only

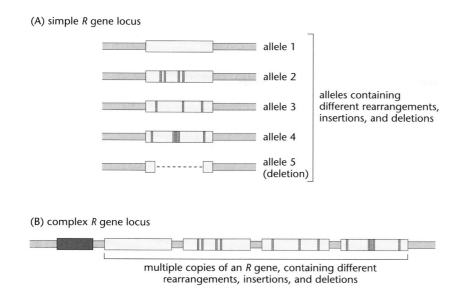

(A) simple *R* gene locus

allele 1
allele 2
allele 3
allele 4
allele 5 (deletion)

alleles containing different rearrangements, insertions, and deletions

(B) complex *R* gene locus

multiple copies of an *R* gene, containing different rearrangements, insertions, and deletions

Figure 8–45.
Simple and complex *R* gene loci.
(A) A simple *R* locus has one copy of the *R* gene. Although any individual plant, if diploid, has only one or two different alleles of the *R* gene, polymorphism can arise because many different alleles exist in the population as a whole. Some alleles may differ in ways that affect only one or a few amino acid residues; others may carry large deletions that prevent expression of the protein.
(B) A complex *R* gene locus consists of multiple, different copies of the *R* gene adjacent to each other on the chromosome. In this case, an individual diploid plant may carry many different copies of the *R* gene family.

two different alleles, the population as a whole may contain large numbers of alleles for that gene locus. At the *L* locus of flax, for example, which confers resistance to rust fungi, at least 10 alleles have been defined, each conferring recognition of a distinct set of rust races. At the *Mla* locus of barley, which confers resistance to powdery mildew, approximately 30 alleles have been defined.

"Complex *R* loci" consist of clusters of *R* genes that constitute **multigene families**. Closely related genes lie adjacent to each other on the chromosome. An individual plant may have many slightly different copies of a particular *R* gene. Loci arise by **gene duplication**, **point mutation**, and **recombination**. Sequence comparisons of *R* gene families suggest that, during evolution, sequences have been exchanged between adjacent family members by recombination. The finding of marked differences among family members, however, suggests that the rate of sequence exchange is not high.

The polymorphisms of *R* genes that encode proteins with LRR domains occur mainly in the sequence encoding the LRR domain, affecting particularly the non-leucine residues that form the interacting surface. Many differences in nucleotide sequence between alleles or family members result in differences in the amino acid sequences of the LRR domains of the proteins they encode (these are called "nonsynonymous" nucleotide changes). The high frequency of amino acid substitutions suggests that these genes are under constant selective pressure for the ability to recognize distinct pathogen effector molecules. This situation is different from that observed for many other types of protein. For an enzyme or transcription factor, for example, changes in amino acid sequence may well result in impairment or loss of function. Selective pressure here is *against* nonsynonymous nucleotide changes. Where polymorphisms exist in genes encoding such proteins, they usually involve "synonymous" nucleotide changes—changes that do not affect the amino acid sequence encoded (Figure 8–46).

Unlike natural populations, modern crops are monocultures of a single genotype. They may be dependent on a single allele of an *R* gene for resistance to a particular pathogen. This places selective pressure on the pathogen carrying the matching *avr* gene for mutations that render its product undetectable by the crop plant. However, although mutations in *avr* genes may allow the pathogen to infect a previously resistant crop plant, this may also have evolutionary costs for the pathogen. Many *avr* genes encode proteins important for the infection process, such as effector proteins that bind host DNA or interact with components of the host's basal defense mechanism. A mutation that prevents an Avr protein from being recognized by a particular host *R* gene may also change

Figure 8–46.

Synonymous and nonsynonymous changes in nucleotide sequence.

Examples of the two sorts of changes are given for the codons for two amino acids, alanine and isoleucine. In both cases, some single-nucleotide changes (synonymous changes; *shaded green*) do not affect the amino acid specified by the codon: thus GCC, GCU, and GCG all specify alanine, and AUU, AUC, and AUA all specify isoleucine. Other single-nucleotide changes (nonsynonymous changes; *red*) alter the amino acid specified by the codon: thus changing GCC to GUC changes the codon specificity from alanine to valine, and changing AUU to AGU changes the codon specificity from isoleucine to serine.

codon	amino acid specified	single nucleotide change	amino acid specified
GCC	alanine	GCU / GCG	alanine / alanine
		GAC / GUC	aspartic acid / valine
AUU	isoleucine	AUC / AUA	isoleucine / isoleucine
		ACU / AGU	threonine / serine

the function of the Avr protein or eliminate its synthesis altogether. Loss of an Avr protein may reduce the effectiveness of the pathogen (a "cost of virulence"). It may now be outcompeted by other pathogen strains carrying different *avr* genes, to which the host plant has no resistance, and its ability to infect plants with no matching *R* gene resistance may be impaired. This imposes the pressure of an evolutionary "snakes and ladders" on the pathogen: every mutation that makes it more fit on one host genotype potentially makes it less fit on another host genotype.

R genes have been selected in crop breeding from the earliest times

It is likely that farmers began selecting for disease resistance in crops with the start of plant domestication, about 10,000 years ago. Early farmers probably selected seed for future sowing from the least-diseased plants in their fields. In the early 1900s, plant breeders started to appreciate that the resistance of crop plants to many important diseases was due to single, dominant genes: the *R* genes. This realization led to searches for new *R* genes to breed into crops to increase disease resistance. Domesticated plants are usually most genetically diverse in the geographic regions where they originated and were first domesticated (see Chapter 9), so expeditions seeking disease-resistant varieties usually concentrate on such regions. The same regions are often also the center of genetic diversity of the pathogens of these plant species.

The deliberate breeding of crops with new *R* genes created its own problems. Within a few years of the large-scale introduction of crop varieties carrying a particular new *R* gene, races of the corresponding pathogen would emerge that could cause disease in this variety. The new races often carried recessive mutations in avirulence (*avr*) genes. This limited period of usefulness of new *R* genes causes what has been referred to as the "boom and bust" cycle: the "boom" period during which a new *R* gene confers effective resistance is almost always followed by a "bust" when the *R* gene–mediated resistance is overcome by new strains of pathogen. Such cycles create a continuous challenge for plant breeders, who are currently wary of introducing varieties that depend on just one new *R* gene.

Some *R* genes have proved to be more durable, such as *Cf-9*, the tomato gene that confers resistance to *C. fulvum* (leaf mold) races carrying the *Avr9* gene. Durable *R* genes are very valuable to the plant breeder. The basis for their durability is not well understood but is likely to involve recognition of *avr* gene products that, when lost or modified through mutation, result in reduced pathogen virulence.

This "plant breeder's treadmill," on which breeders must run very hard to stay in place in terms of maintaining crop productivity in the face of rapid pathogen evolution, is peculiar to the pressure imposed on plant survival by intensive agriculture. In the wild, selective pressures are very different. It seems unlikely that wild populations need to continuously create, at a high rate, new genetic variations for disease resistance in order to match pathogen evolution. Rapidly reproducing pathogens will usually be able to

Figure 8–47.
Advantage of mixing crop plants carrying different *R* genes. When all plants of a crop carry the same *R* gene (*left*), selective pressure for a pathogen that can overcome this single type of resistance is very high. Once this happens, the entire crop will succumb to the pathogen. When individual plants carry one of five different *R* genes (*right*), selective pressure for a pathogen that can overcome any one type of resistance is less than when only one *R* gene is present, and such a pathogen will cause disease only in a relatively small proportion of the total crop (say, in all the *red* plants).

evolve at a higher rate than hosts. As we discussed above, in natural populations extensive polymorphism (multigene families and allelic diversity) at *R* gene loci plays a major role in restricting losses to a specific strain of a particular pathogen. Epidemics in plants are usually limited to the special circumstances of agriculture, in which millions of genetically identical plants are grown in monoculture.

One way for plant breeders to step off the treadmill may be to use mixtures of crop plants that carry different *R* genes (Figure 8–47). Such mixtures of genotypes might be expected to show less disease than a genetically uniform crop, for several reasons. First, plants carrying an *R* gene overcome by any one strain of a pathogen would make up a small proportion of the population, so epidemics affecting the whole crop would be less likely. Second, as we described above, a mutation that enables a pathogen strain to grow in the presence of a particular *R* gene is likely to make the strain less fit in the presence of a different *R* gene, because an effector molecule has been lost or altered. Third, in a mixed planting, each plant is likely to be recurrently infected with avirulent races of the pathogen; this could trigger **systemic acquired resistance (SAR)**, making the plants generally less susceptible to disease (SAR is discussed below).

Insensitivity to toxins is important in plant defense against necrotrophs

Many necrotrophic pathogens produce compounds toxic to the cells of their host plants, and resistance to these toxins is an important component of plant defense mechanisms against such pathogens. Resistance can take the form of detoxification mechanisms that convert the toxin to a less harmful compound, or insensitivity to the toxin due to possession of a modified form of its target molecule or process. We discussed an example of a detoxification mechanism in Section 8.1: the ability of some varieties of maize to resist infection by the fungus *Cochliobolus carbonum* by metabolizing the fungal HC toxin to a harmless compound. Here we describe an agriculturally important example of resistance to a fungus mediated by a host mechanism that prevents the deleterious effects of its toxin.

One of the most spectacular examples of a plant disease caused by a pathogen toxin was provided by the Southern maize leaf blight epidemic in the United States during the 1970s. Hybrid crop plants in general produce higher yields than either parent plant, due to a poorly understood phenomenon called **heterosis** or **hybrid vigor**. To produce hybrid seed, plant breeders exploit plants that exhibit **cytoplasmic male sterility (CMS)** (see Section 9.2 for a discussion of this breeding mechanism). Toward the end of the 1960s maize breeders were using a CMS line that carried a **mitochondrial genome** (see Section 2.6) known as the "T type," and during the 1970s hybrid maize varieties with T-type mitochondrial genotypes were planted on 85% of the maize acreage in the United States.

One race of the fungus (*Cochliobolus heterostrophus*) that causes Southern maize leaf blight makes the compound T-toxin, which is lethal to cells carrying mitochondria with T-type genomes. T-toxin is a mixture of long (35–45 carbons), linear polyketides (Figure 8–48). It interacts directly with a protein encoded only by T-type mitochondrial genomes (see Section 9.2). T-toxin imposes a conformational change on the protein

Figure 8–48.
C. heterostrophus **toxin.** Structure of one of the polyketides in the T-toxin produced by the fungus *Cochliobolus heterostrophus*.

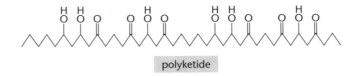

polyketide

Figure 8–49.
Symptoms of Southern maize leaf blight.
Maize plants that carry the T-type mitochondrial genome (*right*) are susceptible to the disease caused by *C. heterostrophus*. Plants that do not carry this mitochondrial genome (*left*) are resistant to the disease. (Courtesy of Jim Holland.)

that makes the inner mitochondrial membrane leaky, preventing ATP synthesis and thus leading to cell death. Southern maize leaf blight was an insignificant disease of maize until the widespread cultivation of hybrid maize carrying T-type mitochondrial genomes in the 1970s. At this point, the T-toxin–synthesizing race of *C. heterostrophus* spread throughout the U.S. corn belt. The leaf blight disease was controlled only when breeders switched to different CMS lines (Figure 8–49).

Plants synthesize antibiotic compounds that confer resistance to some microbes and herbivores

Plants make a range of antibiotic compounds that inhibit or deter pests and pathogens. Some are made constitutively, whereas the synthesis of others is induced or elevated following microbial or herbivore attack. Constitutively synthesized compounds may be present in their biologically active form at all times or may be stored as inactive precursors that are converted to the active forms in response to pathogen attack. We discuss first some examples of the major classes of constitutively synthesized antibiotics, then consider compounds that are synthesized specifically in response to pest or pathogen attack.

Many antibiotic compounds are **terpenoids** (Figure 8–50; synthesis of these compounds is described in Section 4.7). Some plant species produce 10-carbon **monoterpenes**, or monoterpene esters, that kill or repel attacking insects. Monoterpene esters called pyrethroids found in *Chrysanthemum* species are insecticidal, and because they have very low mammalian toxicity they are extracted from plants for use in insect control in homes and gardens. In conifers, monoterpenes such as β-pinene and myrcene are toxic to numerous types of insects. Insect-repellent monoterpenes also include menthol, an essential oil produced in glandular **trichomes** on the surface of mint (*Mentha* spp.) leaves.

Derivatives of the 15-carbon **sesquiterpene** are also implicated in plant defense against pests and pathogens. For example, gossypol is a sesquiterpene derivative in cotton that confers resistance to many insects, deterring insect feeding on the plant. Other deterrents of insect feeding include **phorbols**, derivatives of the 20-carbon diterpenes found in the latex of plants of the spurge family (Euphorbiaceae), and **steroids** and **sterols** synthesized from **squalene**, a 30-carbon triterpene. Azadirachtin, a squalene derivative from the neem tree (*Azadirachta indica*), is one of the most powerful known deterrents to insect feeding. Some plants convert squalene to compounds that mimic the insect molting hormone ecdysone. These "phytoecdysones" disrupt the molting of the larval stages of insects feeding on the plant.

Some pathogens and herbivores have mechanisms that allow them to detoxify or tolerate specific antibiotic compounds produced by plants. Such organisms often feed specifically on plants that make these compounds. Bark beetles, for example, can metabolize and thus tolerate the major monoterpenes in conifer bark that deter attack by other insects. Caterpillars of the monarch butterfly feed only on milkweeds (*Asclepias* spp.); they accumulate the potentially toxic triterpene glycosides from the milkweed leaves in their bodies and as a result become unpalatable to birds and other predators. The roots of oats (*Avena* spp.) constitutively produce avenacin, a toxic triterpene glycoside that prevents attack by many fungi (Figure 8–51). The fungal pathogen *Gaeumannomyces graminis* f. sp. *tritici* causes take-all disease in wheat, but cannot

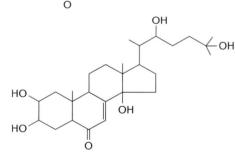

Figure 8–50.
Some terpenoid antibiotic compounds produced by plants.

compounds derived from monoterpene (10 carbons) units

myrcene

menthol

gossypol, a dimer derived from two sesquiterpene (15 carbons) units

phorbol, a diterpene (20 carbons) derivative

azadirachtin A, a squalene (30 carbons) derivative

α-ecdysone, a squalene derivative

infect oat roots. A different form of the fungus, *G. graminis* f. sp. *avenae*, can infect oats, because it can detoxify avenacin by producing the enzyme avenacin glucosidase. If the fungal gene encoding this glucosidase is mutated, *G. graminis* f. sp. *avenae* can no longer grow on oats. When this gene is transferred to *G. graminis* f. sp. *tritici*, the host range of this form of the fungus is extended to oats, proving that the fungal glucosidase is required for pathogenicity on oats. Tomato produces the compound tomatine, which is closely related to avenacin. Pathovars of the fungus *Septoria lycopersici* cause disease on tomato only if they have a glucosidase that can metabolize and thus detoxify tomatine.

Plants produce many nitrogen-containing compounds that are toxic to pests and pathogens. These include alkaloids, cyanogenic glycosides and glucosinolates, and some nonprotein amino acids. **Alkaloids** are derived, through complex biosynthetic pathways, mainly from aspartic acid, lysine, ornithine, tyrosine, or tryptophan. Plants

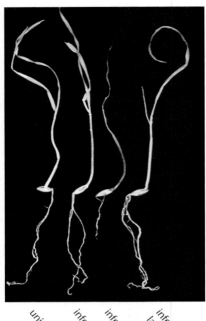

(A)

uninfected
infected with f. sp. tritici
infected with f. sp. avenae
infected with f. sp. avenae lacking avenacin glucosidase
infected with f. sp. avenae

(B)

β-D-glucose

α-L-arabinose-O

β-D-glucose

avenacin

avenacin glucosidase

Figure 8–51.
Action of the oat antibiotic avenacin. (A) Infection of oat seedlings by *Gaeumannomyces graminis*: (*left to right*) uninfected seedling; seedling challenged with *G. graminis* f. sp. *tritici*, which causes take-all disease of wheat but is killed by avenacin produced by oat roots; seedling challenged with *G. graminis* f. sp. *avenae*, which can infect and kill oats because it produces the enzyme avenacin glucosidase that detoxifies avenacin; seedling infected with mutant *G. graminis* f. sp. *avenae* in which avenacin glucosidase is inactivated. (B) Structure of avenacin, showing the action of avenacin glucosidase. (A, courtesy of Anne Osbourn.)

(A)

senecionine

(B)

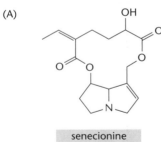

make an extraordinary diversity of alkaloids. Many of these compounds have powerful effects on animals, but they have not been implicated in resistance to microbes. Cocaine, caffeine, morphine, and nicotine are all well-known plant alkaloids that are important because of their effects on human behavior, but many other alkaloids are also of economic significance. For example, quinine from the bark of the tree *Cinchona* was the original antimalarial drug: it is toxic to the apicomplexan *Plasmodium* that causes malaria. Many alkaloids are important because of their adverse effects on farm animals. Grazing animals are poisoned by eating alkaloid-containing plants in pastures, such as ragworts (*Senecio* spp., containing the alkaloid senecionine; Figure 8–52) and lupins (*Lupinus* spp., containing the alkaloid lupinine).

Cyanogenic glycosides and **glucosinolates** are not in themselves toxic, but break down to give volatile poisons when a plant producing them is crushed. Cyanogenic glycosides release hydrogen cyanide (HCN), and glucosinolates release isothiocyanates and nitriles. Both groups of compounds consist of a nitrile (–C≡N) group joined to a carbon that is also linked to alkyl groups and to a sugar. The cyanogenic glycosides and the enzymes that release cyanide from them are stored separately within the plant, so cyanide is released only as a result of mechanical damage that mixes the contents of different compartments together. For example, *Sorghum* makes the cyanogenic glycoside dhurrin (Figure 8–53A), which is stored in the vacuoles of leaf epidermal cells The

Figure 8–52.
The ragwort *Senecio jacobea* as a toxic plant. (A) Structure of senecionine, produced by *S. jacobea*. The toxic alkaloid poisons farm animals that graze in fields containing this ragwort. (B) A plant of *S. jacobea*.

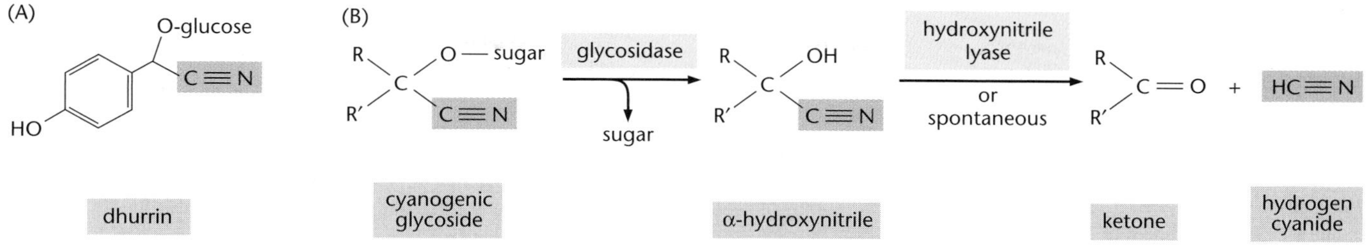

Figure 8–53.
Metabolism of cyanogenic glycosides.
(A) Structure of dhurrin, a cyanogenic glycoside produced in the epidermal cells of the cereal crop sorghum. (B) When plant organs containing cyanogenic glycosides are damaged, the compounds come into contact with enzymes that metabolize them to release toxic derivatives. The sugar group is removed by a glycosidase, and the resulting α-hydroxynitrile is converted to a ketone and hydrogen cyanide, either by the enzyme hydroxynitrile lyase or by spontaneous decay.

hydrolytic enzymes are found in the mesophyll cells, so cyanide is released only after damage that brings the contents of the two cell types together. The initial step in cyanogenic glycoside breakdown is cleavage by a glycosidase to release the sugar (Figure 8–53B). The resulting hydrolysis product (an α-hydroxynitrile) either breaks down spontaneously or is cleaved by the enzyme hydroxynitrile lyase to release cyanide. The presence of cyanogenic glycosides renders some food plants potentially toxic to humans, unless the plants are correctly treated before consumption. The starch-containing roots of cassava, for example, require thorough washing and cooking to prevent cyanide poisoning.

Glucosinolates are primarily found in the Brassicaceae and are responsible for the smell and taste of cabbage, broccoli, radish, and mustard. In these compounds the central carbon is linked to a sugar via a sulfur atom, rather than an oxygen atom as in cyanogenic glycosides (Figure 8–54A). Like cyanogenic glycosides, the glucosinolates are stored in the plant separately from the enzymes that hydrolyze them. Breakdown to release compounds toxic or repellent to herbivores occurs only when the plant is mechanically damaged. A thioglucosidase, also known as a myrosinase, cleaves the sugar–sulfur bond; the resulting compound spontaneously rearranges to produce toxic isothiocyanates and nitriles (Figure 8–54B). Herbivores that can tolerate the breakdown products of glucosinolates often actively seek out plants that make them. For example, the cabbage white butterfly is attracted to lay eggs on *Brassica* spp. leaves by the presence of glucosinolates.

Maize, wheat, and rye synthesize a class of nitrogen-containing compounds, the benzoxazinoid acetal D-glucosides, that are strongly correlated with resistance to the European corn borer (in the case of maize) and other insect pests. The biosynthetic pathway of the benzoxazinoid compounds DIBOA and DIMBOA has been elucidated in maize. Indole is synthesized in the chloroplast, then exported to the cytosol for further metabolism via cytochrome P450 enzymes located in the endoplasmic reticulum. The resulting benzoxazinoids are glucosylated by transfer of glucose from **UDP-glucose**, and the glucosides are stored in the vacuole (Figure 8–55). The glucosides themselves are nontoxic, but a specific glucosidase in the chloroplast can cleave them to release free benzoxazinoids

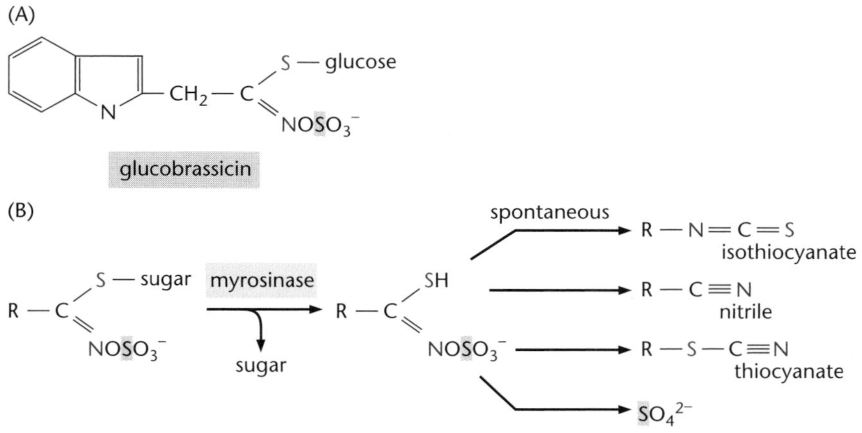

Figure 8–54.
Metabolism of glucosinolates.
(A) Structure of the glucosinolate glucobrassicin. Note the presence of two sulfur atoms, one forming the link between a central carbon atom and the sugar group, the other in a sulfite (SO_3^-) group. (B) When plant organs containing glucosinolates are damaged, the compounds come into contact with the enzyme myrosinase, which cleaves the sugar–sulfur bond to release the sugar. The resulting compound spontaneously decays to produce sulfate (SO_4^{2-}) and toxic isothiocyanates, nitriles, and thiocyanates.

Figure 8–55.
Cellular compartmentation of synthesis and storage of DIMBOA in maize leaves.
Indole is synthesized in chloroplasts from indole-3-glycerol phosphate, an intermediate in the pathway of tryptophan synthesis; this reaction is catalyzed by indole synthase. Indole is exported to the cytosol, where it undergoes reactions catalyzed by three enzymes associated with the endoplasmic reticulum. The resulting benzoxazinoid, DIMBOA, is glucosylated by a cytosolic UDP-glucose glucosyltransferase. The DIMBOA glucoside is stored in the vacuole.

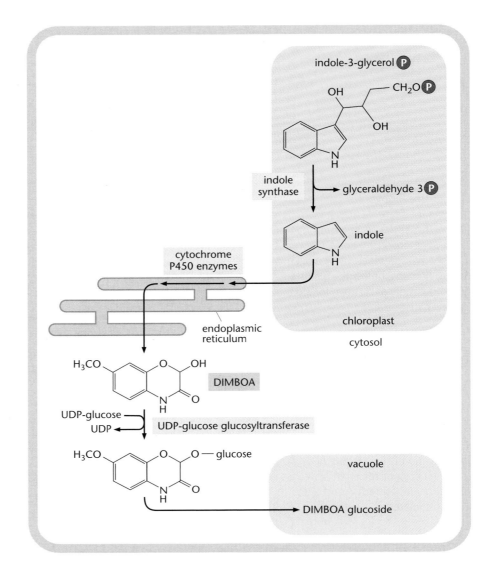

that are toxic to insects. Thus compounds toxic to insects accumulate only when the cell is damaged and the contents of the vacuole and chloroplasts are mixed.

Some plants accumulate high levels of certain nonprotein amino acids that provide protection from herbivores. Some of these compounds closely resemble protein amino acids, and cannot be distinguished from them by animal tRNA synthetases (the enzymes that "charge" **transfer RNAs (tRNAs)** with the correct amino acids in preparation for protein synthesis). The tRNAs of herbivores that consume plants containing these nonprotein amino acids thus become charged with these compounds rather than with the correct amino acids, leading to disruption of protein synthesis or unstable proteins. Canavanine, which accumulates in large amounts in the seeds of jackbean (*Canavallia*), is an example of a toxic nonprotein amino acid (Figure 8–56). It closely resembles arginine, and is mistaken for arginine by the arginyl-tRNA synthetase of animals. However, the arginyl-tRNA synthetase of jackbean can discriminate arginine from canavanine, so the plant is unaffected by its own toxin.

Antimicrobial compounds that are synthesized only in response to microbial attack are known as **phytoalexins** (Figure 8–56B). In many cases, microbial attack triggers the expression of plant genes encoding the enzymes necessary for the synthesis of phytoalexins. For example, expression of genes encoding the enzymes that synthesize capsidiol and rishitin (sesquiterpene phytoalexins) is rapidly induced in response to

(A)

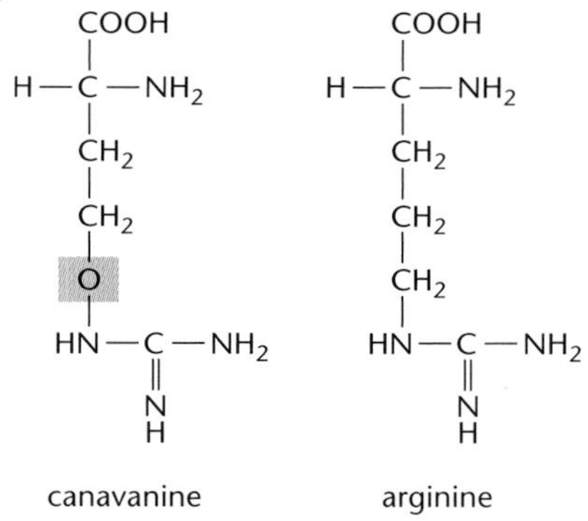

(B)

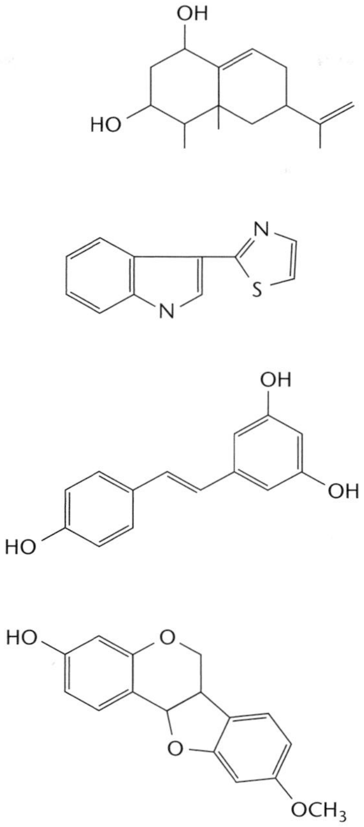

capsidiol, a sesquiterpene from pepper (*Capsicum*) and tobacco

camalexin, a nitrogen- and sulfur-containing compound from Arabidopsis

resveratrol, a grape stilbene derived from phenylpropanoid metabolism

medicarpin, a *Medicago* isoflavonoid derived from phenylpropanoid metabolism

Figure 8–56.
Nonprotein amino acids and phytoalexins. (A) An example of a nonprotein amino acid is canavanine, synthesized by jackbeans; the protein amino acid arginine is shown for comparison. (B) Some phytoalexins.

pathogen attack in tobacco and potato. Different plants make chemically distinct phytoalexins; besides sesquiterpenes, these include simple **phenylpropanoid** derivatives, **flavonoids** and **isoflavonoids**, and polyketides.

There are many reports of correlations between host-plant resistance and phytoalexin accumulation, and some phytoalexins have been shown to be antimicrobial. For example, when genes required for the synthesis of the grapevine phytoalexin resveratrol (Figure 8–56B) were introduced into tobacco, the resulting transgenic plants were more resistant to infection by the fungus *Botrytis*. However, there are relatively few examples of genetic proof that a particular phytoalexin makes a significant difference to a plant's resistance to infection. One such example is the production of medicarpin, an isoflavonoid phytoalexin (Figure 8–56B), by the legume *Medicago* in response to infection by the fungus *Phoma medicaginis* (which causes leaf spot disease). Transgenic *Medicago* plants modified to produce elevated levels of medicarpin exhibit higher levels of resistance to *P. medicaginis* than plants producing normal amounts of medicarpin.

Further genetic evidence for the importance of phytoalexins in disease resistance has come from identification of mutant plants that are deficient in phytoalexin production. Phytoalexins can be detected by biochemical techniques such as **thin-layer chromatography (TLC)** and **high-performance liquid chromatography**. TLC has been used to identify *phytoalexin-deficient* (*pad*) mutants of Arabidopsis. The main phytoalexin synthesized by Arabidopsis in response to microbial attack is camalexin (Figure 8–56B), a fluorescent compound that is visible on TLC plates under UV light. The *pad* mutants deficient in camalexin synthesis are more susceptible to some pathogens, such as the fungus *Alternaria brassicicola*, but unaltered in their response to others, such as the bacterium *Pseudomonas syringae*.

Disease resistance is often associated with the localized death of plant cells

During a plant–pathogen interaction in which the plant exhibits *R* gene–mediated resistance (in other words, an incompatible interaction), a range of defense mechanisms are usually activated in the plant cell or cells at and around the point of attempted invasion. In many cases this culminates rapidly (within a day) in a hypersensitive response: the death of the cell or cells in this region, while cells surrounding the region remain healthy. Localized cell death prevents pathogen invasion and protects the host plant in several ways. First, cell death prevents invasion by biotrophic pathogens, which require living cells for successful colonization. Second, the dead cells often contain high concentrations of antimicrobial compounds. These compounds are synthesized rapidly in response to the pathogen, both in the cells that subsequently die and in the surrounding, living cells. Third, cell death prevents the symplastic spread of toxins or other effector molecules introduced by the pathogen at the site of invasion.

The mechanisms that lead to cell death in the hypersensitive response are not fully understood. It is possible that cells die as a consequence of other aspects of their defense responses, or responses in surrounding cells, such as production of antimicrobial compounds that may also be toxic to the plant. Alternatively, cell death may be a separate defense response rather than an inevitable consequence of other defense responses. So-called **programmed cell death** occurs in other circumstances in plants, such as during formation of gametes, differentiation of the xylem, and formation of **aerenchyma**, the gas-conducting channels formed in response to oxygen deprivation in underground organs (see Section 7.6). In animals, the programmed cell death that occurs as part of organ development is known as **apoptosis**; this is a highly regulated process involving the production of specific proteases called **caspases**. So far, there is little evidence that programmed cell death in plants occurs by the same mechanism.

The hypersensitive response must be tightly regulated, so that the plant prevents pathogen invasion with the minimum loss of plant cells. Important information about the regulation of this process has been gained from the discovery of mutant plants in which hypersensitive responses occur in the absence of a pathogen. These are known as "disease lesion mimic mutants" (Figure 8–57). Their leaves have small patches of dead cells like those seen during an incompatible interaction with a pathogen. Most disease lesion mimic mutants are more resistant to particular pathogens than their wild-type counterparts: they show programmed cell death in response to pathogens to which the wild type is susceptible. For example, a disease lesion mimic mutant of barley, *mlo*, is resistant to powdery mildew, whereas wild-type plants that contain the Mlo protein are not (Figure 8–57A). Mutations in the *LESIONS SIMULATING DISEASE RESISTANCE* (*LSD*) genes of Arabidopsis (Figure 8–57B) and the *ENHANCED DISEASE RESISTANCE 1* (*EDR1*) gene of Arabidopsis result in plants with spontaneous cell death and enhanced disease resistance. Collectively, these mutants indicate that plants have mechanisms that suppress programmed cell death, ensuring that it is triggered only by

Figure 8–57.
Disease lesion mimic mutants. (A) Normal barley plants are susceptible to infection by powdery mildew (*left*); the white areas on the leaf are masses of spore-bearing fungal hyphae. Mutant plants lacking the Mlo protein are resistant to the pathogen (*right*). The Mlo protein is involved in control of the hypersensitive response. Leaves of the *mlo* mutant undergo spontaneous, localized cell death (*not visible*) even in the absence of the pathogen. (B) Disease lesion mimic mutant of Arabidopsis, *lsd4*, showing localized regions of dead cells in the leaves in the absence of a pathogen. (B, courtesy of Jeff Dangl.)

(A)

(B)

highly specific sets of signals. When the suppressor mechanisms are absent, programmed cell death is triggered by a much wider range of signals. Although this results in greater disease resistance, the occurrence of cell death in the absence of pathogens is potentially detrimental to the plant, because resources used for defense become unavailable for growth.

In systemic resistance, plants are "immunized" by biological challenges that lead to cell death

In many host–pathogen interactions, the activation of defense responses is not confined to the particular host cell undergoing attempted invasion. Genes encoding defense-related proteins are activated in surrounding cells, contributing to the limitation of pathogen growth. Some defense genes are also activated throughout the entire plant (systemically). The systemic activation of defense genes provides resistance (called **systemic resistance**) to a broad spectrum of pathogens in uninfected parts of the plant: it can be regarded as a form of "plant immunization."

Two main types of systemic resistance have been recognized, based on the class of signaling molecules involved in the response (Figure 8–58). Systemic acquired resistance (SAR) is triggered when the initial invasion by the pathogen leads to cell death as the result of a hypersensitive response. The establishment of SAR requires production of the signaling molecule **salicylic acid** and is characterized by the accumulation of a specific set of **pathogenesis-related (PR) proteins**, both at the site of initial infection and in uninfected tissues locally and distant to the site of infection. A second type of systemic resistance is triggered by mechanical wounding and by insect attack. Acquisition of this resistance does not require salicylic acid, but instead depends on production of the signaling molecules **jasmonic acid** and **ethylene**.

A typical example of SAR is provided by the response of tobacco plants to inoculation with an avirulent strain of *Tobacco mosaic virus*. When a mature leaf is inoculated with TMV, a hypersensitive response occurs. Groups of cells around the infection sites die, restricting the spread of the virus and giving rise to lesions on the leaf. Subsequent

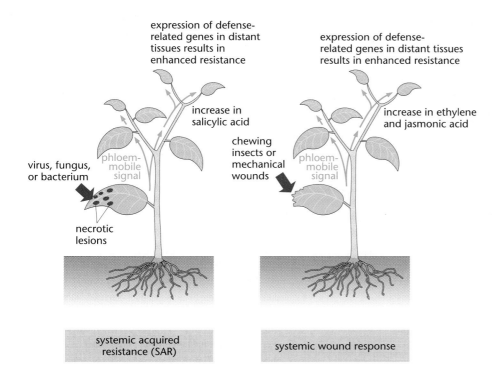

Figure 8–58.
Two types of systemic resistance. In systemic acquired resistance (*left*), a hypersensitive response to invasion by a pathogen results in production of a signaling molecule that moves from the infected leaf via the phloem to other parts of the plant. This triggers the expression of defense-related proteins, and hence promotes enhanced resistance, in these remote parts of the plant. SAR requires production of the signaling molecule salicylic acid, although this is not the signal that moves in the phloem. In the systemic wound response (*right*), cell damage and death caused by necrotrophic pathogens and herbivores similarly results in production of a phloem-mobile signaling molecule that triggers defense responses in remote parts of the plant. In this case, the response requires production of the signaling molecules ethylene and jasmonic acid (discussed later in the text).

inoculation of other leaves with the same or a different strain of TMV also results in cell death, but the lesions are much smaller than those on the first leaf inoculated. This indicates that the defense response has become more effective and stops the virus earlier (Figure 8–59). Thus a signal from the first leaf to be infected must have spread to other leaves, where it triggered increased resistance to viral infection. Tobacco plants in which SAR has been triggered by TMV also show enhanced resistance to other diseases such as blue mold (*Peronospora tabacina*).

Salicylic acid, a compound similar in structure to aspirin (acetylsalicylic acid), plays a central role both in defense responses around the site of infection and in the establishment of SAR. This has been demonstrated by two types of studies: introducing into plants a bacterial gene (*nahG*) encoding an enzyme that degrades salicylic acid (Figure 8–60A), and examining plants with mutations in genes encoding the enzymes required for salicylic acid synthesis. In both cases, disease resistance is found to be reduced and development of SAR compromised. In tobacco plants expressing the bacterial *nahG* gene, for example, TMV inoculation produces much larger lesions, which fail to contain the virus. The virus multiplies throughout the plant, leading to a spreading necrosis as the weakened hypersensitive response is repeatedly triggered.

Although salicylic acid is necessary for SAR, it is not the signal that spreads from the initially infected tissue to trigger SAR elsewhere in the plant. This was shown by grafting experiments in which the younger (upper) parts of normal tobacco shoots were grafted onto the older parts of tobacco plants expressing the *nahG* gene (and so unable

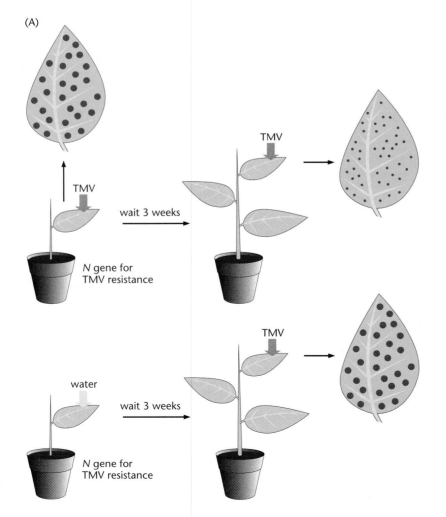

(A)

(B)

Figure 8–59.
Systemic acquired resistance to *Tobacco mosaic virus* in a tobacco plant. (A) When a plant carrying the *N* gene (an *R* gene that confers resistance to TMV) is inoculated with an avirulent strain of TMV, a hypersensitive response occurs on the inoculated leaf (*top*). This triggers SAR. Inoculation of a different leaf three weeks later produces a much more effective defense response, with less cell death. If, in a control experiment, the first inoculation is with water rather than virus (*bottom*), SAR is not triggered; inoculation of a different leaf three weeks later gives an unaltered defense response. (B) Leaves of a tobacco plant infected with TMV without (*left*) and with (*right*) a previous challenge. SAR is triggered only in the plant that was previously challenged (*right*); note the smaller lesions on this leaf. (B, courtesy of Sabg-Wook Park and Dan Klessig.)

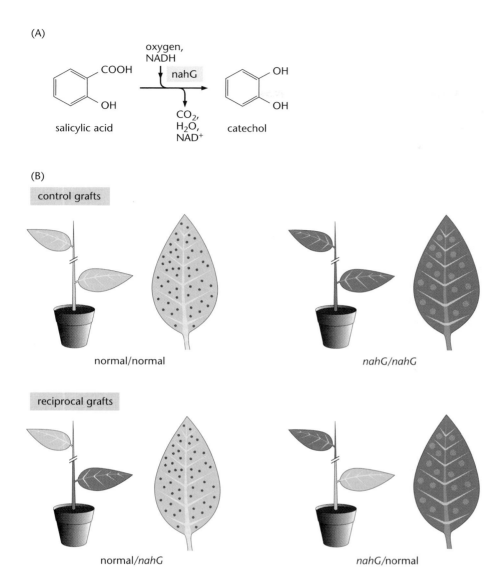

Figure 8–60.
Role of salicylic acid in triggering systemic acquired resistance. (A) Action of the enzyme encoded by the bacterial gene *nahG*. Salicylic acid is degraded to catechol, CO_2, and water. (B) Grafting experiments to determine whether salicylic acid is the phloem-mobile signal that triggers SAR in normal tobacco plants with a strong hypersensitive response (*pale green* leaves with small lesions) and in transgenic plants expressing *nahG* (in which the response is compromised because salicylic acid levels are low; *dark green* leaves with large lesions). After reciprocal grafting (as explained in the text), plants were inoculated with TMV on the lower leaves, then tested for development of SAR by inoculation of the upper leaves seven days later. In grafts where the lower leaves were deficient in salicylic acid (*lower left*), SAR developed normally in the upper leaves. Where the lower leaves were normal but the upper leaves were deficient in salicylic acid (*lower right*), SAR failed to develop in the upper leaves. This shows that production of salicylic acid in the initially infected leaf is not necessary for spread of the SAR signal, but salicylic acid is necessary for the mobile signal to trigger SAR in remote parts of the plant. In the equivalent experiments with the control grafts, the upper and lower parts of the plant were both normal (*upper left*) or both deficient in salicylic acid (*upper right*), showing that the grafting process itself did not affect development of SAR.

to accumulate salicylic acid), and vice versa, with the older parts normal and the younger parts carrying *nahG* and thus lacking salicylic acid (Figure 8–60B). The grafted plants were inoculated with TMV on the older leaves, and then seven days later were inoculated on the younger leaves. In the grafted plants in which salicylic acid was absent from the older leaves but present in the younger leaves, SAR developed normally: the younger leaves were highly resistant to TMV. In the grafts with salicylic acid present in the older leaves but absent from the younger leaves, SAR was *not* induced: the younger leaves were not resistant to TMV. This elegant experiment confirmed that salicylic acid is necessary for SAR to develop in tissues remote from the initial infection, but showed that leaves unable to produce salicylic acid can still produce a signal that elicits SAR in other parts of the plant. The signal that moves from infected to uninfected parts of the plant to trigger SAR is methyl salicylate, which is converted to salicylate in distant leaves by a specific esterase enzyme.

In tissues where SAR has been induced, a specific set of pathogenesis-related genes is expressed. The PR proteins are found either in intercellular spaces (in other words, they are secreted from cells) or in vacuoles. Some PR proteins are enzymes that may have antimicrobial functions, such as **chitinases** and **glucanases**; the functions of other PR proteins are not known. The production of a subset of PR proteins is also seen in systemic responses to mechanical damage and herbivory, but this subset of proteins is different from that seen in SAR.

Wounding and insect feeding induce complex plant defense mechanisms

Wounding is an unavoidable hazard for most plants. Treading and grazing by large animals tears and crushes leaves and shoots. The resulting damaged and dead cells may be more easily colonized by microbes than healthy cells, and they provide a point of entry into the plant body that is no longer protected by thick epidermal cell walls and cuticle.

Plant responses to this type of wounding can be summarized as follows. Wounding triggers the production of reactive oxygen species in surviving cells adjacent to the wound, which in turn triggers a set of defense responses (note the similarity of this sequence of events to the response to microbial challenge described above). Responses include gene expression that leads to the synthesis of **lignin** and other hydrophobic polymers that seal the wound against pathogen invasion. Further protection is given by the production in damaged cells of **polyphenols**, cross-linked phenolic compounds synthesized by polyphenol oxidases when the cells are exposed to an oxidizing environment. Polyphenols are the cause of the characteristic browning on cut surfaces of plant organs such as apples and potatoes. In addition to these localized defenses, wounding also induces some systemic wound responses that render other parts of the plant more resistant to herbivore attack (see Figure 8–58)—for example, through the production of protease inhibitors that render the plant material less digestible by insect proteases.

Invertebrate herbivory has unique features that provoke a more complex set of responses than those induced purely by wounding. Insects such as thrips and spider mites have piercing mouthparts that lacerate leaf cells and suck out the cell contents. These insects can cause considerable mechanical damage to cells and induce wounding responses similar to those seen with larger herbivores. Some other insects, however, are more specialized feeders that cause minimal mechanical damage and little wound response. These include whiteflies, aphids, mealy bugs, and leafhoppers, which have stylets that penetrate and withdraw sap from phloem elements. Resistance to such herbivores can occur through *R* gene–mediated signaling pathways, which may result in localized cell death at the point of stylet contact.

We describe here the responses to wounding and to mechanical damage from herbivory in tomato, in which these responses have been extensively characterized (Figure 8–61). In tomato, mechanical wounding and herbivory induce defense responses (likely to deter herbivores) both in wounded leaves and in undamaged leaves elsewhere on the plant. For example, the protein InhI, a serine protease inhibitor, accumulates adjacent to the wound and also in leaves many centimeters away from wound sites. Thus a signal generated at the site of wounding must move to other parts of the plant, where it initiates defense responses.

An important systemic signal in tomato is **systemin**, a small protein of 18 amino acid residues. Systemin can initiate defense responses at extremely low concentrations—a few femtomoles (10^{-15} moles) per plant. It is synthesized by cleavage from the C-terminus of prosystemin, the 200-residue precursor protein, at the site of wounding. Systemin moves throughout the wounded leaf within 30 minutes and reaches leaves above the wounded leaf within 1 to 2 hours, traveling via the phloem. Other systemic signals can also initiate defense responses distant from the wound site. For example, small, defined fragments of cell wall polysaccharides (oligogalacturonides) generated during wounding are also effective and can potentiate the systemin response.

The importance of systemin in defense against herbivores has been demonstrated using transgenic tomato plants that either produce lower-than-normal levels in response to

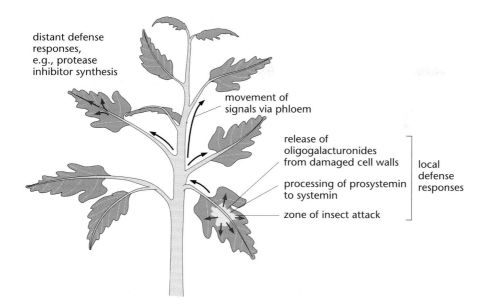

Figure 8–61.
Systemic responses to insect damage in tomato plants. Wounding by herbivorous insects results in local defense responses and production of mobile signals. These responses include systemin, synthesized from a precursor protein in response to wounding, and small cell wall fragments (oligogalacturonides) from damaged cells. Systemin and oligogalacturonides move via the phloem to remote parts of the plant where they trigger responses that enhance resistance to subsequent insect attack.

wounding or produce high levels constantly. When systemin production is lower than normal, plants have severely reduced systemic induction of defense mechanisms and have reduced resistance to herbivory by larvae of the tobacco hornworm *Manduca sexta*. When high levels are permanently present, plants constitutively express defense mechanisms (i.e., whether or not they are wounded).

Systemin activates several signaling pathways that lead to expression of defense mechanisms. One pathway is similar to that elicited by PAMPs, involving ion fluxes across the plasma membrane, an oxidative burst, and the activation of MAP kinases. This pathway produces elevated expression of genes necessary for the production of lignin precursors and the phytohormone ethylene. Ethylene activates further genes involved in defense. Systemin also activates a biochemical pathway responsible for the synthesis of jasmonic acid (JA). JA is an important signaling molecule that activates the expression of further genes involved in defense. JA signaling is also involved in plant responses to stress (see Section 7.7) and in normal plant development. JA is synthesized from the 18-carbon fatty acid **linolenic acid** (Figure 8–62), in reactions catalyzed by enzymes in the chloroplast and the **peroxisome** (the peroxisomal enzymes are part of the **β oxidation** pathway; see Section 4.6). Mutant tomato plants that lack enzymes of this pathway, or are deficient in linolenic acid synthesis, make little or no JA. They are severely deficient in defense responses after wounding; for example, protease inhibitor synthesis is compromised and plants are more sensitive to damage by spider mites and to defoliation by tobacco hornworm larvae (Figure 8–63).

Some parts of the defense response to wounding are conserved across plant species, but others are not. In Arabidopsis, initiation of the wound response resembles that in tomato in that JA plays a crucial role. However, Arabidopsis has no genes that encode proteins resembling prosystemin. The signal that induces systemic wound responses in Arabidopsis is likely to be JA itself.

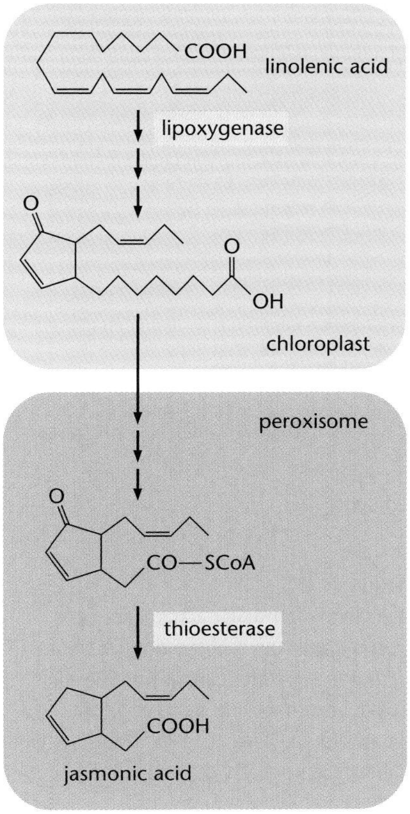

Figure 8–62.
Jasmonic acid synthesis. Jasmonic acid is synthesized from the fatty acid linolenic acid via a pathway in the chloroplast and peroxisome. The first step, in the chloroplast, is catalyzed by the enzyme lipoxygenase. The fatty acid derivative is exported to the peroxisome, where its subsequent metabolism (as an acyl-CoA) involves reduction in the length of the acyl chain by a mechanism analogous to β oxidation. In the final step, a thioesterase removes the coenzyme A to release jasmonic acid.

wild type   spr2

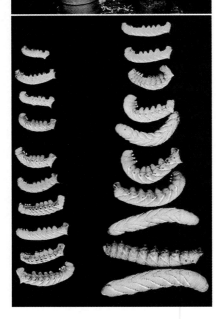

Figure 8–63.

Importance of jasmonic acid in systemic resistance to insect attack. Wild-type tomato plants (*left*) develop systemic resistance to tobacco hornworm caterpillars after an initial attack, limiting damage to the plant and restricting the growth of the caterpillars. An *spr2* mutant of tomato (*right*), with reduced JA synthesis, is unable to develop systemic resistance. The plant is heavily damaged by hornworms and the caterpillars grow larger than on wild-type plants. The wild-type *Spr2* gene encodes a fatty acid desaturase involved in the synthesis of linolenic acid, the precursor of JA (see Figure 8–62), in chloroplasts.

Chewing insects provoke release of volatile compounds that attract other insects

In addition to the direct defense mechanisms triggered by wounding, larval feeding and egg-laying by herbivores can also trigger indirect defense mechanisms. These include the release of volatile organic compounds that both repel the herbivores and attract insects that prey on or parasitize them. The volatile compounds are typically derivatives of the signaling molecules JA and salicylic acid, terpenoids, or indole. When present in the atmosphere surrounding a plant, they can induce defense mechanisms even in the absence of herbivores. Thus attack by a herbivore on one plant can trigger resistance to the herbivore in neighboring, uninfested plants of the same species, via plant-to-plant signaling. We describe two examples of plant defenses mediated by volatile compounds, illustrating some common features of this phenomenon.

When plants of the tobacco species *Nicotiana attenuata* are attacked by hornworm caterpillars (*M. sexta*), JA is synthesized first at the site of attack and later in more distant parts of the plant. JA production triggers the synthesis and release of a cocktail of volatile compounds. Some of these are derived from the same biosynthetic pathway as JA (e.g., hexen-1-ol), while others are terpenoids (e.g., linalool and farnesene) or are derived from salicylic acid (methyl salicylate). These compounds both deter egg-laying by the adult hornworm and attract a parasitic wasp (*Geocoris pallens*) that lays its eggs on hornworm caterpillars. When the wasp eggs hatch, the larvae consume the hornworm caterpillars. Another example of wasp parasitism of a hornworm caterpillar is shown in Figure 8–64. In a similar fashion, maize plants under attack from caterpillars of the beet armyworm (*Spodoptera exigua*) produce volatile compounds, including terpenoids and indole, that attract wasps that parasitize *Spodoptera* larvae.

In both of these cases, the defense reactions are triggered by elicitors in the saliva of the herbivorous caterpillars. One of the elicitors is volicitin, a compound that consists of a fatty acid derivative linked to the amino acid glutamine. The fatty acid component is derived from the breakdown of plant membranes during caterpillar feeding, and the amino acid is produced by the caterpillar itself (Figure 8–65).

Figure 8–64.

Pupae of a parasitic wasp (*Cotesia congregatus*) on a hornworm (*Manduca quinquemaculata*) caterpillar. The adult wasps are attracted to plants on which hornworms are feeding by the plant's systemic production of volatile compounds. (Courtesy of Whitney Cranshaw.)

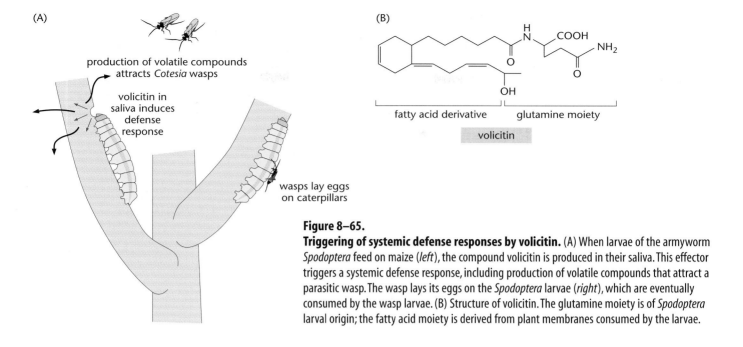

(A)

production of volatile compounds
attracts *Cotesia* wasps

volicitin in
saliva induces
defense
response

wasps lay eggs
on caterpillars

(B)

fatty acid derivative   glutamine moiety

volicitin

Figure 8–65.
Triggering of systemic defense responses by volicitin. (A) When larvae of the armyworm *Spodoptera* feed on maize (*left*), the compound volicitin is produced in their saliva. This effector triggers a systemic defense response, including production of volatile compounds that attract a parasitic wasp. The wasp lays its eggs on the *Spodoptera* larvae (*right*), which are eventually consumed by the wasp larvae. (B) Structure of volicitin. The glutamine moiety is of *Spodoptera* larval origin; the fatty acid moiety is derived from plant membranes consumed by the larvae.

The regulation of production of volatile compounds in response to caterpillar feeding is complex. Although JA is necessary for the response, application of JA to undamaged plants produces a broader, less specific response than that seen in *Manduca* or *Spodoptera* attacks. It seems that insect elicitors trigger the production of more than one signaling molecule in the plant, and the signaling pathways interact in complex ways to control defense responses. For example, in tobacco plants, application of JA and attack by species of insects other than *Manduca* triggers the synthesis of the toxic compound nicotine in the roots. The nicotine is carried to the leaves in the xylem; its synthesis in roots rather than leaves allows a systemic defense to be mounted even when most of the upper parts of a plant have been eaten. However, attack by *Manduca* caterpillars does not trigger nicotine production. *Manduca* caterpillars can tolerate nicotine and their parasitic wasp is repelled by it; thus there has been selective pressure on the plant against nicotine production in response to *Manduca* attack. It seems that an elicitor molecule produced specifically by *Manduca* triggers production of the signaling molecule ethylene, and this suppresses the expression of genes encoding enzymes of nicotine synthesis that would otherwise be induced by JA.

RNA silencing is important in plant resistance to viruses

As has been known for a long time, plants infected with viruses can "recover" and produce new shoots that do not show viral disease symptoms (Figure 8–66). Recent research has revealed that plants can attenuate the replication of many plant viruses by RNA silencing. Although plant viruses vary enormously in morphology, genome organization, replication and protein expression strategies, host range, and so forth, all accumulate viral RNA as part of their life cycle. It is viral RNA, especially when it is double-stranded, that provides the starting point for the silencing mechanism, as described below. The dsRNA can be a component of the replication process, a dsRNA fold within ssRNA, or the RNA product of overlapping bidirectional transcription.

The viral dsRNA is initially degraded into very specific products by a family of plant enzymes called **Dicer-like proteins (DCL)** (or "Dicer"). A DCL cleaves the dsRNA into fragments 21 to 24 nucleotides in length, known as **small interfering RNAs** or **siRNAs** (Figure 8–67). Next, the two strands of the siRNA are separated and one

Figure 8–66.
Recovery from virus infection. The lower leaves of this tobacco plant show strong symptoms of viral infection, but the upper leaves—developed after the plant was infected—are free of symptoms. This photograph is from the first scientific article to describe this phenomenon, published in the *Journal of Agricultural Research* in 1928.

Figure 8–67.
Mechanism of silencing of viral RNA.
Double-stranded RNA produced during replication of the viral (single-stranded) RNA is recognized by the enzyme DCL, or Dicer, which cleaves the dsRNA into sections of 21 to 24 nucleotides. These small interfering RNAs (siRNAs) bind to the RISC complex, which uses the RNA as a template with which to recognize and bind viral RNA molecules. The bound viral RNA molecules are degraded by an RNA nuclease that forms part of the RISC complex.

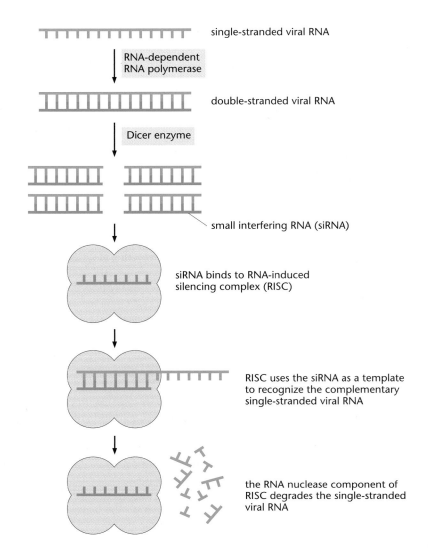

single-stranded viral RNA

RNA-dependent RNA polymerase

double-stranded viral RNA

Dicer enzyme

small interfering RNA (siRNA)

siRNA binds to RNA-induced silencing complex (RISC)

RISC uses the siRNA as a template to recognize the complementary single-stranded viral RNA

the RNA nuclease component of RISC degrades the single-stranded viral RNA

strand is incorporated into a multisubunit complex, the **RNA-induced silencing complex (RISC)**. The complex includes an RNA helicase, necessary for binding the RNA strand, and an RNA nuclease. RISC uses the RNA strand as a template to recognize and bind viral RNA molecules containing the complementary nucleotide sequence. These target viral RNA molecules are then degraded by the RISC RNA nuclease, thus suppressing viral RNA accumulation in the plant.

RNA silencing has three characteristics of an effective defense mechanism. First, it is specific for the viral RNA, because RISC is targeted by siRNAs that are derived from a double-stranded form of the viral RNA. Activation of virus-induced silencing therefore has no effect on host-encoded RNAs. Second, it has the potential for amplification: each viral dsRNA gives rise to multiple siRNAs, and each siRNA incorporated into RISC may catalyze the silencing of multiple targets. This potential for amplification means that silencing can be effective even against a rapidly replicating viral RNA. A third characteristic is the involvement of a mobile signal of silencing that may move either with or ahead of the virus. This signal ensures that the virus cannot escape the effects of silencing by movement between cells or in the phloem.

The sequence-specific nature of RNA silencing is shown by experiments with transgenic plants expressing an introduced gene encoding **green fluorescent protein (GFP)** (Figure 8–68). Under light of appropriate wavelength, these plants fluoresce. If they are subsequently infected with a modified form of *Potato virus X* that also carries the GFP gene, silencing of the GFP gene spreads (as seen by the loss of fluorescence) with the virus. (This phenomenon can also be induced when a second GFP gene is introduced by other methods, such as by infiltration of leaves with *Agrobacterium* containing a T-DNA that carries the gene, as shown in Figure 8–68.) Silencing occurs because RISC complexes primed with GFP-encoding viral RNA are degrading the **messenger RNA (mRNA)** produced from the GFP gene introduced into the plant genome. If the virus does not carry the GFP gene, no silencing of GFP takes place and the plant remains fluorescent. Modified viruses can similarly be used to silence the plant's own genes. If the virus contains part of the sequence of a plant gene, mRNA from that plant gene will be targeted for degradation when the virus is present. This is a valuable method for experimentally reducing or eliminating expression of specific plant genes. (See also Section 2.3.)

Many plant viruses encode proteins that act as suppressors of RNA silencing, allowing the virus to replicate in the plant. Suppressors from different viruses are generally unrelated in sequence and structure. It seems that under the strong selective pressure of RNA silencing, several different viral mechanisms have evolved separately that fulfill the same function—an example of **convergent evolution**. In several cases, the suppressors have additional, unrelated roles in viral biology, and their ability to suppress RNA silencing has evolved as an additional feature. The viral suppressor proteins interfere with various aspects of the plant's RNA silencing machinery. For example, the suppressor P19 encoded in the genome of *Tomato bushy stunt virus* directly binds to short dsRNA molecules, probably thus interfering with their incorporation into RISC complexes. When the suppressor HcPro from *Turnip mosaic virus* is expressed transgenically in turnip plants (i.e., in the absence of virus infection), it produces symptoms similar to those of plant mutants in which the DCL involved in processing regulatory small RNAs (microRNAs) during normal plant growth has been eliminated. This pattern suggests that HcPro interferes with all DCL functions in the plant, not simply the ones involved in suppression of viral replication. It seems likely that some of the viral disease symptoms, particularly stunting and abnormal development, may be the result of interference by viral suppressor proteins in siRNA and microRNA processing.

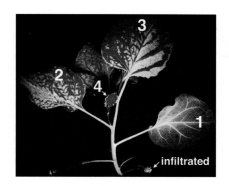

Figure 8–68.
Silencing of GFP expression in tobacco plant by infiltration with *Agrobacterium* carrying a second GFP gene. Transgenic tobacco plants expressing a gene that encodes green fluorescent protein are fluorescent (yellow-green) under light of appropriate wavelength (the red color is due to the natural fluorescence of chlorophyll). Infiltration of a leaf (*arrow*) with *Agrobacterium* carrying the GFP gene triggers the silencing of the plant's GFP gene. The mature leaf above the infiltrated leaf (*leaf 1*) continues to fluoresce because GFP protein was present in this leaf before infiltration. Upper leaves that were still expanding at the time of infiltration are partly silenced (*leaves 2 and 3*), and a very young leaf (*leaf 4*) is completely silenced. (Courtesy of David Baulcombe.)

8.5 COOPERATION

The interactions between plants and microbes or insects described thus far have all been to the detriment of the plant. However, there are many associations between plants and microorganisms, and between plants and insects or other animals, that are mutually beneficial. Associations that benefit both partners are termed "symbioses." In this section we first describe how flowers attract animals for pollination, then discuss two types of symbioses between plants and soil microorganisms: associations between some plant species and bacteria that assimilate atmospheric nitrogen, and more widespread symbioses between plants and fungi.

Many plant species are pollinated by animals

Most flowering plants are pollinated by animals, and most **pollination** is carried out by insects. The evolution of specialized flower forms and the production of floral attractants has occurred in parallel with the evolution of pollinating animals (another example of coevolution). Animal-pollinated flowers attract pollinators through their petal color, shape, and scent, and through rewards for pollinators, including **nectar**. The flowers of plants that are not pollinated by animals (i.e., wind-pollinated flowers) usually lack these features. Pollinators are thought to have visited the earliest flowering plants attracted by pollen itself as a food. The transfer of pollen from one flower to another by feeding insects was probably more efficient than wind pollination, providing a selective pressure for the evolution of other mechanisms that attracted insects and other animals to the flowers.

There is enormous diversity in the shapes of flowers and the attractants they provide, and in the animals that pollinate them (Figure 8–69). Beetle-pollinated flowers tend to have open structures, giving the short mouthparts of these insects access to pollen, nectar, or specialized food structures (clusters of cells within the flower that are eaten by beetles). Beetle- and fly-pollinated flowers are often strongly scented, and may smell of other foods eaten by these insects such as dung and rotting meat. Bee-pollinated flowers are often a yellow, violet, or blue color with strong markings. They typically have a shape that provides a landing platform, and the petals form a tube that allows access to the nectar only for insects with specialized mouthparts. The nectar of flowers pollinated by butterflies, moths, and hummingbirds is also typically at the base of a very long tube or spur, allowing access only to the very long, tubelike mouthparts of these animals. Flowers adapted for moth and bat pollination are often white and strongly scented, and open in the evening or at night.

Some flower shapes are elaborated in ways that maximize the efficiency of pollen transfer. In many bee-pollinated flowers, the shape formed by the petals forces the bee to come in contact initially with the **stigma**, allowing transfer of any pollen it is carrying from another flower. As the bee feeds, it comes into contact with the **stamens** and picks up pollen, which it carries to other flowers. In extreme cases, the structure of the flower actually traps insects until fertilization has been achieved, whereupon the insect is released via a route that ensures that it picks up pollen from the stamens. The flowers of some *Arum* species are examples (see Figure 4–65).

The scents and colors of flowers are created by a wide range of compounds, synthesized via diverse metabolic pathways. Scent is often produced from particular cells (osmophores) in the mesophyll layer of the petals and emitted through specialized regions of the epidermis. Many scent compounds are derived from monoterpene and phenylpropanoid precursors, and the scent of any one flower is usually provided by a complex mixture of different volatile compounds. Flower colors in the yellow, orange, and red range are often due to **carotenoids**, present in plastids of petal cells. These are

Figure 8–69.
Flowers pollinated by animals.
(A) Hoverfly feeding on the open, nectar-producing flowers of the candelabra tree (*Euphorbia ingens*). (B) Flower of *Huernia zebrina*, which, like other members of the family Stapeliaceae, attracts insects because it looks and smells like rotting flesh. (C) Bee feeding from the tubular flower of a *Stachys* species (family Labiatae). (D) Hummingbird visiting the long, hanging tubular flowers of *Tristerix corymbosus* (family Loranthaceae).

(A)

(B)

(C)

(D)

products of a branch of terpenoid metabolism. (Section 4.7 describes the metabolism of these and similar compounds.) Many flower colors, including pinks, reds, blues, and purples, are due to **anthocyanins** (synthesized via a branch of phenylpropanoid metabolism) in the vacuoles of petal cells. The color they convey to the petal depends on their precise molecular structure (in particular, the positions of hydroxyl groups in one of the rings; Figure 8–70A), the pH in the vacuole, and the formation of complexes with metal

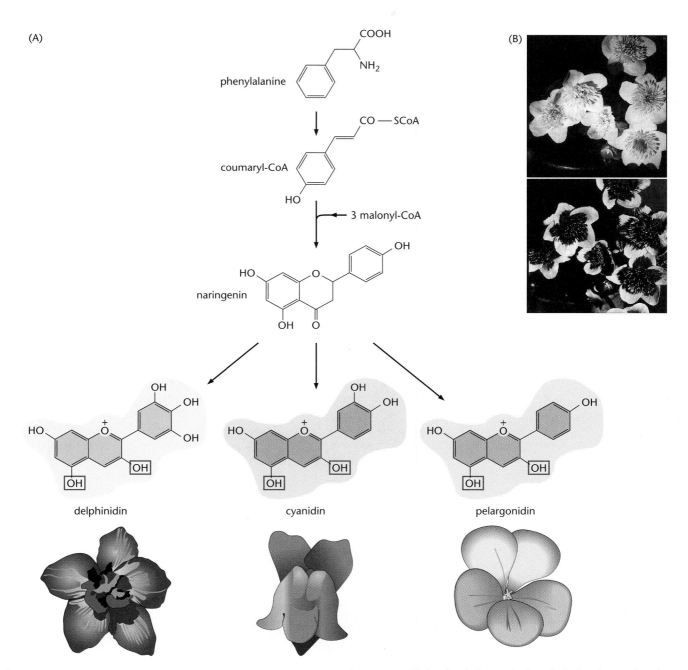

Figure 8–70.
Flower pigments. (A) Synthesis of the anthocyanidins delphinidin, cyanidin, and pelargonidin, which are responsible for the blue, magenta, and orange-pink colors of delphinium, *Antirrhinum*, and geranium flowers respectively. All three compounds are synthesized from the aromatic amino acid phenylalanine. They differ from each other only in the number of hydroxyl groups on the right-hand ring. These anthocyanidins are glycosylated (i.e.,

have sugars added to them) after synthesis, at the hydroxyl groups boxed on the diagram. The glycosylated forms (anthocyanins) are transported into the vacuoles of the petal cells, where they accumulate.
(B) Flowers of *Caltha palustris* in white light (*top*) and in ultraviolet light (*bottom*). Note the dark centers—regions that do not reflect ultraviolet light but are indistinguishable from the rest of the flower in white light.
(B, courtesy of Bob Fosbury.)

ions and with other flavonoid compounds (such as flavones and flavonols) also present in the vacuole. The flavones and flavonols themselves are visible as "color" to some pollinating insects including bees, though not to humans; these compounds absorb light in the ultraviolet range to which bee eyes but not human eyes are sensitive (Figure 8–70B).

The enormous range of colors and patterns of flower petals is the product of complex spatial and temporal regulation of the expression of genes encoding enzymes of flavonoid synthesis. This regulation is a function of multiple transcription factors that target genes encoding particular branches of the biosynthetic pathways. Overlapping patterns of expression of the transcription factors produce precise local variations in flavonoid composition across the petal, seen as shades and patterns of color that attract the attention of particular types of pollinator and guide them into the flower so that pollen is efficiently delivered and dispersed.

Symbiotic nitrogen fixation involves specialized interactions of plants and bacteria

Most plant species obtain their nitrogen as soluble nitrate or ammonium from the soil (see Section 4.8). However, in some groups of plants nitrogen is supplied by symbiotic relationships with **nitrogen-fixing bacteria**. The bacteria reduce atmospheric nitrogen (N_2) to ammonia (NH_3), a process known as **nitrogen fixation**. The bacteria can fix nitrogen only when they are intimately associated with the plant, encased in special structures called **nodules** on the plant roots (Figure 8–71). The ammonia produced by the bacteria in the nodule is used by the plant for amino acid synthesis. Plants capable of forming these symbioses grow better in nitrogen-poor soils than other plants.

Nitrogen-fixing symbiotic associations are formed between plants of the legume family (Leguminosae, also called Fabaceae; an example is shown in Figure 8–71A) and several species of soil bacteria including *Rhizobium* and *Bradyrhizobium*, related species collectively known as **rhizobia**. These symbioses have been extensively studied and we describe them in detail below. Nearly 200 further species of plants, belonging to eight families in the same grouping (clade) of higher-plant families as legumes, form associations with another bacterium, the actinomycete *Frankia* (Figure 8–71B). Although the legume–rhizobia interactions are better understood, the actinomycete symbioses are also of considerable ecological significance. They fix about as much atmospheric nitrogen in the biosphere as do the legume–rhizobia symbioses. Nitrogen-fixing symbioses are also known in several other plant groups. Cyanobacteria (such as *Nostoc* and *Anabaena*) form symbioses with plants as diverse as the water fern (*Azolla*), cycads, and the flowering plant *Gunnera*, although these interactions do not involve the formation of nodules.

Neither rhizobia nor their plant hosts are dependent on the formation of a symbiotic relationship. Rhizobia are free-living soil bacteria that can multiply in the absence of a

Figure 8–71.
Root nodules. (A) Nodules on the root system of a pea plant. The pinkish color is due to high concentrations of the oxygen-carrying protein leghemoglobin. The nodules contain nitrogen-fixing bacteria. (B) Nodules containing the actinomycete *Frankia*, on the root of an alder (*Alnus* sp.) tree. (A, courtesy of J.A. Downie; B, courtesy of David Benson.)

(A)

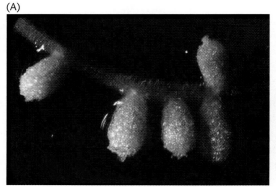

(B)

host plant. Leguminous plants can assimilate nitrogen in the form of nitrate or ammonium in the absence of rhizobia. The symbioses are highly specific: particular species of rhizobia infect only one or a small range of host plant species. The formation of the symbiosis (known as "nodulation") is a remarkable process involving exchange of chemical signals between bacteria and host plant, and large changes occur in gene expression and developmental patterns in both organisms as the nodule develops and the bacteria are encased within it. The process takes place in several steps, which can be outlined as shown in Figure 8–72.

1. Soil rhizobia perceive chemical signals diffusing from the roots of potential host plants; these signals are usually flavonoids, related to the anthocyanin pigments of flowers and leaves.

2. The rhizobia respond to these signals by producing new signal molecules, **Nod factors**, that act on the root hairs of the host plant, causing them to curl over at the tips.

3. Bacteria enter root hairs at a site of curling, and an **infection thread** containing bacteria grows from cell to cell through the cortex of the root.

4. During this process, further secretion of Nod factors promotes new cell divisions in either the cortex or **cambium** that lead to the differentiation of a nodule.

5. As the nodule develops, infection threads enter nodule cells, and rhizobia are released from the threads into these cells, still surrounded by bacterial and host membrane. They differentiate into **bacteroids**, or **symbiosomes**, which are the sites of nitrogen fixation.

The signals released by the roots of a leguminous plant that provoke rhizobia in the soil to make Nod factors are usually (but not always) flavonoids and related compounds—

Figure 8–72.
Stages in the nodulation process.
Uninfected roots of a legume plant release a flavonoid into the soil. (1) This acts as a signal to trigger expression of the *nod* genes of free-living rhizobia in the soil. (2) The *nod* genes encode enzymes that synthesize the Nod factor, which causes root-hair curling. (3) Bacteria enter the curled root hair and grow through the hair cell in an infection thread. (4) Signals from the rhizobia promote cell divisions in the inner cortex of the root, from which the nodule will form (formation of new meristem is discussed later in the text). The infection thread grows through cortical cells to reach the developing nodule. (5) Rhizobia are released from the thread into nodule cells. The nodule enlarges so that it protrudes from the root surface, and becomes vascularized.

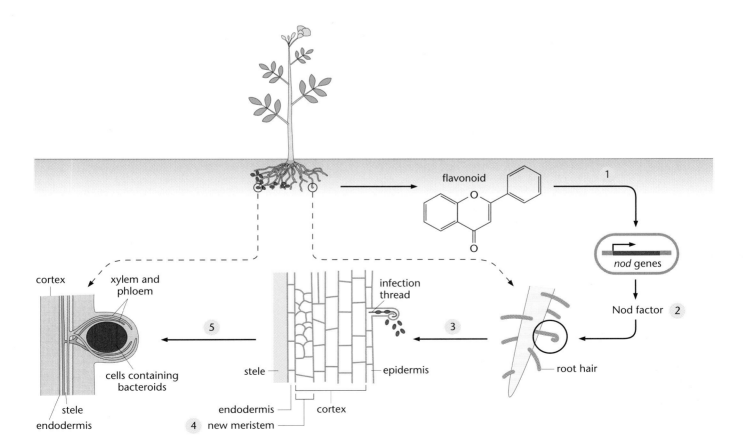

secondary metabolites derived from the phenylpropanoid pathway. Different legume species make a characteristic spectrum of flavonoid molecules; for example, *Medicago* makes luteolin, and other legumes make related but distinct compounds. The host-plant signal is recognized by the appropriate species of rhizobium in the soil. Luteolin, for example, is strongly recognized by *Sinorhizobium meliloti*, which forms the symbiosis with *Medicago*, but not by most other rhizobia species. Recognition of the plant signal causes the bacterium to synthesize Nod factors (Figure 8–73), which trigger nodule formation on the appropriate host-plant species; non-host plants do not recognize and respond to the Nod factors. Thus the bacterium responds only to chemical signals from a plant with which it can form a symbiosis, and the plant responds only to Nod factors from the appropriate rhizobium.

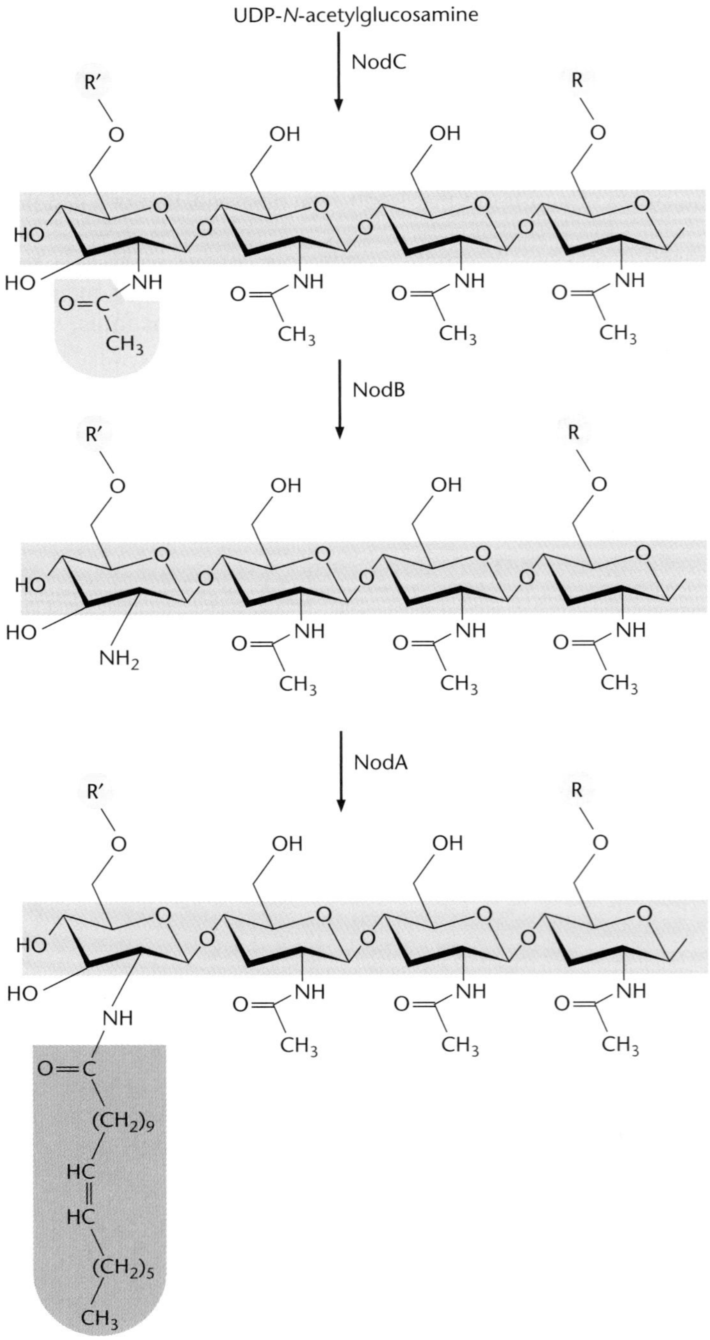

Figure 8–73.
Synthesis of a Nod factor. The sugar nucleotide UDP-*N*-acetylglucosamine is polymerized to form the backbone of the Nod factor, in a reaction catalyzed by the enzyme NodC (encoded by *nodC*). The acetyl group of the end residue (*blue*) is removed by the enzyme NodB, and a fatty acid (*red*) is added in this position by the enzyme NodA. Other enzymes add further groups (R and R′) to the glucosamine backbone. The fatty acid and R groups added vary with the rhizobium species.

Rhizobia capable of forming symbioses are taxonomically very diverse. For example, *Rhizobium* and *Sinorhizobium* are more closely related to *Agrobacterium* than they are to *Bradyrhizobium*. All rhizobia contain two key sets of genes necessary for formation of the symbiosis. The *nod* genes are necessary to induce early nodule formation; some of these are involved in the synthesis of Nod factors (Figure 8–73). The *nif* genes are necessary for the nitrogen-fixing process within the nodule. The *nifK*, *nifD*, and *nifH* genes, for example, encode the three proteins that make up the reductase component of the **nitrogenase complex** that carries out conversion of nitrogen to ammonia. The *nod* and *nif* genes are often clustered together in the bacterial genome. In *Rhizobium* strains, for example, genes required for symbiosis are usually carried on indigenous plasmids (symbiosis plasmids, or sym plasmids) that can range in size from 200 kb to 1200 kb. In free-living rhizobia in the soil, these plasmids can be transferred by conjugation from one strain to another, so that previously nonsymbiotic strains may gain the entire symbiosis "package."

The ability of bacteria to nodulate (i.e., form symbiotic relationships with) legumes correlates with the capacity to produce appropriate signal molecules—the Nod factors—rather than with the relatedness of the bacteria. Nod factors of the different *Rhizobium* species have a common basic structure (Figure 8–73). A β1,4-linked *N*-acetyl-D-glucosamine backbone, usually of four or five units, is linked to a fatty acid on the terminal sugar. Synthesis of the Nod factor backbone is catalyzed by enzymes encoded by the bacterial *nod* genes. The NodC protein catalyzes the polymerization of *N*-acetylglucosamine. NodB removes the acetyl group from the end glucosamine residue of the Nod factor backbone, and then NodA catalyzes the transfer of a fatty acid (acyl) group onto the free amino group. Each rhizobium species makes unique Nod factors. For example, Nod factors from *Rhizobium leguminosarum* and *S. meliloti* differ in the structure of the acyl chain and in the presence of a sulfate group on the *S. meliloti* factor. The *S. meliloti* Nod factor stimulates a response in *Medicago* species but not in other genera of legumes.

Nod factors induce root-hair curling at extremely low concentrations. If purified Nod factor at higher concentrations is applied to roots, it can induce the formation of an entire nodule in the absence of bacteria. In the natural situation, bacteria become entrapped in the fold of the affected root hair (Figure 8–74A). The wall of the root-hair cell adjacent to the bacteria partially degrades and the cell growth is reoriented inward, resulting in invagination of the root-hair cell wall to form a tunnel-like structure. This develops into the infection thread, a structure synthesized by the plant using mechanisms involved in cell wall biosynthesis. Not all legumes initiate symbioses in this way. In peanut (*Arachis*), bacteria enter the root through cracks—such as those that occur when lateral roots emerge—and proliferate in intercellular spaces (Figure 8–74B). This stimulates the plant to make a short infection thread that allows the bacteria to cross the plant cell wall.

Figure 8–74.
Entry of rhizobia into a legume root.
(A) In most legume species, rhizobia enter via root hairs. Root-hair curling induced by a Nod factor encloses the bacteria in a fold of cell wall. Cell wall degradation followed by reorientation of wall synthesis allows the rhizobia to enter the root hair within a tube—the infection thread—which contains cell wall-like material. The infection thread grows down through the root hair, then across cells of the cortex. (B) In some species, including peanut (*Arachis*), rhizobia enter the root through cracks between the main and lateral root walls, rather than via root hairs.

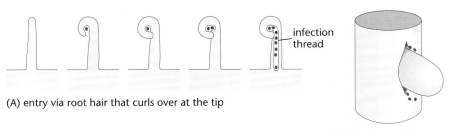

(A) entry via root hair that curls over at the tip

infection thread

(B) "crack entry" - entry via physical gap between main root and lateral root walls

Researchers are beginning to understand the details of the events that lead to root-hair curling. Addition of Nod factor to roots of an appropriate legume species very rapidly triggers a complex signaling pathway in the root-hair cell. Initially, there is a rapid influx of Ca^{2+} ions into the cell, followed by depolarization of the plasma membrane. Membrane depolarization can be induced at Nod factor concentrations as low as 10^{-9} M. About 10 minutes after addition of Nod factor, oscillations in the cytosolic Ca^{2+} concentration (known as "calcium spiking") are induced (Figure 8–75). The calcium spiking probably triggers downstream events that lead to changes in gene expression.

Insight into the signaling pathway upstream and downstream of this event has been gained from the selection of mutant legumes that are either unable to form nodules in the presence of the appropriate rhizobium, or undergo hypernodulation—the formation of many more nodules than normal. Many of the early studies were on pea (*Pisum sativum*), but recently two species have been developed as more tractable models: *Lotus japonicus*, which is nodulated by *Mesorhizobium loti*, and *Medicago truncatula*, nodulated by *S. meliloti*. As in the case of Arabidopsis, genomic and genetic tools are fast being developed for these two species.

Study of *Lotus* mutants that fail to nodulate in the presence of *M. loti* and do not respond to Nod factor has led to the identification of proteins necessary for the initial signaling events caused by Nod factor (Figure 8–76). Several of these are transmembrane protein kinases. One, Nfr, has an extracellular **LysM domain**. LysM domains in other proteins are involved in binding peptidoglycan, the proteinaceous polysaccharide characteristic of bacterial cell walls. Other protein kinases have LRR domains characteristically involved in protein recognition (as in many plant defense proteins; see Section 8.4). These LysM kinases may participate in the recognition of a Nod factor at the plant plasma membrane. They are required for calcium spiking. Proteins involved in the signaling pathway downstream of calcium spiking include protein kinases activated by Ca^{2+} binding, and transcription factors activated by these kinases. Exactly how changes in gene expression are induced and how they result in root-hair curling and initiation of infection threads is not yet understood.

The bacteria grow within the infection thread as it extends along the root-hair cell and down through several cortical cells, often branching in the process. The bacteria in the thread remain topologically outside the plant cells until the thread reaches the nodule

Figure 8–75.
Calcium spiking associated with perception of Nod factor. A root hair was injected with a dye that fluoresces in response to Ca^{2+} ions. After injection of the dye, Nod factor was added and changes in cellular Ca^{2+} ion concentration were monitored as changes in fluorescence. After oscillations in fluorescence were established, images of the cell were taken at four points during a single oscillation (*blue* represents low Ca^{2+} concentration; *red* is the highest concentration). (1, 2) At the start of the oscillation, Ca^{2+} concentration increases rapidly around the nucleus; (3, 4) it then drops more slowly, back to the base level. (ii, courtesy of J.A. Downie.)

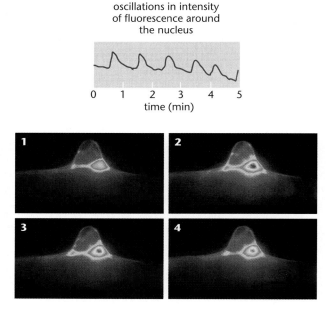

oscillations in intensity of fluorescence around the nucleus

0   1   2   3   4   5
time (min)

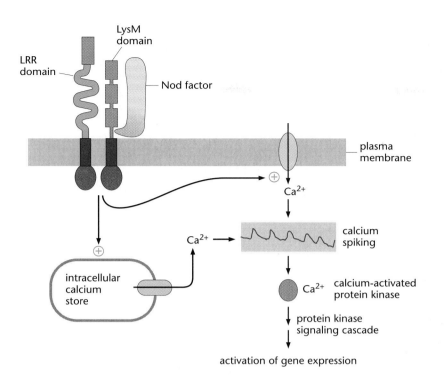

Figure 8–76.
Signaling events triggered in root-hair cells by rhizobia. Two types of transmembrane receptor kinases are necessary for the initial signaling events. One type, with an extracellular LysM domain, may detect Nod factor. The other, with an extracellular LRR domain, hypothetically complexes with the LysM domain protein and is necessary for its activity. Signaling events triggered by the intracellular kinase domains (*red*) of these receptors promote influx of Ca^{2+} ions via plasma membrane calcium channels and release of Ca^{2+} from intracellular stores. Regular oscillations of Ca^{2+} ion concentration activate a protein kinase signaling cascade, which leads to changes in gene expression that result in root-hair curling and formation of infection threads.

cells where nitrogen fixation will occur. At this point no further cell wall material is deposited along the thread, and the bacteria bud off into the plant cell (Figure 8–77).

The nodule **meristem** is generated by cell divisions in the cortex or the cambium of the root at the same time that the infection thread is formed. There is considerable diversity among plant species in nodule structure and differentiation. *Medicago* makes nodules with an **indeterminate** (continually growing) meristem. *L. japonicus* and soybean make nodules with a **determinate** meristem (growth potential is limited). In determinate nodules, the cells mature synchronously so that a mature nodule contains no meristematic (undifferentiated) cells. In indeterminate nodules the opposite is true: newly divided cells are constantly formed at the elongating tip.

The bacteria in the nodule cells are surrounded by a membrane derived from the plant plasma membrane, and these are the bacteroids or symbiosomes. The ability of bacteria to form this intimate association with plant cells depends on the fact that they do not trigger plant defense mechanisms. The composition of the extracellular polysaccharides that surround the bacterial cell is important in this respect. Mutant rhizobia with

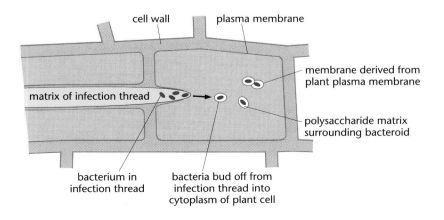

Figure 8–77.
Release of rhizobia into nodule cells. The infection thread, containing bacteria in a matrix of plant-derived cell wall material, grows through cortical cells to cells of the developing nodule. Here, bacteria bud off from the end of the thread into the cytoplasm to become nitrogen-fixing bacteroids. The bacteroids are surrounded by a membrane derived from the plant plasma membrane (the peribacteroid membrane) and a polysaccharide matrix.

Figure 8–78.
Nitrogenase complex. (A) Nitrogenase is a complex of enzymes and cofactors that catalyzes the conversion of nitrogen gas to ammonia, a process that requires large amounts of ATP and reductant (shown here as protons, H^+). (B) The nitrogenase complex consists of dinitrogenase, which catalyzes conversion of nitrogen gas to ammonia, and dinitrogenase reductase, which transfers electrons from reduced ferredoxin (Fd) to dinitrogenase in an ATP-consuming reaction. Electron flow through dinitrogenase reductase and dinitrogenase occurs via an iron-sulfur cluster (shown here as simply Fe) and an iron-sulfur-molybdenum cluster (shown as MoFe), respectively.

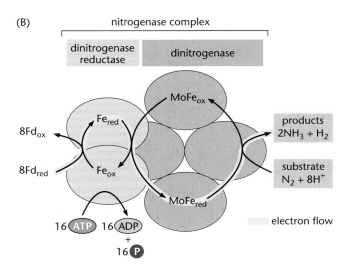

alterations in extracellular polysaccharide structure are unable to infect host plants, perhaps because they trigger host defenses.

The bacteroids utilize carbon sources from the host cell to provide energy for the fixation of nitrogen, and supply the host cell with ammonia for amino acid synthesis (Figure 8–78). The capacity to reduce nitrogen, which constitutes about 80% of the earth's atmosphere, is restricted to prokaryotes, including free-living soil bacteria, cyanobacteria, and the bacteria in symbiotic associations with plants. Nitrogen (N_2), because of the stability of its triple bond, is relatively unreactive and metabolically inaccessible to most organisms. In nitrogen-fixing bacteria the key enzyme, nitrogenase, is made up of two components: a dinitrogenase and a dinitrogenase reductase, neither of which is active on its own. The **dinitrogenase** contains an unusual iron-sulfur-molybdenum cofactor, and is extremely sensitive to (i.e., is destroyed by) oxygen. It reduces a molecule of N_2 to two molecules of NH_3, consuming eight protons and releasing one molecule of hydrogen (Figure 8–78). Electrons for this reduction are transferred to the iron-sulfur-molybdenum cofactor from reduced **ferredoxin** by **dinitrogenase reductase**, an iron-sulfur–containing protein. The mechanism of action of dinitrogenase reductase requires the hydrolysis of two molecules of ATP for every electron transferred from ferredoxin to dinitrogenase. This makes nitrogen fixation energetically expensive, consuming a minimum of 16 mol of ATP to reduce 1 mol of N_2 to 2 mol of NH_3.

The nodule provides an environment in which nitrogen fixation can occur efficiently, because an optimal O_2 concentration is maintained in the region containing bacteroids. Although the nitrogenase can only operate in essentially anaerobic conditions, the ATP required for nitrogen fixation is provided by bacterial **respiration**, a process that typically requires oxygen. Thus there are apparently conflicting requirements for levels of oxygen—the "oxygen paradox"—which are resolved first by maintaining O_2 levels sufficiently low to protect the nitrogenase (microaerobic conditions), and second by changes in the bacterial respiratory **electron transport chain** that enable respiration at very low O_2 levels as the rhizobia differentiate into bacteroids.

Three key factors allow the generation of a microaerobic environment in the nodule, while still allowing aerobic ATP synthesis (Figure 8–79). First, entry of oxygen into the nodule is controlled by a regulated oxygen-permeability barrier around the periphery of the nodule structure. The mechanism by which oxygen entry into the nodule is

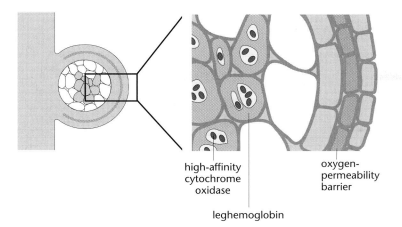

high-affinity
cytochrome
oxidase

oxygen-
permeability
barrier

leghemoglobin

Figure 8–79.
Factors that control oxygen availability in root nodules. Free oxygen inhibits nitrogenase activity, but oxygen is required to generate the ATP (by respiration) that is required for the nitrogenase reaction. This paradox is resolved in three ways: the outer layers of the nodule contain a barrier that restricts oxygen entry; oxygen in the nodule is mostly bound to leghemoglobin protein, so the free O_2 concentration is very low; and the cytochrome oxidase of the bacteroid electron transport chain has an unusually high affinity for oxygen—that is, it operates at its maximum rate at very low O_2 concentrations.

restricted is not understood. Second, the plant synthesizes an oxygen-binding protein called **leghemoglobin**, which reaches millimolar concentrations in the cytosol of nodule cells. Leghemoglobin effectively buffers changes in O_2 concentration that result from fluctuations in the rate of respiration or the operation of the permeability barrier. Oxygen is much more stable in a complex with leghemoglobin than as free oxygen, so rapid flux of oxygen from the air spaces between cells to the bacteroid is facilitated even at low levels of free oxygen. Leghemoglobin-bound oxygen is 70,000-fold more abundant than free oxygen in nodule cells.

The third mechanism to resolve the oxygen paradox is bacterial in origin. Free-living rhizobia have a respiratory electron transport chain similar to that of mitochondria. An additional electron transport chain is expressed when rhizobia differentiate into bacteroids. This enables an increased respiratory rate, making bacteroid respiration a strong sink for oxygen and keeping O_2 concentrations low. In addition, the respiratory chain of bacteroids has a **cytochrome oxidase** with a higher affinity for oxygen than that of the free-living bacterium; in other words, the enzyme of the bacteroid operates at its maximal rate at a lower O_2 concentration than that of the free-living bacterium. The two enzymes have K_m values of 8 nM and 50 nM, respectively (K_m is a measure of the O_2 concentration at which the oxidase-catalyzed reaction occurs at half its maximal rate). The bacteroid form of cytochrome oxidase is expressed only in nodules and is required for nitrogen fixation to occur. The net result of these changes in components of the electron transport chains during bacteroid formation is a very rapid rate of respiration even at O_2 levels sufficiently low to avoid damage to the oxygen-sensitive nitrogenase enzyme.

Fixed nitrogen is released from the nitrogenase complex and the bacteroid into the plant as NH_3. Ammonia is toxic at high levels and cannot be transported through the plant, so it is assimilated into organic compounds in plant cells within the nodule. The pathway by which this occurs is the **glutamine synthetase–GOGAT system**, described in Section 4.8 (Figure 8–80A). The fate of the glutamine synthesized by this pathway varies with the legume species. In alfalfa and pea, for example, glutamine itself, together with asparagine, is exported from the nodule as the nitrogen supply for the plant. In soybean and cowpea, glutamine is converted to **ureides** such as allantoin and allantoic acid, and these are exported from the nodule.

Nitrogen fixation and the production of amino acids in the nodule require a carbon source for bacterial respiration, to produce ATP and reduced ferredoxin for the nitrogenase reaction, and precursor molecules for assimilation of ammonia into amino acids. These are provided by metabolism of **sucrose** imported into nodule cells from the leaves (Figure 8–80B). The carbon sources for the bacteria are the organic acids malate and oxaloacetate, and the main carbon precursor for ammonia assimilation in the plant cells is 2-oxoglutarate.

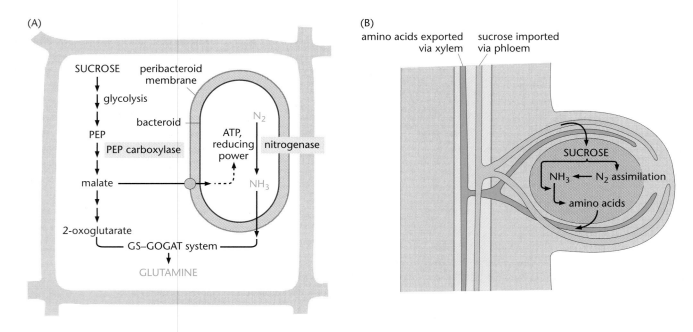

(A)

(B)

amino acids exported via xylem   sucrose imported via phloem

Figure 8–80.
Provision of precursor molecules for bacteroid metabolism and ammonia assimilation. (A) In infected cells, sucrose is metabolized to malate, which serves as a precursor for both bacteroid metabolism and ammonia assimilation. Malate is imported into the bacteroid via a malate transporter in the plant-derived peribacteroid membrane. In the bacteroid, malate is the substrate for generation of reducing power and ATP for the nitrogenase reaction. Malate is also metabolized in the plant cell cytosol to form 2-oxoglutarate, the substrate for the glutamine synthetase (GS)–GOGAT system, in which ammonia generated in the nitrogenase reaction is used to synthesize the amino acid glutamine. (B) Sucrose produced during photosynthesis in the leaves is imported into the nodule via the phloem. Amino acids synthesized by assimilation of ammonia from the nitrogenase reaction are exported from the nodule to other parts of the plant via the xylem. (PEP = phosphoenolpyruvate)

Mycorrhizal fungi form intimate symbioses with plant roots

The roots of many plants form symbiotic associations with fungi, called **mycorrhizae** (derived from words meaning "fungus" and "roots"). The plant benefits from this association because the fungal hyphae exploit a greater soil volume than the plant roots, providing the plant with nutrients from this extended region of the soil. The fungus benefits from access to sugars produced in the plant by photosynthesis. In addition, colonization of the plant by one species of fungus may protect against infection by other fungi. Two major types of mycorrhizal association have been described. Either the fungus surrounds the root as a sheath (**ectomycorrhizae**) or the fungus proliferates inside the plant root and forms haustoria inside root cells (**endomycorrhizae**).

Ectomycorrhizae (Figure 8–81) are formed predominantly on forest trees by toadstool-producing basidiomycete fungi such as the fly agaric *Amanita muscaria*. The fungal hyphae form a tight mesh around the roots, varying in thickness from 1 or 2 to 30 hyphal diameters, depending on the species. Fungal hyphae enter the roots but grow only around the cortical cells rather than penetrating cell walls. Ectomycorrhizal roots are often swollen, short, and more branched than uninfected roots.

Endomycorrhizal associations are established when plant roots are colonized by zygomycete fungi of the genus *Glomus*. Both fossil evidence and DNA-sequence divergence within the Glomales suggest that this symbiosis is about 450 million years old, coinciding with plant colonization of the land (see Section 1.3). Endomycorrhizal fungi are obligate biotrophs—that is, they are strictly dependent on their hosts for survival. In contrast to the restricted clades of plants that enter into symbioses with rhizobia, plant–fungal symbioses are widespread in the plant kingdom, and most plant species (but not Arabidopsis and other Brassicaceae) form mycorrhizae. There are many similarities in plant responses to mycorrhizal and rhizobial symbionts. Some mutants of pea, *Lotus*, and *Medicago* that cannot form nodules with rhizobia are also compromised in the ability to form mycorrhizal associations. This suggests that the association with rhizobia is a refinement of the older and much more widespread association between plants and endomycorrhizal fungi.

The association between the fungus and the host roots starts, as in plant–rhizobia interactions, with the exchange of chemical signals between host and symbiont. These chemical signals are less well defined in mycorrhizal interactions, but exudates from host roots, especially flavonoids, are known to enhance fungal spore germination.

(A)

(B)

Figure 8–81.
Ectomycorrhizal associations. (A) Roots of
the Scots pine (*Pinus sylvestris*) surrounded by
a hyphal sheath of the ectomycorrhizal fungus
Paxillus involutus. (B) Fruiting body of the fly
agaric, a basidiomycete fungus that forms
ectomycorrhizal associations with trees,
including *Betula* spp. (birch).
(A, courtesy of D.J. Read.)

Unlike the rhizobium symbiosis, in which plants produce hydrolytic enzymes that
weaken their own cell walls to enable bacterial colonization, the host plant does not
facilitate fungal entry. At the root surface, the hyphae form appressoria on contact with
the wall of a root epidermal cell. Hyphae develop from the appressoria and enter the
root epidermis, then cross the epidermis by growing either between cells or across cells
in a structure like a rhizobial infection thread. In the inner cortex, the fungus invades
cells. On entry into cortical cells, the hyphae form a highly branched haustorial structure known as an **arbuscule** (Figure 8–82A). This structure creates a large surface area
of contact between the fungus and the host cell. It is separated from the cytosol of the
plant cell by two membranes: the plasma membrane of the plant cell and a perifungal

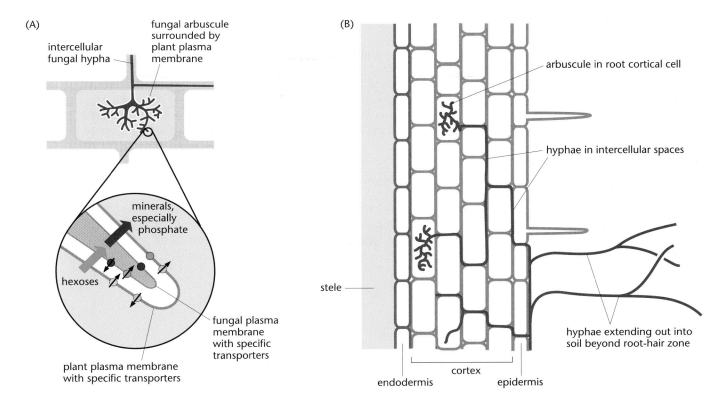

Figure 8–82.
Endomycorrhizal associations. (A) Fungal hyphae penetrate the root and
ramify between the cells, then invade cells to form arbuscules. An arbuscule
remains outside the plant plasma membrane. Plant and fungal plasma
membranes surrounding the arbuscule contain specific transporter proteins.
The arbuscule provides a large area of contact between plant and fungal cells
across which nutrients can be exchanged. The major exchanges are
movement of hexose from the plant to the fungus, as the source of carbon
for fungal growth, and movement of minerals, especially phosphate, from
the fungus to the plant. (B) Fungal hyphae extend out from the surface of an
infected root into soil zones beyond the region that can be exploited by root
hairs. Nutrients taken up by the fungi are transported back to the root, where
they can enter plant cells across the arbuscular membranes.

membrane originating from the fungal plasma membrane. The perifungal membrane is physically separated from the plant plasma membrane and has distinct transport capabilities. The arrangement of the plant–fungus interface is thought to allow tight control over nutrient flow, confining it to the site of fungal penetration.

After infection, hyphae grow out from the root and ramify throughout the soil (Figure 8–82). They take up nutrients from a large volume of the surrounding soil and transport them toward the plant, providing a flow of water and minerals to plant cells via the arbuscules. The fungus receives sugars from the plant in the form of hexoses transported across the perifungal membrane into the arbuscules.

Access to a large volume of soil seems to be of particular importance for plants growing in phosphate-depleted soils. Phosphate is the macronutrient least available to plant roots, in part because a high proportion is adsorbed onto soil particles (see Section 4.9). Endomycorrhizal associations enhance roots' phosphate absorption twofold to sixfold compared with uninfected roots. A specific type of plant phosphate transporter, expressed only in infected cells, is necessary for passage of phosphate from the fungus into the plant. The expression of other phosphate transporters in the root is down-regulated during mycorrhizal symbiosis, presumably because the fungus is providing the bulk of the phosphate requirements of the plant.

The early stages of colonization of the plant by the fungus (mycorrhization) have features in common with the formation of nodules (nodulation) in the rhizobium symbiosis. Expression of several plant genes is rapidly induced during both mycorrhizal and rhizobial symbiotic interactions. The function of many of these genes (early nodulation genes) is not yet defined.

Further common features of nodulation and mycorrhization are being discovered through the study of mutant legumes that cannot form nodules in the presence of rhizobia. Many of these mutants also lose the capacity to support mycorrhizal formation. For example, in pea, three genes are essential for early steps of both interactions (*Sym8*, *Sym9*, and *Sym19*). Plants carrying mutations in these genes cannot form an infection thread with rhizobia, and although endomycorrhizal fungi still form appressoria, they fail to develop intercellular hyphae. On the rare occasions when entry into cells succeeds, the fungus is able to form an arbuscule, showing that the genes are required only for the early stages of fungal penetration.

In comparison with what happens in the plant, and in rhizobial interactions, much less is known about the molecular changes in the fungal symbiont as the mycorrhizal association develops. Genetic analysis is hampered by practical difficulties of working with *Glomus*: it cannot be grown in the absence of a host, and it is extremely heterogeneous genetically.

SUMMARY

Plants are the source of organic carbon—that is, are the food—for all nonphotosynthetic organisms. Many interactions with other organisms are deleterious to the plant. These organisms are classified as pathogens (bacteria, fungi, oomycetes, and viruses) and pests (herbivorous animals). Most plant pathogens and many pests have coevolved with their host plants.

The coevolution of pathogens and host plants can be represented as four phases: (1) pathogens have surface molecules (pathogen-associated molecular patterns, or PAMPs) that are recognized by host-plant receptors, triggering the plant's basal defense mechanisms; (2) pathogens produce effector molecules that suppress these basal defenses; (3) genetic variants of the host plant recognize these effector molecules and respond with further defense mechanisms (*R* gene–mediated defense); (4) genetic variants of the pathogen no longer make the effector recognized by the host, do not trigger a defense mechanism, and can attack and exploit the plant.

Most pathogens are biotrophs or necrotrophs. The three main routes of entry into plants are direct penetration through intact surfaces, entry through natural openings, or opportunistic entry through wound sites. Pathogen infections lead to a broad range of disease symptoms. Effector molecules enhance the pathogens' ability to gain nutrients from their host plant and hence reproduce. Most effector molecules are enzymes, growth regulators (phytohormones), or toxins.

The effector molecules of pathogens can trigger strong defense responses in host plants carrying the appropriate *R* genes. A given species of plant carrying a particular *R* gene can recognize one specific effector molecule produced by a pathogen. This is the basis of the gene-for-gene model of plant–pathogen interactions.

Parasitic nematodes feed on plant roots and produce effector molecules that alter the plant's growth and metabolism. Insects cause extensive losses in crop plants, both directly through feeding and by acting as vectors for microbial pathogens. Some parasitic plants act as plant pathogens. Viruses are major causes of plant disease, transmitted by a variety of vectors. Plant viruses can be classified into several types based on their structures and replication mechanisms. Many viruses also encode proteins that suppress gene silencing in the host plant, which disable host defense mechanisms and interfere with development. RNA silencing is important in plant resistance to viruses.

Basal defense mechanisms include deposition of extra cell wall material, detoxification of toxic compounds produced by pathogens, and production of antimicrobial compounds or other toxic molecules. In *R* gene–mediated defense responses, recognition of pathogen effector molecules triggers defense responses that resemble those of the basal defense response. However, *R* gene–mediated defense responsesare generally stronger than the basal defenses and can result in death of the cell under attack, restricting pathogen invasion.

Cooperative interactions between plants and other organisms include those between flowers and pollinators. The evolution of specialized flowers and floral attractants has occurred in parallel with the evolution of pollinating animals. Symbiotic interactions between plants and nitrogen-fixing bacteria include those between legumes and rhizobia species. The bacteria acquire carbon from the host plant, and the plant receives ammonia for synthesis of amino acids and their derivatives. The roots of many plants form mycorrhizae, symbiotic associations with fungi. Plant nutrition is improved because the fungal hyphae exploit a greater soil volume than the plant roots; the fungus benefits from access to sugars produced by the plant.

FURTHER READING

General

Agrios GN (2005) Plant Pathology, 5th ed. Amsterdam: Elsevier Academic Press.

Chisholm ST, Coaker G, Day B & Staskawicz BJ (2006) Host-microbe interactions: shaping the evolution of the plant immune response. *Cell* 124, 803–814.

Dangl JL & Jones JDG (2001) Plant pathogens and integrated defence responses to infection. *Nature* 411, 826–833.

Jones JDG & Dangl JL (2006) The plant immune system. *Nature* 444, 323–329.

Strange RN & Scott PR (2005) Plant disease: a threat to global food security. *Annu. Rev. Phytopathol.* 43, 83–116.

8.1 Microbial Pathogens

Bent AF & Mackey D (2007) Elicitors, effectors, and R genes: the new paradigm and a lifetime supply of questions. *Annu. Rev. Phytopathol.* 45, 399–436.

Desveaux D, Singer AU & Dangl JL (2006) Type III effector proteins: doppelgangers of bacterial virulence. *Curr. Opin. Plant Biol.* 9, 376–382.

Ellis J, Catanzariti AM & Dodds P (2006) The problem of how fungal and oomycete avirulence proteins enter plant cells. *Trends Plant Sci.* 11, 61–63.

Ellis JG, Dodds PN & Lawrence GJ (2007) The role of secreted proteins in diseases of plants caused by rust, powdery mildew and smut fungi. *Curr. Opin. Microbiol.* 10, 326–331.

Grant SR, Fisher EJ, Chang JH et al. (2006) Subterfuge and manipulation: type III effector proteins of phytopathogenic bacteria. *Annu. Rev. Microbiol.* 60, 425–449.

Kamoun S (2007) Groovy times: filamentous pathogen effectors revealed. *Curr. Opin. Plant Biol.* 10, 358–365.

Koh S & Somerville S (2006) Show and tell: cell biology of pathogen invasion. *Curr. Opin. Plant Biol.* 9, 406–413.

Morgan W & Kamoun S (2007) RXLR effectors of plant pathogenic oomycetes. *Curr. Opin. Microbiol.* 10, 332–338.

Toth IK & Birch PRJ (2005) Rotting softly and stealthily. *Curr. Opin. Plant Biol.* 8, 424–429.

van Kan JAL (2006) Licensed to kill: the lifestyle of a necrotrophic plant pathogen. *Trends Plant Sci.* 11, 247–253.

8.2 Pests and Parasites

Kaloshian I & Walling LL (2005) Hemipterans as plant pathogens. *Annu. Rev. Phytopathol.* 43, 491–521.

Niblack TL, Lambert KN & Tylka GL (2006) A model plant pathogen from the kingdom Animalia: *Heterodera glycines*, the soybean cyst nematode. *Annu. Rev. Phytopathol.* 44, 283–303.

8.3 Viruses and Viroids

Baulcombe D (2004) RNA silencing in plants. *Nature* 431, 356–363.

Baulcombe D (2005) RNA silencing. *Trends Biochem. Sci.* 30, 290–293.

Brodersen P & Voinnet O (2006) The diversity of RNA silencing pathways in plants. *Trends Genet.* 22, 268–280.

8.4 Defenses

Bittel P & Robatzek S (2007) Microbe-associated molecular patterns (MAMPs) probe plant immunity. *Curr. Opin. Plant Biol.* 10, 335–341.

Durrant WE & Dong X (2004) Systemic acquired resistance. *Annu. Rev. Phytopathol.* 42, 185–209.

Engelberth J, Alborn HT, Schmelz EA & Tumlinson JH (2004) Airborne signals prime plants against insect herbivore attack. *Proc. Natl. Acad. Sci. USA* 101, 1781–1785.

Schilmiller AL & Howe GA (2005) Systemic signaling in the wound response. *Curr. Opin. Plant Biol.* 8, 369–377.

Takken FLW, Albrecht M & Tameling WIL (2006) Resistance proteins: molecular switches of plant defence. *Curr. Opin. Plant Biol.* 9, 383–390.

8.5 Cooperation

Brachmann A & Parniske M (2006) The most widespread symbiosis on earth. *PLoS Biol.* 4, 1111–1112.

Chittka L & Raine NE (2006) Recognition of flowers by pollinators. *Curr. Opin. Plant Biol.* 9, 428–435.

Oldroyd GED & Downie JA (2004) Calcium, kinases and nodulation signalling in legumes. *Nat. Rev. Mol. Cell Biol.* 5, 566–576.

Parniske M (2004) Molecular genetics of the arbuscular mycorrhizal symbiosis. *Curr. Opin. Plant Biol.* 7, 414–421.

Paszkowski U (2006) Mutualism and parasitism: the yin and yang of plant symbioses. *Curr. Opin. Plant Biol.* 9, 364–370.

DOMESTICATION AND AGRICULTURE

9

When you have read Chapter 9, you should be able to:

- Describe the origins and history of plant domestication, from wild plants to domesticated plants to scientifically improved crops, focusing on maize, cultivated wheat, and tomato.

- Define "heterosis" and summarize its importance in plant domestication and crop breeding.

- Summarize how crop plant disease resistance is achieved by plant breeding, crop management, and biotechnology, with some examples.

- Describe some achievements of the Green Revolution, and list some uses of transgenic plants in modern agriculture and horticulture.

- Describe the use of the *Agrobacterium* binary vector system and the particle bombardment technique in generating transgenic plants.

- Outline methods for creating herbicide resistance and insect resistance in crop plants.

- Summarize some of the problems and challenges arising from modern agricultural methods.

Plants and people interact everywhere, and these interactions have changed greatly over the course of human history. Many of the plants we encounter in our everyday lives, whether on farmland, in gardens, or in city parks and streets, are deliberately planted by humans, and the types that are planted are generally the products of human selection for qualities such as taste, horticultural utility, enhanced agricultural productivity (yield), or pleasing appearance. In this chapter we review some of the major changes that have occurred in the relationship between humans and plants, from prehistoric times to the present day. We focus in particular on developments that relate to agriculture and horticulture: the initial domestication of plants, the subsequent emergence of locally adapted varieties, the advent of new types of plant breeding in the twentieth century, and the recent development of biotechnological approaches to crop improvement.

We consider the genetic basis of the domestication of some major crop plants—for example, how the difference between wild and cultivated forms of maize depends on variation in the expression of just a few critical genes. We also outline how crosshybridization between cultivated and wild species gave rise to modern bread wheat, and

CHAPTER SUMMARY

9.1 DOMESTICATION

9.2 SCIENTIFIC PLANT BREEDING

9.3 BIOTECHNOLOGY

provide further examples of the ways in which human selection has unknowingly shaped the crop plant species that we are familiar with today. We then turn to the advent of more scientific approaches to the improvement of crop plants, some general concepts that underpin modern plant breeding and agriculture, and some of the ways in which recent advances in understanding of plant biology have contributed to agriculture.

9.1 DOMESTICATION

Selection of crop plants with characteristics desirable to humans (**domestication**) began around 11,000 years ago. One important site of crop domestication was the geographic area known today as the Fertile Crescent (Figure 9–1). The Fertile Crescent was also an early site of the development of various aspects of human civilization that we now take for granted, including commerce, writing, and the creation of social hierarchies. Indeed, it is likely that the development of agriculture promoted the development of civilization: with the greater efficiency of food production in the form of crop cultivation and animal husbandry, some of the population could specialize in tasks other than finding food. Thus the rise of agriculture in the Fertile Crescent has been intensively studied—not only by botanists but also by historians and archaeologists.

The domestication of crop species involved selection by humans

The crop species that first appeared in the Fertile Crescent were derived, by human selection, from wild plant species. For several of these crops the wild ancestors are known, and plant geneticists have identified some of the genetic differences between ancestral and cultivated species. Eight "founder" crops were domesticated in the Fertile Crescent: the cereals emmer wheat, einkorn wheat, and barley; the pulses lentil, pea, chickpea, and bitter vetch; and the fiber crop flax (Figure 9–2).

In addition to the Fertile Crescent, crop domestication arose independently in China, Mesoamerica (central and southern Mexico and adjacent areas of Central America), the Andes and Amazonia, and eastern North America (Table 9–1). Archaeological evidence

Figure 9–1.
The Fertile Crescent. This region lay in the Eastern Mediterranean, covering parts of modern-day Lebanon, Syria, Turkey, Iraq, Iran, Jordan, and Israel.

emmer wheat

einkorn wheat

barley

lentils

chickpeas

bitter vetch

flax

Figure 9–2.
Some crop plants first domesticated in the Fertile Crescent. ('Barley', courtesy of Boye Koch; 'flax' courtesy of Lytton J. Musselman.)

shows that crop cultivation spread to many areas beyond the initial "centers of origin" during prehistoric times (Figure 9–3) and that crops, livestock, and knowledge were passed from one center of origin to another. The main prehistoric spreads were from the Fertile Crescent to Europe, Egypt, North Africa, Ethiopia, and Central Asia; from China to tropical Southeast Asia, the Philippines, Indonesia, Korea, and Japan; and from Mesoamerica to North and South America.

Many domesticated plants are derived from wild species that would have been common around pre-agricultural human settlements: species with the ability to colonize open or disturbed habitats, especially those with exposed soil. Domestication caused some fundamental changes in these plants, changes that made them unable to survive in the wild. For example, early farmers unknowingly selected for **mutations** that facilitated retention (rather than dispersal) of seeds. Since seed dispersal is a favorable characteristic for plants growing under natural conditions, the domesticated forms would be disadvantaged when growing in the wild.

Table 9–1 Some centers of origin of crop domestication

Area of origin	Crops	Approximate date of domestication, BCE
Southwest Asia (Fertile Crescent)	Wheat, pea	8500
China	Rice, millet	7500
Mesoamerica	Maize, beans, squash, tomato	4000
Andes and Amazonia	Potato, cassava	3500
Eastern North America	Sunflower, goosefoot	2500

Figure 9–3.
Progressive spread of crop cultivation from the Fertile Crescent into Europe and western Asia.

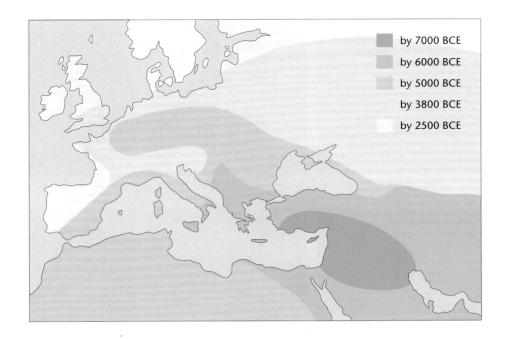

by 7000 BCE
by 6000 BCE
by 5000 BCE
by 3800 BCE
by 2500 BCE

Different plant species were domesticated because of particular individual characteristics. Thus cereals (e.g., barley, wheat, rice, and maize) and pulses (e.g., peas and beans) were domesticated for their seeds, while other plants were domesticated for their leaves (e.g., lettuce, spinach), tubers (e.g., potato), or fruits (e.g., tomato, apple). Domestication often produced large increases in size and changes in shape of the organs harvested from the crop. As we will see in the case of the tomato, domestication has resulted in fruits that differ markedly from their wild ancestors. Figure 9–4 shows some other examples of the wide range of domesticated fruits.

The history of crop plant domestication, then, illustrates how selection changes the form of living organisms. The first chapter of Charles Darwin's *Origin of Species* is devoted not to the evolution of organisms in nature, but to the rapid changes in form that occur during domestication. Darwin recognized that the selection by humans ("artificial selection") that is part of the process of domestication is in some ways similar to the effect of **natural selection** on natural populations. In this section we describe some of the molecular changes associated with crop plant domestication and consider the consequences of these changes for our current understanding of natural selection.

The difference between maize and its wild ancestor, teosinte, can be explained by allelic variation at five different loci

Archaeological evidence suggests that the domestication of maize began around 6000 years ago in Central Mexico. Its wild ancestor has been identified as a large grass—still found in Mexico and Nicaragua—called "teosinte." Teosinte differs so much in appearance from cultivated maize that it was many years before scientists realized that these two plants were closely related. One of the major differences between maize and teosinte concerns the seeds and the structures to which they are attached. Whereas maize seeds (kernels) are permanently attached to a relatively robust structure known as the "cob," teosinte seeds are attached to a fragile stalk-like **rachis** (see Figure 9–5). Teosinte seed dispersal is promoted when the rachis shatters. An additional important difference is that of **dormancy**: teosinte seeds often have variable degrees of dormancy (ensuring that all seeds do not germinate at the same time), whereas maize seeds have

apples

strawberries

melon

apricots

Figure 9–4.
Some domesticated fruits.
(A, courtesy of Blake Winton; B, courtesy of Gemma Cole.)

little or no dormancy. These differences in morphology and dormancy reflect the different **selective pressures** to which maize and teosinte have been exposed. Selection by humans has favored maize traits that minimize seed loss before harvest and permit rapid, even germination. Natural selection, on the other hand, favors efficient seed dispersal and sporadic germination.

Although very different at the morphological level, maize and teosinte differ very little at the genetic level. Both plants have 20 **chromosomes**, with similar morphology and gene order. Crosses between maize and teosinte are successful, resulting in fully fertile **F1 hybrids** (Figure 9–5). In fact, **QTL (quantitative trait locus) analysis** (see Section 2.5) has shown that most of the genetic differences between maize and teosinte can be attributed to just five small chromosomal regions, each affecting several distinct components of the differing traits. In at least two of these five chromosomal regions, the multiple-trait effects observed can be ascribed to the activities of a single gene. It is likely that the other regions also carry single genes of major effect.

As noted above, Darwin used the changes resulting from domestication as one of the mainstays of his argument for evolution by natural selection. We describe below some molecular changes that played a major role in the evolution of maize during domestication from its morphologically distinct progenitor, teosinte. These molecular changes alter the pattern of expression of a gene encoding a regulatory protein. This finding suggests that alterations in patterns of **gene expression** provide an important source of the phenotypic variation on which selection acts during evolution.

Figure 9–5.
Teosinte–maize hybridization. Shown here are ears of teosinte (*left*), maize (*right*), and their F1 hybrid (*middle*). In maize, a female inflorescence (ear) forms at the termini of two to three short lateral branches and is a cob composed of 8 to 24 rows, each row containing approximately 50 kernels. Teosinte has many more lateral branches, each terminating in a spike consisting of 2 rows of 5 to 6 hardened fruit cases each containing a single seed. The maize ear is enclosed by a set of husks, or modified leaves (*not shown*), produced by the lateral branch, whereas the teosinte ear is loosely enclosed within a few husks. Maize kernels adhere to a tough cob or rachis. In contrast, the rachis of the teosinte spike is very fragile, particularly at maturity. (Courtesy of John Doebley.)

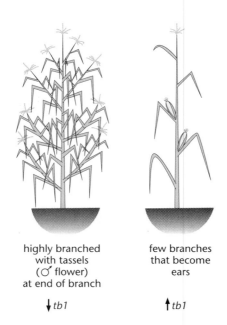

highly branched
with tassels
(♂ flower)
at end of branch

few branches
that become
ears

↓ *tb1*   ↑ *tb1*

Figure 9–6.
Teosinte and maize branching architecture. Teosinte plants (*left*) bear many long axillary branches tipped by male inflorescences (tassels). Maize plants (*right*) have fewer, short axillary branches tipped by female inflorescences (ears). These differences are due to difference in expression of *tb* (here, *tb1*)—relatively low in teosinte, relatively high in maize.

Alterations in the expression of the gene *teosinte branched* played an important role in the domestication of maize

Perhaps the most obvious difference between teosinte and maize is that a teosinte plant is highly branched and a maize plant is not (Figure 9–6). The reduced branching of maize resulted from increased **apical dominance** (see Section 5.4), in which the apex of the main stem inhibits the outgrowth of lateral branches from **lateral meristems**. This reduced branching would have been advantageous to early farmers because it resulted in the production of seeds in relatively large numbers on one or a few lateral shoots (ears), rather than in smaller numbers on many different lateral shoots. This architecture thus increases the ease with which the seeds can be harvested.

The QTL analyses described above found differences in a single chromosomal segment on chromosome 1 that accounted for the differences in branching architecture between maize and teosinte. Furthermore, previous genetic analyses had identified a single mutant gene, *tb* (*teosinte branched*; see also Section 5.4), that confers a teosinte-like architecture on maize plants. The *tb* gene maps to this same segment of chromosome 1 (previously identified by QTL analysis) and is therefore a major determinant of the difference between maize and teosinte shoot-branching architectures.

So what is the difference between the maize and teosinte *tb* genes? The protein encoded by both *tb* genes shares two short regions of sequence similarity with the snapdragon (*Antirrhinum majus*) protein CYCLOIDEA (the *CYCLOIDEA* gene product; see Section 5.5), a transcriptional regulator thought to act as a repressor of organ growth. It is therefore likely that the *tb* gene product also acts as a growth repressor.

There are two possible reasons why the maize and teosinte *tb* genes give rise to plants with such different architectures. First, the TB proteins (the *tb* gene products) might differ in the way they repress growth. This explanation is unlikely, however, because the maize and teosinte TB proteins are virtually identical in amino acid sequence. Second, the maize and teosinte *tb* genes may give rise to different levels of TB protein, and hence different levels of growth repression in the two plants. There is experimental evidence to support this second explanation. **Messenger RNA** transcripts of the maize *tb* gene accumulate to higher levels than transcripts of the teosinte *tb* gene, presumably resulting in higher TB protein levels in maize than in teosinte. Interestingly, in teosinte, changes in environmental conditions can cause changes in *tb* transcript levels. Thus it is likely that, during domestication, humans selected for a variant *tb* gene that is expressed at a higher level in primary axillary meristems than is the original teosinte **allele**. This selection channeled an underlying biological mechanism that previously regulated plant architecture in response to an environmental signal. The result was the formation of a maize-like ear rather than a long teosinte-like branch with a tassel on the end. Analysis of the DNA of maize from archaeological sites indicates that selection for the *tb* alleles associated with domesticated maize occurred very soon after maize cultivation began.

The *teosinte glume architecture* gene regulates glume size and hardness

Another major difference between maize and teosinte is in the development of the **glumes**, floral organs that cover the developing kernel (Figure 9–7). In teosinte, the kernel is contained within a fruit case that is very hard and has a shiny surface. This fruit case confers a selective advantage, because it protects the kernel and reduces the chance of its being eaten by animals. The teosinte fruit case consists of a deep cup-shaped structure called a **cupule**, which contains the kernel, and hardened glumes that curve up and over the kernel, thus covering it and closing the cupule. Maize cupules are much flatter and do not enclose the kernel, and the glumes are soft, thin, and short. Thus maize kernels are not enclosed in a fruit case and are relatively accessible for harvest and for human consumption.

As with *tb*, these different characteristics are largely controlled by a single genetic locus, the gene *teosinte glume architecture* (*tga1*), on chromosome 4. Again as with *tb*, *tga1* is likely to encode a regulatory protein, because it affects several distinct facets of cupule and glume structure. First, *tga1* affects the shininess and hardness of glumes by regulating the deposition of silica in glume epidermal cells. The maize *tga1* gene restricts the deposition of silica to a subset of epidermal cells, whereas the teosinte allele permits deposition throughout the epidermis. Second, *tga1* controls the **lignification** of glume mesophyll cells. Teosinte glumes are tougher than maize glumes because the teosinte allele promotes this lignification. Finally, the rate of growth of glumes and

Figure 9–7.
Teosinte kernels. An immature ear of teosinte (*top*) has kernels completely covered by green glumes. The sectioned immature ear (*middle*) shows kernels within fruit cases. One of the mature fruit cases (*bottom*) is cracked open to show the kernel within. (Courtesy of John Doebley.)

cupules is controlled by *tga1* such that growth is faster in teosinte than in maize. Maize glumes and cupules grow too slowly to completely surround the kernel, thus increasing its accessibility.

So far we have discussed how differences between maize and teosinte (differences that made an important contribution to the domestication and development of cultivated maize) are due to genetic variation at just a few individual genes—genes encoding proteins that regulate multiple aspects of plant morphology and development. It is possible that this reveals a general principle. Evolutionary difference may more often be due to variation in the activity of a few genes having major effects on multiple aspects of the phenotype (i.e., broad **pleiotropy**) than to variation in multiple genetic loci.

Cultivated wheat is polyploid

Unlike maize, cultivated wheats probably did not evolve from a single progenitor. As we will see, modern bread wheat (*Triticum aestivum*) is **polyploid** (hexaploid) and evolved from the hybridization of distinct diploid and tetraploid species. Polyploidization (discussed in Sections 2.4 and 3.1) tends to cause an increase in the size of plant cells and organs, so it is likely that early farmers initially selected for polyploid wheat because it had larger grains than diploid wheats.

Wheat was first domesticated in the Fertile Crescent. Wild progenitors of cultivated wheat, einkorn (*Triticum monococcum*) (Figure 9–2) and emmer (*Triticum turgidum*) (Figure 9–2; Figure 9–8), grew abundantly in this area during the time that wheat was domesticated. Both einkorn and emmer were domesticated individually as cereal crops, and hybrids between emmer and a third, wild grass species are thought to have given rise to the hexaploids that were further domesticated to become wheat.

Einkorn is a diploid member of the genus *Triticum* that includes modern wheats. The **genome** of einkorn, known as the "A genome," consists of seven pairs of chromosomes, 1A to 7A. The major difference between wild and domesticated varieties of einkorn lies in the mechanics of seed dispersal. Wild einkorn has very brittle **spikes** (the stems on which the grains are held). As these spikes dry out during final seed maturation, they tend to fall apart, thus facilitating dispersal of the grain. The first farmers would have preferred plants with nonbrittle spikes and would therefore have (unknowingly) selected for variants with this trait. Thus, cultivated einkorn retains its seeds until harvest.

Although domesticated crop plants are derived from wild species, domestication also often involved further cross-hybridization between cultivated plants and wild relatives. For example, unlike einkorn, emmer is tetraploid. The emmer genome is the product of hybridization between a species with an A genome (*Triticum urartu*) and a second species with a different, "B genome" (Figure 9–8). The exact B-genome donor species is unknown, but is thought to have been a diploid member of a family of goatgrasses (*Aegilops*). Thus the emmer genome contains seven A-genome chromosomes and seven B-genome chromosomes. A pair of A-genome chromosomes is said to be a "homologous" pair (e.g., two 1A chromosomes are **homologs**); the A and B genomes are described as "homoeologous" (e.g., chromosomes 1A and 1B are **homoeologs**).

As with einkorn, the wild forms of emmer have brittle spikes, whereas the cultivated forms do not. Some forms of emmer (*T. turgidum* var. *durum*) are in relatively widespread cultivation today. This "durum" wheat has large, hard grains that are relatively low in gluten (see below) and are particularly suitable as sources of flour for pasta.

Cultivated bread wheat (*T. aestivum*) is hexaploid. It is thought to have arisen from hybridization of emmer (AABB) with another goatgrass, *Aegilops tauschii*. The goatgrass contributed the D genome, resulting in the AABBDD genome of bread wheat

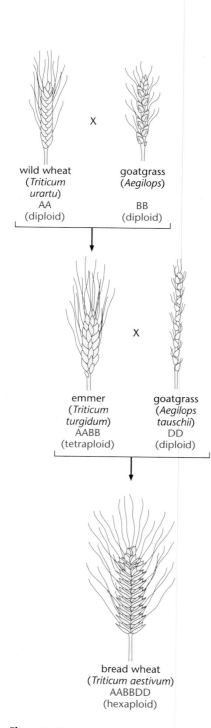

Figure 9–8.
Derivation of hexaploid bread wheat.
Successive rounds of hybridization between diploid and tetraploid species gave rise to hexaploid bread wheat.

wild wheat
(*Triticum urartu*)
AA
(diploid)

X

goatgrass
(*Aegilops*)
BB
(diploid)

emmer
(*Triticum turgidum*)
AABB
(tetraploid)

X

goatgrass
(*Aegilops tauschii*)
DD
(diploid)

bread wheat
(*Triticum aestivum*)
AABBDD
(hexaploid)

(Figure 9–8). The grains of bread wheat contain relatively high levels of gluten proteins (glutenins) in their endosperms (see Section 4.8). Gluten proteins make dough sticky, so the carbon dioxide produced during fermentation with yeast remains trapped in the dough, causing it to rise.

The genome of hexaploid (AABBDD) wheat acts in a diploid manner during **meiosis**. That is to say, chromosome pairing occurs only between homologous chromosomes (for example, between 1A and 1A, but not between 1A and 1B). This enables the stable inheritance of a complete wheat haploid genome (ABD). The restricted pairing in meiosis occurs because the gene *Ph* (on chromosome 5B) prevents the pairing of homoeologs (e.g., of 1A with 1B). In wheat mutants lacking *Ph* function, normal pairing, and thus stable inheritance of the ABD genome, is lost.

Cauliflower arose through mutation of a meristem-identity gene

The domestication of *Brassica* species has resulted in an array of vegetable types that are startlingly different in appearance (Figure 9–9). For example, in cabbage (*Brassica oleracea*) a large number of leaves overlap to surround the terminal bud, thus producing the "head." Brussels sprouts are the enlarged axillary buds of *B. oleracea*, while rutabagas (swedes) consist of the swollen upper root, hypocotyl, and lower stem of *Brassica napus*. The floral organs of *B. oleracea* have also been modified by human selection to produce two other well-known vegetables: broccoli and cauliflower. In most cases, we do not know the genetic nature of the changes from the "original"

Figure 9–9.
Some domesticated brassicas.

cabbage

turnip

kohlrabi

brussels sprouts

B. oleracea (or *B. napus*) to its cultivated forms. We do, however, know something of the molecular changes that led to the development of the cauliflower.

The cauliflower head, or "curd," consists of a mass of flower buds that have failed to undergo normal floral development. Specifically, these buds are arrested following the formation of **inflorescence meristems**. Flower formation can be described as a series of distinct developmental steps (see Section 5.5): **floral induction**, the formation of floral **primordia**, and the initiation of floral organs. In the cauliflower, the apical meristem becomes an inflorescence meristem (floral induction) in the normal way. After this, instead of developing as **floral meristems**, the lateral derivatives of the inflorescence meristem become additional inflorescence meristems. This is a self-generating loop and results in the proliferation of masses of inflorescence meristem tissue, the curd of the cauliflower (Figure 9–10).

Insights into the molecular basis of the cauliflower phenotype came from studies of Arabidopsis mutants with altered flower development. A class of mutants known as *ap1* loss-of-function mutants (see Section 5.5; the *AP1*, or *APETALA1*, gene encodes a **transcription factor** of the **MADS box** family that specifies floral meristem identity and floral organ identity) were originally identified because they lack **petals**, the organs of the second whorl of flower development, and also tend to form inflorescences in the place of flowers. Thus, instead of a single flower, an inflorescence bearing a number of additional flowers is formed. Arabidopsis contains a second gene, *CAL1* (*CAULIFLOWER1*), that encodes a protein of very similar amino acid sequence to the protein encoded by *AP1*. Plants carrying both *ap1* and *cal1* loss-of-function mutations have an extreme phenotype. The new meristems borne on the flanks of the inflorescence meristem are themselves inflorescence meristems, rather than floral meristems (Figure 9–10). Thus the combined genetic and molecular analysis suggests that AP1 and CAL1 proteins have largely overlapping, but not absolutely identical, functions and that they act to promote the transition from an inflorescence to a floral meristem.

The phenotype of Arabidopsis *ap1:cal1* flowers closely resembles that of cauliflower (Figure 9–10), suggesting that the genetic basis of the cauliflower phenotype might be similar to that of Arabidopsis *ap1:cal1* mutants. An **ortholog** of the Arabidopsis *AP1* gene was identified in *B. oleracea* (with normal flowers), and its sequence compared with that found in *B. oleracea* var. *botrytis* (cauliflower). The *B. oleracea AP1* gene encodes a full-length protein very similar to Arabidopsis AP1, and the *B. oleracea* var. *botrytis* allele encodes a truncated and probably nonfunctional AP1 protein, because of a translational **stop codon** that interrupts its **open reading frame**. The structure of the cauliflower curd is therefore due to loss of AP1 function and hence a reduced ability to promote the transition from an inflorescence to a floral meristem.

Figure 9–10.
Comparison of cauliflower and the Arabidopsis *ap1:cal1* double mutant.
(A) Cauliflower. (B) The *ap1* mutant of Arabidopsis; note the lack of petals. (C) The *ap1:cal1* double mutant of Arabidopsis; note the proliferation of inflorescence meristematic tissue and the similarity to the curd of cauliflower (A).
(B and C, courtesy of Martin Yanofsky.)

(A)

(B)

(C)

Increase in fruit size occurred early in the domestication of tomato

Wild fruits would have formed an important part of the human diet before the development of agriculture, and many fruit-bearing crops were domesticated in prehistoric times. Fruits are enormously variable in size, shape, and structure. Domesticated fruits tend to be larger and more palatable to humans than their wild progenitors. Here we consider the tomato as an example and describe a key step in its domestication.

Cultivated tomato (*Solanum lycopersicum*) is one of the world's most important horticultural crops. The first steps in tomato domestication probably occurred in coastal Peru, starting with a wild *Solanum* species like *Solanum pimpinellifolium* that is closely related to tomato and might have been a weed in early maize crops. This "ancestral" tomato would have resembled *Solanum pimpinellifolium* in having small fruit, unlike the cultivated tomato (Figure 9–11). Indeed, it is thought that increases in fruit size (resulting from human selective pressure) were among the earliest of the changes that eventually resulted in the domestication of tomato.

The genetic basis of the difference in size between wild and cultivated tomato species has been explored by QTL analysis. Tomato plants bearing small fruit (*Solanum pimpinellifolium*) were crossed with plants bearing large fruit (the commercially grown Giant Heirloom variety of *S. lycopersicum*). The **F2 generations** of these crosses showed a wide variation in fruit size, and researchers found that the relative size of mature tomato fruits is determined, in part, by the number of cell divisions that occur early in fruit development. QTL analysis revealed that a single gene, *fw2.2*, plays a major role in determining fruit size, and that allelic variants of this gene are substantially responsible for the observed variation in fruit size (Figure 9–11). The *fw2.2* gene encodes a protein with structural similarity to a **GTPase** of the Ras family, an enzyme that regulates cell division in mammalian cells. The protein encoded by *fw2.2* represses cell division in the fertilized **carpels** that constitute the developing tomato fruit. The *S. pimpinellifolium* allele of *fw2.2* causes the formation of small fruit because it produces higher levels of mRNA transcripts and thus more of the division-repressing protein. Since all wild *Solanum* species contain the "small-fruit" allele (as found in *S. pimpinellifolium*) and all cultivated species contain the "large-fruit" allele (as found in Giant Heirloom), it is likely that selection of the large-fruit allele of *fw2.2* was a key step in the domestication of tomato. As in the case of maize *tb* described above, adaptively significant allelic variation is conferred by a difference in the level of expression of a gene encoding a regulatory protein (the protein encoded by *fw2.2*) rather than a difference in the function of the encoded gene product.

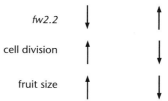

Figure 9–11.
Regulation of tomato fruit size by expression of *fw2.2*. The small fruit (*right*) is from the wild species *Solanum pimpinellifolium*; it is small because higher-level expression of *fw2.2* represses cell division. The much larger fruit (*left*) is from the cultivated *Solanum lycopersicum* (cultivar Giant Red), and is larger (in part) because lower-level expression of *fw2.2* permits increased cell division.
(Courtesy of Steve Tanksley.)

9.2 SCIENTIFIC PLANT BREEDING

The last two to three hundred years have seen profound changes in the genetic makeup of our major crops. The development of wheat in Europe provides a good example, and we begin our discussion there. We then consider other cereal crops and fruits, and describe some of the advances that became known as the "Green Revolution."

Scientific approaches to crop plant improvement have resulted in substantial changes in the genetic structure of many crops

Before modern agriculture and large-scale plant breeding, seed stocks were mostly maintained by individual farmers. This system resulted in the development of thousands of genetically distinct, locally adapted plant varieties that are now known as **landraces**. Many different landraces of wheat were cultivated throughout Europe. The selective pressures that contributed to the development of landraces were from two sources. First, the farmer (largely unconsciously) practiced selection when holding back some of each year's harvest for sowing the following spring. Second, the selection effected by local differences in climate, landscape, and environment also made an important contribution to landrace evolution. Thus, before the advent of modern plant breeding, there was considerable genetic diversity in the wheat that was grown across Europe.

The landraces were themselves relatively heterogeneous in genetic makeup. For example, within any field of landrace wheat there was considerable allelic variation. One of the major changes that has occurred during the development of modern crop varieties is a reduction in genetic heterogeneity. The pace of this reduction has increased over time. Initially, farmers and early plant breeders became more deliberate about the process of selection. Then, as choices were made, fewer (or even single) plants were used as founders for new populations. The result was that the landraces became progressively transformed into "pure-breeding" (inbred) lines. These lines were largely **homozygous** for all genes, and individual plants in a population were very much more like each other than had previously been the case.

At the start of the twentieth century, plant breeders began to apply the new science of genetics to the improvement of crop plants. Deliberate and controlled **cross-pollination** of different pure-breeding lines had a major impact on crop plant development. One of the earliest discoveries was the phenomenon of **heterosis** (also called **hybrid vigor**). This occurs when two low-yielding inbred lines are crossed and the resulting F1 hybrid plants out-yield the parental lines. Heterosis has been of huge importance in the development of modern maize lines, and of many horticultural crops. However, because wheat is a primarily self-pollinating species, heterosis has not contributed significantly to the development of modern wheat varieties.

Twentieth-century wheat breeding generally employed a genetic procedure known as **recurrent backcrossing**. The following imaginary example illustrates the basic principles of this technique. Consider a first wheat variety that is high-yielding and produces good-quality grain, but is susceptible to a particular disease. A second variety is poor-yielding but is disease resistant. Clearly, the development of a new variety that combines the advantages of both initial varieties would be desirable. The plant breeder's approach to this goal would be as follows: cross the first variety with the second, plant out the progeny, and then select (either directly from the F1 or from subsequent generations) those plants that most resemble the first variety in terms of yield but are also disease resistant. These selected disease-resistant plants are then "backcrossed" to the first variety. This backcross is repeated again and again, selecting each time for combined high yield and disease resistance. The end result is a variety that is substantially identical to the first variety, except that it carries some genes (and linked DNA) from the second that confer disease resistance. The effectiveness of recurrent backcrossing and subsequent selection depends on the existence of genetic variability for important traits such as yield and disease resistance in varieties, pure-breeding lines, landraces, and wild, related strains.

The development of crops by modern plant breeding methods has resulted in a substantial narrowing of the genetic base of crops. Each year only a few different varieties are grown over wide areas. This has resulted in the loss of many of the old landraces and of the genetic variation they contained. As we shall see, there is a growing awareness among plant breeders of the potential value of genetic variation from both landraces and their wild relatives in the improvement of crop plants.

One of the aims of recurrent backcross and selection is the introduction of new, beneficial genes with minimal disruption of the original genotype. However, it is sometimes difficult to follow the inheritance of desired traits by observing the phenotypes of just a few plants (particularly in characteristics such as yield, where the traits are quantitative in nature). These difficulties are currently being alleviated by the development of **genome maps** of crop species (see also Section 2.4). Modern genome maps provide molecular **markers** that show tight **genetic linkage** with loci controlling important traits. The inheritance of the markers, rather than of the traits themselves, can be easily followed in breeding programs. Thus genome mapping will make substantial contributions to the effectiveness of plant breeding approaches that depend on the selection and crossing of varieties containing different traits.

Triticale is a synthetic domesticated crop species

For the vast majority of crops grown in the world, domestication from wild relatives began by human selection many thousands of years ago. Triticale is the first crop species deliberately created by human intervention, by the intercrossing of two distinct cereal crop species: wheat (*Triticum*) and rye (*Secale*). The purpose of this interspecific crossing was to combine the desirable nutritional and agronomic properties of wheat (high yield, high protein content in seed, high bread-making quality of flour) with the much greater cold and drought hardiness of rye, thus expanding the available growing area for plants with wheat-like characteristics.

Triticale exists in both hexaploid and octoploid forms. The hexaploid has an AABBRR genome, derived from crosses between a tetraploid (AABB) durum wheat and diploid (RR) rye. The octoploid (AABBDDRR) comes from crosses between hexaploid bread wheat (AABBDD) and rye. These initial crosses were made in the 1940s and resulted in embryos haploid for each of the donor genomes. Resulting haploid plants would have been sterile, but treatment of the embryos with **colchicine** (a drug that blocks chromosome separation during **mitosis**) resulted in the doubling of all chromosomes, allowing normal segregation in subsequent cell generations. These colchicine-treated plants were fully fertile and able to set seed.

Hexaploid triticale is superior to the octoploid form, and has been the focus of most attempts at improvement via plant breeding. Triticale can have yields as high as those of standard bread wheat cultivars, and when used as livestock feed is nutritionally on a par with wheat or barley. In addition, the anticipated increase in hardiness of triticale (relative to the wheat parent) was achieved. As a result, triticale is now grown in substantial amounts in several countries (in eastern Europe, Canada, Mexico, and Brazil; Figure 9–12), where it is used primarily for animal feed.

Figure 9–12.
Triticale growing in a field in Brazil.

Triticale is important from a historical perspective. It represents the first conscious effort by humans to use currently available technology to create a new crop species. In essence it represents the first human-mediated transfer of genetic material from one crop plant species to another. Recent advances in science and technology have made it possible to transfer single genes or small groups of genes, rather than entire genomes, from one plant species to another, as we shall see in Section 9.3.

Disease resistance is an important determinant of yield and can be addressed both by plant breeding and by crop management

Diseases and pests of crops can have a dramatic effect on yield. In addition to losses in the field, post-harvest losses to insects, nematodes, fungi, and bacteria can also reduce the proportion of the harvest that is useful (see Chapter 8). Crops can be protected from diseases and pests in several ways. Some crops are resistant to particular diseases or pests because they have the appropriate capacity to ward off attack. For other crops, changes in farming methods can be used to try to prevent an epidemic from taking hold. Such changes in crop management may involve the use of **herbicides**, fungicides, or pesticides, or may involve alternative methods that are not dependent on the application of chemicals. Here we describe some selected examples of ways of protecting crops against disease; see Section 8.4 for background to the genetics and biochemistry of disease resistance.

As described in Chapter 8, selection favors the evolution of new forms of pests and diseases that can overcome genetically determined disease resistance in crop plants. For this reason, plant breeders are continuously seeking to deploy novel sources of disease resistance. This can be done by crossing crop plant lines that carry desirable traits (e.g., high yield) but are susceptible to a particular strain of disease with disease-resistant lines, and subsequently selecting for plants that retain the desirable traits while also acquiring disease resistance (as described above). More recently, an improved understanding of the mechanisms of disease resistance has enabled scientists to insert cloned disease-resistance genes (***R* genes**) directly into crop plant genomes (see Section 9.3).

To give one example of a conventional genetic-crossing approach, tomato breeders have used disease-resistance genes from wild relatives to improve the resistance of cultivated tomatoes. Tomato (*S. lycopersicum*) is one of nine closely related species within the genus *Solanum*. All of these species are alike in chromosome number (2n = 24) and form interspecific hybrids with *S. lycopersicum* that are fertile enough for further crosses. Given the ease with which tomato species can be crossed, together with a thorough knowledge of their **cytogenetics**, traits from wild relatives have been successfully incorporated by backcrossing and selection into commercially grown varieties of tomato. Such traits include resistance to disease caused by fungi, bacteria, viruses, and nematodes. A major problem with introducing these traits from wild species is known as **linkage drag**: deleterious traits are transferred together with the beneficial traits, due to gene linkage. For example, some potentially useful disease-resistance genes are closely linked to genes that confer reduced fruit size. It may thus be difficult to obtain plants that are both disease resistant and bear large fruit. However, the recent use of precise molecular markers is facilitating the identification of individuals in which the amount of transferred DNA surrounding the resistance gene is as small as possible, thus reducing the linkage-drag problem.

The intensity of the selective pressure driving the evolution of new forms of pests and diseases is reduced when disease-resistance genes are present at low frequency in a crop plant population, rather than being present in every individual. Some breeders therefore advocate the planting of a mixture of crop varieties with resistance to different diseases, thus preventing an epidemic from taking hold. However, this strategy (effectively a reversal of the modern trend toward genetically homogeneous crop plant populations) has the drawback that the different varieties can also vary in certain agronomic and quality traits.

It may eventually become possible to grow crops that are substantially homogeneous, yet heterogeneous for the complement of disease-resistance genes that they carry. This would mirror the way in which these genes work in the wild, since most disease-resistance genes function in natural populations as polymorphic loci (see the discussion of *R* gene **polymorphism** in Section 8.4).

As described above, changes in crop management can also combat disease in crop plants, in ways that are of considerable practical as well as scientific significance. For example, in Africa, maize yields are severely affected by stem borers, particularly the introduced *Chilo partellus* and the indigenous *Busseola fusca*. Other plants, such as molasses grass (*Melinus minutiflora*), chemically repel stem borers and release other chemicals that attract parasites of the borers (see Section 8.4 for more examples of plant-derived insect-repelling chemicals). If molasses grass and maize are grown in alternate rows (**intercropping**), stem borer densities in the maize are significantly lower than when maize is grown alone. This example shows how scientific knowledge about the influence of plant chemistry on the behavior of herbivorous insects and their parasites can lead to increased protection against predation.

The legume silverleaf (*Desmodium uncinatum*) also repels stem borers and can be an effective means of pest control when planted between maize rows. Silverleaf also enhances soil fertility by fixing nitrogen (in *Rhizobium*-containing root nodules; see Section 8.5) and powerfully suppresses the growth of the parasitic weed *Striga hermonthica*, which is prevalent in maize fields in parts of Africa (see Sections 6.1 and 8.2). The mechanism of *Striga* suppression is not yet understood, but seems to involve either suppression of *Striga* seed germination or attraction of the seedlings to (unsuccessful) attempts at colonizing *Desmodium* roots.

Mutations in genes affecting fruit color, fruit ripening, and fruit drop have been used in tomato breeding programs

Tomato breeders have selected for many different shapes, forms, and colors of tomato fruit, so there are considerable differences between tomato varieties (Figure 9–13). Modern tomato breeding focuses on specific adaptations to a wide range of environments (for example, tomatoes are grown in the Mediterranean, in greenhouses in northern Europe, in the humid southeastern United States, and in the arid western United States). In addition, tomato breeders are constantly attempting to improve resistance to a range of pathogens and pests, and have selected for particularly desirable aspects of fruit shape, color, flesh texture, and taste. We present some selected examples of single-gene mutations that have particular effects on the characteristics of the tomato fruit or fruit-bearing structures and have been used in breeding programs.

Color variation in tomato fruit is primarily due to altered quantity and quality of **carotenoids**. All plants make carotenoids, which function in **photosynthesis** (in **light-harvesting complexes** and **reaction centers**; see Section 4.2; see also Section 4.7 for a discussion of carotenoid biosynthesis). Carotenoids accumulate in the tomato fruit in specialized **plastids** called **chromoplasts**. Color changes during tomato fruit ripening (see also below) are associated with increases in the activities of enzymes catalyzing steps in the carotenoid biosynthetic pathway. In many cases, the color differences between tomato varieties are also due to differences in the activities of one or more of the enzymes in this pathway. These varietal differences in enzyme activity (and hence in color) are usually due to allelic variation in the genes that encode the enzymes.

The genetic control of fruit ripening has been exploited by tomato breeders. Tomato and many other fruits undergo a surge of **respiration** and associated metabolic activity at a specific point during ripening, a metabolic event called a **climacteric**. Important aspects of ripening such as color development, softening, and sugar accumulation are associated with the climacteric. This process is triggered by the phytohormone

Figure 9–13.
Heirloom tomatoes. A variety showing a highly distinctive fruit shape and color, just one example of the huge diversity of tomato fruit size, shape and color exhibited by different tomato varieties.

ethylene, and ethylene production increases dramatically just before the onset of the climacteric. Several different mutations affect the responsiveness of the tomato fruit to ethylene. Some of these mutations have proved very useful to plant breeders, because delayed ripening can reduce spoilage of fruit during transport, thus allowing less frequent harvesting of fruit. An example is the semi-dominant *Neverripe* (*Nr*) mutation. *Nr* confers a substantial delay in fruit ripening, by blocking the normal response to ethylene. The wild-type allele encodes an ethylene receptor (and is the tomato ortholog of Arabidopsis *ETR1*). In contrast, the *Nr* allele encodes a mutant ethylene receptor (analogous to that encoded by the Arabidopsis mutant *etr1-1* allele; see Section 6.3 on *ETR1* and other ethylene receptor genes and mutants). This mutant receptor is unable to perceive ethylene, and ethylene responses are therefore blocked. Two other tomato mutations affecting fruit ripening have also been used in tomato breeding: *ripening inhibitor* (*rin*) and *nonripening* (*nor*). The wild-type alleles of these genes encode transcription factors that control virtually all aspects of the ripening process. In plants homozygous for the mutant *rin* or *nor* alleles, the surge in ethylene biosynthesis that triggers the climacteric is inhibited. The *rin* allele has been used widely in the breeding of commercial tomato lines in which ripening is retarded.

Another mutation, *jointless*, has been used by tomato breeders to reduce the problem of crop loss due to premature fruit drop. In wild-type tomato, an **abscission zone** differentiates on the **pedicel** (the stalk at the base of the flower). As the fruit ripens, programmed events coordinated by ethylene (such as the expression of cell wall hydrolytic enzymes in the **middle lamella** between cells of the abscission zone) result in the weakening of the abscission zone, which then breaks, causing the ripening fruit to drop from the plant. Plants homozygous for *jointless* fail to differentiate an abscission zone and thus retain ripening fruit. The wild-type allele of this gene encodes a transcription factor that presumably controls genes involved in the development of abscission zones. Incorporation of *jointless* into commercial tomato breeding **germplasm** has helped to reduce levels of premature fruit drop.

In the "Green Revolution," the use of dwarfing mutations of wheat and rice resulted in major increases in crop yield

The **Green Revolution** was a period during which many countries adopted new varieties and methods of cultivation that resulted in rapid yield increases in the primary cereal crops: wheat and rice. For most of the time that humanity has practiced agriculture, crops have been grown in conditions where nitrogen supplies were limiting and effective competition with weeds for light, water, and nutrients was essential. The application of science and technology to agriculture greatly changed the way in which crop plants are grown. For example, the modern wheat field is a completely different environment from a typical wheat field of the nineteenth century. Use of selective herbicides largely eradicates weeds, and inorganic fertilizers are used to boost soil nitrogen to previously unattainable levels. In particular, the ready availability of inorganic nitrogen (fixed as ammonia in an industrial process, the Haber-Bosch process) has a great effect on crop yields.

In this new environment, a different set of plant characteristics became desirable. Previously, tall plants were advantageous because they could outcompete weeds and avoid being shaded by them; in addition, farmers needed tall straw for bedding for their livestock. In the new environment, tall plants are less desirable. There are fewer weeds to compete against, and tall stems can result in substantial yield losses as they are more vulnerable to being bent by wind or rain than are shorter stems. This vulnerability is exacerbated by the increased tallness of stems that results from increased nitrogen levels in fertilized soils. Wheat ears that are carried on bent stems usually fall to the ground, where they are lost to harvest. Furthermore, the nutrients and energy that go into the growth of tall stems could be more advantageously used in the development of the wheat grain.

Although wheat varieties became progressively shorter during the late nineteenth and early twentieth centuries, the Green Revolution of the 1950s and 1960s saw huge advances in the breeding of dwarf varieties of wheat and rice. We describe first the Green Revolution in wheat breeding.

In the late nineteenth century, Japanese farmers had been growing semi-dwarf varieties of wheat, with stems significantly shorter than those of normal wheats. These varieties contained single genes that conferred **dwarfism** in a dominant fashion. This dwarfism affected stem length but did not significantly affect seed yield per ear. In the 1950s these dwarfing genes were crossed into some relatively high-yielding American varieties, and were then introduced into Mexican wheat breeding programs. It was here that the first true semi-dwarfed Green Revolution varieties were produced. By using testing grounds in two separate regions of Mexico, one cool and one hot, it was possible to select for new dwarf varieties that performed well in both types of environment. Furthermore, these new varieties were "spring" wheats (spring wheats, unlike "winter" wheats, do not require a prolonged period of cold before they will flower). Spring wheats could be planted in spring and harvested in summer in areas where spring and summer are relatively mild, and they could be grown in the winter (planted in autumn, harvested in spring) in areas where winters are mild but spring and summer are too hot.

The new varieties had dramatically higher yields than traditional varieties, for two reasons: first, they increased grain production at the expense of stem growth; second, their shorter stems reduced yield losses from rain or wind damage. The use of these new, Green Revolution wheat varieties quickly spread to many parts of the world. Within a few years, during the late 1960s and early 1970s, some countries, such as India, were transformed from being net importers to net exporters of wheat grain.

The key genes that confer the dwarfism of the Green Revolution wheat varieties are a pair of homoeologous mutant genes, *Rht-B1b* (carried on chromosome 4B) and *Rht-D1b* (carried on chromosome 4D), either of which dominantly confers dwarfism (Figure 9–14). All Green Revolution wheat varieties contain one or other of these genes. *Rht-B1b* and *Rht-D1b* encode mutant versions of the wheat ortholog of the Arabidopsis **gibberellin** signaling protein GAI. GAI was originally identified as a key component of the mechanism by which gibberellin regulates plant growth. GAI is a repressor of growth, whereas gibberellin promotes growth by causing destruction of GAI by the cellular **proteasome** protein-destruction machinery (see Section 5.4). A mutant form of GAI, the gai protein, lacks a highly conserved region at its N-terminus called the **DELLA domain**. Lack of this domain makes the gai protein a constitutive repressor of growth,

Figure 9–14.
Wheat plants showing the effects of "Green Revolution" dwarfing genes.
Shown here are the tall wild type (*left*), semi-dwarf Rht-B1b (*next from left*), semi-dwarf Rht-D1b (*third from left*), and additional dwarfing alleles and combinations (*all others*). (Courtesy of Peter Hedden.)

one that is unaffected by gibberellin. Similarly, the proteins encoded by *Rht-B1b* and *Rht-D1b* lack the DELLA domain and confer gibberellin-insensitive dwarfism in wheat. (See Section 5.4 for more details on gibberellin and the GAI and DELLA proteins.)

The breeding of Green Revolution rice varieties began with the introduction of dwarfing genes into rice breeding programs. The first of these rice varieties (such as IR8) were the progeny of a cross between a dwarf variety from Taiwan (Dee-gee-woo-gen) and a high-yielding variety from Indonesia. Again, the high yields produced by these new dwarf varieties caused a rapid spread in their use throughout many of the rice-growing parts of Asia. The dwarfism of Dee-gee-woo-gen is conferred by a single recessive allele, *sd-1*. Cloning of the wild-type *SD-1* gene showed that it encodes an enzyme of gibberellin (GA) biosynthesis, the GA 20-oxidase. The rice genome contains a family of five different genes encoding GA 20-oxidases, each expressed in a tissue-specific fashion. *SD-1* is expressed in the growing stem and leaves, but not in the inflorescence. Lack of *SD-1* function (conferred by *sd-1*) therefore results in a reduction in stem gibberellin levels, resulting in a dwarfed stem. Seed number (per inflorescence) is unaffected because inflorescence gibberellin levels remain the same, but net crop yields are increased due to the reduction in stem height.

The molecular bases of the Green Revolution wheat and rice dwarfing genes have only recently been established, and the breeders that first used them in the 1950s and 1960s did so without really knowing how the genes conferred their characteristic phenotypes. In retrospect, it is clear that the breeders were using two different methods of altering the gibberellin biology of plants to increase the yields of modern cereal crops. As we have noted, the increased yields resulting from the introduction of the Green Revolution varieties had a worldwide impact, with some countries becoming net exporters rather than importers of food. In addition, the increased grain yields helped reduce the number of deaths from famine. Many people who would otherwise have starved were now fed. During a time when the world population doubled from 3 billion to 6 billion, food production still outpaced population increases.

Heterosis also results in major increases in crop yields

Darwin was the first to notice that when two inbred maize lines were crossed, the F1 progeny were taller and had greater tolerance for cool growing conditions. He concluded that hybrids had "greater height, weight and fertility" than their self-pollinated parents because of "their greater innate constitutional vigor." Plant breeders have exploited this phenomenon—termed "heterosis" (or "hybrid vigor")—to dramatically increase yield.

Heterosis is now used to increase the yield of many agricultural and horticultural crops. Early experiments (performed in 1908) showed that the yield of hybrid maize could be as much as four times greater than that of the parent lines (Figure 9–15). The first commercial hybrid maize became available to U.S. farmers in the early 1920s. By the 1940s the majority of the U.S. maize crop was hybrid, and it remains so to this day.

Between 1930 and 1997 yields increased from an average of 1 metric ton to 8 metric tons per hectare. Improvements in the yield of hybrid lines account for 50 to 70% of this increase, with improved weed control and increased fertilizer application accounting for the rest.

Farmers buy F1 hybrid maize seed from the seed companies, which produce the seed by crossing two inbred lines. The seed (F2) that farmers harvest from the F1 plants segregates for the allelic variation that distinguishes the two original (parental) inbred lines. If this harvested seed were planted, it would generate a range of phenotypically different plants, rather than the phenotypically uniform, high-yielding plants seen in the F1 generation. Thus farmers need to buy new F1 hybrid seed from the seed companies each year.

Figure 9–15.
A modern maize field. Vigorously growing, high-yielding, hybrid maize.

The seed companies generate the F1 seed by controlled cross-pollinations of the parent inbred lines. Since the male and female flowers of maize are physically separated, controlled cross-pollination can be facilitated in several ways: by placing bags over the flowers, by removing the male flowers ("de-tasseling"; Figure 9–16), or by using genetic male sterility (see below).

Heterosis increases the yields of many other agricultural and horticultural crops. For example, commercial tomato varieties are usually F1 hybrids, as are many of the commonly grown varieties of lettuce, spinach, and other leaf crops. Despite its great importance in the improvement of crop plant yield and vigor, the genetic basis of heterosis remains a mystery. This mystery is likely to be unraveled by genetic approaches in the near future.

Figure 9–16.
"De-tasseling" a maize plant in the 1930s. (From D.N. Duvick, *Nature Rev. Genetics* 2:69–74, 2001. With permission from Macmillan Publishers Ltd, courtesy of Pioneer Hi-Bred International, Inc.)

Cytoplasmic male sterility facilitates the production of F1 hybrids

Mutations in mitochondrial DNA can prevent the production of viable pollen and thus result in a failure to produce seed (see Section 2.6 for information on the **mitochondrial genome**). The inability to produce viable pollen is known as "male sterility." In many species of plants, mitochondria are transmitted to the next generation only through the **ovum**, not through the **pollen**. Thus mitochondrial mutations conferring male sterility are inherited only through the maternal parent. This type of inheritance is known as **cytoplasmic male sterility (CMS)**. CMS has been described in more than 150 plant species, and plant breeders have used CMS in crop plants to facilitate the production of F1 hybrids. The use of CMS plants in breeding is often preferable to the laborious alternative: removing **anthers** (emasculation) by hand or cutting away entire male inflorescences (de-tasseling in the case of maize; Figure 9–16) to prevent self-fertilization. In some crops, such as *Sorghum*, the close proximity of numerous, tiny, bisexual flowers makes mechanical emasculation impractical. When CMS plants are fertilized with normal pollen from other cultivars, all of their progeny are F1 hybrids (because the CMS plants do not produce their own viable pollen).

However, the fertility of the F1 hybrid progeny needs to be ensured. For this purpose, CMS plants are crossed with cultivars that contain **nuclear restorer genes**. Nuclear restorer genes suppress the effects of the mutation in the CMS mitochondria, making CMS pollen viable and overcoming the male sterility conferred by the mitochondrial mutation. This ensures the fertility of the F1 hybrids themselves, allowing the production of seeds and fruit when F1 hybrids are grown in agriculture. To produce maize F1 hybrid seed, rows of CMS plants are alternated with rows of restorer plants. Pollination of the CMS plants with pollen from restorer plants results in F1 seed that can germinate to yield fully fertile plants.

The economic importance of CMS has led to a great deal of research to determine its molecular basis. For example, CMS has been well studied in maize, brassicas (crucifers), petunia, rice, sorghum, and sunflower. In many cases, CMS is associated with rearrangements in mitochondrial DNA that result in a mitochondrial genome with novel protein-encoding regions. For example, CMS Texas (T) cytoplasm (CMS-T) was one of the first CMS systems to be used in the production of F1 hybrid maize in the United States. In the mitochondrial genome of CMS-T cytoplasm, a rearrangement has fused the **promoter** of the gene *ATP6* to the partial coding region of another mitochondrial gene, *RRN26*. The result of this fusion is a novel coding region, *T-URF13*, that encodes a novel 13-kDa protein, T-URF13. This protein is an abnormal mutant pore-forming protein; it locates in the mitochondrial inner membrane (Figure 9–17), where its abnormality inhibits normal mitochondrial function and thus causes the formation of nonviable pollen. The products of nuclear restorer genes (*Rf1* and *Rf2*) that were used in combination with CMS-T (to restore fertility to the F1 hybrids generated with the CMS-T system) reduce the effect of T-URF13 on mitochondrial function and restore pollen fertility (Figure 9–18).

The CMS-T system was widely used in the production of F1 hybrid maize in the United States during the late 1960s. However, plants carrying CMS-T cytoplasm are highly susceptible to the disease southern corn blight. An epidemic of this disease resulted in massive destruction of the 1970 U.S. maize crop and severe resultant reductions in harvest yield. For a while, U.S. maize breeders returned to mechanical procedures for making maize F1 hybrids. Fortunately, pathogen susceptibility is not a common feature of other CMS systems.

One of the most intriguing facets of CMS is the question of why the mitochondrial malfunction that causes CMS affects pollen development but not other aspects of plant development. In many cases, CMS is associated with early degeneration of the **tapetum**, the layer of cells in anthers that gives rise to pollen. All of the cells in a CMS plant contain the same mutant mitochondria, so why should this defect manifest only

(A)

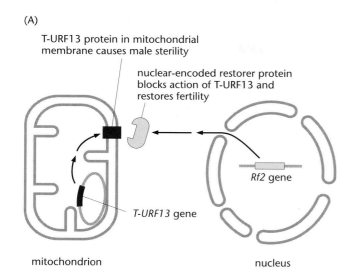

T-URF13 protein in mitochondrial membrane causes male sterility

nuclear-encoded restorer protein blocks action of T-URF13 and restores fertility

Rf2 gene

T-URF13 gene

mitochondrion

nucleus

(B)

during pollen development? The answer to this question remains unclear. However, it is possible that the effects of the mitochondrial defects are observed only in those tissues where mitochondria are functioning at or near their maximal capacity. Perhaps the mutant mitochondria are able to meet the energetic demands of most tissues, but unable to meet the relatively extreme **ATP** requirements of pollen-producing tapetal cells.

9.3 BIOTECHNOLOGY

We now turn to a relatively recent development in plant biology: the ability to insert one or a few genes, carried on a fragment of DNA, into a plant genome. Previously, transfer of novel genetic material into the genome of a plant was only possible via conventional cross-pollination. While such conventional methods have been hugely successful, and are responsible for almost all of the crop plants grown in the world today, the new techniques permit the transfer of selected genes in a more precise way than was previously possible. Selected single genes can be transferred while unwanted genes are not transferred, as they are in conventional breeding. Nor is it necessary to conduct successive rounds of selection (over several generations after the initial cross) to remove unwanted genes from the genome. Besides being increasingly important as a method for crop improvement, the ability to efficiently insert single genes into plant genomes has become important as a technical resource in basic research in plant biology.

In this section we discuss the techniques for gene insertion (the creation of **transgenic plants**; the transferred, foreign genes are referred to as **transgenes**) and then describe some uses of transgenic crops in agriculture. Finally, we consider what may happen in the next few decades: how the application of genetic and biotechnological advances might produce future crop plant improvements.

Agrobacterium-mediated gene transfer is a widely used method for generating transgenic plants

Agrobacterium tumefaciens is a soil-borne bacterium that infects a wide variety of host-plant species. The mechanism by which a segment of *Agrobacterium* DNA is transferred into the nuclear genome of a host is described in Section 8.1, and we recommend that you read that description before continuing here. In brief, the **T-DNA** portion of the *Agrobacterium* **Ti (tumor-inducing) plasmid** is transferred into the nucleus of a host plant, where it integrates into the host's nuclear DNA.

Figure 9–17.

Use of CMS Texas (T) cytoplasm in the production of F1 hybrid maize. (A) T-type male sterility was used in maize breeding programs in the 1960s. Use of a male-sterile line as the female parent in making crosses between different lines ensures that the female parent cannot self-pollinate, and hence all seeds on that parent must be hybrid. The diagram shows the mode of action of T-type male sterility in maize. In lines carrying T-type male sterility, the mitochondrial genome encodes a small protein, T-URF13, located in the inner mitochondrial membrane. This protein renders the pollen of the plant nonviable and thus the plant is male-sterile. A nuclear gene (*Rf2*) that encodes a "restorer" protein can suppress the effect of T-URF13 on pollen viability. Maize breeders used male lines carrying the restorer gene in their breeding programs so that male fertility would be restored in the hybrid offspring. (B) The photographs show tassels of maize plants. T cytoplasm causes male sterility in plants lacking the nuclear restorer genes *Rf1* and *Rf2* (*right*); plants carrying *Rf1* and *Rf2* are male-fertile (*left*).

Transformation is the process of introducing novel DNA into the genome of an organism, and a transgenic plant is a transformed plant, or the descendant of a plant that has been transformed. The natural *Agrobacterium* system has been adapted in the laboratory to create a general method for introducing a specific gene into a recipient plant. The most common *Agrobacterium*-based approach to plant transformation uses **binary vector systems**. In these systems, the *vir* (*Virulence* gene) functions and T-DNA functions of the Ti plasmid are separated into two genetic components (hence the term "binary"). First, the T-DNA is deleted from the Ti plasmid, leaving behind the *vir* genes that are required to accomplish infection (see Figure 8–19). Second, the T-DNA border sequences (LB and RB) (see Figure 8–19) are placed in a plasmid known as a "broad host-range" plasmid because it can grow and be manipulated both in *Escherichia coli* and in *Agrobacterium*. This new plasmid also carries a selectable marker gene that allows antibiotic selection of cells or plants carrying the vector. Novel DNA can be inserted between the two T-DNA border sequences (making a **chimeric** T-DNA) and reintroduced into the *Agrobacterium* strain containing the Ti plasmid from which the T-DNA was initially deleted. Usually this novel DNA consists of a gene and the means of controlling its expression in plant cells: an open reading frame of interest and a promoter that drives expression of the open reading frame.

Agrobacterium carrying the chimeric T-DNA is used to infect plant cells, and hence to insert the chimeric T-DNA into the plant genome. Cells carrying the chimeric T-DNA in their genome are now resistant to the antibiotic and can be selected for during plant regeneration (see Figure 9–19). Alternatively, pollen containing generative nuclei (i.e., nuclei of the **generative cells**; see Section 5.6) transformed with the chimeric T-DNA confer antibiotic resistance on their progeny. Transformant plants identified because they express antibiotic resistance are subsequently tested for expression of the characteristic conferred by expression of the novel gene of interest introduced into the plant genome via insertion of the chimeric T-DNA.

A. tumefaciens usually infects **eudicots**, and not **monocots**. Thus, when *Agrobacterium*-based plant transformation systems were first developed, they were suitable only for eudicots. However, by experimenting with different *Agrobacterium* strains and by manipulating *vir* gene expression, it has been possible to devise transformation systems that are extremely efficient for many monocots, including the important crops maize, rice, wheat, and barley.

The choice of promoter for the chimeric gene construct described above depends on the level or the tissue specificity of gene expression that is required and on the nature of the species that is being transformed. The most widely used promoter in *Agrobacterium*-mediated gene transfer is the *Cauliflower mosaic virus* (CaMV) 35S promoter. This promoter drives high-level gene expression in the tissues of most eudicots, but works only poorly in most monocots. For this reason, a maize ubiquitin gene promoter is commonly used when transferring genes into monocots. Other promoters are expressed at precisely defined developmental stages, in response to external stimuli, or in specific cell types, and these can be used to direct expression of the chimeric gene accordingly.

Particle bombardment–mediated gene transfer is an alternative means of generating transgenic plants

Unlike *Agrobacterium*-mediated gene transfer, the **particle bombardment technique** for plant transformation relies on a physical rather than a biological method for delivering DNA into the plant nucleus. The two techniques are summarized in Figure 9–18. The DNA to be transferred (usually a chimeric gene construct consisting of promoter, open reading frame, and selectable antibiotic-resistance marker) is precipitated and dried onto small gold particles. The particles are accelerated to high velocity—electrostatically, or with a jet of helium, or through the impact of a speeding bullet—so that

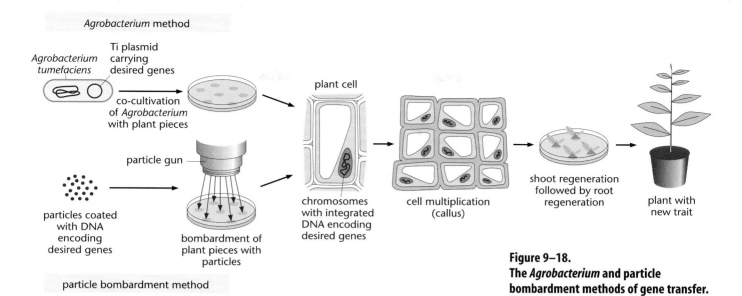

Agrobacterium method

Agrobacterium tumefaciens

Ti plasmid carrying desired genes

co-cultivation of *Agrobacterium* with plant pieces

plant cell

particle gun

particles coated with DNA encoding desired genes

bombardment of plant pieces with particles

chromosomes with integrated DNA encoding desired genes

cell multiplication (callus)

shoot regeneration followed by root regeneration

plant with new trait

particle bombardment method

Figure 9–18.
The *Agrobacterium* and particle bombardment methods of gene transfer.

they penetrate and enter plant cells. Some of the cells become transformed by integration of DNA from the surface of the particle into the genome. Transformed cells can be selected via their antibiotic resistance, and viable plants are regenerated from these cells (these steps are the same for both transformation methods; Figure 9–18).

As mentioned above, it is only recently that researchers have developed methods for *Agrobacterium*-based transformation of species outside the bacterium's normal host range. Accordingly, many of the transgenic maize and soybean varieties currently in commercial use were generated by particle bombardment–mediated gene transfer.

Herbicide resistance in transgenic crops facilitates weed control

Herbicides are chemicals that kill plants, and they are used commercially to kill weeds. The challenge to the farmer is to kill the weeds without killing the crop. Weeds need to be controlled for several reasons: they compete with crops for resources such as light, water, and soil nutrients, leading to losses in crop yields; they can affect the speed and efficiency of harvesting processes; and seeds of poisonous weeds can contaminate crop seeds during harvesting. Specific gene transformation has been used to develop crop plants that are herbicide resistant. These transgenic herbicide-resistant crops increase the flexibility and efficiency with which farmers can use herbicides on their fields.

There are many different classes of herbicide, which kill plants in different ways. Here we focus on one example—**glyphosate**—because transgenes conferring glyphosate resistance have been introduced into many crop species. Glyphosate is a member of a class of herbicides that inhibit amino acid biosynthesis. Specifically, glyphosate (*N*-phosphonomethylglycine) inhibits the activity of 5-enolpyruvylshikimate 3-phosphate (EPSP) synthase, an enzyme of the biosynthetic pathway of the aromatic amino acids, tryptophan, phenylalanine, and tyrosine (the **shikimate pathway** is described in Section 4.8). EPSP synthase converts shikimate 3-phosphate and phosphoenolpyruvate (PEP) to EPSP and inorganic phosphate. Glyphosate binds to the EPSP synthase protein competitively with PEP, which it closely resembles.

Mutations that alter the amino acid sequence of EPSP synthase in the region of the PEP-binding site result in a protein that is insensitive to glyphosate but can still bind PEP. High levels of expression of these glyphosate-insensitive mutant forms of EPSP synthase, particularly in meristems (to which glyphosate is rapidly translocated), result in plants that can tolerate high concentrations of glyphosate. For broad-leaf crops such as soybeans and sugar beet, in which controlling weeds without damaging the crop is extremely difficult, the development of transgenic varieties that express high levels of the glyphosate-insensitive EPSP synthase has made weed-free cultivation much easier. Application of glyphosate to fields of glyphosate-resistant crop varieties kills the weeds but leaves the crop plants unharmed (Figure 9–19).

Transgenic expression of *Bacillus thuringiensis* (Bt) crystal protein in crop plants confers insect resistance and increased yield

Herbicide resistance, particularly glyphosate resistance, is one of two transgenic traits that are currently of great importance in crop plants. The second is resistance to insect attack, conferred by transgenic expression of a gene encoding an insecticidal protein from the bacterium *Bacillus thuringiensis*.

B. thuringiensis is a spore-forming bacterium, and its spores contain crystals formed by a protein of about 130 kDa, the **Bt protein** (or **cry protein**). Following ingestion of the spores by insect larvae, the alkaline environment of the insect midgut causes the crystals to dissolve. Gut proteases cleave the released Bt protein, removing much of the C-terminus and leaving an active, toxic N-terminal domain of 65 to 70 kDa. This active toxin binds specific receptors on the plasma membranes of midgut epithelial cells, forming pores that kill the cells. Insect gut function is thus compromised, and the larva starves.

Once transgenic plants became a reality in 1983, the development of insect-resistant Bt-expressing crops became a major objective. Today, one of the most widely grown Bt-expressing crops is transgenic cotton, in which expression of the cry1Ac Bt protein provides protection against tobacco budworm (*Heliothis virescens*) and, to a lesser extent, cotton bollworm (*Helicoverpa zea*) and pink bollworm (*Pectinophora gossypiella*) (Figure 9–20). In Bt-expressing maize, the cry1Ab Bt protein confers resistance to *Ostrinia nubilalis*, the European corn borer. Cultivation of these transgenic Bt-expressing crops reduces the need for chemical insecticide use in agriculture.

Many different crop plant traits can potentially be improved by transgenesis

The technology for transfer of one or a few genes into crop plants has become relatively straightforward. Many possible applications are apparent, and more are brought into commercial production each year. We focus here on two recent applications that are of particular biological interest: the resistance of papaya to the *Papaya ring spot virus* (PRSV) and the development of rice expressing increased levels of β-carotene.

Figure 9–19.
Glyphosate-resistant soybean.
The soybean crop is resistant to glyphosate treatment, any weeds growing between the crop plants are killed.

Figure 9–20.
Bt cotton growing along the Colorado River in Parker Valley, Arizona. The green field in the foreground has cotton plants genetically modified to produce the Bt toxin cry1Ac. The defoliated, white field in the background contains non-Bt cotton; this planting is a statutory requirement to provide refuge for susceptible pests and thus discourage development of arthropod resistance to the Bt toxin.
(Courtesy of T.J. Dennehy.)

PRSV has been particularly devastating to the papaya crop in recent years. In the 1990s this disease caused severe damage to the papaya industry in Hawaii, and elsewhere. The impact of PRSV on papaya crops has been greatly reduced since the introduction of transgenic papaya engineered to carry a gene construct that expresses the PRSV viral coat protein. Expression of this gene in papaya substantially reduces the effects of PRSV infection, most probably because it imposes homology-dependent gene silencing on PRSV sequences (see the discussion of RNA silencing in Section 8.4). This simple transgenic approach to combating disease has revived the papaya-growing industry in Hawaii and elsewhere (Figure 9–21).

In regions of the world where people subsist primarily on rice, the low vitamin A content of a rice-based diet creates health problems. Large numbers of children die from infections that take hold because of immune systems weakened by vitamin A deficiency. In addition, vitamin A deficiency causes visual problems ranging from mild impairment to total blindness. In Southeast Asia, it is estimated that a quarter of a million children go blind each year because of this nutritional deficiency. The human body synthesizes vitamin A, or retinol, from carotenoids obtained in the diet; β-carotene, for example, is cleaved into two molecules of vitamin A by an enzyme in the liver (Figure 9–22).

Figure 9–21.
PRSV-resistant papaya. Aerial view of a papaya field. PRSV-resistant transgenic papaya plants (the dark green block) are surrounded by (dying) nontransgenic virus-infected papaya trees. (Courtesy of Dennis Gonsalves.)

Figure 9–22.
Formation of vitamin A by cleavage of β-carotene.

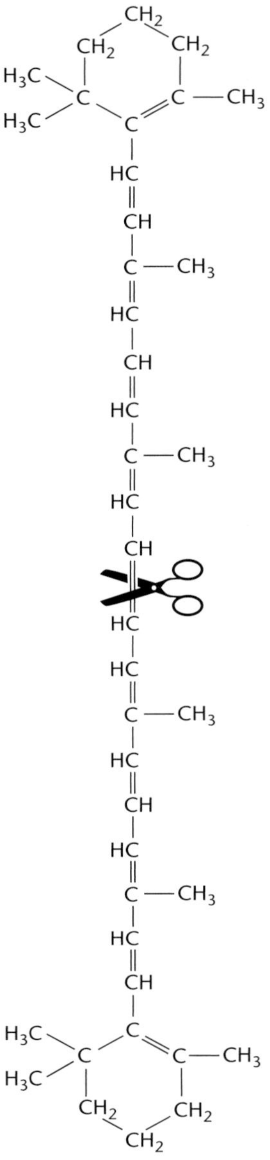

β-carotene

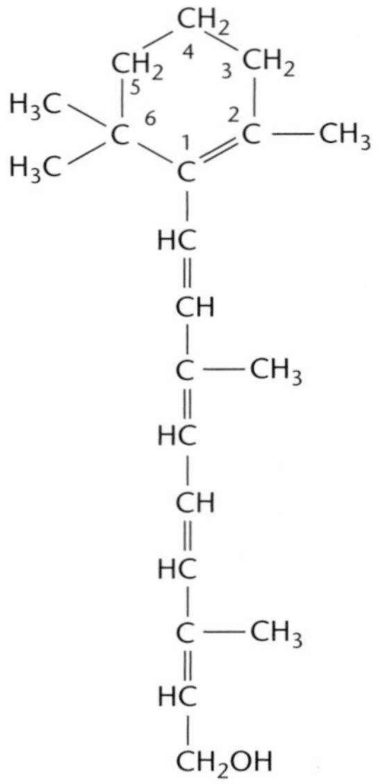

vitamin A$_1$

Transgenic rice expressing enzymes that produce increased levels of β-carotene in **endosperm** provides a means of preventing vitamin A deficiency. Immature rice endosperm synthesizes the early intermediate of the carotene pathway, geranylgeranyl diphosphate (see the discussion of **terpenoid** synthesis in Section 4.7), but lacks the capacity to catalyze significant rates of β-carotene synthesis. Expression in endosperm of the enzyme phytoene synthase (PS) converts the endogenously produced geranylgeranyl diphosphate to phytoene. The synthesis of β-carotene from the phytoene can then be achieved by the expression in rice endosperm of a bacterial carotene desaturase (crtI) and a plant lycopene β-cyclase. Thus, in developing this rice, three genes needed to be transferred to generate the capacity for β-carotene synthesis. Genes expressing PS and lycopene β-cyclase were isolated from *Narcissus pseudonarcissus* (the daffodil: it is the accumulation of β-carotene that gives the flowers their yellow color). The enzymes encoded by these genes carry a chloroplast **transit peptide** and therefore accumulate in plastids. The bacterial *crtI* open reading frame was modified by addition of a sequence encoding a chloroplast transit peptide, so that all three enzymes would locate in the plastids. Rice plants expressing all three enzymes were recovered following *Agrobacterium*-mediated transformation and were found to contain significant levels of endosperm β-carotene. Even higher levels have recently been obtained using a maize (rather than daffodil) version of the PS gene. This β-carotene–synthesizing rice is a potential means of dietary improvement. Experiments are in progress to determine how efficiently the fat-soluble β-carotene of rice endosperm is absorbed into the body and what proportion of dietary requirements this rice can meet.

"The future is green": The relationship between plants and people will continue to develop

In this chapter we have presented a few specific examples to illustrate the changing relationship between plants and people, with special focus on agriculture and horticulture. This relationship is far too broad and complex to document more fully here, but our selected examples clearly illustrate how the processes of plant domestication, plant breeding, and biotechnology have wrought massive changes in the structures and properties of plants and plant populations. These changes have in turn resulted in transformation of agricultural practices, landscape, and ecology. Such changes are certain to continue.

At the present time, the human population faces a challenge. The twentieth century saw unprecedented increases in crop yields, brought about by the development of new varieties and improved methods of cultivation. To a large extent these increases have kept pace with the growing demands of an increasing world population. But in many cases, increased food production has been achieved at a significant cost to the environment. To give just one example, much of the increase in crop productivity has been due to increasing application of nitrogen-based fertilizers. These fertilizers contain chemically reactive forms of nitrogen manufactured from unreactive atmospheric nitrogen gas. Much of the nitrogen fertilizer applied to fields as nitrate and ammonia remains unused by the target plants. As a result, recent decades have seen substantial increases in the amount of chemically reactive nitrogen-containing compounds in the environment. These increases are damaging to natural ecosystems. In lakes and rivers, for example, nitrogen-containing compounds promote the growth of algae, and the increased algal growth depletes the water of oxygen and other nutrients, killing fish and other inhabitants of lakes and rivers.

Modern agriculture also threatens the environment on other fronts. The erosion of natural habitats following land clearance to make space for crops is accelerating the extinction of species. And humans are also responsible for additional changes to the global environment, such as the climate change driven by increased levels of atmospheric CO_2 derived from fossil fuel consumption; these changes are themselves a huge challenge to the maintenance of crop productivity. The realization that the earth provides finite resources, and that ecosystems, habitats, and natural populations must be protected, is growing in many societies around the world. The challenge, for plant biologists and others, is to find ways of sustaining and increasing crop yields while at the same time halting or reversing environmental damage.

SUMMARY

Major changes have occurred in the relationship between plants and people from prehistoric times to the present day. The first domesticated crops were derived by human selection from wild plant species, many of which would have been common around pre-agricultural human settlements. Different plant species were selected for domestication for particular characteristics: cereals and pulses for their seeds, other plants for their leaves, tubers, or fruits.

Many of the differences between maize and its wild ancestor, teosinte, result from allelic variation at just five loci. For example, *teosinte glume architecture* regulates glume size and hardness: the teosinte kernel is contained in a hardened fruit case, maize kernels are not, and are thus more accessible for harvest. Cultivated bread wheat is polyploid and evolved from the hybridization of distinct diploid and tetraploid species. Wild diploid and tetraploid wheats were domesticated as einkorn and emmer, respectively. Hybrids between emmer and a third wild grass species probably gave rise to the hexaploids that were further domesticated to become bread wheat. Increase in fruit size occurred early in the domestication of tomato from a wild *Solanum*.

Later, scientific approaches to crop plant improvement produced substantial changes in the genetic structure of many crops. One result was a reduction in genetic heterogeneity. As plant breeders carried out deliberate and controlled cross-pollination of different pure-breeding plant lines, they discovered heterosis (hybrid vigor). This occurs when crossing of two low-yielding inbred lines results in F1 hybrid plants that out-yield the parental lines. Heterosis has been of huge importance in the development of modern maize and many horticultural crops; most commercial tomato varieties are F1 hybrids, as are many varieties of lettuce, spinach, and other leaf crops. Twentieth-century wheat breeding employed recurrent backcrossing with subsequent selection. One aim of recurrent backcross and selection is the introduction of new, beneficial genes with minimal disruption of the original genotype. The availability of genome maps has greatly eased this process.

Scientific methods also produced triticale by intercrossing wheat and rye; improved tomato varieties through mutations in genes affecting fruit color, fruit ripening, and fruit drop; and introduced disease resistance from wild *Solanum* relatives into tomato. Changes in crop management can also combat crop damage; for example, by intercropping an insect-resistant plant with the crop species (such as molasses grass with maize). In the "Green Revolution," the use of dwarfing mutations of wheat and rice resulted in major increases in crop yield. The new varieties quickly spread to many parts of the world; for example, by the late 1960s and early 1970s, some countries were transformed from net importers to net exporters of wheat grain.

With advances in biotechnology, just one or a few selected genes can be transferred into a plant genome. *Agrobacterium*-mediated and particle bombardment–mediated gene transfer are means of generating transgenic plants. Examples of the use of this technology include introduction of herbicide resistance into transgenic crops to facilitate weed control, and transgenic expression of *Bacillus thuringiensis* (Bt) crystal protein to confer insect resistance.

The twentieth century saw unprecedented increases in crop yields, brought about by the development of new varieties and improved methods of cultivation. To a large extent these increases have kept pace with the growing demands of an increasing world population. But in many cases, increased food production has been achieved at a significant cost to the environment. The challenge for plant biologists and others is to find ways of sustaining and increasing crop yields while halting or reversing environmental damage.

FURTHER READING

General

Chrispeels MJ & Sadava DE (2002) Plants, Genes and Crop Biotechnology, 2nd ed. Sudbury, MA: Jones and Bartlett Publishers, Inc.

Diamond JM (2005) Guns, Germs and Steel, new ed. London: Vintage.

9.1 Domestication

Doebley J (2004) The genetics of maize evolution. *Annu. Rev. Genet.* 38, 37–59.

Doebley J, Stec A & Hubbard L (1997) The evolution of apical dominance in maize. *Nature* 386, 485–488.

Frary A, Nesbitt TC, Grandillo S et al. (2000) *fw2.2*: a quantitative trait locus key to the evolution of tomato fruit size. *Science* 289, 85–88.

Kempin SA, Savidge B & Yanofsky MF (1995) Molecular basis of the cauliflower phenotype in Arabidopsis. *Science* 267, 522–525.

Zohary D & Hopf M (2000). Domestication of Plants in the Old World, 3rd ed. Oxford: Oxford University Press.

9.2 Scientific Plant Breeding

Cui X, Wise RP & Schnable PS (1996) The *rf2* nuclear restorer gene of male-sterile T-cytoplasm in maize. *Science* 272, 1334–1336.

Liu F, Cui X, Horner HT et al. (2001) Mitochondrial aldehyde dehydrogenase activity is required for male fertility in maize. *Plant Cell* 13, 1063–1078.

Mao L, Begum D, Chuang HW et al. (2000) *JOINTLESS* is a MADS-box gene controlling tomato flower abscission zone development. *Nature* 406, 910–913.

Peng J, Richards DE, Hartley NM et al. (1999) "Green Revolution" genes encode mutant gibberellin response modulators. *Nature* 400, 256–261.

Sasaki A, Ashikari M, Ueguchi-Tanaka M et al. (2002) A mutant gibberellin-synthesis gene in rice. *Nature* 416, 701–702.

Vrebalov J, Ruezinsky D, Padmanabhan V et al. (2002) A MADS-box gene necessary for fruit ripening at the tomato *Ripening-inhibitor* (*Rin*) locus. *Science* 296, 343–346.

Wilkinson JQ, Lanahan MB, Yen H-C et al. (1995) An ethylene-inducible component of ethylene signal transduction encoded by *Never-ripe*. *Science* 270, 1807–1809.

9.3 Biotechnology

Comai L, Faccioti D, Hiatt WR et al. (1985) Expression in plants of a mutant aroA gene from *Salmonella typhimurium* confers tolerance to glyphosate. *Nature* 317, 741–744.

Delannay X, Bauman TW, Beighley DH et al. (1995) Yield evaluation of a glyphosate-tolerant soybean line after treatment with glyphosate. *Crop Sci.* 35, 1461–1467.

Feitelson JS, Payne J & Kim L (1992) *Bacillus thuringiensis*: insects and beyond. *Biotechnology* 10, 271–275.

Hoekema A, Hirsch PR, Hooykass PJJ et al. (1983) A binary plant vector strategy based on separation of *vir-* and T-region of the *Agrobacterium tumefaciens* Ti-plasmid. *Nature* 303, 179–180.

Vaeck H, Reynaerts A, Höfte H et al. (1987) Transgenic plants protected from insect attack. *Nature* 328, 33–37.

Ye X, Al-Babili S, Klöti A et al. (2000) Engineering provitamin A (β-carotene) biosynthetic pathway into (carotenoid-free) rice endosperm. *Science* 287, 303–305.

GLOSSARY

14-3-3 proteins Small proteins that bind to specific target proteins, usually in response to phosphorylation of the target, and mediate a change in conformation and/or activity of the target.

α helix A common secondary structure of proteins, characterized by a single, spiral chain of amino acid residues stabilized by hydrogen bonds.

β oxidation The process by which fatty acids are broken down to yield acetyl-CoA.

ABA-response elements (ABREs) Short sequences of nucleotides in the promoters of genes induced by abscisic acid (ABA); they are necessary for specific induction, probably by binding transcription factors induced by ABA.

abaxial Describing the side of an organ oriented away from the apex of the organ or organism; for example, the abaxial (lower) surface of a leaf.

abaxial-adaxial axis An imaginary line running through the leaf blade, from the upper, adaxial side (normally exposed to sunlight) to the lower, abaxial side (normally facing away from sunlight).

ABC model A model summarizing how the different types of floral organs are determined by different combinations of a small group of regulatory genes.

ABC (ATP binding cassette) transporters A superfamily of transmembrane proteins with ATP-binding domains that function in transporting various peptides, sugars, polysaccharides, and ions across membranes.

ablation Elimination, destruction, removal.

abscisic acid (ABA) A phytohormone induced by and regulating responses to stress, particularly dehydration stress.

abscission The shedding by a plant of certain parts, such as leaves, flowers, or fruits.

abscission zone A thin layer of cells that breaks down to cause separation and eventual shedding of a plant part from the plant body.

absorption spectrum A plot of the amount of radiation absorbed versus wavelength of the radiation for specific light-absorbing substances (e.g., solutions of chlorophyll).

accession A genetically uniform line of a species, initially isolated from a wild-type population and subsequently maintained by self-fertilization.

acclimate To undergo a physiological response that allows tolerance to transitory stress conditions.

acetylation The addition of an acetyl group to a molecule.

acetyl-CoA A compound consisting of coenzyme A and an acetyl group derived from pyruvate; it is the substrate for fatty acid synthesis and the Krebs cycle, and the product of fatty acid oxidation and other catabolic processes.

acetyl-CoA carboxylase The enzyme that catalyzes carboxylation of acetyl-CoA to form malonyl-CoA, an intermediate of fatty acid synthesis.

actin A protein that is a key component of the cytoskeleton of eukaryotic cells.

actinomorphy Symmetry around a central point, or radial symmetry. In flowers, this refers to a radially symmetrical arrangement of floral organs.

activation domain The region of a protein that interacts with basal transcription factors to increase the rate of DNA (gene) transcription.

active transport The transport of molecules or ions across a cellular membrane against a concentration gradient (from a region of low concentration to a region of high concentration) with the aid of proteins in the membrane and energy supplied by ATP.

acyl-ACP thioesterase In fatty acid synthesis, the enzyme that catalyzes hydrolysis of the bond between a fatty acid (acyl) chain and acyl carrier protein (ACP).

acyl carrier protein (ACP) The protein to which acyl chains are covalently linked during chain elongation by the fatty acid synthase complex.

acyltransferases Enzymes that catalyze the transfer of an acyl group from one molecule (usually an acyl-CoA) to another.

adapt To undergo modifications that assist in survival of a species in stressful environmental conditions.

adaptation (1) Modifications that assist in survival of a species in stressful environmental conditions. (2) Modifications resulting from selective pressures during evolution.

adaxial Describing the side of an organ oriented toward the apex of the organ or organism; for example, the adaxial (upper) surface of a leaf.

S-**adenosylmethionine** A metabolite used as the source of methyl groups in many types of methylation reactions.

ADP-glucose A sugar nucleotide formed by reaction of glucose 1-phosphate and ATP, catalyzed by ADP-glucose pyrophosphorylase; it is the substrate for starch synthesis.

adventitious embryogeny A type of apomixis in which the haploid embryo sac is invaded by an embryo derived from the surrounding sporophytic tissue.

adventitious roots Roots that develop from a part of the plant other than the embryonic root.

aerenchyma Gas-filled spaces within tissues.

aleurone In seeds, the cell layer that surrounds the endosperm.

algae Photosynthetic eukaryotes that lack stems, leaves, and vascular tissues and generally inhabit wet environments such as seas and rivers. They may be unicellular (e.g., phytoplankton) or multicellular (e.g., seaweeds).

alkaloids A large, diverse class of nitrogen-containing plant compounds, many of which are active against pathogens and have pharmacological properties.

allele One of two or more variant forms of a gene that can be present at a given locus.

allopolyploid Having a set of chromosomes acquired from the combination of chromosome sets of two (or more) different species.

allosteric effector A molecule that regulates the activity of an enzyme or other protein by binding to the protein's allosteric site (a site other than its active site) and increasing (activating) or decreasing (inhibiting) its activity.

allosteric inhibition The inhibition of enzyme activity by binding of a molecule (allosteric inhibitor) to the enzyme's allosteric site.

alternative oxidase pathway An electron transport chain of the inner mitochondrial membrane that transfers electrons to the alternative oxidase, an enzyme that catalyzes transfer of electrons to oxygen to form water.

alternative splicing The production of different messenger RNAs from the same gene by combining exons and introns in different ways.

aminotransferases Enzymes that catalyze the transfer of an amino group from an amino acid to another compound.

amphipathic Having both hydrophilic and hydrophobic parts.

amylopectin A large $\alpha 1,4$, $\alpha 1,6$-linked glucan that makes up 70 to 80% of most starch granules.

amyloplast A nonpigmented starch-storing organelle derived from a plastid, specialized for starch storage.

amylose An $\alpha 1,4$-linked glucan with a small number of $\alpha 1,6$ linkages that makes up 20 to 30% of most starch granules.

anaerobic Completely devoid of oxygen.

anaerobic response elements (AREs) Short sequences of nucleotides in the promoters of genes that respond to anaerobic stress, necessary for specific induction, probably by binding transcription factors induced under anaerobic conditions.

anaphase The stage of mitosis in which sister chromatids separate and daughter chromosomes migrate to opposite poles of the cell; also a stage of meiosis.

anaplerotic pathway A metabolic pathway that replenishes an intermediate of another pathway.

angiosperms Flowering plants, with seeds enclosed in an ovary.

annual growth ring, annual ring A layer of woody growth, including spring and summer wood, formed by a tree during a single growing season; the rings are visible in a cross section of the trunk.

annual plants Plants that complete their entire life cycle within one year.

anoxic Completely devoid of oxygen.

antennae, antenna molecules In chloroplasts, the components of the antenna complex, an array of chlorophyll molecules embedded in the thylakoid membrane that transfer energy to a pair of chlorophyll *a* molecules at a photosystem reaction center.

anthers The regions of a stamen where pollen is produced.

antheridium (pl. **antheridia**) In bryophytes, the organ of the gametophyte that produces male gametes.

anthocyanins A class of red, purple, and blue pigments produced in flowers and leaves.

anticlinal In a direction perpendicular to a nearby surface, especially the outer surface of the plant.

antifreeze proteins Proteins that inhibit the nucleation of ice crystals.

antioxidants Compounds that can scavenge reactive oxygen species by becoming oxidized themselves.

antipodal cells In the ovule, cells located opposite the point where the pollen tube enters.

antipodal end Region of the ovule occupied by the antipodal cells.

antiporters Integral membrane proteins involved in active transport of two or more different molecules or ions across a membrane in opposite directions.

AP2 (APETALA2) family A family of transcription factors that contain the AP2 DNA binding domain, structurally homologous to the prototypic plant transcription factor APETALA2 (AP2). *Also called* ethylene-responsive element binding proteins (EREBPs).

apical-basal axis In the embryo, an imaginary line connecting the regions where the shoot and root apices develop.

apical dominance The ability of the growing shoot apex to inhibit the growth of lateral branches further down the stem.

apical hook The meristem-protecting curved form adopted by the hypocotyl of dark-grown seedlings, resulting from inhibition of cell elongation on the inner side of the curve.

apical meristem A group of actively dividing (meristematic) cells in the growing apex of a root or shoot that gives rise to the primary tissues of the plant body. *See also* **root apical meristem; shoot apical meristem.**

apomixis Reproduction by formation of seeds without fertilization.

apoplast (adj. **apoplastic**) The region of a plant tissue that lies outside the plasma membranes, consisting largely of the cell walls; as distinct from **symplast** (adj. **symplastic**).

apoproteins Native proteins (as defined by open reading frames of messenger RNA transcripts) lacking covalently attached prosthetic groups (e.g., for a phytochrome, the native phytochrome protein lacking the chromophore).

apoptosis One of the main types of genetically programmed cell death in multicellular organisms.

appressorium A flattened, bulblike structure that differentiates from the germ tube of a fungal spore, attaching it to the host-plant surface before formation of the penetration peg.

aquaporins Proteins that serve as water channels and may also selectively transport other small molecules.

arborescence The ability to produce woody tissues; woodiness.

arborescent Having tree-like growth.

arbuscule A highly branched structure formed from the hyphae of mycorrhizal fungi after penetration of host-plant cells.

archegonium (pl. **archegonia**) In bryophytes, the organ of the gametophyte that produces female gametes.

archespore, archesporial cell A cell that undergoes meiosis to produce the haploid cells that eventually give rise to pollen or ovules.

ascorbate–glutathione cycle The metabolic reaction cycle that converts excessive hydrogen peroxide to water in chloroplasts.

asexual reproduction Reproduction that does not involve the union of gametes (fertilization).

assimilate The primary products of carbon dioxide assimilation via the Calvin cycle. *Also called* photoassimilate.

atmosphere The mixture of gases surrounding a planet, organism, organ, or tissue.

ATP (adenosine triphosphate) A high-energy compound formed by photosynthesis and respiration that is required to provide the energy for cellular functions.

ATP synthases Enzyme complexes in the thylakoid membrane of the chloroplast and in the inner mitochondrial membrane that catalyze the synthesis of ATP from ADP and phosphate, in a reaction driven by passage of protons through a membrane pore formed by the complex.

ATP–ADP transporter A membrane protein that allows exchange of ATP and ADP across the membrane.

ATPases Enzymes that catalyze the hydrolysis of ATP to release chemical energy that is used to drive another chemical reaction, a change in protein conformation, or other cellular processes.

atrichoblasts Root epidermal cells that do not form root hairs.

autonomous flowering pathway In Arabidopsis, a pathway that promotes flowering under long or short days. The pathway was defined by a class of mutants that flower later than wild-type plants under long or short days, flower early if vernalized, and show increased expression of the floral repressor FLC (transcription factor encoded by the *FLC* gene).

autophosphorylation Self-phosphorylation of a protein kinase.

autopolyploid Having a set of chromosomes acquired from duplication of the chromosome set of a single ancestral species.

autotrophic organisms, autotrophs Organisms that can make their own organic compounds from simple inorganic precursors, using light or chemical energy.

autotrophy The capacity to synthesize organic compounds from inorganic compounds, using light or chemical energy.

auxin canalization A proposed explanation of how auxin transport determines the paths of developing veins in leaves and stems.

auxins A class of phytohormones involved in several developmental processes, including cell division, cell expansion, apical dominance, root initiation, and flowering; they include indole-3-acetic acid and related compounds.

axil The location where a leaf joins the stem.

axillary buds Lateral buds, found in leaf axils.

axillary meristems A group of rapidly dividing (meristematic) cells located where a leaf joins the stem in growing shoots. *Also called* lateral meristems.

bacteroids Differentiated forms of rhizobial bacteria in the nodule cells of a host plant.

basal angiosperms Angiosperm species on the first branches of an angiosperm phylogenetic tree.

basal defense mechanisms A form of plant immunity activated by the presence of surface components of potential pathogens. *See also* **PAMP-triggered immunity**.

basal meristem A group of rapidly dividing (meristematic) cells at the base of monocot leaves that allows continued leaf growth even after most of the leaf blade has been destroyed (e.g., by grazing).

basal transcriptional machinery A common set of proteins, including RNA polymerase and its cofactors, required for transcription of the majority of genes.

basic helix-loop-helix (bHLH) A protein structural motif that recognizes specific DNA sequences.

basic leucine zipper (bZIP) A protein structural motif that recognizes specific DNA sequences.

betaines A class of sweet-tasting alkaloids found in sugar beets and other plants.

biennial plants, biennials Plants that live for two years, producing vegetative growth in the first year, then flowering, setting seed, and dying in the second year.

binary fission The transverse division of a cell to form two daughter cells, each containing an exact copy of the genetic information in the parent cell.

binary vector systems Experimental plant transformation systems in which the T-DNA region containing the gene of interest is on one vector and the *vir* region is on a separate "disarmed" (without tumor genes) Ti plasmid.

biomass The mass of carbon in living and dead organisms.

biotrophs Pathogens that grow in association with host tissue, without killing the tissue, until they are ready to reproduce.

blade The flat part of a leaf.

blights Plant diseases involving wilting and tissue death; they may be caused by a variety of pathogenic organisms.

bolting Rapid stem elongation during inflorescence formation in rosette plants.

brassinosteroids A family of steroid compounds that act as phytohormones; they regulate many plant processes, including cell expansion, cell division, and vascular development.

Bt protein A protein produced by *Bacillus thuringiensis* that is toxic to insects. *Also called* cry protein.

bundle sheath The cylinder of cells that surrounds the vascular bundle in a leaf.

C3 photosynthesis The standard form of carbon fixation in plants in which carbon dioxide is fixed directly into sugars by the activity of Rubisco and the Calvin cycle.

C3 plants Plants that carry out C3 photosynthesis.

C4 photosynthesis A specialized form of carbon fixation that increases the effective concentration of carbon dioxide in cells that have Rubisco activity, thus limiting carbon loss through photorespiration.

C4 plants Plants that carry out C4 photosynthesis.

Cajal body A spherical sub-organelle in the nucleus of proliferative cells or metabolically active cells, usually one to five per nucleus, the number varying with cell type and phase of the cell cycle. It may be the site of assembly or modification of the nuclear transcriptional machinery.

calcium-dependent protein kinases (CDPKs) Calcium-dependent enzymes, found only in plants and certain parasites, that catalyze the addition of phosphate groups to specific amino acid residues of proteins.

callose A polysaccharide consisting of β1,3-linked glucose units.

Calvin cycle A series of enzyme reactions in chloroplasts in which atmospheric carbon dioxide is reduced and incorporated into triose phosphates and other sugars. *Also called* the reductive pentose phosphate pathway.

CAM *See* **crassulacean acid metabolism**.

CAM cycling A weak form of crassulacean acid metabolism (CAM) in which inorganic acid levels fluctuate daily, but with no net fixation of carbon in the dark.

CAM plants Plants that carry out crassulacean acid metabolism (CAM).

cambium A type of meristem located between the xylem and phloem of stems or roots; it produces new vascular tissue, allowing stems and roots to increase in diameter.

capsid The protein coat surrounding the nucleic acid of a virus.

carbamylation The modification of the amino group of an amino acid residue, or the N-terminus, of a peptide by addition of a carbamyl group.

carbon assimilation The incorporation of carbon dioxide (inorganic carbon) into complex organic molecules by photosynthetic organisms. *Also called* carbon dioxide (CO_2) assimilation.

carbon burial The incorporation of carbon into sediments on the ocean floor; the carbon may be of organic or inorganic origin.

carboxylation The introduction of a carboxyl group into a compound.

carotenoids Fat-soluble pigments found in certain plants; they provide the bright red, orange, or yellow coloration of many vegetables.

carpel The basic unit of the female reproductive structure of a flower.

caryopsis In grasses, a complex structure (the grain) in which maternal layers are fused to the outer layers of the seed.

Casparian strip In roots, deposits of suberin and lignin on the radial and transverse anticlinal walls of the endodermis that limit the flow of water and solutes through the apoplast.

caspases A class of proteases involved in programmed cell death.

cataphylls Scale-like leaves; for example, the leaves that make up the concentric layers of an onion.

caulimoviruses A family of spherical viruses with a genome composed of circular double-stranded DNA.

cauline Describing leaves that arise from the stem, particularly the inflorescence of rosette plants.

caulonema (pl. **caulonemata**) In bryophytes, a multicellular filament with rapidly tip-growing cells that develops from the protonema. The cells have small chloroplasts, and the filaments grow outward to colonize surrounding substrate.

cell cycle The sequence of stages through which a cell passes between one cell division and the next. The four main stages are M phase (nuclear and cytoplasmic division); G1; S phase (DNA replication); and G2.

cell plate The membrane-enclosed space produced during cell division in plants, formed from vesicles of the Golgi apparatus. It fuses with the plasma membrane, dividing the cell into two compartments.

cell wall The outer, rigid layer that surrounds the cells of plants, fungi, and bacteria.

cellularization The subdivision of a syncytium into individual cells.

cellulases Enzymes that catalyze the degradation of cellulose; they function in the degradation of cell walls.

cellulose A polysaccharide consisting of β-glucose units, the major component of the plant cell wall.

cellulose synthase complex The protein complex, organized in a rosette, that catalyzes biosynthesis of cellulose for the plant cell wall.

central cell, central nucleus The product of fusion of two of the nuclei in the female gametophyte; it subsequently fuses with one of the sperm nuclei delivered by the pollen tube, forming the endosperm.

central zone In a shoot meristem, the central region that produces new cells to populate the peripheral zone.

centrioles Barrel-shaped microtubular structures found in most animal cells and in cells of fungi and algae, but not usually found in plants. The spindle apparatus forms from these structures.

centromere The region of the chromosome where the spindle is attached during cell division.

cerebrosides A class of lipids composed of ceramide and a single sugar residue.

chalazal end The region of the ovule opposite the site of entry of the pollen tube.

chaperones Proteins that oversee the correct folding and assembly of polypeptides in cells, but are not components of the final structure.

chelatases Enzymes that catalyze the formation of a complex between a metallic cation and an organic compound.

chelation The formation of a complex between a metallic cation and an organic compound

chemical energy The energy that can be released by a chemical reaction.

chemiosmotic model A model describing how ATP generation in chloroplasts and mitochondria is driven by an electrochemical gradient across a membrane, set up by transfer of electrons along an electron transport chain.

chiasma (pl. **chiasmata**) In meiosis, an observable region in which nonsister chromatids of homologous chromosomes cross over.

chimeric Describing DNA, proteins, cells, or organisms made up of components from different individuals.

chitin A polymer of *N*-acetylglucosamine found in the cell walls of fungi and exoskeletons of insects.

chitinases Enzymes that catalyze the breakdown of chitin.

chloronema (pl. **chloronemata**) In bryophytes, a multicellular filament containing slowly expanding tip-growing cells that develops from the protonema. The cells contain many large chloroplasts.

chlorophylls Green pigments that absorb light energy, one of several types of pigment involved in photosynthesis. In plants, they are found in the thylakoid membranes of chloroplasts, as complexes with proteins and other pigments. Different chlorophylls (e.g., chlorophylls *a*, *b*, *c*, and *d*) have distinct chemical modifications that alter their light-absorbing properties.

chloroplast The specialized, chlorophyll-containing plastid in which photosynthesis takes place.

chlorotic Having a yellowish, pale green, or white color due to reduced amounts of chlorophyll, resulting from the reduced synthesis or accumulation of chlorophyll or from photooxidative damage.

chromatid One-half of a replicated chromosome.

chromatin The complex of DNA and proteins that makes up the chromosomes.

chromophore A molecular moiety that changes conformation when it absorbs light.

chromoplast A specialized plastid in which pigments are synthesizes and stored.

chromosome A single, large molecule of DNA containing an array of genes, and its associated proteins.

circadian Having an approximately 24-hour rhythm.

circadian clock, circadian oscillator A biochemical mechanism that cycles with and controls certain physiological responses, coordinating them with the day–night cycle; the central mechanism that acts as the time-keeper for circadian rhythms.

circadian rhythms Rhythms controlled by the circadian clock. The time between the peak of one cycle and the peak of the next cycle is approximately 24 hours; the rhythm persists in conditions of continuous light or continuous darkness.

circulative viruses Plant viruses that, when ingested by insects, pass through the gut wall and reach the salivary glands via the circulatory system.

cis-element A short, specific DNA sequence in a gene that is recognized by a transcription factor that regulates the activity of the gene.

citrate synthase In the Krebs cycle, the enzyme that catalyzes the first reaction: conversion of acetyl-CoA and oxaloacetate to citrate.

citric acid cycle *See* **Krebs cycle**.

clade A group of species consisting of a common ancestor and all its descendants; it is represented schematically as a branch or branches on an evolutionary tree.

cladogram A tree-like diagram demonstrating evolutionary relatedness.

clathrin A fibrous protein that, together with smaller peptides, coats some vesicles; it may bind to specific receptors on other membranes, such as the plasma membrane.

climacteric The stage in fruit ripening that is associated with a respiratory burst and is stimulated by ethylene.

clonal analysis The experimental use of genetically marked cells to study how their descendants contribute to the makeup of an organism.

clone A set of genes, cells, or organisms with identical genetic composition.

cloning (genetic) An experimental method in which a DNA molecule containing a single gene is isolated and then multiplied in a test tube or in a cell for biochemical study.

co-activator A protein that activates gene transcription and is recruited to its target genes indirectly, by interacting with transcription factors that in turn recognize specific DNA sequences.

codon A three-nucleotide sequence in a protein-coding sequence of DNA (or RNA), corresponding to a single amino acid.

coenzyme A (CoA) A coenzyme important in the metabolism of acyl groups, to which it is covalently linked as a thioester (acyl-S-CoA).

coevolution The evolution of complementary traits in two different species that results from environmental interactions between the species.

cofactor A nonprotein component of an enzyme that is required for the enzyme's activity; for example, a metal ion or an organic molecule (coenzyme).

cohesion-tension theory A hydraulic explanation of the mechanism of water movement in plants, from roots to leaves via the xylem.

colchicine An alkaloid drug that binds tubulin and inhibits the formation of microtubules; it is used experimentally to inhibit the separation of chromosomes in mitosis.

cold acclimation The processes by which plants adjust to transient low temperatures by prior exposure to low, nonfreezing temperatures.

coleoptile In monocots, the protective sheath covering the shoot apex of the embryo.

columella The central region of the root cap.

combinatorial control The combined action of multiple regulatory proteins on a single gene to set its transcription rate.

companion cells Cells adjacent to the sieve tube elements of the phloem, connected by plasmodesmata. Sugars and signaling molecules enter the sieve tube elements from the companion cells.

compensation point The point at which carbon assimilation by photosynthesis equals carbon loss through photorespiration.

complementary DNA (cDNA) A DNA molecule produced by using RNA as a template.

complex I NADH dehydrogenase, part of the electron transport chain in the inner mitochondrial membrane; transfers electrons from NADH to ubiquinone.

complex II A complex of proteins that forms part of the electron transport chain in the inner mitochondrial membrane; transfers electrons from succinate to ubiquinone.

complex III A complex of proteins that forms part of the electron transport chain in the inner mitochondrial membrane; transfers electrons from ubiquinone to cytochrome c.

complex IV Cytochrome c oxidase, part of the electron transport chain in the inner mitochondrial membrane; transfers electrons from cytochrome c to oxygen.

conidia Asexually produced spores of fungi.

conjugation In bacteria, the process by which plasmid DNA is transferred from one bacterium to another.

constitutive Taking place or existing in an organism at all times; as distinct from **inducible**.

continuous variation Gradual phenotypic differences among individuals in a population (e.g., differences in height), in contrast to differences that clearly separate the population into different groups (e.g., plants with white or red flowers).

convergent evolution The independent evolution in different species of processes or structures that fulfill the same function.

core complex A protein–chlorophyll complex in the thylakoid membrane that contains a photosystem reaction center.

cork The material forming the outer bark of woody plants.

cortex, cortical cells The tissue or cells lying between the epidermis and endodermis of a root or stem.

cortical parenchyma A tissue composed of cells with thin cell walls that constitutes most of the cortex of roots and stems.

cotyledons Embryonic leaves; specialized leaves that develop during embryogenesis and often have storage functions.

coumarins Phenylpropanoid compounds derived from cinnamic acid.

covalent modification Modification of a molecule, such as a protein, that involves a change in one or more of its covalent bonds.

crassulacean acid metabolism (CAM) A specialized form of carbon fixation that allows plants to fix carbon at night and keep their stomata closed during the day.

cross-fertilization The fertilization of an egg cell by a sperm cell of another individual of the species.

crossing over In meiosis, the process by which two chromosomes, paired during prophase I, exchange some portion of their DNA. Usually occurs when matching regions on matching chromosomes break and then reconnect to the other chromosome. The result is an exchange of genes (genetic recombination).

crossovers The structures formed when two chromosomes undergo crossing over.

cross-pollination The transfer of pollen from the flower of one individual to the flower of another individual of the species.

cryoprotective Protecting against low temperatures.

cryptochromes A subset of blue-light receptors, a family of flavoproteins that regulate germination, elongation, and photoperiodic responses in higher plants.

cultivar A variety of a crop plant originating and maintained by cultivation (agriculture or horticulture).

cupule A cuplike structure, made of hardened and fused bracts, that holds and protects the kernels of maize and teosinte.

cuticle The waxy layer covering the aerial surfaces of plants.

cuticular waxes Polymers that are components of the cuticle, consisting of esters of fatty acids and long-chain alcohols.

cutin A polymer that is a component of the cuticle, consisting of a network of long, cross-linked fatty acid molecules.

cyanobacteria Oxygen-producing photosynthetic bacteria that can reduce carbon in photosynthesis and can reduce nitrogen in specialized cells such as heterocysts. Photosynthesis resembles that of eukaryotic green algae and higher plants, but differs from that of other photosynthetic bacteria.

cyanogenic glycosides Plant compounds that break down to release hydrogen cyanide when the tissues containing them are damaged.

cyclic electron transport The cycling of electrons around a photosystem. In photosystem I of higher plants, this allows formation of a proton gradient, and hence ATP synthesis, without reduction of $NADP^+$.

cyclin-dependent kinase inhibitors (CKIs) A family of proteins that stop, prevent, or reduce the activity of a cyclin-dependent protein kinase.

cyclin-dependent protein kinases (CDKs) A family of kinases that, once activated by cyclin, regulate the cell cycle by adding phosphate groups to a variety of protein substrates that control processes in the cycle.

cyclins Proteins involved in regulation of the cell cycle. They form a complex with and activate a cyclin-dependent protein kinase (CDK).

cyme An inflorescence in which each branch ends in a flower that opens before the flowers below it.

cytochrome b_6f A complex of proteins that transfers electrons from photosystem II to photosystem I in chloroplasts.

cytochrome c A component of the electron transport chain in the inner mitochondrial membrane.

cytochrome c oxidase A protein complex of the inner mitochondrial membrane (complex IV) that catalyzes transfer of electrons to oxygen to form water. *Also called* cytochrome oxidase.

cytochrome P450 A family of enzymes that catalyze a variety of hydroxylation reactions that use oxygen as substrate.

cytochrome pathway The main electron transport chain in the inner mitochondrial membrane, which transfers electrons to cytochrome *c* oxidase.

cytogenetics The study of the structure and inheritance of chromosomes.

cytokinesis Division of the cytoplasm during cell division, following division of the nucleus.

cytokinins A class of adenine-related phytohormones that control cell division and differentiation.

cytoplasm All the material in the interior of a living cell, bounded by the plasma membrane, excluding the nucleus and vacuole.

cytoplasmic male sterility (CMS) The inability to produce functional pollen due to a mitochondrial mutation.

cytoplasmic sleeve The cytoplasmic content of plasmodesmata that surrounds the desmotubule.

cytoplasmic streaming The movement of cytoplasm in living cells, resulting in the transport of nutrients and enzymes and, for single-cell organisms, the locomotion of the organism.

cytoskeleton The molecular scaffold in the cytoplasm of eukaryotic cells, important in maintaining cell shape and organizing intracellular transport.

cytosol The fluid portion of the cytoplasm, outside the organelles.

day-neutral plants Plants that flower at the same time under all day lengths; flowering time is not responsive to photoperiod.

debranching enzyme An enzyme that catalyzes the hydrolysis of $\alpha1,6$ linkages in $\alpha1,4$, $\alpha1,6$-linked glucans, such as those of starch.

decarboxylation Removal of a carboxyl group from a molecule; the group is often released as carbon dioxide.

de-etiolation The developmental transition of a seedling from a dark-grown (etiolated) to a light-grown phenotype.

DELLA domain A short sequence motif present in regulatory proteins that are degraded in response to gibberellin.

denaturation Loss of the organized three-dimensional structure required for activity (of proteins or nucleic acids).

desmotubule The cylindrical membrane in a plasmodesma that connects the endoplasmic reticulum systems of adjacent cells.

determinate Characterized by a growth pattern that produces a genetically determined number of organs.

diarch In root vasculature, having two strands of protoxylem in cross section.

dicarboxylate transporters Membrane proteins that allow the movement of dicarboxylic acids across the membrane.

Dicer, Dicer-like protein (DCL) The enzyme (an RNase) that catalyzes cleavage of double-stranded RNA at regular intervals to produce small RNAs (21–26 nucleotides).

differentiation zone In roots, the region near the root tip where cell division has ceased and cells develop specialized features.

dinitrogenase The enzyme in rhizobial bacteroids that catalyzes, in several steps, the reduction of a molecule of nitrogen (N_2) to two molecules of ammonia (NH_3).

dinitrogenase reductase The enzyme that catalyzes transfer of electrons to the iron-sulfur-molybdenum cofactor of dinitrogenase.

dioecious Having male and female flowers on different individual plants.

diploid Having two homologous copies of each chromosome.

displacement loop (D-loop) A structure formed during replication of plastid DNA, resulting from replication of only one of the parental DNA strands.

DNA (deoxyribonucleic acid) The double-stranded polymer (nucleic acid) that encodes genetic information. In higher plants, DNA is contained in three distinct parts of the cell: nucleus, mitochondria, and plastids.

DNA binding domain A protein structural motif that has affinity for DNA (in many cases, for specific DNA sequences).

DNA motif A short nucleotide sequence in DNA that is highly conserved and has a specific biological function, such as the binding of a specific transcription factor.

DNA polymerase An enzyme that catalyzes the replication of DNA molecules.

DNA probe A DNA molecule that is experimentally marked in some way (e.g., with a radioactive isotope) and is used to detect DNA with a complementary sequence, to which the probe hybridizes.

domestication (of plants) The process of adaptation that converts wild plant species to plants used in horticulture or agriculture.

dormancy A period of biological rest or inactivity when a plant naturally stops growing and developing and suspends many of its metabolic processes.

double fertilization A fertilization process in angiosperms in which the pollen tube carries two sperm cells to the ovule, one sperm cell fuses with the egg cell to form the (diploid) zygote, and the other fuses with the central cell to form the (triploid) endosperm.

drought deciduous Having leaves that shed in response to adverse weather conditions such as drought.

drought-response elements (DREs) Short sequences of nucleotides in the promoters of genes that are induced by drought. The sequences are necessary for specific induction, probably by binding to transcription factors induced under water deficit.

dwarfism The short stature of plants that have mutations affecting gibberellin metabolism or response (e.g., the semi-dwarf wheat varieties of the Green Revolution).

dynamins Large GTPases (enzymes that hydrolyze GTP) involved in the scission of a wide range of vesicles and organelles.

early wood Secondary xylem (wood) formed first in the growing season and often with characteristics distinct from **late wood** (summer wood). *Also called* spring wood.

ectomycorrhizae Symbiotic associations between fungi and plant roots in which the fungal hyphae surround the root as a sheath.

ectoparasites Parasites living on the outer surface of their hosts.

ectopic Occurring in an unusual location.

effector molecules Pathogen-produced molecules that interfere with and suppress plant defense mechanisms.

effector-triggered immunity (ETI) *See R* gene–mediated defense.

effector-triggered susceptibility The state of a plant that results from suppression of the plant's defense mechanisms by pathogen effector molecules.

egg cell A female gamete, a haploid reproductive cell.

electrochemical gradient A gradient of ion concentration and electrical charge across a membrane; a type of potential energy involving both the concentration difference of the ion and its tendency to move relative to the membrane potential.

electromagnetic radiation Radiation with electrical and magnetic components, including visible light, X-rays, infrared, radio waves, and heat.

electron carrier A molecule that can accept electrons from and donate electrons to other electron carriers or enzymes.

electron transport chain A series of membrane-linked oxidation-reduction reactions in which electrons are transferred from an initial electron donor through a series of intermediates to a final electron acceptor.

elicitor molecules Molecules on the surface of pathogens that trigger plant basal defense mechanisms. *See also* **PAMPs (pathogen-associated molecular patterns)**.

elongase complexes Enzyme complexes in the endoplasmic reticulum that catalyze the elongation of 18-carbon fatty acids by addition of two-carbon units derived from malonyl-CoA.

elongation zone In roots, the region behind the root tip where cells cease dividing and rapidly elongate during root growth.

embryo In seed plants, the part of the seed containing the root and shoot meristems, the hypocotyl (or stem), and one or more cotyledons. The embryo is produced from a fertilized egg cell. On germination, the embryo gives rise to a seedling, which develops into the adult plant.

embryo sac The female gametophyte of angiosperms; a large, thin-walled space in the ovule in which the egg cell and, after fertilization, the embryo develop.

embryogenesis Embryo formation.

embryophytes Plants that develop embryos, form multicellular sporophytes with cuticles, and form gametophytes containing antheridia and archegonia. All land plants are embryophytes.

endoamylases Enzymes that catalyze the hydrolysis of the $\alpha 1,4$ linkages in $\alpha 1,4$, $\alpha 1,6$-linked glucans of starch granules, releasing smaller, soluble glucans.

endodermis, endodermal cells The cell layer surrounding the vascular cylinder of roots or stems that regulates the flow of water and solutes.

endoglucanases Enzymes that catalyze the cleavage of sugar–sugar bonds in glucans (sugar polymers).

endomembrane system A system of membranes in a eukaryotic cell, comprising the plasma membrane, nuclear membrane, endoplasmic reticulum, Golgi apparatus and associated vesicles, lysosomes, and vacuolar membrane (tonoplast).

endomycorrhizae Symbiotic associations between fungi and plant roots in which the fungal hyphae penetrate the root and form arbuscules in the root cells.

endoparasites Parasites living within the bodies of their hosts.

endoplasmic reticulum An membranous organelle of eukaryotic cells that consists of cisternae, vesicles, and tubules in which proteins and complex carbohydrates are processed on their way to other organelles, the cell surface, or the vacuole.

endoreduplication Chromosome replication without cell division, resulting in polyploid cells.

endoribonucleases Enzymes that catalyze the cleavage of RNA at an internal position in the polynucleotide chain.

endosperm In angiosperms, the nutritive tissue of the seed; a triploid tissue formed from fusion of a generative nucleus of the sperm and the central cell (formed from two central nuclei) of the female gametophyte.

endosymbiosis A symbiotic relationship in which one organism lives inside another. During evolution, in endosymbiotic events, one single-cell organism engulfed another to form a stable symbiosis in which the engulfed organism became an organelle (e.g., plastid) in the host cell.

enzymes Proteins (rarely, RNAs) that catalyze biochemical reactions in living organisms.

epialleles Different heritable forms of a gene that have the same nucleotide sequence but different chromatin modifications, such as different patterns of DNA methylation.

epicotyl The region of the embryonic stem above the cotyledons.

epidermis, epidermal cells A cell layer that forms the boundary between an organ or organism and the surrounding environment.

epigenetic Having a difference in gene function that is inherited through cell division but is not caused by a difference in DNA sequence.

epimutation A mutation giving rise to an epiallele.

epinastic Growing more rapidly on the upper surface than on the under surface of an organ.

epiphyte A plant that grows on another plant but does not derive any nourishment from it.

ethanol fermentation Energy-yielding anaerobic breakdown of sugars that produces ethanol.

ethylene A gaseous hydrocarbon that functions as a phytohormone.

etiolation The exaggerated growth and pale color characteristic of dark-grown plants.

etioplast A chloroplast that has not been exposed to light and lacks active photosynthetic pigments.

euchromatin The region of a chromosome that is rich in active genes and has chromatin less densely packed that in other regions; as distinct from **heterochromatin**.

eudicots, eudicotyledons One of the two main groups of angiosperms, with plants typically having broad leaves and having flowers with twofold or fivefold symmetry, and characterized by pollen grains with three or more furrows; as distinct from **monocots, monocotyledons**.

eukaryotes, eukaryotic organisms, eukaryotic cells Organisms of the domain Eukarya, and all cells of these organisms, in which the genetic material (DNA) is contained in a membrane-enclosed nucleus; as distinct from **prokaryotes, prokaryotic organisms**.

exine The tough outer layer of pollen grains.

exoamylases Enzymes that catalyze the cleavage of a two-glucose unit (maltose) from the nonreducing end of a glucan chain.

exocytosis Secretion of substances from a cell by fusion of vesicles with the plasma membrane.

exodermis In roots, the layer of cells (which may be suberinized) that surrounds the cortex on the periphery. *Also called* hypodermis.

exon Part of a gene that is transcribed into an RNA transcript and is present in the mature messenger RNA; as distinct from **intron**.

exoribonucleases Enzymes that catalyze the cleavage of nucleotides from one end of an RNA molecule.

expansins Proteins that regulate the elasticity of the plant cell wall.

expressed sequence tags (ESTs) Partial messenger RNA sequences obtained experimentally by random sequencing of a complementary DNA (cDNA); they are often used to facilitate gene annotation.

expression array An experimental device containing a large number of DNA molecules attached to a solid substrate and used to measure the expression levels of many genes in parallel.

extensins A family of cell wall proteins with a conserved primary sequence, typically rich in hydroxyproline.

external coincidence model of photoperiodic flowering A model proposing that an external signal, light, promotes or represses flowering when it coincides with a particular phase of a circadian rhythm. The model can explain why flowering is induced by long or short days.

F1 hybrids Plants resulting from a cross between two well-defined parental lines, often used to produce plants that are more vigorous than either parent. *See also* **hybrid vigor**.

F2 generation, F2 plants Second generation of a genetic cross, usually derived from self-pollination of first-generation (F1) plants.

facultative Able to grow or perform a specific function in the presence or absence of a particular environmental factor or condition (e.g., a facultative anaerobe can grow in the presence or absence of oxygen); as distinct from **obligate**.

fatty acid A compound consisting of a long hydrocarbon chain with a terminal carboxyl group.

fatty acid desaturases Enzymes that catalyze the introduction of double bonds into fatty acids.

fatty acid synthase complex A complex of proteins that catalyzes the synthesis of fatty acids from acetyl-CoA in plastids.

feedback regulation Regulation by an end product of a pathway or process of an earlier step in the same pathway or process; it may be negative feedback (inhibition) or positive feedback (activation).

feed-forward regulation Regulation by an intermediate in a pathway or process of a later step in the same pathway or process.

fermentation Anaerobic breakdown of sugars or other nutrients that produces energy but no net oxidation.

ferredoxin An iron-sulfur protein that mediates electron transfer in some metabolic reactions.

ferredoxin–thioredoxin system A system of oxygenic photosynthesis composed of ferredoxin, ferredoxin-thioredoxin, and thioredoxin that links light to the regulation of photosynthetic enzymes.

fertilization The fusion of gametes to form a zygote, which develops into a new organism.

flagellin A protein that is the principal constituent of the bacterial flagellum.

flavin adenine dinucleotide (FAD) One form in which flavin is attached to proteins; the flavin group can undergo oxidation-reduction reactions, and can absorb blue light. For example, FAD is one of the chromophores of the cryptochrome photoreceptor.

flavin mononucleotide (FMN) One form in which flavin is attached to proteins; the flavin group can undergo oxidation-reduction reactions, and can absorb blue light. For example, FMN is the chromophore of the phototropin photoreceptor.

flavocytochromes Enzymes with a flavin domain and a cytochrome domain that are involved in transfer of electrons during catalysis of oxidation-reduction reactions.

flavonoids A large class of water-soluble phenylpropanoids (a class of secondary metabolites), plant pigments that include the anthocyanins.

flavonols A class of secondary metabolites in plants.

floral induction The process by which an apical meristem is induced to lose vegetative identity and acquire floral identity.

floral integrator genes In Arabidopsis, genes that promote flowering and/or floral meristem identity and are regulated by signals from more than one flowering pathway.

floral meristem A group of actively dividing (meristematic) cells that gives rise to the organs of the flower.

floral organ identity genes Regulatory genes that control the identity of each of the four types of floral organs: sepals, petals, stamens, and carpels.

floral stimulus A substance or mixture of substances, produced in the leaves and transported through the phloem, that induces flower development. *See also* **florigen**.

florigen A single, phytohormone-like substance, produced in the leaves and transported through the phloem, that induces flower development. *See also* **floral stimulus**.

flower The sexual reproductive structure of angiosperms. Most flowers have showy, leaflike organs such as petals and sepals, which surround the organs bearing the microgametophytes (androecium) and megagametophytes (gynoecium).

fructans Polysaccharides composed of fructose units, sometimes also containing terminal glucose residues.

fructose 1,6-bisphosphatase The enzyme that catalyzes removal of a phosphate group from fructose 1,6-bisphosphate to form fructose 6-phosphate.

fruit In angiosperms, the mature ovary (or ovaries) and associated, fused plant parts; ripened fruits release their seeds by a variety of means.

FtsZ proteins Polymer-forming proteins (GTPases) that drive bacterial cell division and division of chloroplasts and mitochondria; they are structurally and functionally related to eukaryotic tubulins.

fungus A member of the kingdom Fungi, a group of single-cell or filamentous multicellular heterotrophic, eukaryotic organisms.

fusiform initials One of the two cell types of vascular cambium; tall, axially oriented cells. *See also* **ray initials**.

G1 (gap 1), G2 (gap 2) Gap phases in the mitotic cell cycle.

G proteins A family of heterotrimeric proteins (GTPases) that bind activated receptor complexes and, through conformational changes and the cyclic binding and hydrolysis of GTP, directly or indirectly produce alterations in the gating of ion channels, coupling cell surface receptors to intracellular responses.

GA-boxes Short DNA sequences that confer gibberellin-dependent gene transcription.

galactolipids Lipids with fatty acyl groups linked to positions 1 and 2 and a galactose linked to position 3 of the glycerol backbone.

galls Abnormal growths of plant tissue, often caused by microorganism or insect attack.

gamete A specialized haploid cell, male (sperm cell) or female (egg cell), that fuses with another gamete (sperm with egg) to form a diploid zygote.

gametophyte An organism that is the haploid phase of the plant life cycle and produces gametes.

gap junction A gap between the plasma membranes of adjacent cells that contains a network of protein channels and allows the passage of molecules from one cell to another.

G-box A sequence element of DNA (5′–CACGTG–3′) that is present in many plant promoters and is recognized by DNA binding proteins (transcription factors) of the bZIP and bHLH families. It is a common motif in genes whose transcription is induced by light.

geminiviruses A family of plant viruses that consist of two joined, incomplete icosahedra, with a genome of single-stranded DNA.

gene A segment of nucleic acid that encodes a product (RNA or protein) with a distinct biochemical function and is replicated and passed to the next generation, thus passing on the ability to make this same product.

gene annotation An experimental process for identifying individual genes in a genome sequence and adding information about each gene to the sequence database.

gene duplication A cellular process by which additional copies of a gene are inserted in the genome.

gene expression The process by which a gene's functional product (RNA or protein) is formed.

gene-for-gene model A model that explains the relationship between genes that give rise to effector molecules in pathogens and genes that encode defense proteins (R proteins) in plant hosts.

gene silencing The interruption or suppression of gene expression at the transcriptional or translational level.

general transcription factors (GTFs) A common set of proteins required for transcription of the majority of genes; they facilitate the recruitment and function of RNA polymerase.

generative cell A precursor of sperm cells during pollen development.

genetic linkage The preferential inheritance of a particular allele in association with another allele, the physical consequence of the close proximity of the genes on the same chromosome.

genetic map A map showing the relative position of genes and genetic markers on a chromosome.

genetic marker A DNA sequence whose presence in the genome is easily assayed. Experimental use of a genetic marker may involve using DNA sequence polymorphisms to distinguish the inheritance of different alleles and predict the inheritance of linked alleles.

genetic recombination *See* **recombination**.

genetic variants Forms of a species showing genetic variation.

genetic variation A variation in allelic forms or gene sequence in a population of a species.

genome The total genetic content of the haploid set of chromosomes in a organism's nucleus, in an organelle (e.g., chloroplast, mitochondrion), or in a single bacterial chromosome; the DAN or RNA content of a virus.

genome map A representation of the relative order and genetic distances separating genes and markers in an organism's chromosomes.

genomic imprinting A form of gene regulation in which expression of an allele—whether it is or is not expressed—depends on whether it is inherited from the female or male parent.

germ line The cell lineage that gives rise to gametes.

germination The initiation of growth of a seed or spore; in seeds, the emergence of a shoot and root.

germplasm A collection (store) of the extant genetic diversity of a species of crop plant or wild plant.

gibberellic acid (GA) A gibberellin; a phytohormone obtained from the fungus *Gibberella fujikuroi* and used experimentally to promote the growth of plants, especially seedlings.

gibberellins A class of phytohormones that control growth through cell expansion, among other functions.

glaucophytes A group of freshwater algae with plastids having a morphology intermediate between that of a cyanobacterium and a plastid of the green or red algae.

globular embryo A plant embryo at an early stage of development.

globulins A class of storage proteins found primarily in seeds.

glucanases Enzymes that catalyze the hydrolysis (breakdown) of glucans.

glucans Polymers of glucose.

gluconeogenesis Formation of sugars from noncarbohydrate components of the cell, such as fatty acids and amino acids.

glucosinolates Sulfur-containing compounds found in many members of the Brassicaceae, including Arabidopsis. They are broken down to release isothiocyanates and nitriles when the tissues containing them are damaged.

β-glucuronidase (GUS) An enzyme that catalyzes cleavage of glycosidic linkages between glucuronic acid and a hydroxyl group of another molecule. Its gene is frequently used as an experimental tool (reporter gene) in plant biology because the gene product can catalyze cleavage of colorless substrates to yield colored or fluorescent products.

glucuronoarabinoxylans Hemicellulose polysaccharides of plant cell walls, composed of glucuronic acid, arabinose, and xylose residues.

glumes In grasses, bracts in inflorescences that form the husk (chaff) in which cereal grains are enclosed.

glutamine synthetase–GOGAT system In plants, an enzyme system consisting of glutamine synthetase and glutamine–2-oxoglutarate aminotransferase (GOGAT) that is responsible for assimilation of ammonium into amino acids.

glutathione A tripeptide (glutamate–cysteine–glycine) that acts as a coenzyme in some enzyme-catalyzed reactions; its greatest importance may be its role as a nonspecific reducing agent in the cell.

glutathione-*S*-transferases Enzymes that catalyze the reaction of glutathione with an acceptor molecule to form an *S*-substituted glutathione. They play a key role in the detoxification of many compounds.

glutenins Storage proteins found in wheat.

glyceraldehyde 3-phosphate dehydrogenase In glycolysis and the Calvin cycle, the enzyme that catalyzes conversion of 3-phosphoglycerate to glyceraldehyde 3-phosphate.

glycerolipids Lipids with fatty acyl groups linked to positions 1 and 2 and a polar group linked to position 3 of the glycerol backbone.

glycine decarboxylase In the photorespiratory cycle, the enzyme complex that catalyzes conversion of two molecules of glycine to serine, carbon dioxide, and ammonia, with the generation of NADH.

glycogen An α-1,4, α1,6-linked branched polymer of glucose, a storage compound of fungi and animals.

glycolysis The metabolic pathway in which glucose is partially broken down and ATP is produced by a series of enzyme reactions that do not require oxygen.

glycophytes Plants that lack the genetic basis for salt tolerance.

glycoproteins Compounds consisting of carbohydrate and protein.

N-glycosylation The addition of a sugar to a peptide chain, linked through an asparagine residue; it occurs during transfer of the protein into the endoplasmic reticulum.

O-glycosylation The addition of a sugar to a peptide chain, linked through a serine or threonine residue; it occurs in the Golgi apparatus.

glyoxylate cycle The metabolic pathway that converts two molecules of acetyl-CoA to one molecule of succinate during gluconeogenesis.

glyoxysome A specialized type of peroxisome, found predominantly in germinating, fat-storing seeds. Its major function is to convert fatty acids to acetyl-CoA for the glyoxylate cycle, in which two acetyl-CoA molecules are converted to a dicarboxylic acid.

Golgi apparatus A network of stacked vesicles in eukaryotic cells in which proteins and lipids are processed and packaged, often for secretion.

gonidia In a *Volvox* colony, the cell type with reproductive functions.

grana Stacks of thylakoid membranes in chloroplasts.

granal thylakoids Thylakoid membranes that are stacked to form grana; as distinct from **stromal thylakoids**.

gravitropism Reorientation of the direction of plant growth with respect to orientation of the gravitational vector.

green fluorescent protein (GFP) A protein from the jellyfish *Aequorea* that fluoresces green when exposed to blue light; the GFP gene is used as a reporter gene.

Green Revolution Technological, agricultural, and other developments that gave rise to a significant increases in worldwide cereal grain yields between the 1940s and the 1970s.

ground tissue A tissue of the plant body located between the epidermis and the vascular cylinder.

growth regulators *See* **phytohormones**.

GTP (guanosine triphosphate) A high-energy compound involved in signaling and many other cellular processes.

GTPases Enzymes that catalyze the hydrolysis of GTP.

guard cells A pair of specialized epidermal cells that control the pore size of a stoma. *See also* **stoma** (pl. **stomata**).

guttation The exudation of xylem fluid from specialized structures on the margins of leaves.

gymnosperms Seed plants that do not produce flowers; for example, pines and cycads.

gynoecium The female reproductive part of the flower, composed of carpels.

hair cells Specialized root epidermal cells bearing a long, thin outgrowth (root hair) that absorb water and nutrients from the soil.

hairpin structure A double-stranded RNA structure that results from annealing of complementary regions of a single RNA molecule.

halophytes Plants that grow in salty environments.

haploid Containing the same set and same number of chromosomes found in the gametes of a species.

H⁺-ATPase A membrane protein (enzyme) that hydrolyzes ATP and uses the released energy to drive proton movement across the membrane. *See also* **proton pump**.

haustorium (pl. haustoria) A specialized feeding structure formed from a hypha of a biotrophic fungus or oomycete after penetration of a host-plant cell.

heart stage One of the early stages of plant embryogenesis, when the developing cotyledons give the embryo a heart shape.

heartwood The inner core of a woody stem, composed of nonliving cells; it is usually differentiated from the outer wood (sapwood) by its darker color.

heat shock element (HSE) A short sequence of nucleotides in the promoters of genes induced by heat shock, recognized and bound by heat shock factors.

heat shock factor (HSF) A transcription factor that binds to a heat shock element in the promoters of genes encoding heat shock proteins and activates their transcription.

heat shock proteins (HSPs) Proteins that accumulate in response to heat stress.

heat shock response A cellular mechanism to maintain stability when the cell is subjected to stress, often involving the production of heat shock proteins.

helicase An enzyme that catalyzes separation (unwinding) of DNA or RNA strands

heme A porphyrin compound containing an iron atom in the porphyrin ring; a prosthetic group in some classes of proteins involved in electron-transfer reactions.

hemibiotroph A pathogen causing an infection in which host cells initially remain alive but then collapse and die as infection progresses.

hemicelluloses A class of polysaccharides that are more complex than a sugar and less complex than cellulose; they are found in plant cell walls.

herbaceous Nonwoody.

herbicide A substance used to kill unwanted plants.

heterochromatin A region of a chromosome, thought to be genetically inactive, that remains tightly coiled (and stains deeply) during interphase; as distinct from **euchromatin**.

heteroduplex A double-stranded nucleic acid molecule with imperfectly complementary strands.

heterosis The increased strength, size, growth, or fitness of an F1 hybrid in comparison with its purebred parents. *Also called* hybrid vigor.

heterotrophic cell A cell that takes up organic carbon compounds as its main source of carbon (food).

heterotrophy Dependence on organic forms of carbon as food; as distinct from **autotrophy**.

hexokinases Enzymes that catalyze the phosphorylation of a hexose (six-carbon sugar), using ATP, to form the hexose phosphate.

hexose transporters Membrane proteins that allow the movement of hexose sugars (glucose, fructose) across the membrane.

high-performance liquid chromatography (HPLC) A method for separating molecules, based on pumping a solution at high pressure through a column packed with particles that interact differentially with the molecules to be separated.

histidine kinases Enzymes that catalyze the transfer of a phosphate group to specific histidine residues of a protein. Most are cell-surface receptors that transduce an intracellular signal via a two-component pathway, resulting in phosphorylation of an intracellular messenger protein.

histones Nuclear proteins; DNA winds around histones to form chromatin.

holdfast In algae, a multicellular structure that fixes the organism to its substrate (e.g., rock). It initially forms as filaments of tip-growing cells.

holoenzyme An enzyme than contains all the polypeptide subunits and cofactors required for its activity.

homeodomain A protein domain (about 60 amino acids) that recognizes and binds specific DNA sequences in the promoters of target genes.

homeostasis The mechanisms by which an organism regulates its internal processes to maintain a stable, constant state.

homeotic mutant A mutant in which one component of an organism, such as a body segment or a floral whorl, is transformed into a different type of component (e.g., stamens transformed into carpels).

homoeologs Related genes or chromosomes from different species or from different progenitors in allopolyploids.

homogalacturonans Pectins characterized by a backbone of α1,4-linked galacturonic acid residues.

homologs (adj. **homologous**) Chromosomes, genes, nucleic acids, or proteins that have similar sequence, structure, and/or function

homology Similarity in sequence, structure, or function.

homozygous In a diploid organism, having two identical alleles of a gene at a specific locus.

hyaline layer One of the layers of the seed envelope.

hybrid vigor Increased strength, size, growth, or fitness of an F1 hybrid in comparison with its purebred parents. *Also called* heterosis.

hydathodes Specialized organs that develop on leaf margins, particularly at the ends of principal veins. They may be involved in the exudation of water and salts.

hydroids In bryophytes, elongated water-conducting cells with thickened walls. They superficially resemble the water-conducting cells of the xylem of vascular plants, but evolved independently (i.e., by convergent evolution).

hydroquinones Aromatic phenol derivatives.

hydrosphere All water in the earth system: in the atmosphere, in rock (molten and solid), and on the earth's surface.

hydrostatic pressure The pressure exerted on a portion of a column of fluid as a result of the weight of the fluid above it.

hypersensitive response (HR) A mechanism of *R* gene–mediated defense in which plant cells collapse and die when invaded by a pathogen, thus restricting spread of the pathogen to other host tissues.

hypha (pl. **hyphae**) A long, branched, fungal filament; hyphae make up the body of a fungus.

hypocotyl The region of the stem of a eudicot embryo or seedling between the cotyledons and the radicle.

hypodermal layer. In a plant embryo, the cell layer beneath the epidermal layer.

hypoxia Low oxygen conditions.

imbibition The process in which the seed takes up water during germination.

indeterminate Characterized by a growth pattern that produces a number of organs that is not genetically determined.

indoleacetic acid A phytohormone of the auxin class.

inducible Taking place or becoming active in response to defined signals or stimuli; as distinct from **constitutive**.

infection thread During nodule formation on legume roots, a tube that contains rhizobia and extends along the root hair cell and down through several cortical cells to reach the nodule cells.

inflorescence A group of flowers arranged around a stem, which may be single or branched.

inflorescence meristem A group of dividing (meristematic) cells at the growing tip of a stem that gives rise to an inflorescence.

initial, initial cell A meristematic cell that divides continually to renew itself and to contribute a descendant that forms a specialized cell (tissue).

inorganic phosphate An ion of phosphoric acid not combined in an organic molecule; often denoted by P_i.

inorganic pyrophosphate An ion of pyrophosphoric acid not combined in an organic molecule; often denoted by PP_i.

integuments The two layers of cells that envelop the ovule.

intercropping The cultivation of two or more crops in the same space at the same time.

interfascicular regions Regions of ground tissue located between vascular bundles in a stem. *Also called* pith rays.

intermediate-day plants Plants that flower when the day is neither too long nor too short.

internode The segment of a stem between the insertion points of two successive leaves.

intine The innermost of the two cell walls of the pollen grain.

intron Part of a gene that is transcribed into an RNA transcript but is then spliced out (removed) to form the mature messenger RNA; as distinct from **exon**.

invertase The enzyme that hydrolyzes sucrose to glucose and fructose.

ion pump A membrane protein that transports ions against a concentration gradient, using energy from ATP.

iron-sulfur cluster A structure containing iron and inorganic sulfur that mediates electron transfer by some classes of enzymes involved in oxidation-reduction reactions.

isocitrate lyase In the glyoxylate cycle, the enzyme that catalyzes conversion of isocitrate to glyoxylate and succinate.

isoflavonoids A class of secondary metabolites, products of the phenylpropanoid pathway, that includes many phytoalexins.

isoforms Proteins having the same function (i.e., catalyzing the same reaction) but encoded by different genes; they may have small differences in amino acid sequence.

isoprene unit The five-carbon structural unit characteristic of terpenes.

isoprenoid pathway A ubiquitous biosynthetic pathway that synthesizes compounds made up of two or more structural units derived from isoprene.

isozymes Enzymes that differ in amino acid sequence but have the same function (i.e., catalyze the same reaction).

jasmonic acid A volatile phytohormone mainly involved in signaling in response to biotic challenges.

karyokinesis Division of the nucleus during cell division, as distinct from division of the cytoplasm.

ketoacyl-ACP synthase In fatty acid synthesis, the enzyme that catalyzes condensation of malonyl-ACP with an acyl-ACP.

kinases Enzymes that catalyze the transfer of a phosphate group to an organic substrate.

kinesins Motor proteins that use the energy from ATP hydrolysis to move along microtubules, carrying cargo molecules.

kinetochore A chromosomal attachment point for the spindle fibers, located in the centromere.

knockout allele A mutant form of a gene that has completely lost the gene function.

Kranz anatomy The arrangement of photosynthetic cells in concentric cylinders around the vascular bundles in the leaves of C4 plants.

Krebs cycle The cyclic metabolic pathway in mitochondria that oxidizes acetate units derived from glycolysis to carbon dioxide and water, producing reducing power that is used to drive ATP synthesis. *Also called* citric acid cycle; tricarboxylic acid cycle.

lactate fermentation Energy-yielding anaerobic breakdown of sugars that produces lactate.

landraces Pre-modern crop strains and varieties that were farmer-selected and locally specific.

late embryo abundant (LEA) proteins Proteins that accumulate late in embryogenesis and in response to water deficit.

late wood Secondary xylem (wood) formed late in the growing season and often with characteristics distinct from **early wood** (spring wood). *Also called* summer wood.

lateral axis The lateral dimension of a leaf, within the plane of the leaf blade and perpendicular to the main vein.

lateral meristem A group of rapidly dividing (meristematic) cells located where a leaf joins the stem in growing shoots. *Also called* axillary meristem.

leaf The organ of higher plants that is specialized for photosynthesis. Leaves are typically formed as primordia from the shoot apical meristem; when fully expanded, most leaves are flat and thin to maximize light penetration.

leaf primordia Small groups of cells recruited from the shoot meristem to develop into leaves.

leaf trace A vein of vascular tissue in the leaf, connecting the vasculature of stem and leaf.

leaflets The smaller units that make up a compound leaf.

leghemoglobin In leguminous plants, a heme-containing protein that binds molecular oxygen to maintain a reducing environment in root nodules.

legumins A type of storage proteins characteristic of leguminous plants.

lemma In grasses, the lower of two leaflike organs that enclose the flower.

lenticel A crack in bark.

leucine-rich repeat (LRR) A protein structural motif consisting of repeating sequences rich in leucine residues.

leucoplast A nonpigmented plastid.

ligand A molecule or ion that binds specifically to another molecule, such as a receptor.

light compensation point The light intensity at which the rate of carbon dioxide assimilation in photosynthesis equals the rate of loss of carbon dioxide in respiration; net assimilation occurs only at light intensities above the light compensation point.

light-harvesting complex (LHC) The main antenna of a photosystem in chloroplasts; it consists of chlorophyll molecules in association with an abundant integral membrane protein.

lignification Hardening of the cell wall by addition of lignin.

lignin A complex phenolic compound (noncarbohydrate polymer), harder than cellulose and water resistant; it binds to cellulose fibers and strengthens cell walls. It is a main component of wood.

linkage drag The reduction in fitness of a cultivar due to introduction of linked deleterious genes along with the beneficial gene during backcrossing.

linolenic acid A fatty acid with 18 carbons and three double bonds; α-linolenic acid is denoted $18:3^{\Delta 9,12,15}$.

lipases Enzymes that catalyze the hydrolysis of lipids to yield glycerol and fatty acids.

lipid bilayer The basic structure of biological membranes. Two lipid layers are organized so that their hydrocarbon tails face each other and the polar (charged) head groups of one lipid layer are on one side of the membrane and the polar heads of the other layer are on the other side.

lipids A broad class of organic molecules that are insoluble in water but soluble in organic solvents. The term is often used to refer specifically to molecules that contain fatty acids (e.g., triacylglycerols) or are derived from sterols (e.g., cholesterol).

lodicules In grasses, membranous floral organs adjacent to the stamens.

long-day plants Plants that flower when the daylight is longer than a critical length.

long terminal repeats (LTRs) Repeated sequences at the ends of retrotransposons.

luciferase An enzyme found in fireflies that causes emission of light by catalyzing ATP-dependent decarboxylation of the substrate luciferin; the luciferase gene is used as a reporter gene in plants.

LysM domain A widespread protein structural motif, originally identified in enzymes that degrade bacterial cell walls, that is thought to have a general peptidoglycan-binding capacity.

M phase The stage of the eukaryotic cell cycle during which the chromosomes condense and the nucleus and cytoplasm divide.

macronutrients Nutrients essential for plant growth that are required in relatively large amounts: nitrogen, potassium, calcium, magnesium, phosphorus, and sulfur; as distinct from **micronutrients**.

MADS domain The DNA binding domain of proteins belonging to the MADS family of transcription factors.

MADS family The family of transcription factors containing a characteristic structural motif (MADS domain) that recognizes specific DNA sequences.

magnoliid eudicots A monophyletic group consisting of approximately 20 families, including Magnoliaceae (magnolias), Piperaceae (peppers), and Lauraceae (laurels). Many are characterized by large, net-veined leaves and flowers with many spirally arranged tepals, stamens, and carpels.

malate dehydrogenase In the citric acid cycle, the enzyme that catalyzes interconversion of malate and oxaloacetate.

malate synthase In the glyoxylate cycle, the enzyme that catalyzes condensation of acetyl-CoA and glyoxylate to produce malate.

maltases Enzymes that catalyze the hydrolysis of maltose to glucose.

maltose transporters Membrane proteins that allow the movement of maltose across the membrane.

mannans Polysaccharides composed of mannose residues.

MAP kinases *See* **mitogen-activated protein (MAP) kinases**.

map-based cloning Identification of the DNA sequence of a gene by the use of high-resolution genetic mapping to locate the gene to a chromosome region of known sequence.

MAPK cascade *See* **mitogen-activated protein kinase (MAPK) cascade**.

MAPKs *See* **mitogen-activated protein (MAP) kinases**.

mapping line A plant line that is genetically uniform and carries many genetic markers.

marker genes *See* **reporter genes**.

megagametophyte The multicellular haploid structure or organism that contains the egg cell; it develops from a haploid megaspore, which is the product of meiosis.

megaphyll A leaf with multiple leaf traces. Megaphylls may have evolved through the fusion of multiple branches into a single leaflike structure.

megasporangium The tissue containing the cells that undergo meiosis to form megaspores.

megaspore The haploid product of meiosis that gives rise to the megagametophyte.

megasporocyte The diploid cell that undergoes meiosis to produce four haploid megaspores.

meiosis A specialized form of cell division during which a diploid cell produces haploid daughter cells.

membrane potential The charge difference across a membrane, such as across the plasma membrane between the cytosol and the extracellular fluid, caused by a differential distribution of ions.

meristems Tissues that consist of groups of undifferentiated stem cells, located in specific regions of the plant. They are the sources of the differentiated cells of all plant organs.

mesophyll The photosynthetic tissue of a leaf, located under the epidermis.

messenger RNA (mRNA) An RNA molecule transcribed from DNA and translated into a protein sequence by ribosomes.

metabolic control analysis A theoretical framework that allows quantitative analysis of the distribution of control of flux among the enzymes of a metabolic pathway.

metabolic pathway A linked series of enzyme-catalyzed reactions that converts a substrate to final product through a series of intermediates. Various metabolic pathways are interconnected through common intermediates.

metabolism The chemical and physical reactions in the living cell that are responsible for the breakdown of organic molecules to produce energy and the synthesis of new organic molecules.

metabolites Substances produced and consumed in metabolic reactions.

metabolomics The large-scale, parallel study of the many small molecules produced during metabolism.

metaphase The stage of mitosis (cell division) in which condensed chromosomes align in the middle of the cell; also a stage of meiosis.

methylation The attachment of a methyl group to a molecule; for example, a form of covalent modification of proteins and DNA.

methyltransferases Enzymes that catalyze the transfer of a methyl group to an organic substrate.

microgametophyte The haploid gametophyte that gives rise to the mobile sperm.

micronutrients Nutrients essential for plant growth that are required in relatively small amounts; as distinct from **macronutrients**.

microphyll In lycophytes, a small leaf with a single, unbranched vein.

micropyle The opening through which the pollen tube enters the ovule.

microRNAs (miRNAs) Small RNA molecules (usually 21–26 nucleotides) that participate in control of gene expression.

microspore The haploid product of meiosis that gives rise to the microgametophyte.

microsporangium The tissue containing the cells in which microspores develop.

microsporocyte The diploid cell that undergoes meiosis to form four haploid microspores, the precursors of the pollen grain. *Also called* pollen mother cell.

microtubules Long, filamentous protein polymers in the cytoplasm of eukaryotic cells, composed of the protein tubulin. They occur singly or in pairs, triplets, or bundles, as part of the cytoskeleton, and are involved in cell structure and movement.

middle lamella A pectin layer that cements together adjacent plant cells.

midvein The main vein that runs through the center of a leaf.

MIN proteins Proteins involved in the binary fission mechanism of plastids and bacteria.

mitochondrial genome The set of genes present in mitochondrial DNA; as distinct from **nuclear genome**; **plastid genome**.

mitochondrion (pl. **mitochondria**) An organelle of eukaryotic cells. It has a double membrane and contains the enzymes and electron transport chains that generate ATP from the oxidation of pyruvate to carbon dioxide in oxygen-dependent respiration.

mitogen-activated protein (MAP) kinases Serine/threonine protein kinases involved in signaling pathways; the plant MAP kinases have structural and functional homology to the MAP kinases of animals. *Also called* MAPKs.

mitogen-activated protein kinase (MAPK) cascade A cascade of serine/threonine protein kinases. Extracellular stimuli activate a cellular MAP kinase via MAP kinase kinase and MAP kinase kinase kinase.

mitosis The cellular process in which chromosomes are duplicated and separated into two daughter cells.

mixed-linkage glucans Sugar polymers in which the sugar residues are linked by bonds of different types; for example, a mixed-linkage β-glucan has D-glucose units that are either β1,3-linked or β1,4-linked.

molecular clock An analytical method for estimating the time of divergence of two species from a common ancestor by measuring differences in DNA sequence. Because mutations accumulate over time, the number of mutations in a DNA sequence is assumed to be proportional to the time since divergence.

molecular markers (1) Molecules specific to a particular group of organisms or species. (2) Markers in a genome map that show tight genetic linkage with loci controlling important traits. *See also* **genetic markers**.

molybdopterin A molybdenum- and sulfur-containing cofactor required by some classes of enzymes that catalyze oxidation-reduction reactions.

monocistronic Describing a messenger RNA molecule that encodes a single polypeptide; as distinct from **polycistronic**.

monocolpate Having a single opening; monocolpate pollen has a single opening through which the pollen tube emerges.

monocots, monocotyledons One of the two main groups of angiosperms, with plants having a single cotyledon or seed leaf, leaves with parallel veins, and flower parts in multiples of three; as distinct from **eudicots, eudicotyledons**.

monoculture The cultivation of a single variety of a crop in a large area such as an agricultural field.

monolignols Monomeric phenolic compounds that polymerize to form lignin; primarily *para*-coumaryl alcohol, coniferyl alcohol, and sinapyl alcohol.

monophyletic group A group of organism derived from a single common ancestor; as distinct from **polyphyletic group**.

monoterpenes A class of terpenes that consist of two isoprene units, linear or with ring structures.

morphogenesis The development of biological form, involving the coordination of cell division, growth, and differentiation, during the formation of tissues and organs. Cell morphogenesis is the process by which cellular form develops.

motif A domain in a protein or DNA that has recognizable homology to domains with known function. *See also* **DNA motif; protein motif.**

motor proteins Proteins that bind ATP and can move on a suitable substrate with concomitant ATP hydrolysis.

movement proteins Nonstructural proteins encoded by many (if not all) plant viruses that allow movement from an infected cell to adjacent cells.

mRNA transcript *See* **RNA transcript.**

multigene family Several similar but non-identical genes existing in multiple copies, often in the same region of a DNA molecule.

mutation A heritable change in the DNA sequence of a gene.

MYB family A family of transcription factors containing a DNA binding domain (MYB domain) that recognizes specific DNA sequences. Named for structural homology to the prototypic animal protein, c-Myb. *Also called* MYB-like transcription factors.

mycorrhizae Symbiotic associations between a plant and a fungus. They usually involve colonization of the plant root by the fungus, resulting in formation of a novel symbiotic structure. Fungal hyphae grow between the root cells, acquire carbon from the cells, and extend into the surrounding soil; some nutrients (e.g., phosphate) taken up by the hyphae pass into the plant.

mycotoxins Compounds produced by fungi that are toxic to mammals or other organisms.

myosin A motor protein that uses ATP to drive movements along actin filaments.

NAD (nicotinamide adenine dinucleotide) A coenzyme that acts as an electron carrier in many biochemical oxidation-reduction reactions, alternating between its oxidized (NAD^+) and reduced (NADH) state.

NADH (reduced nicotinamide adenine dinucleotide). The reduced form of NAD, a coenzyme that acts as an electron carrier in many biochemical oxidation-reduction reactions.

NADH dehydrogenase An enzyme of mitochondria that transfers electrons from NADH to ubiquinone. *Also called* complex I.

NADP (nicotinamide adenine dinucleotide phosphate) A coenzyme that acts as an electron carrier in many biochemical oxidation-reduction reactions, alternating between its oxidized ($NADP^+$) and reduced (NADPH) state.

NADPH (reduced nicotinamide adenine dinucleotide phosphate) The reduced form of NADP, a coenzyme that acts as an electron carrier in many biochemical reactions.

NADPH oxidase A membrane-bound enzyme complex that catalyzes the transfer of an electron from a molecule inside the cell to oxygen outside the cell, producing superoxide.

natural selection The process whereby particular alleles conferring a selective advantage favor the relative reproductive success of the individuals that contain them.

necrotroph A pathogenic organism that causes the death of host tissues as it invades the cells, such that it is always colonizing dead substrate.

nectar A sweet liquid secreted by some flowers that attracts pollinators such as hummingbirds and insects; it is gathered by bees for making honey.

nectaries Tissues that produce nectar.

nematodes Slender, generally very small, cylindrical worms of the group Nematoda, found in enormous numbers in water, soil, plants, and animals.

nitrate reductase The enzyme that catalyzes reduction of nitrate (NO_3^-) to nitrite (NO_2^-).

nitrate transporters Membrane proteins that allow the movement of nitrate across the membrane.

nitrite reductase The enzyme that catalyzes reduction of nitrite (NO_2^-) to ammonia (NH_3).

nitrogen assimilation In plants, the process by which inorganic forms of nitrogen, such as nitrate and ammonia, are combined into organic molecules, such as amino acids.

nitrogenase complex The enzyme system that catalyzes conversion of molecular nitrogen (N_2) to ammonia (NH_3).

nitrogen fixation In certain bacteria, the biological incorporation of atmospheric nitrogen into organic nitrogen-containing compounds.

nitrogen-fixing bacteria Bacteria that convert nitrogen gas (N_2) to ammonia (NH_3).

Nod (nodulation) factors Signal molecules produced by rhizobia during initiation of nodules on the roots of legumes.

nodules *See* **root nodules.**

nonphotochemical quenching Processes leading to the dissipation of excess energy in excited chlorophyll, usually as heat.

nucellus A diploid tissue surrounding the megagametophyte, located inside the integuments.

nuclear genome The complete set of genes in the chromosomes of the nucleus of a eukaryotic organism; as distinct from **mitochondrial genome; plastid genome.**

nuclear restorer genes Specific mutant alleles in the nucleus that suppress the male-sterility phenotype conferred by mitochondrial genome mutations.

nuclear speckle A nuclear domain rich in components of the messenger RNA splicing machinery and RNA processing.

nuclear-encoded plastidial (NEP) RNA polymerase An RNA polymerase, similar to bacteriophage T7 RNA polymerase, that is encoded in nuclear DNA and imported into plastids.

nucleases Enzymes that catalyze the cleavage of phosphodiester bonds between nucleotide residues in nucleic acids.

nucleic acids Polymers of nucleotides: deoxyribonucleic acid (DNA) and ribonucleic acid (RNA); the nucleotide residues are linked by 3′,5′ phosphodiester bonds.

nucleolar organizer A chromosomal region associated with a nucleolus after nuclear division.

nucleolar organizing region (NOR) A chromosomal region containing many copies of the genes for ribosomal RNA, around which the nucleolus forms.

nucleolus (pl. **nucleoli**) In the nucleus, the site of ribosomal RNA processing and ribosome assembly.

nucleoplasm The nonstaining or slightly chromophilic, liquid or semi-liquid, ground substance of the interphase nucleus; it fills the nuclear space around the chromosomes and nucleoli.

nucleosome A repeating subunit of chromatin, consisting of DNA wrapped around a core of eight histone subunits.

nucleotide A compound consisting of a nitrogen-containing base (adenine, guanine, thymine, cytosine, or uracil), a phosphate group, and a sugar group (deoxyribose or ribose); for example, ATP, ADP, GTP, and the nucleotide residues of DNA and RNA.

nucleus (1) In eukaryotic cells, the membrane-bounded structure that contains the cell's hereditary information and controls cell growth and reproduction. (2) In atoms, the central, positively charged region that contains protons and neutrons.

obligate Restricted to, or absolutely dependent on, a particular condition or process (e.g., an obligate anaerobe can grow only in the absence of oxygen); as distinct from **facultative**.

oil bodies Organelles containing storage oil (triacylglycerols), surrounded by a layer of the protein oleosin.

oleosin A protein with hydrophobic and hydrophilic domains that forms a monolayer around oil bodies.

oligomerization The association of two or more molecules into a multimeric complex.

onium compounds Quaternary ammonium compounds that function as osmolytes.

oomycete A member of the Oomycota, a phylum of filamentous eukaryotic microorganisms that, although fungal-like, are more closely related to the yellow-green algae.

open reading frame The portion of a gene and its RNA transcript that contains a sequence of bases potentially encoding a protein or portion of a protein.

operon A DNA sequence that includes a regulatory region and a region transcribed into a single messenger RNA that may encode several proteins. Operons are typically found in prokaryotes and in organellar (plastid or mitochondrial) DNA.

opines Compounds produced in crown gall tissues by condensation of an amino acid with a keto acid or sugar.

organelles Membrane-bounded structural components of eukaryotic cells; for example, plastids, mitochondria, and peroxisomes.

origin of replication A DNA sequence where replication is initiated.

orthologs Homologous genes that have diverged from a single gene in a common ancestral species.

osmolytes Osmotically active compounds.

osmoprotective Protecting against the adverse effects of low osmotic potential.

osmosis The movement of water through a selectively permeable membrane from a region of high water potential (low solute concentration) to a region of lower water potential (higher solute concentration), tending to equalize water potential and solute concentration on both sides of the membrane.

osmotic potential The potential for water to move through a selectively permeable membrane. The osmotic potential of pure water is 0, and water movement is measured with a negative value.

osmotic pressure The pressure of a solution that can develop when the solution is separated from pure water by a selectively permeable membrane.

ovule In angiosperms, the female gametophyte together with its protective integuments.

oxidases Enzymes that catalyze oxidation reactions, especially enzymes that react with molecular oxygen to catalyze the oxidation.

oxidation An increase in the oxidation state of a molecule, brought about by loss of one or more electrons, loss of hydrogen, or addition of oxygen.

oxidation-reduction (redox) Describing reversible reactions, or the proteins that participate in such reactions, in which substrates undergo oxidation and reduction.

oxidative pentose phosphate pathway A metabolic pathway, linked to glycolysis, that starts with the oxidation of glucose 6-phosphate and produces reducing power in the form of NADPH and four- and five-carbon sugars as carbon precursors for biosynthesis.

oxidative phosphorylation The synthesis of ATP in mitochondria, linked to electron transfer through the electron transport chain from NADH to oxygen. *See also* **ATP synthases**.

P680 The chlorophyll molecule in the reaction center of photosystem II.

P700 The chlorophyll molecule in the reaction center of photosystem I.

palea In grasses, the upper of two leaflike organs that enclose the flower.

paleosols Fossilized soils, preserved intact. The form when the soil is preserved in place by a catastrophic event such as a volcanic eruption that covers the soil and thereby protects it from erosion.

palisade cells, palisade mesophyll Photosynthetic mesophyll cells of the leaf. These elongated cells, with their long axis perpendicular to the plane of the leaf, are found toward the upper (adaxial) surface; they contain numerous chloroplasts.

PAMPs (pathogen-associated molecular patterns) Microbial compounds (e.g., flagellin, chitin) that can be recognized by plants as cues to activate defense mechanisms. *See also* **elicitor molecules**.

PAMP-triggered immunity Defense mechanisms triggered by PAMPs recognized by a plant's surface receptors. *See also* **basal defense mechanisms**.

parallel β sheet A regular secondary structure of proteins consisting of β strands connected laterally by hydrogen bonds to form a pleated-sheet structure.

paralogs Homologous genes that have diverged after gene duplication in a single species.

parenchyma cells, parenchyma The tissue characteristic of an organ, as distinct from the associated connective or supporting tissues; the most abundant type of plant tissue, its functions varying with the organ type. The cells are thin-walled and vacuolated.

particle bombardment technique An experimental method by which DNA is delivered to a plant cell nucleus. The DNA is coated onto accelerated gold particles, which penetrate the cell and the nucleus.

patatin A storage protein of potato tubers.

pathogenesis-related (PR) proteins Proteins, usually secreted, that increase greatly in abundance after pathogen challenge or systemic acquired resistance triggered by salicylic acid.

pathogens Organisms that cause disease in other (host) species.

pathovars Genetic variants of a pathogen.

pectinases Enzymes that catalyze the hydrolysis of pectins, and thus the breakdown of cell walls.

pectins A heterogeneous group of linear and branched sugar polymers (polysaccharides), rich in galacturonic acid but frequently containing several other sugars; they are found in the matrix of plant cell walls.

pedicel A small stalk or stalklike plant part bearing a single flower in an inflorescence.

PEP carboxykinase The enzyme that catalyzes ATP-dependent conversion of oxaloacetate to phosphoenolpyruvate (PEP) and carbon dioxide during gluconeogenesis and in the bundle-sheath cells of some C4 plants.

peptidoglycans Cross-linked polymers of sugars and amino acids that form the wall (outside the plasma membrane) of bacteria. Each monomer contains a pair of *N*-acetylglucosamine and *N*-acetylmuramic acid residues; amino acids are joined as pentapeptides to the *N*-acetylmuramic acid.

perennial plants, perennials Plants that usually take more than one growing season to mature, live for more than two years, and produce seeds more than once during their lifespan.

perianth The outermost, nonreproductive organs (sterile whorls) of the flower; it corresponds to whorls 3 and 4 of the Arabidopsis flower, which develop as petals and sepals, respectively.

periclinal Describing the plane of division or plane of establishment of the cell wall that is parallel to the surface of the organ.

pericycle The cell layer beneath the endodermis of roots and stems.

peripheral zone The region of the shoot meristem where organ primordia are initiated.

peroxisome An organelle of eukaryotic cells, containing catalase and other oxidative enzymes such as peroxidase.

pests Insects, mites, nematodes, and other animals that attack crops.

petals Sterile, showy floral organs.

petiole The stalk that attaches a leaf blade to the stem.

ΔpH pathway A pathway that is dependent on a pH gradient across a membrane. In chloroplasts, this is one pathway by which proteins are translocated from the stroma into the thylakoid lumen.

phenylpropanoids A class of secondary plant metabolites derived from phenylalanine.

phloem The vascular tissue that transports organic nutrients, especially sucrose, through the plant body.

phloem loading The process in which sucrose synthesized in leaf mesophyll cells is transferred into the phloem for export from the leaf.

phorbols A class of diterpenoids involved in defense against insects in some plants.

phosphatidic acid A phospholipid with fatty acyl groups linked to positions 1 and 2 and a phosphate group linked to position 3 of the glycerol backbone.

phosphatidylcholine A phospholipid with fatty acyl groups linked to positions 1 and 2 and choline linked via phosphate to position 3 of the glycerol backbone.

phosphatidylglycerol A phospholipid with fatty acyl groups linked to positions 1 and 2 and glycerol linked via phosphate to position 3 of the glycerol backbone.

phosphoenolpyruvate (PEP) carboxylase The cytosolic enzyme that catalyzes addition of carbon dioxide to phosphoenolpyruvate to form oxaloacetate. This enzyme is involved in the CO_2-concentrating mechanism of C4 plants.

phosphofructokinase In glycolysis, the enzyme that catalyzes ATP-dependent conversion of fructose 6-phosphate to fructose 1,6-bisphosphate.

3-phosphoglycerate kinase In the Calvin cycle and glycolysis, the enzyme that catalyzes conversion of 3-phosphoglycerate to 2,3-bisphosphoglycerate.

phospholipase C An enzyme that catalyzes the hydrolysis of specific ester bonds in phospholipids; it removes a polar head group (such as choline) from the phospholipid.

phospholipids Lipids in which a fatty acyl group of a triacylglycerol is replaced by one of a variety of chemical groups linked via a phosphate group to position 3 of the glycerol backbone; for example, phosphatidylcholine, phosphatidylglycerol.

phosphoribulokinase The enzyme that catalyzes conversion of ribulose 5-phosphate to ribulose bisphosphate, the carbon dioxide acceptor in the Calvin cycle.

phosphorylation/dephosphorylation The addition/removal of phosphate groups through opposing enzyme reactions; for example, phosphorylation/dephosphorylation of specific amino acids residues is a method of regulating some proteins.

photodamage The exposure of plants to higher light intensities than those required for photosynthesis, resulting in the formation of reactive oxygen species in chloroplasts, which cause oxidative stress and damage cellular structures.

photoexcitation The flow of electrons that follows excitation of chlorophyll by light.

photoinhibition Damage to a leaf's capacity for photosynthesis caused by exposure to excess light energy.

photolyases Enzymes that absorb light energy in the visible range and use it to break carbon–carbon bonds in pyrimidine dimers.

photomorphogenesis The light-mediated regulation of plant growth and development.

photon The elementary particle of electromagnetic radiation.

photooxidative damage *See* **photodamage**.

photoperiodic response rhythm A circadian rhythm that controls a photoperiodic response, such as flowering, and is sensitive to exposure to light. This is the circadian rhythm postulated in the external coincidence model.

photoperiodism Plant responses to day length that allow adaptation to the changing seasons.

photophosphorylation The synthesis of ATP in chloroplasts, using light energy from photosynthesis. *See also* **ATP synthases**.

photoreceptor A protein pigment that initiates a signal-transduction cascade when exposed to light of a specific wavelength.

photorespiration The oxidation of carbohydrates with release of carbon dioxide during photosynthesis.

photorespiratory cycle A metabolic pathway that converts the 2-phosphoglycolate produced by oxidation of ribulose 1,5-bisphosphate to 3-phosphoglycerate and carbon dioxide.

photosynthate Any substance synthesized during photosynthesis, particularly sucrose.

photosynthesis The process by which chlorophyll-containing cells absorb light energy and convert it to chemical energy, synthesizing sugars from carbon dioxide and water.

photosystem I, photosystem II Protein–chlorophyll complexes, acting in series and connected by an electron transport chain, that capture energy from light during photosynthesis. Electrons generated by primary charge-separation events in photosystem I (with P700 at its reaction center) reduce $NADP^+$ to NADPH; the reaction center is re-reduced by electrons generated by PSII (P680), transferred by the electron transport chain. The PSII reaction center is re-reduced by electrons donated from water, via a water-splitting complex.

phototropins Blue-light photoreceptors that mediate phototropism responses in higher plants. Along with cryptochromes and phytochromes, they allow plants to alter their growth in response to the light environment. They also regulate intracellular movements of chloroplasts and the opening of stomata.

phototropism Directional growth toward (positive phototropism) or away from (negative phototropism) a light source.

phragmoplast A structure that forms at the site of the spindle equator at the end of mitosis; this is the site where the new primary cell wall grows.

phycobilins Light-absorbing pigments that function in photosynthesis, found in cyanobacteria, red algae, and glaucophytes.

phycobilosomes Protein complexes containing light-absorbing pigments (phycobilins) found in the photosynthetic apparatus of cyanobacteria, red algae, and glaucophytes.

phycocyanobilin A phycobilin, precursor to the linear tetrapyrrole chromophore of phytochrome.

phyllotaxy The arrangement of leaves around a stem.

phylogeny The evolutionary history of an organism, often depicted as a tree (phylogenetic tree) showing the organism's relationship to other species.

phytases Enzymes that catalyze the stepwise hydrolysis of the phosphate groups of phytates.

phytates Phosphate-rich compounds derived from hexose sugars, commonly stored in seeds and used as a source of phosphate on germination.

phytoalexins Compounds produced by plants in response to attack by pathogens; they act to poison or repel the attacking organism.

phytochromes Pigments that act as photoreceptors, sensitive to light in the red and far-red regions of the spectrum. They are involved in the regulation of light-dependent growth processes.

phytoferritin A protein that forms complexes capable of binding large amounts of iron.

phytoglycogen A soluble $\alpha1,4$, $\alpha1,6$-linked glucan produced in place of starch in plants lacking the debranching enzyme isoamylase.

phytohormones Plant substances produced in low concentrations in cells and used as signals to control gene expression or growth in other cells of the same plant. *Also called* growth regulators.

phytosiderophores Plant compounds that chelate iron in the soil and facilitate its uptake.

pilus In bacteria, a tube composed of protein that is used to exchange genetic material during conjugation.

pinocytosis The process by which cells take up soluble substances through the plasma membrane by the formation of vesicles.

pits In water-conducting cells of xylem, perforations in the plates between cells.

pith rays *See* **interfascicular regions**.

plasma membrane The semi-permeable membrane at the boundary of every cell, which acts as a selective barrier; it consists mostly of proteins and phospholipids.

plasmid A small, circular DNA in bacteria that replicates independent of chromosomal replication and can be transferred from one organism to another by conjugation.

plasmodesma (pl. **plasmodesmata**) A microscopic channel that traverses the cell walls and middle lamella between plant cells, linking the symplasts. *See also* **cytoplasmic sleeve**; **desmotubule**.

plastid genome The complete set of genes in an organism's plastid DNA; as distinct from **mitochondrial genome**; **nuclear genome**. *Also called* plastome.

plastid-encoded plastidial (PEP) RNA polymerase The multi-subunit RNA polymerase encoded by the plastid DNA; it is similar to bacterial RNA polymerase.

plastids A class of organelles in some eukaryotes, bounded by a double membrane, often with elaborate internal membrane systems. They perform a variety of functions, including photosynthesis, storage of starch and pigments, and synthesis of key metabolites such as fatty acids. They contain DNA that encodes a limited number of plastidial proteins.

plastocyanin A protein of the thylakoid membrane that is part of the electron transport chain of the light-dependent reactions of photosynthesis.

plastoquinones Quinone compounds that are part of the electron transport chain of the light-dependent reactions of photosynthesis.

pleiotropy The role of a single gene in influencing multiple phenotypic traits.

point mutation A change in DNA sequence involving an alteration of one nucleotide residue.

pollen In seed plants, the microgametophyte (male gametophyte), covered by a hard protective coat and released from specialized organs (in angiosperms, the stamens). It germinates on the surface of the gynoecium in the vicinity of the female gametophyte; a pollen tube (or haustorium in some gymnosperms) delivers the sperm cells (male gametes) to the megagametophyte, where they fertilize the egg cell.

pollen mother cell. *See* **microsporocyte**.

pollen sac The sporophytic (diploid) tissue in which microgametophytes are produced. *Also called* microsporangium.

pollen tube The extension of the male gametophyte that emerges from the pollen grain and delivers the sperm nuclei to the ovule.

pollination In angiosperms, the transfer of pollen from an anther to a stigma of the same (self-pollination) or another (cross-pollination) flower.

poly-A tail A series of adenosine nucleotide residues added to the 3′e end of a newly transcribed messenger RNA.

polyacrylamide gel electrophoresis A method for separating and characterizing DNA or protein molecules on a gel matrix, based on the differential migration of molecules of different size and charge in an electric field.

polyadenylation The addition of a poly-A tail to an RNA molecule.

polycistronic Describing a messenger RNA molecule that contains more than one protein-coding sequence; as distinct from **monocistronic**.

polyketides Compounds produced by plants (as secondary metabolites), bacteria, fungi, and animals; they include some antibiotics and the precursors for a variety of natural products.

polymerase chain reaction (PCR) A method for replicating and amplifying specific DNA sequences *in vitro*.

polymorphism In genetics, the property of a gene that exists in a population in two or more allelic forms.

polyols Polyhydric alcohols produced by hydrogenation or fermentation of various carbohydrates.

polyphenols A group of chemical compounds in plants, characterized by the presence of more than one phenol unit. They include the tannins and phenylpropanoids.

polyphyletic group A group of organisms derived from multiple unrelated ancestors; as distinct from **monophyletic group**.

polyploid Having more than two complete sets of homologous chromosomes.

polyploidy The process of genome doubling that results in organisms with multiple sets of chromosomes; the state of having more than two complete sets of homologous chromosomes.

polysaccharides Polymers consisting of long chains of sugar residues; for example, starch and cellulose.

polyspermy An abnormal condition in which an egg cell is fertilized by more than one sperm cell.

polyubiquitination The addition of chains of ubiquitin units to a protein, marking the protein for destruction by the proteasome.

porphyrins Compounds containing a porphyrin ring (tetrapyrrole ring), often containing a bound metal ion. For example, hemoglobin, chlorophyll, and some enzymes contain porphyrins. *Also called* tetrapyrroles.

post-harvest losses Loss of agricultural crops when pathogens and pests attack plant material after harvest, during transport and storage.

potyviruses A family of plant viruses with a non-enveloped filamentous capsid containing a genome of linear, positive sense (+) single-stranded RNA.

P-proteins Proteins found in large amounts as aggregates or filaments in the sieve elements of phloem.

prenylation The addition of terpenoid (prenyl) units to pigments, electron carriers, and proteins, important in the association of these molecules with membranes

prenyltransferases Enzymes that catalyze the transfer of terpenoid (prenyl) units to pigments, electron carriers, and proteins.

preprophase band A dense band of microtubules that forms just beneath the plasma membrane before the start of cell division.

pressure flow A proposed mechanism by which sugars move through the phloem (sieve tubes) from source to sink organs; it assumes a difference in pressure between the source and sink ends of the phloem.

primary charge separation The event occurring in the reaction center of a photosystem in which the energy of excited electrons in chlorophyll molecules drives the transfer of an electron from a specific chlorophyll molecule to an acceptor molecule.

primary development Plant development that occurs in the vicinity of apical meristems, such as root or shoot apical meristems; as distinct from **secondary development**.

primary growth Plant growth that occurs through the activity of the shoot and root apical meristems, producing the primary plant tissues; as distinct from **secondary growth**.

primary metabolism The metabolic processes occurring in all plant cells and essential to the life of the plant; as distinct from **secondary metabolism**.

primary nitrogen assimilation In plants, the take up and assimilation of inorganic nitrogen, as nitrate or ammonia, from the soil. *See also* **nitrogen assimilation**.

primary root The first root to develop in the seedling, produced by the root apical meristem of the embryo.

primary thickening meristem A meristem that allows stems without secondary growth (such as those of many monocots) to grow in diameter close to the apex. A thick disk of mitotic activity can produce a large number of nodes (and thus leaves) close to the ground.

primordium (pl. **primordia**) In plants, the earliest stage of organ development.

programmed cell death A type of cell death that involves a series of genetically programmed steps.

prokaryotes, prokaryotic organisms Organisms in the domains Bacteria and Archaea. The genetic material (DNA) is located in the cytoplasm, not in a nucleus; as distinct from **eukaryotes, eukaryotic organisms**.

prolamins A class of storage proteins typically found in cereal seeds.

promoter A DNA sequence recognized by regulatory proteins and RNA polymerase to promote transcription of a gene.

propagule Any structure that can give rise to a new plant; for example, a seed or a part of the vegetative body of the plant.

prophase The first stage of mitosis, during which the chromosomes condense but are not yet attached to a mitotic spindle; also a stage of meiosis.

proplastid An organelle that can divide and give rise to plastids.

proteases Enzymes that cleave the peptide bonds between amino acid residues of proteins.

proteasome A multisubunit enzyme complex that degrades proteins specifically targeted for destruction, usually marked by the addition of ubiquitin (polyubiquitination).

protein bodies Organelles packed with storage proteins, found in the cells of storage organs.

protein D1, protein D2 Proteins of the photosystem II reaction center that mediate transfer of electrons from P680 to plastoquinone.

protein domain A region of a protein with a defined structure that is also found in other proteins.

protein kinase/phosphatase cascade A series of reactions, mediated by protein kinases and phosphatases, that results from a single trigger reaction or signal. The cascade involves the phosphorylation and dephosphorylation of specific amino acid residues on specific proteins/enzymes.

protein kinases Enzymes that catalyze the addition of phosphate groups to specific amino acid residues of other proteins, thus changing their activity.

protein motif, protein structural motif A folding pattern (secondary structure) in a protein that is highly conserved and may have a specific biological function.

protein phosphatases Enzymes that catalyze the removal of phosphate groups from specific amino acid residues of other proteins, thus changing their activity.

protein targeting Mechanisms by which a cell transports proteins to the appropriate site for their use, such as insertion into an organelle or secretion from the cell.

protein-import complexes Protein complexes involved in the transfer of proteins across membranes; for example, nuclear pores. *Also called* translocation apparatus or translocases.

proteolysis Cleavage of the peptide bonds of a protein to release peptides or amino acids.

proteomics A set of approaches for analyzing the protein complement of a genome.

prothallus In ferns, a leaflike gametophyte, the haploid stage of the life cycle, which produces sperm and eggs in specialized structures.

protoderm In the plant embryo, the outer cell layer that gives rise to the epidermis.

proto-eukaryotic cell A cell descended from an early endosymbiosis, in which the DNA and replication machinery resembled that of a eukaryotic nucleus, but the cell contained neither mitochondrion nor plastid.

proton pump An integral membrane protein that moves protons across a plasma or organellar membrane, thereby creating a difference in (i.e., a gradient of) both pH and electrical charge across the membrane and tending to establish an electrochemical potential. *See also* **H$^+$-ATPase**.

proton-motive force An electrochemical gradient set up by electron transport chains in the chloroplast thylakoid or inner mitochondrial membrane.

protoplast A plant cell with its cell wall removed, usually by digestion with enzymes that degrade cell walls.

protracheophytes Plants with stems bearing many sporangia on a single stem (polysporangiophytes) and lacking a well-developed vascular system (i.e., lacking tracheids and phloem). Tracheophytes (vascular plants) are probably derived from this group of plants.

provascular strand In plants, the earliest morphologically identifiable stage of a developing vein.

proximo-distal axis An imaginary line running through an organ from its site of attachment to the plant body to its furthest tip.

purines Compounds consisting of a six-membered, pyrimidine ring fused to a five-membered ring; for example, adenine and guanine.

pyrethrins Natural organic compounds with insecticidal activity, produced by *Chrysanthemum* spp.

pyrimidine dimer Two adjacent pyrimidine nucleotides in DNA that are covalently cross-linked following exposure of the DNA to ultraviolet light.

pyrimidines Compounds consisting of a six-membered ring containing two nitrogen atoms; for example, uracil and cytosine.

pyrophosphate–fructose 6-phosphate 1-phosphotransferase (PFP) The enzyme that catalyzes synthesis of fructose 1,6-bisphosphate from fructose 6-phosphate and inorganic pyrophosphate

pyruvate dehydrogenase complex The enzyme complex in mitochondria and chloroplasts that catalyzes conversion of pyruvate to acetyl-CoA.

pyruvate kinase The enzyme that catalyzes synthesis of pyruvate from phosphoenolpyruvate, in a reaction that generates ATP.

pyruvate transporter A membrane protein that allows movement of pyruvate across the membrane.

QTL analysis A method used to map the chromosomal location of quantitative trait loci (QTLs).

quantitative trait locus (QTL) A region of DNA preferentially inherited in association with a particular quantitative phenotypic trait (e.g., high crop yield). Although not necessarily containing the gene (or genes) conferring the phenotype, the QTL region is genetically linked to that gene (or genes).

quiescent center (QC) A region of the root meristem that has little or no mitotic activity.

quorum sensing The sensing of whether a population (e.g., a bacterial population) has reached sufficient size for a particular process or state to occur. For example, some pathogenic bacteria produce enzymes that degrade plant cell walls only when the population of bacteria on the host plant reaches a certain density.

R **gene–mediated defense** A plant defense mechanism activated on detection of pathogen effector molecules. *Also called* effector-triggered immunity (ETI).

R **genes (resistance genes)** In plants, genes that exhibit variation and Mendelian segregation for the capacity to confer host resistance to specific groups of pathogens.

R proteins Plant proteins that recognize pathogen effector molecules; the proteins involved in *R* gene–mediated defense.

rachis The main axis of an inflorescence (spike) of wheat and other cereals.

radicle An embryonic root.

ray cells In trees, elongated cells oriented from the center to the periphery of the trunk.

ray initials One of the two cell types of vascular cambium. *See also* **fusiform initials**.

reaction center The region of a photosystem in which the primary charge-separation event occurs, converting light energy to chemical energy.

reactive oxygen species (ROS) Molecular species such as superoxide, hydrogen peroxide, and hydroxyl radical. At low levels, ROS may function in cellular signaling processes. At higher levels, they may damage cellular macromolecules (e.g., DNA and RNA) and participate in programmed cell death.

reading frame A series of nucleotide triplets in DNA or its RNA transcript that encode a sequence of amino acids, beginning from a specific nucleotide.

receptors Molecules (usually proteins) that detect the presence of other specific molecules and undergo a conformational change that triggers a cellular response.

recessive mutation A mutation with phenotypic effects that are manifest only in the homozygous state.

recombinase An enzyme that catalyzes the exchange of DNA segments between two DNA molecules.

recombination (genetic) The formation in offspring of genetic combinations that are not present in the parents, resulting from crossing over or independent assortment in meiosis during formation of the parental gametes.

recombination frequency The frequency of crossing over between two genes in a chromosome.

recurrent backcrossing The repeated sequential backcrossing of a particular gene (e.g., a disease-resistance gene) to a desirable recipient parent.

reduction A decrease in the oxidation state of a molecule, brought about by the gain of one or more electrons, addition of hydrogen, or loss of oxygen.

reductive activation The activation of an enzyme protein by reduction of the disulfide bond between specific cysteine residues, forming sulfhydryl groups.

reductive pentose phosphate pathway *See* **Calvin cycle**.

replicases Proteins (enzymes) that take part in the replication of nucleic acid sequences.

replication (DNA) The process by which a DNA sequence is copied to form two or more new copies.

replication fork The site at which two DNA strands are separated to allow replication.

reporter genes Genes that encode a protein with an activity that is easily measured and can be introduced into a transgenic organism. These genes are often used to study promoter activity by combining the promoter of interest and an open reading frame encoding the reporter protein. Reporter genes in plant studies typically encode luciferase or β-glucuronidase (GUS), enzymes not normally found in plants. *Also called* marker genes. *See also* **selectable marker genes**.

repression domain A protein region that interacts with basal transcription factors to decrease the rate of gene transcription.

respiration A metabolic process that uses oxygen and releases carbon dioxide and water during the breakdown of carbohydrates to release energy for metabolic activities.

retinoblastoma protein (RBR) A nuclear phosphoprotein that normally acts as an inhibitor of cell proliferation and the cell cycle.

retrotransposon A sequence of DNA that can be inserted into a new location in the genome by a process that involves an RNA intermediate.

retrovirus A virus that has an RNA genome but replicates through a DNA intermediate that is inserted in the host genome.

reverse genetics An experimental approach that begins with a known DNA sequence and proceeds to identify mutations in the corresponding gene to understand its function.

reverse transcriptase An enzyme that uses RNA as a template to produce a complementary DNA (cDNA) molecule.

reverse transcription Replication of a viral DNA via a viral RNA intermediate.

reversion The conversion of a transposon-mutated gene back to the wild-type gene by excision of the transposon.

rhamnogalacturonans A class of pectins with a backbone of rhamnose and galacturonic acid residues and side chains of various sugars.

rhizobia Bacteria that associate with the roots of certain plants (e.g., legumes) to give rise to root nodules in which symbiotic nitrogen fixation takes place.

rhizoid A unicellular or multicellular rootlike structure in algae, bryophytes, lycophytes, and monilophytes, with nutrient-uptake or anchorage functions, or both.

rhizomes Underground stems that send out roots and shoots from their nodes.

rhizosphere The zone of influence of a plant's roots in the soil.

Rho-GTPases, Rho-type GTPases Small monomeric proteins that regulate many cellular processes, including polarization of the cytoskeleton and endocytosis. The GTPase functions as a molecular switch: GTP binding activates interaction between the GTPase and its target protein; hydrolysis of GTP inhibits the interaction.

Ri plasmid A root-inducing plasmid of *Agrobacterium rhizogenes*.

rib meristem The region of the shoot meristem that generates the internal tissues of the stem.

ribonucleases (RNases) Enzymes that catalyze the hydrolysis of RNA molecules.

ribosomal RNA (rRNA) An RNA molecule with a structural role in ribosomes.

ribosomes Cellular organelles consisting of RNA and protein; the site of protein synthesis.

RNA editing A post-transcriptional process that alters an RNA sequence.

RNA interference (RNAi) A process initiated by a double-stranded RNA that leads to degradation of cellular RNA with a similar sequence.

RNA polymerases Enzymes that catalyze the polymerization of ribonucleotides to produce RNAs.

RNA primers Short RNA molecules that hybridize to a larger DNA or RNA molecule and function as the starting point for its replication.

RNA processing The alteration of an RNA transcript to produce the functional, mature RNA; for example, splicing or polyadenylation.

RNA silencing The degradation of RNA with a specific sequence, resulting from RNA interference.

RNA transcript An RNA copy produced by an RNA polymerase from a DNA template; the transcript may be modified to form the mature, active RNA.

RNA-dependent RNA polymerase An enzyme that catalyzes the synthesis of RNA, using another RNA as the template.

rolling-circle replication A type of replication that rapidly synthesizes multiple copies of a circular DNA (e.g., plasmids, some viral DNAs) or circular RNA (e.g., viroidal RNA).

root The organ, usually belowground, that anchors a plant to its substrate (soil) and absorbs water and minerals.

root apical meristem A group of cells originating at the basal (bottom) end of the embryo that gives rise to the primary root.

root cap The structure that covers and protects the root apical meristem.

root conductivity The permeability of a root to water; a measure of the ease with which water can flow from the soil into the xylem.

root hair A specialized cell of the root epidermis, with a long, thin outgrowth that increases the surface area for water and nutrient absorption.

root meristem A group of actively dividing (meristematic) cells in the root tip that provides new cells to support root growth.

root nodules Nodular growths on the root system of certain plants (e.g., legumes) caused by the establishment of symbiotic nitrogen-fixing bacteria in the host tissue.

root pressure The build-up of a positive pressure in root xylem vessels.

rosette A circular cluster of leaves that radiate from a center at or close to the soil surface (e.g., in a dandelion).

Rubisco Ribulose 1,5-bisphosphate carboxylase/oxygenase. In photosynthesis (the Calvin cycle), Rubisco fixes carbon dioxide into ribulose 1,5-bisphosphate to form two molecules of 3-phosphoglycerate. In photorespiration, Rubisco catalyzes the oxygenation of ribulose 1,5-bisphosphate to yield one molecule of 3-phosphoglycerate and one of 2-phosphoglycolate.

S phase The stage of the cell cycle in which DNA synthesis occurs.

salicylic acid A phytohormone that affects plant growth and development; it also serves as an endogenous signal in a plant's defense against pathogens.

salt glands In some plant species, specialized hairs of the aerial epidermis that accumulate and secrete salt.

sapwood The newly formed outer wood located just inside the vascular cambium of a tree trunk and active in the conduction of water.

saturated fatty acids Fatty acids that contain no double bonds in their backbone.

sclerenchymatous cells Thick-walled, usually lignified cells (fiber cells), with a supportive function.

scutellum In grasses, an organ in the caryopsis (grain) that is involved in breakdown of the starchy endosperm; it is thought to be the vestige of a cotyledon.

SEC pathway In chloroplasts, one pathway by which proteins are translocated from the stroma into the thylakoid lumen.

secondary cell wall A plant cell wall that is no longer able to expand and so does not permit growth. Secondary walls consist mostly of cellulose and are strengthened with lignin; they contain less pectin than primary cell walls.

secondary development Plant development derived from cambium or other secondary meristems, such as development of wood and bark; as distinct from **primary development**.

secondary growth The growth of tissues that occurs after completion of primary growth, particularly in woody plants; as distinct from **primary growth**.

secondary metabolism The metabolic processes that produce compounds (secondary metabolites) not essential for basic cell growth, but with specialized roles in plant defense or in development of specialized cell types; as distinct from **primary metabolism**.

secondary metabolites Organic compounds synthesized by the plant that are not immediately essential for growth, development, or reproduction.

secondary thickening meristem In some tree-like monocots, an extension of the primary thickening meristem to form a flanking meristem that produces secondary growth.

seed In gymnosperms and angiosperms, the structure containing the plant embryo and nutrient tissues, enclosed in a protective coat. A seed develops from a fertilized egg, endosperm, and remains of the surrounding megagametophyte tissue, and may be enclosed in maternal tissue.

seed coat The outer protective covering of a seed.

seed plants All plants that produce seeds. *Also called* spermatophytes.

selectable marker genes In genetic studies, genes used to protect cells or organisms from a substance that would otherwise kill them; for example, genes that encode resistance to specific antibiotics or herbicides. In this way, the cells or organisms can be specifically selected from a population. *See also* **reporter genes**.

selective pressure The extent to which environmental demands change the frequency of genes in a population by favoring or eliminating the reproduction of individuals with a particular characteristic (trait).

self-fertilization The process by which pollen fertilizes the ovules of the same flower.

self-incompatibility The inability of pollen to fertilize ovules of the same flower.

self-pollination In angiosperms, the process by which pollen is transferred from the anthers to the stigma of the same flower.

senescence The programmed death of a plant organ, accompanied by regulated breakdown of cells and recovery of nutrients.

sepals The outermost floral organs that envelop the developing bud and the base of a mature flower.

serine hydroxymethyl transferase In the photorespiratory cycle, the enzyme that catalyzes conversion of glycine to serine.

serine/threonine protein kinase An enzyme that catalyzes the addition of a phosphate group to specific serine or threonine residues in another protein.

sesquiterpenes A class of terpenes that consist of three isoprene units, linear or with ring structures; the precursors of the sesquiterpenoids.

sexual reproduction Reproduction that involves the fusion of two haploid gametes to form a diploid zygote, which develops into a diploid organism.

shade avoidance response A growth response of a plant or part of a plant to avoid low-light stress.

sheath In grasses, the part of the leaf that envelopes the stem.

shikimate pathway A series of metabolic reactions that produces amino acids with an aromatic ring (phenylalanine, tyrosine, tryptophan) from carbohydrate precursors.

Shine-Dalgarno sequence A short messenger RNA sequence that precedes the start codon and is complementary to a ribosomal RNA.

shoot The aboveground parts of a plant, including the stem, leaves, and (in angiosperms) flowers.

shoot apical meristem A group of cells originating at the apical (upper) end of the embryo that gives rise to the aboveground plant organs: leaves, sepals, petals, stamens, and ovaries.

shoot meristem A group of rapidly dividing (meristematic) cells in the shoot tip that produces new cells for shoot growth.

short interfering RNAs (siRNAs) Small (20–25 nucleotides) RNA molecules that target complementary messenger RNA during RNA interference.

short-day plants Plants that flower when the day length is shorter than a critical period.

shuttles Exchanges across organellar membranes of metabolites in different oxidation states, enabling the generation and consumption of reducing power in the two compartments separated by the membrane.

sieve elements Elongated cells of the phloem that are involved in sucrose transport.

sieve plates The perforated end walls of sieve elements that connect the lumens of the sieve elements and allow free movement of the cell contents between cells.

sieve tubes Series of specialized transporting cells of the phloem, connected by sieve plates, and with thickened cellulosic walls, that transport sugars throughout the plant.

signal peptide A short (15–60 amino acids) peptide chain that directs post-translational transport of a protein. Some signal peptides are cleaved (removed) by a signal peptidase after the proteins are transported. *Also called* targeting signals or signal sequences.

signal recognition particle A ribonucleoprotein that binds the newly synthesized N-terminus of proteins that are to be transferred to the endoplasmic reticulum.

signaling cascade A sequence of regulatory proteins, such as protein kinases, that relay and amplify a biochemical signal in a cell. *See also* **protein kinase/phosphatase cascade**.

signalosome A protein complex required to repress photomorphogenesis in the dark; it is also necessary for the activity of ubiquitin ligases. *Also called* the COP9 signalosome.

sinapoyl esters Esters of sinapic acid and sugars (e.g., sinapoyl malate, sinapoyl choline, sinapoyl glucose) that accumulate in the Brassicaceae, including Arabidopsis.

sink organs, sink tissues Organs or tissues to which photoassimilate is transported; as distinct from **source organs**, **source tissues**.

siroheme A heme-containing cofactor that catalyzes electron transfers in nitrite and sulfite reductase reactions.

size exclusion limit The upper size limit for a particular process. For example, for plasmodesmata, molecules or ions below this size can pass through but larger molecules cannot. In gel exclusion chromatography, molecules below this size are retained by a column but larger molecules pass through.

skotomorphogenesis Development of the characteristic growth and appearance of plants grown in the absence of light.

small nucleolar RNAs Small RNA transcripts of nuclear DNA, produced in plants by the action of RNA polymerase III.

SNARE proteins (soluble *N*-ethylmaleimide-sensitive factor attachment protein receptors) A conserved family of proteins that are important in membrane and protein trafficking in eukaryotic cells, including the docking and fusion of synaptic vesicles with the plasma membrane.

somatic Describing the vegetative (nonsexual) phase or parts of an organism.

source organs, source tissues Organs or tissues where net assimilation of carbon dioxide occurs and which are the sources of assimilated carbon for other organs of the plant; as distinct from **sink organs**, **sink tissues**.

spadix A thickened, fleshy spike that forms part of the inflorescence of the Araceae (arums).

sperm A male gamete, a haploid reproductive cell.

spermatophytes See **seed plants**.

sphingolipids A class of lipids derived from sphingosine, a long-chain aliphatic amino acid.

spike In cereals, the inflorescence.

spindle An array of microtubules that forms between the two poles of a dividing eukaryotic cell and moves the replicated chromosomes (daughter chromosomes) apart.

spliceosome A nucleoprotein particle that aids in splicing of messenger RNA.

spongy mesophyll An internal leaf tissue specialized for photosynthesis and gas exchange; it has air spaces between loosely arranged, chloroplast-containing cells.

sporangium (pl. **sporangia**) A specialized structure of the diploid plant (sporophyte) that contains the cells that undergo meiosis. The haploid products of meiosis develop as spores.

spores In plants, fungi, bacteria, and other organisms, reproductive (haploid) cells that develop into a new individual without uniting with another cell.

sporophyte The diploid, spore-bearing stage of the plant life cycle; it produces haploid spores by meiosis.

sporopollenin A major component of the outer walls of pollen and spores, very hard, water- resistant, and decay-resistant. It consists of a variety of compounds, including fatty acids and phenolics.

squalene A naturally occurring triterpene produce by all higher organisms.

stamens The male, pollen-producing organs of the flower.

starch An insoluble complex carbohydrate of plants, used for storage of assimilated carbon.

starch phosphorylase An enzyme that catalyzes the conversion of a glucose residue at the nonreducing end of an α1,4-linked glucan to glucose 1-phosphate, a stage in the breakdown of starch.

starch-branching enzyme An enzyme that catalyzes the introduction of α1,6 linkages into α1,4-linked glucose chains during the synthesis of amylopectin.

start codon In a messenger RNA, the three-nucleotide sequence (AUG) that encodes methionine at the beginning (N-terminus) of a newly synthesizing protein.

statoliths Starch-filled plastids that function in gravity perception during gravitropism. *See also* **amyloplasts**.

stele The vascular cylinder of roots, internal to the endodermis.

stem cells Cells that divide to maintain a cell population of their own type and produce precursor cells that form more specialized tissues; in plants, meristematic cells are stem cells.

steroids A large class of terpenoid compounds produced by plants, fungi, and animals, characterized by four fused rings with a variety of attached functional groups. They are produced from sterols derived from the cyclization of squalene (a triterpene).

sterols A family of mainly unsaturated alcohols of the steroid class, present in the fatty tissues of plants and animals; for example, cholesterol and ergosterol.

stigma The specialized part of the carpel where pollen grains attach and germinate.

stilbenes A class of phenylpropanoids that includes some important phytoalexins; for example, resveratrol.

stipe In algae, a structure analogous to the stem of terrestrial plants.

stolons In plants, horizontal stems at or below the ground surface, typically giving rise to tubers or clonal plantlets at their tips.

stoma (pl. **stomata**) A pore on a leaf surface or stem that is involved in gas exchanges.

stop codon A triplet codon (UAA, UGA, or UAG) in messenger RNA that does not specify an amino acid and causes termination of translation.

storage proteins Proteins that accumulate during the development of vegetative and reproductive storage organs, and are hydrolyzed to produce amino acids for protein synthesis during sprouting or germination of the organ.

stratosphere (adj. **stratospheric**) The region of the atmosphere above the troposphere and below the mesosphere.

stroma The contents of a chloroplast outside the photosynthetic membranes.

stromal thylakoids Thylakoid membranes that are not organized into granal stacks.

stromatolites Layered rocky structures thought to comprise fossilized layers of microbes sandwiched between inorganic deposits. The layered structures grow at their upper surface through accretion of new material.

stylet The primitive mouthpart of some nematodes and aphids, adapted for piercing cell walls and providing the animal with access to nutrients contained in the host cell.

suberin A waxy, waterproof substance in the walls of some types of plant cells.

subgenomic RNA An autonomously replicating RNA sequence comprising less than the full sequence of an RNA virus.

subsidiary cell During development of stomata, a cell that derives from a stomatal precursor cell but does not give rise to a guard cell.

substrate (1) Metabolite on which an enzyme acts. (2) Surface or medium on which an organism grows.

succinate dehydrogenase An enzyme complex of the inner mitochondrial membrane that catalyzes the oxidation of succinate to fumarate and passes electrons to the mitochondrial electron transport chain via complex II.

succulent plants Plants that store water.

sucrose A disaccharide of fructose and glucose; the main transport form of carbon in plants.

sucrose phosphate phosphatase The enzyme that catalyzes conversion of sucrose phosphate to sucrose.

sucrose phosphate synthase The enzyme that catalyzes synthesis of sucrose phosphate from fructose 6-phosphate and UDP-glucose.

sucrose synthase The enzyme that catalyzes cleavage of sucrose to form UDP-glucose and fructose.

sucrose transporter A membrane protein that allows movement of sucrose across the membrane.

sulfite reductase The enzyme of the sulfate assimilation pathway that catalyzes reduction of sulfide to sulfite by transfer of six electrons from ferredoxin.

sulfolipids Lipids with fatty acyl groups linked to positions 1 and 2 and a glucose moiety carrying a sulfite group linked to position 3 of the glycerol backbone.

superoxides, superoxide radicals Highly reactive free radicals of oxygen (i.e., with an unpaired electron) that are damaging to biological molecules.

suspensor A structure that connects a developing plant embryo to the mother plant.

symbiont A participant in a symbiotic interaction.

symbiosis *See* **symbiotic interactions.**

symbiosome In endosymbiotic interactions, a double-enveloped cell compartment containing the endosymbiont. The inner envelope is composed of the endosymbiont's plasma membrane; the outer envelope (the perisymbiontic membrane) is derived from the host cell.

symbiotic interactions, symbiotic partnerships, symbiotic relationships Close, often long-term associations between different biological species. It is often defined more narrowly as a relationship from which both species benefit.

symplast (adj. **symplastic**) The intracellular volume bounded by the plasma membrane and interconnected among cells by plasmodesmata, which allow the direct flow of fluid and small molecules such as sugars, amino acids and ions between cells; as distinct from **apoplast** (adj. **apoplastic**).

symporters Membrane proteins that simultaneously transport two different types of substrates.

synapomorphies Derived characteristics common to related organisms.

synaptonemal complex A complex structure that unites homologous chromosomes during prophase of meiosis.

syncytium A continuous region of cytoplasm containing many nuclei.

synergid cells Cells of the female gametophyte that flank the egg cell and have a role in recognizing the pollen tube.

syntaxins Proteins that integrate into membranes and thus allow the membranes to fuse; for example, SNARE proteins

synteny The occurrence of genes in the same order on chromosomes of different species.

systemic acquired resistance (SAR) A "whole plant" resistance response occurring in leaves located above leaves that have undergone an earlier localized exposure to a pathogen that activated resistance. It is associated with induction of pathogenesis-related genes. Activation of SAR involves salicyclic acid.

systemic resistance Resistance activated throughout a plant in response to pests or pathogens. Systemic resistance to chewing insects involves jasmonic acid.

systemin A phytohormone involved in the wound response in the Solanaceae.

tapetum A layer of cells in the anther that encloses and nourishes the developing pollen grains.

TATA box A DNA sequence in the promoter of many genes transcribed by RNA polymerase II that determines the site where transcription is initiated.

taxon The name given to an organism or group of organisms; for example, the name of a species, a class, or a phylum.

taxonomy The classification of species.

T-DNA A DNA segment that is transferred to plant cells by the pathogenic bacterium *Agrobacterium tumefaciens*.

telomerase An enzyme that catalyzes the addition of telomeric sequences to the ends of chromosomes.

telomere An array of short DNA sequences at the end of a eukaryotic chromosome.

telophase The final stage of mitosis, in which the two sets of separated chromosomes decondense and become enclosed by nuclear envelopes; also a stage of meiosis.

tepals Organs of the third and fourth floral whorls that are morphologically similar, as is common in many monocots; as distinct from eudicots, in which the organs of the third and fourth floral whorls are morphologically different.

terminal complexes Multimeric enzyme complexes organized in rosettes, consisting of six hexagonally arranged complexes; for example, the cellulose synthase complex.

terpenoids A large, diverse class of naturally occurring organic compounds similar to terpenes; they are derived from isoprene units assembled and modified in thousands of ways. *Also called* isoprenoids.

testa The outer tissue of the seed coat.

tetraploid Having four copies of the basic set of chromosomes.

tetrapyrroles *See* **porphyrins.**

thallus In algae, a leaflike organ.

thermo-tolerance Tolerance to elevated growth temperatures.

thin-layer chromatography (TLC) A technique for separating molecules based on size and polarity. A thin layer of adsorbent material, such as silica gel or cellulose, is coated on a plate. A liquid phase, consisting of a solution of the molecules to be separated, is drawn up the plate by capillary action.

thioredoxins A class of proteins that donate electrons for the reduction of disulfide bonds between specific cysteine residues (forming sulfhydryl groups) in other proteins (e.g., enzymes of the Calvin cycle), thus activating them.

thylakoids Saclike membrane elements, usually in stacked form, in the stroma of a plastid, usually a chloroplast.

Ti (tumor-inducing) plasmid A plasmid of *Agrobacterium tumefaciens* that is transferred to plant cells and is the causative agent of crown gall disease; used in genetic studies.

TILLING (targeted induced local lesions in genomes) A method for isolating mutant alleles of a gene by screening directly for DNA changes instead of for mutant phenotypes.

tissue culture A method for growing cells or tissues, separated from the organism, on an artificial medium.

tobamoviruses A family of plant viruses consisting of a rigid helical rod containing a genome of linear single-stranded RNA.

tocopherols Vitamin E and its chemical relatives; powerful membrane-associated antioxidants.

tonoplast In plant cells, the membrane that surrounds the vacuole.

topoisomerases Enzymes that catalyze the alteration of supercoiling of double-stranded DNA.

torpedo stage One of the developmental stages of the eudicot embryo.

toxins Poisons produced by certain plants, animals, and bacteria, usually peptides or smaller molecules such as alkaloids or polyketides.

tracheary elements Cells of the xylem that conduct water and provide mechanical support.

tracheids Water-conducting cells of the xylem, with thickened, lignified cell walls.

tracheophytes Plants with a vascular system consisting of xylem and phloem.

transcription The process by which information in a DNA molecule is converted to information in an RNA molecule; that is, RNA synthesis that uses DNA as a template.

transcription factors Proteins that bind to regulatory sequences of a gene to control its expression.

transcriptional activators Transcription factors that have a positive effect on (activate) gene expression.

transcriptional regulators *See* **transcriptional activators**; **transcriptional repressors**.

transcriptional repressors Transcription factors that have a negative effect on (repress) gene expression.

transfer cells Specialized transporting cells, with convoluted wall invaginations and a large surface area of plasma membrane, that take up and carry a variety of ions, molecules, and metabolites.

transfer RNA (tRNA) An RNA molecule that carries a specific amino acid to ribosomes where proteins are synthesized.

transformation The transfer of foreign DNA into an organism's genome.

transgene A foreign gene (DNA segment) artificially introduced into the genome of an organism.

transgenic Containing a transgene.

transit peptide A short amino acid sequence that directs the transport of a protein in a cell.

transketolase In the Calvin cycle, the enzyme that catalyzes transfer of a two-carbon fragment from the five-carbon sugar xylulose 5-phosphate to the five-carbon sugar ribose 5-phosphate, forming the seven-carbon sedoheptulose 7-phosphate and three-carbon glyceraldehyde 3-phosphate.

translation The process in which the information in messenger RNA is converted to the amino acid sequence of a protein.

translocon A complex of proteins associated with the translocation of newly synthesized proteins from the cytosol into the endoplasmic reticulum.

transpiration The movement of water through a plant; in particular, the loss of water from the aerial organs (leaves and stems) through stomata.

transpiration stream A continuous stream of water from the soil via the root, the xylem, and the leaf to the atmosphere, driven primarily by evaporative loss of water (transpiration) through stomata.

transport proteins, transporters Proteins in the cell or in cellular membranes (e.g., plasma membrane) that carry (translocate) specific molecules or ions.

transposase An enzyme that catalyzes the transfer of a transposon from one position to another in the genome.

transposon A DNA segment that can insert itself into other DNA sequences in the same cell.

transposon insertion The movement of a transposon to a new position in the genome.

transposon tagging An experimental method involving gene mutation caused by insertion of a transposon of known sequence; used as a tool to identify the sequence of the mutated gene.

triacylglycerols Storage lipids in which all three positions on the glycerol backbone are linked to fatty acyl groups

tricarboxylic acid cycle *See* **Krebs cycle**.

trichoblast An immature cell that gives rise to a root hair cell.

trichome A hairlike outgrowth of the shoot (stem and leaf) epidermis, consisting of a single modified cell or multiple cells.

tricolpate Having three openings; tricolpate pollen has three furrowed openings in its wall.

triose phosphate transporter A protein in the inner membrane of the chloroplast envelope that allows exchange of triose phosphate for inorganic phosphate across the membrane

triple response A seedling's response to ethylene: inhibition of root and hypocotyl elongation; exaggerated tightening of the apical hook; swelling of the hypocotyl.

triploid Containing three copies of the basic set of chromosomes.

tubers Belowground vegetative storage organs that arise from underground stems (stolons).

tubulin The protein component (monomer) of microtubules.

tunica The outer cell layers of the shoot meristem.

turgor Fullness or tension produced by the fluid content of plant cells; it makes living plant tissue rigid. Loss of turgor, resulting from water loss from plant cells, causes wilting.

type III secretion system A mechanism by which certain (gram-negative) bacterial pathogens deliver effector proteins via a pilus into host cells.

ubiquinone An electron carrier between complex I and complex III in the mitochondrial electron transport chain.

ubiquitin A highly conserved, small regulatory protein, ubiquitous in eukaryotes. Attachment of chains of ubiquitin monomers to a protein (polyubiquitination) marks the protein for destruction by the proteasome.

UDP-glucose A sugar nucleotide with various roles in the cell. For example, it is produced from sucrose by sucrose synthase, is a substrate for sucrose phosphate synthesis by sucrose phosphate synthase, and is a substrate in the reversible reaction catalyzed by UDP-glucose pyrophosphorylase.

uniport Transport of a molecule or ion through a membrane by a carrier mechanism (uniporter) without coupling to transport of another molecule or ion.

ureides Complex chemical derivatives of urea; for example, hydantoin and, more indirectly, guanidine and caffeine.

vacuole A cavity in a plant cell, bounded by a membrane (tonoplast), that stores various plant products and by-products.

vascular bundles Strandlike parts of the vascular system of plants, containing xylem and phloem.

vascular cambium A lateral meristem of woody seed plants that produces secondary xylem to the inside and secondary phloem to the outside.

vascular wilts Plant diseases caused by infection by fungal or bacterial pathogens that results in blockage of xylem vessels and restriction of water transport to the leaves.

vectors Agents of transfer of biological activity. (1) Plasmid vectors are DNA molecules that replicate in bacteria, can be passed between cells during conjugation, and can be used experimentally to insert and propagate DNA or RNA sequences of interest. (2) Insects and nematodes are vectors that carry certain viruses between hosts.

vegetative cell A cell that is part of the pollen grain but does not participate in fertilization.

venation The pattern of veins in a leaf.

vernalization Exposure to a period of cold that is essential for later flowering and reproduction.

vessels In xylem, the water-conducting units.

vicilins A group of storage proteins of legume seeds.

viroids Plant pathogens consisting of a small, highly complementary, circular, single-stranded RNA without a protein coat.

viruses Obligate intracellular parasites that lack the molecular machinery to replicate outside the host. Viral particles usually consist of DNA or RNA contained within a protein coat.

virus-induced gene silencing (VIGS) A method by which the function of a plant gene can be investigated by cloning the corresponding sequence into a viral vector and infecting a plant. Because viral replication involves double-stranded RNA intermediates that trigger silencing of viral sequences, any plant sequence in the virus is also silenced, resulting in loss of function of the corresponding plant gene.

vivipary The germination of a seed while still attached to the mother plant.

voltage-gated channels Ion channels in a membrane that open and close in response to changes in membrane potential.

water potential The chemical potential of water. The water potential of a substance (such as a cell or tissue) is a measure of its ability to absorb or release water relative to another substance.

water-splitting complex The component of photosystem II that is responsible for splitting water in response to harvested light energy.

water–water cycle A cycle of reactions that dissipates the energy of excess electrons generated by photooxidation.

waxes Long-chain aliphatic hydrocarbons of plants, generally water-insoluble.

wood A hard tissue of plant stems, formed mainly of secondary xylem (i.e., xylem produced by the cambium).

xanthophyll cycle In carotenoid metabolism, a cycle of reactions that synthesizes and interconverts xanthophylls.

xanthophylls A class of carotenoids involved in nonphotochemical quenching.

xerophytes Plants that grow in arid environments.

xylem Vascular tissue that transports water from the root to the shoot organs.

xyloglucans Polymers consisting of xylose and glucose residues; a class of hemicelluloses.

zinc finger motif A protein structural motif with regularly spaced cysteine and/or histidine residues that associates with a zinc ion to form a DNA binding domain; many zinc finger motifs are found in transcription factors.

zone of cell division In roots, the region near the root tip, and in leaves, the region at the base of the leaf, where growth occurs primarily by cell division (rather than cell elongation).

zygomorphy Bilateral symmetry.

zygote A diploid cell that results from fertilization (fusion of gametes).

FIGURE ACKNOWLEDGMENTS

Chapter 1

Figure 1-3. A, courtesy of Georgette Douwma/Science Photo Library; B, courtesy of Sinclair Stammers/Science Photo Library. **Figure 1-6.** Cii, courtesy of Michael Abbey/Science Photo Library; Ciii, courtesy of Astrid and Hanns-Frieder Michler/Science Photo Library; Civ, courtesy of Dr. Keith Wheeler/Science Photo Library; Cv, courtesy of Sinclair Stammers/Science Photo Library. **Figure 1-7.** Bii, courtesy of Herve Conge, ISM/Science Photo Library. **Figure 1-8.** A and B, courtesy of Andrew Knoll, Harvard University. **Figure 1-9.** i, courtesy of N. Butterfield, University of Cambridge; ii, courtesy of Lajos Vörös. **Figure 1-10.** From A.H. Knoll, *Proc. Natl Acad. Sci. U.S.A.* 91(15):6743-6750, 1994. With permission from National Academy of Sciences, U.S.A. **Figure 1-13.** A and B, courtesy of Jeremy Pickett-Heaps and Cytographics Pty Ltd. **Figure 1-14.** A, from C.H. Wellman et al., *Nature* 425:282-285, 2003. With permission from Macmillan Publishers Ltd; B, courtesy of Dr. David Polcyn, California State University, San Bernardino. **Figure 1-15.** A and B, courtesy of Hans Kerp and Hagen Hass, Münster; C, courtesy of Hans Steur. **Figure 1-16.** A and B, courtesy of Hans Kerp and Hagen Hass, Münster. **Figure 1-18.** Cii, courtesy of Paul Kenrick. © Natural History Museum, London. **Figure 1-20.** A, courtesy of Vaughan Fleming/Science Photo Library; B, courtesy of Bob Gibbons/Science Photo Library; C, courtesy of John Clegg/Science Photo Library. **Figure 1-21.** A, courtesy of Peter Skelton. **Figure 1-23.** Bi and ii, courtesy of Tobias Kieser; Biii, courtesy of David J Glaves. **Figure 1-26.** A, courtesy of Carleton Ray/Science Photo Library; B, courtesy of Kathy Merrifield/Science Photo Library; C, courtesy of Michael P. Gadomski/Science Photo Library; D, courtesy of Nature's Images/Science Photo Library. **Figure 1-29.** B, courtesy of Philippe Gerrienne. **Figure 1-30.** Ci, courtesy of Dr. Jeremy Burgess/Science Photo Library; Cii, courtesy of Ralf Reski. Previously published in *BMC Plant Biology.* Reproduced under the terms of the BioMed Central Open Access license agreement (http://www.biomedcentral.com/info/about/license) from 'High frequency of phenotypic deviations in Physcomitrella patens plants transformed with a gene-disruption library,' Egener et al. in *BMC Plant Biology* 2002, 2:6. Article available from http://www.biomedcentral.com/1471-2229/2/6. © 2002 Egener et al.; licensee BioMed Central Ltd. This is an Open Access article: verbatim copying and redistribution of this article are permitted in all media for any purpose, provided this notice is preserved along with the article's original URL; Di, courtesy of Claude Nuridsany and Marie Perennou/Science Photo Library, Dii, courtesy of Michael Knee, the Ohio State University. **Figure 1-32.** A, courtesy of Cheryl Power/Science Photo Library; B, courtesy of David Scharf/Science Photo Library. **Figure 1-34.** A and B, from P.K. Endress and A. Igersheim, *Int. J. Plant Sci.* 16(6):5237-5428, 2000. With permission from University of Chicago Press; C, courtesy of Tobias Kieser; D, courtesy of Dr. Nick Kurzenko/Science Photo Library. **Figure 1-35.** A, courtesy of E.R. Degginger/Science Photo Library; B-D, courtesy of Tobias Kieser. **Figure 1-36.** A, courtesy of Tobias Kieser; B, courtesy of Jane Sugarman/Science Photo Library; C, courtesy of Elsa M. Megson/Science Photo Library. **Figure 1-37.** A and B, courtesy of Kim Findlay, John Innes Centre. **Figure 1-38.** A, from E.M. Friis et al., *Nature* 410:357-360, 2001. With permission from Macmillan Publishers Ltd, courtesy of Else Marie Friis; B, from E.M. Friis et al., *Nature* 410:357-360, 2001. With permission from Macmillan Publishers Ltd, courtesy of Pollyanna von Knorring. **Figure 1-39.** B, courtesy of Robert J. Erwin/Science Photo Library; C, courtesy of Chris Knapton/Science Photo Library; D, courtesy of Rod Planck/Science Photo Library. **Figure 1-40.** B-D, courtesy of Tobias Kieser. **Figure 1-41.** B, courtesy of Anthony Cooper/Science Photo Library; C, courtesy of Tobias Kieser; D, taken by Jan Bakker, under the terms of the GNU Free Documentation License, Version 1.2 or any later version published by the Free Software Foundation, from http://commons.wikimedia.org/wiki/File:Teff_(Eragrostis_tef).jpg

Chapter 2

Figure 2-1. Courtesy of N.A. Campbell et al., Biology, 5th ed. Upper Saddle River: Pearson Education, Inc., 1999. **Figure 2-2.** Courtesy of N.A. Campbell et al., Biology, 5th ed. Upper Saddle River: Pearson Education, Inc., 1999. **Figure 2-3.** From T.D. McKnight and D.E. Shippen, *Plant Cell* 16:794-803, 2004. With permission from American Society of Plant Biologists via the Copyright Clearance Center; courtesy of Jack Griffith. **Figure 2-5.** From P.J. Shaw et al., *EMBO J.* 14(12):2896-2906, 1995. With permission from Macmillan Publishers Ltd, courtesy of Peter Shaw. **Figure 2-14.** From P. Cubas et al., *Nature* 401:157-161, 1999. With permission from Macmillan Publishers Ltd; courtesy of Enrico Coen. **Figure 2-18.** Courtesy of Catherine Kidner. **Figure 2-21.** Adapted from The Arabidopsis Genome Initiative, *Nature* 408:796-815, 2000. With permission from Macmillan Publishers Ltd. **Figure 2-22.** Adapted from The Arabidopsis Genome Initiative, *Nature* 408:796-815, 2000. With permission from Macmillan Publishers Ltd. **Figure 2-24.** From S. Liljegren et al., *Nature* 404:766–770, 2000. With permission from Macmillan Publishers Ltd; courtesy of Martin Yanofsky. **Figure 2-25.** A-C, with permission from the Company of Biologists. **Figure 2-27.** Adapted from G. Moore et al., *Curr. Biol.* 5:737-742, 1995, with permission from Elsevier. **Figure 2-31.** From L. Ma et al., *Plant Cell* 13:2589-2607, 2001. With permission from American Society of Plant Biologists via Copyright Clearance Center; courtesy of Xing Wang Deng. **Figure 2-32.** From N. Sato et al., *EMBO J.* 12(2):555–561, 1993. With permission from Macmillan Publishers Ltd; courtesy of Naoki Sato. **Figure 2-33.** Adapted from T. Wakasugi et al., *Plant Mol. Biol. Rep.* 16:231-241, 1998. With permission from Springer Science and Business Media. **Figure 2-34.** With permission from the Company of Biologists.

Chapter 3

Figure 3-1. B, courtesy of Sam Zeeman and Thierry Delatte, ETH, Zurich. **Figure 3-3.** B, courtesy of Herve Conge, ISM/Science Photo Library; C, courtesy of Paul Linstead, John Innes Centre. **Figure 3-12.** Courtesy of Keiko Sugimoto. **Figure 3-15.** B, Courtesy of Sandra McCutcheon and Clive Lloyd from K. Roberts, Handbook of Plant Biology, vol.1, New York: John Wiley and Sons Ltd, 2007. **Figure 3-16.** From C. Lloyd and J. Chan, *Nature Rev. Mol. Cell Biol.* 7:147-152, 2006. With permission from Macmillan Publishers Ltd, courtesy of Jordi Chan, Grant Calder and Clive Lloyd, John Innes Centre. **Figure 3-18.** Courtesy of Henrik Buschmann, John Innes Centre. **Figure 3-24.** A and B, from J.M. Segui-Simarro et al., *Plant Cell* 16:836-856, 2004. With permission from American Society of Plant Biologists via Copyright Clearance Center; courtesy of Jose M. Segui-Simarro. **Figure 3-25.** Courtesy of Roy Brown. **Figure 3-27.** Courtesy of Frederic Berger. **Figure 3-33.** A, from H. Kuroiwa et al., *Planta* 215:185-190, 2002. With permission from Springer Science and Business Media, courtesy of Haruko Kuroiwa; B, from S. Miyagishima et al., *Plant Cell* 15:655-665, 2003. With permission from American Society of Plant Biologists via Copyright Clearance Center, courtesy of Shin-ya Miyagishima; C, from K.W. Osteryoung and R.S. McAndrew, *Annu. Rev. Plant Physiol. Plant Mol. Biol.* 52:315-333, 2001. With permission from Annual Reviews, courtesy of Katherine Osteryoung; D, from J. Marrison, *Plant J.* 18(6):651–662, 1999. With permission from Wiley-Blackwell; courtesy of Joanne Marrison. **Figure 3-41.** A, from B.E.S. Gunning and M.W. Steer, Plant Cell Biology - Structure and Function, Sudbury Mass: Jones and Bartlett Publishers, 1996, courtesy of M.W. Steer; B, from Plant

Cell Biology on DVD, www.plantcellbiologyonDVD.com, micrograph courtesy of B.E.S. Gunning. **Figure 3-44.** B, courtesy of Chris Hawes and Beatrice Satiat-Jeunemaitre from K. Roberts, Handbook of Plant Biology, vol.1, New York: John Wiley and Sons Ltd, 2007. **Figure 3-47.** Courtesy of Charles Delwiche, all rights reserved. **Figure 3-48.** Courtesy of Matilda Crumpton-Taylor, John Innes Centre. **Figure 3-49.** A, courtesy of Marilyn Schaller/Science Photo Library; B, with permission from the Company of Biologists. **Figure 3-52.** A, From T. Arioli et al., *Science* 279(5351):717-720, 1998. With permission from AAAS. **Figure 3-53.** From T. Arioli et al., *Science* 279(5351):717-720, 1998. With permission from AAAS. **Figure 3-56.** Courtesy of Michael Hahn and Glenn Freshour. **Figure 3-57.** From K. Robinson-Beers and R.F. Evert, *Planta* 184:307-318, 1991. With permission from Springer Science and Business Media. Courtesy of Ray Evert. **Figure 3-58.** A and B, with permission from the Company of Biologists. **Figure 3-64.** Courtesy of Gary Meszaros/Science Photo Library. **Figure 3-68.** From K. Schumacher et al., *Genes & Development* 13:3259-3270, 1999. With permission from Cold Spring Harbor Laboratory Press, courtesy of Karin Schumacher. **Figure 3-69.** A and B, from I. Shomer et al., *Can. J. Bot.* 67:625-632, 1989, courtesy of NRC Research Press and Ilan Shomer. **Figure 3-71.** C, image taken by Liming Zhao. From *Am. J. Bot.* 86(7):929-939, 1999. With permission from the American Journal of Botany. **Figure 3-74.** A and B, with permission from the Company of Biologists. C, from Whittington et al., *Nature* 411:610-613, 2001. With permission from Macmillan Publishers Ltd, courtesy of Geoffrey Wasteneys. **Figure 3-75.** C, from T. Murata et al., JSPP, *Plant Cell Physiol.* 38(2):201-209, 1997. With permission from Oxford University Press, courtesy of Takashi Muarata. **Figure 3-76.** From J. Chan et al., *Proc. Natl Acad. Sci. U.S.A.* 96(26):14931-14936, 1999. With permission from National Academy of Sciences, U.S.A., courtesy of Jordi Chan and Clive Lloyd. **Figure 3-77.** A and B, courtesy of David Oppenheimer; C and D, from D.G. Oppenheimer et al., *Proc. Natl Acad. Sci. U.S.A.* 94(12):6261-6266, 1997. With permission from National Academy of Sciences, U.S.A., courtesy of Jordi Chan and Clive Lloyd. **Figure 3-78.** A, courtesy of Dr. Seiji Takeda. B, courtesy of N. Moreno and J. Feijo. **Figure 3-79.** B, Ca2+ image courtesy of N. Moreno and J. Feijo, taken by two-photon raciometric imaging of Oregon-BAPTA/Rhodamine. **Figure 3-80.** From C. Ringli et al., *Plant Physiol.* 129:1-9, 2002. With permission from American Society of Plant Biologists via Copyright Clearance Center, courtesy of Christoph Ringli. **Figure 3-82.** A and B, courtesy of Simon Turner. **Figure 3-83.** B, courtesy of John Sij; D, from R. Franke et al., *Phytochemistry* 66:2643–2658, 2005. With permission from Elsevier, courtesy of Christiane Nawrath. **Figure 3-84.** From R. Zhong et al., *Plant Physiol.* 123:59-69, 2000. With permission from the American Society of Plant Biologists via Copyright Clearance Center; courtesy of Zheng-Hua Ye. **Figure 3-87.** Courtesy of Isabelle His-Mauger. **Figure 3-88.** Ai, courtesy of Eye of Science/Science Photo Library; Aii, courtesy of Eye of Science/Science Photo Library; C, courtesy of P. Dayanandan/Science Photo Library. **Figure 3-89.** Courtesy of J.C. Revy/Science Photo Library. **Figure 3-90.** B, courtesy of Andrew Syred/Science Photo Library; C, courtesy of Dr. Jeremy Burgess/Science Photo Library. **Figure 3-91.** Courtesy of David T. Webb. **Figure 3-92.** Courtesy of Eye of Science/Science Photo Library. **Figure 3-93.** Courtesy of Long Ashton Research Station/Science Photo Library. **Figure 3-94.** From A.A. Millar et al., *Plant Cell* 11:825-839, 1999. With permission from the American Society of Plant Biologists via Copyright Clearance Center; courtesy of Ljerka Kunst. **Figure 3-95.** A and B, from S.J. Lolle et al., *Genetics* 149:607-619, 1998. With permission from the Genetics Society of America via Copyright Clearance Center, courtesy of Robert E. Pruitt. **Figure B3-1.** Courtesy of Peter Shaw.

Chapter 4

Figure 4-4. B, courtesy of Eldon Newcomb and Department of Botany, University of Wisconsin Madison. C, courtesy of K. Plaskitt. **Figure 4-29.** B, from S.C. Zeeman et al., *New Phytologist* 163(2):247-261, 2004. With permission from Wiley-Blackwell. **Figure 4-32.** A, courtesy of Eldon Newcomb and Department of Botany, University of Wisconsin Madison. Originally published in *J. Cell Biol.* 43:343–353, 1969. With permission from Rockefeller University Press. **Figure 4-35.** From M. Lancien et al., *Plant J.* 29(3):347–358, 2002. With permission from

Wiley-Blackwell. **Figure 4-38.** C, courtesy of Ray F. Evert. **Figure 4-40.** C, courtesy of Ray F. Evert. **Figure 4-41.** B, courtesy of Tobias Kieser; C, courtesy of Martyn F. Chillmaid/Science Photo Library. **Figure 4-42.** A, courtesy of Bob Gibbons/Science Photo Library; B, courtesy of Keir Morse. **Figure 4-44.** A, courtesy of William A. Russin; B, from G. Hoffman-Thoma et al., *Planta* 212:231-242, 2001. With permission from Springer Science and Business Media, courtesy of Gudrun Hoffmann-Thoma and Katrin Ehlers. **Figure 4-46.** A, From R. Kempers et al., *Plant Physiol.* 116:271-278, 1998. With permission from the American Society of Plant Biologists via Copyright Clearance Center. B, from G. Hoffmann-Thoma et al., *Planta* 212:231-242, 2001. With permission from Springer Science and Business Media, courtesy of Gudrun Hoffman-Thoma and Katrin Ehlers. **Figure 4-49.** A and B, courtesy of Tobias Kieser; D, courtesy of Adrian Thomas/Science Photo Library. **Figure 4-50.** A, from L. Borisjuk et al., *Plant Biol.* 6:375–386, 2001. With permission from Georg Thieme Verlag; courtesy of Hans Weber; B, from C.E. Offler et al., from *Annu. Rev. Plant Biol.* 54:431-454, 2003. With permission from Annual Reviews. **Figure 4-65.** D, courtesy of Tobias Kieser. **Figure 4-69.** Courtesy of Nigel Cattlin/Science Photo Library. **Figure 4-70.** A, courtesy of FAO/18293/P. Cenini; B, courtesy of ICARDA Photo. **Figure 4-72.** From M.G. Neuffer et al., Mutants of Maize, Cold Spring Harbor: Cold Spring Harbor Laboratory Press, 1997. **Figure 4-76.** B, From S.C. Zeeman et al., *New Phytologist* 163(2):247-261, 2004. With permission of Wiley-Blackwell. **Figure 4-78.** B, courtesy from Ray F. Evert. **Figure 4-80.** Ci and ii, courtesy of Tobias Kieser; Ciii, courtesy of David G. Smith; Civ, courtesy of Thomas Kramer/Stadionwelt. **Figure 4-83.** Courtesy of Sue Bunnewell and Vasiolos Andriotis, John Innes Centre. **Figure 4-88.** iii, courtesy of Tobias Kieser; iv, courtesy of A. Fiedler (www.nativeplants.msu.edu). **Figure 4-90.** Ai-iv, from H. Fu et al., *Plant Cell* 7:1387-1394, 1995. With permission from the American Society of Plant Biologists via Copyright Clearance Center. **Figure 4-91.** Bi and ii, from N. Appeldoorn et al., *Physiologia Plantarum* 115:303–310, 2002. With permission from Wiley-Blackwell. **Figure 4-92.** From E. Lopez-Juez and K.A. Pyke, *Int. J. Dev. Biol.* 49:557-577, 2005. With permission from UBC Press. **Figure 4-102.** Bi, courtesy of Dennis W. Woodland, Andrews University; Biii, courtesy of Andrew Syred/Science Photo Library; Bii, from G. Turner et al., *Plant Physiol.* 124:665-679, 2000. With permission from the American Society of Plant Biologists via Copyright Clearance Center; Biii, courtesy of Andrew Syred/Science Photo Library. **Figure 4-126.** B, courtesy of Dr. Jeremy Burgess/Science Photo Library. **Figure 4-128.** E, courtesy of Tobias Kieser; F, courtesy of Steve Gschmeissner/Science Photo Library. **Figure 4-136.** B, courtesy of Keith Roberts, John Innes Centre. **Figure 4-138.** A, courtesy of Mike Murphy. **Figure 4-140.** Courtesy of H. Reinhard/www.bciusa.com.

Chapter 5

Figure 5-1. A, courtesy of Judy Jernstedt, Botanical Society of America Image Collection, used with permission, photograph by Judy Jernstedt; B, courtesy of Sally Bensusen/Science Photo Library. **Figure 5-3.** From M. Drew et al., *J. Exp. Bot.* 24(83):1189-1202, 1973. With permission from Oxford University Press, courtesy of Malcolm Drew. **Figure 5-4.** From J.A. Banks, *Annu. Rev. Plant Physiol. Plant Mol. Biol.* 50:163-186, 1999. With permission from Annual Reviews, courtesy of Jo Ann Banks. **Figure 5-5.** Courtesy of David Kirk. **Figure 5-6.** A, from D.L. Kirk, *BioEssays*, 27:299–310, 2005. With permission from John Wiley and Sons, Inc., courtesy of David Kirk. **Figure 5-7.** A, from B. Goodner and R.S. Quatrano, *Plant Cell* 5(10):1471–1481, 1993. With permission from the American Society of Plant Biologists via Copyright Clearance Center. **Figure 5-9.** From F. Berger et al., *Science* 263:1421-1423, 1994. With permission from AAAS. **Figure 5-10.** Courtesy of Gerd Jürgens. With permission from the Company of Biologists. **Figure 5-11.** Courtesy of Gerd Jürgens. **Figure 5-12.** From T. Steinmann et al., *Science* 286(5438):316-318, 1999. With permission from AAAS. **Figure 5-13.** Courtesy of John Schiefelbein. With permission from the Company of Biologists and the author. **Figure 5-14.** A-F, from K. Nakajima et al., *Nature* 413:307-311, 2001. With permission from Macmillan Publishers Ltd. **Figure 5-15.** From K. Nakajima et al., *Nature* 413:307–311, 2001. With permission from Macmillan Publishers Ltd, courtesy of Keiji Nakajima. **Figure 5-17.** A and B, with permission from the Company of Biologists. **Figure 5-18.** A-F, with permission from the Company of

Biologists. **Figure 5-19.** From C. Boisnard-Lorig et al., *Plant Cell* 13:495-509, 2001. With permission from the American Society of Plant Biologists via Copyright Clearance Center; courtesy of Frederic Berger. **Figure 5-20.** A and B, from T. Kiyose et al., *Proc. Natl. Acad. Sci. U.S.A.* 96:4186–4191, 1999. With permission from National Academy of Sciences, U.S.A, courtesy of Bob Fischer. **Figure 5-21.** From M.G. Neuffer et al., Mutants of Maize, Cold Spring Harbor: Cold Spring Harbor Laboratory Press, 1997. **Figure 5-25.** A, courtesy of Frank Hochholdinger, University of Tuebingen; B, from W. Hetz et al., *Plant J.* 10(5):845-857. With permission from Wiley-Blackwell, courtesy of Frank Hochholdinger, University of Tuebingen. **Figure 5-26.** Courtesy of Dr. Seiji Takeda. **Figure 5-29.** A and B, courtesy of John Schiefelbein. **Figure 5-31.** A-C, with permission from the Company of Biologists. **Figure 5-32.** ii, from P. Stieger et al., *Plant J.* 32:509–517, 2002. With permission from Wiley-Blackwell, courtesy of Pia Stieger. **Figure 5-33.** i, from J. Fletcher et al., *Curr. Op. Plant Biol.* 3(1):23-30, 2000. With permission from Elsevier. **Figure 5-34.** A-D, from S. Satina et al., *Am. J. Bot.* 27(10):895-905, 1940. With permission from the Botanical Society of America. **Figure 5-35.** From Clark, *Curr. Op. Plant Biol.* 4(1):28-32, 2001. With permission from Elsevier. **Figure 5-37.** Courtesy of James Mauseth. **Figure 5-40.** A and B, courtesy of Sarah Hake. **Figure 5-44.** A and B, courtesy of Tobias Kieser. **Figure 5-45.** A-C, courtesy of Tobias Kieser. **Figure 5-47.** D, courtesy of David T. Webb. **Figure 5-49.** A and B, courtesy of John Bowman. With permission from the Company of Biologists. **Figure 5-50.** D, courtesy of Andrew Hudson. With permission from the Company of Biologists. **Figure 5-52.** Courtesy of Tomohiko Tsuge. With permission from the Company of Biologists. **Figure 5-53.** Courtesy of James Mauseth. **Figure 5-54.** A and B, From T. Nelson and N. Dengler, *Plant Cell* 9:1125-1135, 1997. With permission from the American Society of Plant Biologists via Copyright Clearance Center, courtesy of Nancy Dengler. **Figure 5-56.** A and B, with permission from the Company of Biologists. **Figure 5-57.** Courtesy of David T. Webb. **Figure 5-58.** From A. Schnittger et al., *Plant Cell* 11:1105-1116, 1999. With permission from American Society of Plant Biologists via Copyright Clearance Center, courtesy of Arp Schnittger. **Figure 5-59.** A and B, from A. Schnittger et al., *Plant Cell* 11:1105-1116, 1999. With permission from American Society of Plant Biologists via Copyright Clearance Center, courtesy of Arp Schnittger. C and D, from A. Schnittger et al., *Plant Cell* 11:1105-1116, 1999. With permission from American Society of Plant Biologists via Copyright Clearance Center, courtesy of Arp Schnittger. **Figure 5-60.** A and B, from M. Geisler et al., *Plant Cell* 12:2075–2086, 2000. With permission from American Society of Plant Biologists via Copyright Clearance Center, courtesy of Fred Sack. **Figure 5-62.** From Quirino et al., *Trends Plant Sci.* 5:278-282, 2000. With permission from Elsevier. **Figure 5-64.** A and B, courtesy of Dörte Müller and Klaus Theres. **Figure 5-65.** From White et al., *Trends Gen.* 14(8):327-332, 1998. With permission from Elsevier. **Figure 5-69.** From Sterky et al., *Proc. Natl Acad. Sci. U.S.A.* 95:13330-13335, 1998. With permission from National Academy of Sciences, U.S.A. **Figure 5-70.** C, courtesy of Leonardo Alves Jr. **Figure 5-71.** From Coen et al., *Cell* 63(1):1311-1322, 1990. With permission from Elsevier. **Figure 5-72.** Bi, courtesy of Enrico Cohen; Ci, from Komatsu et al., *Dev. Biol.* 231:364–373, 2001. With permission from Elsevier, courtesy of Junko Kyozuka. **Figure 5-73.** Courtesy of Desmond Bradley. With permission from the Company of Biologists. **Figure 5-74.** Ai and Bi, courtesy of Eliezer Lifschitz. With permission from the Company of Biologists. **Figure 5-75.** A-C, courtesy of Tobias Kieser. **Figure 5-76.** A and B, courtesy of Tobias Kieser. **Figure 5-77.** Ai, from E. Coen, *EMBO J.* 15(24):6777–6788, 1996. With permission from Macmillan Publishers Ltd, courtesy of Enrico Cohen; Aii, from R. Sablowski, *J. Exp. Bot.* 58(5):899-907, 2007. With permission from Oxford University Press; Bi, from E. Coen, *EMBO J.* 15(24):6777–6788, 1996. With permission from Macmillan Publishers Ltd, courtesy of Enrico Cohen; Bii, from Sablowski et al., *Cell* 92:93-103, 1998. With permission from Elsevier; Ci, from E. Coen, *EMBO J.* 15(24):6777–6788, 1996. With permission from Macmillan Publishers Ltd, courtesy of Enrico Cohen; Cii, from R. Sablowski, *J. Exp. Bot.* 58(5):899-907, 2007. With permission from Oxford University Press. **Figure 5-78.** A, courtesy of Leonardo Alves Jr.; B, from H.G. Dickinson, *Can. J. Bot.* 72:384-401, 1994. With permission from NRC Research Press, courtesy of Elizabeth Lord. **Figure 5-80.** Ai, from *J. Exp. Bot.*

58(5):899-907, 2007. With permission from Oxford University Press. **Figure 5-82.** A, courtesy of C. Whipple and R. Schmidt. B, from Ambrose et al., *Mol. Cell* 5(1):569-579, 2000. With permission from Elsevier. **Figure 5-83.** A-C, courtesy of Tobias Kieser. **Figure 5-84.** A and B, from E. Coen, *EMBO J.* 15(24):6777–6788, 1996. With permission from Macmillan Publishers Ltd, courtesy of Enrico Coen. **Figure 5-87.** A and B, from H.A. Owen and C.A. Makaroff, *Protoplasma* 85:7-21, 1995. With permission from Springer Science and Business Media, courtesy of Chris Makaroff. **Figure 5-88.** A and B, from Yang et al., *Genes & Development* 13:2108-2117, 1999. With permission from Cold Spring Harbour Laboratory Press, courtesy of Venkatesan Sundaresan. **Figure 5-90.** A-C, adapted from L. Reisler and R.L. Fischer, *Plant Cell* 5:1291-1301, 1993. With permission from American Society of Plant Biologists via Copyright Clearance Center; D, from G. Drews et al., *Plant Cell* 10:5-17, 1998. With permission from American Society of Plant Biologists via Copyright Clearance Center; E and F, courtesy of Gary Drews. **Figure 5-91.** From G.C. Angenent et al., *Plant Cell* 7:1569-1582, 1995. With permission from American Society of Plant Biologists. **Figure 5-92.** From V.E. Franklin-Tong, *Plant Cell* 11:727-738, 1999. With permission from American Society of Plant Biologists via Copyright Clearance Center. **Figure 5-93.** From T. Higashiyama et al., *Plant Cell* 10:2019-2031, 1998. With permission from American Society of Plant Biologists via Copyright Clearance Center, courtesy of Tetsuya Higashiyama. **Figure 5-97.** From C. Boisnard-Lorig et al., *Plant Cell* 13:495-509, 2001. With permission from American Society of Plant Biologists via Copyright Clearance Center, courtesy of Frederic Berger. **Figure B5-2.** Courtesy of Samantha Fox, John Innes Centre.

Chapter 6

Figure 6-1. Courtesy of Barbara J. Collins. **Figure 6-2.** Courtesy of Dr. Zeyaur R. Khan. **Figure 6-32.** B and C, from J. Li et al., *Science* 272:398-401, 1996. With permission from AAAS, courtesy of Jianming Li. **Figure 6-36.** A-C, courtesy of Rod King, CSIRO; D, courtesy of Jim Lewis and Lucy Rubino, Fordham University. **Figure 6-40.** Courtesy of Tobias Kieser. **Figure 6-47.** iv and v, from J. Colosanti et al., *Cell* 93:593-603, 1998. With permission from Elsevier, courtesy of Joe Colosanti; vi, from J. Colosanti et al., *Cell* 93:593-603, 1998. With permission from Elsevier, courtesy of Joe Colosanti. **Figure 6-53.** From I.R. Henderson et al., *Annu. Rev. Gen.* 37:371-392, 2003. With permission from Annual Review of Genetics, courtesy of Caroline Dean, John Innes Centre, and Josh Mylne, University of Queensland. **Figure 6-61.** From H. Zhang and B.G. Forde, *Science* 279:407-409, 1998. With permission from AAAS, courtesy of Brian G. Forde.

Chapter 7

Figure 7-1. A and B, courtesy of Tobias Kieser. **Figure 7-3.** From K.K. Niyogi et al., *Plant Cell* 10(7):1121-1134, 1998. With permission from the American Society of Plant Biologists. **Figure 7-7.** From S. Munne Bosch and L. Alegre, *Crit. Rev. Plant Sci.*, 21(1):31-57, 2002. With permission from Taylor and Francis Ltd. **Figure 7-8.** From M.H. Montane and K. Kloppstech, *Gene* 258:1-8, 2000. With permission from Elsevier. **Figure 7-9.** From B. Osmond et al., *Trends Plant Sci.* 2:119-121, 1997. With permission from Elsevier. **Figure 7-10.** A and B, from S. Yano and I. Terashima, *Plant Cell Physiol.* 42:1303–1310, 2001. With permission from Oxford University Press, courtesy of Satoshi Yano; C, from Y. Park et al., *Plant Physiol.* 111:867-875, 1996. With permission from the American Society of Plant Biologists. **Figure 7-11.** A and B, from T. Feild et al., *Int. J. Plant Sci.* 162(5):999-1008, 2001. With permission from the University of Chicago Press, courtesy of Taylor Feild. **Figure 7-17.** From M.A.K. Jansen et al., *Trends Plant Sci.* 3:131-135, 1998. With permission from Elsevier. **Figure 7-20.** A-C, courtesy of Tobias Kieser. **Figure 7-23.** Adapted from F. Schöffl et al., *Plant Physiol.* 117:1135-1141, 1998. With permission from the American Society of Plant Biologists. **Figure 7-24.** Courtesy of Pete Lowry, Missouri Botanical Garden. **Figure 7-25.** Courtesy of Arthur Gibson. **Figure 7-28.** From K. Shinozaki and K. Yamaguchi-Shinozaki, *Plant Physiol.* 115:327-334, 1997. With permission from the American Society of Plant Biologists. **Figure 7-29.** From M.R. McAinsh et al., *Physiologica Plantarum,* 100:16-29, 1997. With permission from Blackwells Publishing Ltd. **Figure 7-33.** From R.L. Jones et al., Biochemistry and Molecular Biology of Plants. Somerset:

American Society of Plant Biologists, 2000. With permission from the American Society of Plant Biologists. **Figure 7-34.** Courtesy of Tobias Kieser. **Figure 7-35.** From H.G. Nimmo, *Trends Plant Sci.* 5:75-80, 2000. With permission from Elsevier. **Figure 7-36.** B, courtesy of Dorothea Bartels. **Figure 7-37.** A-C, courtesy of Tobias Kieser. **Figure 7-38.** B, courtesy of Ruth MacCormack. D, © Guenther Eichhorn (geichhorn@aerobaticsweb.org) **Figure 7-39.** Courtesy of R.F. Daubenmire. **Figure 7-40.** A, from M.O. Dillon, *Plant Systematics and Evolution*, 212:261-278, 1998. With permission from Springer Science and Business Media, courtesy of Michael O. Dillon; B, courtesy of Dr. Morley Read/Science Photo Library; C, courtesy of Julian Collins and H. A. Sadaqat; D, courtesy of Bromeliad Society International. **Figure 7-41.** Courtesy of Dierk Wanke. **Figure 7-42.** From R.L. Jones et al., Biochemistry and Molecular Biology of Plants. Somerset: American Society of Plant Biologists, 2000. With permission from the American Society of Plant Biologists. **Figure 7-46.** Courtesy of Erick Dronnet. **Figure 7-48.** From D.L. Hallahan et al., Advances in Botanical Research: Plant Trichomes, vol. 31, 2000. With permission from Elsevier. **Figure 7-49.** From D.L. Hallahan et al., Advances in Botanical Research: Plant Trichomes, vol. 31, 2000. With permission from Elsevier. **Figure 7-50.** From D.L. Hallahan et al., Advances in Botanical Research: Plant Trichomes, vol. 31, 2000. With permission from Elsevier. **Figure 7-51.** A and B, courtesy of Gavin W. Maneveldt, Department of Biodiversity and Conservation Biology, University of the Western Cape, South Africa. **Figure 7-55.** A, courtesy of John A. Lawrence, Jr.; B, courtesy of Brian Thomas, TRC. **Figure 7-56.** From R. Dolferus et al., *Ann. Bot.* 79:21-31, 1997. With permission from Dr. Rudy Dolferus and Oxford University Press. **Figure 7-57.** From M. Sauter, *Naturwissenschaften* 87:289-303, 2000. With permission from Springer-Verlag. **Figure 7-58.** A, courtesy of Dr. Jean Armstrong, Department of Biological Sciences, University of Hull; B, from A.H.L.A.N. Gunawardena et al., *Planta* 212(2):205-214, 2001. With permission from Springer-Verlag. **Figure 7-60.** A and B, courtesy of Tobias Kieser. **Figure 7-61.** A, courtesy of Petr Krasa, Botany.cz; B, courtesy of David T. Webb. **Figure 7-62.** A, courtesy of Bob Gibbons/Science Photo Library; B and C, courtesy of Wayne Armstrong. **Figure 7-66.** From G. Noctor and C.H. Foyer, *Annu. Rev. Plant Physiol. Plant Mol. Biol.* 49:249-279, 1998. With permission from Annual Reviews. **Figure 7-67.** From G. Noctor and C.H. Foyer, *Annu. Rev. Plant Physiol. Plant Mol. Biol.* 49:249-279, 1998. With permission from Annual Reviews. **Figure 7-68.** Adapted from P. Conklin, *Plant, Cell and Environment*, 24:383-394, 2001. With permission from Blackwell Publishing.

Chapter 8

Figure 8-1. Courtesy of Willmer Perez (CIP). **Figure 8-3.** A, from M. Schuster, Simon and Schuster's Beginner's Guide to Understanding Wine. London: Mitchell Beazley Ltd, 1990. With permission from the author and publisher, courtesy of M. Schuster; B, from J.R. Alfano and A. Collmer, Mechanisms of Bacterial Pathogenesis in Plants: Familiar Foes in a Foreign Kingdom. Oxford: Elsevier, 2001. With permission of Elsevier, courtesy of Allan Collmer; C, from T. Pryor et al., *Trends Gen.* 3(1):157-161, 1987. With permission from Elsevier, courtesy of Tony Pryor; D, courtesy of Michael Shintaku and Rick Nelson, Samuel Roberts Noble Foundation. **Figure 8-6.** A, courtesy of Nick Talbot; B-D, from R.J. Howard and B. Valent, *Annu. Rev. Microbiol.* 50:491-512, 1996. With permission from Annual Reviews, courtesy of Richard J. Howard. **Figure 8-7.** B, micrograph from MycoAlbum CD by George Barron. **Figure 8-10.** A, courtesy of Geoff Kidd/Science Photo Library; B, courtesy of J. Pscheidt, from Compendium of Stone Fruit Diseases, St. Paul: American Phytopathological Society, 1995; C, courtesy of Marc Cubeta. **Figure 8-11.** Courtesy of Scubazoo/Science Photo Library. **Figure 8-13.** A, courtesy of Guri Johal. **Figure 8-14.** A, courtesy of David Barker, Ohio State University. **Figure 8-15.** B, courtesy of G.N. Agrios, Plant Pathology, 5th ed. New York: Elsevier Ltd, 2005. **Figure 8-16.** Courtesy of Yuji Kamiya. **Figure 8-17.** A, courtesy of Halvor Aarnes; B, from Csekse et al., *Plant Cell Rep.* 26:1529-1538, 2007. With permission from Springer Science and Business Media, courtesy of G.K. Podila. **Figure 8-26.** Courtesy of AVRDC, the World Vegetable Center, www.avrdc.org **Figure 8-27.** B, with permission from the Society of Nematologists; C, courtesy of Victor Dropkin and Mactode Publications. **Figure 8-28.** A, courtesy of Bob O'Neill, Purdue University. B, from the Agricultural Research Service, the United States Department of Agriculture, taken by Scott Bauer; C, courtesy of Norm Thomas/Science Photo Library; D, from D.M. Shah, C.M.T. Rommens, R.N. Beachy, *Trends Biotechnol.* 13:362-368, 1995. With permission from Elsevier. **Figure 8-30.** A, courtesy of Lea Paterson/Science Photo Library; B, courtesy of Bob Gibbons/Science Photo Library; C, courtesy of Troy Davies. **Figure 8-31.** A, courtesy of Rothamsted Research; D and E, from I. Dorr, *Ann. Bot.* 79:463-472, 1997. With permission from Oxford University Press. **Figure 8-32.** A, courtesy of Dr. John Finch/Science Photo Library. **Figure 8-34.** A, courtesy of Margaret Boulton. **Figure 8-40.** From R. Hückelhoven, *Annu. Rev. Phytopathol.* 45:101-127, 2007. With permission from Annual Reviews. **Figure 8-41.** From A. Di Matteo et al., *Proc. Natl Acad. Sci. U.S.A.* 100(17):10124-10128, 2003. With permission from National Academy of Sciences, U.S.A., courtesy of Felice Cervone. **Figure 8-49.** Courtesy of Jim Holland. **Figure 8-51.** A, courtesy of Anne Osbourn. **Figure 8-52.** B, courtesy of Geoff Kidd/Science Photo Library. **Figure 8-57.** A, from R. Panstruga and P. Schulze-Lefert, *Microbes and Infection* 5:429-437, 2003. With permission from Editions scientifiques et medicales Elsevier SAS, courtesy of Ralph Panstruga; B, courtesy of Jeff Dangl. **Figure 8-59.** B, courtesy of Sabg-Wook Park and Dan Klessig. **Figure 8-63.** i and ii, from C. Li et al., *Plant Cell* 15:1646-1661, 2003. With permission from American Society of Plant Biologists via Copyright Clearance Center. **Figure 8-64.** Courtesy of Whitney Cranshaw. **Figure 8-68.** From O. Voinnet et al., *Cell* 95:177-187, 1998. With permission from Elsevier, courtesy of David Baulcombe. **Figure 8-69.** A, courtesy of M.F. Merlet/Science Photo Library; B, courtesy of Steve Taylor/Science Photo Library. **Figure 8-70.** B, courtesy of Bob Fosbury. **Figure 8-71.** A, from B.P. Surin et al., *Mol. Microbiol.* 4:245-252, 1990. With permission from Wiley-Blackwell, courtesy of J.A. Downie; B, courtesy of David Benson. **Figure 8-75.** ii, from H. Miwa, J. Sun, G.E.D. Oldroyd and J.A. Downie, *Mol. Plant-Microbe Interactions* 19: 914-923, 2006. With permission from the American Phytopathological Society, courtesy of J.A. Downie. **Figure 8-81.** A, courtesy of D.J. Read; B, courtesy of High Hall Nursery www.highhallnursery.co.uk

Chapter 9

Figure 9-2. A and B, courtesy of Butser Ancient Farm, Chalton, Hampshire; C, courtesy of Boye Koch, Riso National Laboratory, Denmark; G, courtesy of Carl Farmer; H, courtesy of Lytton J. Musselman. **Figure 9-4.** A, courtesy of Blake Winton; B, courtesy of Gemma Cole; D, courtesy of Western Fertiliser Technology Pty. Ltd. **Figure 9-5.** Courtesy of John Doebley. **Figure 9-7.** Courtesy of John Doebley. **Figure 9-9.** A, courtesy of Rosie Lerner and the Purdue University Horticulture and Landscape Architecture Department; C, courtesy of Enza Zaden Beheer B.V. **Figure 9-10.** A, courtesy of Enza Zaden Beheer B.V.; B and C, courtesy of Martin Yanofsky. **Figure 9-11.** Courtesy of Steve Tanksley. **Figure 9-12.** Courtesy of Sven Freydank, Institute of Agronomy and Crop Science Halle/Saale. **Figure 9-13.** Courtesy of David Gubernick/AgstockUSA/Science Photo Library. **Figure 9-14.** Courtesy of Peter Hedden. **Figure 9-15.** Courtesy of Alex Bartel/Science Photo Library. **Figure 9-16.** From D.N. Duvick, *Nature Rev. Gen.* 2:69–74, 2001. With permission from Macmillan Publishers Ltd., courtesy of Pioneer Hi-Bred International, Inc. **Figure 9-17.** A, from R. Wise et al., *Genetics* 143:1383-1394, 1996. With permission from Genetics Society of America via Copyright Clearance Center, courtesy of Roger Wise. **Figure 9-19.** Courtesy of Bill Barksdale/AgstockUSA/Science Photo Library. **Figure 9-20.** Courtesy of T.J. Dennehy, University of Arizona. **Figure 9-21.** Courtesy of Dennis Gonsalves with permission from the American Phytopathological Society.

INDEX